Graduate Texts in Mathematics 125

Graduate Texts in Mathematics

1 TAKEUTI/ZARING. Introduction to Axiomatic Set Theory. 2nd. ed.
2 OXTOBY. Measure and Category. 2nd ed.
3 SCHAEFFER. Topological Vector Spaces.
4 HILTON/STAMMBACH. A Course in Homological Algebra.
5 MAC LANE. Categories for the Working Mathematician.
6 HUGHES/PIPER. Projective Planes.
7 SERRE. A Course in Arithmetic.
8 TAKEUTI/ZARING. Axiometic Set Theory.
9 HUMPHREYS. Introduction to Lie Algebras and Representation Theory.
10 COHEN. A Course in Simple Homotopy Theory.
11 CONWAY. Functions of One Complex Variable. 2nd ed.
12 BEALS. Advanced Mathematical Analysis.
13 ANDERSON/FULLER. Rings and Categories of Modules.
14 GOLUBITSKY GUILEMIN. Stable Mappings and Their Singularities.
15 BERBERIAN. Lectures in Functional Analysis and Operator Theory.
16 WINTER. The Structure of Fields.
17 ROSENBLATT. Random Processes. 2nd ed.
18 HALMOS. Measure Theory.
19 HALMOS. A Hilbert Space Problem Book. 2nd ed., revised.
20 HUSEMOLLER. Fibre Bundles 2nd ed.
21 HUMPHREYS. Linear Algebraic Groups.
22 BARNES/MACK. An Algebraic Introduction to Mathematical Logic.
23 GREUB. Linear Algebra. 4th ed.
24 HOLMES. Geometric Functional Analysis and its Applications.
25 HEWITT/STROMBERG. Real and Abstract Analysis.
26 MANES. Algebraic Theories.
27 KELLEY. General Topology.
28 ZARISKI/SAMUEL. Commutative Algebra. Vol. I.
29 ZARISKI/SAMUEL. Commutative Algebra. Vol. II.
30 JACOBSON. Lectures in Abstract Algebra I. Basic Concepts.
31 JACOBSON. Lectures in Abstract Algebra II. Linear Algebra.
32 JACOBSON. Lectures in Abstract Algebra III. Theory of Fields and Galois Theory.
33 HIRSCH. Differential Topology.
34 SPITZER. Principles of Random Walk. 2nd ed.
35 WERMER. Banach Algebras and Several Complex Variables. 2nd ed.
36 KELLEY/NAMIOKA et al. Linear Topological Spaces.
37 MONK. Mathematical Logic.
38 GRAUERT/FRITZSCHE. Several Complex Variables.
39 ARVESON. An Invitation to C^*-Algebras.
40 KEMENY/SNELL/KNAPP. Denumerable Markov Chains. 2nd ed.
41 APOSTOL. Modular Functions and Dirichlet Series in Number Theory.
42 SERRE. Linear Representations of Finite Groups.
43 GILLMAN/JERISON. Rings of Continuous Functions.
44 KENDIG. Elementary Algebraic Geometry.
45 LOÈVE. Probability Theory I. 4th ed.
46 LOÈVE. Probability Theory II. 4th ed.
47 MOISE. Geometric Topology in Dimentions 2 and 3.

continued after index

Carlos A. Berenstein Roger Gay

Complex Variables

An Introduction

With 68 Illustrations

Springer-Verlag
New York Berlin Heidelberg London
Paris Tokyo Hong Kong Barcelona

Carlos A. Berenstein
Department of Mathematics
University of Maryland
College Park, MD 20742
USA

Roger Gay
Department of Mathematics
University of Bordeaux
351 Cours de la Liberation
33504 Talence, France

Library of Congress Cataloging-in-Publication Data
Berenstein, Carlos A.
Complex variables: an introduction / Carlos A. Berenstein, Roger Gay.
p. cm. — (Graduate texts in mathematics ; 125)
Includes bibliographical references (p.) and index.
ISBN 978-0-387-97349-4 (alk. paper)
1. Functions of complex variables. I. Gay, R. (Roger), 1934– II. Title. III. Series.
QA331.7.B46 1991
515'.9—dc20 90-39311
CIP

Printed on acid-free paper

Typeset by Asco Trade Typesetting Ltd., Hong Kong.

9 8 7 6 5 4 3 2 1

ISBN 978-0-387-97349-4 Springer-Verlag New York Berlin Heidelberg

To our families for their patience and encouragement

To Lars V. Ahlfors, with our admiration

Nadie puede escribir un libro. Para
Que un libro sea verdaderamente,
Se requieren la aurora y el poniente,
Siglos, armas y el mar que une y separa.

Jorge Luis Borges

Preface

Textbooks, even excellent ones, are a reflection of their times. Form and content of books depend on what the students know already, what they are expected to learn, how the subject matter is regarded in relation to other divisions of mathematics, and even how fashionable the subject matter is. It is thus not surprising that we no longer use such masterpieces as Hurwitz and Courant's *Funktionentheorie* or Jordan's *Cours d'Analyse* in our courses.

The last two decades have seen a significant change in the techniques used in the theory of functions of one complex variable. The important role played by the inhomogeneous Cauchy–Riemann equation in the current research has led to the reunification, at least in their spirit, of complex analysis in one and in several variables. We say reunification since we think that Weierstrass, Poincaré, and others (in contrast to many of our students) did not consider them to be entirely separate subjects. Indeed, not only complex analysis in several variables, but also number theory, harmonic analysis, and other branches of mathematics, both pure and applied, have required a reconsideration of analytic continuation, ordinary differential equations in the complex domain, asymptotic analysis, iteration of holomorphic functions, and many other subjects from the classic theory of functions of one complex variable. This ongoing reconsideration led us to think that a textbook incorporating some of these new perspectives and techniques had to be written. In particular, we felt that introducing ideas from homological algebra, algebraic topology, sheaf theory, and the theory of distributions, together with the systematic use of the Cauchy–Riemann $\bar{\partial}$-operator, were essential to a complete understanding of the properties and applications of the holomorphic functions of one variable.

The idea that function theory can be integrated into other branches of mathematics is not unknown to our students. It is our experience that undergraduates see many applications of complex analysis, such as the use of partial fractions, the Laplace transform, and the explicit computation of integrals and series which could not be done otherwise. Graduate students thus have a powerful motivation to understand the foundation of the theory of functions of one complex variable.

The present book evolved out of graduate courses given at the universities of Maryland and Bordeaux, where we have attempted to give the students a sense of the importance of new developments and the continuing vitality of the theory of functions. Because of the amount of material covered, we are presenting our work in two volumes.

We have tried to make this book self-contained and accessible to graduate students, while at the same time to reach quite far into the topics considered. For that reason we assume mainly knowledge that is found in the undergraduate curriculum, such as elementary linear algebra, calculus, and point set topology for the complex plane and the two-dimensional sphere S^2. Beyond this, we assume familiarity with metric spaces, the Hahn–Banach theorem, and the theory of integration as it can be found in many introductory texts of real analysis. Whenever we felt a subject was not universally known, we have given a short review of it.

Almost every section contains a large number of exercises of different levels of difficulty. Those that are not altogether elementary have been starred. Many starred exercises came from graduate qualifying examinations. Some exercises provide an insight into a subject that is explained in detail later in the text.

In the same vein, we have made each chapter, and sometimes each section, as independent as possible of the previous ones. If an argument was worth repeating, we did so. This is one of the reasons the formulas have not been numbered; when absolutely essential, they have been marked for ease of reference in their immediate neighborhood. There are some propositions and proofs that have also been starred, and the reader can safely skip them the first time around without loss of continuity. Finally, we have left for the second volume some subjects that require a somewhat better acquaintance with functional analysis.

Let us give a short overview of this volume. Some of the basic properties of holomorphic functions of one complex variable are really topological in nature. For instance, Cauchy's theorem and the theory of residues have a homotopy and a homology form. In the first chapter, we give a detailed description of differential forms (including a proof of the Stokes formula), homotopy theory, homology theory, and other parts of topology pertinent to the theory of functions in the complex plane. Later chapters introduce the reader to sheaf theory and its applications. We conclude Chapter 1 with the definition of holomorphic functions and with the properties of those functions that are immediate from the preceding topological considerations.

In the second chapter we study analytic properties of holomorphic functions, with emphasis on the notion of compact families. This permits an early proof of the Riemann mapping theorem, and we explore some of its consequences and extensions. The class S of normalized univalent functions is introduced as an example of a compact family. A one-semester course in complex analysis could very well start in this chapter and refer the student back to selected topics as necessary.

In the third chapter we consider the solvability of the inhomogeneous Cauchy–Riemann equation. As a corollary we obtain a simple exposition of ideal theory and corresponding interpolation theorems in the algebra of holomorphic functions. We also study the boundary values of holomorphic functions in the sense of distributions, showing that every distribution on $\mathbb{R}$ can be obtained as boundary value of a holomorphic function in $\mathbb{C}\setminus\mathbb{R}$. (An appendix to this chapter gives a short introduction to the concepts of distribution theory.) The Edge-of-the-Wedge theorem, an important generalization of the Schwarz reflection principle, is proven. These ideas lead directly to the theory of hyperfunctions to be considered in the second volume. We conclude this chapter with a totally new approach to the theory of residues.

In the fourth chapter we develop the theory of growth of subharmonic functions in such a way that Hadamard's infinite product expansion for entire functions of finite order is generalized to subharmonic functions. We give a proof due to Bell and Krantz of the fact that a biholomorphic mapping between smooth domains extends smoothly to the boundary. This is used to prove simply and rigorously properties of the Green function of a domain.

In order to develop fully the concept of analytic continuation, Chapter 5 has a short introduction to the theory of sheaves, covering spaces and Riemann surfaces. Among the applications of these ideas we give the index theorem for linear differential operators in the complex plane. This chapter also contains an introduction to the theory of Dirichlet series.

In the second volume the reader will find the application of the ideas and methods developed in the present volume to harmonic analysis, functional equations, and number theory. For instance, elliptic functions, mean-periodic functions, the corona theorem, the Bezout equation in spaces of entire functions, and the Leroy–Lindelöf theory of analytic continuation and its relation to functional equations and overconvergence of Dirichlet series.

This being a textbook, it is impossible to be entirely original, and we have benefited from the existence of many excellent monographs and even unpublished lecture notes, too many to give credit to all of them in every instance. The list of references contains their titles as well as those of a number of research articles relevant to the subjects we touched upon. In a few places we have also tried to steer the reader into further lines of study that were naturally related to the subject at hand, but that, due to the desire to keep this book within manageable limits, we were compelled to leave aside.

Almost everything that the reader will find in our book can be traced in one way or another to Ahlfors' *Complex Analysis*. When it appeared, it changed entirely the way the subject was taught. Although we do not aspire to such achievement, our sincere hope is that we have not let him down.

Finally, we would like to thank Virginia Vargas for the excellent typing and her infinite patience. A number of our friends and students, among them, F. Colonna, D. Pascuas, A. Sebbar, A. Vidras, and A. Yger, have gladly played the role of guinea pigs, reading different portions of the manuscript and offering excellent advice. Our heartfelt thanks to all of them.

Carlos A. Berenstein
Bethesda, Maryland

Roger Gay
Saucats, La Brede

Contents

CHAPTER 1

Topology of the Complex Plane and Holomorphic Functions

§1. Some Linear Algebra and Differential Calculus

The complex plane $\mathbb{C}$ coincides with $\mathbb{R}^2$ by the usual identification of a complex number $z = x + iy$, $x = \operatorname{Re} z$, $y = \operatorname{Im} z$, with the vector (x, y). As such it has two vector space structures, one as a two-dimensional vector space over $\mathbb{R}$ and the other as a one-dimensional vector space over $\mathbb{C}$. The relations between them lead to the classical Cauchy-Riemann equations.

Let $\mathscr{L}_{\mathbb{R}}(\mathbb{C}, \mathbb{R})$ be the space of all $\mathbb{R}$-linear maps from $\mathbb{C}$ into $\mathbb{R}$. These maps are also called (real) linear forms. It is clear that $\mathscr{L}_{\mathbb{R}}(\mathbb{C}, \mathbb{R})$ is an $\mathbb{R}$-vector space. Moreover, since $\{1, i\}$ forms an $\mathbb{R}$-basis for $\mathbb{C}$, the pair of linear forms

$$dx : h \mapsto \operatorname{Re} h \quad \text{and} \quad dy : h \mapsto \operatorname{Im} h$$

constitutes a basis for $\mathscr{L}_{\mathbb{R}}(\mathbb{C}, \mathbb{R})$.

Let $\mathscr{L}_{\mathbb{R}}(\mathbb{C})$ denote the space of all $\mathbb{R}$-linear maps of $\mathbb{C}$ into itself. It is a vector space of dimension 4 over $\mathbb{R}$ and dimension 2 over $\mathbb{C}$. One way to see this is the following. The inclusion $\mathbb{R} \subseteq \mathbb{C}$ allows us to consider $\mathscr{L}_{\mathbb{R}}(\mathbb{C}, \mathbb{R})$ as an $\mathbb{R}$-linear subspace of $\mathscr{L}_{\mathbb{R}}(\mathbb{C})$. We can decompose a form $L \in \mathscr{L}_{\mathbb{R}}(\mathbb{C})$ as $L = \operatorname{Re} L + i \operatorname{Im} L$. Hence, as real vector spaces

$$\mathscr{L}_{\mathbb{R}}(\mathbb{C}) = \mathscr{L}_{\mathbb{R}}(\mathbb{C}, \mathbb{R}) \oplus i\mathscr{L}_{\mathbb{R}}(\mathbb{C}, \mathbb{R}),$$

and we see immediately that $\dim_{\mathbb{R}} \mathscr{L}_{\mathbb{R}}(\mathbb{C}) = 4$. Moreover, any $\mathbb{R}$-basis of $\mathscr{L}_{\mathbb{R}}(\mathbb{C}, \mathbb{R})$ is a $\mathbb{C}$-basis of $\mathscr{L}_{\mathbb{R}}(\mathbb{C})$ and, conversely, any $\mathbb{C}$-basis of $\mathscr{L}_{\mathbb{R}}(\mathbb{C})$ consisting of real-valued mappings is an $\mathbb{R}$-basis for $\mathscr{L}_{\mathbb{R}}(\mathbb{C}, \mathbb{R})$. In particular, the pair $\{dx, dy\}$ is a $\mathbb{C}$-basis for $\mathscr{L}_{\mathbb{R}}(\mathbb{C})$.

We shall consider now the complex subspace $\mathscr{L}_{\mathbb{C}}(\mathbb{C})$ of $\mathscr{L}_{\mathbb{R}}(\mathbb{C})$ consisting of those linear forms which are $\mathbb{C}$-linear. Observe that a linear form $L = P\,dx + Q\,dy \in \mathscr{L}_{\mathbb{R}}(\mathbb{C})$ is $\mathbb{C}$-linear if and only if

$$L(ih) = iL(h)$$

for every $h \in \mathbb{C}$. Writing $h = h_1 + ih_2, h_1, h_2 \in \mathbb{R}$, we find $ih = -h_2 + ih_1$ and

$$L(ih) = -Ph_2 + Qh_1,$$

while

$$iL(h) = i(Ph_1 + Qh_2) = iPh_1 + iQh_2.$$

Therefore, $L \in \mathscr{L}_{\mathbb{C}}(\mathbb{C})$ if and only if $Q = iP$. Define the linear form

$$dz := dx + i\,dy,$$

$dz \in \mathscr{L}_{\mathbb{C}}(\mathbb{C})$ and $L = P\,dz$ whenever $Q = iP$. In particular, $\mathscr{L}_{\mathbb{C}}(\mathbb{C})$ has complex dimension 1 and real dimension 2.

Finally, let us denote by $\overline{\mathscr{L}_{\mathbb{C}}(\mathbb{C})}$ the subspace of $\mathscr{L}_{\mathbb{R}}(\mathbb{C})$ of $\mathbb{C}$-antilinear transformations. That is, $L(\alpha h) = \bar{\alpha}L(h)$ for every $\alpha, h \in \mathbb{C}$. The involution $z \to \bar{z}$ can be extended from $\mathbb{C}$ to $\mathscr{L}_{\mathbb{R}}(\mathbb{C})$, and it exchanges the subspaces $\mathscr{L}_{\mathbb{C}}(\mathbb{C})$ and $\overline{\mathscr{L}_{\mathbb{C}}(\mathbb{C})}$. It provides a direct sum decomposition (as real vector spaces):

$$\mathscr{L}_{\mathbb{R}}(\mathbb{C}) = \mathscr{L}_{\mathbb{C}}(\mathbb{C}) \oplus \overline{\mathscr{L}_{\mathbb{C}}(\mathbb{C})}.$$

The linear form $\overline{dz} = dx - i\,dy$ is in $\overline{\mathscr{L}_{\mathbb{C}}(\mathbb{C}}$ and it is usually denoted $d\bar{z}$. It is immediate to vertify that $\{dz, d\bar{z}\}$ is also a $\mathbb{C}$-basis of $\mathscr{L}_{\mathbb{R}}(\mathbb{C})$.

As an illustration of this, let us consider the formulas for the change of basis. When we write an element $L \in \mathscr{L}_{\mathbb{R}}(\mathbb{C})$ in terms of those two bases we have

$$L = P\,dx + Q\,dy = A\,dz + B\,d\bar{z},$$

where $P, Q, A, B \in \mathbb{C}$ are related by the equations

$$A = \tfrac{1}{2}(P - iQ), \qquad B = \tfrac{1}{2}(P + iQ)$$

$$P = A + B, \qquad Q = i(A - B).$$

The transformation L is $\mathbb{C}$-linear, i.e., $L \in \mathscr{L}_{\mathbb{C}}(\mathbb{C})$, if and only if $B = 0$. This is the familiar Cauchy-Riemann condition found earlier:

$$P = \frac{1}{i}Q.$$

When we identify $\mathbb{C}$ to $\mathbb{R}^2$, then L correpsonds to a 2×2 real matrix $\begin{bmatrix} a & c \\ b & d \end{bmatrix}$, related to the preceding representation by

$$P = a + ib, \qquad Q = c + id.$$

The Cauchy-Rieman condition takes the more familiar form of the pair of equations:

$$\begin{cases} a = d \\ b = -c. \end{cases}$$

Thus, the $\mathbb{C}$-linear transformation of multiplication by $P = a + ib \in \mathbb{C}$ has the matrix representation $\begin{bmatrix} a & -b \\ b & a \end{bmatrix}$.

It is also clear from these computations that $\mathscr{L}_{\mathbb{C}}(\mathbb{C}) \cap \mathscr{L}_{\mathbb{R}}(\mathbb{C}, \mathbb{R}) = \{0\}$, i.e., the only real-valued $\mathbb{C}$-linear transformation of $\mathbb{C}$ is the identically zero map.

We denote by $\mathscr{B}(\mathbb{R}^2 \times \mathbb{R}^2, \mathbb{C})$ the complex vector space of the alternating $\mathbb{R}$-bilinear mappings from $\mathbb{R}^2 \times \mathbb{R}^2$ into $\mathbb{C}$.

Recall that if $h = (h_1, h_2) \in \mathbb{R}^2$, $k = (k_1, k_2) \in \mathbb{R}^2$, and $B \in \mathscr{B}(\mathbb{R}^2 \times \mathbb{R}^2, \mathbb{C})$ then $h \to B(h, k)$ and $k \to B(h, k)$ are $\mathbb{R}$-linear and $B(h, k) = -B(k, h)$. An example of such a map is:

$$B(h, k) = \det\begin{pmatrix} h_1 & k_1 \\ h_2 & k_2 \end{pmatrix} = h_1 k_2 - h_2 k_1.$$

We can generate other $\mathbb{R}$-bilinear maps by the following procedure: If $\phi, \theta \in \mathscr{L}_{\mathbb{R}}(\mathbb{C})$, then we define the ***wedge product*** (or ***exterior product***) $\phi \wedge \theta$, as the element in $\mathscr{B}(\mathbb{R}^2 \times \mathbb{R}^2, \mathbb{C})$ given by

$$(\phi \wedge \theta)(h, k) = \det\begin{pmatrix} \phi(h) & \theta(h) \\ \phi(k) & \theta(k) \end{pmatrix} = \phi(h)\theta(k) - \phi(k)\theta(h).$$

In this notation the previous example is simply $dx \wedge dy$.

Let us see that $\{dx \wedge dy\}$ is a $\mathbb{C}$-basis for $\mathscr{B}(\mathbb{R}^2 \times \mathbb{R}^2, \mathbb{C})$ and hence, $\dim_{\mathbb{C}} \mathscr{B}(\mathbb{R}^2 \times \mathbb{R}^2, \mathbb{C}) = 1$. It is evident that $dx \wedge dy \neq 0$. Moreover, elementary calculation shows that for any $B \in \mathscr{B}(\mathbb{R}^2 \times \mathbb{R}^2, \mathbb{C})$ we have

$$B = B(e_1, e_2)\, dx \wedge dy,$$

where $e_1 = (1, 0)$ and $e_2 = (0, 1)$.

One verifies that the mapping

$$\mathscr{L}_{\mathbb{R}}(\mathbb{C}) \times \mathscr{L}_{\mathbb{R}}(\mathbb{C}) \to \mathscr{B}(\mathbb{R}^2 \times \mathbb{R}^2, \mathbb{C})$$

$$(\phi, \theta) \mapsto \phi \wedge \theta$$

is also $\mathbb{R}$-bilinear and alternating. This proves the distributivity of the wedge product with respect to the sum and shows $\phi \wedge \phi = 0$ for every $\phi \in \mathscr{L}_{\mathbb{R}}(\mathbb{C})$. In particular,

$$dx \wedge dx = dy \wedge dy = dz \wedge dz = d\bar{z} \wedge d\bar{z} = 0,$$

$$dx \wedge dy = -dy \wedge dx,$$

and

$$dz \wedge d\bar{z} = -d\bar{z} \wedge dz = -2i\, dx \wedge dy,$$

which shows that $\{dz \wedge d\bar{z}\}$ is also a $\mathbb{C}$-basis for $\mathscr{B}(\mathbb{R}^2 \times \mathbb{R}^2, \mathbb{C})$.

Let Ω be an open subset of $\mathbb{C}$ and E a normed space defined over $\mathbb{R}$. A mapping $f: \Omega \to E$ is said to be ***differentiable*** at $a \in \Omega$ if there is a linear transformation $L \in \mathscr{L}_{\mathbb{R}}(\mathbb{C}, E)$ (the space $\mathbb{R}$-linear transformations from $\mathbb{C}$ into E) such that for every $h \in \mathbb{C}$ of absolute value sufficiently small we have

$$f(a+h) = f(a) + L(h) + |h|\varepsilon(h),$$

where $\lim_{h\to 0} \varepsilon(h) = 0$. If such an L exists, it is unique and one calls it the derivative of f at the point a. It is denoted $Df(a)$.

If f is differentiable at every point $a \in \Omega$ we can define a function $Df: \Omega \to \mathscr{L}_{\mathbb{R}}(\mathbb{C}, E)$ by $a \mapsto Df(a)$.

Recall that $\mathscr{L}_{\mathbb{R}}(\mathbb{C}, E)$ is also a normed space with the norm

$$\|u\| = \sup_{|h|\le 1} \|u(h)\|_E,$$

where $\|\cdot\|_E$ is the norm in E. Since every $\mathbb{R}$-linear map $u: \mathbb{C} \to E$ is continuous due to the finite dimensionality of $\mathbb{C}$, it follows that $\|u\|$ is well defined for every $u \in \mathscr{L}_{\mathbb{R}}(\mathbb{C}, E)$.

It is easy to see that if $f: \Omega \to E$ is differentiable (everywhere in Ω) it is continuous. One says that f is continuously differentiable, or f is C^1 (or $f \in C^1(\Omega)$), if f is differentiable and Df is continuous. One says that f is twice differentiable if f is differentiable and its derivative $Df: \Omega \to \mathscr{L}_{\mathbb{R}}(\mathbb{C}, E)$ is also differentiable. It is clear that if f is twice differentiable then it is C^1. One says that f is twice continuously differentiable, or f is C^2, if Df is itself C^1. These notions can be recursively extended to any integer $k \ge 1$, and even $k \ge 0$ if one agrees to say f is C^0 when f is continuous. One says f is C^∞, or infinitely differentiable, if f is C^k for every $k \ge 0$.

Let $f: \Omega \to E$, $a = a_1 + ia_2 \in \Omega$. We say f has a partial derivative with respect to x at the point a if the function $f^{a_2}(x) := f(x, a_2)$ is differentiable, as a function of the real variable x, at a_1. Denote by $\frac{\partial f}{\partial x}(a)$ the $\mathbb{R}$-linear transformation $D(f^{a_2})(a_1)$. In the same way we can define the partial derivative with respect to y. One has: $\frac{\partial f}{\partial x}(a) \in \mathscr{L}_{\mathbb{R}}(\mathbb{R}, E)$ and $\frac{\partial f}{\partial y}(a) \in \mathscr{L}_{\mathbb{R}}(\mathbb{R}, E)$. We can identify $\mathscr{L}_{\mathbb{R}}(\mathbb{R}, E)$ to E, and with this identification in mind, one can verify that if f is differentiable at the point a, then it admits both partials at a and

$$\begin{aligned} Df(a) &= \frac{\partial f}{\partial x}(a) \circ dx + \frac{\partial f}{\partial y}(a) \circ dy \\ &= \frac{\partial f}{\partial x}(a)\, dx + \frac{\partial f}{\partial y}(a)\, dy, \end{aligned}$$

where we allow the multiplication of the vectors in E by (real) scalars to take place also on the right. The reader should recall that a sufficient condition for the function to be differentiable at a is that both partial derivatives are continuous at the point a.

Exercises 1.1

1. Write down the 2×2 real matrix corresponding to the $\mathbb{C}$-linear transformation $z \mapsto e^{i\theta} z$ ($\theta \in \mathbb{R}$). Compute its determinant.

2. Let $f: \mathbb{C} \to \mathbb{C}$ be defined by $f(0) = 0$, $f(z) = \dfrac{z^3}{\bar{z}}$ for $z \neq 0$. Show f is C^1 and find the points where Df is $\mathbb{C}$-linear.

3. Is the function $f(z) = z^5|z|^{-4}$ $(z \neq 0)$, $f(0) = 0$, continuous at $z = 0$? Is it differentiable at $z = 0$?

§2. Differential Forms on an Open Subset Ω of $\mathbb{C}$

1.2.1. Definitions

1. A ***differential form of degree* 0 *and class* C^k** in Ω ($k \in \mathbb{N} \cup \{\infty\}$) is a function $f: \Omega \to \mathbb{C}$ which is C^k in Ω. We will denote by $\mathscr{E}_k^0(\Omega)$, for simplicity $\mathscr{E}_k(\Omega)$, the set of all differential forms of degree 0. If $k = \infty$ we will omit it from the notation.

2. A ***differential form of degree* 1 *and class* C^k** in Ω is a function $\omega: \Omega \to \mathscr{L}_{\mathbb{R}}(\mathbb{C})$ of class C^k in Ω. We denote by $\mathscr{E}_k^1(\Omega)$ the set of these differential forms, omitting the index k if $k = \infty$.

A differential form of degree 1 in Ω can be written in a unique way as

$$\omega = P\,dx + Q\,dy = A\,dz + B\,d\bar{z}$$

where P, Q, A, and B are complex-valued functions in Ω of the same class as ω.

3. Let $f: \Omega \to \mathbb{C}$ be a differentiable function. One denotes by df, the ***differential*** of f, the differential form of degree 1 given by

$$df = \frac{\partial f}{\partial x}dx + \frac{\partial f}{\partial y}dy.$$

If we express df in terms of the basis dz, $d\bar{z}$, its coefficients will be denoted $\dfrac{\partial f}{\partial z}$ and $\dfrac{\partial f}{\partial \bar{z}}$ by analogy with the previous expression:

$$df = \frac{\partial f}{\partial z}dz + \frac{\partial f}{\partial \bar{z}}d\bar{z}.$$

The elementary calculation of change of basis mentioned in §1 gives the relations

$$\frac{\partial f}{\partial z} = \frac{1}{2}\left(\frac{\partial f}{\partial x} + \frac{1}{i}\frac{\partial f}{\partial y}\right)$$

$$\frac{\partial f}{\partial \bar{z}} = \frac{1}{2}\left(\frac{\partial f}{\partial x} - \frac{1}{i}\frac{\partial f}{\partial y}\right).$$

Note that $\dfrac{\partial f}{\partial z}$ and $\dfrac{\partial f}{\partial \bar{z}}$ are not partial derivatives of f with respect to the "variables" z and $\bar{z}$, but rather the result of applying to f the differential operators.

$$\frac{\partial}{\partial z} = \frac{1}{2}\left(\frac{\partial}{\partial x} + \frac{1}{i}\frac{\partial}{\partial y}\right), \qquad \frac{\partial}{\partial \bar{z}} = \frac{1}{2}\left(\frac{\partial}{\partial x} - \frac{1}{i}\frac{\partial}{\partial y}\right).$$

One verifies easily the following relations (all the functions appearing here are assumed to be differentiable):

$$\frac{\partial}{\partial z}(u+v) = \frac{\partial u}{\partial z} + \frac{\partial v}{\partial z}, \qquad \frac{\partial(\lambda u)}{\partial z} = \lambda\frac{\partial u}{\partial z} \qquad (\lambda \in \mathbb{C}),$$

$$\frac{\partial}{\partial z}(u\cdot v) = u\frac{\partial v}{\partial z} + v\frac{\partial u}{\partial z}, \qquad \frac{\partial}{\partial z}(u^n) = nu^{n-1}\frac{\partial u}{\partial z} \qquad (n \in \mathbb{Z}),$$

$$\frac{\partial}{\partial z}\left(\frac{u}{v}\right) = \frac{\dfrac{\partial u}{\partial z}v - u\dfrac{\partial v}{\partial z}}{v^2},$$

all of which hold when $\dfrac{\partial}{\partial z}$ is replaced by $\dfrac{\partial}{\partial \bar{z}}$. Similarly,

$$d(u+v) = du + dv, \qquad d(u\cdot v) = u\,dv + v\,du,$$

$$d(u^n) = nu^{n-1}\,du, \qquad d\left(\frac{u}{v}\right) = \frac{v\,du - u\,dv}{v^2}, \quad \text{and}$$

$$d(f\circ g) = \left(\frac{\partial f}{\partial x}\circ g\right)dg_1 + \left(\frac{\partial f}{\partial y}\circ g\right)dg_2,$$

where $f:\Omega \to \mathbb{C}$ and $g:\Omega' \to \Omega$ are differentiable, Ω and Ω' are open subsets of $\mathbb{C}$, and $g = g_1 + ig_2$. Furthermore, one has

$$d(f\circ g) = \left(\frac{\partial f}{\partial z}\circ g\right)dg + \left(\frac{\partial f}{\partial \bar{z}}\circ g\right)d\bar{g}.$$

Writing $\zeta = \xi + i\eta$ as the variable in Ω' and $z = x + iy$ the variable in Ω, and using $dg_1 = \dfrac{\partial g_1}{\partial \xi}d\xi + \dfrac{\partial g_1}{\partial \eta}d\eta$ and a similar expression for g_2, one can also write

$$d(f\circ g) = \left(\left(\frac{\partial f}{\partial x}\circ g\right)\frac{\partial g_1}{\partial \xi} + \left(\frac{\partial f}{\partial y}\circ g\right)\frac{\partial g_2}{\partial \xi}\right)d\xi + \left(\left(\frac{\partial f}{\partial x}\circ g\right)\frac{\partial g_1}{\partial \eta} + \left(\frac{\partial f}{\partial y}\circ g\right)\frac{\partial g_2}{\partial \eta}\right)d\eta$$

and

$$d(f\circ g) = \left(\left(\frac{\partial f}{\partial z}\circ g\right)\frac{\partial g}{\partial \zeta} + \left(\frac{\partial f}{\partial \bar{z}}\circ g\right)\frac{\partial \bar{g}}{\partial \zeta}\right)d\zeta + \left(\left(\frac{\partial f}{\partial z}\circ g\right)\frac{\partial g}{\partial \bar{\zeta}} + \left(\frac{\partial f}{\partial \bar{z}}\circ g\right)\frac{\partial \bar{g}}{\partial \bar{\zeta}}\right)d\bar{\zeta}.$$

Occasionally we will need to use the relations

$$\overline{\frac{\partial u}{\partial z}} = \frac{\partial \bar{u}}{\partial \bar{z}} \quad \text{and} \quad \overline{\frac{\partial u}{\partial \bar{z}}} = \frac{\partial \bar{u}}{\partial z}.$$

4. A ***differential form of degree* 1, *type* (1, 0), *and class*** C^k ***in*** Ω, is a function $\omega : \Omega \to \mathscr{L}_{\mathbb{C}}(\mathbb{C})$ of class C^k. The space of these forms is denoted $\mathscr{E}_k^{1,0}(\Omega)$. The differential forms of type (0, 1) are the functions $\omega : \Omega \to \overline{\mathscr{L}_{\mathbb{C}}(\mathbb{C})}$. The corresponding space is denoted $\mathscr{E}_k^{0,1}(\Omega)$. We evidently have

$$\mathscr{E}_k^1(\Omega) = \mathscr{E}_k^{1,0}(\Omega) \oplus \mathscr{E}_k^{0,1}(\Omega).$$

Every element $\omega \in \mathscr{E}_k^{1,0}(\Omega)$ (resp. $\mathscr{E}_k^{0,1}(\Omega)$) can be written as $\omega = A\,dz$ (resp. $\omega = B\,d\bar{z}$) with A, B complex-valued functions of class C^k in Ω.

5. A ***differential form of degree* 2 *and class*** C^k ***in*** Ω is a mapping $\omega : \Omega \to \mathscr{B}(\mathbb{R}^2 \times \mathbb{R}^2, \mathbb{C})$ of class C^k. The space of all these forms will be denoted $\mathscr{E}_k^2(\Omega)$ (omitting k when it is ∞). By the preceding remarks, a differential form of degree 2 and class C^k can be written in a unique way

$$\omega = C\,dx \wedge dy = D\,dz \wedge d\bar{z},$$

where $C, D : \Omega \to \mathbb{C}$ are functions of class C^k.

Later on we will consider differential forms with coefficients less regular than continuous. For instance, one can speak about differential forms with coefficients that are locally integrable (with respect to the Lebesgue measure) in Ω. We will denote these spaces $(L^1_{\text{loc}}(\Omega))^0$, $(L^1_{\text{loc}}(\Omega))^1$, $(L^1_{\text{loc}}(\Omega))^{1,0}$, $(L^1_{\text{loc}}(\Omega))^{0,1}$ $(L^1_{\text{loc}}(\Omega))^2$. In other cases we will use corresponding notations without further comments.

We have already introduced the differential form df, the differential of a differentiable function f. The differential defines, for $k \geq 1$, a mapping:

$$d : \mathscr{E}_k^0(\Omega) \to \mathscr{E}_{k-1}^1(\Omega)$$
$$f \mapsto df,$$

which can be decomposed into the sum of two mappings

$$\partial : \mathscr{E}_k^0(\Omega) \to \mathscr{E}_{k-1}^{1,0}(\Omega)$$
$$f \mapsto \partial f := \frac{\partial f}{\partial z}\,dz$$

and

$$\bar{\partial} : \mathscr{E}_k^0(\Omega) \to \mathscr{E}_{k-1}^{0,1}(\Omega)$$
$$f \mapsto \bar{\partial} f := \frac{\partial f}{\partial \bar{z}}\,d\bar{z}$$

so that $d = \partial + \bar{\partial}$.

One extends to the space of 1-forms of class C^k the operation of wedge product

$$\mathscr{E}_k^1(\Omega) \times \mathscr{E}_k^1(\Omega) \to \mathscr{E}_k^2(\Omega)$$

$$(\omega_1, \omega_2) \mapsto \omega_1 \wedge \omega_2,$$

where $(\omega_1 \wedge \omega_2)(z) := \omega_1(z) \wedge \omega_2(z)$ for all $z \in \Omega$. If $\omega_1 = P_1\,dx + Q_1\,dy = A_1\,dz + B_1\,d\bar{z}$ and $\omega_2 = P_2\,dx + Q_2\,dy = A_2\,dz + B_2\,d\bar{z}$ then

$$\omega_1 \wedge \omega_2 = (P_1Q_2 - P_2Q_1)dx \wedge dy = (A_1B_2 - A_2B_1)dz \wedge d\bar{z}.$$

One can also introduce the ***exterior differential***, or ***differential*** for short. It is a mapping, still denoted d, $d : \mathscr{E}_k^1(\Omega) \to \mathscr{E}_{k-1}^2(\Omega)$ $(k \geq 1)$ defined by

$$d\omega := dP \wedge dx + dQ \wedge dy,$$

if $\omega = P\,dx + Q\,dy$. Since ω can also be written as $A\,dz + B\,d\bar{z}$ we have the relations

$$d\omega = dA \wedge dz + dB \wedge d\bar{z} = \left(\frac{\partial Q}{\partial x} - \frac{\partial P}{\partial y}\right)dx \wedge dy = \left(\frac{\partial B}{\partial z} - \frac{\partial A}{\partial \bar{z}}\right)dz \wedge d\bar{z}$$

and, for any function f of class C^k,

$$d(f\omega) = df \wedge \omega + f\,d\omega.$$

Let us also consider here operators ∂, $\bar{\partial}$ for which $d = \partial + \bar{\partial}$, where:

$$\partial : \mathscr{E}_k^1(\Omega) \to \mathscr{E}_{k-1}^2(\Omega)$$

$$\omega \mapsto \partial A \wedge dz + \partial B \wedge d\bar{z} = \frac{\partial B}{\partial z}dz \wedge d\bar{z}$$

and

$$\bar{\partial} : \mathscr{E}_k^1(\Omega) \to \mathscr{E}_{k-1}^2(\Omega)$$

$$\omega \mapsto \bar{\partial} A \wedge dz + \bar{\partial} B \wedge d\bar{z} = -\frac{\partial A}{\partial \bar{z}}dz \wedge d\bar{z}.$$

For $k \geq 2$ it makes sense to consider the composition of the mappings:

$$\mathscr{E}_k^0(\Omega) \xrightarrow{d} \mathscr{E}_{k-1}^1(\Omega) \xrightarrow{d} \mathscr{E}_{k-2}^2(\Omega)$$

One has $d^2 = d \circ d = 0$ since

$$(d \circ d)(f) = d(df) = d\left(\frac{\partial f}{\partial x}dx + \frac{\partial f}{\partial y}dy\right) = \left(\frac{\partial^2 f}{\partial x \partial y} - \frac{\partial^2 f}{\partial y \partial x}\right)dx \wedge dy$$

which is zero by the theorem of Schwarz on the identity of the mixed partials.

Therefore, a necesssary condition for $\omega = P\,dx + Q\,dy \in \mathscr{E}_{k-1}^1(\Omega)$ $(k \geq 2)$ to be of the form $\omega = df$ for some $f \in \mathscr{E}_k^0(\Omega)$ is that $d\omega = 0$ (that is, $\frac{\partial Q}{\partial x} - \frac{\partial P}{\partial y} = 0$). This condition is not sufficient as shown by the well-known example: $\Omega = \mathbb{C}\setminus\{0\}$ and $\omega = \dfrac{x\,dy - y\,dx}{x^2 + y^2}$ $\left(\text{or } \omega = \dfrac{dz}{z}\right)$. See Exercise 1.2.6 herein.

6. Let $\omega \in \mathscr{E}^j_k(\Omega)$ ($j = 1, 2$, and $k \geq 0$). A differential form $\alpha \in \mathscr{E}^{j-1}_{k+1}(\Omega)$ such that $\omega = d\alpha$ is called a ***primitive*** of ω.

7. A differential form $\omega \in \mathscr{E}^1_k(\Omega)$ ($k \geq 1$) such that $d\omega = 0$ is called a ***closed form***. A form admitting a primitive is called on ***exact form***. The previous remarks indicate that every exact 1-form of class C^1 is closed.

Recall that an open set Ω is called ***star-shaped*** (with respect to the origin) if for every $z \in \Omega$ the line segment $[0, z] = \{tz : t \in \mathbb{R}, 0 \leq t \leq 1\}$ is completely contained in Ω.

1.2.2. Propositon (Poincaré's lemma). *Let Ω be a star-shaped open set. We have:*
(a) *every $\omega \in \mathscr{E}^1_k(\Omega)$ ($k \geq 1$) that is closed is exact.*
(b) *every $\alpha \in \mathscr{E}^2_k(\Omega)$ ($k \geq 0$) is exact.*

PROOF. (a) Let $\omega = P\,dx + Q\,dy$ and define

$$f(x, y) = x \int_0^1 P(tx, ty)\,dt + y \int_0^1 Q(tx, ty)\,dt.$$

It is legitimate to take derivatives under the integral sign due to the differentiability hypothesis assumed on ω (hence on P and Q). One obtains

$$\frac{\partial f}{\partial x}(x, y) = \int_0^1 P(tx, ty)\,dt + x \int_0^1 \frac{\partial P}{\partial x}(tx, ty)t\,dt + y \int_0^1 \frac{\partial Q}{\partial x}(tx, ty)t\,dt.$$

Since $d\omega = 0$ means that $\dfrac{\partial Q}{\partial x} = \dfrac{\partial P}{\partial y}$, we have

$$\frac{\partial f}{\partial x}(x, y) = \int_0^1 P(tx, ty)\,dt + \int_0^1 \left(x\frac{\partial P}{\partial x}(tx, ty) + y\frac{\partial P}{\partial y}(tx, ty) \right) t\,dt.$$

The expression in brackets can be rewritten as $\dfrac{d}{dt}(P(tx, ty))$. Therefore

$$\frac{\partial f}{\partial x}(x, y) = \int_0^1 P(tx, ty)\,dt + \int_0^1 t\frac{d}{dt}(P(tx, ty))\,dt,$$

which we can simplify by integrating by parts the second integral, so

$$\frac{\partial f}{\partial x}(x, y) = \int_0^1 P(tx, ty)\,dt + (tP(tx, ty))_0^1 - \int_0^1 P(tx, ty)\,dt = P(x, y).$$

One shows in the same way that $\dfrac{\partial f}{\partial y} = Q$. This proves (a). Note the essential way the geometric hypothesis on Ω was used to define f.

(b) Let $\alpha = C\,dx \wedge dy$, set

$$\theta(x, y) = \left(-y \int_0^1 tC(tx, ty)\,dt \right) dx + \left(x \int_0^1 tC(tx, ty)\,dt \right) dy.$$

We can again differentiate under the integral sign and obtain

$$
\begin{aligned}
d\theta(x, y) &= \left(2\int_0^1 tC(tx, ty)\,dt + x\int_0^1 t^2\frac{\partial C}{\partial x}(tx, ty)\,dt \right. \\
&\quad \left. + y\int_0^1 t^2\frac{\partial C}{\partial y}(tx, ty)\,dt\right) dx \wedge dy \\
&= \left(2\int_0^1 tC(tx, ty)\,dt + \int_0^1 t^2\frac{d}{dt}(C(tx, ty))\,dt\right) dx \wedge dy \\
&= \left(2\int_0^1 tC(tx, ty)\,dt + (t^2C(tx, ty))_0^1 - 2\int_0^1 tC(tx, ty)\,dt\right) dx \wedge dy \\
&= C(x, y)\,dx \wedge dy = \alpha. \qquad \square
\end{aligned}
$$

We will return to the problem of deciding when a closed 1-form is exact in §1.9 later.

We shall now define the ***pull-back*** or ***inverse image*** of a differential form by a differentiable mapping. For simplicity, we will always assume the mapping to be of class C^∞. In order to proceed we need to consider the notion of a differential form on an open subset U of $\mathbb{R}$. A differential form of degree 0 and class C^k is simply a function $f: U \to \mathbb{C}$ of class C^k. A differential form of degree 1 and class C^k is a function $\omega: U \to \mathscr{L}_{\mathbb{R}}(\mathbb{R}, \mathbb{C})$ of class C^k, where $\mathscr{L}_{\mathbb{R}}(\mathbb{R}, \mathbb{C})$ is the space of $\mathbb{R}$-linear transformations of $\mathbb{R}$ into $\mathbb{C}$. The space $\mathscr{L}_{\mathbb{R}}(\mathbb{R}, \mathbb{C})$ is a complex vector space isomorphic to $\mathbb{C}$. In fact, it has the basis $\{dt\}$, $dt(s) = s$ for $s \in \mathbb{R}$. A differential form ω of degree 1 can be written $\omega = g\,dt$, $g: U \to \mathbb{C}$ of class C^k. An example is the differential $df = f'(t)\,dt$ of a differentiable function f on U.

We consider three separate cases in order to define the inverse image of a differential form.

1.2.3. Inverse Image: The case of a Mapping $\gamma: \Omega_1 \to \Omega_2$ of Class C^∞ Between Two Open Subsets of $\mathbb{R}$

The inverse image by γ of a differential form of degree 0, $g: \Omega_2 \to \mathbb{C}$, is the form of degree 0:

$$
\gamma^* g: \Omega_1 \to \mathbb{C}
$$

$$
\gamma^* g := g \circ \gamma.
$$

The inverse image by γ of a differential form of degree 1, $\omega = g\,dt$, is the differential form of degree 1:

$$
\gamma^*\omega := (g \circ \gamma)\,d\gamma = (g \circ \gamma)\cdot\gamma'\,dt
$$

defined in Ω_1. Note that if $\langle \alpha, h\rangle$ denotes the action of the map α on the vector h, then for $h \in \mathbb{R}$,

$$\langle \gamma^*\omega(t), h\rangle = \langle \omega(\gamma(t)), \gamma'(t)h\rangle.$$

If g is of degree 0 and class C^k, $k \geq 1$, one verifies that

$$\gamma^*(dg) = d(\gamma^* g).$$

1.2.4. Inverse Image: The Case of a Mapping $\gamma : \Omega_1 \to \Omega_2$, Ω_1 Open Set in $\mathbb{R}$, Ω_2 Open in $\mathbb{C}$, γ of Class C^∞

For a differential form of degree 0, $g : \Omega_2 \to \mathbb{C}$, the definition is the same as in §1.2.3, $\gamma^* g = g \circ \gamma$. If $\omega = P\,dx + Q\,dy$, then its inverse image by γ is the differential form of degree 1 in Ω_1 given by

$$\gamma^*\omega := (P \circ \gamma) d(x \circ \gamma) + (Q \circ \gamma) d(y \circ \gamma).$$

If $\gamma = (\gamma_1, \gamma_2) = \gamma_1 + i\gamma_2$ we then have

$$\gamma^*\omega = ((P \circ \gamma)\gamma_1' + (Q \circ \gamma)\gamma_2')\,dt.$$

One can also verify that $d(\gamma^* g) = \gamma^*(dg)$ for $g : \Omega_2 \to \mathbb{C}$ of degree 0 and class C^k ($k \geq 1$). One also has for $h \in \mathbb{R}$

$$\langle \gamma^*\omega(t), h\rangle = \langle \omega(\gamma(t)), \gamma'(t)(h)\rangle.$$

1.2.5. Inverse Image: The Case of a Mapping $\gamma : \Omega_1 \to \Omega_2$ of Class C^∞, Where Ω_1, Ω_2 are Open Subsets of $\mathbb{C}$

We continue defining $\gamma^* g = g \circ \gamma$ for $g : \Omega_2 \to \mathbb{C}$ a differential form of degree 0. For a 1-form $\omega = P\,dx + Q\,dy$ we have

$$\begin{aligned}\gamma^*\omega &:= (P \circ \gamma) d(x \circ \gamma) + (Q \circ \gamma) d(y \circ \gamma) \\ &= \left((P \circ \gamma)\frac{\partial \gamma_1}{\partial \xi} + (Q \circ \gamma)\frac{\partial \gamma_2}{\partial \xi}\right) d\xi + \left((P \circ \gamma)\frac{\partial \gamma_1}{\partial \eta} + (Q \circ \gamma)\frac{\partial \gamma_2}{\partial \eta}\right) d\eta\end{aligned}$$

where $\gamma = \gamma_1 + i\gamma_2$ and $\zeta = \xi + i\eta$ denotes the variable in Ω_1. A formula similar to that in §1.2.4 can be obtained if we represent ω by $A\,dz + B\,d\bar{z}$ (see Exercise 1.2.3).

For a differential form ω of degree 2, $\omega = A\,dx \wedge dy$, $\gamma^*\omega$ is defined by

$$\gamma^*\omega := (A \circ \gamma)\,d\gamma_1 \wedge d\gamma_2 = (A \circ \gamma)\left(\frac{\partial \gamma_1}{\partial \xi}\frac{\partial \gamma_2}{\partial \eta} - \frac{\partial \gamma_1}{\partial \eta}\frac{\partial \gamma_2}{\partial \xi}\right) d\xi \wedge d\eta.$$

One can see that if $J(\gamma)$ denotes the Jacobian determinant of γ, as a map $\mathbb{R}^2 \to \mathbb{R}^2$, then

$$\gamma^*\omega = (A \circ \gamma) J(\gamma)\, d\xi \wedge d\eta.$$

It is now possible to verify that if the degree of ω is 0 or 1 then

$$d(\gamma^*\omega) = \gamma^*(d\omega).$$

It is also true that if ω has degree 1 and h, k are two vectors in $\mathbb{R}^2$, then

$$\langle \gamma^*\omega(\zeta), h\rangle = \langle \omega(\gamma(\zeta)), \gamma'(\zeta)(h)\rangle,$$

where γ' is the derivative of γ. If ω has degree 2, then $\langle \gamma^*\omega(\zeta), (h,k)\rangle = \langle \omega(\gamma(\zeta)), (\gamma'(\zeta)(h), \gamma'(\zeta)(k))\rangle$. In the same vein, it is not hard to see that

$$\gamma^*(\omega_1 \wedge \omega_2) = \gamma^*\omega_1 \wedge \gamma^*\omega_2.$$

As an example, let us verify $d(\gamma^*\omega) = \gamma^*(d\omega)$ for a form $\omega = P\,dx + Q\,dy$ of degree 1. We assume the distributivity of the inverse image with respect to the wedge product and that the formula has been verified for degree 0. We have:

$$\begin{aligned} d(\gamma^*w) &= d((P\circ\gamma)\,d\gamma_1 + (Q\circ\gamma)\,d\gamma_2) \\ &= d(P\circ\gamma)\wedge d\gamma_1 + (P\circ\gamma)\,d^2\gamma_1 + d(Q\circ\gamma)\wedge d\gamma_2 + (Q\circ\gamma)\,d^2\gamma_2 \\ &= d(\gamma^*P)\wedge d(\gamma^*x) + d(\gamma^*Q)\wedge d(\gamma^*y) \\ &= \gamma^*(dP)\wedge\gamma^*(dx) + \gamma^*(dQ)\wedge\gamma^*(dy) \\ &= \gamma^*(dP\wedge dx + dQ\wedge dy) = \gamma^*(d\omega). \end{aligned}$$

In summary, the inverse mapping γ^* is a linear transformation between differential forms of the same degree, which commutes with the operations of exterior derivative d and wedge product.

1.2.6. Example (Polar Coordinates). Consider $\Omega_1 =]0,\infty[\times \mathbb{R}$ as an open subset of $\mathbb{R}^2$ with variables denoted (ρ,θ), $\Omega_2 = \mathbb{C}^* = \mathbb{C}\setminus\{0\}$. Let $\gamma : \Omega_1 \to \Omega_2$ be given by

$$\gamma_1(\rho,\theta) = \rho\cos\theta \qquad \text{and} \qquad \gamma_2(\rho,\theta) = \rho\sin\theta,$$

hence $\gamma(\rho,\theta) = \rho e^{i\theta}$. We remind the reader that if $z = x + iy = \rho e^{i\theta}$, then $\rho = |z| = \sqrt{x^2+y^2} = \sqrt{z\bar{z}}$ is the **absolute value** of z and $\theta = \arg z$ is the **argument** of z. When we can choose $\theta \in]-\pi,\pi[$, we denote it by $\operatorname{Arg} z$, the **principal value** of the argument.

For $\omega = \dfrac{x\,dy - y\,dx}{x^2+y^2}$, we have

$$\gamma^*\omega = d\theta \qquad \text{and} \qquad d(\gamma^*\omega) = 0.$$

1.2.7. Remark. The notion of inverse image makes sense for forms of class C^k if γ is of class C^j, $j \geq k$.

1.2.8. Definitions. 1. Let Ω be an open subset of $\mathbb{C}$. A ***path*** in Ω is a continuous function $c : [0,1] \to \Omega$. The point $c(0)$ is called the ***starting point*** of the path. The point $c(1)$ is the ***endpoint***.

2. The set of all paths in Ω is the set $\mathscr{C}([0,1],\Omega)$ of all continuous functions in $[0,1]$ with values in Ω.

3. We say a path c is **piecewise-C^j** ($j \geq 1$) if there is a partition $\sigma: 0 = t_0 < t_1 < \cdots < t_n = 1$ of the segment $[0, 1]$ such that all the functions $c_k := c|[t_{k-1}, t_k]$ are of class C^j. (This means that all the derivatives in $]t_{k-1}, t_k[$ extend continuously to $[t_{k-1}, t_k]$. Equivalently, there is a C^j function in an open interval which restricts to c_k in the closed interval.)

4. Let $\omega \in \mathscr{E}_0^1(\Omega)$ and let c be a piecewise-C^j path in Ω ($j \geq 1$). Then we can define the integral of ω along c by:

$$\int_c \omega := \sum_{1 \leq k \leq n} \int_{t_{k-1}}^{t_k} c_k^*(\omega) = \sum_{1 \leq k \leq n} \int_{t_{k-1}}^{t_k} \langle \omega(c(t)), c_k'(t) \rangle \, dt.$$

One can verify without any difficulty that the value thus obtained is independent of the chosen partition σ of c. In fact, by introducing an extra point $\tau \in]t_{k-1}, t_k[$ we have

$$\int_{t_{k-1}}^{t_k} c_k^*(\omega) = \int_{t_{k-1}}^{\tau} c_k^*(\omega) + \int_{\tau}^{t_k} c_k^*(\omega).$$

If σ_1 and σ_2 are two partitions associated to c, one compares the value associated to each of them with the value corresponding to the partition $\sigma_1 \cup \sigma_2$.

5. A path can also be defined as a continuous (or piecewise-C^j) map $c: [a, b] \to \Omega$. A ***change of parameterization*** is a strictly increasing C^1 map $\varphi: [c, d] \to [a, b]$. We obtain a new path $\varphi^* c$ whose image in Ω, starting point, and endpoint coincide with those of c. Clearly we can define the integral of a form ω along c in the same way as earlier and we find without difficulty that

$$\int_c \omega = \int_{\varphi^* c} \omega,$$

that is, the value of the integral is independent of the parameterization.

1.2.9. Proposition (Barrow's Rule). *Let c be a piecewise-C^1 path in Ω and $f \in \mathscr{E}_1^0(\Omega)$. Then*

$$\int_c df = f(c(1)) - f(c(0)).$$

PROOF. Let us choose a partition $\sigma: 0 = t_0 < \cdots < t_n = 1$ such that the corresponding $c_k = c|[t_{k-1}, t_k]$ are continuously differentiable. We have

$$\begin{aligned} \int_c df &= \sum_{1 \leq k \leq n} \int_{t_{k-1}}^{t_k} c_k^*(df) = \sum_{1 \leq k \leq n} \int_{t_{k-1}}^{t_k} d(f \circ c_k) \\ &= \sum_{1 \leq k \leq n} \int_{t_{k-1}}^{t_k} (f \circ c_k)'(t) \, dt \\ &= \sum_{1 \leq k \leq n} \{f(c(t_k)) - f(c(t_{k-1}))\} = f(c(1)) - f(c(0)). \end{aligned}$$

□

Let c be a piecewise-C^1 path in Ω. We recall that such a path is rectifiable and its length $\ell(c)$ is given by

$$\ell(c) = \int_0^1 |c'(t)|\,dt.$$

The parameterization of c such that $|c'(s)| \equiv 1$, $s \in [0, \ell(c)]$, is called the **arc length parameterization**. The integral of a function g defined on the image of c using arc length parameterization is denoted indistinctly by

$$\int_c g\,ds = \int_c g\,|dz| := \int_0^{\ell(c)} g(c(s))\,ds.$$

With this notation we have $\ell(c) = \int_c |dz|$.

Let now $\omega \in \mathscr{E}_0^1(\Omega)$. For each $z \in \Omega$ denote $\|\omega(z)\|$ the norm of the linear map $\omega(z) \in \mathscr{L}_{\mathbb{R}}(\mathbb{C})$. The following simple inequality will have very important applications:

$$\left|\int_c \omega\right| \leq \ell(c)\cdot \sup_{0\leq t\leq 1} \|\omega(c(t))\|.$$

In fact, let $\sigma : 0 = t_0 < t_1 < \cdots < t_n = 1$ be a suitable partition for the path c. We have

$$\left|\int_c \omega\right| = \left|\sum_{1\leq k\leq n} \int_{t_{k-1}}^{t_k} c_k^*(\omega)\right| \leq \sum_{1\leq k\leq n} \int_{t_{k-1}}^{t_k} |\langle \omega(c(t)), c'(t)\rangle|\,dt$$

$$\leq \sup_{0\leq t\leq 1} \|\omega(c(t))\| \cdot \sum_{1\leq k\leq n} \int_{t_{k-1}}^{t_k} |c'(t)|\,dt$$

$$= \sup_{0\leq t\leq 1} \|\omega(c(t))\| \cdot \ell(c).$$

EXERCISES 1.2

1. Let Ω be an open connected subset of $\mathbb{R}^2$, $f \in C^1(\Omega)$ such that $df = 0$. Show that f is a constant.
2. In the situation of §1.2.4, verify that if $\omega = A\,dz + B\,d\bar{z}$, then
$$\gamma^*\omega = (A\circ\gamma)\,d\gamma + (B\circ\gamma)d\bar{\gamma} = ((A\circ\gamma)(\gamma_1' + i\gamma_2') + (B\circ\gamma)(\gamma_1' - i\gamma_2'))\,dt.$$
3. In the situation of 1.2.5, compute $\gamma^*\omega$ when $\omega = A\,dz + B\,d\bar{z}$.
4. In the situation of 1.2.5, verify that if $\omega = B\,dz \wedge d\bar{z}$, then
$$\gamma^*\omega = (B\circ\gamma)\left(\left|\frac{\partial\gamma}{\partial\zeta}\right|^2 - \left|\frac{\partial\gamma}{\partial\bar{\zeta}}\right|^2\right)d\zeta \wedge d\bar{\zeta}.$$
5. Let $\gamma(\rho, \theta) = (\rho\cos\theta, \rho\sin\theta)$ as in Example 1.2.6. Show that if f is a differentiable function in $\mathbb{C}^*$, then

$$\frac{\partial(f\circ\gamma)}{\partial\rho}=\frac{1}{\rho}\left(x\frac{\partial f}{\partial x}+y\frac{\partial f}{\partial y}\right)\circ\gamma,$$

$$\frac{\partial(f\circ\gamma)}{\partial\theta}=\left(x\frac{\partial f}{\partial y}-y\frac{\partial f}{\partial x}\right)\circ\gamma,$$

$$\frac{\partial f}{\partial x}(\rho\cos\theta,\rho\sin\theta)=\cos\theta\frac{\partial(f\circ\gamma)}{\partial\rho}(\rho,\theta)-\frac{\sin\theta}{\rho}\frac{\partial(f\circ\gamma)}{\partial\theta}(\rho,\theta),$$

and find corresponding formulas for $\frac{\partial f}{\partial y}\circ\gamma$, $\frac{\partial f}{\partial z}\circ\gamma$, and $\frac{\partial f}{\partial \bar{z}}\circ\gamma$.

6. In this exercise we show that the form $\omega=\dfrac{x\,dy-y\,dx}{x^2+y^2}$ is not exact in $\mathbb{C}^*$. In fact, if $\omega=df$ for some $f\in C^1(\mathbb{C}^*)$, show that $f\circ\gamma=\theta+c$, for some constant $c\in\mathbb{C}$. Conclude from this that f cannot be continuous. Find another proof that ω is not exact using Proposition 1.2.9.

7. Let $\alpha>0$ be a parameter, $\Omega=\mathbb{C}^*$ and $\omega=\dfrac{(x-y)\,dx+(x+y)\,dy}{(x^2+y^2)^\alpha}$. For which values of α is ω closed? For which values of α is ω exact?

8. Let Ω be an open subset of $\mathbb{C}$, $f\in\mathscr{E}^0(\Omega)$, $\alpha=df$, and suppose $\alpha(z)\neq 0$ for every $z\in\Omega$. For $\omega\in\mathscr{E}^2(\Omega)$ solve the equation

$$\alpha\wedge\beta=\omega$$

with $\beta\in\mathscr{E}^1(\Omega)$.

9. Let $n\in\mathbb{Z}$, compute $\frac{\partial}{\partial z}(z^n)$ and $\frac{\partial}{\partial\bar{z}}(z^n)$, where $z\neq 0$ if $n<0$.

10. Compute $d|z|^2$ and $d\log^2|z|$ (where $z\neq 0$).

11. Let $m,n\in\mathbb{N}$, compute $\frac{\partial}{\partial z}(z^m\bar{z}^n)$ and $\frac{\partial}{\partial\bar{z}}(z^m\bar{z}^n)$.

§3. Partitions of Unity

One can easily verify that the function φ defined on the real line by

$$\varphi(t):=\begin{cases}\exp(-1/(1-t^2)) & \text{if } |t|<1\\ 0 & \text{if } |t|\geq 1,\end{cases}$$

is of class C^∞ in $\mathbb{R}$, even, and strictly positive on $]-1,1[$.

Let us denote supp f, the ***support*** of a continuous function f (i.e., $\operatorname{supp} f=\overline{\{x: f(x)\neq 0\}}$ = the closure of the set of points where f is different from zero). Then we see that supp φ is exactly the compact set $[-1,1]$.

Consider the function θ defined in $\mathbb{R}^n$ by

$$\theta(x) := \begin{cases} k_n \exp(-1/(1 - \|x\|^2)) & \text{if } \|x\| < 1 \\ 0 & \text{if } \|x\| \geq 1, \end{cases}$$

where k_n is chosen so that $\int_{\mathbb{R}^n} \theta(x)\,dx = 1$. (Here $x = (x_1, \ldots, x_n)$, $\|x\| = \left(\sum_1^n |x_j|^2\right)^{1/2}$ is the Euclidean norm.) It is a C^∞ function in $\mathbb{R}^n$, radial (i.e., depends only on $\|x\|$), nonnegative, has integral 1, and its support is contained in the closed unit ball $\bar{B}(0, 1)$. A function θ with these properties will be called ***standard***. In that case, for $\varepsilon > 0$ we will denote by θ_ε the function given by $\theta_\varepsilon(x) = \varepsilon^{-n}\theta(x/\varepsilon)$. This new function will have the same properties as θ except that its support will be contained in $\bar{B}(0, \varepsilon)$.

Let Ω be an open set in $\mathbb{R}^n$, we will denote $\mathscr{D}(\Omega)$ the complex vector space of all C^∞ functions in $\mathbb{R}^n$ with compact support contained in Ω. The existence of standard functions shows this space is nontrivial.

1.3.1. Proposition. *Let Ω be an open set of $\mathbb{R}^n$ and $\mathscr{B}$ a basis of open sets in Ω. There is a sequence $(U_j)_{j \geq 1}$ of open sets in $\mathscr{B}$ such that*

(1) $\bigcup_{j \geq 1} U_j = \Omega$

(2) *For every compact set K in Ω the set $\{j : K \cap U_j \neq \emptyset\}$ is finite.*

The first condition means that $(U_j)_{j \geq 1}$ is an ***open covering*** of Ω. The second means that this covering is ***locally finite***.

PROOF. Let $(K_j)_{j \geq -1}$ be an exhaustion of Ω by compact sets, where $K_{-1} = K_0 = \emptyset$ for convenience. That is

(i) $K_j \subseteq \mathring{K}_{j+1}$ for $j \geq 1$,

(ii) $\Omega = \bigcup_{j \geq 1} K_j$.

Consider $W_r := \mathring{K}_{r+1} \backslash K_{r-2}$, $V_r := K_r \backslash \mathring{K}_{r-1}$ for $r \geq 1$. Hence, each W_r is open, each V_r is compact, $V_r \subseteq W_r$, and $\Omega = \bigcup_{r \geq 1} V_r$.

For every $x \in V_r$ there is $U_{x,r} \in \mathscr{B}$ such that $x \in U_{x,r} \subseteq W_r$. Since V_r is compact, there exist finitely many points $x_{r,1}, \ldots, x_{r,k_r} \in V_r$ such that

$$V_r \subseteq \bigcup_{1 \leq i \leq k_r} U_{x_{r,i},r} \subseteq W_r.$$

The collection $(U_{x_{r,i},r})_{r \geq 1, 1 \leq i \leq k_r}$ is countable and satisfies (1) and (2) since any compact K in Ω intersects only a finite number of W_r. □

1.3.2. Proposition (C^∞ Partition of Unity, I). *Let Ω be a nonempty open subset of $\mathbb{R}^n$. Let $(\Omega_i)_{i \in I}$ be an open covering of Ω. There exists a sequence $(\alpha_j)_{j \geq 1}$ of elements $\alpha_j \in \mathscr{D}(\Omega)$ such that*

(1) *for every* $j \geq 1$ *there is an* $i = i(j) \in I$ *such that* $\operatorname{supp} \alpha_j \subseteq \Omega_i$. *The family* $(\operatorname{supp} \alpha_j)_{j \geq 1}$ *is locally finite.*
(2) $0 \leq \alpha_j \leq 1$ *for every* $j \geq 1$.
(3) $\sum_{j \geq 1} \alpha_j(x) = 1$ *for every* $x \in \Omega$.

This sequence $(\alpha_j)_{j \geq 1}$ *is said to be a* C^∞ ***partition of unity subordinate to the covering*** $(\Omega_i)_{i \in I}$.

PROOF. For every $x \in \Omega$ there is $r_x > 0$ such that $\bar{B}(x, r_x) \subseteq \Omega_{i_x}$ for some $i_x \in I$. The family $\mathscr{B}$ of all the balls $B(x, r)$, $x \in \Omega$, $0 < r \leq r_x$, is a basis of open sets of Ω. Therefore, there is a sequence $B(x_j, r_j)$, $j \geq 1$, satisfying the properties (1) and (2) of Proposition 1.3.1. For $j \geq 1$ we have

$$B(x_j, r_j) \subseteq \bar{B}(x_j, r_j) \subseteq \Omega_{i(j)},$$

where we have set $i(j) = i_{x_j}$. Let θ be a standard function and define functions $\beta_j \in \mathscr{D}(\Omega)$ by $\beta_j(x) = \theta_{r_j}(x - x_j)$. The family $(\operatorname{supp} \beta_j)_{j \geq 1}$ is locally finite by construction. Hence the function

$$s(x) = \sum_{j \geq 1} \beta_j(x)$$

is a C^∞ function in Ω. Furthermore, $s(x) > 0$ everywhere in Ω. Let $\alpha_j = \beta_j / s$. This sequence has the desired properties. □

1.3.3. Corollary. *Let* K *be a compact subset of an open set* Ω *in* $\mathbb{R}^n$. *Let* V *be an open neighborhood of* K, $V \subseteq \Omega$. *There is a function* $\varphi \in \mathscr{D}(V)$ *such that*

(1) $0 \leq \varphi \leq 1$,
(2) $\varphi \equiv 1$ *in a neighborhood of* K.

PROOF. For $\varepsilon > 0$, denote $V(K, \varepsilon) = \{x \in \mathbb{R}^n : \operatorname{dist}(x, K) < \varepsilon\}$. Choose $\varepsilon > 0$ so that $K \subseteq V(K, \varepsilon) \subseteq \bar{V}(K, 2\varepsilon) \subseteq V$. We apply §1.3.2 to the covering of Ω consisting of the two open sets $\Omega_1 = V(K, 2\varepsilon)$ and $\Omega_2 = \Omega \setminus \bar{V}(K, \varepsilon)$, and define

$$\varphi := \sum_j{}' \alpha_j,$$

where the prime indicates the sum takes place over only those indices j for which $\operatorname{supp} \alpha_j \subseteq \Omega_1$. The function φ is clearly in $\mathscr{D}(\Omega)$, and its support is contained in $V(K, 2\varepsilon)$. It is also identically equal to one on a neighborhood of K, since if the index k does not appear in the sum defining φ we must have $\operatorname{supp} \alpha_k \nsubseteq \Omega_1$. Therefore $\operatorname{supp} \alpha_k \subseteq \Omega_2$. It follows that $\alpha_k = 0$ on $\bar{V}(K, \varepsilon)$. Hence

$$\varphi | V(K, \varepsilon) = \left(\sum_{j \geq 1} \alpha_j \right) | V(K, \varepsilon) = 1.$$

This ends the proof of the corollary. □

Such a function φ is called a ***plateau*** function.

1.3.4. Proposition (C^∞ Partition of Unity, II). *Let $(U_i)_{i\in I}$ be a covering of an open set Ω by nonempty open sets. For every $i \in I$ there is $\alpha_i \in \mathscr{E}(\Omega)$ such that $0 \le \alpha_i \le 1$ and $\operatorname{supp}\alpha_i$ is a relatively closed subset of U_i. Furthermore, the family $(\operatorname{supp}\alpha_i)_{i\in I}$ is locally finite and $\sum_{i\in I} \alpha_i \equiv 1$.*

PROOF. We already know that there is a sequence $(\beta_j)_{j\ge1} \subseteq \mathscr{D}(\Omega)$ with the properties (1), (2), and (3) of §1.3.2. For every $i \in I$ let $I_i = \{j : i(j) = i\}$ and define

$$\alpha_i = \sum_{j\in I_i} \beta_j.$$

The family $(\operatorname{supp}\beta_j)_{j\ge1}$ is locally finite, hence it follows that $\alpha_i \in \mathscr{E}(\Omega)$, $0 \le \alpha_i \le 1$, and $\operatorname{supp}\alpha_i \subseteq \bigcup_{j\in I_i} \operatorname{supp}\beta_j$ is a relatively closed subset of Ω contained in U_i.

We need to show that $(\operatorname{supp}\alpha_i)_{i\in I}$ is also a locally finite family. In fact, each $z \in \Omega$ has a neighborhood V_z such that

$$E = \{j \in \mathbb{N}^* : \operatorname{supp}\beta_j \cap V_z \ne \emptyset\}$$

is finite. Let $i(E) = \{i(j) : j \in E\}$. If $i \notin i(E)$, then one must have $\operatorname{supp}\alpha_1 \cap V_z = \emptyset$, otherwise there is a j with $i(j) = i$ such that $\operatorname{supp}\beta_j \cap V_z \ne \emptyset$. This implies that $j \in E$ and hence $i \in i(E)$. Therefore, $\#\{i : \operatorname{supp}\alpha_i \cap V_z \ne \emptyset\} \le \#(i(E)) \le \#(E) < \infty$.

Finally, it is clear that

$$\sum_{i\in I} \alpha_i = \sum_{i\in I}\left(\sum_{j\in I_i} \beta_j\right) = \sum_{j\ge1} \beta_j \equiv 1. \qquad \square$$

In the sequel to this volume, we will need more precision on the behavior of the derivatives of the function φ obtained in Corollary 1.3.3. This precision is given by the following proposition originally due to H. Whitney. The reader can safely skip its proof for the moment.

***1.3.5. Proposition.** *Let $\emptyset \ne F \subseteq \Omega \subseteq \mathbb{R}^n$, F closed and Ω open. Define $d(x) := \max\{d(x, F), d(x, \Omega^c)\}$. There is a C^∞ function φ in $\mathbb{R}^n$ such that*

(i) *$\varphi \equiv 1$ on F,*
(ii) *$\operatorname{supp}\varphi \subseteq \Omega$, and*
(iii) *for some constants $c_k > 0$ (independent of F and Ω), any derivative of φ of order k satisfies the estimate*

$$\left|\frac{\partial^k}{\partial x^\alpha}\varphi(x)\right| \le c_k d(x)^{-k}$$

everywhere. (Here $\alpha \in \mathbb{N}^k$, $\alpha = (\alpha_1, \ldots, \alpha_n)$, $|\alpha| = \alpha_1 + \cdots + \alpha_n = k$.)

Moreover, if F is compact then φ can be taked in $\mathscr{D}(\mathbb{R}^n)$.

If $\Omega = \mathbb{R}^n$, replacing $d(x)$ by a positive constant, the same statements hold.

PROOF. First note that one can easily verify that $d(x) > 0$ everywhere and $|d(x) - d(y)| \leq \|x - y\|$.

Using Zorn's lemma we can now construct a maximal sequence of points x_m in $\mathbb{R}^n$ such that the balls $B(x_m, \frac{1}{10}d(x_m))$ are pairwise disjoint. We claim that the balls $(B(x_m, \frac{1}{4}d(x_m)))_{m\geq 1}$ form an open covering of $\mathbb{R}^n$, i.e., $\bigcup_{m\geq 1} B(x_m, \frac{1}{4}d(x_m)) = \mathbb{R}^n$. In fact, if $x_0 \notin \bigcup_{m\geq 1} B(x_m, \frac{1}{10}d(x_m))$ then there must exist an integer m such that

$$B\left(x_m, \frac{1}{10}d(x_m)\right) \cap B\left(x_0, \frac{1}{10}d(x_0)\right) \neq \varnothing.$$

Otherwise the sequence $(x_m)_m$ would not be maximal. Let y be a point in the intersection. Then

$$\|x_0 - x_m\| \leq \|x_0 - y\| + \|y - x_m\| < \frac{1}{10}d(x_0) + \frac{1}{10}d(x_m)$$

$$= \frac{1}{10}(d(x_0) - d(x_m)) + \frac{2}{10}d(x_m) \leq \frac{1}{10}\|x_0 - x_m\| + \frac{2}{10}d(x_m).$$

Hence $\frac{9}{10}\|x_0 - x_m\| \leq \frac{2}{10}d(x_m)$ and $x_0 \in B(x_m, \frac{2}{9}d(x_m)) \subseteq B(x_m, \frac{1}{4}d(x_m))$. A fortiori, the balls of center x_m and radius $\frac{1}{2}d(x_m)$ also form a covering of $\mathbb{R}^n$. Let us verify that the number of such balls that can have a common point x_0 is bounded by a constant that depends only on the dimension n. Namely, let us consider $M = \{m : x_0 \in B(x_m, \frac{1}{2}d(x_m))\}$ be nonempty. Then for $m \in M$ we have

$$d(x_m) = d(x_m) - d(x_0) + d(x_0) \leq \|x_0 - x_m\| + d(x_0) \leq \frac{1}{2}d(x_m) + d(x_0),$$

hence

$$d(x_m) \leq 2d(x_0),$$

and therefore

$$B\left(x_m, \frac{1}{2}d(x_m)\right) \subseteq B(x_0, 2d(x_0)).$$

The same reasoning shows that

$$d(x_0) \leq \frac{3}{2}d(x_m), \tag{$*$}$$

and therefore

$$B\left(x_m, \frac{1}{10}d(x_m)\right) \supseteq B\left(x_m, \frac{2}{30}d(x_0)\right).$$

The disjointness of the balls $B(x_m, \frac{1}{10}d(x_m))$ $(m \in M)$ implies that the sum of their volumes cannot be bigger than the volume of $B(x_0, 2d(x_0))$, whence the inequality

$$\#(M)\left(\frac{2}{30}d(x_0)\right)^n \le (2d(x_0))^n,$$

or

$$\#(M) \le 30^n.$$

The last property of these balls that we will need is that no $\bar{B}(x_m, \frac{1}{2}d(x_m))$ can simultaneously intersect F and Ω^c. In fact, if there are $x_0 \in \bar{B}(x_m, \frac{1}{2}d(x_m)) \cap F$ and $x_0' \in \bar{B}(x_m, \frac{1}{2}d(x_m)) \cap \Omega^c$, then

$$d(x_m, F) \le \|x_0 - x_m\| \le \frac{1}{2}d(x_m),$$

and

$$d(x_m, \Omega^c) \le \|x_0' - x_m\| \le \frac{1}{2}d(x_m),$$

which contradicts the definition of $d(x_m)$.

Let θ be a standard function and consider the function

$$\psi(x) := \int_{|y| \le 1.5} \theta_{1/2}(x - y)\,dy.$$

(This function is in fact the convolution product of the characteristic function of the ball $\bar{B}(0, 1.5)$ and $\theta_{1/2}$.) It is easy to verify that $\psi \in \mathscr{D}(\mathbb{R}^n)$, $\psi \equiv 1$ on $\bar{B}(0, 1)$, $\operatorname{supp}\psi \subseteq \bar{B}(0, 2)$, $0 \le \psi \le 1$, and we have some constants $c_k' > 0$ such that

$$\left|\frac{\partial^{|\alpha|}}{\partial x^\alpha}\psi(x)\right| \le c_k' \qquad (k = |\alpha|).$$

Now we adapt the function ψ to the balls $B(x_m, \frac{1}{4}d(x_m))$ introducing the functions

$$\psi_m(x) = \psi\left(\frac{4(x - x_m)}{d(x_m)}\right), \qquad m \in \mathbb{N}^*.$$

It is clear that $0 \le \psi_m \le 1$, $\psi_m \equiv 1$ on $B(x_m, \frac{1}{4}d(x_m))$, $\operatorname{supp}\psi_m \subseteq \bar{B}(x_m, \frac{1}{2}d(x_m))$, and that for $x \in \operatorname{supp}\psi_m$, the following inequalities hold:

$$\left|\frac{\partial^{|\alpha|}}{\partial x^\alpha}\psi_m(x)\right| \le c_k' 4^k d(x_m)^{-k} \le c_k' 12^k d(x)^{-k}.$$

The last inequality is a consequence of the above inequality (∗). Finally, for every $x \in \mathbb{R}^n$,

$$1 \le \Psi(x) := \sum_{m \ge 1} \psi_m(x) \le 30^n.$$

The lower bound is a consequence of the fact that $(B(x_m, \frac{1}{4}d(x_m)))_{m \ge 1}$ is a covering of $\mathbb{R}^n$; the upper bound, from the bound on the number of balls of radius $\frac{1}{2}d(x_m)$ intersecting at a single point.

Let $M_0 = \{m : \bar{B}(x_m, \frac{1}{2}d(x_m)) \cap F \neq \varnothing\}$. We define the required function φ by the formula

$$\varphi(x) := \left(\sum_{m \in M_0} \psi_m(x)\right) / \Psi(x).$$

It is easy to verify that φ has all the properties stated in the proposition. □

EXERCISES 1.3

1. Let φ, ψ be functions in $\mathscr{D}(\mathbb{R})$, with supp φ and supp ψ contained the interval $[a, b]$. Let $\alpha = \int_{-\infty}^{\infty} \varphi(x)\,dx$, $\beta = \int_{-\infty}^{\infty} \psi(x)\,dx$. Show that the function

$$\chi(x) = \alpha \int_{-\infty}^{x} \psi(t)\,dt - \beta \int_{-\infty}^{x} \varphi(t)\,dt$$

is also in $\mathscr{D}(\mathbb{R})$, $\mathrm{supp}(\chi) \subseteq [a, b]$.

2. Let $f \in L^1_{\mathrm{loc}}(\Omega)$, Ω open in $\mathbb{R}^n$. Show that if

$$\int_\Omega f\varphi\,dx = 0$$

for every $\varphi \in \mathscr{D}(\Omega)$, then $f = 0$ a.e. in Ω. (Hint: Show first that for every hypercube Q, $\bar{Q} \subseteq \Omega$, one has $\int_Q f\,dx = 0$.)

*3 (*Borel's Lemma*). Let $(a_n)_{n \geq 0}$ be an arbitrary sequence of complex numbers. Show that there is a C^∞ function f in $\mathbb{R}$ such that $f^{(n)}(0) = a_n$, $n = 0, 1, 2, \ldots$. (Hint: Let $\varphi \in \mathscr{D}(]-1, 1[)$ such that $\varphi(x) \equiv 1$ in a neighborhood of 0. Let α_n be a conveniently chosen increasing sequence, $\alpha_n \to \infty$. Define $f(x) = \sum_{n=0}^{\infty} a_n \frac{x^n}{n!} \varphi(\alpha_n x)$.)

4. Let f be a C^∞ function defined in an open set $\Omega \subseteq \mathbb{R}^2$ such that the differential $df(z) \neq 0$ for every $z \in \Omega$. Let $S = \{z \in \Omega : f(z) = 0\}$. Show that
   (i) For every $z \in S$ there is an open set $V_z \subseteq \Omega$ and a C^∞ diffeomorphism $\varphi : V_z \to]-1, 1[\times]-1, 1[$ such that $(f \circ \varphi^{-1})(u, v) = v$.
   (ii) If g is a C^∞ function in the square $]-1, 1[\times]-1, 1[$ which vanishes on the axis $v = 0$, then there is a C^∞ function h such that $g(u, v) = vh(u, v)$ in the square.
   (iii) If $G \in \mathscr{E}(\Omega)$, $G = 0$ on S, then there is $H \in \mathscr{E}(\Omega)$ such that $G = fH$ in Ω.

*5. The goal of this exercise is to construct, without appeal to Zorn's lemma, the maximal sequence $\{x_m\}_{m \geq 1}$ found in the proof of Proposition 1.3.5. We keep the notation from that proposition. Pick an arbitrary point $x_1 \in \mathbb{R}^n$ and proceed by induction. Assume you have already found the first $m - 1$ points $x_1, \ldots, x_{m-1}$ of the sequence, and let us try to find the point x_m as follows:
   (i) Show that the set
$$E_m := \{x \in \mathbb{R}^n : B(x, d(x)/10) \cap B(x_j, d(x_j)/10) = \varnothing, j = 1, \ldots, m - 1\}$$
   is not empty.

(ii) Let $r_m := \inf\{\|x - x_1\| : x \in E_m\}$. Prove there is a point $x_m \in E_m$ such that $\|x_m - x_1\| = r_m$.

(iii) Show that that sequence of positive real numbers $\{r_m\}_{m\geq 2}$ is unbounded.

Deduce that the sequence $\{x_m\}_{m\geq 1}$ is maximal for the property that the balls $B(x_m, d(x_m)/10)$ are pairwise disjoint.

6. Show that a compact subset K of an open set Ω is the support of a function $\varphi \in \mathscr{D}(\Omega)$ if and only if $\overline{\mathring{K}} = K$.

*7. Show that every closed subset F of an open set $\Omega \subseteq \mathbb{C}$ coincides with the set of zeros of a function of class C^∞ in Ω.

§4. Regular Boundaries

1.4.1. Definition. Let Ω be an open subset of $\mathbb{R}^2$. We say that Ω has a ***regular boundary*** of class C^k $(k \geq 1)$ if for every $p \in \partial\Omega$ there is a neighborhood U_p of p and a diffeomorphism φ_p of class C^k from U_p onto a neighborhood V_p of 0 in $\mathbb{R}^2$ such that $\varphi_p(p) = 0$, $\varphi_p(U_p \cap \bar{\Omega}) = V_p \cap \{(x, y) \in \mathbb{R}^2 : x \leq 0\}$, and the Jacobian determinant $J(\varphi_p)$ is > 0 in U_p.

One can assume without loss of generality that $V_p =]-1, 1[\times]-1, 1[$ and that φ_p is still a diffeomorphism in a neighborhood of U_p.

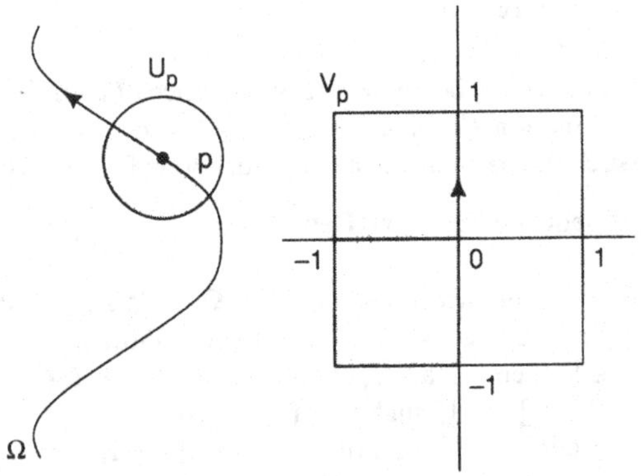

Figure 1.1

1.4.2. Remarks. (1) If $\varphi_p = (\rho_p, \sigma_p)$ then the function $\rho_p : U_p \to \mathbb{R}$ is such that $U_p \cap \Omega = \{\zeta \in U_p : \rho_p(\zeta) < 0\}$ and $U_p \cap \partial\Omega = \{\zeta \in U_p : \rho_p(\zeta) = 0\}$. Furthermore, $\psi_p := \sigma_p|(U_p \cap \partial\Omega) : U_p \cap \partial\Omega \to V_p \cap \{(x, y) \in \mathbb{R}^2 : x = 0\}$ is a homeomorphism onto $]-1, 1[$ such that $\psi_p^{-1} :]-1, 1[\to \Omega$ is of class C^k. Let $\zeta = (\xi, \eta)$ and $y = \psi_q(\zeta)$. Since $\rho_p \circ \psi_p^{-1} \equiv 0$ in $]-1, 1[$ one sees that the vector $\operatorname{grad} \rho_p(\zeta) = \left(\frac{\partial \rho_p}{\partial \xi}(\zeta), \frac{\partial \rho_p}{\partial \eta}(\zeta)\right)$ is orthogonal to $(\psi_p^{-1})'(y)$, the tangent vector to the curve $U_p \cap \partial\Omega$ parameterized by ψ_p^{-1}. As we mentioned earlier, we can suppose that ψ_p^{-1} is a homeomorphism of the closed interval $[-1, 1]$ onto $\psi_p^{-1}([-1, 1]) \subseteq \partial\Omega$. Moreover, the conditon $\det J(\varphi_p) > 0$ means that the basis of $\mathbb{R}^2$ given by $\{\operatorname{grad} \rho_p(\zeta), (\psi_p^{-1})'(y)\}$ is a basis defining the usual orientation of $\mathbb{R}^2$.

(2) If $U_p \cap U_q \neq \emptyset$ for two points $p, q \in \partial\Omega$, then the map

$$\varphi_p \circ \varphi_q^{-1} | \varphi_q(U_p \cap U_q) : \varphi_q(U_p \cap U_q) \to \varphi_p(U_p \cap U_q)$$

is a diffeomorphism of class C^k.

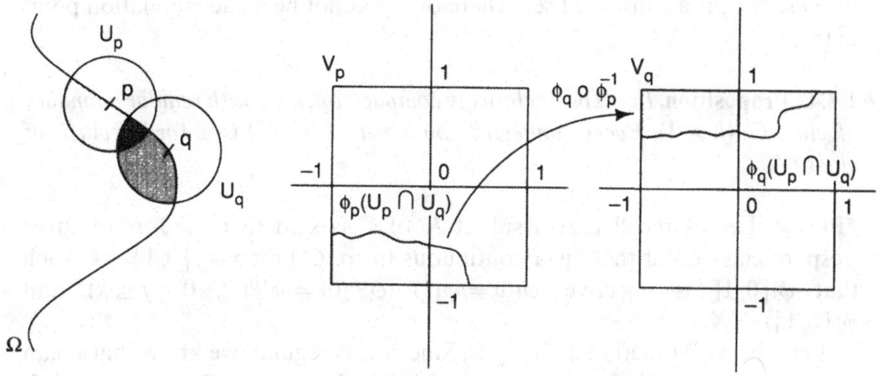

Figure 1.2

(3) There is a map $\rho : \mathbb{R}^2 \to \mathbb{R}$ of class C^k such that $\Omega = \{\zeta : \rho(\zeta) < 0\}$, $\partial\Omega = \{\zeta : \rho(\zeta) = 0\}$ and, furthermore, $d\rho(\zeta) \neq 0$ if $\zeta \in \partial\Omega$. To see this, consider a covering of $\partial\Omega$ by open sets U_p ($p \in \partial\Omega$) as those obtained earlier and let $(\beta_0, \beta_1, \alpha_p : p \in \partial\Omega)$ be a C^∞ partition of unity subordinate to the covering $\{\Omega, \mathbb{R}^2 \backslash \bar{\Omega}, U_p (p \in \partial\Omega)\}$ of $\mathbb{R}^2$, as given by Proposition 1.3.4. Hence, $\operatorname{supp} \beta_0 \subseteq \Omega$, $\operatorname{supp} \beta_1 \subseteq \mathbb{R}^2 \backslash \bar{\Omega}$, and $\operatorname{supp} \alpha_p \subseteq U_p$. Set

$$\rho(\zeta) := -\beta_0(\zeta) + \beta_1(\zeta) + \sum_p \alpha_p(\zeta)\rho_p(\zeta),$$

which is evidently of class C^k. By (1) if $\zeta \in \partial\Omega$ all the $d\rho_p(\zeta)$ for which $\alpha_p(\zeta) \neq 0$ are different from zero and proportional to each other with a positive constant of proportionality. It follows that $d\rho(\zeta) \neq 0$ for every $\zeta \in \partial\Omega$. It is clear that if $\zeta \in \Omega$ then $\rho(\zeta) < 0$. Conversely, if $\zeta \notin \bar{\Omega}$ one can easily see $\rho(\zeta) > 0$.

One can also see without difficulty that, given $\rho : \mathbb{R}^2 \to \mathbb{R}$ of class C^k ($k \geq 1$) such that when Ω is defined by $\Omega := \{\zeta : \rho(\zeta) < 0\}$ one has $d\rho(\zeta) \neq 0$ for $\zeta \in \partial\Omega$, then Ω is an open set with regular boundary of class C^k. We remark that if ρ_1 and ρ_2 define the same Ω, then $\rho_1 = h\rho_2$ for some strictly positive function h (see Exercise 1.3.4).

Finally, we note that one can also consider an open subset ω of an open set Ω, and say that ω has a regular boundary relative to Ω, if the relative boundary $\partial_\Omega \omega$ is regular in the preceding sense. Since we are only going to use this for ω relatively compact in Ω, the two notions coincide.

1.4.3. Proposition. *Let Ω be a relatively compact, open set with regular boundary. Then, the number of connected components of $\partial\Omega$ is finite.*

PROOF. Let us suppose that the set $(C_i)_{i \in I}$ of connected components of $\partial\Omega$ is infinite. Choose $z_i \in C_i$. The family $(z_i)_{i \in I}$ admits an accumulation point $z \in \partial\Omega$. By definition of regular boundary, we can choose a neighborhood U_z of z such that $\partial\Omega \cap U_z$ is a connected set. Therefore z cannot be an accumulation point of $(z_i)_{i \in I}$. □

1.4.4. Proposition. *Let Ω be a relatively compact, open set with regular boundary of class C^k ($k \geq 1$). Every connected component Γ of $\partial\Omega$ is a Jordan curve of class C^k.*

*PROOF. Let us recall that a subset K of $\mathbb{C}$ is said to be a Jordan curve (resp. of class C^k) if there is a continuous (resp. C^k) map $\varphi: [0, 1] \to \mathbb{C}$ such that $\varphi|[0, 1[$ is injective, $\varphi(0) = \varphi(1)$ $(\varphi^{(j)}(0) = \varphi^{(j)}(1),\ 0 \leq j \leq k)$, and $\varphi([0, 1]) = K$.

Let ρ be a C^k function defining Ω. Since $\partial\Omega$ is regular, we know that Γ can be represented as the finite union of a family of open arcs $\Gamma_1, \ldots, \Gamma_N$, $N \geq 2$, of the form $\Gamma_i = \varphi_i(]-1, 1[)$, where each φ_i is the restriction to $]-1, 1[$ of an injective function, still denoted φ_i, $\varphi_i: [-1, 1] \to \Gamma$ which is not surjective. Moreover, for any $t \in [-1, 1]$, the pair (grad $\rho(\varphi_i(t)), \varphi_i'(t))\}$ is an orthogonal basis of $\mathbb{R}^2$ with the canonical orientation. We can also assume that $\Gamma_i \not\subseteq \bigcup_{j \neq i} \Gamma_j$ $(1 \leq i \leq N)$ and that every $p \in \Gamma$ has a neighborhood U_p such that if $p \in \Gamma_i$ then

$$U_p \cap \Gamma_i = U_p \cap \Gamma. \qquad (*)$$

We shall show by induction on N that Γ is a Jordan curve of class C^k.

Consider first the case $N = 2$. Then $\Gamma = \Gamma_1 \cup \Gamma_2$; the connectedness of Γ implies $\Gamma_1 \cap \Gamma_2 \neq \varnothing$. Thus $\varphi_1^{-1}(\Gamma_2)$ is a nonempty open proper subset of $]-1, 1[$. Therefore $\varphi_1^{-1}(\Gamma_2)$ is a countable union of disjoint, nonempty open intervals, $\varphi_1^{-1}(\Gamma_2) = \bigcup_{n \geq 1}]a_n, b_n[$.

Suppose $a_n > -1$. We shall show that $b_n = 1$. Since neither $\varphi_1(a_n)$ nor $\varphi_1(b_n)$ belongs to Γ_2, we have that $\varphi_1(]a_n, b_n[) = \varphi_1([a_n, b_n]) \cap \Gamma_2$. Hence $\varphi_1(]a_n, b_n[)$ is a closed subarc of Γ_2. If $b_n < 1$, then $\varphi_1(]a_n, b_n[)$ would also be open in Γ_2, whence $\Gamma_2 = \varphi_1(]a_n, b_n[)$. It would follow that $\Gamma_2 \subseteq \Gamma_1$, which is impossible. This argument shows that $b_n = 1$ when $a_n > -1$. A similar argument shows that if $b_n < 1$, then $a_n = -1$. We conclude that $\varphi_1^{-1}(\Gamma_2)$ is the union of at most two disjoint nonempty intervals $]-1, b[$, $]a, 1[$, with $b \leq a$. The set $\varphi_1^{-1}(\Gamma_2)$ is relatively open in $[-1, 1]$ and both points $\varphi_1(-1)$ and $\varphi_1(1)$ belong to Γ_2. Therefore $\varphi_1^{-1}(\Gamma_2)$ must have exactly two components.

Clearly we also have that $\varphi_2^{-1}(\Gamma_1)$ is the union of two disjoint intervals $]-1,\beta[$ and $]\alpha,1[$, with $-1<\beta\le\alpha<1$. Therefore, $\Gamma_1\cap\Gamma_2$ is the disjoint union of the arcs $\varphi_1(]-1,b[)$ and $\varphi_1(]a,1[)$, and it is also the disjoint union of the arcs $\varphi_2(]-1,\beta[)$ and $\varphi_2(]\alpha,1[)$. We claim that

$$\varphi_1(]-1,b[)=\varphi_2(]\alpha,1[)\qquad\text{and}\qquad\varphi_1(]a,1[)=\varphi_2(]-1,\beta[).$$

Otherwise, $\varphi_1(]-1,b[)=\varphi_2(]-1,\beta[)$ and $\varphi_1(]a,1[)=\varphi_2(]\alpha,1[)$. Hence $\varphi_2^{-1}\circ\varphi_1$ is a diffeomorphism from $]-1,b[$ onto $]-1,\beta[$, and from $]a,1[$ onto $]\alpha,1[$. This function is strictly increasing because its derivative is positive. In fact, for $t\in]-1,b[\cup]a,1[$ and $s=\varphi_2^{-1}(\varphi_1(t))$, we have that $\{\operatorname{grad}\rho(\varphi_1(t)),\varphi_1'(t)\}$ and $\{\operatorname{grad}\rho(\varphi_2(s)),\varphi_2'(s)\}$ are bases of $\mathbb{R}^2$ with the same orientation. Since $\varphi_1(t)=\varphi_2(s)$, this implies that $\varphi_1'(t)=(\varphi_2^{-1}\circ\varphi_1)'(t)\varphi_2'(s)$ is a positive scalar multiple of $\varphi_2'(s)$, i.e., $(\varphi_2^{-1}\circ\varphi_1)'(t)>0$.

By continuity, we obtain $\varphi_1(-1)=\varphi_2(-1)$, $\varphi_1(1)=\varphi_2(1)$, $\varphi_1(b)=\varphi_2(\beta)$, $\varphi_1(a)=\varphi_2(\alpha)$. A quick look at Figure 1.3 will convince the reader that this is a contradiction with the assumption (∗) at the point $p=\varphi_1(b)$.

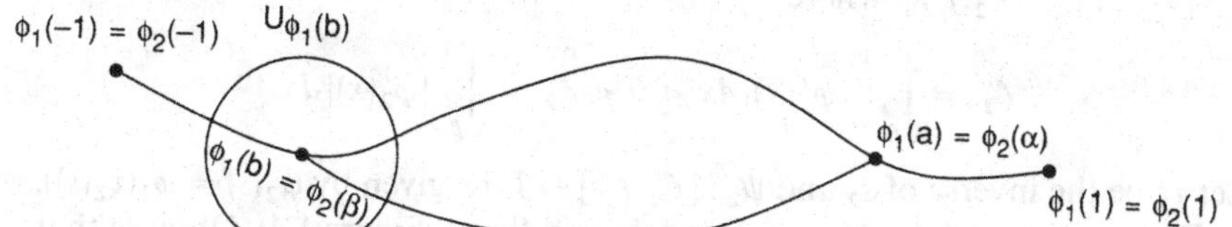

Figure 1.3

In conclusion, $\varphi_1(]-1,b[)=\varphi_2(]\alpha,1[)$ and $\varphi_1(]a,1[)=\varphi_2(]-1,\beta[)$. Moreover, since $\varphi_2^{-1}\circ\varphi_1$ is strictly increasing, $\varphi_1(-1)=\varphi_2(\alpha)$, $\varphi_1(b)=\varphi_2(1)$, $\varphi_1(a)=\varphi_2(-1)$, and $\varphi_1(1)=\varphi_2(\beta)$. From Figure 1.4 we see now that Γ is a Jordan curve.

Now it is not difficult to show that Γ is a C^k Jordan curve. For that purpose

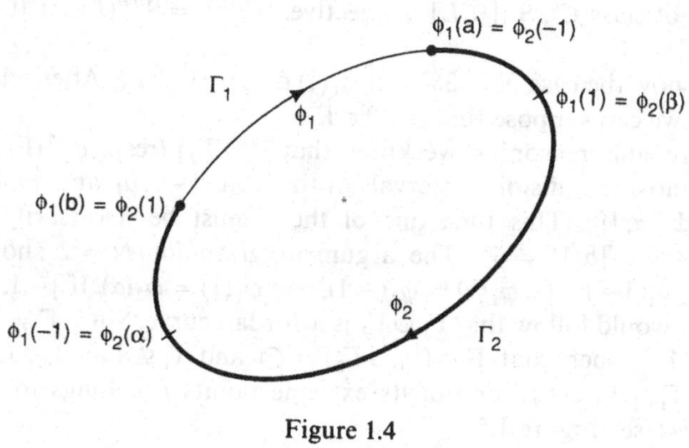

Figure 1.4

we will reparameterize Γ_1 and Γ_2 using arc length parameters in such a way that we can patch these parameterizations into a C^k map defining a C^k Jordan curve.

First consider s_1, the arc length parameter for Γ_1, defined by

$$s_1(t) = \int_1^t |\varphi_1'(x)|\,dx.$$

Then $s_1 : [-1, 1] \to [0, \ell_1]$, where

$$\ell_1 := \int_1^1 |\varphi_1'(x)|\,dx = \ell(\Gamma_1).$$

Let τ_1 be the inverse of s_1 and $\psi_1 : [0, \ell_1] \to \Gamma$ given by $\psi_1(s) = \varphi_1(\tau_1(s))$.

We also consider the arc length parameter for Γ_2, defined by

$$s_2(t) = \int_\beta^t |\varphi_2'(x)|\,dx;$$

$s_2 : [-1, 1] \to [\ell_2', \ell_2'']$, where

$$\ell_2' := \int_\beta^{-1} |\varphi_2'(x)|\,dx < 0 < \ell_2'' := \int_\beta^1 |\varphi_2'(x)|\,dx.$$

Let τ_2 be the inverse of s_2 and $\psi_2 : [\ell_2', \ell_2''] \to \Gamma$ be given by $\psi_2(s) = \varphi_2(\tau_2(s))$.

Denote $\ell_1' := s_1(a)$, $\ell_2 = s_2(\alpha)$, and $\ell_0 = s_1(b)$ (cf. Figure 1.4). Observe that $\psi_1(\ell_1 + s)$, $\ell_1' - \ell_1 \le s \le 0$, and $\psi_2(s)$, $\ell_2' \le s \le 0$, are both arc length parameterizations of the arc $\varphi_1([a, 1]) = \varphi_2([-1, \beta])$, with $\psi_1(\ell_1) = \varphi_1(1) = \varphi_2(\beta) = \psi_2(0)$. Therefore, these two parameterizations must coincide. Hence $\ell_1' - \ell_1 = \ell_2'$ and $\psi_1(\ell_1 + s) = \psi_2(s)$, for $\ell_2' \le s \le 0$. A similar argument shows that $\ell_0 = \ell_2'' - \ell_2$ and $\psi_1(s) = \psi_2(\ell_2 + s)$, for $0 \le s \le \ell_0$.

Define now $L = \ell_1 + \ell_2$ and a map $\Psi : [0, L] \to \mathbb{C}$ by

$$\Psi(s) = \begin{cases} \psi_1(s) & \text{if } 0 \le s \le \ell_1 \\ \psi_2(s - \ell_1) & \text{if } \ell_1 \le s \le L. \end{cases}$$

This map is of class C^k, $\Psi|[0, L]$ is injective, $\Psi^{(j)}(0) = \Psi^{(j)}(L)$ for $0 \le j \le k$, and $\Psi([0, L]) = \Gamma$.

Consider now the case $N \ge 3$. Then $\varphi_1(1) \in \Gamma_2 \cup \cdots \cup \Gamma_N$. After relabeling, if necessary, we can suppose that $\varphi_1(1) \in \Gamma_2$.

By the previous reasoning we know that $\varphi_1^{-1}(\Gamma_2)$ (resp. $\varphi_2^{-1}(\Gamma_1)$) is the union of at most two disjoint intervals of the form $]-1, b[$ and $]a, 1[$ (resp. $]-1, \beta[$ and $]\alpha, 1[$). This time one of them must be necessarily empty. Since $\varphi_1(1) \in \Gamma_2$, $]b, 1[\ne \varnothing$. The argument given for $N = 2$ shows that $\varphi_1(]b, 1[) = \varphi_2(]-1, \alpha[)$, $\varphi_1(b) = \varphi_2(-1)$, and $\varphi_1(1) = \varphi_2(\alpha)$. If $]-1, a[$ were not empty, it would follow that $\Gamma_1 \cup \Gamma_2$ is a Jordan curve. Since $\Gamma \ne \Gamma_1 \cup \Gamma_2$, there is $i \ne 1, 2$ such that $\Gamma_i \cap (\Gamma_1 \cup \Gamma_2) \ne \varnothing$ and $\Gamma_i \nsubseteq \Gamma_1 \cup \Gamma_2$. Consider an arc $\gamma_i \subseteq \Gamma_i \cap (\Gamma_1 \cup \Gamma_2)$, one of its extreme points p belongs to Γ_i. This contradicts $(*)$ (see Figure 1.5).

Therefore, using the same method as in the case $N = 2$, we can construct

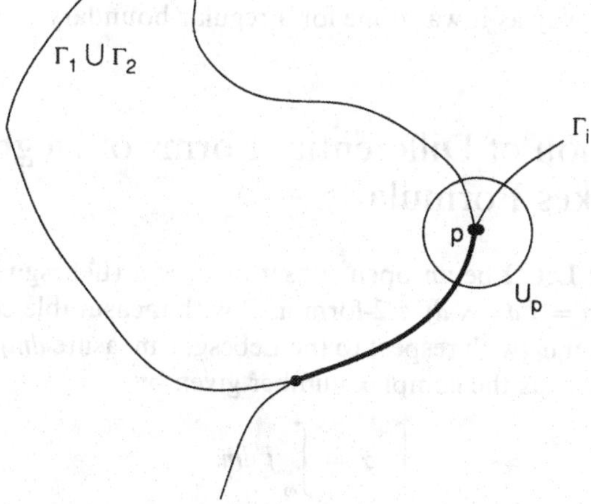

Figure 1.5

an injective C^k function $\Psi_{1,2}:[-1,1]\to\Gamma$ such that $\Psi_{1,2}(]-1,1[)=\Gamma_1\cup\Gamma_2$ and $\{\operatorname{grad}\rho(\Psi_{1,2}(t)),\Psi'_{1,2}(t)\}$ is an orthogonal basis of $\mathbb{R}^2$ with the canonical orientation for every $t\in[-1,1]$. Applying now the inductive hypothesis to $\Gamma_1\cup\Gamma_2,\Gamma_3,\ldots,\Gamma_N$, we conclude that Γ is a C^k Jordan curve. □

1.4.5. Remark. If $\rho:\mathbb{C}\to\mathbb{R}$ is a function of class C^k defining Ω and if $\varphi:[0,1]\to\mathbb{C}$ is a parameterization of a component Γ of $\partial\Omega$ of class C^k (i.e., $\varphi|[0,1[$ injective, $\varphi^{(j)}(0)=\varphi^{(j)}(1)$, $0\le j\le k$), then one can choose φ so that for every point $p\in\Gamma$ the pair $\{\operatorname{grad}\rho(p),\varphi'(\varphi^{-1}(p))\}$ is an orthogonal basis of $\mathbb{R}^2$ with the canonical orientation. This determines an orientation for Γ, independent of the choice of ρ and φ.

We also note that Proposition 1.4.4 does not really use that Ω is relatively compact.

1.4.6. Definition. An open subset Ω has a ***piecewise regular boundary*** (of class C^k, $k\ge1$) if, for every point $p\in\partial\Omega$, there is a neighborhood U_p of p and a diffeomorphism φ_p of class C^k from U_p onto $]-1,1[\times]-1,1[$ such that $\varphi_p(p)=0$, $J(\varphi_p)>0$, and $\varphi_p(U_p\cap\bar\Omega)$ is one of the following sets:

(1) $\varphi_p(U_p\cap\bar\Omega)=]-1,0]\times]-1,1[$
(2) $\varphi_p(U_p\cap\bar\Omega)=]-1,0]\times]-1,0]$
(3) $\varphi_p(U_p\cap\bar\Omega)=(]-1,1[\times]-1,1[)\backslash(]0,1[\times]0,1[)$

1.4.7. Proposition. *Let Ω be a relatively compact, open subset of $\mathbb{C}$ with piecewise regular boundary (of class $C^k\ge1$). There is only a finite number of connected components of $\partial\Omega$ and each of them is a Jordan curve (piecewise C^k).*

PROOF. The proof is analogous to the proofs of 1.4.3 and 1.4.4. □

1.4.8. Remark. One can give an orientation to $\partial\Omega$ in the case of Proposition 1.4.7 in the same way as it was done for a regular boundary.

§5. Integration of Differential Forms of Degree 2: The Stokes Formula

1.5.1. Definition. Let Ω be an open subset of $\mathbb{C}$, ω a (Lebesgue) measurable subset of Ω and $\alpha = f\,dx \wedge dy$ a 2-form in Ω with measurable coefficient f. If f is integrable over ω (with respect to the Lebesgue measure dm) we define the integral of α over ω as the complex number given by

$$\int_\omega \alpha := \int_\omega f\,dm.$$

1.5.2. Proposition. *Let $\varphi : \Omega_1 \to \Omega_2$ be a diffeomorphism of class C^1 between two open subsets of $\mathbb{C}$ such that $J(\varphi) > 0$, let ω_i be open subsets of $\Omega_i (i = 1, 2)$ such that $\varphi(\omega_1) = \omega_2$, and let $\alpha = f\,dx \wedge dy$ be a 2-form with f measurable on Ω_2 and integrable over ω_2. Let $\varphi(\zeta, \eta) = (x, y)$, then $\varphi^*(\alpha) = (f \circ \varphi)J(\varphi)\,d\xi \wedge d\eta$, $(f \circ \varphi)J(\varphi)$ is measurable on Ω_1, integrable over ω_1, and $\int_{\omega_1} \varphi^*(\alpha) = \int_{\omega_2} \alpha$.*

Proof. In fact, the formula of change of variables for the Lebesgue integral becomes here

$$\int_{\omega_2} f\,dm = \int_{\omega_1} (f \circ \varphi)J(\varphi)\,dm,$$

which can be translated into

$$\int_{\omega_2} \alpha = \int_{\omega_1} \varphi^*(\alpha).$$ □

1.5.3. Proposition (The Stokes Formula). *Let Ω be an open subset of $\mathbb{C}$, and ω a relatively compact, open subset of Ω with piecewise regular boundary. Let γ be a 1-form of class C^1 in Ω. We have the relation*

$$\int_\omega d\gamma = \int_{\partial\omega} \gamma,$$

where $\int_{\partial\omega} \gamma$ represents $\sum_{1 \le i \le n} \int_{\Gamma_i} \gamma$, the Γ_i being the connected components of $\partial\omega$, canonically oriented.

Proof. For each $p \in \partial\omega$ we can find an open neighborhood U_p of p in Ω and a diffeomorphism φ_p of U_p onto $]-1, 1[\times]-1, 1[= V$ such that $\varphi_p(U_p \cap \bar{\omega})$

is of one of three types indicated in §1.4.6. We can find a finite number of points $p_1, \ldots, p_N$ so that $U_{p_1}, \ldots, U_{p_N}$ is a covering of $\partial\omega$. Hence the family of open sets $\Omega\backslash\bar{\omega}, \omega, U_{p_1}, \ldots, U_{p_N}$ is a covering of Ω. Let $(\alpha_j)_{j\geq 1}$ be a C^∞ partition of unity subordinate to that covering. We have $\gamma = \sum_{j\geq 1} \alpha_j\gamma$ and $d\gamma = \sum_{j\geq 1} d(\alpha_j\gamma)$.

In order to compute $\int_\omega d(\alpha_j\gamma)$ we need to consider three cases.

(1) First case: $\operatorname{supp}\alpha_j \subseteq \omega$. We have here $\int_\omega d(\alpha_j\gamma) = \int_C d(\alpha_j\gamma)$. If $\alpha_j\gamma = P\,dx + Q\,dy$, one can write

$$\int_C d(\alpha_j\gamma) = \int_C \left(\frac{\partial Q}{\partial x} - \frac{\partial P}{\partial y}\right) dx\,dy$$
$$= \int_{-\infty}^{\infty}\left(\int_{-\infty}^{\infty} \frac{\partial Q}{\partial x}dx\right)dy - \int_{-\infty}^{\infty}\left(\int_{-\infty}^{\infty} \frac{\partial P}{\partial y}dy\right)dx = 0,$$

since P, Q are C^1 functions of compact support. Therefore, in this case, have

$$0 = \int_\omega d(\alpha_j\gamma) = \int_{\partial\omega} \alpha_j\gamma,$$

since the form $\alpha_j\gamma$ vanishes on $\partial\omega$.

(2) Second case: there is an index k such that $\operatorname{supp}\alpha_j \subseteq U_{p_k}$.

There are three subcases depending on the type of $\varphi_{p_k}(U_{p_k} \cap \bar{\omega})$ according to §1.4.6.

(1) $\varphi_{p_k}(U_{p_k} \cap \bar{\omega}) =]-1,0] \times]-1,1[= V'$ (see Figure 1.6).

We have

$$\int_\omega d(\alpha_j\gamma) = \int_{U_{p_k}\cap\omega} d(\alpha_j\gamma) = \int_{V'} (\varphi_{p_k}^{-1})^*(d(\alpha_j\gamma)) = \int_{V'} d[(\varphi_{p_k}^{-1})^*(\alpha_j\gamma)].$$

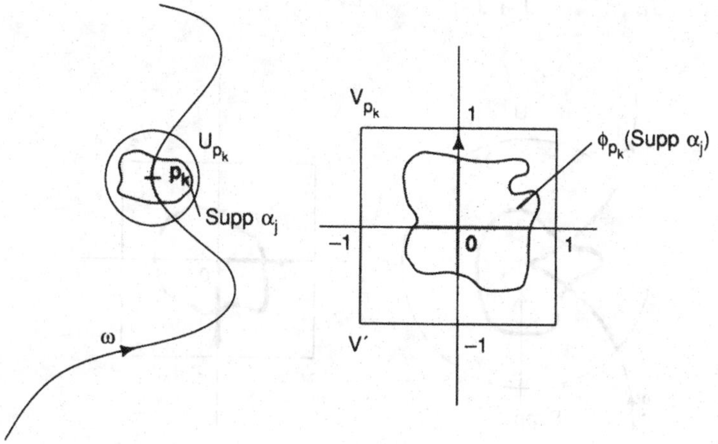

Figure 1.6

Writing $(\varphi_{p_k}^{-1})^*(\alpha_j\gamma) = P\,dx + Q\,dy$, the preceding expression becomes

$$\int_\omega d(\alpha_j\gamma) = \int_{-1}^{1}\left\{\int_{-1}^{0}\left(\frac{\partial Q}{\partial x} - \frac{\partial P}{\partial y}\right)dx\right\}dy$$
$$= \int_{-1}^{1}\left\{\int_{-1}^{0}\frac{\partial Q}{\partial x}dx\right\}dy - \int_{-1}^{0}\left\{\int_{-1}^{1}\frac{\partial P}{\partial y}dy\right\}dx$$
$$= \int_{-1}^{1} Q(0,y)\,dy.$$

If ψ denotes $\varphi_{p_k}^{-1}$ restricted to $\{0\} \times]-1,1[$, one has

$$\psi^*(\alpha_j\gamma|(U_{p_k} \cap \partial\omega)) = Q(0,y)\,dy$$

and hence

$$\int_{U_{p_k}\cap\partial\omega} \alpha_j\gamma = \int_{-1}^{1} \psi^*(\alpha_j\gamma|(U_{p_k} \cap \partial\omega)) = \int_{-1}^{1} Q(0,y)\,dy.$$

Therefore one has

$$\int_\omega d(\alpha_j\omega) = \int_{U_{p_k}\cap\partial\omega} \alpha_j\gamma = \int_{\partial\omega} \alpha_j\gamma.$$

(2) $\varphi_{p_k}(U_{p_k} \cap \bar{\omega}) =]-1,0] \times]-1,0] = V''$ (see Figure 1.7).
Here we have

$$\int_\omega d(\alpha_j\gamma) = \int_{U_{p_k}\cap\omega} d(\alpha_j\gamma) = \int_{V''} (\varphi_{p_k}^{-1})^*(d(\alpha_j\gamma)) = \int_{V''} d[(\varphi_{p_k}^{-1})^*(\alpha_j\gamma)].$$

Again writing $(\varphi_{p_k}^{-1})^*(\alpha_j\gamma) = P\,dx + Q\,dy$, this expression becomes

$$\int_\omega d(\alpha_j\gamma) = \int_{-1}^{0}\int_{-1}^{0}\left(\frac{\partial Q}{\partial x} - \frac{\partial P}{\partial y}\right)dx\,dy$$

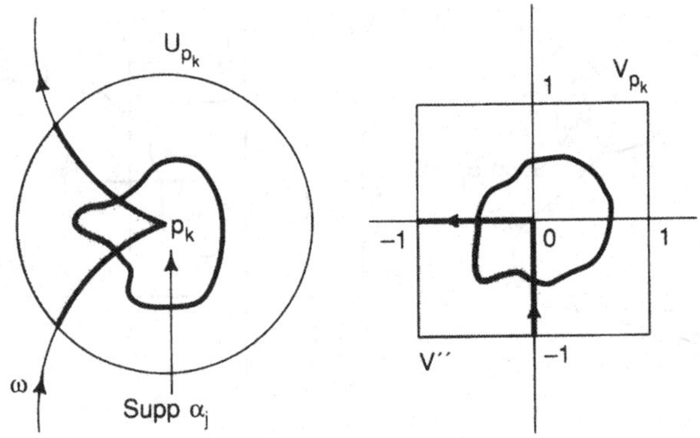

Figure 1.7

$$= \int_{-1}^{0}\left(\int_{-1}^{0}\frac{\partial Q}{\partial x}\,dx\right)dy - \int_{-1}^{0}\left(\int_{-1}^{0}\frac{\partial P}{\partial y}\,dy\right)dx$$

$$= \int_{-1}^{0} Q(0,y)\,dy - \int_{-1}^{0} P(x,0)\,dx.$$

If ψ denotes $\varphi_{p_k}^{-1}|\{0\}\times]-1,1[$ and θ denotes $\varphi_{p_k}^{-1}|]-1,1[\times\{0\}$, then $\psi^*(\alpha_j\gamma|(U_{p_k}\cap\partial\omega)) = Q(0,y)\,dy$ and $\theta^*(\alpha_j\gamma|(U_{p_k}\cap\partial\omega)) = P(x,0)\,dx$. Therefore

$$\int_{-1}^{0} Q(0,y)\,dy - \int_{-1}^{0} P(x,0)\,dx = \int_{\partial\omega}\alpha_j\gamma,$$

as shown earlier.

(3) $\varphi_{p_k}(U_{p_k}\cap\bar{\omega}) = V\backslash(]0,1[\times]0,1[) = V'''$ (see Figure 1.8). We leave this subcase to the reader.

(3) Third case: $\operatorname{supp}\alpha_j \subseteq \Omega\backslash\bar{\omega}$. Here we immediately have

$$\int_{\partial\omega}\alpha_j\gamma = \int_{\omega} d(\alpha_j\gamma) = 0.$$

Therefore, in every case we have the identity $\int_{\omega} d(\alpha_j\gamma) = \int_{\partial\omega}\alpha_j\gamma$. The proposition follows by summation over j. □

1.5.6. Corollary (Ostrogradski's Formula). *Let $z\mapsto A(z) = (A_1(z), A_2(z))$ be a C^1-vector field on an open set Ω in $\mathbb{C}$ (i.e., the coordinates $A_j(z)$ of the vector $A(z)$ are functions of class C^1). Let ω be a relatively compact open subset of Ω with piecewise regular boundary of class C^1. One has the relation*

$$\int_{\omega}\operatorname{div} A\,dx\,dy = \int_{\partial\omega}(A(z)|n(z))\,|dz|,$$

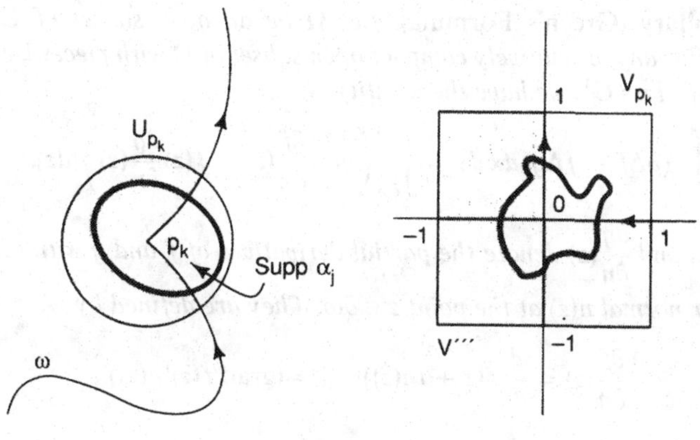

Figure 1.8

where $\operatorname{div} A = \dfrac{\partial A_1}{\partial x} + \dfrac{\partial A_2}{\partial y}$ *is the divergence of the vector field* A, $n(z)$ *denotes the exterior unit normal to* $\partial\omega$ *at the point* $z \in \partial\omega$, *and* $(A|n)$ *denotes the scalar product of the two vectors.*

If the component of $\partial\omega$ is parameterized near z by $\varphi : [0, 1] \to \Omega$, then one has that $n(\varphi(t))$ can be identified to the complex number $-i\varphi'(t)/|\varphi'(t)|$, $|dz| = |\varphi'(t)|\, dt$, and

$$(A(\varphi(t))|n(\varphi(t))) = \frac{A_1(\varphi(t)) \cdot \operatorname{Im} \varphi'(t) - A_2(\varphi(t)) \cdot \operatorname{Re} \varphi'(t)}{|\varphi'(t)|}.$$

Also note that $n(z)$ is well defined with the possible exception of finitely many $z \in \partial\omega$.

Proof. If $(\Gamma_j)_{1 \le j \le n}$ is the family of connected components of $\partial\omega$, and $\varphi_j = x_j + iy_j$ is their piecewise-C^1 parametric representation (with the canonical orientation), one has

$$\begin{aligned}\int_{\partial\omega} (A(z)|n(z))\,|dz| &= \sum_{1 \le j \le n} \int_{\Gamma_j} (A(\varphi_j(t))|n(\varphi_j(t)))|\, dz(\varphi_j(t))| \\ &= \sum_{1 \le j \le n} \int_0^1 (A_1(\varphi_j(t))y_j'(t) - A_2(\varphi_j(t))x_j'(t))\, dt \\ &= \sum_{1 \le j \le n} \int_0^1 (A_1\, dy - A_2\, dx) = \int_{\partial\omega} \gamma,\end{aligned}$$

where we have introduced the 1-form $\gamma = A_1\, dy - A_2\, dx$. This form has the property that $d\gamma = \left(\dfrac{\partial A_1}{\partial x} + \dfrac{\partial A_2}{\partial y}\right) dx \wedge dy$. Therefore, by Stokes' formula,

$$\int_\omega \operatorname{div} A\, dx\, dy = \int_\omega d\gamma = \int_{\partial\omega} \gamma = \int_{\partial\omega} (A(z)|n(z))\,|dz|. \qquad \square$$

1.5.7. Corollary (Green's Formula). *Let* Ω *be an open subset of* $\mathbb{C}$ *and* f, $g \in \mathscr{E}_2(\Omega)$. *For any* ω *relatively compact open subset of* Ω *with piecewise regular boundary of class* C^1, *we have the identity*

$$\int_\omega (g\Delta f - f\Delta g)\, dx\, dy = \int_{\partial\omega} \left(g(z)\frac{\partial f}{\partial n}(z) - f(z)\frac{\partial g}{\partial n}(z)\right)|dz|,$$

where $\dfrac{\partial f}{\partial n}(z)$ *and* $\dfrac{\partial g}{\partial n}(z)$ *denote the partial derivatives of* f *and* g *with respect to the exterior normal* $n(z)$ *at the point* $z \in \partial\omega$. *They are defined by*

$$\frac{\partial f}{\partial n}(z) = \frac{d}{dt} f(z + tn(z))|_{t=0} = (\operatorname{grad} f(z)|n(z))$$

$$\frac{\partial g}{\partial n}(z) = \frac{d}{dt} g(z + tn(z))|_{t=0} = (\operatorname{grad} g(z)|n(z)).$$

The symbol Δ *represents the Laplace (differential) operator,*

$$\Delta = \frac{\partial^2}{\partial x^2} + \frac{\partial^2}{\partial y^2} = 4\frac{\partial^2}{\partial z\,\partial \bar{z}}.$$

PROOF. Use §1.5.6 with $A = g\cdot \operatorname{grad} f - f\cdot \operatorname{grad} g$. □

1.5.8. Corollary (Gauss' Formula). *With the same hypotheses as in* §1.5.7, *we have*

$$\int_\omega \Delta f\,dx\,dy = \int_{\partial\omega} \frac{\partial f}{\partial n}(z)|dz|.$$

1.5.9. Corollary. *Let* Ω *be an open set in* $\mathbb{C}$, $\alpha \in \mathscr{E}_1^1(\Omega)$ *a* 1*-form of class* C^1 *with compact support in* Ω. *Then*

$$\int_\Omega d\alpha = 0.$$

PROOF. The form α can be considered as an element of $\mathscr{E}_1^1(\mathbb{C})$ (extended to be zero outside Ω, as we have done before). For $R > 0$ sufficiently large we have

$$\int_\Omega d\alpha = \int_{B(0,R)} d\alpha = \int_{\partial B(0,R)} \alpha = 0. \qquad \square$$

We will denote by $\mathscr{D}_k^0(\Omega)$, or for simplicity $\mathscr{D}_k(\Omega)$, the set of 0-forms (i.e., functions) of class C^k and compact support in Ω. Similarly one denotes $\mathscr{D}_k^1(\Omega)$ (and $\mathscr{D}_k^2(\Omega)$) the set of 1-forms (and 2-forms) of class C^k and compact suport in Ω. We omit the subscript k when $k = \infty$.

We are going to consider the following two sequences of $\mathbb{C}$-linear mappings:

$$\mathscr{E}^0(\Omega) \xrightarrow{d} \mathscr{E}^1(\Omega) \xrightarrow{d} \mathscr{E}^2(\Omega),$$
$$\mathscr{D}^0(\Omega) \xrightarrow{d} \mathscr{D}^1(\Omega) \xrightarrow{d} \mathscr{D}^2(\Omega),$$

where d denotes the differential of functions or 1-forms, according to case. Recall that $d^2 = 0$, hence the image of the first d is contained in the kernel of the next one. One says the sequence is ***exact*** if these two spaces coincide. To measure how much these sequences deviate from exactness one introduces the following vector spaces:

$Z^1(\Omega) := \operatorname{Ker}[d: \mathscr{E}^1(\Omega) \to \mathscr{E}^2(\Omega)]$, space of the ***1-cocycles.***
$B^1(\Omega) := \operatorname{Im}[d: \mathscr{E}^0(\Omega) \to \mathscr{E}^1(\Omega)]$, space of the ***1-coboundaries.***
$H^1(\Omega) := Z^1(\Omega)/B^1(\Omega)$, the ***first de Rham cohomology vector space*** *of* Ω.
$Z^2(\Omega) := \mathscr{E}^2(\Omega)$.
$B^2(\Omega) := \operatorname{Im}[d: \mathscr{E}^1(\Omega) \to \mathscr{E}^2(\Omega)]$, space of the ***2-coboundaries.***
$H^2(\Omega) := Z^2(\Omega)/B^2(\Omega)$, the ***second de Rham cohomology vector space*** *of* Ω.
$H^0(\Omega) := Z^0(\Omega) := \operatorname{Ker}[d: \mathscr{E}^0(\Omega) \to \mathscr{E}^1(\Omega)]$, the ***zeroth de Rham cohomology vector space*** *of* Ω.

Note that the elements of $H^0(\Omega)$ are locally constant functions, therefore $H^0(\Omega)$ can be identified with the Cartesian product $\prod_{i \in J} \mathbb{C}_i$, where J is the set of connected components of Ω and $\mathbb{C}_i$ is a copy of $\mathbb{C}$.

One defines analogously the spaces $Z_c^j(\Omega)$, $B_c^j(\Omega)$, $H_c^j(\Omega)$, by replacing $\mathscr{E}$ with $\mathscr{D}$ everywhere. One calls them the same way, adding "with compact support." Note that $H_c^0(\Omega) = 0$.

The dimension (as a vector space over $\mathbb{C}$) of $H^0(\Omega)$ is denoted by b_0 and called the ***zeroth Betti number*** of Ω. It is exactly the number of connected components of Ω. Similarly, the ***jth Betti number*** of Ω is defined to be $b_j := \dim_{\mathbb{C}} H^j(\Omega)$ $(j = 1, 2)$.

It is standard to call ***domain*** a nonempty connected open set in $\mathbb{C}$ (later we will use this name for subsets of the sphere S^2.)

It is easy to see that if $\varphi : \Omega \to \Omega'$ is a C^∞ diffeomorphism between two open subsets of $\mathbb{C}$, the inverse image maps φ^* induce isomorphisms

$$\tilde{\varphi}^* : H^j(\Omega') \to H^j(\Omega) \qquad (j = 0, 1, 2),$$

by passage to the quotient. One needs only to recall the commutation relations $\varphi^* d = d\varphi^*$ established in §1.2.

We shall see later that the space $H^1(\Omega)$ plays a fundamental role in the theory of holomorphic functions. For the time being though, we shall concentrate on studying the spaces $H_c^2(\Omega)$ and $H^2(\Omega)$.

As we have seen in Corollary 1.5.9 the linear map "integration over Ω"

$$I : \mathscr{D}^2(\Omega) \to \mathbb{C}$$

is zero on $d(\mathscr{D}^1(\Omega))$ and hence induces a linear map $\tilde{I}$ in the quotient space, $\tilde{I} : H_c^2(\Omega) \to \mathbb{C}$. This map is surjective: if $\lambda \in \mathbb{C}$, and $B(z_0, R) \subseteq \Omega$, one can easily find a 2-form $\omega = f\,dx \wedge dy$ with $f \in \mathscr{D}(B(z_0, R))$ such that

$$\int_\Omega \omega = \int_{B(z_0, R)} f\,dx\,dy = \lambda.$$

One can make this statement more precise.

1.5.10 Proposition. *Let Ω be a connected open set of $\mathbb{C}$. The map $\tilde{I}$ is an isomorphism. In other words, $\dim_{\mathbb{C}} H_c^2(\Omega) = 1$. Equivalently, in order for a 2-form of class C^∞ with compact support to be exact, it is necessary and sufficient that its integral vanishes.*

PROOF. We already know that if $\alpha \in d(\mathscr{D}^1(\Omega))$ then $\int_\Omega \alpha = 0$. To show the converse we first need the following lemmas:

1.5.11. Lemma. *Let $\alpha, \beta \in \mathscr{D}^2(B(z_0, R))$. In order that α and β be cohomologous (i.e., $\alpha - \beta \in d(\mathscr{D}^1(B(z_0, R)))$) it is necessary and sufficient that*

$$\int_{B(z_0, R)} \alpha = \int_{B(z_0, R)} \beta.$$

PROOF. Since $B(0,1)$ is diffeomorphic to $\mathbb{C}$ by the orientation preserving diffeomorphism

$$\varphi(z) := \frac{z}{1-|z|^2}$$

and $B(z_0, R)$ is diffeomorphic to $B(0,1)$ by the orientation preserving diffeomorphism

$$\theta(z) := \frac{z-z_0}{R},$$

it is clear that we only need to prove the following:

1.5.12. Lemma. *Let α, $\beta \in \mathscr{D}^2(\mathbb{C})$. Then $\alpha - \beta \in d(\mathscr{D}^1(\mathbb{C}))$ if and only if $\int_{\mathbb{C}} \alpha = \int_{\mathbb{C}} \beta$.*

PROOF. It is enough to show that if $\gamma \in \mathscr{D}^2(\mathbb{C})$, $\gamma = g\,dx \wedge dy$, is such that $I(\gamma) = \int_{\mathbb{C}} \gamma = \int_{\mathbb{C}} g\,dx\,dy = 0$ then $\gamma \in d(\mathscr{D}^1(\mathbb{C}))$. For that, it is enough to show that for every $\varphi \in \mathscr{D}^2(\mathbb{C})$ there exists $A(\varphi) \in \mathscr{D}^1(\mathbb{C})$ such that

$$d(A(\varphi)) = \varphi - I(\varphi)\rho,$$

where $\rho \in \mathscr{D}^2(\mathbb{C})$ is a conveniently fixed 2-form. In fact, if we take $\varphi = \gamma$ in this formula it follows that $d(A(\gamma)) = \gamma$.

Let $k \in \mathscr{D}(\mathbb{R})$ be such that $\int_{-\infty}^{\infty} k(t)\,dt = 1$ and set $\rho := k(x)k(y)\,dx \wedge dy$. For $\varphi = f\,dx \wedge dy$, we define the 1-form $A(\varphi)$ by:

$$A(\varphi) := \left\{\int_{-\infty}^{x} \left[f(t,y) - k(t)\int_{-\infty}^{\infty} f(s,y)\,ds\right] dt\right\} dy$$

$$- k(x)\left\{\int_{-\infty}^{y} \left[\int_{-\infty}^{\infty} f(s,u)\,ds - I(\varphi)k(u)\right] du\right\} dx.$$

One verifies easily that $A(\varphi) \in \mathscr{D}^1(\mathbb{C})$ and $d(A(\varphi)) = \varphi - I(\varphi)\rho$. □

1.5.13. Lemma. *For every $\omega \in \mathscr{D}^2(\Omega)$ and $B(z_0, R) \subseteq \Omega$ there is $\alpha \in \mathscr{D}^2(B(z_0, R))$ such that $\omega - \alpha \in d(\mathscr{D}^1(\Omega))$.*

PROOF. Let $(B(z_i, r_i))_{1 \le i \le n}$ be a covering of $\operatorname{supp} \omega$. For a C^∞ partition of unity $(\alpha_j)_{j \ge 1}$ subordinate to the covering $\{(B(z_i, r_i))_{1 \le i \le n}, \Omega \setminus \operatorname{supp} \omega\}$ of Ω, we can write $\omega = \sum_{j \ge 1} \alpha_j \omega$, with $\operatorname{supp}(\alpha_j \omega)$ contained in one of the disks $B(z_i, r_i)$ if $\alpha_j \omega \neq 0$. Since the set of j for which $\alpha_j \omega \neq 0$ is finite, it sufficies to prove the lemma for a 2-form ω with compact support in a disk $B(\zeta, r) \subseteq \Omega$.

Since Ω is connected, we can find a finite family of disks $(B(\zeta_j, R_j))_{1 \le j \le k}$

such that: (1) $B(\zeta_1, R_1) = B(z_0, R)$, (2) $B(\zeta_k, R_k) = B(\zeta, r)$, and (3) $B(\zeta_j, R_j) \cap B(\zeta_{j+1}, R_{j+1}) \neq \varnothing$ for $1 \leq j \leq k-1$.

If we can find differential forms $\omega_j \in \mathscr{D}^2(B(\zeta_j, R_j) \cap B(\zeta_{j+1}, R_{j+1}))$, $\beta_j \in \mathscr{D}^1(B(\zeta_{j+1}, R_{j+1}))$ such that

(i) $\omega = \omega_{k-1} + d\beta_{k-1}$, and
(ii) $\omega_{j+1} = \omega_j + d\beta_j$, $j = 1, 2, \ldots, k-2$,

then we will have $\omega = \omega_1 + \sum_{1 \leq j \leq k-1} d\beta_j$. Setting $\alpha = \omega_1$ will prove the lemma.

Let us show how to find ω_{k-1}, $d\beta_{k-1}$; the rest of the proof is a repetition of this step.

We know we can find $\omega_{k-1} \in \mathscr{D}^2(B(\zeta_k, R_k) \cap B(\zeta_{k-1}, R_{k-1}))$ such that

$$\int_{B(\zeta_k, R_k) \cap B(\zeta_{k-1}, R_{k-1})} \omega_{k-1} = \int_{B(\varepsilon_k, R_k)} \omega.$$

Hence $\omega - \omega_{k-1} = d\beta_{k-1}$ for some $\beta_{k-1} \in \mathscr{D}^1(B(\zeta_k, R_k))$ by Lemma 1.5.11.

□

Let us now go back to the proof of Proposition 1.5.10. Given $\omega \in \mathscr{D}^2(\Omega)$ with $\int_\Omega \omega = 0$ we want to show that ω is of the form $d\gamma$ with $\gamma \in \mathscr{D}^1(\Omega)$. Choose $\nu \in \mathscr{D}^2(\Omega)$ such that $\int_\Omega \nu = 1$ and the support of ν is contained in some disk $B(z_0, R)$. The form α given by Lemma 1.5.13 is cohomologous to $\mu := \left(\int_{B(z_0, R)} \alpha\right)\nu$, since both have the same integral. Therefore $\alpha - \mu = d\psi$ for some $\psi \in \mathscr{D}^1(B(z_0, R)) \subseteq \mathscr{D}^1(\Omega)$, and ω is also cohomologous to μ in Ω, i.e., $\omega - \mu = d\gamma$ for some $\gamma \in \mathscr{D}^1(\Omega)$. The hypothesis $\int_\Omega \omega = 0$ now gives $\int_{B(z_0, R)} \alpha = 0$, hence $\mu = 0$ and $\omega = d\gamma$. □

1.5.14. Proposition. *For every open set Ω in $\mathbb{C}$ one has $H^2(\Omega) = 0$.*

PROOF. We can assume without loss of generality that Ω is connected. Let $(B(z_i, r_i))_{i \geq 1}$ be a countable, locally finite covering of Ω by disks $B(z_i, r_i) \subset\subset \Omega$ such that $B(z_i, r_i) \cap B(z_{i+1}, r_{i+1}) \neq \varnothing$ for every i. Let $(\varphi_i)_{i \geq 1}$ be a C^∞ partition of unity subordinate to the covering. We can assume $\varphi_i \in \mathscr{D}(B(z_i, r_i))$.

If $\alpha \in Z^2(\Omega)$, then every $\alpha_i = \varphi_i \alpha$ has compact support inside $B(z_i, r_i)$.

Using Lemma 1.5.11 and induction, one can see that for every $i \geq 1$ there exist forms $\alpha_{i,j} \in \mathscr{D}^2(B(z_j, r_j) \cap B(z_{j-1}, r_{j-1}))$ and $\beta_{i,j} \in \mathscr{D}^1(B(z_{j-1}, r_{j-1}))$, $j = i+1$, $i+2, \ldots$ such that

(A) $\alpha_i = \alpha_{i,i+1} + d\beta_{i,i+1}$
(B) $\alpha_{i,j} = \alpha_{i,j+1} + d\beta_{i,j+1}$, $j \geq i+1$.

The family $(\operatorname{supp} \beta_{i,j})_{j \geq i+1}$ is locally finite and one has $\alpha_i = d\gamma_i$, with $\gamma_i =$

$\sum_{j \geq i+1} \beta_{i,j}$, which is a 1-form of class C^∞ having its support contained in $\bigcup_{j \geq i+1} B(z_j, r_j)$. It follows that the family $(\operatorname{supp} \gamma_i)_{i \geq 1}$ is also locally finite and $\alpha = d\gamma$ with $\gamma = \sum_{i \geq 1} \gamma_i$. This proves the proposition. □

1.5.15. Proposition (Mayer-Viétoris). *Let U and V be nonempty open subsets of an open subset Ω in $\mathbb{C}$ such that $\Omega = U \cup V$. Denote λ_j the linear maps $\lambda_j : \mathscr{E}^j(\Omega) \to \mathscr{E}^j(U) \oplus \mathscr{E}^j(V)$, $j = 0, 1, 2$, defined by $\lambda_j(\omega) := (\omega|U, \omega|V)$. Let μ_j be the linear maps $\mu_j : \mathscr{E}^j(U) \oplus \mathscr{E}^j(V) \to \mathscr{E}^j(U \cap V)$,*

$$\mu_j(\alpha, \beta) := (\alpha|(U \cap V)) - (\beta|(U \cap V)).$$

Then:

(1) *The sequences*

$$0 \to \mathscr{E}^j(\Omega) \xrightarrow{\lambda_j} \mathscr{E}^j(U) \oplus \mathscr{E}^j(V) \xrightarrow{\mu_j} \mathscr{E}^j(U \cap V) \to 0$$

are exact.

(2) *If $U \cap V$ is connected, passing from λ_1 and μ_1 to the quotient maps $\tilde{\lambda}_1$, $\tilde{\mu}_1$ induces an exact sequence*

$$0 \to H^1(\Omega) \xrightarrow{\tilde{\lambda}_1} H^1(U) \oplus H^1(V) \xrightarrow{\tilde{\mu}_1} H^1(U \cap V) \to 0.$$

PROOF. (1) The only thing that needs to be shown is the surjectivity of μ_j. Let $\{\varphi_U, \varphi_V\}$ be a C^∞ partition of unity subordinate to the covering $\{U, V\}$ of Ω. If $\alpha \in \mathscr{E}^j(U \cap V)$ then $\varphi_V \alpha \in \mathscr{E}^j(U)$ and $\varphi_U \alpha \in \mathscr{E}^j(V)$. For example, $\varphi_V \alpha \in \mathscr{E}^j(U)$ since it is obtained by putting together the form identically zero in $U \setminus \operatorname{supp} \varphi_V$ and the form $\varphi_V \alpha$, which is C^∞ in $U \cap V$ (recall $U = (U \setminus \operatorname{supp} \varphi_V) \cup (U \cap V)$). It is clear that $\alpha = \mu_j(\varphi_V \alpha, -\varphi_U \alpha)$.

(2) If $U \cap V$ is connected let us show that the map

$$H^1(\Omega) \xrightarrow{\tilde{\lambda}_1} H^1(U) \oplus H^1(V)$$

induced by $\lambda_1 : Z^1(\Omega) \to Z^1(U) \oplus Z^1(V)$ (which passes to the quotient since $\lambda_1(B^1(\Omega)) \subseteq B^1(U) \oplus B^1(V)$) is injective. Denote by $\tilde{\omega}$ the class of a closed form ω. If $\tilde{\lambda}_1(\tilde{\omega}) = 0$ then $\omega|U = df$, for some $f \in \mathscr{E}(U)$, and $\omega|V = dg$, for some $g \in \mathscr{E}(V)$. Therefore $f - g$ is constant in the connected open set $U \cap V$, say $f - g = c$. It follows that the function h defined by $h = f$ in U and $h = g + c$ in V is in $\mathscr{E}(\Omega)$ and $dh = \omega$. Hence $\tilde{\omega} = 0$.

Let now $(\alpha, \beta) \in Z^1(U) \oplus Z^1(V)$ be such that $\tilde{\mu}_1(\tilde{\alpha}, \tilde{\beta}) = 0$. This means that $(\alpha|(U \cap V)) - (\beta|(U \cap V)) = d\gamma$, for some $\gamma \in \mathscr{E}((U \cap V))$. Since μ_0 is surjective, $\gamma = \gamma_U - \gamma_V$ with $\gamma_U \in \mathscr{E}(U)$, $\gamma_V \in \mathscr{E}(V)$. It follows that

$$(\alpha|(U \cap V)) - (\beta|(U \cap V)) = d(\gamma_U|(U \cap V)) - d(\gamma_V|(U \cap V)),$$

hence $(\alpha|(U \cap V)) - d(\gamma_U|(U \cap V)) = (\beta|(U \cap V)) - d(\gamma_V|(U \cap V))$. In other words, $\alpha - d\gamma_U$ and $\beta - d\gamma_V$ define a single 1-form $\delta \in \mathscr{E}^1(\Omega)$ such that $\delta \in Z^1(\Omega)$ and $\tilde{\lambda}_1(\tilde{\delta}) = (\tilde{\alpha}, \tilde{\beta})$. This shows the exactness at $H^1(U) \oplus H^1(V)$.

Finally, let us prove that $\tilde{\mu}_1$ is surjective. Let $\omega \in Z^1(U \cap V)$, set $\omega_U = \varphi_V \omega \in \mathscr{E}^1(U)$ and $\omega_V = \varphi_U \omega \in \mathscr{E}^1(V)$ as earlier, so that $\mu_1(\omega_U, -\omega_V) = \omega$.

We have $d\omega_U = d\varphi_V \wedge \omega$, $d(-\omega_V) = -d\varphi_U \wedge \omega$, but since $\varphi_U + \varphi_V = 1$ we also have $d\varphi_U = -d\varphi_V$. Therefore $d\omega_U = d(-\omega_V)$ in $U \cap V$, which implies that the pair $d\omega_U$, $d(-\omega_V)$ defines a global form $\eta \in \mathscr{E}^2(\Omega)$. Since $H^2(\Omega) = 0$, there is $\theta \in \mathscr{E}^1(\Omega)$ such that $\eta = d\theta$. Hence $\omega_U - (\theta|U) \in Z^1(U)$, $\omega_V - (\theta|V) \in Z^1(V)$ and

$$\tilde{\mu}_1((\omega_U - (\theta|U))^{\sim}, -(\omega_V - (\theta|V))^{\sim}) = \tilde{\omega}.$$

This ends the proof of Proposition 1.5.15. □

1.5.16. Corollary. *Let Ω be an open subset of $\mathbb{C}$ which is C^∞ diffeomorphic to $\mathbb{C}$, $p_1, \ldots, p_n$, distinct points of Ω, and $\Omega' = \Omega \setminus \{p_1, \ldots, p_n\}$. Then $H^1(\Omega')$ is isomorphic to $\mathbb{C}^n$ and one can take as a basis the classes $\tilde{\omega}_j$, corresponding to the forms $\omega_j = \frac{1}{2\pi i} \cdot \frac{dz}{z - p_j}$, $1 \le j \le n$.*

Proof. Let us first note that an elementary computation shows that $\omega_j \in Z^1(\Omega')$. Let $\varphi : \mathbb{C} \to \Omega$ be a C^∞ diffeomorphism, $a_j \in \mathbb{C}$ such that $\varphi(a_j) = p_j$ and denote by $\alpha_1, \ldots, \alpha_n$ the 1-forms $\varphi^*(\omega_1), \ldots, \varphi^*(\omega_n)$. Recall that φ^* induces an isomorphism between $H^1(\Omega')$ and $H^1(\mathbb{C} \setminus \{a_1, \ldots, a_n\})$.

We are going to argue by induction on n.

Case $n = 1$. We need to show that $H^1(\mathbb{C} \setminus \{a_1\}) \simeq \mathbb{C}$ and that the 1-form α_1, which is certainly a nonzero cocycle, gives in fact a generator $\tilde{\alpha}_1$ of $H^1(\mathbb{C} \setminus \{a_1\})$. We can clearly assume that $a_1 = 0$ and drop the index for α_1.

For $\beta \in Z^1(\mathbb{C} \setminus \{0\})$ let $I(\beta) = \int_\gamma \beta$, where γ is the circle $t \mapsto e^{2\pi i t}$ $(0 \le t \le 1)$. Consider the auxiliary expression

$$v(\beta) := \beta - I(\beta)\alpha.$$

We are going to show that $I(\alpha) = 1$, and hence $I(v(\beta)) = 0$. We have

$$I(\alpha) = \int_\gamma \alpha = \int_{\varphi \circ \gamma} (\varphi^{-1})^*(\alpha) = \frac{1}{2\pi i} \int_{\varphi \circ \gamma} \frac{dz}{z - p_1}.$$

Let us choose $R > 0$ sufficiently large so that $\varphi \circ \gamma([0, 1]) \subseteq B(p_1, R)$, and the open set $D = B(p_1, R) \setminus \overline{\varphi(B(0, 1))}$ has a regular boundary. Recall $p_1 = \varphi(0)$, hence $p_1 \notin \bar{D}$. Therefore $d\left(\frac{dz}{z - p_1}\right) = 0$ in a neighborhood of $\bar{D}$ and ∂D is composed of $\partial B(p_1, R)$ and of $\varphi \circ \gamma$ traced in the opposite sense. Using the Stokes formula we obtain

$$\frac{1}{2\pi i} \int_{\varphi \circ \gamma} \frac{dz}{z - p_1} = \frac{1}{2\pi i} \int_{\partial B(p_1, R)} \frac{dz}{z - p_1} = 1.$$

Hence $I(\alpha) = 1$.

Let us now introduce $U_1 = \mathbb{C}\setminus]-\infty,0]$ and $U_2 = \mathbb{C}\setminus[0,\infty[$. For $z \in U_1$ consider the segment σ_z which joins 1 to z ($\sigma_z(t) = 1 + t(z-1), 0 \le t \le 1$). For $z \in U_2$, let τ_z be the segment that joins -1 to z ($\tau_z(t) = -1 + t(z+1), 0 \le t \le 1$). Define two functions F_1, F_2 by

$$F_1(z) := \int_{\sigma_z} \nu(\beta), \qquad z \in U_1,$$

$$F_2(z) := \int_{\tau_z} \nu(\beta), \qquad z \in U_2,$$

which one can verify are of class C^∞ and satisfy $dF_j = \nu(\beta)$ in their respective domains of definition. Since $U_1 \cap U_2$ is disconnected it is not at all clear that $G(z) = F_1(z) - F_2(z)$ is independent of z. This is evident though, for the upper half-plane and lower half-plane, respectively, since $dG = 0$. (One can also obtain this result applying the Stokes formula to convenient quadrilaterals.) Let us now take any two points, one in each half-plane: to fix ideas let them be $+i$ and $-i$. Then the definition of G, F_1, F_2 indicates that

$$G(i) - G(-i) = \int_\Gamma \nu(\beta),$$

where Γ is the quadrilateral of vertices $1, i, -1, -i$ with the counterclockwise orientation. Since $d\nu(\beta) = 0$ outside the origin, we can apply Stokes' formula twice and obtain

$$\int_\Gamma \nu(\beta) = \int_{\partial B(0,2)} \nu(\beta) = \int_\gamma \nu(\beta) = I(\nu(\beta)) = 0$$

(recall $\gamma = \partial B(0,1)$). Hence, $G(z) \equiv c \in \mathbb{C}$ throughout $U_1 \cap U_2$, and the functions F_1 in U_1 and $F_2 + c$ in U_2 define jointly a C^∞ function F_β in $\mathbb{C}\setminus\{0\}$ such that $dF_\beta = \nu(\beta)$. It follows that

$$\beta = dF_\beta + I(\beta)\alpha$$

and $\tilde{\beta} = I(\beta)\tilde{\alpha}$. This ends the proof for $n = 1$.

Case $n \ge 2$. It is clear that one can find an index j_0 and an open strip S whose boundary is formed by two parallel lines L_0 and L_1, such that a_{j_0} is in one of the components of $\mathbb{C}\setminus\bar{S}$ and all the other a_j are in the other one. Let us call Π the open half-plane defined by the line L_0 (closest to a_{j_0}) which does not contain a_{j_0} (see Figure 1.9).

Similarly, let Π' be the open half-plane defined by L_1 which contains a_{j_0}. Hence, $S = \Pi \cap \Pi'$ and $\{a_j\}_{j \ne j_0} \subseteq \Pi$. By induction we see that:

(1) $H^1(\Pi\setminus\{a_j\}_{j\ne j_0}) \simeq \mathbb{C}^{n-1}$ and $\{\tilde{\omega}_j\}_{j\ne j_0}$ is a basis for this vector space.
(2) $H^1(\Pi'\setminus\{a_{j_0}\}) \simeq \mathbb{C}$ and $\{\tilde{\omega}_{j_0}\}$ is a basis for this vector space.

Let $U = \Pi\setminus\{a_j)_{j\ne j_0}$, $V = \Pi'\setminus\{a_{j_0}\}$. Then $U \cap V = S$, which is connected, and $H^1(S) = 0$ by Poincare's lemma 1.2.2(a) (since S is star shaped). By Mayer-Viétoris we have

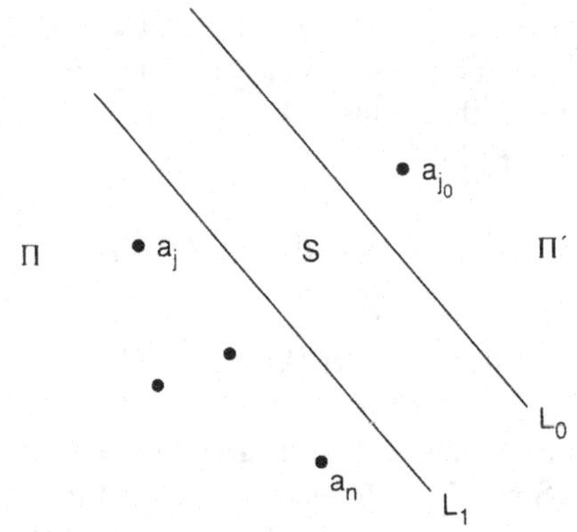

Figure 1.9

$$H^1(\mathbb{C}\backslash\{a_1,\ldots,a_n\}) \simeq H^1(U) \oplus H^1(V) \simeq \mathbb{C}^n.$$

In this isomorphism the class $\tilde{\omega}_j$ in the left-hand side of ω_j corresponds to the class of ω_j in $H^1(U)$ $(j \neq j_0)$ and the class $\tilde{\omega}_{j_0}$ to that of ω_{j_0} in $H^1(V)$. This proves the last statement of the corollary. □

Exercises 1.5

1. Let γ be a piecewise C^1 path in $\mathbb{C}$ and γ^* its complex conjugate, i.e., $\gamma^*(t) = \overline{\gamma(t)}$. Let f be continuous in a neighborhood of $\operatorname{Im}\gamma$ and $f^*(z) = \overline{f(\bar{z})}$. Show that

$$\overline{\int_\gamma f(z)\,dz} = \int_{\gamma^*} f^*(z)\,dz.$$

Conclude that if $\gamma(t) = e^{2\pi i t}$ $(0 \leq t \leq 1)$, i.e., the unit circle traversed in the positive sense, then

$$\overline{\int_\gamma f(z)\,dz} = -\int_\gamma \overline{f(z)}\frac{dz}{z^2}.$$

2. Prove the formula $\operatorname{area}(\Omega) = \int_{\partial\Omega} x\,dy$, for an open set Ω with piecewise regular boundary. Compute the area of the ellipse $\dfrac{x^2}{a^2} + \dfrac{y^2}{b^2} = 1$.

3. Let Ω be a domain with piecewise regular boundary, symmetric with respect to the origin in $\mathbb{C}$. Compute

$$\int_{\partial\Omega} (yx + e^y)\,dx + (xy^3 + xe^y - 2y)\,dy.$$

4. Let Ω be a domain as in the previous section. Let $\theta(z)$ be the angle between the exterior normal $n(z)$ to $\partial\Omega$ at the point z, and the positive real axis. Compute

$$\int_{\partial\Omega} (x\cos\theta(z) + y\sin\theta(z))|dz|,$$

where $z = x + iy$.

5. Let $m, n \in \mathbb{N}$. Use the Stokes formula to compute $\int_{B(0,1)} z^m \bar{z}^n \, dx\, dy$.

6. Show that if Ω is a star-shaped open subset of $\mathbb{C}$, then its Betti numbers are $b_0 = 1$, $b_1 = b_2 = 0$.

7. Show that, if Ω is a connected open set in $\mathbb{C}$, there is a covering of Ω, $(B(z_i, r_i))_{i \in \mathbb{N}}$, by disks such that $B(z_i, r_i) \cap B(z_{i+1}, r_{i+1}) \neq \varnothing$ (cf. §5.1.14).

8. Show that $\frac{dz}{z}$ is a closed form in $\mathbb{C}^*$.

9. Let Ω be a connected open set in $\mathbb{C}$, Ω diffeomorphic to $\mathbb{C}$, and $a_1, \ldots, a_n$, distinct points in Ω. Given a form $\omega \in Z^1(\Omega \backslash \{a_1, \ldots, a_n\})$, find a procedure to compute the coefficients λ_j in the decomposition $\tilde{\omega} = \sum_{j=1}^{n} \lambda_j \tilde{\omega}_j$, with $\omega_j = \frac{1}{2\pi i} \frac{dz}{z - a_j}$.

10. Let Ω be a relatively compact, open set with piecewise regular boundary, and $u, v \in C^2(\bar{\Omega})$. Show that

$$\int_{\Omega} (\operatorname{grad} u | \operatorname{grad} v)\, dx\, dy + \int_{\Omega} u \Delta v \, dx\, dy = \int_{\partial\Omega} u \frac{\partial v}{\partial n} |dz|.$$

11. Let $g \in C^2(B(0, r))$ for some $r > 1$, $h = \Delta g$. Show that if $B = B(0, 1)$, then

$$2\pi g(0) = \int_B h(z) \log|z|\, dx\, dy + \int_{\partial B} g(z) |dz|.$$

(Hint: use Green's formula with $\omega = B \backslash \bar{B}(0, \varepsilon)$, $0 < \varepsilon < 1$.)

12. Show that the function $f(z) = \frac{1}{z}$ is locally integrable in $\mathbb{C}$.

13. Let Ω be an open set in $\mathbb{C}$, $a_1, \ldots, a_n$ distinct points in Ω, $\Omega' = \Omega \backslash \{a_1, \ldots, a_n\}$. Let $f \in \mathscr{E}(\Omega')$ be such that $f(z) \neq 0$ if $z \in \Omega'$ and there are disjoint disks B_j centered at a_j, nonvanishing functions $\psi_j \in \mathscr{E}(B_j)$ and integers k_j with the property that $f(z) = (z - a_j)^{k_j} \psi_j(z)$ in $B_j\{a_j\}$. Show that

(i) $\frac{df}{f} = \frac{k_j\, dz}{z - a_j} + \frac{d\psi_j}{\psi_j}$ in $B_j \backslash \{a_j\}$.

(ii) Given any $g \in \mathscr{D}(\Omega)$, we can find $g_0 \in \mathscr{D}(\Omega')$, $g_j \in \mathscr{D}(B_j)$ with $g_j = g$ near a_j such that $g = g_0 + g_1 + \cdots + g_n$.

(iii) $\int_{\Omega} \frac{df}{f} \wedge dg_0 = -\int_{\Omega} d\left(g_0 \frac{df}{f}\right) = 0$. (Why do the integrals make sense?)

(iv) $\int_{B_j} \frac{d\psi_j}{\psi_j} \wedge dg_j = -\int_{B_j} d\left(g_j \frac{d\psi_j}{\psi_j}\right) = 0.$

(v) $\frac{1}{2\pi i} \int_{B_j} \frac{dz}{z - a_j} \wedge dg_j = g(a_j).$

Conclude that for any $g \in \mathscr{D}(\Omega)$:

$$\frac{1}{2\pi i} \int_{\Omega} \frac{df}{f} \wedge dg = \sum_{j=1}^{n} k_j g(a_j).$$

14. Let $\eta = \frac{x\, dy - y\, dx}{x^2 + y^2}$ in $\mathbb{R}^2 \backslash \{0\}$. Verify that $d\eta = 0$. Compute $(x\, dx + y\, dy) \wedge \eta$.

15. Let Ω be an open subset of $\mathbb{R}^2$ with regular boundary of class C^1, f a function C^1 in a neighborhood of $\bar{\Omega}$, (x_0, y_0) a point in Ω. Define

$$\tau = \tau_{x_0, y_0} : (x, y) \mapsto (x - x_0, y - y_0)$$

and let η be the 1-form from the previous exercise. Show that

$$f(x_0, y_0) = \frac{1}{2\pi}\left[\int_{\partial\Omega} f(\tau^*\eta) - \int_{\Omega} df \wedge \tau^*\eta\right].$$

Use this formula to compute $\int_{\partial\Omega} \tau^*\eta$.

*16. Let $\rho : \mathbb{C}^* \to \mathbb{R}$ of class C^∞ be such that $\lim_{|z|\to\infty} \rho(z) = +\infty$ and $d\rho(z) \neq 0$ for every $z \in \mathbb{C}^*$. Define $S_t = \{z \in \mathbb{C}^* : \rho(z) = t\}$ for $t \in \mathbb{R}$.
(i) Show that for every $\omega \in \mathscr{D}^2(\mathbb{C}^*)$ there is $\alpha \in \mathscr{D}^1(\mathbb{C}^*)$ such that $\omega = d\rho \wedge \alpha$.
(ii) Let $\alpha_1, \alpha_2 \in \mathscr{D}^1(\mathbb{C}^*)$ be such that $d\rho \wedge \alpha_1 = d\rho \wedge \alpha_2$. Show that there is $\beta \in \mathscr{D}^0(\mathbb{C}^*)$ such that $\alpha_1 = \alpha_2 + \beta\, d\rho$. Show, moreover, that for every $t \in \mathbb{R}$,

$$\int_{S_t} \alpha_1 = \int_{S_t} \alpha_2.$$

(iii) Let $\alpha \in \mathscr{D}^1(\mathbb{C}^*)$, $\omega = d\rho \wedge \alpha$. Show that if $g(t) = \int_{S_t} \alpha$ then

$$\int_{\mathbb{C}^*} \omega = \int_{-\infty}^{\infty} g(t)\, dt.$$

Study in particular the case $\rho(z) = |z|$.

17. Let $B = B(0, 1) \subseteq \mathbb{R}^2$ and f a C^∞ function from a neighborhood of $\bar{B}$ into $\mathbb{R}^2$ such that $f(\bar{B}) \subseteq \partial B$, $f | B = id_{\partial B}$ (i.e., $f(x, y) = (x, y)$ if $(x, y) \in \partial B$).
(i) Show that if $f = (f_1, f_2)$ then $df_1 \wedge df_2 = 0$ in $\bar{B}$.
(ii) Compute $\int_{\partial B} f_1\, df_2$ in two different ways to show that such function f cannot exist.
*(iii) Show that the conclusion from (ii) still holds if we remove the assumption that $f(z) = z$ for all $z \in \partial B$. (Hint: Consider the auxiliary function $z + (1 - |z|^2)f(z)$.)
*(iv) Conclude that there is no C^∞ map g from a neighborhood of $\bar{B}$ into $\mathbb{R}^2$ such that $g(\bar{B}) \subseteq \bar{B}$ and $g(x, y) \neq (x, y)$ for every $(x, y) \in \bar{B}$.
*(v) Prove statement (iii) under the hypothesis that g is only continuous.

§6. Homotopy: Fundamental Group

We recall from §1.2.8 that a path γ in an open set Ω in $\mathbb{C}$ (or more generally a topological space, e.g., Ω a closed subset of $\mathbb{C}$) is simply a continuous map $\gamma : [0, 1] \to \Omega$. We call $a = \gamma(0)$ the ***starting point*** of the path and $b = \gamma(1)$ the ***endpoint***. Let us denote by $\mathscr{C}(\Omega; a, b)$ the collection of paths in Ω starting at a

and ending at b. If $a = b$ we say the path is ***closed*** and a is called the ***base point***. $\mathscr{C}(\Omega; a)$ is then the family of all closed paths (***loops*** in Ω with *base point* a). If $z \in \Omega$ we denote by ε_z the constant path with base point z, $\varepsilon_z(t) = z$ $(0 \leq t \leq 1)$. If $\gamma \in \mathscr{C}(\Omega; a, b)$ we denote by $\overline{\gamma}$ the inverse path to γ, $\overline{\gamma} \in \mathscr{C}(\Omega; b, a)$ and is defined by $\overline{\gamma}(t) = \gamma(1 - t)$ $(0 \leq t \leq 1)$. If $\gamma \in \mathscr{C}(\Omega; a)$ then $\overline{\gamma} \in \mathscr{C}(\Omega; a)$ also. Furthermore, $\overline{\varepsilon}_a = \varepsilon_a$.

If $\alpha \in \mathscr{C}(\Omega; a, b)$ and $\beta \in \mathscr{C}(\Omega; b, c)$ we denote by $\alpha\beta$ the path in $C(\Omega; a, c)$ defined by:

$$\alpha\beta(t) := \begin{cases} \alpha(2t) & \text{if } 0 \leq t \leq \frac{1}{2} \\ \beta(2t - 1) & \text{if } \frac{1}{2} \leq t \leq 1. \end{cases}$$

The path $\alpha\beta$ is called the ***composition*** of α and β. It is easy to see that $\overline{\alpha\beta} = \overline{\beta}\overline{\alpha}$. On the other hand, when $(\alpha\beta)\gamma$ is defined, the $\alpha(\beta\gamma)$ is also defined but, in general, they do not coincide.

1.6.1. Definition. Two paths $\gamma_0, \gamma_1 \in \mathscr{C}(\Omega; a, b)$ are said to be ***homotopic with fixed endpoints*** if there is a continuous map $H : [0, 1] \times [0, 1] \to \Omega$ such that $H(t, 0) = \gamma_0(t)$, $H(t, 1) = \gamma_1(t)$ $(0 \leq t \leq 1)$ and, $H(0, s) = a$, $H(1, s) = b$ $(0 \leq s \leq 1)$.

One says the homotopy H carries γ_0 into γ_1. (See Figure 1.10.)

1.6.2. Proposition. *Homotopy is an equivalence relation in* $\mathscr{C}(\Omega; a, b)$.

Proof. Reflexivity and symmetry are clear. Let us see the transitivity. Let $\gamma_0, \gamma_1, \gamma_2 \in \mathscr{C}(\Omega; a, b)$, γ_0, γ_1 homotopic by H and γ_1, γ_2 homotopic by K. One obtains a homotopy L carrying γ_0 into γ_2 by:

$$L(t, s) = \begin{cases} H(t, 2s) & 0 \leq s \leq \frac{1}{2} \\ K(t, 2s - 1) & \frac{1}{2} \leq s \leq 1. \end{cases}$$

□

Denote by $[\alpha]$ the class of α under homotopy.

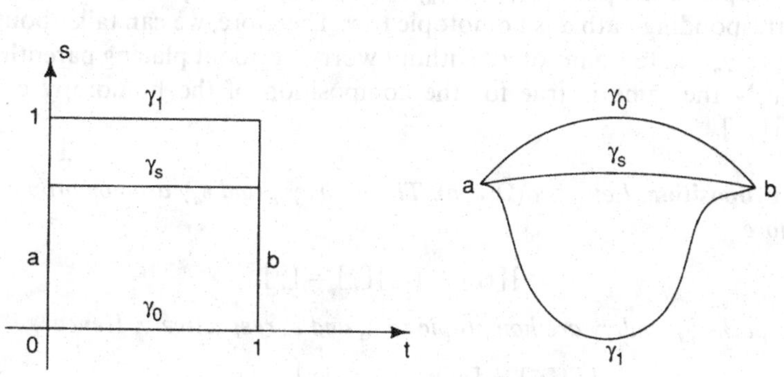

Figure 1.10

1.6.3. Proposition (Change of Parameters). *Let $\alpha \in \mathscr{C}(\Omega; a, b)$ and $f: [0, 1] \to [0, 1]$ continuous, $f(0) = 0$, $f(1) = 1$. Then the paths α and $f^*(\alpha) = \alpha \circ f$ are homotopic with fixed endpoints.*

PROOF. It is enough to define the homotopy H,

$$H(t, s) := \alpha((1 - s)t + sf(t)). \qquad \square$$

1.6.4. Proposition. *The composition of paths and the inversion of paths are compatible with homotopies. The composition thus defined on the equivalence classes of paths is associative.*

PROOF. Let F be a homotopy carrying α_0 into α_1, $\alpha_0, \alpha_1 \in \mathscr{C}(\Omega; a, b)$, and G a homotopy carying β_0 into β_1, $\beta_0, \beta_1 \in \mathscr{C}(\Omega; b, c)$. Then

$$H(t, s) := \begin{cases} F(2t, s), & 0 \le t \le \frac{1}{2} \\ G(2t - 1, s), & \frac{1}{2} \le t \le 1, \end{cases}$$

is a homotopy carrying $\alpha_0\beta_0$ into $\alpha_1\beta_1$.

If F is a homotopy carrying γ_0 into γ_1 then $(t, s) \mapsto F(1 - t, s)$ is a homotopy carrying $\overline{\gamma}_0$ into $\overline{\gamma}_1$.

Finally, let $\alpha \in \mathscr{C}(\Omega; a, b)$, $\beta \in \mathscr{C}(\Omega; b, c)$ and $\gamma \in \mathscr{C}(\Omega; c, d)$. We want to show that $\delta_0 = (\alpha\beta)\gamma$ and $\delta_1 = \alpha(\beta\gamma)$ are homotopic. Let f be the unique continuous piecewise affine map such that $f(0) = 0$, $f(1/2) = 1/4$, $f(3/4) = 1/2$, $f(1) = 1$. Then $\delta_1 = f^*(\delta_0)$, whence δ_0 and δ_1 are homotopic. $\square$

1.6.5. Remark. More generally, let $(\gamma_i)_{1 \le i \le n}$ be a family of paths such that $\gamma_i(1) = \gamma_{i+1}(0)$ $(1 \le i \le n - 1)$ and let $0 = t_0 < t_1 < \cdots < t_n = 1$ be a partition of $[0, 1]$. The path γ defined by

$$\gamma(t) := \gamma_i\left(\frac{t - t_{i-1}}{t_i - t_{i-1}}\right) \qquad (t_{i-1} \le t \le t_i, 1 \le i \le n)$$

is homotopic to the path $\gamma_1(\gamma_2(\dots\gamma_n)\dots)$. Moreover, if α_i is homotopic to γ_i, the corresponding path α is homotopic to γ. Therefore, we can talk about the path $\gamma_1 \dots \gamma_n$ "up to homotopy," without worrying about placing parentheses. Obviously the same is true for the composition of the homotopy classes $[\gamma_1]\dots[\gamma_n]$.

1.6.6. Proposition. *Let $\gamma \in \mathscr{C}(\Omega; a, b)$. The paths $\gamma\varepsilon_b$ and $\varepsilon_a\gamma$ are homotopic to γ. Therefore*

$$[\gamma][\varepsilon_b] = [\varepsilon_a][\gamma] = [\gamma].$$

The paths $\gamma\overline{\gamma}$ and $\overline{\gamma}\gamma$ are homotopic to ε_a and ε_b respectively. Hence

$$[\gamma][\overline{\gamma}] = [\varepsilon_a], \qquad [\overline{\gamma}][\gamma] = [\varepsilon_b].$$

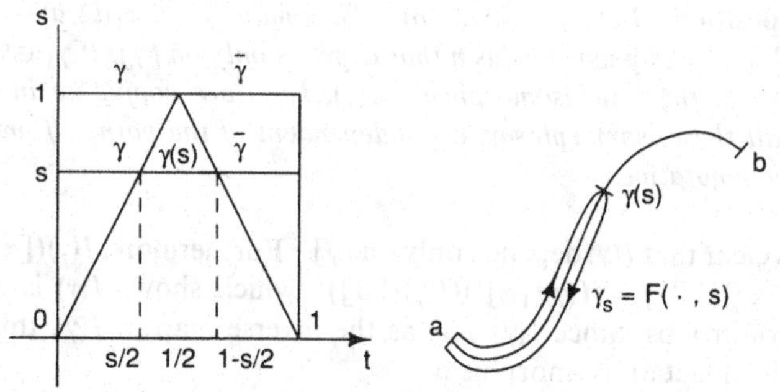

Figure 1.11

PROOF. One has $\gamma\varepsilon_b = \gamma \circ f$ with $f(t) = 2t$ if $0 \le t \le 1/2$ and $f(t) = 1$ for $1/2 \le t \le 1$. A similar formula holds for $\varepsilon_a\gamma$. Therefore, by Proposition 1.6.3, $\gamma\varepsilon_b$ and $\gamma\varepsilon_a$ are homotopic to γ.

The function

$$F(t,s) = \begin{cases} \gamma(2t) & 0 \le t \le s/2 \\ \gamma(s) & s/2 \le t \le 1 - s/2 \\ \bar{\gamma}(2t-1) & 1 - s/2 \le t \le 1 \end{cases}$$

is a homotopy carrying ε_a into $\gamma\bar{\gamma}$. (See Figure 1.11.) □

Let us denote by $\pi_1(\Omega; a)$ the set of equivalence classes (under homotopy) of $\mathscr{C}(\Omega; a)$. $\pi_1(\Omega; a, b)$ is defined similarly. The following theorem summarizes several of the previous propositions.

1.6.7. Theorem. *For the composition law*

$$\pi_1(\Omega; a) \times \pi_1(\Omega; a) \to \pi_1(\Omega; a)$$
$$([\alpha], [\beta]) \mapsto [\alpha][\beta] = [\alpha\beta],$$

the set $\pi_1(\Omega; a)$ is a group whose identity element is $[\varepsilon_a]$. The inverse $[\gamma]^{-1}$ of $[\gamma]$ is given by $[\bar{\gamma}]$.

1.6.8. Definition. The group $\pi_1(\Omega; a)$ will be called the ***fundamental group*** of Ω (with base point a). It is also known as the ***first homotopy group*** of Ω with base point a.

1.6.9. Remark. The group $\pi_1(\Omega; a)$ is, in general, nonabelian, as will be seen in later examples.

We want to compare now the fundamental groups with base points a, b in the same connected component of Ω.

1.6.10. Proposition. *Let $\gamma \in \mathscr{C}(\Omega; a, b)$. The map $I(\gamma): \pi_1(\Omega; a) \to \pi_1(\Omega; b)$, $[\alpha] \mapsto [\bar{\gamma}\alpha\gamma]$, is a group isomorphism that depends only on $[\gamma]$. If $\gamma_1 \in \mathscr{C}(\Omega; a, b)$ is another path, then the isomorphisms $I(\gamma)$, $I(\gamma_1)$ are conjugate in $\pi_1(\Omega; b)$. Moreover, all these isomorphisms are independent of the path γ if and only if $\pi_1(\Omega; a)$ is commutative.*

PROOF. It is clear that $I(\gamma)$ depends only on $[\gamma]$. Furthermore, $I(\gamma)([\alpha][\beta]) = [\bar{\gamma}\alpha\beta\gamma] = [\bar{\gamma}\alpha\gamma][\bar{\gamma}\beta\gamma] = (I(\gamma)([\alpha]))(I(\gamma)([\beta]))$, which shows $I(\gamma)$ is a homomorphism of groups. Since $I(\bar{\gamma})$ acts as the inverse map to $I(\gamma)$, this homomorphism is in fact an isomorphism.

Given now $\gamma, \gamma_1 \in \mathscr{C}(\Omega; a, b)$ we have $c = \bar{\gamma}_1\gamma \in \mathscr{C}(\Omega; b)$ and

$$I(\gamma_1)([\alpha]) = [\bar{\gamma}_1\alpha\gamma_1] = [\bar{\gamma}_1\gamma][\bar{\gamma}\alpha\gamma][\bar{\gamma}\gamma_1] = [c]I(\gamma)([\alpha])[c]^{-1}.$$

Finally, if $\pi_1(\Omega; a)$ is abelian, $\pi_1(\Omega; b)$ is also abelian and $I(\gamma) = I(\gamma_1)$. Conversely, let us assume $I(\gamma) = I(\gamma_1)$ for every pair of paths $\gamma, \gamma_1 \in \mathscr{C}(\Omega; a, b)$. We want to show that every inner automorphism $[\alpha] \mapsto [c][\alpha][c]^{-1}$ of $\pi_1(\Omega; a)$ is the identity. Pick γ as done earlier. Given $c \in \mathscr{C}(\Omega; a)$ let $\gamma_1 = \bar{c}\gamma \in \mathscr{C}(\Omega; a, b)$. Hence we have $[c] = [\gamma][\bar{\gamma}c] = [\gamma][\bar{\gamma}_1]$. On the other hand, the identity $I(\gamma) = I(\gamma_1)$ means that

$$[\bar{\gamma}][\alpha][\gamma] = [\bar{\gamma}_1][\alpha][\gamma_1].$$

Multiplying on the left by $[\gamma]$ and on the right by $[\bar{\gamma}]$, this identity leads to

$$[\alpha] = [\gamma\bar{\gamma}_1][\alpha][\gamma_1\bar{\gamma}] = [c][\alpha][c]^{-1}. \qquad \square$$

If Ω is connected we denote by $\pi_1(\Omega)$ one of the groups $\pi_1(\Omega; a)$, with $a \in \Omega$ chosen arbitrarily.

1.6.11. Definition. An open subset Ω of $\mathbb{C}$ is called ***simply connected*** if Ω is connected and $\pi_1(\Omega) = 0$ (i.e., $\pi_1(\Omega; a) = \{[\varepsilon_a]\}$ for every $a \in \Omega$).

1.6.12. Proposition. *If Ω is simply connected, the set $\pi_1(\Omega; a, b)$ contains a single element for every pair $a, b \in \Omega$.*

PROOF. If for some $a \neq b$, $\pi_1(\Omega; a, b)$ contains $[\gamma] \neq [\gamma_1]$, then $\pi_1(\Omega; a)$ contains two different elements $[\varepsilon_a]$ and $[\gamma_1\bar{\gamma}]$. $\square$

1.6.13. Examples

(1) A star-shaped open subset of $\mathbb{C}$ (with respect to a point $a \in \Omega$) is simply connected. If $\gamma \in \mathscr{C}(\Omega; a)$ then $H(t, s) := sa + (1 - s)\gamma(t)$ is a homotopy carrying γ into ε_a.

(2) Let U_1, U_2 be simply connected open subsets of $\mathbb{C}$ such that $U_1 \cap U_2$ is nonempty and connected. Then $\Omega = U_1 \cup U_2$ is simply connected.

It is clear that Ω is connected. Let $a \in U_1 \cap U_2$, $\gamma \in \mathscr{C}(\Omega; a)$. We need to show

$[\gamma] = [\varepsilon_a]$. There is a partition of the interval $[0, 1]$, $0 = t_0 < t_1 < \cdots t_n = 1$, such that $\gamma(t_j) \in U_1 \cap U_2$ and $\gamma_k := \gamma|[t_{k-1}, t_k]$ $(1 \leq k \leq n)$ is a path entirely contained in one of the U_i, which we will denote $i(k)$.

Choose $\alpha_j \in \mathscr{C}(U_1 \cap U_2;\ a,\ \gamma(t_j))$ $(1 \leq j \leq n-1)$. It is clear that $[\gamma] = [\gamma_1\bar{\alpha}_1][\alpha_1\gamma_2\bar{\alpha}_2]\ldots[\alpha_{n-1}\gamma_n]$.

Now, we can consider that $\gamma_1\bar{\alpha}_1 \in \mathscr{C}(U_{i(1)}; a)$, $\alpha_1\alpha_2\bar{\alpha}_2 \in \mathscr{C}(U_{i(2)}; a)$, etc. Therefore (with an obvious abuse of notation, which is justified by considering the canonical injections of $U_1 \cap U_2$, U_1, and U_2 into Ω) we have

$$[\gamma_1\bar{\alpha}_1] = [\alpha_1\gamma_2\bar{\alpha}_2] = \cdots = [\alpha_{n-1}\gamma_n] = [\varepsilon_a].$$

Hence $[\gamma] = [\varepsilon_a]$ and Ω is simply connected.

1.6.14. Definition. Two loops $\gamma_0 \in \mathscr{C}(\Omega; a)$ and $\gamma_1 \in \mathscr{C}(\Omega; b)$ are called ***free homotopic*** if there is a continuous map $H: [0, 1] \times [0, 1] \to \Omega$—still called homotopy—with the following properties:

(1) $H(t, 0) = \gamma_0(t)$, $0 \leq t \leq 1$
(2) $H(t, 1) = \gamma_1(t)$, $0 \leq t \leq 1$
(3) $H(0, s) = H(1, s)$, $0 \leq s \leq 1$

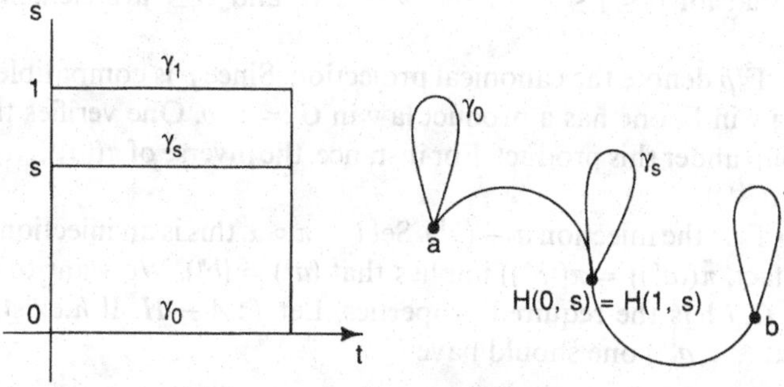

Figure 1.12

1.6.15. Proposition. *A connected open set Ω is simply connected if and only if any two loops in Ω are free homotopic.*

PROOF. Exercise. □

We would like now to compute the fundamental group of an open subset of $\mathbb{C}$ which is homeomorphic to $\mathbb{C}\setminus\{p_1, \ldots, p_n\}$. For that purpose we need the notion of a free group generated by a set.

1.6.16. Proposition *Let A be a nonempty set. There is a group G and an injection $i: A \to G$ with the following two properties:*

(1) *G is generated by* $i(A)$.
(2) *If* f *is a mapping from* A *into a group* H, *there is a unique group homomorphism* $h: G \to H$ *such that* $h \circ i = f$.

PROOF. Let Γ be the set of finite sequences (or "words") of elements of $A \times \{-1, 1\}$. We denote by $(a_1^{\varepsilon_1}, \ldots, a_m^{\varepsilon_m})$ a word in Γ, where $a_j \in A$, $\varepsilon_j \in \{-1, 1\}$. If $m = 0$, the empty word is denoted by 1. One can consider an associative product in Γ, which has 1 as its neutral element, by means of:

$$(a_1^{\varepsilon_1}, \ldots, a_m^{\varepsilon_m})(b_1^{\eta_1}, \ldots, b_n^{\eta_m}) := (c_1^{\theta_1}, \ldots, c_{n+m}^{\theta_{n+m}}),$$

where $c_j^{\theta_j} = a_j^{\varepsilon_j}$ if $1 \le j \le m$ and, $c_j^{\theta_j} = b_{j-m}^{\eta_{j-m}}$ if $m + 1 \le j \le n + m$. That is, Γ is a monoid.

Consider on Γ the equivalence relation ρ compatible with this product and generated by the identification of $(a^\varepsilon)(a^{-\varepsilon})$ with 1 for every $a \in A$. (If $B = \{((a^\varepsilon)(a^{-\varepsilon}), 1): a \in A, \varepsilon = \pm 1\} \subseteq \Gamma \times \Gamma$, then ρ is the intersection of all the equivalence relations in Γ containing B.) We say that two words M, N are elementarily equivalent if $M = N$, or $M = m(a^\varepsilon)(a^{-\varepsilon})n$ and $N = mn$, or $M = mn$ and $N = m(a^\varepsilon)(a^{-\varepsilon})n$, for some $a \in A$, $\varepsilon \in \{-1, 1\}$, m, $n \in \Gamma$, and juxtaposition denotes product. One can verify that two words U, V are ρ-equivalent if and only if there is a finite collection of words $U_0 = U, U_1, \ldots, U_r = V$ such that for $0 \le j \le r - 1$ the words U_j and U_{j+1} are elementarily equivalent.

Let $\pi: \Gamma \to \Gamma/\rho$ denote the canonical projection. Since ρ is compatible with the product law in Γ, one has a product law in $G := \Gamma/\rho$. One verifies that G is in fact a group under this product. For instance, the inverse of $\pi((a_1^{\varepsilon_1}, \ldots, a_n^{\varepsilon_n}))$ is $\pi((a_n^{-\varepsilon_n}, \ldots, a_1^{-\varepsilon_1}))$.

Let $\alpha: A \to \Gamma$ be the injection $a \mapsto (a^1)$. Set $i = \pi \circ \alpha$, this is an injection from A into G. In fact, $\pi((a^\varepsilon)) = \pi((b^\eta))$ implies that $(a^\varepsilon) = (b^\eta)$. We want to show that the pair G, i has the required properties. Let $f: A \to H$. If h exists and $m \in \Gamma$, $m = (a_1^{\varepsilon_1}, \ldots, a_k^{\varepsilon_k})$ one should have

$$\begin{aligned} h(\pi(m)) &= (h \circ \pi)((a_1^{\varepsilon_1})) \times \cdots \times (h \circ \pi)((a_k^{\varepsilon_k})) \\ &= ((h \circ \pi)(a_1))^{\varepsilon_1} \times \cdots \times ((h \circ \pi)(a_k))^{\varepsilon_k} \\ &= (f(a_1))^{\varepsilon_1} \times \cdots \times (f(a_k))^{\varepsilon_k}, \end{aligned}$$

where $\times$ denotes the product in H. This identity indicates how to define h and ensures the uniqueness of h. Namely, the condition $j((a^\varepsilon)) = (f(a))^\varepsilon$ determines a unique monoidal homomorphism $j: \Gamma \to H$ by

$$j((a_1^{\varepsilon_1}, \ldots a_k^{\varepsilon_k})) = (f(a_1))^{\varepsilon_1} \times \cdots \times (f(a_k))^{\varepsilon_k},$$

since every element of Γ is written in a unique way as $(a_1^{\varepsilon_1}, \ldots, a_k^{\varepsilon_k})$. This map j is constant on the equivalence classes with respect to ρ: for instance, if $M = m(a^\varepsilon)(a^{-\varepsilon})n$ and $N = mn$, one has

$$j(M) = j(m)(f(a))^\varepsilon(f(a))^{-\varepsilon}j(n) = j(m)j(n) = j(N).$$

Therefore, there is a unique group homomorphism $h: G \to H$ such that $h \circ \pi = j$. □

What one really does in practice is to restrict oneself to "reduced" words. Given a word $(a_1^{\varepsilon_1}, \ldots, a_m^{\varepsilon_m})$ we reduce it by suppressing two consecutive entries of the form $a^{\varepsilon}, a^{-\varepsilon}$. A reduced word is one such that for every $j (1 \leq j \leq m-1)$ $a_{j+1}^{\varepsilon_{j+1}} \neq a_j^{-\varepsilon_j}$. One verifies easily that every word is equivalent to a reduced word. Every element of $G \backslash \{1\}$ can be written in a unique way in the form $x_1^{n_1} \ldots x_k^{n_k}$ where $x_j \in i(A)$, $n_j \in \mathbb{Z} \backslash \{0\}$ and $x_j \neq x_{j+1}$ for $1 \leq j \leq k-1$.

The pair (G, i) is unique up to isomorphisms, cf. Exercise 1.6.12.

1.6.17. Definition. One says that the group G constructed in 1.6.16 is the ***free group generated by*** A. We denote it by $L(A)$. We will identify A to $i(A)$. If $A = \emptyset$ we denote by $L(A)$ the trivial group $G = \{1\}$. If $\# A = n \geq 1$ we say G is a free group with n generators.

For an arbitrary group G, one calls ***commutator*** $[\alpha, \beta]$ of two elements $\alpha, \beta \in G$ the element

$$[\alpha, \beta] := \alpha\beta\alpha^{-1}\beta^{-1}.$$

The subgroup generated by the commutators is denoted $[G, G]$ and is called the ***commutator subgroup*** of G.

1.6.18. Proposition. *The subgroup $[G, G]$ is normal in G and the quotient group $G/[G, G]$ is abelian.*

PROOF. In fact, $\gamma[\alpha, \beta]\gamma^{-1} = (\gamma\alpha\gamma^{-1})(\gamma\beta\gamma^{-1})(\gamma\alpha\gamma^{-1})^{-1}(\gamma\beta\gamma^{-1})^{-1}$ and $\beta\alpha = \alpha\beta[\beta^{-1}, \alpha^{-1}]$. □

The subgroup $[G, G]$ is the smallest normal subgroup H of G such that G/H is abelian. In fact, the projection $\theta: G \to G/H$ sends $[\alpha, \beta]$ into the neutral element of G/H and $[G, G] \subseteq \operatorname{Ker} \theta = H$. The group $G/[G, G]$ is sometimes called the ***abelianized*** version of G.

It follows that any homomorphism of G into a commutative group K induces a homomorphism from $G/[G, G]$ into K (cf. Exercise 1.6.13).

Let us denote by $\mathbb{Z}^{(A)}$ the set of all functions $\varphi: A \to \mathbb{Z}$ that are zero except at finitely many points in A. It is an abelian group under the operation of sum of functions.

1.6.19. Proposition. *Let $G = L(A)$. Then $G/[G, G]$ is isomorphic to the group $\mathbb{Z}^{(A)}$.*

PROOF. Let us denote δ_a the Kronecker function $A \to \mathbb{Z}$, $\delta_a(b) = 1$ or 0 according to whether $a = b$ or not. We have our associated map $f: A \to \mathbb{Z}^{(A)}$ given by $f(a) = \delta_a$. The homomorphism $h: G \to \mathbb{Z}^{(A)}$ that extends f is given by

$$h(a_1^{n_1}, \ldots, a_k^{n_k}) = \sum_{1 \le j \le k} n_j \delta_{a_j},$$

for the reduced element $(a_1^{n_1}, \ldots, a_k^{n_k})$. It is clear that h is surjective and $\operatorname{Ker} h \supseteq [G, G]$.

We claim that $\operatorname{Ker} h \subseteq [G, G]$. Assume $h((a_1^{n_1}, \ldots, a_k^{n_k})) = 0$. Let $b_1, \ldots, b_r$ be distinct points of A such that $\{b_1, \ldots, b_r\} = \{a_1, \ldots, a_k\}$. Therefore, for every i one has $\Sigma^i n_j = 0$ where Σ^i denotes the sum over those j for which $a_j = b_i$. Let $\theta: G \to G/[G, G]$ be the canonical projection. We also have

$$\theta((a_1^{n_1}, \ldots, a_k^{n_k})) = \prod_i \prod_{j : a_j = b_i} \theta(a_j^{n_j}) = \prod_i \theta(b_i)^{\Sigma^i n_j} = 1.$$

Hence $(a_1^{n_1}, \ldots, a_k^{n_k}) \in [G, G]$ and the proposition has been proved. □

1.6.20. Proposition. *Let Ω be an open subset of $\mathbb{C}$, homeomorphic to $\mathbb{C}$. The fundamental group of $\Omega \backslash \{p\}$, for $p \in \Omega$, is isomorphic to $\mathbb{Z}$. A generator can be taken to be the homotopy class of the loop γ given by $\gamma: t \mapsto p + re^{2\pi i t}$ $(0 \le t \le 1)$ and $r > 0$ is sufficiently small in order that $\bar{B}(p, r) \subseteq \Omega$.*

PROOF. It will depend on several lemmas, the first one being:

1.6.21. Lemma (Existence of a Continuous Determination of the Logarithm over any Path in $\mathbb{C}^*$). *Let γ be a path in $\mathbb{C}^*$, and let $z_0 \in \mathbb{C}$ be such that $e^{z_0} = \gamma(0)$. There is a unique path c in $\mathbb{C}$ such that $c(0) = z_0$ and $\exp(c(t)) = \gamma(t)$ for $0 \le t \le 1$.*

PROOF. Let us first prove the uniqueness of the ***lifting*** c of γ. Let $x \in [0, 1]$ and c_1, c_2 two liftings of γ defined over $[0, x]$ such that $c_1(0) = c_2(0)$. Then we have $1 = e^{c_1(t) - c_2(t)}$ for $0 \le t \le x$. Hence $c_1 - c_2$ takes values in $2\pi i \mathbb{Z}$, and since $c_1(0) - c_2(0) = 0$, we have $c_1 = c_2$ in $[0, x]$.

Let us now prove the existence of a lifting c of γ. Let $I = \{x \in [0, 1] : \exists\, c_x : [0, x] \to \mathbb{C}$, continuous, $c_x(0) = z_0$, and $\exp c_x = \gamma | [0, x]\}$. The set I is not empty because $0 \in I$. Let $x \in I$ and $0 \le x' \le x$, then $c_{x'} = c_x | [0, x']$ has all the necessary properties, hence $x' \in I$. Therefore I is an interval with endpoints 0 and b $(0 \le b \le 1)$. We want to show that $b \in I$ and $b = 1$. By the proof of uniqueness given at the beginning we have a well-defined continuous map $c_b : [0, b[\to \mathbb{C}$ such that $\exp c_b = \gamma | [0, b[$ and $c_b(0) = z_0$. Namely, for $x < b$ define $c_b(x) = c_x(x)$.

Choose $a \in \mathbb{C}$ satisfying $e^a = \gamma(b)$ but otherwise arbitrary. For $0 < \eta < 2\pi$ the disks $B(a + 2\pi i k, \eta)$ $(k \in \mathbb{Z})$ are disjoint and exp is a C^∞ diffeomorphism of each of them into a neighborhood V of $\gamma(b)$ in $\mathbb{C}^*$. There exist $\delta > 0$ and $k_0 \in \mathbb{Z}$ such that for $t \in [b - \delta, b + \delta] \cap [0, 1]$ one has $\gamma(t) \in V$, and for $t \in [b - \delta, b[\cap [0, 1]$ one has $c_b(t) \in B(a + 2\pi i k_0, \eta)$. Define a continuous function c in $[0, b + \delta] \cap [0, 1]$ by

$$c(t) := \begin{cases} c_b(t) & \text{if } t < b \\ (\exp | B(a + 2\pi i k_0, \eta))^{-1} \circ \gamma(t) & \text{if } t \in]b - \delta, b + \delta] \cap [0, 1]. \end{cases}$$

The function c is continuous since the two definitions coincide in $]b-\delta, b[\cap[0,1]$. Therefore c is a lifting of γ over $[0, b+\delta]\cap[0,1]$ such that $c(0)=0$. This shows $I=[0,1]$. □

1.6.22. Lemma. *Let $H:[0,1]\times[0,1]\to\mathbb{C}^*$ be a homotopy with fixed endpoints betwen two paths γ_0 and γ_1 in $\mathbb{C}^*$. Let $z_0=\gamma_0(0)=\gamma_1(0)$, $z_1=\gamma_0(1)=\gamma_1(1)$, and $a\in\mathbb{C}$ be such that $e^a=z_0$. For every $s\in[0,1]$ let c_s be the unique path starting at a which lifts the path $\gamma_s: t\mapsto\gamma_s(t)=H(t,s)$. Define $\tilde{H}$ by $\tilde{H}(t,s)=c_s(t)$. Then every c_s has the same endpoint, $\tilde{H}$ is a homotopy with fixed endpoints between c_0 and c_1, and $\exp\tilde{H}=H$.*

PROOF. Note that $\exp\tilde{H}=H$ follows from the definition (as was done in the previous lemma). Choose $\eta>0$ such that exp is a diffeomorphism from every $B(a+2\pi ik,\eta)$ onto a neighborhood V of z_0. We claim:

(i) There is an $\varepsilon_0>0$ such that $\tilde{H}$ is continuous in $[0,\varepsilon_0[\times[0,1]$. Since $H(0,s)=z_0$, $0\le s\le 1$, then there is ε_0 such that $H([0,\varepsilon_0[\times[0,1])\subseteq V$. The uniqueness of the lifting of the paths γ_s over $[0,\varepsilon_0[$ shows that $c_s|[0,\varepsilon_0]=(\exp|B(a,\eta))^{-1}\circ\gamma_s|[0,\varepsilon_0[$. Hence $\tilde{H}=(\exp|B(a,\eta))^{-1}\circ H$ on $[0,\varepsilon_0[\times[0,1]$, which shows that (i) holds.

(ii) $\tilde{H}$ is continuous in $[0,1]\times[0,1]$. Assume $\tilde{H}$ is not continuous at the point (t_0,σ) Define τ as $\tau=\inf\{t:\tilde{H}$ discontinuous at the point $(t,\sigma)\}$. Then $\varepsilon_0\le\tau$. Let $x=H(\tau,\sigma)$, $y=\tilde{H}(\tau,\sigma)$. We have by definition, $\gamma_\sigma(\tau)=x=e^y=\exp(c_\sigma(\tau))$. For some $\rho>0$, the exponential map is a diffeomorphism of $B(y,\rho)$ onto an open neighborhood W of x in $\mathbb{C}^*$. To simplify the notation we call it φ.

By the continuity of H at (τ,σ) we can find ε, $0<\varepsilon<\varepsilon_0$ such that

$$H(]\tau-\varepsilon,\tau+\varepsilon[\times]\sigma-\varepsilon,\sigma+\varepsilon[\cap[0,1]\times[0,1])\subseteq W.$$

In particular, $\gamma_\sigma(]\tau-\varepsilon,\tau+\varepsilon[)\subseteq W$. (We replace $\tau+\varepsilon$ by 1 if $\tau+\varepsilon>1$.) Hence $c_\sigma|]\tau-\varepsilon,\tau+\varepsilon[=\varphi^{-1}\circ\gamma_\sigma|]\tau-\varepsilon,\tau+\varepsilon[$ takes its values in $B(y,\rho)$.

Let t_1 be arbitrary in $]\tau-\varepsilon,\tau[$. We have $\tilde{H}(t_1,\sigma)=c_\sigma(t_1)\in B(y,\rho)$. The continuity of $\tilde{H}$ at (t_1,σ) allows us to find δ, $0<\delta<\varepsilon$, such that $\tilde{H}(t_1,s)=c_s(t_1)\in B(y,\rho)$ for $s\in]\sigma-\delta,\sigma+\delta[$. (Again we replace $\sigma+\delta$ by 1 if $\sigma+\delta>1$.) The uniqueness of the lifting gives $c_s|]\tau-\varepsilon,\tau+\varepsilon[=\varphi^{-1}\circ\gamma_s|]\tau-\varepsilon,\tau+\varepsilon[$ for every $s\in]\sigma-\delta,\sigma+\delta[$ (since both sides coincide at $\sigma=t_1$). Hence $\tilde{H}=\varphi^{-1}\circ H$ in $]\tau-\varepsilon,\tau+\varepsilon[\times]\sigma-\delta,\sigma+\delta[$, showing that $\tilde{H}$ is continuous at the points (t,σ) with $\tau\le t\le\tau+\varepsilon$. This contradiction with the choice of τ shows that $\tilde{H}$ is continuous over the whole square $[0,1]\times[0,1]$ as asserted by (ii).

Since $\exp(\tilde{H})=H$ and $H(\{1\}\times[0,1])=\{z_1\}$, we have $\tilde{H}(\{1\}\times[0,1])\subseteq\exp^{-1}(z_1)\in b+2\pi i\mathbb{Z}$ ($e^b=z_1$). By continuity there is a unique $k_0\in\mathbb{Z}$ such that $\tilde{H}(\{1\}\times[0,1])=\{b+2\pi ik_0\}$. Therefore, c_0 and c_1 are homotopic by $\tilde{H}$ with fixed endpoints a and $c_0(1)=c_1(1)=b+2\pi ik_0$. □

1.6.23. Definition. Let γ be a closed loop in $\mathbb{C}^*$ with base point z_0. We call ***degree*** of γ the integer $(c(1)-c(0))/2\pi i$, where c is a lifting of γ. Denote by $d(\gamma)$ this integer.

This definition presupposes that $(c(1) - c(0))/2\pi i$ does not depend on the liefting chosen. But if c_1 and c_2 are two liftings of γ and if $n = (c_1(0) - c_2(0))/2\pi i$, then $c_2' = c_1 - 2\pi in$ is another lifting of γ starting at $c_2'(0) = c_1(0) - 2\pi in = c_2(0)$. Hence $c_2' = c_2$ and it follows that $c_2(1) = c_2'(1) = c_1(1) - 2\pi in$, implying $c_2(1) - c_2(0) = c_1(1) - c_1(0)$.

For example, if γ_n is the loop in $\mathbb{C}^*$ with base point 1, $\gamma_n(t) = \exp(2\pi int)$, $n \in \mathbb{Z}$, then $d(\gamma_n) = n$ since $t \mapsto 2\pi int$ is a lifting of γ_n.

1.6.24. Lemma *Two loops γ_0, γ_1 in $\mathbb{C}^*$, with base point z_0, are homotopic in $\mathbb{C}^*$ with fixed endpoint if and only if $d(\gamma_0) = d(\gamma_1)$.*

Proof. If $d(\gamma_0) = d(\gamma_1) = d$, the liftings c_0 and c_1 of γ_0 and γ_1 respectively, which are defined by $c_0(0) = c_1(0) = a$ $(e^a = z_0)$, must satisfy $c_0(1) = c_1(1) = a + 2\pi id$. Since $\mathbb{C}$ is simply connected (being star-shaped) it follows that c_0 and c_1 are homotopic by a means of a homotopy $\tilde{H}$ with fixed endpoints. Hence γ_0 and γ_1 are homotopic by $H := \exp(\tilde{H})$.

The converse is valid by 1.6.22. □

1.6.25. Lemma. *Let γ_0, γ_1 be two loops in $\mathbb{C}^*$ with base point z_0. We have*

$$d(\gamma_0\gamma_1) = d(\gamma_0) + d(\gamma_1).$$

Proof. In fact, if $e^a = z_0$ and c_0, c_1 are the respective liftings of γ_0, γ_1 starting at a, then

$$c(t) := \begin{cases} c_0(2t) & 0 \le t \le \frac{1}{2} \\ c_1(2t-1) + c_0(1) - a & \frac{1}{2} \le t \le 1 \end{cases}$$

is a lifting of $\gamma_0\gamma_1$ starting at a. Therefore $d(\gamma_0\gamma_1) = (c(1) - c(0))/2\pi i$, hence

$$\begin{aligned} d(\gamma_0\gamma_1) &= (c_1(1) + c_0(1) - a - c_0(0))/2\pi i \\ &= ((c_1(1) - c_1(0))/2\pi i) + ((c_0(1) - c_0(0))/2\pi i) \\ &= d(\gamma_0) + d(\gamma_1). \end{aligned}$$

□

1.6.26. Lemma. *Let Ω_1, Ω_2 be two open subsets of $\mathbb{C}$ homeomorphic by $f : \Omega_1 \to \Omega_2$ and $z_0 \in \Omega_1$. Then $\pi_1(\Omega_1; z_0)$ and $\pi_1(\Omega_2; f(z_0))$ are isomorphic by $f_* : [\alpha] \to [f \circ \alpha]$.*

Proof. Exercise. □

Let us go back to the proof of Proposition 1.6.20. By Lemma 1.6.26 we can assume $\Omega = \mathbb{C}$ and $p = 0$. Hence $\Omega \backslash \{p\} = \mathbb{C}^*$. If $z_0 \in \mathbb{C}^*$, the map $d : \mathscr{C}(\mathbb{C}^*; z_0) \to \mathbb{Z}$ which associates to γ its degree $d(\gamma)$ passes by 1.6.24 to the quotient and induces $\bar{d} : \pi_1(\mathbb{C}^*; z_0) \to \mathbb{Z}$, $\bar{d}([\gamma]) = d(\gamma)$. This map is a group isomorphism. Its inverse is $n \mapsto [\gamma_n]$, $\gamma_n(t) = z_0 e^{2\pi int}$. A generator of $\pi_1(\mathbb{C}^*; z_0)$ is then $[\gamma_1]$. This was the result we wanted to prove (cf. Exercise 1.6.14). □

1.6.27. Proposition. *Let Ω be an open subset of $\mathbb{C}$ homeomorphic to $\mathbb{C}$, $p_1, \ldots, p_n$ distinct points in Ω. Let $\gamma_j : t \mapsto p_j + r_j e^{2\pi i t}$ $(0 \le t \le 1)$, where $r_j > 0$ are so small that $\bar{B}(p_j, r_j) \subseteq \Omega \setminus \{p_1, \ldots, p_{j-1}, p_{j+1}, \ldots, p_n\}$ $(1 \le j \le n)$. Then the fundamental group of $\Omega \setminus \{p_1, \ldots, p_n\}$ is isomorphic to the free group generated by $A_n = \{\gamma_1, \ldots, \gamma_n\}$.*

PROOF. Note that γ_j is free homotopic in $\Omega \setminus \{p_1, \ldots, p_n\}$ to $t \mapsto p_j + \varepsilon e^{2\pi i t}$ for any $0 < \varepsilon < r_j$, in fact to any loop γ in $B(p_j, r_j)$ with $d(\gamma) = 1$ in $\mathbb{C} \setminus \{p_j\}$. This follows from the previous proposition. With this remark in mind, we can now suppose $\Omega = \mathbb{C}$. After a possible renumbering we can suppose there are two parallel lines λ_1, λ_2 such that in the open half-plane V determined by λ_1 and containing p_n there are no other p_j in $\bar{V}$ and $\lambda_2 \subseteq V$. Similarly, we require that $p_1, \ldots, p_{n-1}$ belong to the half-plane U determined by λ_2 which contains λ_1 and that $p_n \notin \bar{U}$. (See Figure 1.9 for a similar situation.) Then $U \cap V = S$ is a strip with $\partial S = \lambda_1 \cup \lambda_2$ and no p_j belongs to $\bar{S}$. Note that we also have $U \cup V = \mathbb{C}$. Due to the preceding remark we can assume that the γ_j are in fact circles as in the statement of the proposition, but with radii so small that they do not intersect $\bar{S}$ and that the corresponding closed disks are disjoint. We can take the base point $z_0 \in S$ and choose paths α_j starting at z_0, ending at a point $\zeta_j \in \gamma_j([0, 1])$ and, furthermore, α_n is entirely contained in V and the others are entirely contained in U.

Let $\alpha \in \mathscr{C}(\mathbb{C} \setminus \{p_1, \ldots, p_n\}; z_0)$. There is a partition $0 = t_0 < t_1 < \cdots < t_N = 1$ of the unit interval such that

(1) $\alpha(t_i) \in S$, $0 \le i \le N$;
(2) for every i $(0 \le i \le N - 1)$ either $\alpha([t_i, t_{i+1}]) \subseteq U$ or $\alpha([t_i, t_{i+1}]) \subseteq V$;
(3) in successive intervals $[t_i, t_{i+1}]$ and $[t_{i+1}, t_{i+2}]$ the images do not lie in the same half-plane.

Such a partition exists: if M is an integer sufficiently large so that $1/M$ is smaller than the Lebesgue number of the covering $(\alpha^{-1}(U), \alpha^{-1}(V))$ of $[0, 1]$, then one has $\alpha\left(\left[\frac{k-1}{M}, \frac{k}{M}\right]\right)$ contained entirely in U or V $(1 \le k \le M)$. If $\alpha\left(\frac{k}{M}\right) \notin S$, then one can remove the point k/M since then either both $\alpha\left(\left[\frac{k-1}{M}, \frac{k}{M}\right]\right)$ and $\alpha\left(\left[\frac{k}{M}, \frac{k+1}{M}\right]\right)$ are in U or both are in V. We can eliminate more points if necessary to satisfy (3). The t_i are the remaining points.

Let now $\beta_i = \alpha | [t_i, t_{i+1}]$ and pick δ_i path in S from z_0 to $\alpha(t_i)$ (with the choice $\delta_0 = \delta_N = \varepsilon_{z_0}$). The loops $\eta_i = \delta_i \beta_i \bar{\delta}_{i+1}$ have base point z_0 and lie entirely in U or V.

We are going to continue the proof by induction on n, the case $n = 1$ being the previous proposition. We can assume more precise knowledge of the isomorphism between $\pi_1(U \setminus \{p_1, \ldots, p_{n-1}\}; z_0)$ and $L(A_{n-1})$. (It is just a question of reformulating the induction hypothesis and verifying that the case $n = 1$ still holds.) Namely, let c_j be the loops in $\mathbb{C} \setminus \{p_1, \ldots, p_{n-1}\}$ with base point z_0, $c_j := \alpha_j \gamma_j \bar{\alpha}_j$. We require:

Every loop σ with base point z_0 in $U\setminus\{p_1,\ldots,p_{n-1}\}$ is homotopic to a unique loop which is ε_{z_0} or can be written in the reduced form $c_{l_1}^{q_1}\ldots c_{l_k}^{q_k}$, where $q_i \in \mathbb{Z}^*$ and $l_i \in \{1,\ldots,n-1\}$. The isomorphism is given by $[\sigma]\mapsto \gamma_{l_1}^{q_1}\ldots\gamma_{l_k}^{q_k}$ (and $[\varepsilon_{z_0}]\mapsto 1$).

Returning now to the loop α, we remark that it is homotopic to the loop $\beta = \eta_0\eta_1\ldots\eta_{N-1}$. The case $n=1$ and the preceding induction hypothesis ensure that every η_i is either

homotopic to some $c_{l_1}^{q_1}\ldots c_{l_k}^{q_k}$ written in reduced form, $l_i \in \{1,\ldots,n-1\}$ if η_i lies in $U\setminus\{p_1,\ldots,p_{n-1}\}$, or
homotopic to c_n^q if η_i lies in $V\setminus\{p_n\}$.

It follows that α is homotopic to an expression of the form $c_{m_1}^{r_1}\ldots c_{m_s}^{r_s}$, $m_j \in \{1,\ldots,n\}$ directly juxtaposing the reduced forms. Condition (3) imposed on the subdivision points t_i implies this is a reduced expression already.

The map $\theta:\pi_1(\mathbb{C}\setminus\{p_1,\ldots,p_n\};z_0)\to L(A_n)$, given by $[\alpha]\mapsto\gamma_{m_1}^{r_1}\ldots\gamma_{m_s}^{r_s}$ is also a group homomorphism. In fact, if α_1 is homotopic to $c_{l_1}^{q_1}\ldots c_{l_k}^{q_k}$ and α_2 is homotopic to $c_{m_1}^{r_1}\ldots c_{m_k}^{r_n}$ then $\alpha_1\alpha_2$ is homotopic to $\gamma_{l_1}^{q_1}\ldots\gamma_{l_k}^{q_k}\cdot\gamma_{m_1}^{r_1}\ldots\gamma_{m_k}^{r_n}$, which is exactly the product $\theta([\alpha_1])\theta([\alpha_2])$.

This homomorphism is clearly surjective. Let us see it is also injective. If $\theta([\alpha]) = \gamma_{l_1}^{q_1}\ldots\gamma_{l_k}^{q_k}$ is the neutral element in $L(A_n)$ then $q_1 = \cdots = q_k = 0$, since the reduced form in $L(A_n)$ is unique. This says that $[\alpha] = [\varepsilon_{z_0}]$. □

1.6.28 Proposition (Existence of a Continuous Logarithm). *Let Ω be a simply connected open subset of $\mathbb{C}$, $f:\Omega\to\mathbb{C}^*$ continuous, $z_0\in\Omega$, $w_0\in\mathbb{C}$ such that $e^{w_0} = f(z_0)$. There is a unique continuous function $g:\Omega\to\mathbb{C}$ such that $g(z_0)=w_0$ and* $\exp(g) = f$ (i.e., $g = \log f$).

PROOF. *Uniqueness*: If g_1, g_2 are two such liftings of f, then we have $g_1(z) - g_2(z) \in 2\pi i\mathbb{Z}$ for every $z\in\Omega$, but $g_1(z_0) - g_2(z_0) = 0$. Hence $g_1 \equiv g_2$.

Existence: Let $z\in\Omega$. For every path γ in Ω starting at z_0 and ending at z, the path $f\circ\gamma$ admits a unique lifting g_γ such that $\exp(g_\gamma) = f\circ\gamma$ and $g_\gamma(0) = w_0$. If γ_1, γ_2 are two such paths, then they are homotopic and it follows $g_{\gamma_0}(1) = g_{\gamma_1}(1)$. Define g by $g(z) := g_\gamma(1)$.

Let us show that g is continuous in Ω. Let $z_1\in\Omega$. There is $\eta > 0$ such that exp is a C^∞-diffeomorphism of $B(g(z_1),\eta)$ onto a neighborhood V of $f(z_1)$ in $\mathbb{C}^*$. There is $\delta > 0$ such that $B(z_1,\delta)\subseteq\Omega$ and $f(B(z_1,\delta))\subseteq V$.

If $\zeta\in B(z_1,\delta)$ one can find a path γ from z_0 to ζ of the form αs_ζ, where α is a fixed path from z_0 to z_1 in Ω and s_ζ is the line segment $s_\zeta(t) = t(\zeta - z_1) + z_1$. A lifting σ of $f\circ\gamma$ is therefore the following. If c is a lifting of $f\circ\alpha$ starting at w_0, let $\sigma(t) = c(2t)$ $(0\le t\le 1/2)$ and

$$\sigma(t) = ((\exp|B(g(z_1),\eta))^{-1}\circ f)((2t-1)(\zeta - z_1) + z_1)$$

$(1/2\le t\le 1)$. It follows that $g(\zeta) = \sigma(1) = ((\exp|B(g(z_1),\eta))^{-1}\circ f)(\zeta)$ for $\zeta\in B(z_1,\delta)$ and hence g is continuous. □

1.6.29. Remarks (1) If f is of class C^k then g is of less C^k.

(2) The function f has n-roots of any order $n \in \mathbb{N}^*$ (as differentiable as f), and, more generally, f^α is defined for any $\alpha \in \mathbb{C}$.

1.6.30. Remark. Given a loop γ in $\mathbb{C}^*$, we can write $\gamma(t) = |\gamma(t)| e^{i\alpha(t)}$, where α is a continuous path in $\mathbb{R}$. This follows from 1.6.21 by lifting the loop $\beta(t) = \gamma(t)/|\gamma(t)|$. By Exercise 1.6.2, $d(\gamma) = d(\beta) = \frac{1}{2\pi}(\alpha(1) - \alpha(0))$. Since $\alpha(t)$ is the argument of the complex number $\gamma(t)$, this formula is usually stated as saying "the degree of the loop γ equals the variation of the argument along γ," and it is written as

$$d(\gamma) = \frac{1}{2\pi} \underset{z \in \gamma}{\Delta} \arg z = \frac{1}{2\pi i} \underset{z \in \gamma}{\Delta} \log z.$$

This formula is also known as the ***argument principle.***

EXERCISE 1.6

1. Identify $\pi_1(\Omega, z_0)$ $(z_0 \in \Omega)$ in the following cases:
   (i) $\Omega =]-2, 2[^2 \backslash [-1, 1]^2$
   (ii) $\Omega = (]-4, 4[\times]-2, 2[) \backslash ([-1, 1]^2 \cup \bar{B}(2, \frac{1}{2}))$
   (iii) $\Omega = \mathbb{C} \backslash \left(\bigcup_{j=1}^{p} R_j \cup \{a_1, \ldots, a_q\} \right)$, where the R_j are closed half-lines, pairwise disjoint, and containing none of the points $a_1, \ldots, a_q$.
2. Show that $\mathbb{C}^*$ every loop is free homotopic to a loop contained in the unit circle $C := \{z \in \mathbb{C} : |z| = 1\}$. Show they have the same degree.
3. Show that two loops in $\mathbb{C}^*$ which have the same degree are free homotopic.
4. Show that for any loop γ in $\mathbb{C}^*$, $d(\gamma) = -d(\bar{\gamma})$.
5. Let $f : \Omega \to \mathbb{C}^*$ have a continuous logarithm g, $f = e^g$. Show that for any loop γ in Ω, $d(f \circ \gamma) = 0$.

*6. Let $f : \Omega \to \mathbb{C}^*$ be a continuous function such that $d(f \circ \gamma) = 0$ for every loop γ in Ω. Show that f admits a continuous logarithm in Ω. (Hint: Follow the proof of Proposition 1.6.28.)

7. Let $f : C \to \mathbb{C}^*$ be continuous, then there is an integer n and a continuous function $g : C \to \mathbb{C}$ such that $f(z) = z^n e^{g(z)}$.

*8. Let Ω be an open set homeomorphic to $\mathbb{C}$, $\Omega' = \Omega \backslash \{p_1, \ldots, p_n\}$, p_j distinct points of Ω. Let $f : \Omega' \to \mathbb{C}^*$ be continuous. Show there exist integers $k_1, \ldots, k_n$ and a continuous function $g : \Omega' \to \mathbb{C}$ such that

$$f(z) = (z - p_1)^{k_1} \ldots (z - p_n)^{k_n} e^{g(z)}$$

for every $z \in \Omega'$. (Hint: Let $\gamma_1, \ldots, \gamma_n$ be loops in Ω' as in Proposition 1.6.27 and $k_j = d(f \circ \gamma_j)$.)

9. Let $\alpha : [0, 1] \to \Omega$ be a continuous path in an open set $\Omega \subseteq \mathbb{C}$. Show there is a $\delta > 0$ such that if $\beta : [0, 1] \to \mathbb{C}$ is continuous and

$$\sup\{|\alpha(t) - \beta(t)| : t \in [0, 1]\} < \delta,$$

then β is another path in Ω and β is free homotopic to α in Ω. (We leave to the reader the task of supplying the definition of free homotopy for paths, similar to 1.6.14.)

10. Every path in an open subset of $\mathbb{C}$ is free homotopic in Ω to a polygonal path in Ω whose segments are parallel to one of the coordinate axes. Show that this is also true for homotopy with fixed endpoints.

*11. Let $P(z) = z^n + a_1 z^{n-1} + \cdots + a_n$, $a_j \in \mathbb{C}$ be a polynomial of degree $n > 0$ in $\mathbb{C}$ and, for $r \in]0, \infty[$, consider the loop $f_r : t \mapsto P(re^{2\pi i t})$, $t \in [0, 1]$.
   (i) Show that for r sufficiently large, $d(f_r) = n$
   (ii) Show that if P has no roots in $r_1 \leq |z| \leq r_2$ then $d(f_r)$ is constant for $r \in [r_1, r_2]$.
   (iii) Assume $a_n \neq 0$, what is $d(f_0)$?
   (iv) (Fundamental theorem of algebra) Show that P has a root in $\mathbb{C}$.

12. Let $A \neq \varnothing$, G_1, G_2 two groups, and $i_1 : A \to G_1$, $i_2 : A \to G_2$ two injective maps such that both pairs (G_1, i_1) and (G_2, i_2) satisfy the conditions 1 and 2 of Proposition 1.6.16. Show there is a unique group isomorphism $k : G_1 \to G_2$ such that $k \circ i_1 = i_2$.

13. Let G be a group, K a commutative group, and $f : G \to K$ a homomorphism. Show there is a unique homomorphism $\tilde{f} : G/[G, G] \to K$ such that $f = \tilde{f} \circ \pi$, where π is the canonical projection of G onto $G/[G, G]$. Moreover, the map $f \mapsto \tilde{f}$ is a group isomorphism from $\mathrm{Hom}(G, K)$ onto $\mathrm{Hom}(G/[G, G], K)$.

14. Let $f : \Omega \to \mathbb{C}$ be a homeomorphism, $f(p) = 0$, $r > 0$ be such that $\bar{B}(p, r) \subseteq \Omega$, and $\gamma(t) = p + re^{2\pi i t}$ $(0 \leq t \leq 1.)$ Let $z_0 \in \mathbb{C}^*$ and $\gamma_1(t) = z_0 e^{2\pi i t}$ $(0 \leq t \leq 1)$. Show that $[f \circ \gamma] = [\gamma_1]$.

§7. Integration of Closed 1-Forms Along Continuous Paths

To study better the homotopy of continuous paths one needs to extend the notion of integration of 1-forms (of class C^1). Such extension is possible if we limit ourselves to the case of closed forms.

1.7.1. Lemma. *Let Ω be an open subset of $[0, 1]$. The map $\dfrac{d}{dx} : \mathscr{E}(\Omega) \to \mathscr{E}(\Omega)$ is surjective and its kernel, denoted $\mathbb{C}(\Omega)$, is the space of functions which are locally constant on Ω.*

PROOF. Exercise. □

1.7.2. Lemma. *Let $(V_i)_{i \in I}$ be an open covering of $[0, 1]$ and for every pair $(i, j) \in I \times I$ such that $V_i \cap V_j \neq \varnothing$ let $c_{i,j} \in \mathbb{C}(V_i \cap V_j)$ be given so that for any triplet $(i, j, k) \in I \times I \times I$ such that $V_i \cap V_j \cap V_k \neq \varnothing$ they satisfy $c_{i,j} + c_{j,k} +$*

$c_{k,i} = 0$ on $V_i \cap V_j \cap V_k$. Then, for every i there is $c_i \in \mathbb{C}(V_i)$ such that $c_{i,j} = c_i - c_j$ for every (i, j) such that $V_i \cap V_j \neq \emptyset$.

PROOF. Note that by a judicious choice of (i, j, k) the hypotheses imply $c_{i,i} \equiv 0$ and $c_{i,j} = -c_{j,i}$. If neither 0 nor 1 belong to V_i we set $\tilde{V}_i = V_i$. If V_i contains only one of them, say 0, we set $\tilde{V}_i =]-\infty, 0[\cup V_i$. (Similarly, in the other case, $\tilde{V}_i = V_i \cup [1, \infty[$.) Finally, if both $0, 1 \in V_i$, set $\tilde{V}_i =]-\infty, 0] \cup V_i \cup [1, \infty[$. Now the family $(\tilde{V}_i)_{i \in I}$ is an open covering of $\mathbb{R}$. There exists a C^∞ partition of unity $(\alpha_\nu)_{\nu \geq 1}$ given by elements of $\mathcal{D}(\mathbb{R})$ subordinate to this covering: for every $\nu \geq 1$ there is $j(\nu) \in I$ such that $\alpha_\nu \in \mathcal{D}(\tilde{V}_{j(\nu)})$. Restrict this partition of unity to $[0, 1]$.

Since the family $(\operatorname{supp} \alpha_\nu)_{\nu \geq 1}$ is locally finite in $\mathbb{R}$, only finitely many α_ν are not identically zero on $[0, 1]$. Let $f_i = \sum_{\nu \geq 1} \alpha_\nu c_{i,j(\nu)} \in \mathcal{E}(V_i)$. If $V_i \cap V_j \neq \emptyset$, in $V_i \cap V_j$ we have

$$f_i - f_j = \sum_{\nu \geq 1} \alpha_\nu (c_{i,j(\nu)} - c_{j,j(\nu)}) = \sum_{\nu \geq 1} \alpha_\nu c_{i,j} = c_{i,j},$$

since, if $\alpha_\nu(x) \neq 0$ then $V_i \cap V_j \cap V_{j(\nu)} \neq \emptyset$ and hence

$$c_{i,j(\nu)} - c_{j,j(\nu)} = c_{i,j(\nu)} + c_{j(\nu),j} = -c_{j,i} = c_{i,j}.$$

Differentiating this expression we obtain $f_i' - f_j' = 0$ on $V_i \cap V_j$. Therefore, the f_i' induce a globally defined function $g \in \mathcal{E}([0, 1])$. There exists $h \in \mathcal{E}([0, 1])$ such that $h' = g$. Let $c_i = f_i - (h|V_i)$. A priori, c_i is just a C^∞ function in V_i but $c_i' = f_i' - (h'|V_i) = f_i' - g = 0$, hence $c_i \in \mathbb{C}(V_i)$. It is immediate that $c_{i,j} = c_i - c_j$. □

Recall that $\mathscr{C}([0, 1], \Omega)$ denotes the space of continuous paths $\alpha : [0, 1] \to \Omega$. We now have the following result.

1.7.3. Proposition. *Let $(U_i)_{i \in I}$ be an open covering of an open subset Ω of $\mathbb{C}$. For every $i \in I$ let $f_i : U_i \to \mathbb{C}$ be given such that if $U_i \cap U_j \neq \emptyset$ then $f_i - f_j$ is locally constant in $U_i \cap U_j$. Then there exists a unique map*

$$I : \mathscr{C}([0, 1], \Omega) \to \mathbb{C}$$

such that

(1) *If the path $\alpha \in \mathscr{C}([0, 1], \Omega)$ has its image in U_i, then*

$$I(\alpha) = f_i(\alpha(1)) - f_i(\alpha(0)).$$

(2) *If β is obtained from α by a parameter change* (i.e., $\beta = \alpha \circ \varphi$, $\varphi : [0, 1] \to [0, 1]$ *continuous*, $\varphi(0) = 0$, $\varphi(1) = 1$) *then $I(\beta) = I(\alpha)$.*

(3) *If α and β are consecutive* ($\alpha(1) = \beta(0)$) *then*

$$I(\alpha\beta) = I(\alpha) + I(\beta).$$

PROOF. *Uniqueness*: We say that a partition σ of $[0, 1]$, $\sigma : 0 = t_0 < t_1 < \cdots < t_n = 1$, is ***adapted*** to a path α in Ω if for every k, $0 \leq k \leq n - 1$, there is $i_k \in I$

such that $\alpha([t_k, t_{k+1}) \subseteq U_{i_k}$. As we have already shown a few times, e.g., in §1.6.27, such adapted partitions exist. Let $\alpha_k(t) = (\alpha|[t_k, t_{k+1}])((1-t)t_k + t_{k+1})$. Hence α is obtained from $\alpha_0\alpha_1 \ldots \alpha_{n-1}$ by a parameter change, therefore $I(\alpha) = I(\alpha_0) + \cdots + I(\alpha_{n-1})$ by (2) and (3). By (1) we now have

$$I(\alpha) = \sum_{0 \le k \le n-1} (f_{i_k}(\alpha(t_{k+1})) - f_{i_k}(\alpha(t_k))).$$

This proves the uniqueness of $I(\alpha)$. This reasoning proves in fact a little bit more: under hypotheses (1), (2), and (3), the right-hand side is independent of the subidivision σ adapted to α and of the choice of indices i_k such that $\alpha([t_k, t_{k+1}]) \subseteq U_{i_k}$.

Existence. Let $\alpha \in \mathscr{C}([0,1],\Omega)$. The family $(\alpha^{-1}(U_i))_{i\in I}$ is an open covering of $[0,1]$. For every $i \in I$, $f_i \circ \alpha$ maps $\alpha^{-1}(U_i)$ into $\mathbb{C}$. If $U_i \cap U_j \neq \varnothing$ then $\alpha^{-1}(U_i) \cap \alpha^{-1}(U_j) = \alpha^{-1}(U_i \cap U_j) \neq \varnothing$ and the function $c_{i,j} := f_i \circ \alpha - f_j \circ \alpha = (f_i - f_j) \circ \alpha$ is locally constant. On the nonempty intersection of three sets of the covering one has $c_{i,j} + c_{j,k} + c_{k,i} = 0$. By §1.7.2 there exist $g_i \in \mathbb{C}(\alpha^{-1}(U_i))$ such that

$$c_{i,j} = g_i - g_j \quad \text{on} \quad \alpha^{-1}(U_i \cap U_j) \neq \varnothing.$$

Therefore,

$$f_i \circ \alpha - g_i = f_j \circ \alpha - g_j \quad \text{on} \quad \alpha^{-1}(U_i) \cap \alpha^{-1}(U_j),$$

and this defines $h: [0,1] \to \mathbb{C}$ by $h|\alpha^{-1}(U_i) := f_i \circ \alpha - g_i$. Furthermore, $h - f_i \circ \alpha$ is locally constant on $\alpha^{-1}(U_i)$.

Let $I(\alpha) = h(1) - h(0)$. The complex number $I(\alpha)$ thus defined does not depend on the family $(g_i)_{i\in I}$. Namely, let $(\tilde{g}_i)_{i\in I}$ be a second family of functions with the same properties. Define $\tilde{h}$ by patching together $f_i \circ \alpha - \tilde{g}_i$ as before. Then the functions $\tilde{h} - f_i \circ \alpha$ are also locally constant on $\alpha^{-1}(U_i)$. Therefore, $h - \tilde{h}$ is locally constant on $[0,1]$, hence constant. It follows that $\tilde{h}(1) - \tilde{h}(0) = h(1) - h(0)$, as we wanted to show.

This function I has the desired properties on $\mathscr{C}([0,1],\Omega)$:

(a) If α is a path with $\alpha([0,1]) \subseteq U_i$, then $h - f_i \circ \alpha$ is locally constant on $[0,1]$, hence constant and

$$I(\alpha) = h(1) - h(0) = f_i(\alpha(1)) - f_i(\alpha(0)).$$

(b) Let β be obtained from α by a change of parameters, $\beta = \alpha \circ \varphi$. Since $h - f_i \circ \alpha$ is locally constant on $\alpha^{-1}(U_i)$, then

$$h \circ \varphi - f_i \circ \beta = h \circ \varphi - f_i \circ \alpha \circ \varphi$$

is locally constant on $\varphi^{-1}(\alpha^{-1}(U_i)) = \beta^{-1}(U_i)$. Moreover, if we start with $g_{i,j} = f_i \circ \beta - f_j \circ \beta$ then $g_{i,j} = c_{i,j} \circ \varphi = g_i \circ \varphi - g_j \circ \varphi$ and then, to define $I(\beta)$, we can consider $\tilde{h} = f_i \circ \beta - g_i \circ \varphi = h \circ \varphi$. Therefore,

$$I(\beta) = \tilde{h}(1) - \tilde{h}(0) = h(\varphi(1)) - h(\varphi(0)) = h(1) - h(0) = I(\alpha).$$

(c) Let $\alpha, \beta \in \mathscr{C}([0,1],\Omega)$ be such that $\alpha(1) = \beta(0)$. Denote by ℓ_1, ℓ_2 the maps

$t \mapsto 2t$ $(0 \leq t \leq \frac{1}{2})$, $t \mapsto 2t - 1$. The set $A = \{i \in I : \alpha(1) = \beta(0) = \alpha\beta(1/2) \in U_i\}$ is not empty. The covering sets are given by

$$(\alpha\beta)^{-1}(U_i) = \ell_1^{-1}(\alpha^{-1}(U_i)) \cup \ell_2^{-1}(\beta^{-1}(U_i)).$$

Let us denote by $g_i^\alpha, g_i^\beta, g_i^{\alpha\beta}, h^\alpha, h^\beta, h^{\alpha\beta}$ the functions associated to the paths $\alpha, \beta, \alpha\beta$. Let now $i, j \in A$. We have the relations

$$(f_i - f_j) \circ \alpha = g_i^\alpha - g_j^\alpha,$$

$$(f_i - f_j) \circ \beta = g_i^\beta - g_j^\beta.$$

Since $f_i(\alpha(1)) - f_j(\alpha(1)) = f_i(\beta(0)) - f_j(\beta(0))$, we obtain $g_i^\alpha(1) - g_j^\alpha(1) = g_i^\beta(0) - g_j^\beta(0)$. Hence, the quantity $s = g_i^\alpha(1) - g_i^\beta(0)$ does not depend on the index $i \in A$. We can therefore define functions g_i by:

$$g_i(t) := \begin{cases} g_i^\alpha(2t) & \text{if } t \in \ell_1^{-1}(\alpha^{-1}(U_i)) \\ g_i^\beta(2t - 1) + s & \text{if } t \in \ell_2^{-1}(\beta^{-1}(U_i)). \end{cases}$$

These functions are well defined on $(\alpha\beta)^{-1}(U_i)$ since if

$$t \in \ell_1^{-1}(\alpha^{-1}(U_i)) \cap \ell_2^{-1}(\beta^{-1}(U_i))$$

it must be the case that $t = 1/2$ and $i \in A$ (recall $0 \leq t \leq 1$ always holds). One verifies without difficulty that they are locally constant. If $0 \leq t \leq 1/2$ and t is in $(\alpha\beta)^{-1}(U_i) \cap (\alpha\beta)^{-1}(U_j)$ $(= \ell_1^{-1}(\alpha^{-1}(U_i \cap U_j)))$ then

$$g_i(t) - g_j(t) = g_i^\alpha(2t) - g_j^\alpha(2t) = (f_i \circ \alpha)(2t) - (f_j \circ \alpha)(2t) = ((f_i - f_j) \circ (\alpha\beta))(t).$$

If $1/2 \leq t \leq 1$ and t is in $(\alpha\beta)^{-1}(U_i) \cap (\alpha\beta)^{-1}(U_j)$ $(= \ell_2^{-1}(\beta^{-1}((U_i \cap U_j)))$ then we have

$$\begin{aligned} g_i(t) - g_j(t) &= (g_i^\beta(2t - 1) + s) - (g_j^\beta(2t - 1) + s) \\ &= g_i^\beta(2t - 1) - g_j^\beta(2t - 1) = ((f_i - f_j) \circ \beta)(2t - 1) \\ &= ((f_i - f_j) \circ (\alpha\beta))(t). \end{aligned}$$

Therefore, the functions $f_i \circ (\alpha\beta) - g_i$ determine a global function $h^{\alpha\beta}$. We have

$$I(\alpha\beta) = h^{\alpha\beta}(1) - h^{\alpha\beta}(0) = h^{\alpha\beta}(1) - h^{\alpha\beta}(1/2) + h^{\alpha\beta}(1/2) - h^{\alpha\beta}(0).$$

On the other hand,

$$h^{\alpha\beta}(1) - h^{\alpha\beta}(1/2) = h^\beta(1) - h^\beta(0) = I(\beta).$$

Namely, for $t \geq 1/2$, $t \in (\alpha\beta)^{-1}(U_i)$ we have

$$h^{\alpha\beta}(t) = (f_i \circ (\alpha\beta))(t) - g_i(t) = f_i(\beta(2t - 1)) - (g_i^\beta(2t - 1) + s).$$

Hence, for $t = 1$, if $1 \in (\alpha\beta)^{-1}(U_i)$, we get

$$h^{\alpha\beta}(1) = f_i(\beta(1)) - g_i^\beta(1) - s = h^\beta(1) - s.$$

Similarly, for $t = 1/2$, we have

$$h^{\alpha\beta}(1/2) = f_i(\beta(0)) - g_i^\beta(0) - s = h^\beta(0) - s.$$

This proves the claim. In the same way we obtain

$$h^{\alpha\beta}(1/2) - h^{\alpha\beta}(0) = h^{\alpha}(1) - h^{\alpha}(0) = I(\beta).$$

This ends the proof of the proposition. □

1.7.4. Proposition. *Under the same conditions as in §1.7.3, the map I defined there has the following additional properties:*

(1) $I(\varepsilon_a) = 0$ *for every* $a \in \Omega$.
(2) *If* α, β *are two paths which are homotopic with fixed endpoints, then* $I(\alpha) = I(\beta)$.
(3) *For every path* α *in* Ω, $I(\bar{\alpha}) = -I(\alpha)$.
(4) *If* α, β *are two free homotopic loops in* Ω, *then* $I(\alpha) = I(\beta)$.

PROOF. (1) There is some i for which $a \in U_i$, that is, $\varepsilon_a([0,1]) \subseteq U_i$. Hence $I(\varepsilon_a) = f_i(\varepsilon_a(1)) - f_i(\varepsilon_a(0)) = f_i(a) - f_i(a) = 0$.

(2) Let H be a homotopy with fixed endpoints carrying α into β. Let us denote $a = \alpha(0) = \beta(0)$, $b = \alpha(1) = \beta(1)$, and H_s is the path $t \mapsto H(t,s)$. We are going to show that the function $s \mapsto I(H_s)$ is locally constant. This will suffice to prove (2) since $H_0 = \alpha$, $H_1 = \beta$.

Given $s \in [0,1]$ there is a partition $0 = t_0 < t_1 < \cdots < t_n = 1$ and open sets $U_{i_0}, \ldots, U_{i_{n-1}}$ of the covering such that $H_s([t_k, t_{k+1}]) \subseteq U_{i_k}$. By the uniform continuity of H, there is $\delta > 0$ such that if $|s - s'| < \delta$ and $s' \in [0,1]$ then $H_{s'}([t_k, t_{k+1}]) \subseteq U_{i_k}$ is also valid. As observed in the proof of the uniqueness in §1.7.3 we have

$$\begin{aligned} I(H_s) &= f_{i_0}(H_s(t_1)) - f_{i_0}(H_s(0)) + \cdots + f_{i_{n-1}}(H_s(1)) - f_{i_{n-1}}(H_s(t_{n-1})) \\ &= -f_{i_0}(H_s(0)) + (f_{i_0}(H_s(t_1)) - f_{i_1}(H_s(t_1))) + \cdots \\ &\quad + (f_{i_{n-2}}(H_s(t_{n-1})) - f_{i_{n-1}}(H_s(t_{n-1}))) + f_{i_{n-1}}(H_s(1)) \end{aligned}$$

and the same identity holds for $H_{s'}$.

Since $f_i - f_j$ is locally constant in $U_i \cap U_j$, there is a neighborhood of $H_s(t_k)$ where $f_{i_{k-1}} - f_{i_k}$ is constant. By reducing the size of δ if necessary, we can assume $H_{s'}(t_k)$ belongs to the same neighborhood for $0 \le k \le n$. Hence, for $1 \le k \le n$ we have

$$(f_{i_{k-1}} - f_{i_k})(H_s(t_k)) = (f_{i_{k-1}} - f_{i_k})(H_{s'}(t_k)).$$

It follows that

$$\begin{aligned} I(H_s) - I(H_{s'}) &= (f_{i_0}(H_{s'}(0)) - f_{i_0}(H_s(0))) \\ &\quad + (f_{i_{n-1}}(H_s(1)) - f_{i_{n-1}}(H_{s'}(1))) = 0. \end{aligned}$$

(3) One remarks that $\alpha\bar{\alpha}$ is homotopic to ε_a with fixed endpoints. Hence $0 = I(\varepsilon_a) = I(\alpha\bar{\alpha}) = I(\alpha) + I(\bar{\alpha})$.

(4) Let α, β be two free homotopic loops, H the corresponding homotopy. The starting point of the path $H_s(t) = H(t,s)$ traverses a path γ joining

$a = \alpha(0) = \alpha(1)$ to $b = \beta(0) = \beta(1)$. The loop $\gamma\beta\bar{\gamma}$ is homotopic with fixed endpoints to α (up to a change of parameters) by the homotopy

$$\Phi(t, s) = \begin{cases} 0 \leq t \leq 1/3 & : H(0, 3ts) \\ 1/3 \leq t \leq 2/3 : & H(3t - 1, s) \\ 2/3 \leq t \leq 1 & : H(1, 3s - 3ts). \end{cases}$$

Hence

$$I(\alpha) = I(\gamma\beta\bar{\gamma}) = I(\gamma) + I(\beta) - I(\gamma) = I(\beta). \qquad \square$$

1.7.5. Proposition. *Let $(U_i, f_i)_{i \in I}$, $(V_j, g_j)_{j \in J}$ be two families satisfying the conditions of* §1.7.3. *Assume further that for any pair $(i, j) \in I \times J$ such that $U_i \cap U_j \neq \varnothing$, the difference $f_i - g_j$ is locally constant on $U_i \cap V_j$. Then the integration functionals I_1, I_2 associated to the two families by* §1.7.3 coincide.

Proof. It is enough to prove the uniqueness in the case in which one of the coverings consists of a single set, since then, for every U_i and every path α in U_i one can compute $f_i(\alpha(1)) - f_i(\alpha(0))$ in terms of the covering $(U_i \cap V_j)_{j \in J}$ and the functions $(g_j|(U_i \cap V_j))_{j \in J}$.

Let us assume then that I contains a single element Ω, and a corresponding function f. Let $\sigma : 0 = t_0 < t_1 < \cdots < t_n = 1$ be a subdivision adapted to α and the covering $(V_j)_{j \in J}$, so that for $0 \leq k \leq n - 1$ there exists $j_k \in J$ such that $\alpha([t_k, t_{k+1}]) \in V_{j_k}$. One has

$$I_2(\alpha) = \sum_{0 \leq k \leq n-1} (g_{j_k}(\alpha(t_{k+1})) - g_{j_k}(\alpha(t_k))),$$

and

$$I_1(\alpha) = f(\alpha(1)) - f(\alpha(0)) = \sum_{0 \leq k \leq n-1} (f(\alpha(t_{k+1})) - f(\alpha(t_k))).$$

Now for every k, the function $x \mapsto f(x) - g_{j_k}(x)$ is constant over the connected set $\alpha([t_k, t_{k+1}]) \subseteq U_{j_k}$. Hence

$$I_1(\alpha) - I_2(\alpha) = \sum_{0 \leq k \leq n-1} [(f - g_{j_k})(\alpha(t_{k+1})) - (f - g_{j_k})(\alpha(t_k))] = 0.$$

The proposition has therefore been proved. $\square$

Given $\omega \in \mathscr{E}_1^1(\Omega)$, ω closed, let S be the collection of open subsets V of Ω for which there is a primitive g_V of ω on V. Poincaré's lemma (Proposition 1.2.2.a) ensures that S is a covering of Ω. Furthermore, if U, $V \in S$ and $U \cap V \neq \varnothing$ it follows that $g_U - g_V$ is locally constant on $U \cap V$. There is hence a unique functional

$$I_\omega : \mathscr{C}([0, 1], \Omega) \to \mathbb{C}$$

which verifies the properties stated in §1.7.3 and §1.7.4. By Proposition 1.7.5 it does not depend on the choice of subcovering or of primitives.

The relation between this functional and the integration of ω along a piecewise-C^1 path is given by the following lemma.

1.7.6. Lemma. *If α is a piecewise-C^1 path in Ω and $\omega \in \mathscr{E}_1^1(\Omega)$ is a closed form, then*

$$I_\omega(\alpha) = \int_\alpha \omega.$$

PROOF. If $\alpha([0,1]) \subseteq V$ *for some* $V \in S$, then

$$I_\omega(\alpha) = g_V(\alpha(1)) - g_V(\alpha(0)) = \int_\alpha dg_V = \int_\alpha \omega.$$

If not, let $0 = t_0 < t_1 < \cdots < t_n = 1$ be a partition adapted to α. Then $\alpha([t_k, t_{k+1}]) \subseteq V_k$, $0 \le k \le n-1$ and

$$I_\omega(\alpha) = \sum_{0 \le k \le n-1} (g_{V_k}(\alpha(t_{k+1})) - g_{V_k}(\alpha(t_k)))$$

$$= \sum_{0 \le k \le n-1} \int_{\alpha|[t_k, t_{k+1}]} \omega = \int_\alpha \omega. \qquad \square$$

1.7.7. Proposition. *Let $\omega \in \mathscr{E}_1^1(\Omega)$ be a closed form. Then:*

(1) If α, β are two piecewise-C^1 paths in Ω, homotopic with fixed endpoints or, if they are piecewise-C^1 loops, free homotopic, then

$$\int_\alpha \omega = \int_\beta \omega.$$

(2) *If α, β are two consecutive piecewise-C^1 paths then*

$$\int_{\alpha\beta} \omega = \int_\alpha \omega + \int_\beta \omega.$$

(3) *If α is a piecewise-C^1 path, then*

$$\int_{\bar{\alpha}} \omega = -\int_\alpha \omega.$$

PROOF. A consequence of §1.7.4 and §1.7.6. $\square$

1.7.8. Remark. Let Ω_1, Ω_2 be two open subsets of $\mathbb{C}$ and $h : \Omega_1 \to \Omega_2$ a C^1 map with positive Jacobian. If α is a piecewise-C^1 path in Ω_1 and $\omega \in \mathscr{E}_1^1(\Omega_2)$ then we know that

$$\int_{h \circ \alpha} \omega = \int_\alpha h^*(\omega).$$

If, moreover, ω is closed and $(U_i)_{i \in I}$ is a covering of Ω_2 by open sets on which ω admits a primitive g_i, then $h^*(\omega)$ is closed in Ω_1 and has a primitive $h^*(g_i) = g_i \circ h = f_i$ in the open sets $h^{-1}(U_i)$ of the covering $(h^{-1}(U_i))_{i \in I}$ of Ω_1.

One concludes that

$$I_\omega(h \circ \alpha) = I_{h^*(\omega)}(\alpha).$$

Let's consider now a connected open set Ω, $a \in \Omega$, $\omega \in \mathscr{E}_1^1(\Omega)$ a closed form.

The map $I_\omega : \mathscr{C}([0,1], \Omega) \to \mathbb{C}$ restricted to $\mathscr{C}(\Omega; a)$ can be factored through a map $j_\omega : \pi_1(\Omega; a) \to \mathbb{C}$ which makes the following diagram commutative ($q(\alpha) = [\alpha]$):

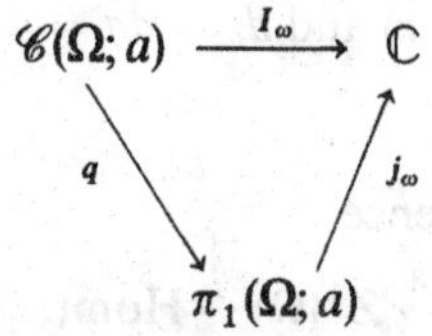

Since $I_\omega(\alpha\beta) = I_\omega(\alpha) + I_\omega(\beta)$, the map j_ω is a group homomorphism. The values $j_\omega([\alpha])$ are called the ***periods of*** ω and, correspondingly, the subgroup $\operatorname{Im} j_\omega$ of $\mathbb{C}$ is called the ***group of periods*** of ω (or the **lattice of periods** of ω).

Consider now the set $\operatorname{Hom}(\pi_1(\Omega; a), \mathbb{C})$ of group homomorphisms from $\pi_1(\Omega; a)$ into $\mathbb{C}$. It is clearly a complex vector space. The map

$$j : Z^1(\Omega) \to \operatorname{Hom}(\pi_1(\Omega; a), \mathbb{C})$$

$$\omega \mapsto j_\omega$$

vanishes on $B^1(\Omega)$, hence passes to the quotient $H^1(\Omega) = Z^1(\Omega)/B^1(\Omega)$ as shown by the following proposition.

1.7.9. Proposition. *Let Ω be a connected open set in $\mathbb{C}$. The kernel of the map j defined earlier is exactly $B^1(\Omega)$. Furthermore, this map induces an isomorphism $\bar{j}$ of complex vector spaces*

$$\bar{j} : H^1(\Omega) \to \operatorname{Hom}(\pi_1(\Omega; a), \mathbb{C}).$$

In particular, two closed 1-forms are cohomologous if and only if they have the same periods.

PROOF. We will admit for the moment the surjectivity of the map j. It will be proved in Chapter 5, Proposition 5.14.14.

If $\omega = df$, one needs a single set in the covering to construct I_ω. Then if α is a loop we have

$$I_\omega(\alpha) = f(\alpha(1)) - f(\alpha(0)) = 0.$$

Therefore j vanishes on $B^1(\Omega)$ and induces $\bar{j}$.

Let us show that $\bar{j}$ is injective. Let $\omega \in Z^1(\Omega)$ be such that $j_\omega = 0$. Let c_z be an arbitrary path in Ω joining a to z, then $I_\omega(c_z)$ does not depend on the path c_z. Namely, if c_1, c_2 are two such paths, then $[c_1\bar{c}_2] \in \pi_1(\Omega; a)$ and

$$0 = I_\omega(c_1\bar{c}_2) = I_\omega(c_1) - I_\omega(c_2).$$

Let us now show that the (well-defined) function

$$f(z) := I_\omega(c_z)$$

is C^∞ in Ω. Let $z_0 \in \Omega$ and $r > 0$ be such that $B(z_0, r) \subseteq \Omega$. There is a C^∞ primitive g of ω in $B(z_0, r)$. For $z \in B(z_0, r)$ let α_z be the segment $\alpha_z(t) = z_0 + t(z - z_0)$. The path $c_{z_0}\alpha_z\bar{c}_z$ is a loop with base point a. Hence $I_\omega(c_{z_0}\alpha_z\bar{c}_z) = 0$. In other words, $f(z_0) + g(z) - g(z_0) - f(z) = 0$, which means that in $B(z_0, r)$, f can be represented as

$$f(z) = g(z) + f(z_0) - g(z_0).$$

It follows that f is C^∞ in $B(z_0, r)$ and $df = dg = \omega$ in $B(z_0, r)$. Hence $f \in \mathscr{E}(\Omega)$ and $df = \omega$, or $\omega \in B^1(\Omega)$. □

1.7.10. Remarks. (a) The sequence

$$0 \to \mathbb{C} \to \mathscr{E}^1(\Omega) \xrightarrow{d} Z^1(\Omega) \xrightarrow{j} \mathrm{Hom}(\pi_1(\Omega; a), \mathbb{C}) \to 0$$

is exact.

(b) If U is an open set homeomorphic to $\mathbb{C}$ and $p_1, \ldots, p_n$ are distinct points in U, then, for $\Omega = U \setminus \{p_1, \ldots, p_n\}$, we know that $H^1(\Omega)$ is isomorphic to $\mathbb{C}^n$ and every element $\tilde{\omega} \in H^1(\Omega)$ can be written in a unique way as

$$\tilde{\omega} = \sum_{1 \le j \le n} \lambda_j \tilde{\omega}_j,$$

where $\lambda_j \in \mathbb{C}$ and $\omega_j = \dfrac{1}{2\pi i} \dfrac{dz}{z - p_j}$.

We know also that every loop α in Ω with base point a is homotopic to a loop of the form

$$c_{l_1}^{q_1} \ldots c_{l_k}^{q_k},$$

where $q_j \in \mathbb{Z}$, $c_j = \gamma_j \alpha_j \bar{\gamma}_j$, α_j is a circle centered at p_j and radius r_j (i.e., $\alpha_j(t) = p_j + r_j e^{2\pi i t}$), r_j so small that $\bar{B}(p_j, r_j) \subseteq U$ and $\bar{B}(p_j, r_j) \cap \bar{B}(p_k, r_k) = \varnothing$ if $j \neq k$. We also choose γ_j to be a path in Ω from a to $\zeta_j \in \alpha_j([0, 1])$.

We have

$$\tilde{j}(\tilde{\omega}_k)([c_j]) = I_{\omega_k}(c_j) = I_{\omega_k}(\gamma_j \alpha_j \bar{\gamma}_j) = I_{\omega_k}(\alpha_j) = \frac{1}{2\pi i} \int_{\alpha_j} \frac{dz}{z - p_k} = \delta_{j,k}.$$

Therefore

$$\tilde{j}(\tilde{\omega})([\alpha]) = \sum_{j,k} \lambda_j q_k \delta_{j,l_k}.$$

If $E_j = \{k : l_k = j\}$ and $m_j = \sum_{k \in E_j} q_k$ then

$$\tilde{j}(\tilde{\omega})([\alpha]) = \sum_j \lambda_j m_j.$$

It follows that the group of periods of ω is the set $\sum_j \lambda_j m_j$, where $(m_1, \ldots, m_n) \in \mathbb{Z}^n$. Hence, one finds in this case that

$$\mathrm{Hom}(\pi_1(\Omega; a), \mathbb{C}) = \mathrm{Hom}\left(\frac{\pi_1(\Omega; a)}{[\pi_1(\Omega; a), \pi_1(\Omega, a)]}, \mathbb{C}\right) \simeq \mathrm{Hom}(\mathbb{Z}^n, \mathbb{C}) = \mathbb{C}^n.$$

For the first isomorphism, see Exercise 1.6.13. The next one is due to the fact that a free group of n generators modulo its subgroup of commutators is isomorphic to $\mathbb{Z}^n$. This shows, in this particular case, that j is indeed surjective.

1.7.11. Proposition. *Let Ω be a simply connected open subset of $\mathbb{C}$, then every closed 1-form is exact.*

PROOF. $\pi_1(\Omega; a) = \{[\varepsilon_a]\}$, hence j is the zero map and $\operatorname{Ker} j = Z^1(\Omega)$. It follows that $B^1(\Omega) = Z^1(\Omega)$. □

Note that this proposition did not use the fact that j is surjective. Neither does the following one.

1.7.12. Proposition. *Let Ω be an open subset of $\mathbb{C}$. In order for a closed 1-form to admit a primitive it is necessary and sufficient that all its periods vanish.*

PROOF. It is also a corollary of §1.7.9. □

From now on, whenever ω is a closed 1-form in Ω and γ is a loop in Ω, we will write $\int_\gamma \omega$ to represent $I_\omega(\gamma)$. □

EXERCISES 1.7

1. Let $\omega = \frac{1}{2\pi i}\frac{dz}{z}$, γ a loop in $\mathbb{C}^*$, $d(\gamma) = 1$, show that $\int_\gamma \omega = 1$. What happens if $d(\gamma) = n$?

2. Compute $\int_0^{2\pi} \frac{dt}{a^2\cos^2 t + b^2 \sin^2 t}$ $(a, b > 0)$ by comparison with $\int_{\partial\Omega} \frac{dz}{z}$, Ω being the domain $\frac{x^2}{a^2} + \frac{y^2}{b^2} < 1$.

3. Let $P(z) = a_0 z^n + a_1 z^{n-1} + \cdots + a_n$ be a complex polynomial of degree n. Show that for $r > 0$ sufficiently large $\int_{\partial B(0,r)} \frac{dP}{P} = 2\pi i d(P \circ \gamma_r) = 2\pi i n$, ($\gamma_r = \partial B(0, r)$ with the positive orientation).

4. Recall that the Peano curve is a continuous surjective map $\gamma_0 : [0, 1] \to [0, 1]^2$ such that $\gamma_0(0) = (0, 0)$ and $\gamma_0(1) = (1, 1)$. Let Ω_+ (resp. Ω_-) be the closed set in $\mathbb{C}$ defined by $1 \le |z| \le 2$, $\operatorname{Re} z \ge 0$ (resp. $\operatorname{Re} z \le 0$). Using γ_0, constant a path γ_+ starting at the point $z = 2$ and ending at the point $z = -1$, whose image is the whole Ω_+. Similarly, let γ_- fill Ω_- starting at $z = -1$ and ending at $z = 2$. Let γ be the loop $\gamma = \gamma_+\gamma_-$, compute $d(\gamma)$.

5. Let $\omega = \frac{x\,dx}{(x-1)^2 + ((x+1)^2 + y^2)} + \frac{y(x^2 + y^2 + 1)\,dy}{((x-1)^2 + y^2)^2((x+1)^2 + y^2)^2}$. Show $d\omega = 0$ in $\mathbb{R}^2 \backslash \{1, -1\}$. Compute its periods. Does it admit a primitive?

§8. Index of a Loop

If α is a loop in $\mathbb{C}$ we will define its index with respect to points $a \in \mathbb{C}\setminus\alpha([0,1])$. It is an integer that describes how many times the loop α turns around the point a.

1.8.1. Definition. We call ***index*** *of the loop α with respect to the point* $a \in \mathbb{C}\setminus\alpha([0,1])$ the degree of the loop γ, $\gamma(t) = \alpha(t) - a$, in $\mathbb{C}^*$. We denote it $\operatorname{Ind}_\alpha(a)$.

Some properties of the index follow.

1.8.2. Proposition. *Let α be a loop in $\mathbb{C}$ and $a \in \mathbb{C}\setminus\alpha([0,1])$. The condition $\operatorname{Ind}_\alpha(a) = 0$ is equivalent to α being homotopic to $\varepsilon_{\alpha(0)}$ in $\mathbb{C}\setminus\{a\}$.*

PROOF. It is an immediate consequence of 1.6.24. □

1.8.3. Proposition. *If α is piecewise-C^1 and $a \in \mathbb{C}\setminus\alpha([0,1])$ we have*

$$\operatorname{Ind}_\alpha(a) = \frac{1}{2\pi i}\int_\alpha \frac{dz}{z-a}.$$

In particular, if $a = 0$, the degree of α is

$$d(\alpha) = \frac{1}{2\pi i}\int_\alpha \frac{dz}{z}.$$

PROOF. We can assume $a = 0$. Let c be a lifting of α, it is also piecewise-C^1 since exp is a local diffeomorphism. We have

$$\frac{1}{2\pi i}\int_\alpha \frac{dz}{z} = \frac{1}{2\pi i}\int_c (\exp)^*\left(\frac{dz}{z}\right) = \frac{1}{2\pi i}\int_c \frac{d(e^u)}{e^u}$$

$$= \frac{1}{2\pi i}\int_c du = \frac{1}{2\pi i}(c(1) - c(0)) = d(\alpha) = \operatorname{Ind}_\alpha(0). \qquad \square$$

1.8.4. Remarks. (1) Let ω_a be the closed form $\omega_a = \dfrac{1}{2\pi i}\dfrac{dz}{z-a}$. We have

$$\operatorname{Ind}_\alpha(a) = I_{\omega_a}(\alpha).$$

Namely, by Remark 1.7.8, $I_{\omega_a}(\alpha) = I_{h^*(\omega_a)}(c)$ with $h = \exp$ and c a lifting of α. It follows that if α, β are two loops free homotopic in $\mathbb{C}\setminus\{a\}$, then $\operatorname{Ind}_\alpha(a) = \operatorname{Ind}_\beta(a)$. For the same reason, if α, β are two loops with the same base point in $\mathbb{C}\setminus\{a\}$, we have $\operatorname{Ind}_{\alpha\beta}(a) = \operatorname{Ind}_\alpha(a) + \operatorname{Ind}_\beta(a)$.

In fact, these properties were already used when we discussed the degree of a loop.

(2) The map sending $\alpha \in \mathscr{C}(\mathbb{C}\setminus\{a\}; z_0)$ to $\operatorname{Ind}_\alpha(a) \in \mathbb{Z}$ factors through

$\pi_1(\mathbb{C}\setminus\{a\};z_0)$. One obtains this way j_{ω_a} and $\operatorname{Im} j_{\omega_a}$, the group of the periods of ω_a.

(3) The function $a \mapsto \operatorname{Ind}_\alpha(a)$ is locally constant in $\mathbb{C}\setminus\alpha([0,1])$. In fact, $\operatorname{Ind}_\alpha(a+b) = \operatorname{Ind}_{\alpha-b}(a) = \operatorname{Ind}_\alpha(a)$ if $b \in \mathbb{C}$, $|b|$ very small, since then α, $\alpha - b$ are free homotopic in $\mathbb{C}\setminus\{a\}$. (Here we define $(\alpha - b)(t) := \alpha(t) - b$.) It follows that $a \mapsto \operatorname{Ind}_\alpha(a)$ is constant on each connected component of $\mathbb{C}\setminus\alpha([0,1])$. Furthermore it is zero on the unbounded component, since if $|a|$ is very large then $\alpha([0,1])$ is contained in a half-plane $\subseteq \mathbb{C}\setminus\{a\}$. Hence α is homotopic to a constant loop in $\mathbb{C}\setminus\{a\}$.

(4) *Effective way of computing the index*: Assume α is a piecewise-C^1 loop in $\mathbb{C}$ and $z \notin \operatorname{Im}(\alpha) = \alpha([0,1])$. Assume that a half-line R of direction u, starting at the point z, does not intersect $\operatorname{Im}(\alpha)$ except for a finite number of simple points $\alpha(t_1), \ldots, \alpha(t_n)$, $0 < t_1 < \cdots < t_n < 1$, where the tangent vectors $\dot\alpha(t_j)$ are defined and are not parallel to u. Let t_j be such that the pair $\{u, \dot\alpha(t_j)\}$ is a basis of $\mathbb{R}^2$ with the canonical orientation. A judicious use of Sard's theorem and the inverse function theorem shows that one can find $\varepsilon > 0$ sufficiently small so that for $B = B(\alpha(t_j), \varepsilon)$, $\operatorname{Im}(\alpha) \cap B$ has only one connected component, it contains no other intersection point with R, and $B\setminus\operatorname{Im}(\alpha)$ has exactly two components, Ω_1, Ω_2. Let Ω_1 be the component that intersects the straight line segment $[z, \alpha(t_j)]$. Reducing ε further, if necessary, we can assume that the connected component of $\alpha^{-1}(B)$ containing t_j is $]t_j - \varepsilon_1, t_j + \varepsilon_2[$ and $\alpha(]t_j - \varepsilon_1, t_j + \varepsilon_2[) = B \cap \operatorname{Im}(\alpha)$. We can assume further α is C^1 in $[t_j - \varepsilon_1, t_j + \varepsilon_2]$. We claim that for any $w_1 \in \Omega_1$ and $w_2 \in \Omega_2$ one has the relation

$$\operatorname{Ind}_\alpha(w_1) = \operatorname{Ind}_\alpha(w_2) + 1.$$

In fact, let β be the arc of ∂B which does not intersect $[z, \alpha(t_j)]$, starts at $\alpha(t_j - \varepsilon_1)$ and ends at $\alpha(t_j + \varepsilon_2)$. Let $\gamma = \beta(\alpha|[t_j - \varepsilon_1, t_j + \varepsilon_2])^-$ (the $-$ denotes the inverse path). It is clear, by continuing R in the direction $-u$, that z is in the unbounded component of $\operatorname{Im}(\gamma)$. Hence, the same is true for w_1. Therefore, $\operatorname{Ind}_\gamma(w_1) = 0$.

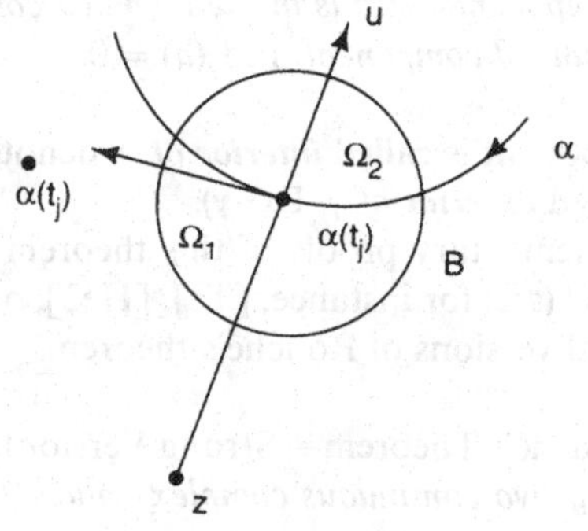

Figure 1.13

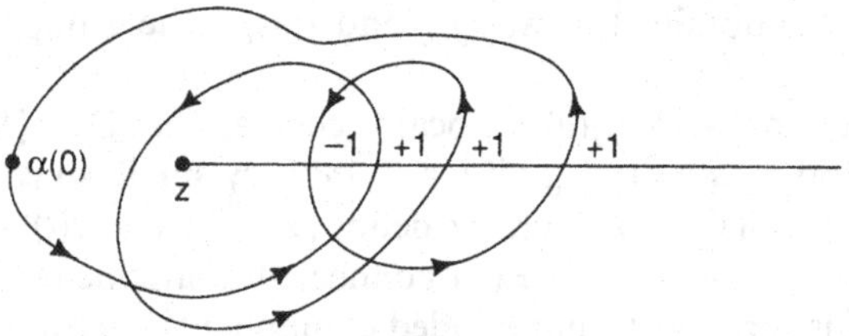

Figure 1.14

It follows, by homotopy, that if δ represents the loop obtained from α by replacing $\alpha|[t_j - \varepsilon_1, t_j + \varepsilon_2]$ by the arc of circle β, then $\mathrm{Ind}_\alpha(w_1) = \mathrm{Ind}_\delta(w_1)$. The same reasoning shows that if β' represents the arc of ∂B that starts at $\alpha(t_j - \varepsilon_1)$, ends at $\alpha(t_j + \varepsilon_2)$ and intersects $[z, \alpha(t_j)]$, and δ' is the loop obtained from α replacing $\alpha|[t_j - \varepsilon_1, t_j + \varepsilon_2]$ by β', then $\mathrm{Ind}_\alpha(w_2) = \mathrm{Ind}_{\delta'}(w_2)$. Moreover, it is clear that $\mathrm{Ind}_\delta(w_1) = \mathrm{Ind}_\delta(\alpha(t_j))$ and $\mathrm{Ind}_{\delta'}(w_2) = \mathrm{Ind}_{\delta'}(\alpha(t_j))$. Therefore,

$$\mathrm{Ind}_\alpha(w_1) - \mathrm{Ind}_\alpha(w_2) = \mathrm{Ind}_\delta(\alpha(t_j)) - \mathrm{Ind}_{\delta'}(\alpha(t_j)) = \mathrm{Ind}_{\partial B}(\alpha(t_j)) = 1.$$

Similarly, if the orientation of the pair $\{u, \dot{\alpha}(t_k)\}$ is negative one has

$$\mathrm{Ind}_\alpha(w_1) = \mathrm{Ind}_\alpha(w_2) - 1,$$

with the same definitions of Ω_1 and Ω_2.

One can now see that $\mathrm{Ind}_\alpha(z)$ is the sum $\sum_{j=1}^{n} \sigma_j$, where $\sigma_j = +1$ if orientation of $\{u, \dot{\alpha}(t_j))\}$ is positive, $\sigma_j = -1$ in the other case. For instance, in Figure 1.14, we have $\mathrm{Ind}_\alpha(z) = 2$.

Let us recall here that a ***Jordan curve*** is a loop γ in $\mathbb{C}$ such that $\gamma|[0, 1[$ is injective. The computation of the function Ind_γ is given in this case by the following famous theorem.

1.8.5. Jordan Curve Theorem. *Let γ be a Jordan curve, then $\mathbb{C}\backslash\gamma([0,1])$ has only two connected components. If a is in the bounded component,* $\mathrm{Ind}_\gamma(a) = 1$ *and, if a is in the unbounded component,* $\mathrm{Ind}_\gamma(a) = 0$.

The bounded component is called ***interior** of* γ, denoted by $\mathrm{Int}(\gamma)$, and the unbounded one is called ***exterior** of* γ, $\mathrm{Ext}(\gamma)$.

There are several elementary proofs of this theorem originally stated by Camille Jordan in 1887 (see, for instance, [Tr], [He2], or [Bu]).

We give now several versions of Rouché's theorem.

1.8.6. Proposition (Rouché's Theorem—Strong Version). *Let Ω be an open set in $\mathbb{C}$, α a loop in Ω, f, g two continuous complex-valued functions on $\alpha([0, 1])$ such that*

$$|f(\alpha(t)) - g(\alpha(t))| < |f(\alpha(t))| + |g(\alpha(t))|$$

for every $t \in [0, 1]$. *Then* $f \circ \alpha$ *and* $g \circ \alpha$ *are two loops in* $\mathbb{C}^*$ *of the same degree, i.e.,*

$$\mathrm{Ind}_{f \circ \alpha}(0) = \mathrm{Ind}_{g \circ \alpha}(0).$$

PROOF. The strict inequality in the hypotheses shows that neither f nor g can vanish on $\mathrm{Im}(\alpha)$. Moreover, the function $\lambda(t) = f(\alpha(t))/g(\alpha(t))$ takes values in $\mathbb{C} \setminus]-\infty, 0]$. Namely, $\lambda = 0$ is impossible and if $\lambda < 0$ then the inequality of the statement implies $1 + |\lambda| < 1 + |\lambda|$.

Let $H(t, s) := (1 - s) + s\lambda(t)$. This establishes a homotopy between the constant loop ε_1 and the loop λ. Hence $K(t, s) := H(t, s)g(\alpha(t))$ is a homotopy between $g \circ \alpha$ and $f \circ \alpha$. □

1.8.7. Corollary (Usual Version of Rouché's Theorem). *Let* Ω *be an open subset of* $\mathbb{C}$, α *a loop in* Ω, f, g *two continuous complex-valued functions on* $\alpha([0, 1])$ *such that*

$$|f(\alpha(t)) - g(\alpha(t))| < |g(\alpha(t))]$$

for every $t \in [0, 1]$. *Then* $f \circ \alpha$ *and* $g \circ \alpha$ *are two loops in* $\mathbb{C}^*$ *with the same degree*:

$$\mathrm{Ind}_{f \circ \alpha}(0) = \mathrm{Ind}_{g \circ \alpha}(0).$$

EXERCISES 1.8

1. Compute the index in each connected component of $\mathbb{C} \setminus \mathrm{Im}(\alpha)$ for the following loops:

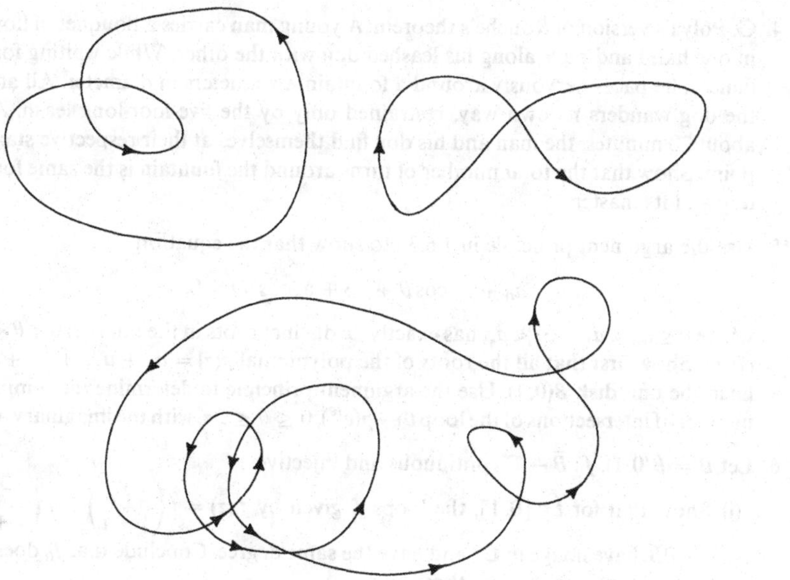

Figure 1.15

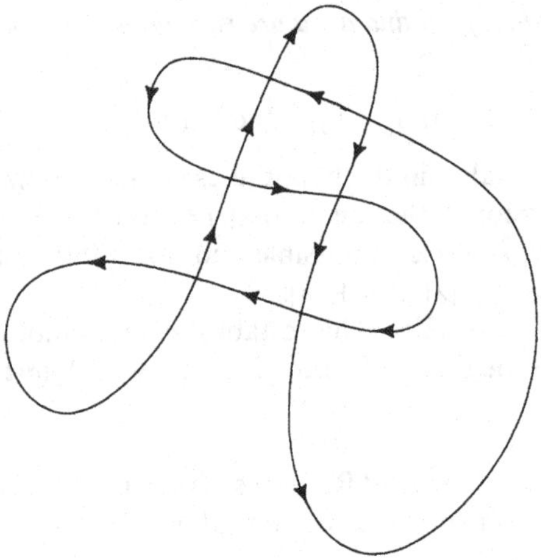

Figure 1.15 (*cont.*)

2. Let $a \in B(0, 1)$, and α the loop given by

$$\alpha : t \mapsto \frac{e^{2\pi i t} - a}{1 - \bar{a}e^{2\pi i t}} \qquad (0 \le t \le 1).$$

Show that $\Delta \arg \alpha(t) = 2\pi$. Compute $\mathrm{Ind}_\alpha(z)$ for $|z| \ne 1$.

3. Let $B = B(0, 1)$, $a_1, \ldots, a_n \in B$, $f(z) = \prod_{j=1}^{n} \left(\frac{z - a_j}{1 - \bar{a}_j z}\right)$. Compute $\underset{z \in \partial B}{\Delta} \arg f(z)$.

4. G. Pólya's version of Rouche's theorem: A young man carries a bouquet of flowers in one hand and pulls along his leashed dog with the other. While waiting for his fiancée, he paces nervously around a fountain, two meters in diameter. All along, the dog wanders its own way, restrained only by the five-foot-long leash. After about 20 minutes, the man and his dog find themselves at their respective starting point. Show that the total number of turns around the fountain is the same for the dog and its master.

*5. Use the argument principle in 1.6.30 to show that the equation

$$a_0 + a_1 \cos\theta + \cdots + a_n \cos n\theta = 0,$$

where $0 \le a_0 < a_1 < \cdots < a_n$, has exactly $2n$ distinct roots in the interval $0 < \theta < 2\pi$. (Hint: Show first that all the roots of the polynomial $p(z) = a_0 + a_1 z + \cdots + a_n z^n$ lie in the unit disk $B(0, 1)$. Use the argument principle to determine the minimum number of intersections of the loop $\theta \mapsto p(e^{i\theta})$, $0 \le \theta \le 2\pi$, with the imaginary axis.)

*6. Let $B = B(0, 1)$, $f: \bar{B} \to \mathbb{C}$ continuous and injective.

(i) Show that for $t \in [0, 1]$, the loops f_t given by $f_t(z) = f\left(\frac{z}{1 + t}\right) - f\left(\frac{-tz}{1 + t}\right)$, $z \in \partial B$, have image in $\mathbb{C}^*$ and have the same degree. Conclude that f_0 does not have a continuous logarithm.

(ii) Show that if $f(0)$ is not an interior point of $f(\bar{B})$ then there are values w so that $|f(0) - w|$ is very small and $z \mapsto f(z) - w$ admits a continuous logarithm in $\bar{B}$.

(iii) Using Rouche's theorem, show that (i) and (ii) are contradictory. Conclude that if $f: \Omega \to \mathbb{C}$, Ω open in $\mathbb{C}$, f continuous and injective, then $f(\Omega)$ is open.

§9. Homology

Let Ω be an open subset of $\mathbb{C}$. In this section we will denote $S_0(\Omega)$ the set Ω itself, $S_1(\Omega)$ the set $\mathscr{C}([0,1], \Omega)$ of paths in Ω, and by $S_2(\Omega)$ the set $\mathscr{C}(\Delta_2, \Omega)$ of continuous maps $\Delta_2 \to \Omega$, $\Delta_2 := \{(x, y) \in \mathbb{R}^2 : x \geq 0, y \geq 0, 0 \leq x + y \leq 1\}$. If $\Delta_0 = \{0\}$, $\Delta_1 = [0,1]$ then $S_i(\Omega) = \mathscr{C}(\Delta_i, \Omega)$ for $i = 0, 1, 2$. An element from $S_i(\Omega)$ will be called an ***elementary i-chain***. We denote by $\mathscr{C}_i(\Omega; \mathbb{Z})$ the free $\mathbb{Z}$-module $\mathbb{Z}^{(S_i(\Omega))}$ of all maps from $S_i(\Omega)$ into $\mathbb{Z}$ that are supported by a finite number of points. The elements of $\mathscr{C}_i(\Omega; \mathbb{Z})$ are called ***i-chains with integral coefficients***. In other words, every i-chain is a (formal) linear combination $\sum v_\sigma \cdot \sigma$ where $\sigma \in S_i(\Omega)$, $v_\sigma \in \mathbb{Z}$, and only finitely many v_σ are distinct from zero. Note that $\sum v_\sigma \cdot \sigma = 0$ is equivalent to $v_\sigma = 0$ for every σ. The ***boundary*** of an elementary 1-chain (i.e., of a path) $\gamma: [0,1] \to \Omega$, is the 0-chain $\gamma(1) - \gamma(0)$, denoted $\partial\gamma$ or $\partial_1\gamma$. The boundary $\partial\sigma$ (or $\partial_2\sigma$) of an elementary 2-chain $\sigma: \Delta_2 \to \Omega$ is the 1-chain $\gamma_0 - \gamma_1 + \gamma_2$, where $\gamma_0(t) := \sigma(1 - t, t)$, $\gamma_1(t) := \sigma(0, t)$, $\gamma_2(t) := \sigma(t, 0)$.

We can extend the definition of the boundary operators to $\mathscr{C}_1(\Omega; \mathbb{Z})$ and $\mathscr{C}_2(\Omega; \mathbb{Z})$ by making them $\mathbb{Z}$-linear:

$$\partial\left(\sum_{\sigma \in S_1(\Omega)} v_\sigma \cdot \sigma\right) = \sum_{\sigma \in S_1(\Omega)} v_\sigma \cdot \partial(\sigma) \qquad (i = 1, 2).$$

This way we obtain two homomorphisms

$$\partial_1 : \mathscr{C}_1(\Omega; \mathbb{Z}) \to \mathscr{C}_0(\Omega; \mathbb{Z})$$

$$\partial_2 : \mathscr{C}_2(\Omega; \mathbb{Z}) \to \mathscr{C}_1(\Omega; \mathbb{Z}),$$

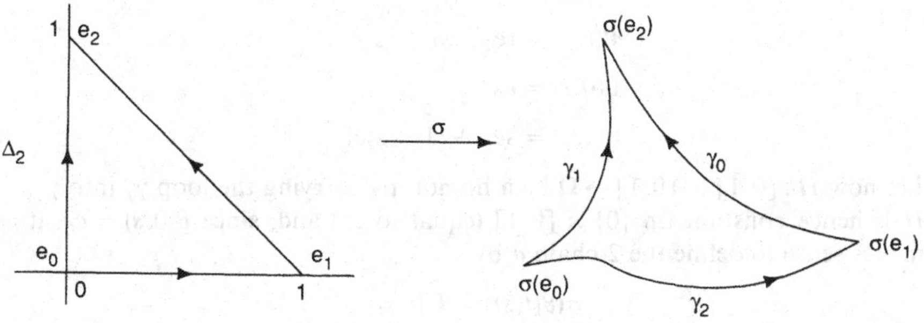

Figure 1.16

and one verifies easily that $\partial_1 \circ \partial_2 = 0$ (sometimes also written $\partial^2 = 0$). If we introduce the homomorphism

$$\varepsilon : \mathscr{C}_0(\Omega; \mathbb{Z}) \to \mathbb{Z}$$

$$\varepsilon\left(\sum_{\sigma \in S_0(\Omega)} v_\sigma \cdot \sigma\right) = \sum_{\sigma \in S_0(\Omega)} v_\sigma,$$

we can also verify $\varepsilon \circ \partial_1 = 0$. It is then natural to consider the sequence

$$\mathscr{C}_2(\Omega; \mathbb{Z}) \xrightarrow{\partial_2} \mathscr{C}_1(\Omega; \mathbb{Z}) \xrightarrow{\partial_1} \mathscr{C}_0(\Omega; \mathbb{Z}) \xrightarrow{\varepsilon} \mathbb{Z} \to 0,$$

where $\operatorname{Im} \partial_2 \subseteq \operatorname{Ker} \partial_1$ and $\operatorname{Im} \partial_1 \subseteq \operatorname{Ker} \varepsilon$. Denote by $Z_1(\Omega; \mathbb{Z})$ the set $\operatorname{Ker} \partial_1$, its elements are called ***1-cycles***. The **1-*boundaries*** are the elements of $B_1(\Omega; \mathbb{Z}) = \operatorname{Im} \partial_2$. The quotient group $H_1(\Omega; \mathbb{Z}) = Z_1(\Omega; \mathbb{Z})/B_1(\Omega; \mathbb{Z})$ is called the ***first homology group*** *of* Ω (***with integral coefficients***). Two elements of $Z_1(\Omega; \mathbb{Z})$ which are congruent modulo $B_1(\Omega; \mathbb{Z})$ are said to be ***homologous***.

We are going to show now that if Ω is connected then $H_1(\Omega; \mathbb{Z})$ can be identified to $\pi_1(\Omega)/[\pi_1(\Omega), \pi_1(\Omega)]$. We have really shown this already in the case of an open set homeomorphic to $\mathbb{C}$ punctured by a finite number of points (i.e., $\mathbb{C}\setminus\{p_1, \ldots, p_n\}$).

A loop $\gamma \in \mathscr{C}(\Omega; z_0)$ appears in the languange of homology as an elementary 1-chain σ_γ such that $\partial\sigma_\gamma = \gamma(1) - \gamma(0) = 0$. We have therefore a canonical injection

$$\chi : \mathscr{C}(\Omega; z_0) \to Z_1(\Omega; \mathbb{Z}).$$

1.9.1 Lemma. *If two loops γ_1, γ_2 with base point z_0 are homotopic in Ω then $\sigma_{\gamma_1} = \chi(\gamma_1)$ and $\sigma_{\gamma_2} = \chi(\gamma_2)$ are homologous. (That is, χ induces a map from $\pi_1(\Omega; z_0)$ into $H_1(\Omega; \mathbb{Z})$.)*

PROOF. Let us denote (as in Figure 1.16) $e_0 = (0, 0)$, $e_1 = (1, 0)$, $e_2 = (0, 1)$, the vertices of Δ_2. Let $Q(s) := se_2 + (1 - s)e_1$ $(0 \le s \le 1)$ and $\theta(t, s) := tQ(s) + (1 - t)e_0 = t(1 - s)e_1 + tse_2$ $(0 \le t \le 1)$. Now, θ is a continuous surjection from $[0, 1] \times [0, 1]$ to Δ_2 such that

$$\theta(t, 0) = te_1$$

$$\theta(t, 1) = te_2$$

$$\theta(0, s) = e_0$$

$$\theta(1, s) = se_2 + (1 - s)e_1.$$

Let now $H : [0, 1] \times [0, 1] \to \Omega$ be a homotopy carrying the loop γ_1 into γ_2. H is hence constant on $\{0\} \times [0, 1]$ (equal to z_0) and, since $\theta(0, s) = e_0$, it makes sense to define the 2-chain σ by

$$\sigma(\theta(t, s)) = H(t, s).$$

(Note that θ is injective except on $\{0\} \times [0, 1]$ and Δ_2 can be considered as the quotient space of $[0, 1] \times [0, 1]$ by the equivalence $z \sim z'$ if and only if

$\theta(z) = \theta(z')$.) We have now

$$\partial(\sigma) = (s \mapsto H(1,s)) - (t \mapsto H(t,1)) + (t \mapsto H(t,0)) = \sigma_{\varepsilon_{z_0}} - \sigma_{\gamma_2} + \sigma_{\gamma_1}.$$

On the other hand, if σ' is the elementary 2-chain $\sigma'(t,s) := z_0$ then $\partial(\sigma') = \sigma_{\varepsilon_{z_0}}$, therefore $\partial(\sigma' - \sigma) = \sigma_{\gamma_2} - \sigma_{\gamma_1} = \chi(\gamma_2) - \chi(\gamma_1)$, which is what we wanted to prove. □

1.9.2. Lemma. *Let γ_1, γ_2 be two paths in Ω such that $\gamma_1(1) = \gamma_2(0)$. Let $\gamma_0 = \gamma_1\gamma_2$. Then:*

(a) *The 1-chain $\sigma_{\gamma_1} + \sigma_{\gamma_2} - \sigma_{\gamma_0}$ is a boundary* (i.e., *$\sigma_{\gamma_1} + \sigma_{\gamma_2}$ is homologous to σ_{γ_0}*).
(b) *The map $\chi : \mathscr{C}(\Omega; z_0) \to Z_1(\Omega, \mathbb{Z})$ induces a quotient map*

$$\bar{\chi} : \pi_1(\Omega; z_0) \to H_1(\Omega, \mathbb{Z})$$

which is a group homomorphism.

PROOF. (a) Define an elementary 2-chain σ as follows (see Figure 1.17):

$$\begin{cases} \sigma(x,y) = \gamma_1(x + 2y) & \text{if } x + 2y \le 1 \\ \sigma(x,y) = \gamma_2(x + 2y - 1) & \text{if } x + 2y \ge 1. \end{cases}$$

It is immediate to check that σ is a 2-chain in Ω satisfying $\sigma|[e_0, e_1] = \gamma_1$, $\sigma|[e_1, e_2] = \gamma_2$, $\sigma|[e_2, e_0] = \gamma_1\gamma_2 = \gamma_0$. That is, $\partial(\sigma) = \sigma_{\gamma_2} - \sigma_{\gamma_0} + \sigma_{\gamma_1}$. This proves (a).

(b). We have $\chi(\gamma_1\gamma_2)$ homologous to $\chi(\gamma_1) + \chi(\gamma_2)$, hence $\bar{\chi}([\gamma_1][\gamma_2]) = \bar{\chi}([\gamma_1]) + \bar{\chi}([\gamma_2])$. □

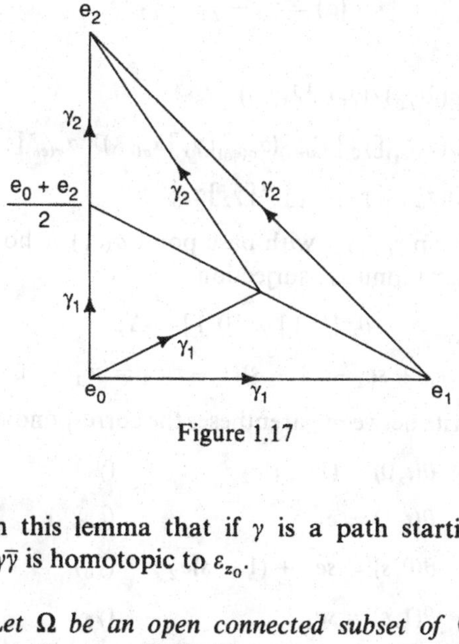

Figure 1.17

It follows from this lemma that if γ is a path starting at z_0 one has $\bar{\chi}([\gamma\bar{\gamma}]) = 0$ since $\gamma\bar{\gamma}$ is homotopic to ε_{z_0}.

1.9.3. Theorem. *Let Ω be an open connected subset of $\mathbb{C}$, then the group homomorphism $\bar{\chi} : \pi_1(\Omega; z_0) \to H_1(\Omega; \mathbb{Z})$ is surjective. Its kernel $\operatorname{Ker} \bar{\chi}$ is the* *subgroup of commutators of* $\pi_1(\Omega; z_0)$ *and* $H_1(\Omega; \mathbb{Z})$ *can be identified to*

$$\pi_1(\Omega; z_0)/[\pi_1(\Omega; z_0), \pi_1(\Omega; z_0)].$$

PROOF. Since $\bar{\chi}$ takes values in a commutative group it follows that $\operatorname{Ker} \bar{\chi} \supseteq [\pi_1(\Omega; z_0), \pi_1(\Omega; z_0)]$ and, if φ is the quotient map of $\pi_1(\Omega, z_0)$ onto its abelianized $\pi_1(\Omega; z_0)/[\pi_1(\Omega; z_0), \pi_1(\Omega; z_0)]$, then there is a unique homomorphism χ' of this last group into $H_1(\Omega; \mathbb{Z})$ such that

$$\bar{\chi} = \chi' \circ \varphi.$$

We want to define a homomorphism κ from $H_1(\Omega; \mathbb{Z})$ into this quotient group such that $\kappa \circ \chi' = id$, $\chi' \circ \kappa = id$. This will prove the theorem.

For each $z \in \Omega$, let us choose a homotopy class α_z of a path joining z_0 to z, with the convention that $\alpha_{z_0} = [\varepsilon_{z_0}]$. Let now $u : [0, 1] \to \Omega$ be an arbitrary path, set

$$\lambda(u) := \alpha_{u(0)}[u]\alpha_{u(1)}^{-1} \in \pi_1(\Omega; z_0).$$

This defines a map $\lambda : S_1(\Omega) \to \pi_1(\Omega; z_0)$ by $\lambda(\sigma_u) = \lambda(u)$. By $\mathbb{Z}$-linearly extending to $\mathscr{C}_1(\Omega; \mathbb{Z})$ the map $\varphi \circ \lambda$, we obtain a homomorphism:

$$\mu : \mathscr{C}_1(\Omega; \mathbb{Z}) \to \pi_1(\Omega; z_0)/[\pi_1(\Omega; z_0), \pi_1(\Omega; z_0)].$$

1.9.4. Lemma. *The map* μ *vanishes on* $B_1(\Omega; \mathbb{Z})$.

PROOF. Let $\sigma : \Delta_2 \to \Omega$ be an elementary 2-chain. Using the previous notation (see Figure 1.16), we have

$$\partial(\sigma) = \gamma_0 - \gamma_1 + \gamma_2.$$

Hence,

$$\begin{aligned}\mu(\partial(\sigma)) &= \varphi(\lambda(\gamma_0)\lambda(\gamma_1)^{-1}\lambda(\gamma_2))\\ &= \varphi(\alpha_{\sigma(e_1)}[\gamma_0]\alpha_{\sigma(e_2)}^{-1}(\alpha_{\sigma(e_0)}[\gamma_1]\alpha_{\sigma(e_2)}^{-1})^{-1}\alpha_{\sigma(e_0)}[\gamma_2]\alpha_{\sigma(e_1)}^{-1})\\ &= \varphi(\alpha_{\sigma(e_1)}[\gamma_0][\gamma_1]^{-1}[\gamma_2]\alpha_{\sigma(e_1)}^{-1}).\end{aligned}$$

We claim that the loop $\gamma_0\bar{\gamma}_1\gamma_2$, with base point $\sigma(e_1)$, is homotopic to $\varepsilon_{\sigma(e_1)}$. Namely, let θ be the continuous surjection

$$\theta : [0, 1] \times [0, 1] \to \Delta_2$$

$$\theta(t, s) := t(1 - s)e_0 + se_1 + (1 - s)(1 - t)e_2 = se_1 + (1 - s)(1 - t)e_2.$$

It is clear that (we write between parentheses the corresponding paths under σ)

$$\begin{aligned}\theta(t, 0) &= (1 - t)e_2 &&(\bar{\gamma}_1)\\ \theta(t, 1) &= e_1 &&(\varepsilon_{\sigma(e_1)})\\ \theta(0, s) &= se_1 + (1 - s)e_2 &&(\bar{\gamma}_0)\\ \theta(1, s) &= se_1 &&(\gamma_2).\end{aligned}$$

Hence $\tau := \sigma \circ \theta : [0,1] \times [0,1] \to \Omega$ is a continuous map such that

$$\tau|[0,1] \times \{0\} = \bar{\gamma}_1$$
$$\tau|\{1\} \times [0,1] = \gamma_2$$
$$\tau|[0,1] \times \{1\} = \varepsilon_{\sigma(e_1)}$$
$$\tau|\{0\} \times [0,1] = \bar{\gamma}_0.$$

Let us consider the following figure:

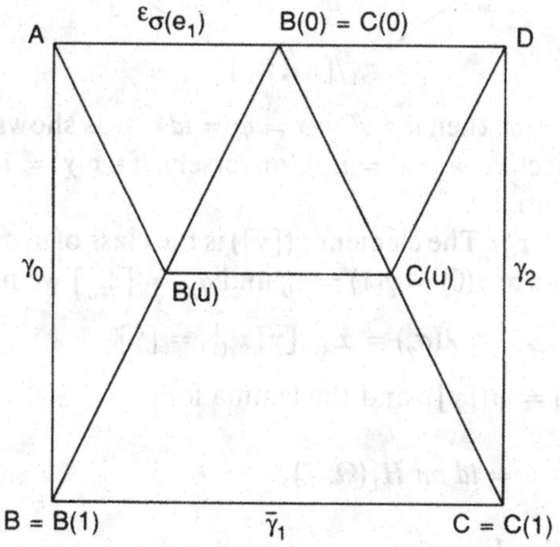

Figure 1.18

It shows that $\gamma_0\bar{\gamma}_1\gamma_2$ is homotopic to $\varepsilon_{\sigma(e_1)}$ as follows: For $u \in [0,1]$ as the homotopy parameter, it traverses the segment $[A, B(u)]$ during the first third of the time, $[B(u), C(u)]$ during the second third, and $[C(u), D]$ the last third. More precisely, let

$$H(t,u) := \begin{cases} \tau\left(\dfrac{3t(1-u)}{2}, 1 - 3tu\right) & 0 \le t \le 1/3 \\ \tau\left(\dfrac{1-u}{2} + (3t-1)u, 1-u\right) & 1/3 \le t \le 2/3 \\ \tau\left(\dfrac{1+u}{2} + \dfrac{(3t-2)(1-u)}{2}, 1 - u + u(3t-2)\right) & 2/3 \le t \le 1 \end{cases}$$

Hence, $\mu(\partial(\sigma)) = \varphi([\varepsilon_{z_0}]) = 0$. □

Therefore, the restriction of μ to $Z_1(\Omega; \mathbb{Z})$ induces a homomorphism

$$\kappa : H_1(\Omega; \mathbb{Z}) \to \pi_1(\Omega; z_0)/[\pi_1(\Omega; z_0), \pi_1(\Omega; z_0)].$$

1.9.5. Lemma. $\kappa \circ \chi' = id$ *on* $\pi_1/[\pi_1, \pi_1]$.

PROOF. It is enough to show that $\kappa \circ \bar{\chi} = \varphi$.

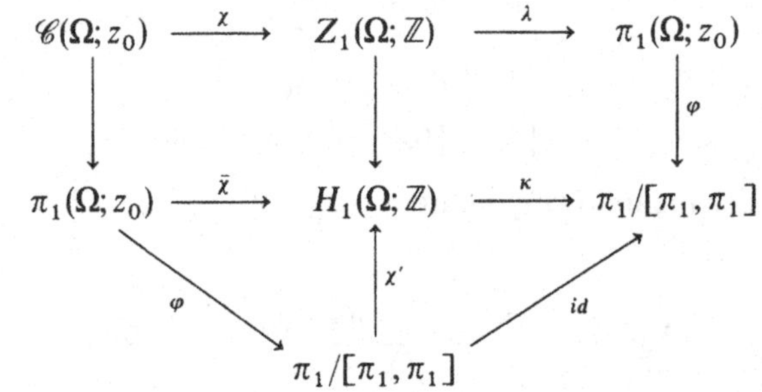

Indeed, if $\kappa \circ \bar{\chi} = \varphi$, then $\kappa \circ \chi' \circ \varphi = \varphi = id \circ \varphi$ as shown by this diagram and, since φ is surjective, $\kappa \circ \chi' = id$. Conversely, if $\kappa \circ \chi' = id$, then $\kappa \circ \bar{\chi} = \varphi$, by the same diagram.

Let now $\gamma \in \mathscr{C}(\Omega; z_0)$. The element $\bar{\chi}([\gamma])$ is the class of σ_γ and $(\kappa \circ \bar{\chi})([\gamma]) = \mu(\sigma_\gamma) = \varphi(\lambda(\sigma_\gamma))$. Since $\gamma(0) = \gamma(1) = z_0$ and $\alpha_{z_0} = [\varepsilon_{z_0}]$ we have

$$\lambda(\sigma_\gamma) = \alpha_{\gamma(0)}[\gamma]\alpha_{\gamma(1)}^{-1} = [\gamma].$$

Hence $(\kappa \circ \bar{\chi})([\gamma]) = \varphi([\gamma])$, and the lemma follows. $\square$

1.9.6. Lemma. $\chi' \circ \kappa = id$ *on* $H_1(\Omega; \mathbb{Z})$.

PROOF. Consider the diagram:

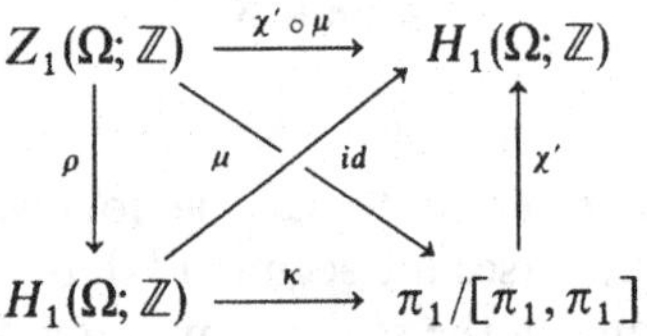

The statement is equivalent to $\chi' \circ \mu = \rho =$ canonical projection $Z_1(\Omega; \mathbb{Z}) \to H_1(\Omega; \mathbb{Z})$. Namely, if $\chi' \circ \kappa = id$ then $\chi' \circ \mu = \chi' \circ \kappa \circ \rho = \rho$. On the other hand, if $\chi' \circ \mu = \rho$ then $\chi' \circ \kappa \circ \rho = id \circ \rho$ and, since ρ is surjective, $\chi' \circ \kappa = id$.

Let $\sigma = \sum_i v_i \sigma_{u_i}$ be a 1-cycle ($v_i \in \mathbb{Z}$, u_i path in Ω). We have

$$\mu(\sigma) = \sum_1 v_i \mu(\sigma_{u_i}) = \sum_i v_i \varphi(\lambda(\sigma_{u_i})).$$

with $\lambda(\sigma_{u_i}) = \alpha_{u_i}(0)[u_i]\alpha_{u_i(1)}^{-1}$ (note that we use additive notation because μ takes values in an abelian group). Let c_ζ be a path in the class α_ζ for every $\zeta \in \Omega$. The loop $c_{u_i(0)} u_i \bar{c}_{u_i(1)}$ defines a cycle homologous to the cycle $\tilde{\sigma}_i = \sigma_{c_{u_i(0)}} + \sigma_{u_i} - \sigma_{c_{u_i(1)}}$. Hence $\chi'(\mu(\sigma_{u_i})) = \chi'(\varphi(\lambda\sigma_{u_i})))$ is the homology class of the cycle $\tilde{\sigma}_i$. By linearity, $\chi'(\mu(\sigma))$ is the class of the cycle

$$\sum_i v_i \tilde{\sigma}_i = \sum_i v_i \sigma_{u_i} + \sum_i v_i \left(\sigma_{c_{u_i(0)}} - \sigma_{c_{u_i(1)}} \right) = \sigma + \sum_i v_i (\sigma_{c_{u_i(0)}} - \sigma_{c_{u_i(1)}}).$$

We need therefore to show that the last sum vanishes. We have not yet really used that σ is a cycle; this means precisely that

$$\partial(\sigma) = \sum_i v_i (u_i(1) - u_i(0)) = 0$$

in $\mathscr{C}_0(\Omega; \mathbb{Z})$. For $\zeta \in \Omega$, let

$$I(\zeta) := \{ i \in I : u_i(0) = \zeta \}$$
$$J(\zeta) := \{ i \in I : u_i(1) = \zeta \}.$$

Recall that I is finite, so for only finitely many $\zeta \in \Omega$ we can have $I(\zeta) \neq \varnothing$ or $J(\zeta) \neq \varnothing$. The equality $\partial(\sigma) = 0$ leads to

$$0 = \sum_{\zeta \in \Omega} \left(\sum_{i \in I(\zeta)} v_i - \sum_{i \in J(\zeta)} v_i \right) \zeta,$$

hence $\sum_{i \in I(\zeta)} v_i = \sum_{i \in J(\zeta)} v_i$ for every $\zeta \in \Omega$. Therefore

$$\sum_i v_i (\sigma_{c_{u_i(0)}} - \sigma_{c_{u_i(1)}}) = \sum_{\zeta \in \Omega} \left(\sum_{i \in I(\zeta)} v_i - \sum_{i \in J(\zeta)} v_i \right) \sigma_{c_\zeta} = 0.$$

This concludes the proofs of Lemma 1.9.6 and Theorem 1.9.3. □

1.9.7. Corollary. *If Ω is a simply connected open subset of $\mathbb{C}$, then $H_1(\Omega; \mathbb{Z}) = 0$.*

1.9.8. Example. In $\mathbb{C} \setminus \{p_1, p_2\}$ the loop with base point z_0 of Figure 1.19 induces a 1-boundary σ_γ.

In fact, γ is homotopic to the commutator $\alpha\beta\alpha^{-1}\beta^{-1}$, where α, β are shown in Figure 1.20.

More generally, if Ω is homeomorphic to $\mathbb{C}$, then $H_1(\Omega \setminus \{p_1, \ldots, p_n\}; \mathbb{Z})$ is isomorphic to $\mathbb{Z}^n$ and generated by the loops $\alpha_j : t \mapsto p_j + r_j e^{2\pi i t}$, $r_j > 0$ small.

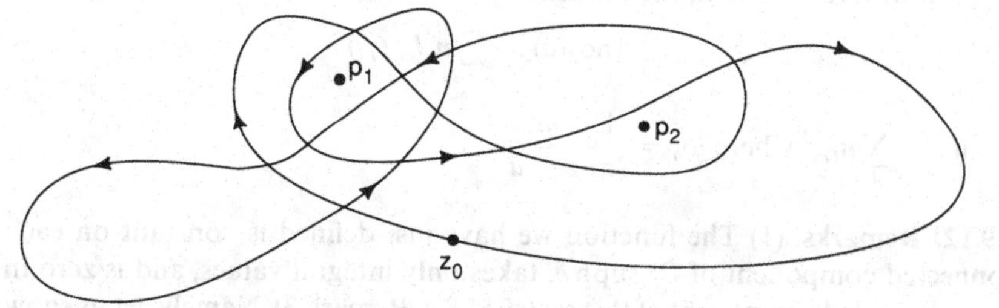

Figure 1.19

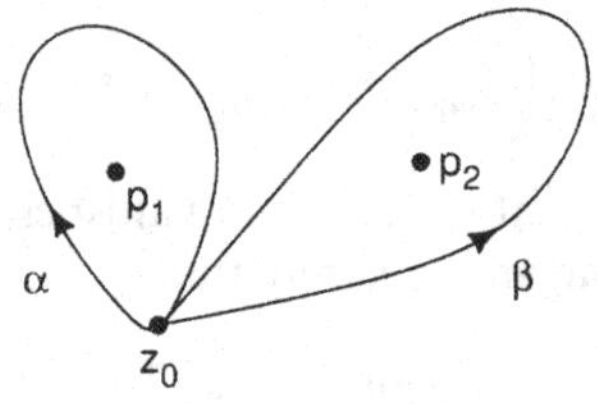

Figure 1.20

1.9.9. Definition. We call a ***special chain*** in Ω a 1-chain $c = \sum_i n_i \gamma_i$, where $n_i \in \mathbb{Z}$ and γ_i is either a horizontal or a vertical path (i.e., parallel to one of the coordinates axes) of the form $\gamma_i(t) = (ta + (1-t)b, c)$ or $\gamma_i(t) = (c, ta + (1-t)b)$.

1.9.10. Proposition. *Let Ω be an open subset of $\mathbb{C}$. Every cycle $\sigma \in Z_1(\Omega; \mathbb{Z})$ is homologous to a special cycle.*

PROOF. It is enough to show that every path γ is homotopic in Ω to a special path (i.e., composed of horizontal and vertical line segments). This is a consequence of the following observation: if γ_1, γ_2 are two paths in Ω with the same endpoints and $\sup_{0 \le t \le 1} |\gamma_1(t) - \gamma_2(t)|$ is sufficiently small, then γ_1, γ_2 are homotopic by $H(t, s) = s\gamma_1(t) + (1-s)\gamma_2(t)$ which is a homotopy in Ω with fixed endpoints. The uniform continuity of a path γ now shows that it is homotopic to a piecewise linear path in Ω composed of very small segments. Moreover, this piecewise linear path is itself homotopic to a special path. □

1.9.11 Definitions.

(1) We call ***support*** of an i-chain ($i = 0, 1, 2$) $\sigma = \sum_i v_i \sigma_i$, the set

$$\operatorname{supp} \sigma := \bigcup_i \operatorname{Image}(\sigma_i).$$

(2) If Ω is a connected open set, $\delta \in Z_1(\Omega; \mathbb{Z})$ and $a \in \mathbb{C} \setminus \operatorname{supp} \delta$, we call ***index*** *of δ with respect to a*, the number

$$\operatorname{Ind}_\delta(a) := \sum_i n_i I_{\omega_a}(\gamma_i)$$

if $\delta = \sum_i n_i \gamma_i$, where $\omega_a = \dfrac{1}{2\pi i} \dfrac{dz}{z-a}$.

1.9.12. Remarks. (1) The function we have just defined is constant on each connected component of $\mathbb{C} \setminus \operatorname{supp} \delta$, takes only integral values, and is zero in the unbounded component of this set (cf. §1.8.4, Remark 3). Namely, let us show that $\operatorname{Ind}_\delta(a) = (\bar{j}_{\omega_a} \circ \mu)(\delta)$, where $\bar{j}_{\omega_a} : \pi_1/[\pi_1, \pi_1] \to \mathbb{Z}$ is the quotient map corresponding to $j_{\omega_a} : \pi_1(\Omega; z_0) \to \mathbb{Z}$ (this can be done since $\operatorname{Ker} j_{\omega_a} \supseteq [\pi_1, \pi_1]$, because $\mathbb{Z}$ is commutative). This identity will then prove the remark by the properties of j_{ω_a}.

If $\delta = \sum_i n_i\gamma_i$ with $\sum_i n_i(\gamma_i(1) - \gamma_i(0)) = \partial(\delta) = 0$, then $\mu\left(\sum_i n_i\gamma_i\right)$ is the class in $\pi_1/[\pi_1, \pi_1]$ of $\prod_i (\alpha_{i(0)}[\gamma_i]\alpha_{i(1)}^{-1})^{n_i}$. It follows that

$$\bar{j}_{\omega_a}\left(\varphi\left(\prod_i (\alpha_{i(0)}[\gamma_i]\alpha_{i(1)}^{-1})^{n_i}\right)\right) = \sum_i n_i I_{\omega_a}(c_{a_{i(0)}}\gamma_i\bar{c}_{a_{i(1)}})$$
$$= \sum_i n_i I_{\omega_a}(c_{a_{i(0)}}) + \sum_i n_i I_{\omega_a}(\gamma_i) - \sum_i n_i I_{\omega_a}(c_{a_{i(1)}})$$

(we have kept the notations from the previous theorem and corresponding lemmas). As before, $I(z) = \{i \in I : \gamma_i(0) = \text{starting point of } \gamma_i = z\}$, and $J(z) = \{i \in I : \gamma_i(1) = \text{endpoint of } \gamma_i = z\}$, hence we have $\sum_{i \in I(z)} n_i = \sum_{i \in J(z)} n_i$ for every $z \in \Omega$, because δ is a cycle. Let us note now that

$$\sum_i n_i I_{\omega_a}(c_{a_{i(0)}}) = \sum_{z \in \Omega}\left(\sum_{i \in I(z)} n_i\right) I_{\omega_a}(c_z)$$
$$\sum_i n_i I_{\omega_a}(c_{a_{i(1)}}) = \sum_{z \in \Omega}\left(\sum_{i \in J(z)} n_i\right) I_{\omega_a}(c_z).$$

It follows that

$$\mathrm{Ind}_\gamma(a) = \sum_i n_i I_{\omega_a}(\gamma_i) = (\bar{j}_{\omega_a} \circ \mu)(\delta).$$

(2) What we have just done is to "integrate" the form ω_a along the cycle δ. More generally:

(a) We call ***integration of* 0*-forms*** the operation

$$\mathscr{C}_0(\Omega; \mathbb{Z}) \times \mathscr{E}_0^0(\Omega) \to \mathbb{C}$$
$$(c, f) \mapsto \langle c, f\rangle = \sum_{z \in \Omega} v_z f(z),$$

where c is the 0-chain $c = \sum_{z \in \Omega} v_z \cdot z$.

(b) ***Integration of the closed* 1*-forms*** *of class* C^1, the operation

$$\mathscr{C}_1(\Omega; \mathbb{Z}) \times Z_1^1(\Omega) \to \mathbb{C}$$
$$(c, \omega) \mapsto \langle c, \omega\rangle = \sum_{\sigma \in S_1(\Omega)} v_\sigma I_\omega(\sigma),$$

where $Z_1^1(\Omega)$ is the set of closed 1-forms of class C^1 and $c = \sum_{\sigma \in S_1(\Omega)} v_\sigma \sigma$. If the σ appearing in c are piecewise C^1 then one can also integrate along c a 1-form ω which could eventually be not closed, by setting $\langle c, \omega\rangle = \sum_{\sigma \in S_1(\Omega)} v_\sigma \int_\sigma \omega$.

For instance, if U is an open set with piecewise regular boundary of class

C^1, then the boundary ∂U of U is a 1-cycle $\delta := \partial U = \sum_{1 \le i \le n} \gamma_i$, where the γ_i are the piecewise-C^1 Jordan curves that are the connected components of ∂U, parameterized following the canonical orientation. If ω is a 1-form of class C^1 on a neighborhood of $\bar{U}$, the Stokes formula can be written as

$$\int_U d\omega = \int_{\partial U} \omega = \sum_i \int_{\gamma_i} \omega = \langle \delta, \omega \rangle.$$

(c) Let us show that if ω is a closed 1-form of class C^1 and $\delta \in B_1(\Omega; \mathbb{Z})$ then $\langle \delta, \omega \rangle = 0$. Without loss of generality we can assume $\delta = \partial\sigma$, $\sigma \in S_2(\Omega)$. Then, with the notation of Figure 1.16, we have $\delta = \gamma_0 - \gamma_1 + \gamma_2$. Therefore

$$\langle \delta, \omega \rangle = I_\omega(\gamma_0) - I_\omega(\gamma_1) + I_\omega(\gamma_2) = I_\omega(\gamma_0 \bar{\gamma}_1 \gamma_2) = 0,$$

because, as we have shown in the proof of 1.9.4, the loop $\gamma_0 \bar{\gamma}_1 \gamma_2$ is homotopic to $\varepsilon_{\sigma(e_1)}$. In particular the index of a 1-boundary δ with respect to any point $a \in \mathbb{C} \setminus \operatorname{supp} \delta$ is zero.

More generally, we have the following proposition.

1.9.13. Proposition. *Let Ω be a connected open set in $\mathbb{C}$. A 1-cycle $\delta \in Z_1(\Omega; \mathbb{Z})$ is a 1-boundary if and only if* $\operatorname{Ind}_\delta(a) = 0$ *for every $a \in \mathbb{C} \setminus \Omega$.*

PROOF. We can assume δ is a special 1-cycle $\Sigma v_k \gamma_k$. Consider two (finite) increasing sequences of real numbers $\{a_i\}$, $\{b_j\}$ where the $\{a_i\}$ contains all the projections of the endpoints of the γ_k on the x-axis, $\{b_j\}$ those on the y-axis. We can, in fact, suppose that every γ_k is either of the form $[a_i, a_{i+1}] \times \{b_j\}$ or $\{a_i\} \times [b_j, b_{j+1}]$. Denote by $Q_{i,j}$ the rectangle $[a_i, a_{i+1}] \times [b_j, b_{j+1}]$, which will

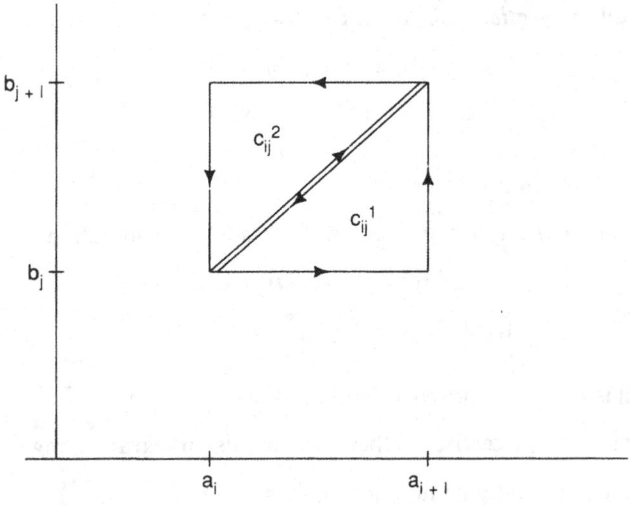

Figure 1.21

be considered as the support of the 2-chain $c_{i,j} = c^1_{i,j} + c^2_{i,j}$ as shown by Figure 1.21.

Let $z_{i,j}$ be a point in the interior of $Q_{i,j}$. Consider the 1-chain

$$\delta' := \sum_{i,j} \operatorname{Ind}_\delta(z_{i,j})\partial(c_{i,j}),$$

which is evidently a 1-cycle in $\mathbb{C}$. We want to show that δ' is in fact a 1-boundary in Ω.

If $\operatorname{Ind}_\delta(z_{i,j}) \neq 0$ we claim that $Q_{i,j} \subseteq \Omega$. Assume this is not the case, let $z \in Q_{i,j} \backslash \Omega$. We have then $\operatorname{Ind}_\delta(z) = 0$ by hypothesis. The segment $[z_{i,j}, z]$ does not intersect $\operatorname{supp} \delta$ since $[z_{i,j}, z[\subseteq \mathring{Q}_{i,j}$ and $z \notin \Omega$. Therefore, $\operatorname{Ind}_\delta$ is constant on that segment and $\operatorname{Ind}_\delta(z_{i,j}) = \operatorname{Ind}_\delta(z) = 0$, a contradiction.

Therefore, δ' is the boundary of the 2-chain $c \in \mathscr{C}_2(\Omega; \mathbb{Z})$,

$$c := \sum_{i,j}{}' \operatorname{Ind}_\delta(z_{i,j})c_{i,j},$$

where the sum takes place only over the indices i, j, for which $\operatorname{Ind}_\delta(z_{i,j}) \neq 0$.

Let us now show that $\delta = \delta'$. One can write $\delta - \delta' = \sum_p n_p\sigma_p$, where $n_p \in \mathbb{Z}$, and σ_p are segments, either vertical or horizontal, and have disjoint interiors. Let us assume $n_{p_0} \neq 0$, and to fix ideas, assume σ_{p_0} is a vertical segment located on the left side of Q_{i_0,j_0}.

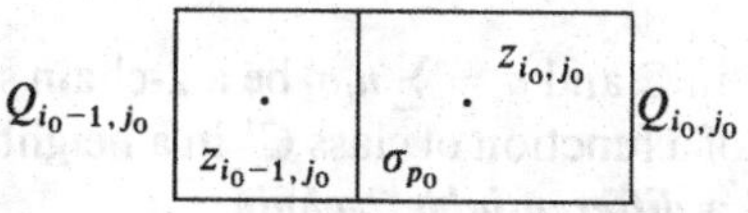

Figure 1.22

Define $\delta'' := \delta - \delta' + n_{p_0}\partial(c_{i_0,j_0})$. We now have $\operatorname{Ind}_{\delta-\delta'}(z_{i,j}) = 0$ for every i, j and $\operatorname{Ind}_{\partial(c_{i_0,j_0})}(z_{i_0,j_0}) = 1$. Therefore $\operatorname{Ind}_{\delta''}(z_{i_0,j_0}) = n_{p_0}$. For the same reason $\operatorname{Ind}_{\delta''}(z_{i_0-1,j_0}) = 0$. On the other hand, $n_{p_0}\partial(c_{i_0,j_0})$ contains the term $-n_{p_0}\sigma_{p_0}$, hence $\operatorname{supp} \delta''$ does not intersect $[z_{i_0-1,j_0}, z_{i_0,j_0}]$. From here we conclude that $n_{p_0} = 0$. This contradiction shows that $\delta = \delta'$. Therefore we have obtained $\delta = \partial(c)$. □

1.9.14. Corollary. *Let Ω be a connected open subset of $\mathbb{C}$ and α a loop in Ω. The following conditions are equivalent:*

(1) *For every $\omega \in Z^1(\Omega)$ we have $I_\omega(\alpha) = j_\omega([\alpha]) = 0$.*
(2) *For every $a \in \mathbb{C} \backslash \Omega$ we have $\operatorname{Ind}_\alpha(a) = 0$.*
(3) *The class $[\alpha]$ of α in $\pi_1(\Omega; \alpha(0))$ belongs to the commutator subgroup.*

Proof. (1) implies (2) since ω_a is closed. (2) implies (3) since, by Proposition 1.9.13, $\chi(\alpha) = \sigma_\alpha$ is a 1-boundary in Ω. This means that $[\alpha] \in [\pi_1, \pi_1]$ because in the isomorphism between $H_1(\Omega; \mathbb{Z})$ and $\pi_1/[\pi_1, \pi_1]$ to the class of σ_α corresponds $[\alpha]$.

Finally, (3) implies (1) since j_ω vanishes on $[\pi_1, \pi_1]$. □

1.9.15. Proposition. *Let ω be a closed 1-form of class C^1 in Ω, Ω connected open set in $\mathbb{C}$. The differential form ω is exact if and only if for every*

$$\delta = \sum_{i \in 1} n_i \gamma_i \in Z_1(\Omega; \mathbb{Z})$$

it satisfies

$$\langle \delta, \omega \rangle = \sum_{i \in I} n_i I_\omega(\gamma_i) = 0.$$

PROOF. If $\omega = df$ with $f \in \mathscr{E}_2^0(\Omega)$ then

$$\langle \delta, \omega \rangle = \sum_{i \in I} n_i(f(\gamma_i(1)) - f(\gamma_i(0))) = \sum_{z \in \Omega} \left(\sum_{i \in J(z)} n_i - \sum_{i \in I(z)} n_i \right) f(z),$$

with $I(z) = \{i \in J : \gamma_i(0) = z\}$, $J(z) = \{i \in I : \gamma_i(1) = z\}$. Since δ is a 1-cycle we know that $\sum_{i \in I(z)} n_i = \sum_{i \in J(z)} n_i$, hence $\langle \delta, \omega \rangle = 0$.

Conversely, the condition $\langle \delta, \omega \rangle = 0$ for every 1-cycle implies that all the periods of ω vanish, hence the function $f(z) = \int_{c_z} \omega$ is well defined and satisfies $df = \omega$, as in Proposition 1.7.8. □

Let Ω be an open set in $\mathbb{C}$ and $\sigma = \sum n_i \sigma_i$ be a 2-chain such that $\sigma_i : \Delta_2 \to \Omega$ be the restriction to Δ_2 of a function of class C^1 in a neighborhood of Δ_2. Such a 2-chain will be called a ***differentiable 2-chain***.

If ω is a 2-form in Ω with coefficients in $L^1_{\text{loc}}(\Omega)$, we will call the ***integral of ω over the differentiable 2-chain*** σ the complex number

$$\langle \sigma, \omega \rangle := \sum_i n_i \int_{\sigma_i} \omega := \sum_i n_i \int_{\Delta_2} \sigma_i^*(\omega).$$

A form $\omega \in \mathscr{E}_0^1(\Omega)$ is said to be ***locally exact*** if for every $a \in \Omega$ there is a disk $B = B(a, r) \subseteq \Omega$ and a C^1-function f in B such that $\omega = df$ in B. If Ω' is an open subset of Ω, Ω' simply connected, then the proof of Proposition 1.7.9 shows that ω has a primitive f in Ω' (i.e., $f \in \mathscr{E}_1(\Omega')$, $\omega = df$). If ω were of class C^1 we could have applied directly Proposition 1.7.11.

1.9.16. Proposition. (1) *Let ω be a 1-form of class C^1 in Ω. The differential form ω is closed if and only if $\langle \delta, \omega \rangle = 0$ for every $\delta \in B_1(\Omega; \mathbb{Z})$ which is of the form $\delta = \partial(\sigma)$, σ a differentiable 2-chain in Ω.*

(2) *If ω is a continuous 1-form, we have the following equivalence: The form ω is locally exact if and only if $\int_\alpha \omega = 0$ for every loop α in Ω which is of class piecewise-C^1 and homotopic to a point Ω. It is enough to let α be a boundary of a rctangle of sides parallel to the axes which is homotopic to a point in Ω.*

PROOF. (1) If ω is C^1 and closed, and $\delta = \partial(\Sigma n_i \sigma_i)$, with σ_i differentiable, we have $\langle \delta, \omega \rangle = 0$ by §1.9.12, 2(c). Conversely, if $d\omega \neq 0$ and $\omega = P\,dx + Q\,dy$, we can assume $\dfrac{\partial Q}{\partial x} - \dfrac{\partial P}{\partial y}$ does not vanish in some triangle T. Hence $\displaystyle\int_T \left(\frac{\partial Q}{\partial x} - \frac{\partial P}{\partial y}\right) \times dx\,dy \neq 0$. If σ is the affine 2-chain that parameterizes T, then

$$\langle \partial(\sigma), \omega \rangle = \int_{\partial T} \omega = \int_T d\omega \neq 0.$$

(2) Every $a \in \Omega$ has neighborhood $B(a, r) \subseteq \Omega$ where the function

$$F(z) = \int_{\gamma_z} \omega$$

is a primitive of ω such that $\dfrac{\partial F}{\partial x} = P$, $\dfrac{\partial F}{\partial y} = Q$ when $\omega = P\,dx + Q\,dy$. (γ_z is the path indicated in Figure 1.23.)

Let us verify that $\dfrac{\partial F}{\partial x} = P$. Assume $a = 0$ for simplicity. For $z \in B(0, r)$ h real, we have (say $h > 0$)

$$F(z + h) - F(z) = \int_x^{x+h} P(t)\,dt + \int_0^y Q(x + h + it)\,dt - \int_0^y Q(x + it)\,dt$$

$$= \int_x^{x+h} P(t + iy)\,dt$$

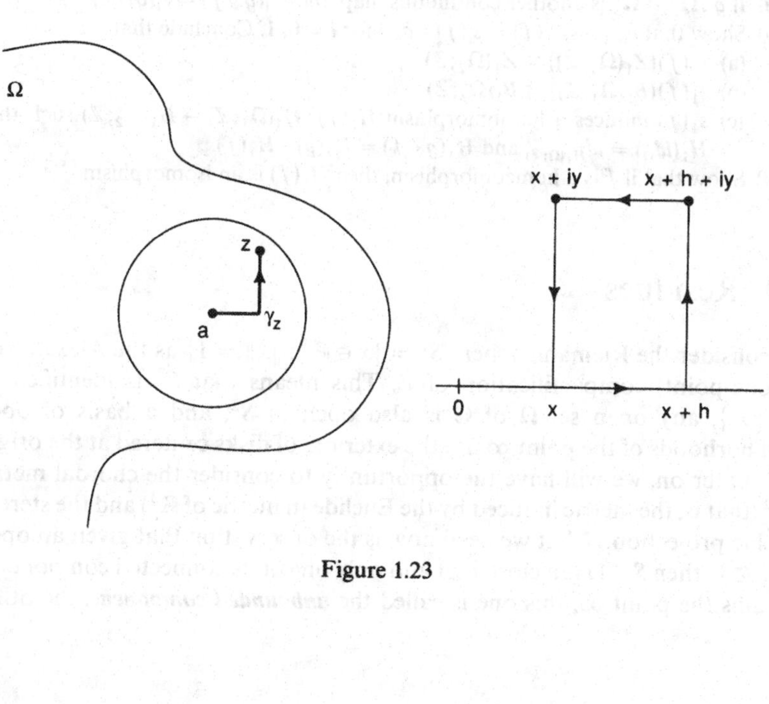

Figure 1.23

by the hypotheses (Figure 1.23). It follows that $\frac{\partial F}{\partial x}(z)$ exists and equals $P(z)$.

□

These results allow us to define on open connected subsets Ω of $\mathbb{C}$, a duality bracket

$$H_1(\Omega; \mathbb{Z}) \times H^1(\Omega) \to \mathbb{C}$$

which to the class $\bar{\delta}$ of the 1-cycle $\delta = \sum n_i \gamma_i$, and to the class $\bar{\omega}$ of the closed form ω associates $\langle \delta, \omega \rangle$. This mapping is well defined, $\mathbb{Z}$-linear in $\bar{\delta}$, $\mathbb{C}$-linear in $\bar{\omega}$, and nondegenerate, namely,

$$\begin{cases} \langle \delta, \omega \rangle = 0 \quad \forall \delta \Rightarrow \bar{\omega} = 0 \quad (1.9.15), \\ \langle \delta, \omega \rangle = 0 \quad \forall \omega \Rightarrow \bar{\delta} = 0 \quad (1.9.13). \end{cases}$$

EXERCISES 1.9

1. Let $a_1, \ldots, a_{n+1}$ be a collection of distinct points in $\mathbb{C}$, γ_j paths starting at a_j and ending at a_{j+1} $(1 \le j \le n)$, $\sigma_j = \sigma_{\gamma_j}$ the corresponding elementary 1-chains. Show that if $\sigma = \sum_{j=1}^{n} \lambda_j \sigma_j$ is a 1-cycle then $\lambda_j = 0$ for all j.

2. Compute $H_1(\Omega; \mathbb{Z})$ for the Ω given in Exercise 1.6.1.

3. Let $f: \Omega_1 \to \Omega_2$ be a continuous transformation between two open subsets of $\mathbb{C}$.
   (i) Show that the map $s_i(f)(\sigma) = f \circ \sigma$, $s_i(f): S_i(\Omega_1) \to S_i(\Omega_2)$ $(i = 0, 1, 2)$, extends to a $\mathbb{Z}$-linear map from $\mathscr{C}_i(\Omega_1, \mathbb{Z})$ into $\mathscr{C}_i(\Omega_2, \mathbb{Z})$.
   (ii) Prove that if $\Omega_1 = \Omega_2$, $f = id_{\Omega_1}$ then $s_i(f) = id_{\mathscr{C}_i(\Omega_1, \mathbb{Z})}$.
   (iii) If $g: \Omega_2 \to \Omega_3$ is another continuous map, then $s_i(g \circ f) = s_i(g) \circ s_i(f)$.
   (iv) Show that $\partial_{i+1} \circ s_{i+1}(f) = s_i(f) \circ \partial_{i+1}$ for $i = 0, 1$. Conclude that
      (a) $s_1(f)(Z_1(\Omega_1; \mathbb{Z})) \subseteq Z_1(\Omega_2; \mathbb{Z})$
      (b) $s_1(f)(B_1(\Omega_1; \mathbb{Z})) \subseteq B_1(\Omega_2; \mathbb{Z})$
      (c) $s_1(f)$ induces a homomorphism $H_1(f): H_1(\Omega_1; \mathbb{Z}) \to H_1(\Omega_2; \mathbb{Z})$ such that $H_1(id_\Omega) = id_{H_1(\Omega; \mathbb{Z})}$ and $H_1(g \circ f) = H_1(g) \circ H_1(f)$.
   (v) Show that if f is a homeomorphism, then $H_1(f)$ is an isomorphism.

§10. Residues

We consider the Riemann sphere $S^2 = \{x \in \mathbb{R}^3 : \|x\| = 1\}$ as the Alexandrov (or one-point) compactification of $\mathbb{C}$. This means that S^2 is identified to $\mathbb{C} \dot{\cup} \{\infty\}$, any open set Ω of $\mathbb{C}$ is also open in S^2, and a basis of open neighborhoods of the point ∞ are the exteriors of disks centered at the origin in $\mathbb{C}$. Later on, we will have the opportunity to consider the chordal metric in S^2 (that is, the metric induced by the Euclidean metric of $\mathbb{R}^3$) and the stereographic projection. What we need now is the observation that given an open set $\Omega \subseteq \mathbb{C}$ then $S^2 \setminus \Omega$ is a closed set and only one of its connected components contains the point ∞, this one is called the ***unbounded component***; the other ones are closed sets, bounded in $\mathbb{C}$ (hence compact), and are called the ***holes*** of Ω.

1.10.1. Proposition. *Let Ω be a connected open set in $\mathbb{C}$. The following statements are equivalent*:

(1) *Ω does not have holes (i.e., $S^2 \setminus \Omega$ is connected).*
(2) *$H_1(\Omega; \mathbb{Z}) = 0$ (i.e., every 1-cycle is a boundary).*
(3) *$H^1(\Omega) = 0$ (i.e., every closed 1-form is exact).*
(4) *Ω is simply connected.*

PROOF. (1) $\Rightarrow$ (2): Let δ be a 1-cycle and $a \notin \Omega$. Then $\operatorname{Ind}_\delta(a) = 0$ since a is in the unbounded component of $\mathbb{C} \setminus \operatorname{supp} \delta$. Hence δ is a boundary by Proposition 1.9.3.

(2) $\Rightarrow$ (1): We prove first an important lemma.

1.10.2. Lemma. *Let K be a compact subset of $\mathbb{C}$, V an open neighborhood of K, and $a \in K$. There is $\delta \in Z_1(V \setminus K; \mathbb{Z})$, which can be taken to be a special cycle, such that $\operatorname{Ind}_\delta(a) = 1$ and $\delta \in B_1(V, \mathbb{Z})$.*

PROOF. If $V = \mathbb{C}$ the proof is obvious. If not, let $\rho = d(K, V^c) > 0$. Let us consider the ***tiling*** of $\mathbb{C}$ by (closed) squares of center $a + \frac{\rho}{2}(p + iq)$, $p, q \in \mathbb{Z}$, and side $\rho/2$ (called the size of these tiles).

Let $Q_0, Q_1, \ldots, Q_n$ be those squares intersecting K, numbered in such a way that $a \in Q_0$. We consider Q_j as the support of the 2-chain $c_j = c_j^1 + c_j^2$, where c_j^k ($k = 1, 2$) are triangles defined as in Figure 1.21.

Let $\delta = \sum_{0 \le j \le n} \partial(c_j)$, which is clearly a 1-boundary in V. We claim that δ is also a 1-chain in $V \setminus K$ (and hence a 1-cycle). In fact, if one of the sides γ of Q_j is not entirely in $V \setminus K$, γ must intersect K. Then the square of the paving contiguous to Q_j along γ intersects K, hence it is one from the preceding list, say Q_k, $0 \le k \le n$, $k \ne j$. It follows that the segment γ is not in $\operatorname{supp} \delta$ since it appears exactly two times in δ, once in $\partial(c_j)$ and once in $\partial(c_k)$, but with opposite signs.

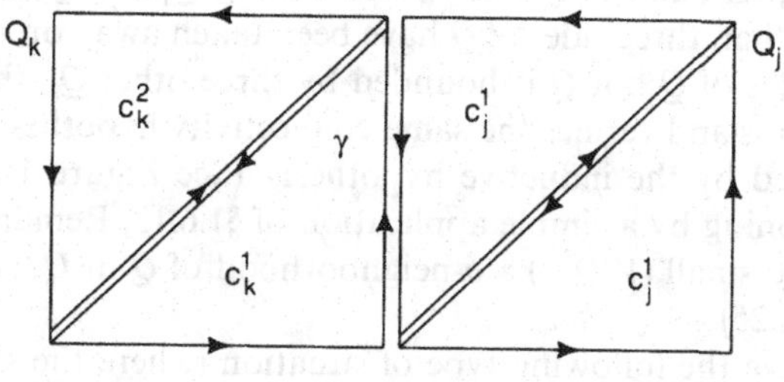

Figure 1.24

The chain δ is then a cycle in $V \backslash K$ since its boundary, as a chain in $V \backslash K$, has the same formal expression as the boundary of δ, considered as a chain in V.

Finally, $\mathrm{Ind}_\delta(a) = 1$ since $\mathrm{Ind}_{\partial(c_0)}(a) = 1$ and $\mathrm{Ind}_{\partial(c_j)}(a) = 0$ for $j \geq 1$. This concludes the lemma. □

Let us go back to the proof of (2) ⇒ (1). We argue by contraposition. Assume that $H_1(\Omega; \mathbb{Z}) = 0$ but there are holes in Ω. Since now $S^2 \backslash \Omega$ is not connected, then there are two closed sets $A, B \subseteq S^2 \backslash \Omega$, not empty, disjoint, and such that $S^2 \backslash \Omega = A \cup B$. Assume $\infty \in A$, then B is compact. Let us apply the preceding lemma to $V = \Omega \cup B$ $(= S^2 \backslash A)$, $K = B$, $a \in B$. There is then $\delta \in Z_1(V \backslash K; \mathbb{Z}) = Z_1(\Omega; \mathbb{Z})$ such that $\mathrm{Ind}_\delta(a) = 1$. Hence $\delta \notin B_1(\Omega; \mathbb{Z})$, otherwise $\mathrm{Ind}_\delta(a) = 0$. This contradicts $H_1(\Omega; \mathbb{Z}) = 0$.

(2) ⇔ (3) It follows from the duality between $H_1(\Omega; \mathbb{Z})$ and $H^1(\Omega)$ established at the end of the previous section.

(4) ⇒ (1), (2), (3): if $\pi_1(\Omega; z_0) = \{[\varepsilon_{z_0}]\}$ for $z_0 \in \Omega$, then $H_1(\Omega; \mathbb{Z}) = 0$.

(1), (2), (3) ⇒ (4): If α is a loop in Ω there is an $\varepsilon > 0$ such that if Q is a square of a tiling of size ε such that $\mathring{Q} \cap \partial\Omega \neq \varnothing$ then $Q \cap \mathrm{Image}(\alpha) \neq \varnothing$. Let us add to $S^2 \backslash \Omega$ those squares of the tiling whose interiors intersect $\partial\Omega$. One obtains this way a closed set F_ε in S^2 such that $\Omega_\varepsilon = S^2 \backslash F_\varepsilon$ is an open set in Ω and α is a loop in Ω_ε. Since Ω has no holes, F_ε is connected. Let $\{\Omega_{\varepsilon,i}\}$ be the family of connected components of Ω_ε. For each i, $F_\varepsilon \cap \bar{\Omega}_{\varepsilon,i}$ is not empty by construction, hence $F_\varepsilon \cup \Omega_{\varepsilon,i}$ is connected. (Both F_ε, $\Omega_{\varepsilon,i}$ are connected.) Therefore, for each i, $S^2 \backslash \Omega_{\varepsilon,i}$ is connected since it is the union of connected sets, $F_\varepsilon \cup \Omega_{\varepsilon,j}$ $(j \neq i)$, with nonempty intersection F_ε. Therefore, since α is contained in one of the $\Omega_{\varepsilon,i}$ and $\Omega_{\varepsilon,i}$ has no holes, we are reduced to prove, by induction on N, the following:

Let U be an open set of $\mathbb{C}$ obtained as union of N squares $Q_1, \ldots, Q_N$ of a tiling (possibly taking away some of their sides).

Assume U and $S^2 \backslash U$ are connected. Then U is simply connected.

The case $N = 1$ being obvious, we consider the case $N \geq 2$. Among the squares in U, we have one which can be singled out by being "further to the right and up" than any other square in U (for instance, the point $q_0 = (x_0, y_0) \in \bar{U}$ which satisfies $x_0 = \max\{x : (x, y) \in \bar{U}\}$ and $y_0 = \max\{y : (x_0, y) \in \bar{U}\}$ is a vertex of this square and only this square among $Q_1, \ldots, Q_N$.) Let us call Q this square. If either, three sides of Q have been taken away or, this happened for only two sides of Q but Q is bounded by three other Q_i, then $V = U \backslash Q$ has $N - 1$ squares and verifies the same connectivity hypotheses. Hence V is simply connected by the inductive hypothesis. (See Figure 1.25.) One concludes the reasoning by a simple application of §1.6.13, Remark 2, to V and $V(Q, \delta)$ for $\delta > 0$ small ($V(Q, \delta)$ a δ-neighborhood of Q in U, i.e., the shaded area in Figure 1.25).

If not, we have the following type of situation (where the shaded area in Figure 1.26 corresponds to $S^2 \backslash U$):

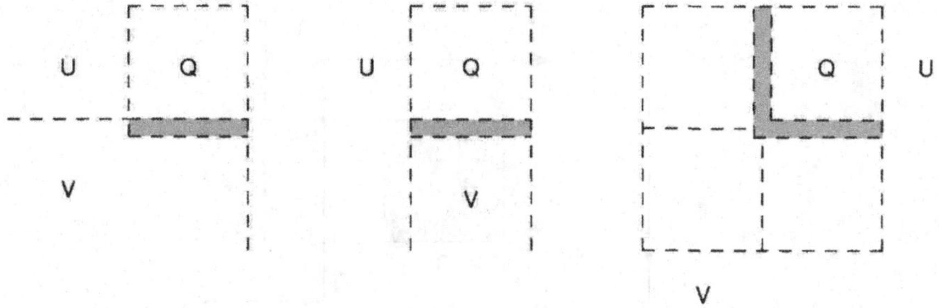

Figure 1.25

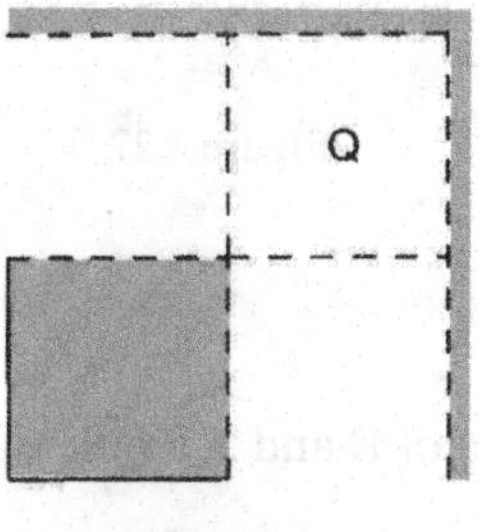

Figure 1.26

The problem here is that if we remove Q then we may disconnect U or not. The set $U \backslash Q$ has at most two connected components by the choice of Q. If it has exactly two, U_1 and U_2, then they are simply connected by the induction hypothesis (for instance, $S^2 \backslash U_1 = Q \cup U_2 \cup (S^2 \backslash U)$ is connected). Applying §1.6.13 once gives $Q \cup U_1$ simply connected (with the correct sides of Q taken away), reapplying it shows that $U = (Q \cup U_1) \cup U_2$ is also simply connected.

The troublesome case arises when $U \backslash Q$ has a single component. Then $V = U \backslash Q$ is again simply connected by induction, but we cannot say anything about U. We are going to see that this case does not really arise. Let a and b be the midpoints of the sides of Q which are not exterior to U. Since V is connected one can find a special path γ in V joining a to b. One can further assume that γ has no self-intersections (i.e., the continuous map $\gamma : [0, 1] \to V$ is injective). Let α be the special path joining b to a in Q and passing through the center of Q. The path $\beta = \gamma\alpha$ is hence a Jordan curve in U (not homotopic to a point!). One can see that this situation is impossible because it shows that $S^2 \backslash U$ is not connected. Namely, the square C from Figure 1.27 is in the bounded component of $S^2 \backslash U$ as one can see computing by continuity the index with respect to β. The index changes from 0 to 1 when one traverses α. This concludes the proof of Proposition 1.10.1. □

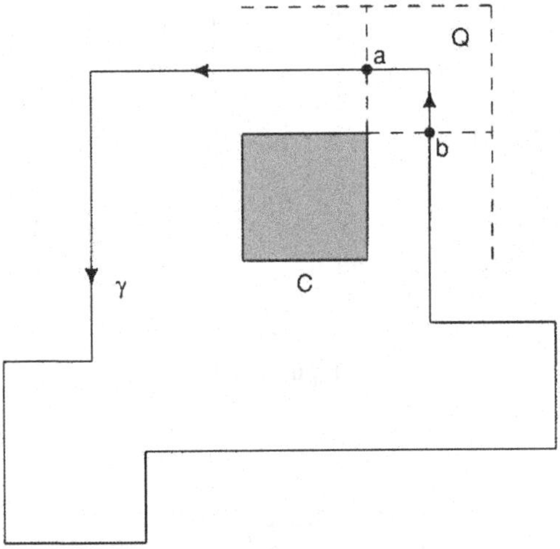

Figure 1.27

1.10.3. Definitions

(1) Let Ω be an open subset of $\mathbb{C}$ and T a hole of Ω. T is said to be ***admissible*** if $\Omega \cup T$ is open in $\mathbb{C}$.
(2) An open subset of $\mathbb{C}$ is called ***admissible*** if all its holes are admissible.

For example, if $\Omega = \mathbb{C}\setminus\{0, 1, 1/2, 1/3, \ldots, 1/n, \ldots\}$ then $T_n = \{1/n\}$ is an admissible hole for $n \geq 1$. $T_0 = \{0\}$ is not. Hence Ω is inadmissible.

1.10.4. Proposition. *An open set Ω in $\mathbb{C}$ is admissible if and only if the family $(T_i)_{i\in I}$ of its holes is locally finite in the open set $U = S^2\setminus T_\infty$, where T_∞ is the component of ∞ in $S^2\setminus\Omega$.*

PROOF. If $(T_i)_{i\in I}$ is locally finite in U then for every $i_0 \in I$ we have $\Omega \cup T_{i_0}$ is open. In fact, since T_{i_0} is compact and the family is locally finite we can find a neighborhood of T_{i_0} which meets only finitely many other T_i, by reducing the size of this neighborhood one finds an open set V that doesn't intersect any T_i, $i \neq i_0$ and $V \supseteq T_{i_0}$, hence $V \cup \Omega = \Omega \cup T_{i_0}$ is open.

Conversely, if Ω is admissible, then every compact subset K of U must intersect only a finite number of holes T_i. If not, let $(T_j)_{j\in J}$ be a countably infinite family of holes intersecting K. Pick distinct $z_j \in T_j \cap K$. The sequence $\{z_j\}_{j\in J}$ admits an accumulation point $z_0 \in K \cap (S^2\setminus\Omega)$. Therefore z_0 itself belongs to some T_{i_0}. It follows $\Omega \cup T_{i_0}$ cannot be open, since every neighborhood of z_0 (in particular, $\Omega \cup T_{i_0}$, if open) must contain an infinite number of z_j, $j \in J$, which is clearly impossible. □

For example, if Ω has only finitely many holes or if the set of holes is a discrete set of points, then Ω is admissible.

1.10.5. Proposition. *Let Ω be an open set in $\mathbb{C}$, T an admissible hole. There exists $\delta_T \in Z_1(\Omega; \mathbb{Z}) \cap B_1(\Omega \cup T; \mathbb{Z})$ such that*

(1) *If $b \in T$, then $\operatorname{Ind}_{\delta_T}(b) = 1$.*
(2) *If $c \notin \Omega \cup T$, then $\operatorname{Ind}_{\delta_T}(c) = 0$.*

Furthermore, if $\delta'_T \in Z_1(\Omega; \mathbb{Z}) \cap B_1(\Omega \cup T; \mathbb{Z})$ verifies also (1) and (2), then $\delta_T - \delta'_T \in B_1(\Omega; \mathbb{Z})$.

Proof. One applies Lemma 1.10.2 to $V = \Omega \cup T$, $K = T$ and $a \in T$, this lemma furnishes δ_T.

If δ'_T has the same properties as δ_T then $\delta_T - \delta'_T$ is a 1-boundary in Ω since its index with respect to any $z \in S^2 \setminus \Omega$ is zero. □

1.10.6. Definition. Let T be a hole of an open set Ω and $\delta \in Z_1(\Omega; \mathbb{Z})$. We denote by $\operatorname{Ind}_\delta(T)$, ***index** of δ with respect to T*, the integer $\operatorname{Ind}_\delta(a)$ for $a \in T$ (which is independent of the choice of a).

1.10.7. Proposition. *Let Ω be an admissible open set. For every 1-cycle $\delta \in Z_1(\Omega; \mathbb{Z})$ one has $\operatorname{Ind}_\delta(T) = 0$ for every hole T, with the exception of at most a finite number of holes.*

Proof. The connected component of ∞ in $U = S^2 \setminus \operatorname{supp} \delta$ contains all the holes of Ω except possibly for a finite number (cf. §1.10.4), and $\operatorname{Ind}_\delta(a) = 0$ for $a \in U$. □

1.10.7. Residue Theorem (First Version). *Let Ω be an admissible open set. For every hole T of Ω choose $\delta_T \in Z_1(\Omega; \mathbb{Z}) \cap B_1(\Omega \cup T; \mathbb{Z})$ such that $\operatorname{Ind}_{\delta_T}(a) = 1$ if $a \in T$, $\operatorname{Ind}_{\delta_T}(a) = 0$ if $a \notin \Omega \cup T$. Then, for every $\delta \in Z_1(\Omega; \mathbb{Z})$ one has*

$$\delta' := \delta - \sum_T \operatorname{Ind}_\delta(T)\delta_T \in B_1(\Omega; \mathbb{Z}).$$

Proof. Let us consider the index of δ' with respect to $a \notin \Omega$. If a is in the unbounded component, then $\operatorname{Ind}_\delta(a)$, $\operatorname{Ind}_{\delta_T}(a)$ are all zero, hence $\operatorname{Ind}_{\delta'}(a) = 0$.

If $a \in T_0$, T_0 hole of Ω, $\operatorname{Ind}_\delta(a) = \operatorname{Ind}_\delta(T_0)$ and

$$\sum_T \operatorname{Ind}_\delta(T) \operatorname{Ind}_{\delta_T}(a) = \operatorname{Ind}_\delta(T_0) \operatorname{Ind}_{\delta_{T_0}}(a) = \operatorname{Ind}_\delta(T_0).$$

Hence $\operatorname{Ind}_{\delta'}(a) = 0$. It follows that δ' is a 1-boundary. □

1.10.8. Remarks

(1) If δ is homologous to $\sum_T n_T \delta_T$, then $n_T = \operatorname{Ind}_\delta(T)$.

(2) Let I be the set of holes in Ω. The first version of the residue theorem means that $H_1(\Omega; \mathbb{Z})$ is isomorphic to $\mathbb{Z}^{(I)}$ (the set of maps $I \to \mathbb{Z}$ with finite support) via the mapping $\bar{\delta} \mapsto (\operatorname{Ind}_\delta(T))_{T \in I}$.

1.10.9. Definition. Let Ω be an open set in $\mathbb{C}$, T an admissible hole, ω a closed 1-form of class C^1 in Ω, and δ_T a 1-cycle corresponding to T. We denote $\operatorname{Res}_\omega(T)$, the ***residue** of ω with respect to T*, the number

$$\operatorname{Res}_\omega(T) = \frac{1}{2\pi i} I_\omega(\delta_T).$$

We remark that this value does not depend on the choice of δ_T, since if δ'_T is another cycle satisfying the same condition, $\delta = \delta'_T - \delta_T$ is a 1-boundary in Ω and hence, $I_\omega(\delta) = 0$. Moreover, two closed 1-forms which are cohomologous have the same residue.

1.10.10. Residue Theorem (Second Version). *Let Ω be an admissible open subset of $\mathbb{C}$, ω a closed 1-form of class C^1 in Ω and $\delta \in Z_1(\Omega; \mathbb{Z})$. One has the relation ("residue formula")*

$$\frac{1}{2\pi i}\int_\delta \omega = \sum_T \operatorname{Ind}_\delta(T)\operatorname{Res}_\omega(T),$$

where $\int_\delta \omega = \langle \delta, \omega\rangle$, i.e., $\int_\delta \omega = \sum_i n_i I_\omega(\gamma_i)$ if $\delta = \sum_i n_i\gamma_i$. The sum takes place over all the holes of Ω.

Proof. We already known that $\delta' = \delta - \sum_T \operatorname{Ind}_\delta(T)\delta_T$ is a 1-boundary. Furthermore, $\int_{\delta'} \omega = 0$ since ω is closed. □

1.10.11. Remark. With I the set of holes in Ω as in §1.10.8, one can consider the map

$$H^1(\Omega) \to \mathbb{C}^I$$

$$\bar{\omega} \mapsto (\operatorname{Res}_\omega(T))_{T \in I}.$$

This map is $\mathbb{C}$-linear and injective. If $\operatorname{Res}_\omega(T) = 0$ for every hole T, the residue theorem shows that $\langle \delta, \omega\rangle = 0$ for every $\delta \in Z_1(\Omega; \mathbb{Z})$. We already know that this implies that ω is exact, hence $\bar{\omega} = 0$.

Therefore, the residue theorem states that the "duality" $H_1(\Omega; \mathbb{Z}) \times H^1(\Omega)$, $(\bar{\delta}, \bar{\omega}) \mapsto \langle \delta, \omega\rangle$, is exactly the usual duality bracket of $\mathbb{Z}^{(I)} \times \mathbb{C}^I$ via the injection $H^1(\Omega) \to \mathbb{C}^I$ given earlier. If I is finite, the map $H^1(\Omega) \to \mathbb{C}^I$ is clearly bijective.

Exercises 1.10

1. A ***Jordan region*** is defined to be $\operatorname{Int}(\gamma)$, where γ is a Jordan curve. Show any Jordan region is simply connected.

2. (a) Let Ω_1, Ω_2 be two connected open sets whose intersection is also connected. Show that if $H^1(\Omega_1) = H^1(\Omega_2) = 0$ then $H^1(\Omega_1 \cup \Omega_2) = 0$.

(b) Under the hypothesis of part (a), it follows that if ω is a 1-form which admits a primitive in Ω_1 and in Ω_2, then it admits a primitive in $\Omega_1 \cup \Omega_2$. Is it true in general that if a 1-form ω admits a primitive in two connected open sets Ω_1, Ω_2 with $\Omega_1 \cap \Omega_2$ connected, then ω admits a primitive in $\Omega_1 \cup \Omega_2$?

3. Let $\Omega = \mathbb{C}\backslash(T_1 \cup T_2)$, $T_1 = \bar{B}(0, 1)$, $T_2 = \bar{B}(3, 1)$. Let $\omega = \frac{dz}{z}$, compute $\mathrm{Res}_\omega(T_j)$, $j = 1, 2$.

4. Let Ω be an admissible open set in $\mathbb{C}$, possibly having infinitely many holes. Let $T_1, \ldots, T_n$ be holes of Ω, and $\lambda_1, \ldots, \lambda_n$ arbitrary complex numbers. Construct a closed 1-form ω in Ω such that $\mathrm{Res}_\omega(T_j) = \lambda_j$ for $j = 1, \ldots, n$. Could this be done when $n = \infty$?

5. Let Ω be an open connected set in $\mathbb{C}$, show it is simply connected if and only if, for every Jordan curve γ in Ω, one has $\mathrm{Int}(\gamma) \subseteq \Omega$.

§11. Holomorphic Functions

In this section we introduce the subject of study of this book, the holomorphic functions, and point out some properties that are immediate consequences of the topological considerations in this chapter. The deeper properties of holomorphic functions will be considered in the following chapters.

1.11.1. Definition. Let Ω be an open set in $\mathbb{C}$. A function $h \in \mathscr{E}_1(\Omega)$ is said to be ***holomorphic*** if $\bar{\partial} h = 0$, in other words, if its differential is of type $(1, 0)$. We denote by $\mathscr{H}(\Omega)$ the set of functions holomorphic in Ω. A function in $\mathscr{H}(\mathbb{C})$ is said to be an ***entire function***.

It is convenient to say that f is holomorphic in a set K if there is an open set Ω, $K \subseteq \Omega$, such that f is defined in Ω and $f \in \mathscr{H}(\Omega)$.

The condition required for a function h to be holomorphic is that it satisfies the differential equation $\frac{\partial h}{\partial \bar{z}} = 0$, which is known as the ***homogeneous Cauchy-Riemann equation***.

If $h = P + iQ$, the equation $\frac{\partial h}{\partial \bar{z}} = 0$ is equivalent to the system

$$\begin{cases} \dfrac{\partial P}{\partial x} - \dfrac{\partial Q}{\partial y} = 0 \\ \dfrac{\partial Q}{\partial x} + \dfrac{\partial P}{\partial y} = 0 \end{cases}.$$

It follows from these equations that if h is holomorphic

$$\frac{\partial h}{\partial z} = \frac{\partial h}{\partial x} \quad \text{and} \quad \frac{\partial h}{\partial z} = \frac{1}{i} \frac{\partial h}{\partial y}.$$

These identities and the observation that when h is holomorphic, the differential form $\omega = h\,dz$ is closed, play a very important role in computations involving holomorphic functions.

We recall from §1 that the condition of holomorphicity is exactly that the differential $dh(z_0)$ be a $\mathbb{C}$-linear map at each point $z_0 \in \Omega$. One can rephrase this observation as follows: since h is differentiable, if z is close to z_0 we have

$$h(z) - h(z_0) = dh(z_0)(z - z_0) + |z - z_0|\varepsilon(z - z_0)$$

with $\lim_{|w|\to 0} \varepsilon(w) = 0$. Since $dh = \dfrac{\partial h}{\partial z} dz$, the action of the map $dh(z_0)$ on the vector $z - z_0$ becomes multiplication by the complex number $\dfrac{\partial h}{\partial z}(z_0)$:

$$h(z) - h(z_0) = \frac{\partial h}{\partial z}(z_0)\cdot(z - z_0) + |z - z_0|\varepsilon(z - z_0).$$

We profit from the fact that we can divide by the nonzero complex number $z - z_0$ (if $z \neq z_0$) to obtain the existence of the following limit

$$\lim_{z\to z_0} \frac{h(z) - h(z_0)}{z - z_0} = \frac{\partial h}{\partial z}(z_0).$$

This formula is analogous to the usual definition in one real variable of the derivative of a function; for that reason we denote this limit also $h'(z_0)$, or sometimes $\dfrac{dh}{dz}(z_0)$ and call it the ***derivative*** of the holomorphic function h at the point z_0. The derivative is hence a new (continuous) function in Ω defined by

$$h'(z) := \frac{\partial h}{\partial z}(z).$$

A natural question arises here. Assume $h : \Omega \to \mathbb{C}$ has a "complex derivative" at each point in the following sense: the limit

$$h'(z_0) = \lim_{z\to z_0} \frac{h(z) - h(z)}{z - z_0}$$

exists for every $z_0 \in \Omega$. Is h a holomorphic function? The answer is affirmative, as we will see in the next chapter. Meanwhile we will describe some elementary properties of holomorphic functions.

From now on, unless otherwise mentioned, a polynomial is an expression of the form $a_0 + a_1 z + \cdots + a_n z^n$. That is, it is a polynomial in the single variable z (as opposed to the more general polynomials in the two real variables x, y, $\sum a_{j,k} x^j y^k$, where $z = x + iy$).

1.11.2. Proposition. *The set $\mathscr{H}(\Omega)$ is an algebra containing the polynomials.*

PROOF. Exercise for the reader. □

An entire function which is not a polynomial is said to be ***transcendental.***

1.11.3. Examples. (1) The following functions are entire:

$$e^z := e^x(\cos y + i \sin y) \qquad \left(\frac{d}{dz} e^z = e^z\right)$$

$$\sin z := \frac{e^{iz} - e^{-iz}}{2i} \qquad \left(\frac{d}{dz} \sin z = \cos z\right)$$

$$\cos z := \frac{e^{iz} + e^{-iz}}{2} \qquad \left(\frac{d}{dz} \cos z = -\sin z\right)$$

$$\sinh z := \frac{e^z - e^{-z}}{2} \qquad \left(\frac{d}{dz} \sinh z = -\cosh z\right)$$

$$\cosh z := \frac{e^z + e^{-z}}{2} \qquad \left(\frac{d}{dz} \cosh z = \sinh z\right).$$

(2) In $\Omega = \mathbb{C} \setminus]-\infty, 0]$, the function

$$\operatorname{Log} z = \log|z| + i \operatorname{Arg} z,$$

$-\pi < \operatorname{Arg} z < \pi$, is holomorphic and satisfies $\exp(\operatorname{Log} z) = z$. It is the ***principal branch of the logarithm.*** One can verify that $\operatorname{Log} z$ is a primitive of the differential form $\omega = \dfrac{dz}{z}$ in Ω. The function $z^\alpha := \exp(\alpha \operatorname{Log} z)$ ($\alpha \in \mathbb{C}$) is holomorphic in the same set.

Let us show, for instance, that $\operatorname{Log} z$ is holomorphic in Ω. The exponential function is a C^∞ bijection of the strip $S := \{w : -\pi < \operatorname{Im} w < \pi\}$ onto Ω with Jacobian $e^{2\operatorname{Re} w} > 0$. It follows that there is a unique inverse φ of class C^∞, $\varphi : \Omega \to S$, $\exp \circ \varphi = id_\Omega$, $\varphi \circ \exp = id_S$. Assume $\tilde{\varphi}$ is a continuous function in Ω satisfying $\exp \circ \tilde{\varphi} = id_\Omega$. Then, for $z \in \Omega$, one has $\varphi(z) - \tilde{\varphi}(z) = 2\pi i n(z)$, where $n : \Omega \to \mathbb{Z}$ is a continuous function, hence a constant. Since the function $\operatorname{Log} z = \log|z| + i \operatorname{Arg} z$ is a continuous function such that $\exp(\operatorname{Log} z) = z$ and $\operatorname{Log} : \Omega \to S$, it follows that $\operatorname{Log} z \equiv \varphi(z)$. Therefore Log is a C^∞ function.

Let us now compute $\dfrac{\partial \varphi}{\partial z}$. After differentiating the identity $z = \exp(\varphi(z))$, the chain rule allows us to conclude that $0 = \exp(\varphi(z)) \cdot \dfrac{\partial \varphi}{\partial \bar{z}}$. Hence $\dfrac{\partial \varphi}{\partial \bar{z}} = 0$ and $\varphi(z)$ is holomorphic. In the same way we find that $1 = \exp(\varphi) \cdot \dfrac{\partial \varphi}{\partial z}$. Therefore $\dfrac{\partial \varphi}{\partial z} = \dfrac{1}{z}$ and $\operatorname{Log} z$ is the primitive in Ω of the closed differential form $\omega = \dfrac{dz}{z}$, which vanishes at $z = 1$,

$$\operatorname{Log} z = \int_{\sigma_z} \frac{dt}{t},$$

where σ_z is the segment joining 1 to z in Ω. (We can take any other path since Ω is simply connected.)

(3) If $c \neq 0$ the rational function $z \mapsto \dfrac{az+b}{cz+d}$ from $\mathbb{C}\setminus\{-d/c\}$ into $\mathbb{C}$ is holomorphic. In particular $\dfrac{1}{z}$ is holomorphic in $\mathbb{C}^* = \mathbb{C}\setminus\{0\}$.

(4) $z \mapsto |z|^2$ is not holomorphic.

(5) Let Ω_1, Ω_2 be open sets in $\mathbb{C}$, $f \in \mathscr{H}(\Omega_1)$, $g \in \mathscr{H}(\Omega_2)$, and $f(\Omega_1) \subseteq \Omega_2$. Then $g \circ f \in \mathscr{H}(\Omega_1)$.

This follows immediately from the chain rule applied to the differential operator $\dfrac{\partial}{\partial \bar{z}}$.

As we have already pointed out, when $f \in \mathscr{H}(\Omega)$, the differential form $\omega = f\,dz$ is of class C^1 and closed. Therefore, Proposition 1.7.7, Proposition 1.9.15, and Theorem 1.10.10 can be translated into the following.

1.11.4. Cauchy's Theorem. *Let Ω be an open set in $\mathbb{C}$, $f \in \mathscr{H}(\Omega)$.*

(1) *If α, β are two paths in Ω, homotopic with fixed endpoints, then*

$$\int_\alpha f\,dz = \int_\beta f\,dz.$$

(2) *If α, β are two loops which are free homotopic in Ω, then*

$$\int_\alpha f\,dz = \int_\beta f\,dz.$$

In particular, if α is a loop free homotopic to a point

$$\int_\alpha f\,dz = 0.$$

If Ω is simply connected, then one has $\int_\alpha f\,dz = 0$ for every loop α.

(3) *If δ_1 and δ_2 are two 1-cycles that are homologous in Ω then*

$$\int_{\delta_1} f\,dz = \int_{\delta_2} f\,dz.$$

If δ is a 1-boundary in Ω, then $\int_\delta f\,dz = 0$.

(4) *If Ω is an admissible open set and δ is a 1-cycle in Ω, then the residue formula is the expression*

$$\frac{1}{2\pi i}\int_\delta f\,dz = \sum_T \operatorname{Ind}_\delta(T)\operatorname{Res}_{f\,dz}(T).$$

(5) *If $\omega = f\,dz$ admits a primitive F in Ω, then F is holomorphic in Ω.*

Example. If Ω is a simply connected set and $f \in \mathscr{H}(\Omega)$, then $\omega = f\,dz$ admits primitives. All of them are holomorphic.

We conclude this chapter with a discussion of two generalizations of Example 1.11.3, (2). If Ω is a simply connected open set and $f \in \mathscr{H}(\Omega)$, $f(z) \neq 0$ for every $z \in \Omega$, then we can define in a unique way a C^1 function $g(z) = \log f(z)$ once a value for $\log f(z_0)$ has been fixed for some $z_0 \in \Omega$. This follows from §1.6.29 (1). Since $f(z) = \exp(g(z))$, we can apply the same reasoning as in Example 1.11.3, (2) to show that $\dfrac{\partial g}{\partial \bar{z}} = 0$, hence $g \in \mathscr{H}(\Omega)$. Furthermore, $g'(z) = \dfrac{f'(z)}{f(z)}$, hence g is a primitive of the form $\dfrac{f'(z)}{f(z)}\,dz$. A priori this form is only of class C^0, it follows from §2.1.6 that, in fact, it is a closed C^∞ form. Note that in particular we can define other branches $\log z$ of the logarithm in any Ω simply connected such that $0 \notin \Omega$.

What happens when f omits not only the value 0 but also another value, say the value 1, in Ω? We have the following lemma of Landau.

1.11.5. Lemma. *Let Ω be a simply connected open set in $\mathbb{C}$, $f \in \mathscr{H}(\Omega)$, $f(z) \not\equiv 0, 1$ for every $z \in \Omega$. Fix a point $z_0 \in \Omega$. There is then a function $h \in \mathscr{H}(\Omega)$ such that*

$$f(z) = \exp(2\pi i \cosh(h(z)))$$

and $-\pi \leq \operatorname{Im} h(z_0) < \pi$. This function satisfies the estimate

$$|h(z_0)| \leq \log\left(2 + \frac{|\log|f(z_0)||}{\pi}\right) + \pi.$$

Proof. Since f does not vanish, we can uniquely determine $\log f$ once we normalize its imaginary part at the point z_0. Let $g \in \mathscr{H}(\Omega)$ be such that $-\frac{1}{2} \leq \operatorname{Re} g(z_0) < \frac{1}{2}$ and $f(z) = \exp(2\pi i g(z))$ in Ω. It is then clear that $g(\Omega) \cap \mathbb{Z} = \varnothing$, otherwise $f(z) = 1$ would have a solution in Ω. In particular, $g(z)^2 - 1$ does not vanish in Ω and has a holomorphic square root G in Ω. We have then $g^2 = 1 + G^2$, or

$$(g - G)(g + G) = 1.$$

Let H be one of the functions $g - G$, $g + G$, chosen so that $|H(z_0)| \geq 1$. It follows that H does not vanish and we choose

$$h = \log H, \qquad -\pi \leq \operatorname{Im} h(z_0) < \pi.$$

Let us verify that $\cosh(h(z)) \equiv g(z)$. Assume that $H(z) \equiv g(z) - G(z)$, $(H(z))^{-1} \equiv g(z) + G(z)$. We have

$$\cosh(h(z)) = \frac{e^{h(z)} + e^{-h(z)}}{2} = \frac{H(z) + (H(z))^{-1}}{2}$$

$$= \frac{g(z) + G(z) + g(z) - G(z)}{2} = g(z).$$

The other case leads to the same result. From the definition of g we see that

$$f(z) = \exp(2\pi i \cosh h(z))).$$

The estimate of $|h(z_0)|$ is left for the exercises. □

EXERCISES 1.11

Ω represents a domain in $\mathbb{C}$, $B = B(0, 1)$. We collect here some geometric properties of the complex plane that will be useful in the study of holomorphic functions.

1. Let $z \in \mathbb{C}\setminus]-\infty, 0]$, $\sqrt[n]{z}$ the principal branch of the nth root. Compute

$$\lim_{n\to\infty} (n\sqrt[n]{z} - 1).$$

2. Show that if $f: \Omega \to \mathbb{C}$ has a complex derivative at every point, i.e., $\lim_{h\to 0} \dfrac{f(z+h)-f(z)}{h}$ exists for every $z \in \Omega$, then f is continuous.

3. (i) Write down the Cauchy-Riemann equations in polar coordinates. That is, show that if $f(z) = u(x, y) + iv(x, y)$ and (r, θ) are the corresponding polar coordinates, the Cauchy-Riemann equations become

$$\frac{\partial u}{\partial r} = \frac{1}{r}\frac{\partial v}{\partial \theta}, \qquad \frac{\partial v}{\partial r} = -\frac{1}{r}\frac{\partial u}{\partial \theta},$$

with some abuse of notation (see Exercise 1.2.5 for a precise version).

(ii) Let $\{s, n\}$ be an orthonormal basis of $\mathbb{R}^2$ defining the usual orientation. Let $\dfrac{\partial}{\partial n}$ and $\dfrac{\partial}{\partial s}$ represent differentiation directions n and s, respectively, at a point z_0 in Ω. Then the Cauchy-Riemann equations become

$$\frac{\partial u}{\partial s} = \frac{\partial v}{\partial n}, \qquad \frac{\partial u}{\partial n} = -\frac{\partial v}{\partial s}.$$

4. Let $f, g \in \mathscr{H}(\Omega)$, $g(z) \neq 0$ for every $z \in \Omega$. Show that $f/g \in \mathscr{H}(\Omega)$.

5. Show that if $f = u + iv$ is holomorphic in Ω, then for $p \geq 2$ one has (Δ is the Laplace operator):

(i) $\Delta u = \Delta v = \Delta f = 0$

(ii) $\Delta |f(z)|^p = p^2 |f(z)|^{p-1} |f'(z)|^2$

(iii) $\Delta |u(z)|^p = p(p-1)|u(z)|^{p-2}|f'(z)|^2$.

(In proving (ii) and (iii) you should be careful about the points where $f(z) = 0$ or $u(z) = 0$. Formulas (ii) and (iii) hold for every p real if $f(z) \neq 0$ or $u(z) \neq 0$.)

(iv) $\dfrac{\partial}{\partial z}|f(z)| = \dfrac{1}{2}|f(z)|\dfrac{f'(z)}{f(z)}$ (whenever $f(z) \neq 0$)

(v) $\dfrac{\partial}{\partial z}u(z) = \dfrac{1}{2}f'(z)$ and $\dfrac{\partial v}{\partial z}(z) = \dfrac{1}{2i}f'(z)$

(vi) $\Delta e^{p|f(z)|} = p^2 e^{p|f(z)|}\left(p + \dfrac{1}{|f(z)|}\right)|f'(z)|^2 \quad (p \in \mathbb{C},\ f(z) \neq 0)$

(vii) $\Delta \log(1 + |f(z)|^2) = \dfrac{4|f'(z)|^2}{1 + |f(z)|^2}$.

6. Consider $z \mapsto z^2$ as a transformation of $\mathbb{C}$ into $\mathbb{C}$:
   (i) What is the image of an angle $\theta_0 < \arg z < \theta_1, \theta_1 - \theta_0 < \pi$?
   (ii) What is the image of a family of parallel lines not passing through the origin?

7. Consider the transformation $z \mapsto e^z$:
   (i) What is the image of a family of lines parallel to one of the coordinate axes? (The answer depends on the axis.)
   (ii) What is the image of a line not parallel to any of the coordinate axes?

8. Consider the function $f: z \mapsto \cos z$.
   (i) What is the image of the segment $y = y_0, 0 < x < \pi$, in the three cases $y_0 > 0$, $y_0 = 0$, $y_0 < 0$?
   (ii) Same question for the straight line $x = x_0$ in the two cases $x_0 \neq k\frac{\pi}{2}$ $(k \in \mathbb{Z})$ and $x_0 = k\frac{\pi}{2}$.
   (iii) Let D be the strip $0 < x < \pi$. Show f takes every complex value when z is in $\bar{D}$, and it is injective in D.

9. Show that the map $f(z) = \frac{1}{2}\left(z + \frac{1}{z}\right)$ transforms every circle $\partial B(0, r)$, $r > 0$, into an ellipse. Find the image of the segments $z = te^{i\alpha}$, $0 < t < 1$ ($\alpha \in [0, 2\pi[$ fixed).

10. Let $f_r(z)$ be defined for $z \in B(0, r)$ $(r > 0)$ by the formula
$$f_r(z) = \int_{|\zeta|=r} \exp\left(\zeta + \frac{1}{\zeta}\right)\frac{d\zeta}{\zeta - z}.$$
   (i) Show that f_r is C^1, by taking derivatives as a function of a parameter, and it satisfies $\frac{\partial f_r}{\partial \bar{z}} = 0$ in $B(0, r)$.
   (ii) Show that $f_R | B(0, r) = f_r$ when $R > r$, using Theorem 1.11.4.
   (iii) Conclude there is an entire function f_∞ such that $f_\infty | B(0, r) = f_r$ for every $r > 0$.

11. Prove the inequality in the statement of Lemma 1.11.5. (Hint: Use that $H + \frac{1}{H} = 2g$ and $|H(z_0)| \geq 1$.)

12. Show that the function h from Lemma 1.11.5 does not take any of the values
$$\pm\cosh^{-1}(n + 1) + 2m\pi i \qquad (n \in \mathbb{N}, m \in \mathbb{Z}).$$
Draw a picture of these points in the plane to convince yourself that any disk of radius ≥ 4 contains one of these omitted values (actually 3.22 works).

Notes to Chapter 1

The reader will find [Gre] and [Go] as very reasonable introductions to algebraic topology, which is the subject of most of this chapter. The material on differential calculus can be found in [Ca] and [Sp]. We also recommend [Bur] and [Fo] for further insights into the relation between topology and complex analysis.

CHAPTER 2

Analytic Properties of Holomorphic Functions

§1. Integral Representation Formulas

At the end of Chapter 1 we introduced the holomorphic functions, that is, those functions $f \in C^1(\Omega)$ that satisfy the Cauchy-Riemann differential equation $\dfrac{\partial f}{\partial \bar{z}} = 0$ throughout an open set $\Omega \subseteq \mathbb{C}$. As an immediate consequence of the topological tools developed in that chapter we found that the holomorphic functions enjoyed the following remarkable property (Cauchy's theorem 1.11.4).

2.1.1. Proposition. *If α is a loop homotopic to a point in an open set Ω and $f \in \mathscr{H}(\Omega)$, then*

$$\int_\alpha f(z)\,dz = 0.$$

In order to study further the holomorphic functions we need to pursue their analytic properties, that is, those properties that depend on Stokes' formula, integral representations, power series, etc. The first step in that direction is the following integral representation formula, valid for arbitrary C^1 functions.

2.1.2. Proposition (Pompeiu's Formula). *Let Ω be an open set in $\mathbb{C}$, D an open relatively compact subset of Ω with piecewise regular boundary of class C^1. Let f be a C^1 function in Ω. Then, for every $\zeta \in D$ we have*:

$$f(\zeta) = \frac{1}{2\pi i}\int_{\partial D}\frac{f(z)}{z-\zeta}\,dz + \frac{1}{2\pi i}\int_D \frac{\partial f}{\partial \bar{z}}(z)\frac{1}{z-\zeta}\,dz \wedge d\bar{z}.$$

PROOF. Let $D_\varepsilon := \{z \in D : |z - \zeta| > \varepsilon\}$ $(0 < \varepsilon < d(\zeta, D^c))$. We apply Stokes' formula to the 1-form $\frac{f(z)}{z-\zeta}dz$ and the open set D_ε, whose boundary is piecewise regular and it is the 1-cycle $\partial D - \partial B(\zeta, \varepsilon)$ of Ω. Parameterizing $\partial B(\varepsilon, \zeta)$ by $\theta \in [0, 2\pi] \mapsto \zeta + \varepsilon e^{i\theta}$, we obtain:

$$-\int_{D_\varepsilon} \frac{\partial f}{\partial \bar{z}}(z)\frac{dz \wedge d\bar{z}}{z-\zeta} = \int_{\partial D} \frac{f(z)}{z-\zeta}dz - i\int_0^{2\pi} f(\zeta + \varepsilon e^{i\theta})\, d\theta,$$

since $d\left(\frac{f(z)}{z-\zeta}dz\right) = -\frac{\partial f}{\partial \bar{z}}(z)\frac{dz \wedge d\bar{z}}{z-\zeta}$. Let $\varepsilon \to 0$ to obtain the result. □

As a consequence of the last proposition we obtain *Cauchy's integral representation* for holomorphic functions.

2.1.3. Corollary. *Under the same hypotheses as in Proposition* 2.1.2, *we have*

$$f(\zeta) = \frac{1}{2\pi i}\int_{\partial D} \frac{f(z)}{z-\zeta}dz \qquad (\zeta \in D)$$

for any $f \in \mathscr{H}(\Omega)$.

2.1.4. Corollary. *If $u \in \mathscr{D}(\mathbb{C})$ we have*

$$u(\zeta) = \frac{1}{2\pi i}\int_{\mathbb{C}} \frac{\partial u}{\partial \bar{z}}(z)\frac{1}{z-\zeta}dz \wedge d\bar{z} \qquad (\zeta \in \mathbb{C}).$$

PROOF OF THE COROLLARIES. The first one is immediate from §2.1.2. For the proof of the second, let $D = B(0, R)$ in §2.1.2 with R so large that $\operatorname{supp} u \subset B(0, R)$. Then let $R \to +\infty$. □

We are now going to consider a kind of converse to these corollaries.

2.1.5. Proposition. *Let μ be a complex measure of compact support in $\mathbb{C}$. The integral*

$$\hat{\mu}(\zeta) := -\frac{1}{\pi}\int \frac{d\mu(z)}{z-\zeta}$$

defines a function $\hat{\mu}$ holomorphic outside $\operatorname{supp}\mu$. *If in an open set ω, the measure μ has the form $\varphi\, dx\, dy$, $\varphi \in \mathscr{E}_k(\omega)$, $k \geq 0$, we also have $\hat{\mu} \in \mathscr{E}_k(\omega)$, moreover, $\frac{\partial \hat{\mu}}{\partial \bar{z}} = \varphi$ if $k \geq 1$. The function $\hat{\mu}$ is called the* ***Cauchy transform*** *of μ.*

PROOF. The function $\hat{\mu}$ is of class C^∞ in $\mathbb{C}\setminus\operatorname{supp}\mu$ as one sees by differentiation under the integral sign. We find thus in that open set

$$\frac{\partial \hat{\mu}}{\partial \bar{\zeta}}(\zeta) = -\frac{1}{\pi}\int \frac{\partial}{\partial \bar{\zeta}}\left(\frac{1}{z-\zeta}\right) d\mu(z) = 0,$$

since $\dfrac{1}{z-\zeta}$ is a holomorphic function for $\zeta \in \mathbb{C}\setminus\operatorname{supp}\mu$ if $z \in \operatorname{supp}\mu$. Therefore $\hat{\mu}$ is holomorphic on $\mathbb{C}\setminus\operatorname{supp}\mu$.

To prove the second part, let us first assume $\omega = \mathbb{C}$ and $\mu = \varphi\, dx\, dy$, $\varphi \in \mathscr{D}_k(\mathbb{C})$. We have then:

$$\hat{\mu}(\zeta) = \frac{1}{2\pi i}\int_{\mathbb{C}} \frac{\varphi(z)}{z-\zeta} dz \wedge d\bar{z}.$$

Let us introduce the change of variables $z = \zeta - u = f(u)$, then

$$f^*\left(\frac{\varphi(z)}{z-\zeta} dz \wedge d\bar{z}\right) = -\frac{\varphi(\zeta-u)}{u} du \wedge d\bar{u}.$$

The function $1/u \in L^1_{\mathrm{loc}}(\mathbb{C})$ and $\operatorname{supp}\varphi$ is compact, it follows that $\hat{\mu}$ is at least a continuous function of ζ. Furthermore, if φ is of class C^k $(k \geq 1)$ then one can take derivatives under the integral sign and $\hat{\mu} \in C^k$ also. Hence

$$\begin{aligned}\frac{\partial \hat{\mu}}{\partial \bar{\zeta}}(\zeta) &= -\frac{1}{2\pi i}\int_{\mathbb{C}} \frac{\partial \varphi}{\partial \bar{\zeta}}(\zeta-u)\frac{du \wedge d\bar{u}}{u} \\ &= \frac{1}{2\pi i}\int_{\mathbb{C}} \frac{1}{z-\zeta}\frac{\partial \varphi}{\partial \bar{z}}(z)\, dz \wedge d\bar{z} \\ &= \varphi(z) \qquad \text{(by §2.1.4).}\end{aligned}$$

In the general case, when $\omega \neq \mathbb{C}$, let $z_0 \in \omega$ and take $\psi \in \mathscr{D}(\omega)$, $\psi \equiv 1$ on a neighborhood V of z_0. Let $\mu_1 = \psi\mu$, $\mu_2 = (1-\psi)\mu$. Since $\mu_1 = \varphi\psi\, dm$, $\varphi\psi \in \mathscr{D}_k(\mathbb{C})$ (after it is defined by zero outside ω), we have $\hat{\mu}_1 \in C^k$, and if $k \geq 1$, $\dfrac{\partial \hat{\mu}_1}{\partial \bar{\zeta}} = \varphi\psi$. Since $V \subseteq \mathbb{C}\setminus\operatorname{supp}\mu_2$, $\hat{\mu}_2 \in C^\infty$ and $\dfrac{\partial \hat{\mu}_2}{\partial \bar{\zeta}} = 0$ in V. The proposition follows. □

2.1.6. Corollary. *Every $h \in \mathscr{H}(\Omega)$ is of class C^∞ in Ω. Furthermore, the derivative h' is also a holomorphic function in Ω.*

Proof. For $z_0 \in \Omega$, let $\omega = B(z_0, R) \subset\subset \Omega$. Cauchy's integral representation formula is then

$$h(\zeta) = \frac{1}{2\pi i}\int_{\partial B(z_0,R)} \frac{h(z)}{z-\zeta} dz, \qquad \zeta \in B(z_0, R).$$

In $B(z_0, R)$, h appears then as the Cauchy transform of a measure of support contained in the compact $\partial B(z_0, R)$. By §2.1.5, h is of class C^∞ in $B(z_0, R)$. Therefore $h \in \mathscr{E}(\Omega)$.

We have $h' = \dfrac{\partial h}{\partial z}$ and, since h is of class C^∞, we have

$$\frac{\partial h'}{\partial \bar{z}} = \frac{\partial}{\partial \bar{z}}\left(\frac{\partial h}{\partial z}\right) = \frac{\partial}{\partial z}\left(\frac{\partial h}{\partial \bar{z}}\right) = 0.$$

Therefore h' is also holomorphic in Ω. □

2.1.7. Remark. Let Ω be a nonempty open set in $\mathbb{C}$. The operator $\dfrac{d}{d\bar{z}}: \mathscr{D}(\Omega) \to \mathscr{D}(\Omega)$ is not surjective. In fact, if $f = \dfrac{\partial u}{\partial \bar{z}}$, $u \in \mathscr{D}(\Omega)$, then $f\,dz \wedge d\bar{z} = -d(u\,dz)$. It follows that $\displaystyle\int_\Omega f\,dz \wedge d\bar{z} = -\int_\Omega d(u\,dz) = 0$. On the other hand, there are $f \in \mathscr{D}(\Omega)$ such that $\displaystyle\int_\Omega f\,dz \wedge d\bar{z} \neq 0$.

The following is an important result; in fact, it is the converse to Cauchy's theorem 2.1.1.

2.1.8. Proposition (Morera's Theorem). *Let f be a continuous function in an open connected subset Ω of $\mathbb{C}$. Assume that for every piecewise-C^1 loop α which is free-homotopic to a point in Ω we have $\displaystyle\int_\alpha f\,dz = 0$. Then f is holomorphic in Ω. Furthermore, it is enough to consider only rectangles with sides parallel to the axes and homotopic to a point in Ω.*

PROOF. By §1.9.16, item (2), the differential form $f\,dz$ admits a primitive F of class C^1. Since F is a primitive, $\dfrac{\partial F}{\partial x} = f$, $\dfrac{\partial F}{\partial y} = if$, hence $\dfrac{\partial F}{\partial \bar{z}} = 0$. F is then holomorphic, and so is its derivative F'. But $F' = \dfrac{\partial F}{\partial z} = f$, which shows that locally f is C^∞ and satisfies the Cauchy-Riemann equation. Since f is defined in the whole set Ω, it follows that $f \in \mathscr{H}(\Omega)$. □

2.1.9. Remarks. (1) A first application of Morera's theorem is the following: Let (X, μ) be a σ-finite measure space and $\mu \geq 0$ (or even just a complex-valued measure on X). Let f be a function $\Omega \times X \to \mathbb{C}$ such that

(i) For every $z \in \Omega$, $t \mapsto f(z,t)$ is measurable and defined a.e. in X.
(ii) For every $z \in \Omega$ there is a closed disk $\bar{B}(z,r) \subseteq \Omega$ and a function g on X integrable with respect to μ (respectively $|\mu|$, if μ is a complex measure) such that $|f(\zeta,t)| \leq g(t)$ a.e. for every $\zeta \in \bar{B}(z,r)$.
(iii) For a.e. t, $z \mapsto f(z,t)$ is holomorphic in Ω.

Then, the function

$$F(z) := \int_X f(z,t)\,d\mu(t)$$

is holomorphic in Ω.

The hypotheses (i) and (ii) show that F is continuous. Moreover, the hypothesis (ii) implies that f is integrable on $\partial R \times X$ for every closed rectangle contained in $B(z, r) \subset\subset \Omega$. Thus we can apply Fubini's theorem and obtain

$$\int_{\partial R} F(z)\,dz = \int_{\partial R}\int_X f(z,t)\,d\mu(t)\,dz = \int_X\int_{\partial R} f(z,t)\,dz\,d\mu(t) = 0,$$

since by (iii), for almost every t we have $\displaystyle\int_{\partial R} f(z,t)\,dz = 0$.

(2) As other immediate corollaries of Morera's theorem, let us mention here:

- If f is continuous in Ω and holomorphic in $\Omega \backslash L$, where L is an affine real line then f is holomorphic in Ω.
- If f is continuous in Ω, A a discrete subset of Ω, and f holomorphic in $\Omega \backslash A$, then f is holomorphic in Ω.

2.1.10. Proposition (The Schwarz Reflection Principle). *Let f be a function holomorphic in the half-disk $D = \{z \in \mathbb{C} : |z| < r, \operatorname{Im} z > 0\}$, such that f is continuous on $\bar{D}$ and real valued on the real axis. Define a function F in* $B(0, r)$ by

$$F(z) := \begin{cases} f(z) & \text{if } \operatorname{Im} z \geq 0 \\ \overline{f(\bar{z})} & \text{if } \operatorname{Im} z < 0. \end{cases}$$

Then F is holomorphic in $B(0, r)$.

Proof. One verifies easily that F is holomorphic outside the real axis and F is continuous in $B(0, r)$. Apply then the previous remark. □

We will find stronger versions of this principle in other parts of this text, e.g., the Edge-of-the-Wedge theorem in §3.6.

We are now ready to prove that a function having a complex derivative at every point is holomorphic, thus answering a question raised at the end of Chapter 1.

2.1.11. Proposition (Cauchy-Goursat's Theorem). *Let f be a function whose complex derivative exists at every point of an open set Ω. Then f is holomorphic in Ω.*

Proof. It is clear that f is continuous in Ω. By Morera's theorem it is enough to show that $\displaystyle\int_{\partial R} f\,dz = 0$ for every closed rectangle R in Ω. For that purpose, denote

$$\alpha(R) := \int_{\partial R} f(z)\,dz.$$

Let us now subdivide the rectangle in four parts by dividing each side in two equal parts. Let γ_i be the oriented boundaries of the four rectangles R_i thus obtained. We have

$$\int_{\partial R} f(z)\,dz = \sum_{1\leq i\leq 4} \int_{\gamma_i} f(z)\,dz = \sum_{1\leq i\leq 4} \alpha(R_i).$$

Among these four rectangles there must be one such that $|\alpha(R_i)| \geq |\alpha(R)|/4$. Let us call R_1 this rectangle. Divide R_1 into four rectangles by the same procedure. At least one of them, denoted R_2, verifies

$$|\alpha(R_2)| \geq |\alpha(R)|/4^2.$$

Iterating this procedure we get a sequence of closed rectangles $\{R_k\}$, $R_{k+1} \subseteq R_k$,

$$|\alpha(R_k)| \geq |\alpha(R)|/4^k,$$

$$d_k := \text{diameter}(R_k) = \frac{1}{2^k}\text{diameter}(R) = d/2^k;$$

$$L_k := \text{length}(\partial R_k) = \frac{1}{2^k}\text{length}(\partial R) = L/2^k.$$

Let z_0 be the unique common point to all these rectangles. Since $z_0 \in R \subseteq \Omega$ and f has a complex derivative at z_0, we have

$$f(z) = f(z_0) + f'(z_0)(z - z_0) + |z - z_0|\varepsilon(z - z_0),$$

with $\varepsilon(t) \to 0$ as $t \to 0$. We have therefore

$$\int_{\partial R_k} f(z)\,dz = f(z_0)\int_{\partial R_k} dz + f'(z_0)\int_{\partial R_k} (z - z_0)\,dz$$
$$+ \int_{\partial R_k} |z - z_0|\varepsilon(z - z_0)\,dz.$$

Since dz and $(z - z_0)\,dz$ are closed differential forms, the first two terms on the right-hand side vanish, leading to

$$|\alpha(R_k)| = \left|\int_{\partial R_k} f(z)\,dz\right| \leq d_k L_k \sup\{|\varepsilon(z - z_0)| : z \in \partial R_k\}.$$

From here it follows that if $\delta(k)$ indicates the supremum in the last formula, we have

$$|\alpha(R)| \leq dL\,\delta(k).$$

Since $\delta(k) \to 0$ as $k \to \infty$ we obtain $\alpha(R) = 0$. □

2.1.12. Proposition (Cauchy's Integral Formulas). *Let f be a holomorphic function in the open set $\Omega \subseteq \mathbb{C}$ and let $a \in \Omega$. The following two formulas hold.*

(1) *Let γ be a loop in $\Omega\backslash\{a\}$ homotopic to a point in Ω. Then:*

$$\frac{1}{2\pi i}\int_\gamma \frac{f(z)}{z - a}\,dz = \text{Ind}_\gamma(a)\cdot f(a).$$

(2) *Let* $\delta = \sum_j v_j\gamma_j$ *be a* 1*-boundary in* Ω *such that* $\operatorname{supp}\delta \subseteq \Omega\backslash\{a\}$. *Then*:

$$\frac{1}{2\pi i}\int_\delta \frac{f(z)}{z-a}\,dz = \operatorname{Ind}_\delta(a)\cdot f(a).$$

In particular, if $\delta = \sum_{1\le j\le n} \gamma_j$, *where the* γ_j *are loops in* $\Omega\backslash\{a\}$ *such that for every* $z \in \mathbb{C}\backslash\Omega$ *one has* $\sum_j \operatorname{Ind}_{\gamma_j}(z) = 0$, *then*

$$\frac{1}{2\pi i}\int_\delta \frac{f(z)}{z-a}\,dz = \left(\sum_{1\le j\le n} \operatorname{Ind}_{\gamma_j}(a)\right) f(a).$$

PROOF. Let us note first that we are dealing with integrals of the closed form $(f(z)/(z-a))\,dz$ in $\Omega/\{a\}$ along continuous paths.

The function $g(z) := (f(z) - f(a))/(z-a)$ if $z \neq a$, $g(a) := f'(a)$ is holomorphic in $\Omega\backslash\{a\}$ and continuous in Ω. It follows from §2.1.9, (2) that g is holomorphic in Ω. Therefore $\int_\gamma g(z)\,dz = 0$ if γ is a loop as in (1). One obtains the first formula recalling that $\operatorname{Ind}_\gamma(a) = \frac{1}{2\pi i}\int_\gamma \frac{dz}{z-a}$.

The other relations are obtained in the same way. □

2.1.13. Definition. The holomorphic function $h^{(n)}$ is the **nth derivative of the holomorphic function** h and is defined by recurrence as follows:

$$h^{(n)} := (h^{(n-1)})', \quad h^{(1)} = h', \quad h^{(0)} = h.$$

2.1.14. Proposition (Cauchy's Formula for the Derivatives). *Let h be a holomorphic function in an open set* $\Omega \subseteq \mathbb{C}$, $\omega \subset\subset \Omega$ *an open subset with piecewise regular boundary. The following relation holds*:

$$h^{(j)}(\zeta) = \frac{j!}{2\pi i}\int_{\partial\omega} \frac{h(z)}{(z-\zeta)^{j+1}}\,dz \qquad (\zeta \in \omega).$$

PROOF. Let us show that

$$h'(\zeta) = \frac{1}{2\pi i}\int_{\partial\omega} \frac{h(z)}{(z-\zeta)^2}\,dz \qquad (\zeta \in \omega).$$

Consider the differential of the function $F(z) = \dfrac{h(z)}{z-\zeta}$, holomorphic in $\Omega\backslash\{\zeta\}$. We have $dF = \left(\dfrac{h'(z)}{z-\zeta} - \dfrac{h(z)}{(z-\zeta)^2}\right)dz$. We know that $I_{dF}(\gamma) = 0$ for every loop γ in $\Omega\backslash\{\zeta\}$, hence the integral of dF over $\partial\omega$ vanishes. It follows that

$$h'(\zeta) = \frac{1}{2\pi i}\int_{\partial\omega} \frac{h'(z)}{(z-\zeta)}\,dz = \frac{1}{2\pi i}\int_{\partial\omega} \frac{h(z)}{(z-\zeta)^2}\,dz.$$

The proof continues by recurrence. The relation

$$h^{(n+1)}(\zeta) = \frac{n!}{2\pi i} \int_{\partial\omega} \frac{h'(z)}{(z-\zeta)^{n+1}}\,dz$$

holds by induction applied to the nth derivative of the holomorphic function h'. Using now the differential of the auxiliary function $F(z) = \dfrac{h(z)}{(z-\zeta)^{n+1}}$, as done earlier, leads to the Cauchy formulas. □

A slightly more general version of the Cauchy formulas for the derivatives is the following.

2.1.15. Proposition. *Let f be a holomorphic function in the open set Ω, $a \in \Omega$. Then:*

(1) *Let γ be a loop in $\Omega\setminus\{a\}$, homotopic to a point in Ω. We have*

$$\operatorname{Ind}_\gamma(a)\cdot f^{(j)}(a) = \frac{j!}{2\pi i}\int_\gamma \frac{f(z)}{(z-a)^{j+1}}\,dz \qquad (j \geq 0).$$

(2) *Let δ be a 1-boundary in Ω such that* $\operatorname{supp}\delta \subseteq \Omega\setminus\{a\}$, *then*

$$\operatorname{Ind}_\delta(a)\cdot f^{(j)}(a) = \frac{j!}{2\pi i}\int_\delta \frac{f(z)}{(z-a)^{j+1}}\,dz \qquad (j \geq 0).$$

In particular, if $\delta = \sum_{1\leq i\leq n} \gamma_i$, where the γ_i are loops in $\Omega\setminus\{a\}$ such that $\sum_i \operatorname{Ind}_{\gamma_i}(z) = 0$ for every $z \in \mathbb{C}\setminus\Omega$, we have

$$\frac{1}{2\pi i}\int_\delta \frac{f(z)}{(z-a)^{j+1}}\,dz = \left(\sum_i \operatorname{Ind}_{\gamma_i}(a)\right)\frac{f^{(j)}(a)}{j!}.$$

Proof. It is analogous to the proof of the preceding proposition taking into account that integration over a 1-cycle annihilates the exact differential forms of class C^1 and degree 1. We also use the fact that a 1-boundary of Ω with support in $\Omega\setminus\{a\}$ is a 1-cycle in $\Omega\setminus\{a\}$. □

We are going to show now that the holomorphic functions in an open set Ω can be characterized as those functions that are locally representable by convergent power series.

2.1.16. Definition. Given a sequence $(a_n)_{n\geq 0}$ of complex numbers, the ***power series*** about z_0 ($z_0 \in \mathbb{C}$) with coefficients $(a_n)_{n\in\mathbb{N}}$ is the formal series whose general term is $a_n(z-z_0)^n$, that is, $\sum_{n\geq 0} a_n(z-z_0)^n$.

We recall here that if a power series $\sum_{n\geq 0} a_n z^n$ converges at a point $z_1 \neq 0$, then it converges for every z in the disk $B(0, |z_1|)$. It follows that if a power series converges at some point $z \neq 0$, there is a largest R (possibly $R = \infty$) such that the power series converges in $B(0, R)$. This value R is called the ***radius of convergence*** of $\sum_{n\geq 0} a_n z^n$. We declare it to be $R = 0$ if the series converges only at $z = 0$, and $R = \infty$ if it converges everywhere. There is a very simple way of obtaining R, namely,

$$R = \sup\{r \geq 0 : \exists M(r) \text{ such that } |a_n| r^n \leq M(r) < \infty \ (n \geq 0)\}.$$

2.1.17. Proposition. *Given a power series* $\sum_{n\geq 0} a_n(z - z_0)^n$, *its radius of convergence is given by the formula due to Hadamard*:

$$\frac{1}{R} = \limsup_{n\to\infty} \sqrt[n]{|a_n|}.$$

The series converges absolutely in $B(z_0, R)$, *and absolutely and uniformly in every compact subset of* $B(z_0, R)$. *Moreover, it does not converge anywhere in the exterior of* $\bar{B}(z_0, R)$.

PROOF. It is left as an exercise to the reader (cf. [Mar], [Ah1]). □

2.1.18. Proposition. *Let* Ω *be an open subset of* $\mathbb{C}$. *For every* $f \in \mathscr{H}(\Omega)$ *and* $z_0 \in \Omega$, *the power series*

$$\sum_{n\geq 0} \frac{1}{n!} f^{(n)}(z_0)(z - z_0)^n$$

converges uniformly and absolutely over any compact subset of $B(z_0, d(z_0, \Omega^c))$. *Its sum equals* $f(z)$ *at every point of that disk.*

Furthermore, this series (called the ***Taylor series*** *of* f *at* z_0) *is the unique power series* $\sum_{n\geq 0} a_n(z - z_0)^n$ *whose sum equals* f *in* $B(a_0, d(z_0, \Omega^c))$. *(Here we have set* $d(z_0, \Omega^c) = +\infty$ *if* $\Omega = \mathbb{C}$ *and* $d(z_0, \Omega^c) := \inf\{|\zeta - z_0| : \zeta \in \Omega^c\}$, if not.)

PROOF. Let $0 < r < d(z_0, \Omega^c)$. For $z \in \partial B(z_0, r)$ and $\zeta \in B(z_0, r)$ we have

$$\frac{1}{z - \zeta} = \frac{1}{(z - z_0) - (\zeta - z_0)} = \frac{1}{(z - z_0)} \frac{1}{\left(1 - \dfrac{\zeta - z_0}{z - z_0}\right)} = \sum_{n\geq 0} \frac{(\zeta - z_0)^n}{(z - z_0)^{n+1}}.$$

This series is uniformly and absolutely convergent over any set of the form $K \times \partial B(z_0, r)$, where $K \subset\subset B(z_0, r)$.

We can therefore apply Cauchy's formula with $\omega = B(z_0, r)$ and interchange the order of integration and summation. For fixed $\zeta \in B(z_0, r)$ this gives:

$$f(\zeta) = \frac{1}{2\pi i} \int_{\partial B(z_0, r)} \frac{f(z)}{z - \zeta} dz = \frac{1}{2\pi i} \int_{\partial B(z_0, r)} f(z) \sum_{n\geq 0} \frac{(\zeta - z_0)^n}{(z - z_0)^{n+1}} dz$$

$$= \sum_{n\geq 0}\left(\frac{1}{2\pi i}\int_{\partial B(z_0,r)}\frac{f(z)}{(z-z_0)^{n+1}}\,dz\right)(\zeta - z_0)^n$$

$$= \sum_{n\geq 0}\frac{f^{(n)}(z_0)}{n!}(\zeta - z_0)^n.$$

The last step is a consequence of the Cauchy formula for the derivatives. By choosing $r \in \,]|\zeta - z_0|, d(z_0, \Omega^c)[$ we can prove the same representation holds for any $\zeta \in B(z_0, d(z_0, \Omega^c))$.

This reasoning yields the uniform convergence of the series over any compact subset of $B(z_0, d(z_0, \Omega^c))$. This can also be obtained from the general theory of power series since the argument shows that the radius of convergence is at least $d(z_0, \Omega^c)$. (One can also estimate the coefficients using their integral representation and Hadamard's formula.) Finally, the uniqueness of the series follows from the Cauchy formula for the derivatives. □

We will show later on that, conversely, a function representable locally by convergent power series is holomorphic. For the time being, let us recall some familiar examples of series expansions.

2.1.19. Examples

$$e^z = \sum_{n\geq 0}\frac{z^n}{n!} \qquad (z \in \mathbb{C}),$$

$$\sin z = \sum_{n\geq 0}(-1)^n\frac{z^{2n+1}}{(2n+1)!} \qquad (z \in \mathbb{C}),$$

$$\cos z = \sum_{n\geq 0}(-1)^n\frac{z^{2n}}{(2n)!} \qquad (z \in \mathbb{C}),$$

$$\cosh z = \frac{e^z + e^{-z}}{2} = \sum_{n\geq 0}\frac{z^{2n}}{(2n)!} \qquad (z \in \mathbb{C}),$$

$$\sinh z = \frac{e^z - e^{-z}}{2} = \sum_{n\geq 0}\frac{z^{2n+1}}{(2n+1)!} \qquad (z \in \mathbb{C}),$$

$$\frac{1}{(1-z)^p} = \sum_{k\geq 0}\binom{k+p-1}{p-1}z^k \qquad (|z| < 1), \qquad p \in \mathbb{N}^*.$$

2.1.20. Proposition (Cauchy's Inequalities). *Let Ω be an open set in $\mathbb{C}$, $f \in \mathscr{H}(\Omega)$:*

(1) *For $z_0 \in \Omega$, $0 < r < d(z_0, \Omega^c)$ we have*

$$|f^{(n)}(z_0)| \leq n!\frac{M(|f|, r)}{r^n},$$

with $M(|f|, r) = \max\limits_{|z-z_0|=r}|f(z)|$. In particular, if f is bounded in Ω we have

$$|f^{(n)}(z_0)| \le n!\frac{\|f\|_\infty}{(d(z_0,\Omega^c))^n},$$

with $\|f\|_\infty = \sup_{z\in\Omega}|f(z)|$.

As a corollary, if f is a bounded entire function, then f is a constant (Liouville's theorem).

(2) *If ω is a relatively compact, open subset of Ω, with piecewise regular boundary of class C^1, and if $z_0 \in \omega$, then:*

$$|f^{(n)}(z_0)| \le \frac{n!\int_{\partial\omega}|f(z)||dz|}{2\pi(d(z_0,\partial\omega))^{n+1}},$$

which leads to

$$|f^{(n)}(z_0)| \le \frac{n!\ell(\partial\omega)\max\{|f(z)| : z\in\partial\omega\}}{2\pi(d(z_0,\partial\omega))^{n+1}},$$

where, we recall, $\ell(\partial\omega)$ denotes the length of the boundary of ω.

(3) *If K is a compact subset of Ω and U a neighborhood of K which is relatively compact in Ω, there are positive constants $C_n (n \in \mathbb{N})$ such that for every $f \in \mathscr{H}(\Omega)$*

$$\sup_{z\in K}|f^{(n)}(z)| \le C_n\|f\|_{L^1(U)}.$$

PROOF. The statements (1) and (2) are immediate consequences of the Cauchy formulas for the derivatives.

Let us prove (3). Let $\psi \in \mathscr{D}(U)$ be identically equal to 1 on a neighborhood of K. We have

$$\frac{\partial}{\partial\bar{z}}(\psi f) = f\frac{\partial\psi}{\partial\bar{z}},$$

and that, if $K_1 := \operatorname{supp}\frac{\partial\psi}{\partial\bar{z}}$, K_1 is a compact subset of U and $d(K,K_1) > 0$. Therefore, applying Pompeiu's formula (cf. §2.1.4) to ψf and differentiating under the integral sign gives, for $\zeta \in K$,

$$f^{(n)}(\zeta) = \frac{n!}{2\pi i}\int_{K_1} f(z)\frac{\partial\psi}{\partial\bar{z}}(z)\frac{dz\wedge d\bar{z}}{(z-\zeta)^{n+1}}.$$

The constants C_n are obtained by estimating directly this integral. □

2.1.21. Proposition (Principle of Analytic Continuation). *Let Ω be a connected open subset of $\mathbb{C}$ and $h \in \mathscr{H}(\Omega)$ be a function such that for some $z_0 \in \Omega$ verifies $h^{(n)}(z_0) = 0$ for every $n \ge 0$. Then $h \equiv 0$.*

PROOF. The set $E := \{z \in \Omega : h^{(n)}(z) = 0 \text{ for every } n \ge 0\}$ is closed in Ω since all the functions $h^{(n)}$ are continuous. It is also open by the convergence of the

Taylor expansion

$$h(\zeta) = \sum_{n\geq 0} \frac{h^{(n)}(z)}{n!}(\zeta - z)^n$$

in $B(z, d(z, \Omega^c))$ for every $z \in \Omega$. This expansion implies that if $z \in E$, then $h \equiv 0$ in that disk and hence, $E \supseteq B(z, d(z, \Omega^c))$. The hypothesis ensures that E is not empty, hence $E = \Omega$ and the proposition follows. □

2.1.22. Corollary. *If h is holomorphic in a connected open set* Ω *and* $h \not\equiv 0$, *then for every* $z_0 \in \Omega$ *there is a unique integer* $k \geq 0$ *such that*

$$h(z) = (z - z_0)^k g(z) \qquad (z \in \Omega),$$

with $g \in \mathscr{H}(\Omega)$ *and* $g(z_0) \neq 0$. *In particular, the zeros of h are isolated.*

PROOF. We know that h is represented in $B(z_0, d(z_0, \Omega^c))$ by the Taylor series

$$h(z) = \sum_{n\geq 0} \frac{1}{n!} h^{(n)}(z_0)(z - z_0)^n$$

and that, by §2.1.21, the set $\{n \in \mathbb{N} : h^{(n)}(z_0) \neq 0\}$ is not empty, hence it has a smallest element k. Therefore, we can define g by $g(z) = h(z)/(z - z_0)^k$ in $\Omega \setminus \{z_0\}$ and

$$g(z) = \sum_{n\geq k} \frac{1}{n!} h^{(n)}(z_0)(z - z_0)^{n-k}$$

in $B(z_0, d(z_0, \Omega^c))$. This shows that g is holomorphic in $\Omega \setminus \{z_0\}$ and at least continuous at z_0, hence $g \in \mathscr{H}(\Omega)$. Furthermore, $g(z_0) = \dfrac{h^{(k)}(z_0)}{k!} \neq 0$. □

2.1.23. Corollary. *If* h_1, $h_2 \in \mathscr{H}(\Omega)$, Ω *open connected and* $h_1|E = h_2|E$ *for some nonempty subset E of* Ω *which has an accumulation point in* Ω, *then* $h_1 \equiv h_2$.

PROOF. $h_1 - h_2$ would have some nonisolated zeros if it were not identically zero. □

2.1.24. Example. We have the relation

$$\sin^2 z + \cos^2 z = 1$$

in $\mathbb{C}$. (Since we already know it to be true in $\mathbb{R}$!)

2.1.25. Proposition (Open Mapping Property). *Let* Ω *be a nonempty open subset of* $\mathbb{C}$ *and* $h \in \mathscr{H}(\Omega)$ *a holomorphic function which is not constant in any connected component of* Ω. *Then* $h(\Omega)$ *is an open subset of* $\mathbb{C}$.

PROOF. We can assume that $0 \in \Omega$ and $h(0) = 0$. It is enough then to prove that $h(\Omega)$ is a neighborhood of the origin. The zeros of h are isolated, hence there is $r > 0$ such that $\bar{B}(0, r) \subseteq \Omega$ and $h(z) \neq 0$ for $0 < |z| \leq r$. Let

$$\delta := \inf_{|z|=r} |h(z)| > 0.$$

We want to show that $B(0, \delta/2) \subseteq h(\Omega)$.

Either $B(0, \delta) \subseteq h(\Omega)$, and we are done, or there is w, $|w| < \delta$ and $w \notin h(\Omega)$. Hence, the function

$$\varphi(z) := \frac{1}{h(z) - w}$$

is holomorphic in Ω. The Cauchy inequalities show

$$\frac{1}{|w|} = |\varphi(0)| \leq \sup_{|z|=r} \frac{1}{|h(z) - w|} \leq \frac{1}{\delta - |w|}.$$

Therefore $|w| \geq \delta/2$. It follows that $B(0, \delta/2) \subseteq h(\Omega)$ as we wanted to show. □

2.1.26. Corollary. *Let Ω be a connected open set in $\mathbb{C}$ and $f \in \mathscr{H}(\Omega)$, if either of* $\operatorname{Re} f$, $\operatorname{Im} f$, *or* $|f|$ *is constant, then f is also constant.*

PROOF. $f(\Omega)$ would be a subset of the x-axis, y-axis, or a circle respectively. None of them could be open. □

2.1.27. Corollary (Maximum Principle). *Let Ω be a bounded connected open set in $\mathbb{C}$, let $h \in \mathscr{H}(\Omega)$ be nonconstant, and*

$$M = \sup_{\zeta \in \partial\Omega} \left(\limsup_{\substack{z \to \zeta \\ z \in \Omega}} |h(z)| \right).$$

Then the inequality

$$|h(z)| < M$$

holds for every $z \in \Omega$.

PROOF. We can assume $M < \infty$. The function φ defined on the compact set $\bar{\Omega}$ by

$$\varphi(\zeta) := \begin{cases} |h(\zeta)| & \text{if } \zeta \in \Omega \\ \limsup\limits_{\substack{z \to \zeta \\ z \in \Omega}} |h(z)| & \text{if } \zeta \in \partial\Omega \end{cases}$$

is upper semicontinuous in $\bar{\Omega}$. Therefore, φ is bounded. It follows that $U = h(\Omega)$ is a bounded open subset of $\mathbb{C}$. Since h is an open mapping, for every $w \in \partial U$ there is a sequence $(z_n)_{n \geq 1}$ of points in Ω such that $w = \lim_{n \to \infty} h(z_n)$ and $(z_n)_{n \geq 1}$ is itself convergent to a point in $\partial\Omega$. We conclude that

$$\partial U \subseteq \{w \in \mathbb{C} : |w| \leq M\},$$

hence

$$U \subseteq \{w \in \mathbb{C} : |w| < M\}.$$

This proves the corollary. □

2.1.28. Corollary. *Let Ω be a connected open subset of $\mathbb{C}$ and $h \in \mathscr{H}(\Omega)$ such that for some $a \in \Omega$ we have $|h(z)| \leq |h(a)|$ for every z in a neighborhood of a. Then h is constant.*

PROOF. The proof is left to the reader as an exercise. □

2.1.29. Proposition (Schwarz's Lemma). *Let $g : B(0,1) \to B(0,1)$ be a holomorphic function such that $g(0) = 0$. Then*

$$|g(z)| \leq |z|$$

and

$$|g'(0)| \leq 1.$$

If either inequality becomes an equality (even at a single point, except for $z = 0$, for the first one) then $g(z) = Cz$ for some constant C, $|C| = 1$.

PROOF. The function $h(z) := g(z)/z (z \neq 0)$, $h(0) := g'(0)$ is holomorphic in $B(0,1)$ and, for any r, $0 < r < 1$,

$$\max_{|z| \leq r} |h(z)| = \max_{|z|=r} |h(z)| = \frac{\max_{|z|=r} |g(z)|}{r} \leq \frac{1}{r}.$$

Since r is arbitrary, it follows that $|h(z)| \leq 1$ for every $z \in B(0,1)$. If there is some $z_0 \in B(0,1) \backslash \{0\}$ such that $|h(z_0)| = 1$, then h is a constant C of absolute value 1 by §2.1.28. The same argument holds if $|g'(0)| = |h(0)| = 1$. □

EXERCISES 2.1

(Ω represents a connected open set in $\mathbb{C}$ and $B = B(0,1)$. It is understood that a contour like $|z| = r$ is traversed only once and counterclockwise.)

1. With the help of Cauchy's integral representation formulas compute the following integrals:

   (i) $\displaystyle\int_{|z+i|=3} \sin z \frac{dz}{z+i}$

   (ii) $\displaystyle\int_{|z|=2} \frac{e^z}{z-1} dz$

   (iii) $\displaystyle\int_{|z|=4} \frac{\cos z}{z^2 - \pi^2} dz$

   (iv) $\displaystyle\int_{|z|=2} \frac{dz}{(z-1)^n(z-3)}$ $\quad (n = 1, 2, 3, \ldots)$

2. Let $f \in \mathscr{H}(B) \cap C^0(\bar{B})$, show that

$$\int_{|\zeta|=1} f(\zeta)\, d\zeta = 0$$

and

$$f(z) = \frac{1}{2\pi i} \int_{|\zeta|=1} f(\zeta) \frac{d\zeta}{\zeta - z} \qquad (z \in B).$$

Conclude that

$$f(0) = \frac{1}{2\pi} \int_{-\pi}^{\pi} f(e^{i\theta})\, d\theta.$$

Write down the corresponding mean-value property for $f(a)$, when $B(a, r) \subseteq \Omega$ and $f \in \mathscr{H}(\Omega)$.

3. Let $f, g \in \mathscr{H}(B) \cap C^0(\bar{B})$, show that

$$\frac{1}{2\pi i} \int_{|\zeta|=1} \left(\frac{f(\zeta)}{\zeta - z} + \frac{z g(\zeta)}{z\zeta - 1} \right) d\zeta = \begin{cases} f(z) & \text{if } |z| < 1 \\ g(1/z) & \text{if } |z| > 1. \end{cases}$$

4. Let $f \in \mathscr{H}(\Omega)$, $\bar{B}(z_0, R) \subseteq \Omega$, $0 \le r < R$. Compute

$$\int_{r < |z - z_0| < R} f(z)\, dx\, dy.$$

5. Let $p > 0$ and f be an entire function such that for some constant $M > 0$,

$$|f(z)| \le M(1 + |z|)^p$$

for every $z \in \mathbb{C}$. Show that f is a polynomial of degree at most p.

6. Let f be an entire function satisfying $|f(z)| \le M e^{|z|}$ everywhere. Show that $|f(0)| \le M$ and

$$\frac{|f^{(n)}(0)|}{n!} \le M \left(\frac{e}{n} \right)^n \qquad (n \in \mathbb{N}^*).$$

7. Let γ be a piecewise C^1 Jordan curve, φ a continuous function on $\operatorname{Im}(\gamma)$, and $0 \in \operatorname{Int}(\gamma)$. Show that

$$\int_\gamma \frac{\varphi(t)}{t - z}\, dt = 0 \qquad \text{for every } z \in \operatorname{Int}(\gamma)$$

if and only if

$$\int_\gamma t^n \varphi(t)\, dt = 0 \qquad \text{for } n = -1, -2, -3, \ldots.$$

*8. Let $S = \{z \in \mathbb{C} : |\operatorname{Im} z| < a < \infty\}$ and $f \in \mathscr{H}(S)$ satisfying the two conditions

(i) $\displaystyle\lim_{|z| \to \infty} \frac{f(z)}{z} = 0,$

(ii) $\displaystyle\int_{-\infty}^{\infty} \frac{|f(x + iy)|}{1 + |x|}\, dx < \infty,$ for every $y \in\,]-a, a[$.

Show that there exist functions f_1 (resp. f_2) holomorphic in the half-plane $\operatorname{Im} z > -a$ (resp. $\operatorname{Im} z < a$) satisfying the conditions: for any $\varepsilon > 0$, $f_1(z) \to 0$ when $|z| \to \infty$ while $\operatorname{Im} z \geq -a + \varepsilon$ (resp. $f_2(z) \to 0$ when $|z| \to \infty$ while $\operatorname{Im} z \leq a - \varepsilon$), and

$$f(z) = f_1(z) - f_2(z) \qquad \text{for every } z \in S.$$

(Hint: For $0 < b < a$, define f_1 in $\operatorname{Im} z > -b$ by

$$f_1(z) = \frac{1}{2\pi i} \int_{-\infty - ib}^{\infty - ib} \frac{f(\zeta)}{\zeta - z} d\zeta.$$

Use a similar definition for f_2 and Cauchy's theorems 2.1.3 and 1.11.4.)

9. Show that

(a) For any $\alpha \in \mathbb{C}$ we have $\int_{-\infty}^{\infty} e^{-t^2} dt = \int_{-\infty}^{\infty} e^{-(t+\alpha)^2} dt$. (Why is this not an immediate consequence of the translation invariance of the Lebesgue measure?)

(b) Consider the integral of the function $e^{i\pi z^2} \tan \pi z$ around the parallelogram in Figure 2.1. Let $R \to \infty$ to show that

$$\int_{-\infty}^{\infty} e^{-x^2} dx = \sqrt{\pi}.$$

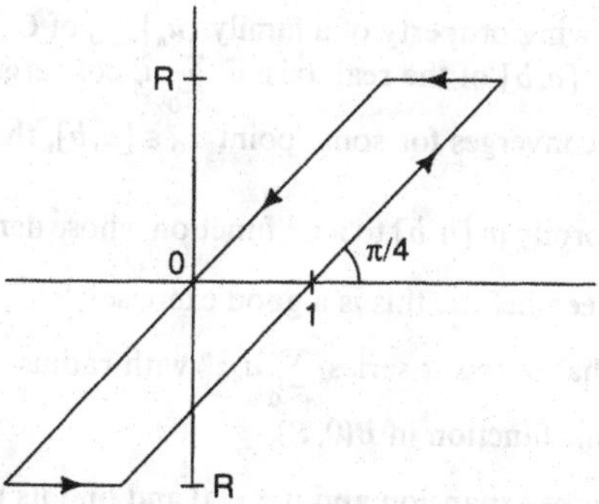

Figure 2.1

(c) Integrate e^{-z^2} along the contour of Figure 2.2 to show that

$$\int_0^{\infty} \sin(x^2)\, dx = \int_0^{\infty} \cos(x^2)\, dx = \frac{1}{2}\sqrt{\frac{\pi}{2}}.$$

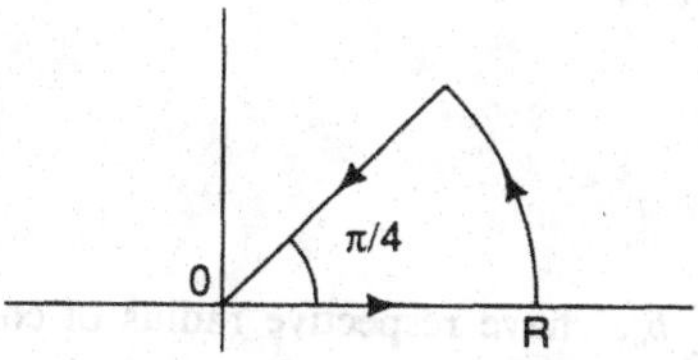

Figure 2.2

10. (a) Prove Proposition 2.1.17 and the preceding statements about the radius of convergence.
(b) Show that the radius of convergence of a power series $\sum_{n\geq 0} a_n z^n$ and that of the series of its formal derivative, $\sum_{n\geq 1} na_n z^{n-1}$, coincide.
(c) Show that the function (defined for $h \neq 0$, $n \in \mathbb{N}^*$)

$$\frac{(z+h)^n - z^n}{h} - nz^{n-1}$$

is a polynomial in z and h with positive coefficients. Use this fact to prove the inequality

$$\left|\frac{(z+h)^n - z^n}{h} - nz^{n-1}\right| \leq \frac{(|z|+|h|)^n - |z|^n}{|h|} - n|z|^{n-1}.$$

Conclude that

$$\left|\frac{1}{h}\left(\sum_{n\geq 0} a_n(z+h)^n - \sum_{n\geq 0} a_n z^n\right) - \sum_{n\geq 0} na_n z^{n-1}\right|$$
$$\leq \sum_{n\geq 0} |a_n|\left(\frac{(|z|+|h|)^n - |z|^n}{|h|}\right) - \sum_{n\geq 0} n|a_n||z|^{n-1}$$

(d) Recall the following property of a family $\{u_n\}_{n\geq 0}$ of C^1 functions defined on a finite interval $[a,b]$ of the real axis: if $\sum_{n\geq 0} u_n'$ converges uniformly in $[a,b]$ and $\sum_{n\geq 0} u_n(x_0)$ converges for some point $x_0 \in [a,b]$, then the series $\sum_{n\geq 0} u_n(x)$ converges uniformly in $[a,b]$ to a C^1 function whose derivative is $\left(\sum_{n\geq 0} u_n\right)' = \sum_{n\geq 0} u_n'$. (If not seen before, this is a good exercise).

Conclude that a power series $\sum_{n\geq 0} a_n z^n$ with radius of convergence $R > 0$ is a holomorphic function in $B(0,R)$.

11. Give the Taylor series expansion about $z = 0$ and find its radius of convergence for the following functions:

(i) $\dfrac{e^z + e^{-z} + 2\cos z}{4}$

(ii) $\dfrac{z^2 + 4z^4 + z^6}{(1-z^2)^4}$

(iii) $(1-z)^{-m-1}$, $m \in \mathbb{N}$

(iv) $(1-z^6)^{-3}$

(v) $\dfrac{z^5}{(z^2+1)(z-1)}$

(vi) Let $\sum_{n\geq 0} a_n z^n$, $\sum_{n\geq 0} b_n z^n$ have respective radius of convergence $\alpha > 0$, $\beta > 0$.

What can be said about the radius of convergence of the series $\sum_{n\geq 0} c_n z^n$, where $c_n = \sum_{k=0}^{n} a_{n-k} b_k$?

12. (a) Let $\omega \subset\subset \Omega$, $\partial\omega$ piecewise regular, $f \in \mathscr{H}(\Omega)$. Apply the Cauchy inequality 2.1.20, (2)

$$|f(z_0)| \leq \frac{\ell(\partial\omega)\max\{|f(z)| : z \in \partial\omega\}}{2\pi d(z_0, \partial\omega)},$$

which is valid for $z_0 \in \omega$, to the holomorphic functions $(f(z))^n$, $n = 1, 2, \ldots$, to show that $|f(z_0)| \leq \max\{|f(z)| : z \in \partial\omega\}$. Obtain a proof of the *weak maximum principle*:

If Ω bounded, $f \in \mathscr{H}(\Omega) \cap C^0(\bar{\Omega})$ then $\|f\|_{L^\infty(\Omega)} \leq \|f\|_{L^\infty(\partial\Omega)}$.

(By comparison with this exercise, the statement 2.1.27 is sometimes called the strong maximum principle.)

(b) Show that the open mapping property 2.1.25 implies the weak maximum principle.

13. Show that the maximum principle 2.1.27 fails if Ω is not bounded.
14. Assume $f \in \mathscr{H}(\Omega)$ is not constant. Show that $|f|$ cannot have a minimum at a point $z_0 \in \Omega$, unless $f(z_0) = 0$.
15. Suppose $\varphi \in C^1(\Omega)$, $N := \{z \in \Omega : \varphi(z) = 0\}$, and grad $\varphi(z) \neq 0$ if $z \in N$. Show that if $f \in \mathscr{H}(\Omega)$ and $f(\Omega) \subseteq N$ then f is constant.
16. Let p be a polynomial of degree n and $a > 0$. How many connected components can the open set $\{z \in \mathbb{C} : |p(z)| < a\}$ have?
17. Let f be an entire function such that $|f(z)| \leq e^{\operatorname{Re} z}$ everywhere. What can you say about f?
18. Assume $f \in \mathscr{H}(B)$ and it satisfies $f''\left(\frac{1}{2^n}\right) = f\left(\frac{1}{2^n}\right)$ for every $n \in \mathbb{N}^*$, show that f is an entire function.
19. Is there a function holomorphic in B such that it satisfies any of these conditions for every $n \in \mathbb{N}^*$:

(a) $f\left(\frac{1}{n}\right) = f\left(-\frac{1}{n}\right) = \frac{1}{n^2}$;

(b) $f\left(\frac{1}{n}\right) = f\left(-\frac{1}{n}\right) = \frac{1}{2n+1}$;

(c) $\left|f\left(\frac{1}{n}\right)\right| < 2^{-n}$;

(d) $\frac{1}{\sqrt{n}} < \left|f\left(\frac{1}{n}\right)\right| < \frac{2}{\sqrt{n}}$.

20. Let p be a polynomial of degree n. $M(r) = M(|p|, r) = \max\{|p(z)| : |z| = r\}$. Show that if $0 < r < R < \infty$ then

$$\frac{M(r)}{r^n} \geq \frac{M(R)}{R^n}.$$

What can you say about p if the equality holds for some r, R?

21. Let $B_+ = B \cap \{z \in \mathbb{C} : \operatorname{Im} z > 0\}$, $f \in \mathscr{H}(B_+) \cap \mathscr{C}(\bar{B}_+)$. Show that if $f(x) = 0$ for $-1 < x < 1$, then $f \equiv 0$.

*22. Let B_+ be as in the previous exercise, $f \in \mathscr{H}(B_+) \cap \mathscr{C}(\bar{B}_+)$ such that f is real valued on the real axis, $\operatorname{Im} f(z) \neq 0$ in B_+ and f is injective in B_+. Show that there is a function $F \in \mathscr{H}(B)$, F injective and $F|B_+ = f$.

23. Let f be holomorphic in the upper-half-plane and continuous up to the real axis. Assume f is real valued on the real axis and $\|f\|_\infty \leq 1$. Show $f \equiv c \in [-1, 1]$.

24. Show that the function $z \mapsto \cos(\sqrt{z})$ which is a priori defined only in $-\pi < \operatorname{Arg} z < \pi$, can be extended to be an entire function. Give its Taylor series expansion about $z = 0$. Does it depend on the determination of $\sqrt{z}$?

25. Show that the Γ function defined here is holomorphic in the right half-plane $\{z \in \mathbb{C} : \operatorname{Re} z > 0\}$,

$$\Gamma(z) := \int_0^\infty e^{-t} t^{z-1}\, dt.$$

Verify that

$$\Gamma(z + 1) = z\Gamma(z).$$

Conclude that for $n \in \mathbb{N}$,

$$\Gamma(n + 1) = n!.$$

26. Show that the function $z \mapsto \displaystyle\int_0^\infty \frac{e^{-tz}}{1 + t^2}\, dt$ is holomorphic for $\operatorname{Re} z > 0$.

27. Let f be holomorphic in $\bar{B}$, $f \in \mathscr{H}(\bar{B})$ (see §1.11.1). Compute

$$\int_B \bar{z}^{n-1} f(z)\, dx\, dy \qquad (n \geq 1).$$

28. Let $f \in \mathscr{H}(B)$ and k a radial continuous function of compact support in B such that $\displaystyle\int_{\mathbb{C}} k(z)\, dm(z) = 2\pi \int_0^1 rk(r)\, dr = 1$. Show that

$$f(0) = \int_{\mathbb{C}} f(z)k(z)\, dm(z).$$

(as always in this book, dm stands for the Lebesgue measure.)

29. Let $f_1, \ldots, f_n \in \mathscr{H}(\Omega)$ be such that $|f_1(z)|^2 + \cdots + |f_n(z)|^2 \equiv 1$ in Ω. Show that $f_1, \ldots, f_n$ are constants.

30. Let $f : \Omega \to \Omega$ be holomorphic and satisfy $f \circ f = f$. Show that either f is constant or $f = id_\Omega$.

31. (a) Let $f \in \mathscr{H}(\Omega)$, $\Omega \supseteq \bar{B}(0, r)$, $f(z) = \sum_{n \geq 0} a_n z^n$ be its Taylor expansion about

$z = 0$. Show that

$$\sum_{n\geq 0} |a_n|^2 r^{2n} = \int_{-\pi}^{\pi} |f(re^{i\theta})|^2 \frac{d\theta}{2\pi}.$$

(b) When $r = 1$ in part (a) show that

$$\sum_{n\geq 0} \frac{|a_n|^2}{n+1} = \frac{1}{\pi} \int_B |f(z)|^2 \, dm(z).$$

32. Apply exercise 31 to show that if f is a polynomial of degree n such that $f(\bar{B}) \subseteq \bar{B}$, then

$$\frac{1}{\pi} \int_B |f'(z)|^2 \, dm(z) \leq n.$$

Find all the polynomials for which this is an identity.

33. Let $f \in \mathscr{H}(B)$, $f(z) = \sum_{n\geq 0} a_n z^n$. Write $u = \operatorname{Re} f$, $v = \operatorname{Im} f$. Show that for any $r \in]0, 1[$, $n \in \mathbb{N}^*$,

$$a_n = \frac{1}{\pi r^n} \int_0^{2\pi} u(re^{i\theta}) e^{-in\theta} \, d\theta = \frac{i}{\pi r^n} \int_0^{2\pi} v(re^{i\theta}) e^{-in\theta} \, d\theta.$$

Use the first identity to show that if $f(0) = 1$, $u \geq 0$, then $|a_n| \leq 2$, $n = 1, 2, \ldots$. Conclude that

$$\frac{1-|z|}{1+|z|} \leq |f(z)| \leq \frac{1+|z|}{1-|z|}.$$

(Hint: Use that $f(z) \neq 0$ to prove the lower bound.)

34. Use Liouville's theorem (cf. §2.1.20) to prove the Fundamental Theorem of Algebra, i.e., every nonconstant polynomial has a zero.

§2. The Fréchet Space $\mathscr{H}(\Omega)$

We recall that the space $\mathscr{C}(\Omega)$ of all the continuous complex-valued functions in an open set $\Omega \subseteq \mathbb{C}$ admits the following metric:

$$d(f, g) := \sum_{n\geq 1} \frac{1}{2^n} \frac{p_n(f-g)}{1 + p_n(f-g)},$$

where $p_n(h) = \sup_{z\in K_n} |h(z)|$ and $(K_n)_{n\geq 1}$ is an exhaustive sequence of compacts in Ω. $\left(\text{That is, } K_n \subseteq \mathring{K}_{n+1} \text{ for every } n \geq 1 \text{ and } \Omega = \bigcup_{n\geq 1} K_n. \text{ An example of such a sequence is } K_n = \bar{B}(0, n) \cap \left\{z \in \Omega : d(z, \Omega^c) \geq \frac{1}{n}\right\}.\right)$ This distance is invariant under translations, i.e., $d(f + h, g + h) = d(f, g)$, and it induces the topology of uniform convergence on every compact subset of Ω. $\mathscr{C}(\Omega)$ is then a complete metric space, i.e., a Fréchet space.

We will always consider $\mathscr{H}(\Omega)$ with the topology induced from $\mathscr{C}(\Omega)$.

2.2.1. Proposition. *The space $\mathscr{H}(\Omega)$ is a Fréchet space.*

PROOF. It is enough to show that $\mathscr{H}(\Omega)$ is a closed subspace of $\mathscr{C}(\Omega)$. Let $(f_n)_{n\geq 1}$ be a sequence of holomorphic functions that converges uniformly over every compact subset of Ω towards a continuous function f. We have to show that f is holomorphic. If R is a closed rectangle contained in Ω, then

$$\int_{\partial R} f(z)\,dz = \lim_{n\to\infty} \int_{\partial R} f_n(z)\,dz = 0,$$

by the uniform convergence on the compact ∂R. Morera's theorem implies that f is indeed holomorphic in Ω. □

2.2.2. Corollaries

(1) *If a power series $\sum_{n\geq 0} a_n z^n$ is convergent in a disk $B(0, R)$, its sum is a holomorphic function in that disk.*
(2) *The space $\mathscr{H}(\Omega)$ is identical to the collection of all functions in Ω that admit locally an expansion in power series.*

PROOF. The proofs are left to the reader. □

2.2.3. Proposition. *The mapping $\dfrac{d}{dz}: f \mapsto f'$ from $\mathscr{H}(\Omega)$ into itself is continuous.*

PROOF. Let K be a compact subset of Ω and ω a relatively compact neighborhood of K in Ω. The Cauchy inequalities assert that there is a constant C such that

$$\sup_{z\in K} |f'(z)| \leq C \sup_{z\in\omega} |f(z)|.$$

This proves the continuity of the operator $\dfrac{d}{dz}$. □

2.2.4. Corollaries

(1) *If $s(z) = \sum_{n\geq 1} a_n z^n$ in the disk $B(0, R)$ then $s'(z) = \sum_{n\geq 1} n a_n z^{n-1}$ in the same disk.*
(2) *If $(f_n)_{n\geq 1}$ is a convergent sequence in $\mathscr{H}(\Omega)$ to a function f, then $(f_n')_{n\geq 1}$ converges in $\mathscr{H}(\Omega)$ to f'.*
(3) *A similar statement holds for series: If $\sum_{n\geq 1} f_n$ converges to f in $\mathscr{H}(\Omega)$, then $\sum_{n\geq 1} f_n'$ converges to f' in this space.*

2.2.5. Remark. If $f(z) = \sum a_n z^n$ in $B(0, R)$, then a primitive for f in the same disk is

$$F(z) = \sum_{n \geq 0} \frac{a_n}{n+1} z^{n+1}.$$

2.2.6. Examples. (1) Let $f(z) = \dfrac{1}{1-z}$, its expansion in $B(0, 1)$ is $\sum_{n \geq 0} z^n$. The function $F(z) = \int_0^z f(\zeta)\, d\zeta$ is a primitive of f and admits the expansion

$$F(z) = \sum_{n \geq 0} \frac{z^{n+1}}{n+1}.$$

Consider now the holomorphic function $g(z) := e^{-F(z)}$. Its derivative can be readily computed by the chain rule

$$g' = \frac{\partial g}{\partial z} = -\frac{\partial F}{\partial z} g = -fg.$$

The product function fg is also holomorphic in $B(0, 1)$ and

$$(fg)' = fg' + f'g.$$

Since $g' = -fg$ and $f' = 1/(1-z)^2 = f^2$ we have

$$(fg)' = -f^2 g + f^2 g = 0.$$

Therefore fg is a constant equal to $f(0)g(0) = 1$. It follows that $-F$ is a determination of the logarithm of $1 - z$. Moreover, if $|z| < 1$ we have $1 - z \in \mathbb{C} \setminus]-\infty, 0]$ and we have already defined the principal value of its logarithm, $\operatorname{Log}(1-z)$. The two continuous functions $\operatorname{Log}(1-z)$ and $-F(z)$ are determinations of the logarithm of $1 - z$ in $B(0, 1)$ and coincide for $z = 0$. Therefore $F(z) = -\operatorname{Log}(1-z)$ throughout $B(0, 1)$ and we have

$$\operatorname{Log}(1-z) = -\sum_{n \geq 0} \frac{z^{n+1}}{n+1} \qquad (|z| < 1).$$

(2) Consider the function $(1-x)^\alpha$ ($\alpha \in \mathbb{C}$) defined for $x \in]-1, 1[$ by

$$(1-x)^\alpha = e^{\alpha \operatorname{Log}(1-x)}.$$

From calculus we know that this function has the Taylor expansion convergent in $]-1, 1[$,

$$(1-x)^\alpha = \sum_{n \geq 0} (-1)^n \frac{\alpha(\alpha-1)\ldots(\alpha-n+1)}{n!} x^n.$$

The power series obtained by replacing x with z converges in $B(0, 1)$ and hence defines a holomorphic function there, also denoted $(1-z)^\alpha$. One can verify that $(1-z)^\alpha$ coincides in $B(0, 1)$ with $e^{\alpha \operatorname{Log}(1-z)}$. In fact, these two holomorphic functions already coincide on $]-1, 1[$.

(3) Let $f(z) = \dfrac{1}{1+z^2} = \sum_{n\geq 0}(-1)^n z^{2n}$ in $B(0,1)$. The function $F(z) = \int_0^z f(\zeta)\,d\zeta$ is a primitive of f whose Taylor series expansion about 0 is

$$F(z) = \sum_{n\geq 0}(-1)^n \frac{z^{2n+1}}{2n+1} \qquad (|z|<1).$$

Let us recall that the entire function $\cos w$ vanishes exactly at the points $\pi/2 + k\pi (k \in \mathbb{Z})$. Therefore, the function $\tan w = \sin w/\cos w$ is holomorphic in $\mathbb{C}\setminus\{\pi/2 + k\mathbb{Z}\}$. It is well known that $\tan x$ is strictly increasing in the segment $]-\pi/2, \pi/2[$ of the real axis, where it admits an inverse function $\arctan x$ whose derivative is $\dfrac{1}{1+x^2}$.

There is a connected open set ω in $\mathbb{C}$ containing $]-\pi/4, \pi/4[$ such that $g(z) := F(\tan z)$ is well defined and holomorphic in ω. One has $g(0) = 0$ and

$$g'(z) = F'(\tan z)\frac{d}{dz}\tan z = \frac{\sec^2 z}{1+\tan^2 z} = 1,$$

hence $g(z) = z$. It follows that in $]-1, 1[$ the function $\arctan x$ is the restriction of the holomorphic function F.

2.2.7. Definition. A subset $A \subseteq \mathscr{H}(\Omega)$, $A \neq \varnothing$, is said to be ***bounded*** if for every compact $K \subseteq \Omega$ there is a constant $M(K) < \infty$ such that

$$\sup_{f\in A}\left(\sup_{z\in K}|f(z)|\right) = \sup_{f\in A}\|f\|_K \leq M(K).$$

This is a particular case of the usual definition of bounded subsets of a topological vector space. The reader should note that the sets that are bounded for the metric defining the topology of $\mathscr{H}(\Omega)$ are not generally bounded in the sense of Definition 2.2.7. For instance, $\mathscr{H}(\Omega)$ is itself bounded for this distance being contained in the ball centered at the zero function and radius one. Evidently, $\mathscr{H}(\Omega)$ is not bounded in the sense of §2.2.7.

2.2.8. Theorem (Montel). *The bounded sets of $\mathscr{H}(\Omega)$ are precisely the relative compact subsets of $\mathscr{H}(\Omega)$. (In other words, $\mathscr{H}(\Omega)$ is a Fréchet-Montel space* (cf. [Sch]).)

Proof. Since for every $K \subset\subset \Omega$, the numerical function $f \mapsto \|f\|_K$ is continuous in $\mathscr{H}(\Omega)$, it is clear that every relatively compact subset of $\mathscr{H}(\Omega)$ is bounded in the sense of §2.2.7.

Conversely, if A is a bounded set in $\mathscr{H}(\Omega)$, it is also equicontinuous by the Cauchy inequalities. The theorem of Arzelá-Ascoli now shows that A is relatively compact in $\mathscr{C}(\Omega)$. Since $\mathscr{H}(\Omega)$ is a closed subspace of $\mathscr{C}(\Omega)$, A is also relatively compact in $\mathscr{H}(\Omega)$. □

2.2.9. Examples. (1) Let Ω be a connected open set in $\mathbb{C}$, $(f_n)_{n\geq 1}$ a bounded sequence in $\mathscr{H}(\Omega)$, and E a subset of Ω which has an accumulation point in Ω. If for every $z \in E$, $(f_n(z))_{n\geq 1}$ is a convergent sequence, then the sequence $(f_n)_{n\geq 1}$ converges in $\mathscr{H}(\Omega)$.

In fact, let $(f_{n_k})_{k\geq 1}$ be a convergent subsequence, its limit g does not depend on the choice of convergent subsequence since it is completely determined by its values on E. Therefore the sequence $(f_n)_{n\geq 1}$ converges to g in $\mathscr{H}(\Omega)$. (If not, there is $\varepsilon > 0$ and a subsequence $(f_{n_k})_{k\geq 1}$ such that $d(f_{n_k}, g) \geq \varepsilon$. This subsequence will have itself a convergent subsequence $(f_{n_{k_j}})_{j\geq 1}$ and its limit will coincide with g on E, hence everywhere. This is an obvious contradiction with the choice of ε and (f_{n_k}).)

(2) Let $\Omega = B(0,1)$ and $(f_n)_{n\geq 0}$ be a bounded sequence of holomorphic functions in Ω. Let $f_n(z) = \sum_{k\geq 0} c_{n,k}z^k$ be their Taylor series expansion about the origin. Then $(f_n)_{n\geq 1}$ converges in $\mathscr{H}(\Omega)$ to f_0 if and only if for every $k \geq 0$, $\lim_{n\to\infty} c_{n,k} = c_{0,k}$.

If $\lim_{n\to\infty} f_n = f_0$ then it is clear that $(c_{n,k})_{n\geq 1}$ converges to $c_{0,k}$ for every $k \geq 0$. Conversely, since $c_{n,k} = f_n^{(k)}(0)/k!$, if $g = \lim_{j\to\infty} f_{n_j}$ is the limit of a subsequence, then $g^{(k)}(0)/k! = \lim_{j\to\infty} c_{n_j,k} = c_{0,k}$ and hence g is independent of the subsequence. The rest of the proof is the same as in Example 1.

(3) Let U be an open set in $\mathbb{C}$, $f \in \mathscr{H}(U)$ such that $f(U) \subseteq U$, and $z_0 \in U$ be such that $f(z_0) = z_0$ and $|f'(z_0)| < 1$. Under these conditions, there is a disk $\bar{B}(z_0, r) \subseteq U$ such that $\lim_{n\to\infty} f^{[n]}(z) = z_0$ uniformly in that disk. (Here $f^{[n]}$ denotes the nth iterate of f, $f^{[1]} = f$, $f^{[n]} = f \circ f^{[n-1]}$ for $n \geq 2$.) If U is bounded and connected, it is also true that $\lim_{n\to\infty} f^{[n]} = z_0$ (the constant function) in $\mathscr{H}(U)$.

Set $R = \frac{1}{2}(1 + |f'(z_0)|) < 1$. From the definition of $f'(z_0)$ it follows that there is $r > 0$ such that $\bar{B}(z_0, r) \subseteq U$ and

$$\left|\frac{f(z) - f(z_0)}{z - z_0}\right| \leq R \qquad \text{if } 0 < |z - z_0| \leq r.$$

Therefore, since $f(z_0) = z_0$,

$$|f(z) - z_0| \leq R|z - z_0| \qquad \text{if } 0 \leq |z - z_0| \leq r.$$

This implies that $f(B(z_0, r)) \subseteq B(z_0, r)$ and hence we can iterate the last inequality, whence

$$|f^{[n]}(z) - z_0| \leq R^n|z - z_0| \leq R^n r \qquad (z \in \bar{B}(z_0, r)).$$

From this inequality the uniform convergence of $f^{[n]}$ to the constant z_0 in $\bar{B}(z_0, r)$ is evident. Finally, if U is bounded and connected, the sequence $(f^{[n]})_{n\geq 1}$ is bounded in $\mathscr{H}(U)$. From (1) it follows that $f^{[n]}$ converges to z_0 in $\mathscr{H}(U)$.

(4) Let Ω be a connected open subset of $\mathbb{C}$ and $f \in \mathscr{H}(\Omega)$ be such that $\overline{f(\Omega)}$ is a compact subset of Ω. The iterates $f^{[n]}$ of f converge to a constant $z_0 \in \Omega$.

In order to see this, let $\Omega_n = f^{[n]}(\Omega)$ and $K = \bigcap_{n \geq 1} \overline{\Omega_n}$. We have $\overline{\Omega_{n+1}} = \overline{f(\Omega_n)} \subseteq f(\bar{\Omega}_n) \subseteq f(\Omega_{n-1}) = \Omega_n$ by induction. It follows that K is also given by $K = \bigcap_{n \geq 2} \bar{\Omega}_n$, hence it is compact and nonempty. Let $(f^{[n_k]})_{k \geq 1}$ be a subsequence of iterates of f converging to g in $\mathscr{H}(\Omega)$. It is clear that $g(\Omega) \subseteq K$. We claim that $g(\Omega) = K$. If $w \in K$ then $w \in \Omega_{n+1} = f^{[n+1]}(\Omega) = f^{[n]}(\Omega_1)$, hence for every $n \geq 1$ there is $w_n \in \Omega_1$ such that $w = f^{[n]}(w_n)$. Since $\bar{\Omega}_1$ is compact, we can find a subsequence $(w_{n_{k_j}})_{j \geq 1}$ of the sequence $(w_{n_k})_{k \geq 1}$ converging to $w_0 \in \bar{\Omega}_1$. The uniform convergence of $(f^{[n_{k_j}]})_{j \geq 1}$ to g on $\bar{\Omega}_1$ implies that $\lim_{j \to \infty} f^{[n_{k_j}]}(w_{n_{k_j}}) = g(w_0)$, so that $w = g(w_0) \in g(\bar{\Omega}_1) \subseteq g(\Omega)$.

We can conclude that g must be a constant function z_0, $z_0 \in \Omega$. If not, g will be an open mapping, $K = g(\Omega)$ will be both open and closed, hence $K = \Omega$ which is impossible.

We recommend [Ab] and [Bl] as introductions to the iteration theory of holomorphic maps.

(5) Let $f \in L^1_{\text{loc}}(\Omega)$ be such that for every $\varphi \in \mathscr{D}(\Omega)$ one has

$$\int_\Omega f \frac{\partial \varphi}{\partial \bar{z}} \, dm = 0.$$

We claim there is a holomorphic function $\tilde{f}$ in Ω such that $f = \tilde{f}$ a.e. Let φ be a standard function and set $\varphi_n(z) = n^2 \varphi(nz)$. Let $\Omega_n = \left\{ z \in \Omega : d(z, \Omega^c) > \frac{1}{n} \right\}$. The C^∞ function

$$F_n(z) = \int_{\bar{B}(0, 1/n)} f(z - \zeta) \varphi_n(\zeta) \, dm(\zeta)$$

is then holomorphic in Ω_n. In fact, we can write

$$F_n(z) = \int_{\bar{B}(0, 1/n)} f(u) \varphi_n(z - u) \, dm(u),$$

and hence, by differentiation under the integral sign,

$$\frac{\partial F_n}{\partial \bar{z}} = \int_\Omega f(u) \frac{\partial}{\partial \bar{z}} (\varphi_n(z - u)) \, dm(u) = -\int_\Omega f(u) \frac{\partial}{\partial \bar{u}} (\varphi_n(z - u)) \, dm(u) = 0,$$

since $u \mapsto \varphi_n(z - u) \in \mathscr{D}(\Omega)$ when $z \in \Omega_n$.

For n_0 fixed, the sequence $(F_n)_{n \geq n_0}$ is bounded in Ω_{n_0}: Let $K \subset\subset \Omega_{n_0}$, $r > 0$ such that $\bar{V}\left(K, r + \frac{1}{n_0}\right) \subseteq \Omega$. (We recall that for $\varepsilon > 0$, $\bar{V}(K, \varepsilon) = \{z \in \mathbb{C} : d(z, K) < \varepsilon\}$.) For $n \geq n_0$ and $z \in K$ we can write

$$F_n(z) = \frac{1}{\pi r^2} \int_{B(z, r)} F_n(w) \, dm(w).$$

(In fact, we have, for $0 < \rho \le r$,

$$F_n(z) = \frac{1}{2\pi i}\int_{|\zeta - z| = \rho} \frac{F_n(\zeta)}{\zeta - z}\,d\zeta = \frac{1}{2\pi}\int_0^{2\pi} F_n(z + \rho e^{i\theta})\,d\theta.$$

Integrating this identity with respect to $\rho\,d\rho$, we obtain

$$\frac{1}{\pi r^2}\int_0^r \left(\int_0^{2\pi} F_n(z + \rho e^{i\theta})\,d\theta\right)\rho\,d\rho = \frac{1}{\pi r^2}\int_0^r F_n(z)\rho\,d\rho = F_n(z),$$

which yields the preceding claim.)

Writing now $w = u + iv$, $\zeta = \xi + i\eta$, we obtain

$$\begin{aligned}|F_n(z)| &\le \frac{1}{\pi r^2}\int_{B(z,r)} |F_n(w)|\,du\,dv\\ &\le \frac{1}{\pi r^2}\int_{B(z,r)} du\,dv \int_{\bar{B}(0,1/n)} |f(w - \zeta)|\varphi_n(\zeta)\,d\xi\,d\eta,\\ &\le \frac{1}{\pi r^2}\left(\int_{\bar{V}(K, r+1/n_0)} |f(w)|\,du\,dv\right)\int_{B(0,1/n)} \varphi_n(\zeta)\,d\xi\,d\eta.\end{aligned}$$

That is, for $z \in K$ and $n \ge n_0$ we have

$$|F_n(z)| \le \frac{1}{\pi r^2}\int_{\bar{V}(K, r+1/n_0)} |f(w)|\,du\,dv.$$

This shows that the sequence $(F_n)_{n \ge n_0}$ is bounded in $\mathscr{H}(\Omega_{n_0})$.

The sequence $(F_n)_{n \ge 1}$ has the following property: for every $\psi \in \mathscr{D}(\Omega)$,

$$\lim_{n \to \infty}\int_\Omega F_n\psi\,dx\,dy = \int_\Omega f\psi\,dx\,dy.$$

To verify this identity, let n be such that $\Omega_n \supseteq \operatorname{supp}\psi$. Then

$$\int_\Omega (f - F_n)\psi\,dx\,dy = -\int_\Omega\left[\int_{B(0,1/n)} (f(z - \zeta) - f(z))\varphi_n(\zeta)\,dx\,dy\right]\psi(z)\,dx\,dy.$$

The function $(z, \zeta) \mapsto (f(z - \zeta) - f(z))\psi(z)\varphi_n(\zeta)$ is integrable on $\bar{V}\left(\operatorname{supp}\psi, \frac{1}{n}\right) \times \bar{B}\left(0, \frac{1}{n}\right)$, hence we can apply Fubini's theorem and obtain

$$\int_\Omega (f - F_n)\psi\,dx\,dy = \int_{B(0,1/n)}\left(\int_\Omega (f(z - \zeta) - f(z))\psi(z)\,dx\,dy\right)\varphi_n(\zeta)\,d\xi\,d\eta.$$

Let $z - \zeta = w = u + iv$, the identity becomes

$$\int_\Omega (f - F_n)\psi\,dx\,dy = \int_{B(0,1/n)}\left(\int_\Omega f(w)(\psi(\zeta + w) - \psi(w))\,du\,dv\right)\varphi_n(\zeta)\,d\xi\,d\eta.$$

By the uniform continuity of ψ, given $\varepsilon > 0$ there is N such that if $n > N$

$$\left|\int_{\Omega}(f - F_n)\psi \, dx\, dy\right| \leq \varepsilon \int_{V(\operatorname{supp}\psi,\, 1/N)} |f(w)| \, du\, dv,$$

which shows the limit of $\int_{\Omega} F_n\psi$ exists and equals $\int_{\Omega} f\psi$.

If n_0 is such that $\Omega_{n_0} \supseteq \operatorname{supp}\psi$, and if $(F_{n_k})_{k\geq 1}$ is a subsequence of $(F_n)_{n\geq n_0}$ which converges in $\mathscr{H}(\Omega_{n_0})$ to g, then

$$\lim_{k\to\infty}\int_{\Omega} F_{n_k}\psi \, dx\, dy = \int_{\Omega_{n_0}}\left(\lim_{k\to\infty} F_{n_k}\right)\psi \, dx\, dy = \int_{\Omega_{n_0}} g\psi \, dx\, dy,$$

Therefore

$$\int_{\Omega_{n_0}} g\psi \, dx\, dy = \int_{\Omega} f\psi \, dx\, dy, \qquad \text{for every } \psi \in \mathscr{D}(\Omega_{n_0}),$$

which shows that the holomorphic function g coincides with f a.e. in Ω_n, and hence g does not depend on the subsequence $(F_{n_k})_{k\geq 1}$. We can conclude that the condition $f \in L^1_{\mathrm{loc}}(\Omega)$ and $\int_{\Omega} f \frac{\partial\varphi}{\partial\bar{z}} \, dx\, dy = 0$ for every $\varphi \in \mathscr{D}(\Omega)$ implies that there is a function $\tilde{f} \in \mathscr{H}(\Omega)$ such that $f = \tilde{f}$ a.e.

(6) Let $(f_n)_{n\geq 1}$ be a sequence of holomorphic functions in Ω such that $(f_n)_{n\geq 1}$ converges in $L^1_{\mathrm{loc}}(\Omega)$ to a function $f \in L^1_{\mathrm{loc}}(\Omega)$ (i.e., for every $K \subset\subset \Omega$, $f_n \to f$ in $L^1(K)$). Then there exists $\tilde{f} \in \mathscr{H}(\Omega)$ such that $f = \tilde{f}$ a.e. and $f_n \to \tilde{f}$ in $\mathscr{H}(\Omega)$.

In fact, one verifies easily that for every $\varphi \in \mathscr{D}(\Omega)$,

$$\int_{\Omega} f \frac{\partial\varphi}{\partial\bar{z}} \, dx\, dy = \lim_{n\to\infty}\int_{\Omega} f_n \frac{\partial\varphi}{\partial\bar{z}} \, dx\, dy = 0$$

and then one can use (5) and inequality 2.1.20, (3).

(7) Let $0 < \delta < \alpha \leq \pi$, Ω the angular region given by $\Omega := \{z \in \mathbb{C} : 0 < |z| < 1, -\alpha < \operatorname{Arg} z < \alpha\}$, and $f \in \mathscr{H}(\Omega)$, be a bounded function such that $\lim_{x\to 0+} f(x) = L \in \mathbb{C}$ exists. It follows that $\lim_{z\to 0} f(z) = L$ holds uniformly within $-\alpha + \delta \leq \operatorname{Arg} z \leq \alpha - \delta$ (i.e., only $|z| \to 0$ counts, as long as z lies in the smaller angle).

To see this, let us introduce the bounded sequence of holomorphic functions in Ω, $f_n(z) := f(z/2^n)$. This sequence verifies, for every $x \in\,]0, 1[$,

$$\lim_{n\to\infty} f_n(x) = L.$$

It follows from (1) that $(f_n)_{n\geq 1}$ converges uniformly on every compact subset of Ω to the constant L.

Let $K = \{z \in \Omega : \frac{1}{4} \leq |z| \leq \frac{1}{2}, -\alpha + \delta \leq \operatorname{Arg} z \leq \alpha - \delta\}$. For $\varepsilon > 0$ there is $n_0 \in \mathbb{N}$ such that if $n > n_0$ we have $|f_n(z) - L| < \varepsilon$ for every $z \in K$. Let z be such that $0 < |z| \leq 2^{-n_0}$, $-\alpha + \delta \leq \operatorname{Arg} z \leq \alpha + \delta$, and let $j > n_0$ such that $1/2^{j+2} < |z| \leq 1/2^{j+1}$ (which means that $2^j z \in K$), then $f(z) = f_j(2^j z)$ and

$$|f(z) - L| = |f_j(2^j z) - L| < \varepsilon.$$

Therefore, $-\alpha + \delta \leq \operatorname{Arg} z \leq \alpha - \delta$ and $0 < |z| \leq 1/2^{n_0}$ implies that $|f(z) - L| < \varepsilon$. This means that the uniform convergence to the constant L holds in the smaller angle.

2.2.10. Remark. A different generalization of the Cauchy-Goursat Proposition 2.1.11 than the one given in 2.2.9, (5) is the Loman-Menchoff theorem: Let $f \in \mathscr{C}(\Omega)$ be such that the partial derivatives $\frac{\partial f}{\partial x}, \frac{\partial f}{\partial y}$ exist at every point $z \in \Omega$ and satisfy $\frac{\partial}{\partial \bar{z}} f(z) = 0$ everywhere, then $f \in \mathscr{H}(\Omega)$.

For a proof see [SZ] or [Na1]. This theorem is not so frequently used as Example 5 in §2.2.9.

2.2.11. Definition** (Analytic Functionals). An ***analytic functional T ***defined on the open set*** Ω of $\mathbb{C}$ is a continuous linear function $T : \mathscr{H}(\Omega) \to \mathbb{C}$. The space of all those functionals is the topological dual of the Fréchet space $\mathscr{H}(\Omega)$ and it is denoted $\mathscr{H}'(\Omega)$ (cf. [Sch]).

Since $\mathscr{H}(\Omega)$ is a closed subspace of $\mathscr{C}(\Omega)$, the Hahn-Banach and Riesz theorems allow us to conclude that for any $T \in \mathscr{H}'(\Omega)$, there is at least one complex-valued measure μ with compact support in Ω representing T, that means:

$$\langle T, f \rangle = \int_\Omega f \, d\mu \qquad (f \in \mathscr{H}(\Omega)).$$

It is easy to see that in general the measure μ is not unique. For instance, if T is the evaluation at the point $a \in \Omega$, the measures μ_r $(0 < r < d(a, \Omega^c))$ defined by

$$\int f \, d\mu_r = \frac{1}{2\pi i} \int_0^{2\pi} f(a + re^{i\theta}) \, d\theta,$$

as well as the Dirac measure δ_a represent T. (They are not the only ones!)

One can define a topology in $\mathscr{H}'(\Omega)$ that makes it a Hausdorff locally convex topological vector space. To every bounded set $A \subseteq \mathscr{H}(\Omega)$ we associate the seminorm p_A on $\mathscr{H}'(\Omega)$ given by

$$p_A(T) = \sup_{f \in A} |\langle T, f \rangle|.$$

With this topology, one can consider the topological dual $\mathscr{H}''(\Omega) = (\mathscr{H}'(\Omega))'$ of $\mathscr{H}'(\Omega)$. Montel's theorem 2.2.8 has as a consequence that $\mathscr{H}''(\Omega)$ can be identified to $\mathscr{H}(\Omega)$. This is a classical result in the theory of topological vector spaces (cf. [Sch]). We will return to the study of $\mathscr{H}'(\Omega)$ in the sequel to this volume.

EXERCISES 2.2

($B = B(0, 1)$ and Ω is a domain in $\mathbb{C}$).

1. Let $f \in \mathscr{H}(B)$ be holomorphic and nonconstant, $\|f\|_\infty \le 1$. Show that $g(z) := \sum_{n=0}^{\infty} (f(z))^n$ is holomorphic in B. Is g bounded?
2. Show that the family $\mathscr{F} \subseteq \mathscr{H}(B), \mathscr{F} = \{f : \|f\|_{L^2} \le 1\}$ is a compact family in $\mathscr{H}(B)$.
3. Let $(f_n)_{n\ge1} \subseteq \mathscr{H}(B)$, $\|f_n\|_{L^\infty} \le 1$ for all n, and $\lim_{n\to\infty} f_n(z)$ exists for every $z \in B$. Show that $(f_n)_{n\ge1}$ has a limit in $\mathscr{H}(B)$. Show that it is enough to assume that $\lim_{n\to\infty} f_n(z)$ exists for $z \in A$, where A is a set with an accumulation point $z_0 \in B$.

*4. Let T be an analytic functional in $\mathscr{H}'(\mathbb{C})$ (cf. §2.2.11), let $a_n = \langle T, z^n \rangle$, $n \ge 0$. Show that
   (a) The function f defined by
$$f(\zeta) := \sum_{n\ge0} \frac{a_n}{n!} \zeta^n$$
   is an entire function which satisfies for every $\zeta \in \mathbb{C}$,
$$|f(\zeta)| \le Ae^{B|\zeta|}$$
   for some constants $A, B > 0$.
   (b) Let $\varphi \in \mathscr{H}(\mathbb{C})$ be arbitrary, $\varphi(z) = \sum_{n\ge0} b_n z^n$, then
$$\langle T, \varphi \rangle = \sum_{n\ge0} a_n b_n.$$
   Formally this can be interpreted as showing that
$$T = \sum_{n\ge0} \frac{a_n}{n!} \frac{d^n}{dz^n}\bigg|_{z=0} = \sum_{n\ge0} (-1)^n \frac{a_n}{n!} \delta_0^{(n)},$$
   where δ_0 is the Dirac mass at $z = 0$.
   (c) Can you find a function Φ defined in $\mathbb{C} \times \mathbb{C}$ which is entire holomorphic in each variable separately, and such that for fixed $\zeta \in \mathbb{C}$
$$\langle T, \Phi(\cdot, \zeta) \rangle = f(\zeta)$$
   with f as in part (a)?

5. Let $0 < \lambda_1 < \lambda_2 < \cdots$ be a sequence such that
$$0 < \alpha = \limsup_{n\to\infty} \frac{n}{\lambda_n} < \infty,$$
   and let $\{a_n\}_{n\ge1}$ be a sequence of complex numbers such that
$$1 < \rho = \limsup_{n\to\infty} \sqrt[n]{|a_n|} < \infty.$$
   Show that the series

$$\sum_{n\geq 1} a_n e^{-\lambda_n z}$$

converges to a function holomorphic in the half-plane $\operatorname{Re} z > \alpha \log \rho$.

6. Show that the following series of holomorphic functions converge to a holomorphic function in the indicated domains:

(a) $\displaystyle\sum_{n\geq 0} \frac{\cos nz}{n!} \quad (z \in \mathbb{C});$

(b) $\displaystyle\sum_{n\geq 1} \frac{1}{n(z-n)} \quad (z \notin \mathbb{N}^*);$

(c) $\displaystyle\sum_{n\geq 1} \frac{n!}{n^n} \sin nz \quad (|\operatorname{Im} z| < 1)$ (use Stirling's formula for $n!$);

(d) $\displaystyle\sum_{n\geq 0} e^{-z^2\sqrt{n}} \quad \left(|\operatorname{Arg} z| < \frac{\pi}{4}\right).$

7. Show that the function $z \mapsto \displaystyle\int_0^\infty \frac{t \sin t}{t^2 + z^2}\, dt$ is holomorphic when $\operatorname{Re} z > 0$. (Note the integral has to be considered as an improper integral.)

8. (Abel summation theorem).
   (i) Let $u_1, \ldots, u_n, v_1, \ldots, v_n$ be a collection of $2n$ (not necessarily distinct) complex numbers. Let $s_k = u_1 + \cdots + u_k$ $(1 \leq k \leq n)$. Show that
   $$\sum_{k=1}^{n} u_k v_k = s_n v_n + \sum_{k=1}^{n-1} s_k (v_k - v_{k+1}).$$
   (ii) Show that if $\sum_{n\geq 1} u_n$ is a convergent numerical series and $v_n(z)$ is a sequence of functions on a set K such that $|v_1(z)| + \sum_{n=1}^{\infty} |v_n(z) - v_{n+1}(z)|$ is uniformly bounded on K, then the series $\sum_{n\geq 1} u_n v_n(z)$ is uniformly convergent on K.
   (iii) Consider now a convergent numerical series $A = \sum_{n\geq 0} a_n$. Show that the function $f(z) = \sum_{n\geq 0} a_n z^n$ is holomorphic in B and, moreover, for any $\alpha > 0$ if $z \in B$ remains in the angle $|1 - z| \leq \alpha(1 - |z|)$, the limit $\lim_{|z|\to 1} f(z) = A$ is uniform.

9. Suppose $\mathscr{F}$ is a relatively compact family in $\mathscr{H}(\Omega)$ such that for some open set D one has $f(\Omega) \subseteq D$ for every $f \in \mathscr{F}$. Let $g \in \mathscr{H}(D)$, which is bounded on bounded sets. Show that the family of $\{g \circ f : f \in \mathscr{F}\}$ is also relatively compact in $\mathscr{H}(\Omega)$.

10. Let $\mathscr{F} \subseteq \mathscr{H}(\Omega)$, suppose that $\mathscr{F}$ is a bounded family, show that $\mathscr{F}' = \{f' : f \in \mathscr{F}\}$ is also a bounded family in $\mathscr{H}(\Omega)$. Is the converse true?

11. Let $(f_n)_{n\geq 1} \subseteq \mathscr{H}(B)$, $f_n(0) = 0$. Suppose $\operatorname{Re} f_n$ converges to zero locally uniformly. Show that $f_n \to 0$ in $\mathscr{H}(B)$ (Hint: consider $g_n = e^{f_n}$.)

12. Let Ω be a bounded domain in $\mathbb{C}$ and $\mathscr{F} = \{f \in \mathscr{H}(\Omega) : f(\Omega) \subseteq \Omega\}$.
    (a) Show that $\mathscr{F}$ is a relatively compact family in $\mathscr{H}(\Omega)$ and find its closure $\bar{\mathscr{F}}$.

(b) Show that under the operation of composition of maps, $\mathscr{F}$ is a semigroup.
(c) Show that if $(f_n)_{n\geq 1}$, $(g_n)_{n\geq 1}$ are sequences in $\mathscr{F}$, $f_n \to f$, $g_n \to g$ in $\mathscr{H}(\Omega)$ and f, $g \in \mathscr{F}$, then $f_n \circ g_n \to f \circ g$ in $\mathscr{H}(\Omega)$.

§3. Holomorphic Maps

A holomorphic function $f = u + iv \in \mathscr{H}(\Omega)$ can also be interpreted as a transformation from Ω into $\mathbb{C}$. It is evidently a C^∞ transformation. One can compute the Jacobian matrix $D(f)$ of the transformation

$$\Omega \xrightarrow{f} \mathbb{R}^2$$

$$(x, y) \longmapsto (u, v)$$

$$(z = x + iy \longmapsto w = f(z) = u + iv)$$

$$D(f) = \frac{\partial(u, v)}{\partial(x, y)} = \begin{pmatrix} \dfrac{\partial u}{\partial x} & \dfrac{\partial u}{\partial y} \\ \dfrac{\partial v}{\partial x} & \dfrac{\partial v}{\partial y} \end{pmatrix}.$$

The Jacobian $J(f) = \det D(f)$ of this transformation equals $|f'|^2$, hence it is greater than or equal to 0 and f preserves the orientation. This is a direct consequence of the Cauchy-Riemann equations but it can also be seen as follows.

From $dw \wedge d\bar{w} = -2i\, du \wedge dv$, $dz \wedge d\bar{z} = -2i\, dx \wedge dy$ and the relation

$$f^*(du \wedge dv) = J(f)\, dx \wedge dy,$$

one deduces

$$J(f)\, dx \wedge dy = f^*\left(\frac{i}{2} dw \wedge d\bar{w}\right) = \frac{i}{2} f'\, dz \wedge \overline{f'}\, d\bar{z} = |f'|^2\, dx \wedge dy,$$

hence,

$$J(f) = |f'|^2.$$

One concludes from the inverse function theorem that if $f'(z_0) \neq 0$ then f is locally invertible at the point z_0 as a C^∞ map. Before proving that this local inverse is also a holomorphic map, let us reobtain a result already mentioned.

2.3.1. Proposition (Composition of Holomorphic Maps). *Let Ω_1, Ω_2 be two open sets in $\mathbb{C}$, $f \in \mathscr{H}(\Omega_1)$ taking values in Ω_2 and $g \in \mathscr{H}(\Omega_2)$. The function $h = g \circ f$ is holomorphic in Ω_1 and*

$$h' = (g' \circ f)f'.$$

PROOF. We will prove first:

2.3.2. Lemma. *Let $\varphi : \Omega_1 \to \Omega_2$ be a C^1 map, φ is holomorphic if and only if the inverse image of every form $A\,dw$ of type $(1,0)$ (in w) is a form of type $(1,0)$ (in z).*

PROOF. For a C^1 map φ we have

$$\varphi^*(A\,dw) = (A \circ \varphi)\,d\varphi = (A \circ \varphi)\left\{\frac{\partial\varphi}{\partial z}dz + \frac{\partial\varphi}{\partial\bar{z}}d\bar{z}\right\}.$$

If φ is holomorphic, then $\dfrac{\partial\varphi}{\partial\bar{z}} = 0$ and hence

$$\varphi^*(A\,dw) = (A \circ \varphi)\frac{\partial\varphi}{\partial z}dz = (A \circ \varphi)\varphi'\,dz,$$

which is clearly of type $(1,0)$.

Conversely, if $\varphi^*(A\,dw)$ is of type $(1,0)$ for every C^∞ function A, then necessarily $\dfrac{\partial\varphi}{\partial\bar{z}} = 0$ and φ is holomorphic. □

2.3.3. Remark. If φ is holomorphic one has

$$\partial \circ \varphi^* = \varphi^* \circ \partial \quad\text{and}\quad \bar{\partial} \circ \varphi^* = \varphi^* \circ \bar{\partial},$$

so that φ^* preserves also the forms of type $(0,1)$.

PROOF OF 2.3.1. Since $h^* = (g \circ f)^* = f^* \circ g^*$, and both f^* and g^* preserve the $(1,0)$ forms, then h^* also preserves the $(1,0)$ forms and h is holomorphic by Lemma 2.3.2. Besides, using $d(g \circ f) = f^*(dg)$, one sees that

$$h' = \frac{\partial h}{\partial z} = \frac{\partial}{\partial z}(g \circ f) = \left(\frac{\partial g}{\partial w} \circ f\right)\frac{\partial f}{\partial z} = (g' \circ f)f',$$

which concludes the proof. □

2.3.4. Corollaries

(1) *If $f : \Omega_1 \to \Omega_2$ is a holomorphic map which is also a C^1-diffeomorphism, then the inverse transformation f^{-1} is holomorphic.*

(2) *If $g : \Omega_1 \to \Omega_2$ is a holomorphic map such that for some $z_0 \in \Omega_1$, $g'(z_0) \neq 0$, the local inverse is holomorphic in a neighborhood of $g(z_0)$.*

PROOF. They are an easy consequence of Lemma 2.3.2; we encourage the reader to prove them in that framework. A direct proof of (2) can be obtained as follows.

Let $f : \Omega \to \mathbb{C}$ be holomorphic and such that $f'(z) \neq 0$ for every $z \in \Omega$. If we fix $z_0 \in \Omega$ and consider z, ζ near z_0 we obtain

$$f(z) = f(z_0) + f'(z_0)(z - z_0) + O((z - z_0)^2)$$
$$f(\zeta) = f(z_0) + f'(z_0)(\zeta - z_0) + O((\zeta - z_0)^2),$$

hence,

$$f(z) - f(\zeta) = f'(z_0)(z - \zeta) + (z - \zeta)O(|z - z_0| + |\zeta - z_0|).$$

Therefore,

$$|f(z) - f(\zeta)| \geq \frac{1}{2}|f'(z_0)||z - \zeta|$$

if both z and ζ are close to z_0. Hence f is injective near z_0 and the inverse f^{-1} can be defined near $f(z_0)$, which is continuous since f is an open map.

Let now $w = f(z)$, $\omega = f(\zeta)$ for z, ζ near z_0, then

$$\frac{f^{-1}(w) - f^{-1}(\omega)}{w - \omega} = \frac{z - \zeta}{f(z) - f(\zeta)} \to \frac{1}{f'(\zeta)} \qquad \text{when } w \to \omega,$$

which shows that f^{-1} is holomorphic with derivative $1/f'(\zeta)$ at ω. □

2.3.5. Remark. The inverse of a differentiable homeomorphism is not differentiable in general as shown by the example $x \mapsto x^3$ from $\mathbb{R}$ to $\mathbb{R}$. Nevertheless, for holomorphic maps we have the following.

2.3.6. Proposition. *A bijective holomorphic map $f: \Omega_1 \to \Omega_2$ has a holomorphic inverse.*

PROOF. We already know f is a homeomorphism since f is an open map. It suffices to prove that f' does not vanish. Let us assume that for some $z_0 \in \Omega_1$ we have $f'(z_0) = 0$. We might as well suppose that $z_0 = f(z_0) = 0$.

There is then an integer $k \geq 2$, a function g holomorphic near 0, $g(0) = 1$, and a number $c \in \mathbb{C}\setminus\{0\}$ such that

$$f(z) = c^k z^k g(z).$$

There is a $\delta > 0$ such that $|g(z) - 1| < 1$ if $|z| < \delta$. The function $w \mapsto w^{1/k} = e^{1/k \log w}$ is holomorphic in $B(1, 1)$. By composition, the function $h(z) = (g(z))^{1/k}$ is holomorphic in $|z| < \delta$. This implies that f cannot be injective in a neighborhood of zero. In fact, the map $\varphi: z \mapsto czh(z)$ is a local diffeomorphism about $z = 0$ by §2.3.4 since its derivative at the origin is c. Hence, there is an open neighborhood V ($V \subseteq \{|z| < \delta\}$) of the origin and $\varepsilon > 0$ such that $\varphi: V \to B(0, \varepsilon)$ is bijective. In V, $f = \varphi^k$, and since $w \mapsto w^k$ is a k-to-1 map on any circle centered at the origin, f cannot be injective. □

2.3.7. Corollary. *Let f be a nonconstant holomorphic function in a connected open set Ω, $z_0 \in \Omega$, and $f(z_0) = w_0$. There is an integer $k \geq 1$ and a holomorphic* *function φ defined in a neighborhood of z_0 such that $\varphi(z_0) = 0$, $\varphi'(z_0) \neq 0$, and*

$$f(z) = w_0 + (\varphi(z))^k.$$

PROOF. It is a consequence of the proof of §2.3.6. □

2.3.8. Remarks. (1) From the local representation 2.3.7, one can give a new proof of the fact that a nonconstant holomorphic map is an open map. This is left to the reader as an exercise.

(2) Let Ω be open and connected, $f : \Omega \to \mathbb{C}$ holomorphic and nonconstant. The set U of points in Ω where f is locally injective is open and its complementary $D = \Omega \setminus U$ is a discrete set. Namely, D is the set of zeros of f'. A point $z_0 \in \Omega$ is in U if and only if the integer k in §2.3.7 equals 1.

For a point $z_0 \in \Omega$ there is a neighborhood W of $f(z_0) = w_0$, and a neighborhood V of z_0 such that $f(V) = W$, $f^{-1}(w_0) \cap V = \{z_0\}$, and, for every $\zeta \in W \setminus \{w_0\}$, the equation $f(z) = \zeta$ has exactly k distinct solutions in $V \setminus \{z_0\}$.

The integer k is sometimes called the ***ramification index*** (or ***branching order***) of f at z_0. If $w_0 = 0$ it is also called the ***multiplicity*** of z_0 as a zero of f. The proof of characterizes k as the smallest positive integer such that $f^{(k)}(z_0) \neq 0$.

2.3.9. Definition. A holomorphic map $f : \Omega_1 \to \Omega_2$ between two open subsets of $\mathbb{C}$ which is bijective is called a ***biholomorphism*** or a ***conformal map***.

If $\Omega_1 = \Omega_2$, it is customary to say that f is an ***automorphism*** of Ω_1. The collection of all automorphisms of an open set Ω is a group under composition denoted $\mathrm{Aut}(\Omega)$.

We would like to explain here where the name "conformal map" comes from. Given two vectors $z_1, z_2 \in \mathbb{C}^*$, let us recall that the oriented angle $\theta = \theta(z_1, z_2)$ between them is defined to be the value of θ, $0 \leq \theta < 2\pi$, such that

$$e^{i\theta} = \frac{z_2 \bar{z}_1}{|z_2||z_1|}.$$

Given a differentiable map $\varphi : \Omega_1 \to \Omega_2$ and a C^1 curve $\gamma :]-\varepsilon, \varepsilon[\to \Omega_1$, recall that the tangent vector $\dot{\alpha}(0)$ of the curve $\alpha = \varphi \circ \gamma$ at $\alpha(0)$ is given by

$$\dot{\alpha}(0) = D(\varphi)(\gamma(0))\dot{\gamma}(0) = \frac{\partial \varphi}{\partial z}(\gamma(0))\dot{\gamma}(0) + \frac{\partial \varphi}{\partial \bar{z}}(\gamma(0))\overline{\dot{\gamma}(0)},$$

where $\dot{\gamma}(0)$ is the tangent vector to the curve γ at $\gamma(0)$.

Let now γ_1, γ_2 be two curves of class C^1 in Ω_1 such that $\gamma_1(0) = \gamma_2(0) = z_0 \in \Omega_1$, and let $\alpha_1 = \varphi \circ \gamma_1$, $\alpha_2 = \varphi \circ \gamma_2$.

Assume φ is holomorphic in a neighborhood of z_0 and $\varphi'(z_0) \neq 0$. If $\theta = \theta(\dot{\gamma}_1(0), \dot{\gamma}_2(0))$ and $\tilde{\theta} = \theta(\dot{\alpha}_1(0), \dot{\alpha}_2(0))$ then we have

$$e^{i\tilde{\theta}} = \frac{\dot{\alpha}_2(0)\overline{\dot{\alpha}_1(0)}}{|\dot{\alpha}_2(0)||\dot{\alpha}_1(0)|} = \frac{\varphi'(z_0)\dot{\gamma}_2(0)}{|\varphi'(z_0)\dot{\gamma}_2(0)|} \frac{\overline{\varphi'(z_0)\dot{\gamma}_1(0)}}{|\varphi'(z_0)\dot{\gamma}_1(0)|} = e^{i\theta}.$$

Therefore $\tilde{\theta} = \theta$ and we see that φ preserves the oriented angles between the two tangent vectors at z_0.

Conversely, suppose the differentiable map φ preserves the oriented angles between any two tangent vectors at z_0 and $J(\varphi)(z_0) \neq 0$. Consider the two curves

$$\gamma_1(t) = z_0 + te^{i\theta_1}, \qquad \gamma_2(t) = z_0 + te^{i\theta_2},$$

$0 \le \theta_j < 2\pi$. For them $\dot{\gamma}_1(0) = e^{i\theta_1}$ and $\dot{\gamma}_2(0) = e^{i\theta_2}$. Also,

$$\dot{\alpha}_1(0) = e^{i\theta_1}\frac{\partial \varphi}{\partial z}(z_0) + e^{-i\theta_1}\frac{\partial \varphi}{\partial \bar{z}}(z_0)$$

$$\dot{\alpha}_2(0) = e^{i\theta_2}\frac{\partial \varphi}{\partial z}(z_0) + e^{-i\theta_2}\frac{\partial \varphi}{\partial \bar{z}}(z_0).$$

The condition $e^{i\tilde{\theta}} = e^{i\theta}$ implies that the quadratic form

$$Z(e^{i\theta_1}, e^{i\theta_2}) = \left(\frac{\partial \varphi}{\partial z}(z_0) + \frac{\partial \varphi}{\partial \bar{z}}(z_0)e^{-2i\theta_2}\right)\left(\frac{\overline{\partial \varphi}}{\partial z}(z_0) + \frac{\overline{\partial \varphi}}{\partial \bar{z}}(z_0)e^{-2i\theta_1}\right) > 0$$

for all choices of θ_1, θ_2. This immediately implies that $\dfrac{\partial \varphi}{\partial \bar{z}}(z_0) = 0$. If not, taking $\theta_1 = 0$ and varying θ_2 one sees that the point $Z(1, e^{i\theta_2})$ describes a circle of positive radius.

We have therefore shown the following.

2.3.10. Proposition. *Let $\varphi : \Omega_1 \to \Omega_2$ be a differentiable map with nonvanishing Jacobian $J(\varphi)$. Then φ is holomorphic if and only if φ preserves the oriented angles.*

2.3.11. Remarks

(1) Let φ be holomorphic in a neighborhood of z_0, if $\varphi'(z_0) = 0$, then the angles between tangent vectors are *not* preserved but amplified by the factor $k \ge 2$, ramification index of φ at z_0.

(2) There are differentiable maps φ which are not holomorphic but $J(\varphi)(z_0) = 0$ and φ preserves the angles between tangent vectors at z_0. For example, $z_0 = 0$ and $\varphi(z) = z|z|$. This transformation preserves lines in a wide sense. For instance, the image of a ray through the origin is exactly the same ray. Only the parameterization changes.

(3) A useful corollary of Proposition 2.3.10 is the following: Let $\varphi : \Omega_1 \to \Omega_2$ be a differentiable homeomorphism between two open subsets of $\mathbb{C}$, then φ is a biholomorphic map if and only if $J(\varphi)$ never vanishes and φ preserves the oriented angles.

2.3.12. Examples. (1) A transformation σ of $S^2 = \mathbb{C} \cup \{\infty\}$ of the form

$$\sigma(z) = \frac{az + b}{cz + d},$$

with $ad - bc \neq 0$ is called a **Moebius transformation**. If $c = 0$ it is a **similarity** (an invertible affine transformation of $\mathbb{C}$ onto itself) and we have $\sigma(\infty) = \infty$. If $c \neq 0$ we have $\sigma(-d/c) = \infty$ and $\sigma(\infty) = a/c$. If σ is a Moebius transformation, the new transformation σ^{-1} defined by $\sigma^{-1}(z) = \dfrac{dz - b}{-cz + a}$, is also a Moebius transformation and $\sigma(\sigma^{-1}(z)) = \sigma^{-1}(\sigma(z)) = z$. If σ and τ are two Moebius transformations one can see without difficulty that the same holds for the composition $\sigma \circ \tau$. Hence the family $\mathscr{M}$ of Moebius transformations is a group. It is clear that if $\lambda \in \mathbb{C}^*$ and σ is the transformation just described it is also true that $\sigma(z) = \dfrac{\lambda az + \lambda b}{\lambda cz + \lambda d}$, therefore the coefficients a, b, c, d are not uniquely determined. But this is really the extent of the indeterminacy, one can say this more precisely as follows.

Let $GL(2, \mathbb{C})$ be the group of invertible 2×2 matrices. The map $\Phi : GL(2, \mathbb{C}) \to \mathscr{M}$ given by $\Phi\left(\begin{pmatrix} a & b \\ c & d \end{pmatrix}\right) = \sigma$, where σ is the transformation just described, is a group homomorphism whose kernel is the subgroup of $GL(2, \mathbb{C})$ of diagonal matrices.

(2) If $a \in \mathbb{C}$, $|a| < 1$, consider the Moebius transformation

$$\varphi_a(z) = \frac{z - a}{1 - \bar{a}z}.$$

Note that φ_a is holomorphic in the disk $B(0, 1/|a|)$ and that $\varphi_a^{-1} = \varphi_{-a}$. We claim that φ_a is an automorphism of $B(0, 1)$. It is enough to show that $|\varphi_a(z)| < 1$ if $|z| < 1$. But the condition $|\varphi_a(z)| < 1$ is equivalent to $|\varphi_a(z)|^2 < 1$, which itself is equivalent to

$$|z|^2 + |a|^2 < 1 + |a|^2|z|^2.$$

This last inequality is a consequence of the identity

$$|z|^2 + |a|^2 - |a|^2|z|^2 - 1 = (|z|^2 - 1)(1 - |a|^2) < 0.$$

It is also easy to see directly that $\varphi_a(\partial B(0, 1)) = \partial B(0, 1)$; it is only necessary to observe that for $\theta \in \mathbb{R}$

$$|\varphi_a(e^{i\theta})| = \frac{|e^{i\theta} - a|}{|1 - \bar{a}e^{i\theta}|} = \frac{|e^{i\theta} - a|}{|e^{-i\theta} - \bar{a}|} = 1.$$

Let us note also that $\varphi_a(0) = -a$, $\varphi_a(a) = 0$, $\varphi_a'(0) = 1 - |a|^2$ and $\varphi_a'(a) = 1/(1 - |a|^2)$. More generally, the following relation holds

$$|\varphi_a'(z)|(1 - |z|^2) = 1 - |\varphi_a(z)|^2.$$

Let now $f : B(0, 1) \to B(0, 1)$ be a holomorphic map. Let $a \in B(0, 1)$ and $b = f(a)$. Consider the holomorphic map

$$g := \varphi_b \circ f \circ \varphi_{-a} : B(0, 1) \to B(0, 1)$$

which satisfies $g(0) = 0$. By Schwarz's lemma we have $|g'(0)| < 1$, unless $g(z) = Cz$, $|C| = 1$. Computing this derivative we find

$$g'(0) = (\varphi_b \circ f)'(\varphi_{-a}(0)) \cdot \varphi'_{-a}(0) = (\varphi_b \circ f)'(a) \cdot (1 - |a|^2)$$

$$= \varphi'_b(b) f'(a)(1 - |a|^2) = \frac{1 - |a|^2}{1 - |b|^2} f'(a).$$

Therefore,

$$|f'(a)| \leq \frac{1 - |b|^2}{1 - |a|^2} = \frac{1 - |f(a)|^2}{1 - |a|^2},$$

and the first inequality is an equality only if $|g'(0)| = 1$, in which case, we will have

$$f(z) = \varphi_{-b}(C\varphi_a(z)), \qquad |z| < 1.$$

Since $|C| = 1$, this says that f is an automorphism of $B(0, 1)$ (as a composition of three automorphisms: φ_a, multiplication by C, φ_{-b}). One can find by direct computation that f is then of the form $\tilde{C}\varphi_{\tilde{a}}$ for some $|\tilde{a}| < 1$, $|\tilde{C}| = 1$. In particular, we can show that every biholomorphism f of $B(0, 1)$ is of the form $f = C\varphi_a$ for a convenient constant C, $|C| = 1$ and point $a \in B(0, 1)$. In fact, let $a = f^{-1}(0)$, then $g = f \circ \varphi_{-a}$ is another biholomorphic map of $B(0, 1)$ with $g(0) = 0$. If h is its inverse, then by Schwarz's lemma $|g'(0)| \leq 1$ and $|h'(0)| \leq 1$ but $g'(0)h'(0) = 1$, hence $g(z) = Cz$ and, hence $f = C\varphi_a$.

Summarizing, we have shown that every holomorphic map of $B(0, 1)$ into itself satisfies

$$|f'(z)| \leq \frac{1 - |f(z)|^2}{1 - |z|^2} \qquad (|z| < 1)$$

and that equality at a single point occurs if and only if f is a biholomorphic map, hence of the form $C\varphi_a(|a| < 1, |C| = 1)$. In this last case equality holds everywhere.

Exercises 2.3

Ω is an open connected set in $\mathbb{C}$, $B = B(0, 1)$.

1. Let us recall that the explicit identification of unit sphere $S^2 = \{p = (x_1, x_2, x_3) \in \mathbb{R}^3 : x_1^2 + x_2^2 + x_3^2 = 1\}$ with $\mathbb{C} \cup \{\infty\}$ is given by the ***stereographic projection:*** $\mathbb{C}$ is considered as the plane $x_3 = 0$. The north pole $N = (0, 0, 1)$ is identified to ∞ and a point $z = x + iy \in \mathbb{C}$ corresponds to the unique point $p \in S^2$ such that the straight line through N and p passes also through $(x, y, 0)$. The ***chordal distance*** of two points $z_1, z_2 \in \mathbb{C}$ corresponding to $p_1, p_2 \in S^2$ is $\sigma(z_1, z_2) = |p_1 - p_2| =$ Euclidean distance in $\mathbb{R}^3$. Show that:

(a) $$x_1 = \frac{z + \bar{z}}{1 + |z|^2}, \qquad x_2 = \frac{z - \bar{z}}{i(1 + |z|^2)}, \qquad x_3 = \frac{|z|^2 - 1}{1 + |z|^2}$$

and

$$z = \frac{x_1 + ix_2}{1 - x_3}.$$

What is the image of a circle in S^2 under stereographic projection?

(b) The topology in S^2 induced by the chordal distance and that considered in the text coincide. Moreover,

$$\sigma(z_1, z_2) = \frac{2|z_1 - z_2|}{((1 + |z_1|^2)(1 + |z_2|^2))^{1/2}}.$$

Find $\sigma(z, \infty)$ for $z \in \mathbb{C}$. Show also that $\sigma\left(\frac{1}{z_1}, \frac{1}{z_2}\right) = \sigma(z_1, z_2)$, for every pair z_1, $z_2 \in \mathbb{C}^*$.

(c) Let Γ be a piecewise-C^1 path in S^2 and γ the piecewise C^1-path in $\mathbb{C} \cup \{\infty\}$ obtained by stereographic projection. Then, if we consider the length of Γ in the metric induced by σ, we have

$$\text{length}(\Gamma) = \int_\gamma \frac{2}{1 + |z|^2} |dz|.$$

This length induces the usual metric on the sphere S^2.

The aim of the following questions is to find out what kind of transformation of $\mathbb{C} \cup \{\infty\}$ corresponds to a rotation of S^2 under stereographic projection.

(d) Show that every rotation of $\mathbb{R}^3$ admits 1 as an eigenvalue, hence a fixed axis.

(e) Show that every rotation is the composition of rotations about the coordinate axes. Show that in the complex plane, these rotations correspond via stereographic projection to a Moebius transformation.

(f) Compute the Moebius transformation corresponding to the matrix $\begin{pmatrix} 1 & 0 & 0 \\ 0 & -1 & 0 \\ 0 & 0 & -1 \end{pmatrix}$.

*(g) Does every Moebius transformation correspond to a rotation of S^2? (Show that only those $z \mapsto \dfrac{az + b}{cz + d}$ such that $|a|^2 + |b|^2 = 1, c = -b, d = \bar{a}$ have this property. The corresponding group of matrices is denoted $PSU(2, 2)$, cf. [JS].)

2. Let $f \in \mathscr{H}(\Omega)$ be an isometry for the chordal distance, that is, $\sigma(f(z_1), f(z_2)) = \sigma(z_1, z_2)$ for every $z_1, z_2 \in \Omega$. We want to find f.

(i) Show that f satisfies the differential equation

$$|f'(z)|^2 = \left(\frac{1 + |f(z)|^2}{1 + |z|^2}\right)^2.$$

(ii) Show that by composition with Moebius transformations one can assume $0 \in \Omega$, $f(0) = 0$, $f'(0) = 1$.

(iii) Show that under the conditions of (ii) one has $|f(z)| = |z|$ in a neighborhood of 0. Conclude that $f(z) = z$ in this case.

(iv) (Alternate proof of (iii)). Differentiating the differential equation obtained in (i) with respect to $\dfrac{\partial}{\partial z}$, show that $f^{(n)}(0) = 0$ for $n \geq 2$. Conclude that every local holomorphic isometry for the chordal metric is a Moebius transformation. (In fact, it is possible to show that if f preserves the metric (locally) and the orientation, it is differentiable and conformal, hence f is automatically holomorphic.)

3. Let $f \in \mathscr{H}(\Omega)$. Show that, if σ is the chordal distance,

   (a) $f^{\#}(z) := \lim_{\zeta \to z} \dfrac{\sigma(f(\zeta), f(z))}{|\zeta - z|} = \dfrac{2|f'(z)|}{1 + |f(z)|^2}$.

   (b) If $f(z) \neq 0$, show that $f^{\#}(z) = \left(\dfrac{1}{f}\right)^{\#}(z)$.

   (c) The length in S^2 (in the metric induced by the chordal distance) of $f \circ \gamma$, γ piecewise-C^1 path in Ω, is
   $$\int_\gamma f^{\#}(z)|dz|.$$
   The function $f^{\#}$ is called the ***spherical derivative*** of f.

   (d) Let γ be a piecewise-C^1 path in Ω, starting at a and ending at b. Then
   $$\sigma(f(a), f(b)) \leq \int_\gamma f^{\#}(z)|dz|.$$

4. (a) Show that any Moebius transformation can be obtained as a composition of the following three kinds of simpler Moebius transformations: (i) translations: $z \mapsto z + a$, $a \in \mathbb{C}$; (ii) dilations: $z \mapsto bz$, $b \in \mathbb{C}^*$, and (iii) inversions: $z \mapsto \dfrac{1}{z}$.

   (b) What does this say about the group $GL(2, \mathbb{C})$?

   (c) Conclude that any Moebius transformation preserves the family composed of all circles and lines in $\mathbb{C}$ (you can think of the lines as being circles through $\infty \in S^2$).

   (d) Show that a Moebius transformation is determined by the image of three different points.

   (e) The quantity
   $$(z_1, z_2, z_3, z_4) := \left(\frac{z_1 - z_3}{z_1 - z_4}\right) \Big/ \left(\frac{z_2 - z_3}{z_2 - z_4}\right)$$
   is called the ***cross ratio*** of the four points $z_1, z_2, z_3, z_4 \in \mathbb{C}$ when at least three of them are different. If one of them is ∞ it can be defined by limits, e.g.,
   $$(\infty, z_2, z_3, z_4) = \frac{z_2 - z_4}{z_2 - z_3}.$$
   Show that Moebius transformations preserve the cross ratio. Moreover, show that if φ is a Moebius transformation and $z_2 = \varphi^{-1}(1)$, $z_3 = \varphi^{-1}(0)$, $z_4 = \varphi^{-1}(\infty)$, then
   $$\varphi(z) = (z, z_2, z_3, z_4).$$
   Conclude that given two triples of distinct points $\{z_1, z_2, z_3\}$ and $\{w_1, w_2, w_3\}$ in S^2, there is a Moebius transformation ψ such that $\psi(z_j) = w_j$ $(j = 1, 2, 3)$.

   (f) Show that the cross ratio $(z_1, z_2, z_3, z_4) \in \mathbb{R}$ if and only if the four points lie on a circle or on a line.

5. Show that the Moebius transformations which map the unit disk onto itself can also be identified to the subgroup $SU(1,1) := \left\{\begin{pmatrix} a & b \\ \bar{b} & \bar{a} \end{pmatrix} := |a|^2 - |b|^2 = 1\right\}$ of $GL(2,\mathbb{C})$. More precisely, $SU(1,1)/\{\pm I\}$, I = identity matrix. Conclude that $\operatorname{Aut}(B)$ is isomorphic to $SU(1,1)/\{\pm I\}$.

6. Show that if φ is a Moebius transformation of B onto itself, it preserves the ***hyperbolic*** length of curves, defined by

$$L(\gamma) := \int_\gamma \frac{|dz|}{1 - |z|^2},$$

where γ is a piecewise-C^1 path in B. It also preserves the hyperbolic distance between points

$$\rho(z_1, z_2) := \operatorname{arctanh}\left(\frac{|z_1 - z_2|}{|1 - z_1\bar{z}_2|}\right).$$

Taking for granted that the geodesic, with respect to this metric, joining the origin to the point r in the segment $]0,1[$ is the segment $[0,r]$, find all geodesics.

7. Let $\Omega \subseteq B$, $f \in \mathscr{H}(\Omega)$, $f(\Omega) \subseteq B$. Assume f is an isometry for the hyperbolic metric. Show that f is a Moebius transformation from B onto B. (Hint: Compare with Exercise 2.3.2).

8. Let $f \in \mathscr{H}(B)$, $f(B) \subseteq B$, $f(z) \not\equiv z$. Show that it can have at most one fixed point.

9. Let $f: B \to B$ be holomorphic; let $m \in \mathbb{N}^*$ be the multiplicity of the origin as a zero of f. Show that

$$|f(z)| \le |z|^m \qquad (z \in B).$$

10. Let $f: B \to B$ be holomorphic and vanish at $z_1, \ldots, z_m$. Show that

$$|f(0)| \le |z_1 \ldots z_m|.$$

*11. Suppose that $f \in \mathscr{H}(B)$ is injective, $f(0) = 0$, $|f'(0)| \le 1$, and $f(B) \supseteq B$. Show that $f(z) = cz$, for some constant c, $|c| = 1$.

*12. Let $a \in B$, $\psi(z) = z\varphi_a(z) = z(z-a)/(1 - \bar{a}z)$. Consider the sequence of functions $\psi^{[n]} = \psi \circ \cdots \circ \psi$. Show that
   (i) if $z \in B$ then $\psi^{[n]}(z) \to 0$ as $n \to \infty$ (uniformly on compacts of B);
   (ii) if $z \in \bar{B}^c$ then $\psi^{[n]}(z) \to \infty$ as $n \to \infty$ (uniformly on compacts of $\bar{B}^c$);
   (iii) if $|z| = 1$, then either $\psi^{[n]}(z)$ does not have a limit or the limit is the solution z_0 of $\varphi_a(z_0) = 1$.

13. Let $f \in \mathscr{H}(\Omega)$ be injective, $D = f(\Omega)$. Show that

$$\operatorname{area}(D) = \int_\Omega |f'(z)|^2 \, dx\, dy.$$

14. Let $0 < \alpha < 2$. Show that the map $z \mapsto z^\alpha$ is a biholomorphic mapping of the upper half plane H onto the angular region $\{z \in \mathbb{C} : 0 < \operatorname{Arg} z < \alpha\pi\}$.

15. (a) Show that the map $z \mapsto w = \dfrac{1}{i}\dfrac{z+1}{z-1}$ is a biholomorphism of B onto $H = \{w \in \mathbb{C} : \operatorname{Im} w > 0\}$.

(b) Use part (a) and Example 2.3.12(2) to show that every biholomorphism of H onto itself is given by a Moebius transformation

$$w \mapsto \frac{aw+b}{cw+d}, \qquad a, b, c, d \in \mathbb{R}, \qquad ad - bc = 1.$$

This identifies the group $\mathrm{Aut}(H)$ with $SL(2, \mathbb{R})/\{\pm I\}$, where $SL(2, \mathbb{R})$ is the group of 2×2 real matrices with determinant equal to 1.

(c) Show that if $f \in \mathrm{Aut}(H)$ fixes the points 0 and ∞ in S^2, then $f(w) = \alpha w$, $\alpha \in]0, \infty[$. When is $\alpha = 1$?

(d) Show that every conformal map of B onto H is a Moebius transformation.

(e) Show that if $f \in \mathrm{Aut}(H)$ is such that $w = 0$ is a double zero of $f(w) - w = 0$ then $f(w) = \dfrac{w}{cw+1}$, $c \in \mathbb{R}$. When is $c = 0$? What is the form of f when $w = \infty$ is a double fixed point?

16. (a) Let $\varphi \in \mathrm{Aut}(B)$. Show that φ has either exactly one or exactly two fixed points in $\bar{B}$ or $\varphi = id_B$. If the fixed point z_0 is in B, show $|\varphi'(z_0)| = 1$.

(b) If φ has exactly two fixed points, show there is a Moebius transformation ψ from B onto H so that $\psi \circ \varphi \circ \psi^{-1} \in \mathrm{Aut}(H)$ and fixes 0 and ∞.

(c) Show that if φ has exactly two fixed points then there is one of them, say z_0, such that the iterates $\varphi^{[n]}$ converge in $\mathscr{H}(B)$ to the constant function z_0.

(d) Show that if φ has exactly one fixed point z_0 in ∂B then the conclusion of (c) also holds. (Hint: cf. Exercise 2.3.15, (e).)

17. Find a biholomorphic map of B onto the first quadrant $\{z \in \mathbb{C} : \mathrm{Re}\, z > 0 \text{ and } \mathrm{Im}\, z > 0\}$.

18. Construct a biholomorphic mapping from $B_+ = \{z \in B : \mathrm{Im}\, z > 0\}$ onto B.

19. Construct a biholomorphic mapping from $\Omega = \{z \in \mathbb{C} : |z| > 1 \text{ and } \mathrm{Im}\, z > 0\}$ onto B.

20. Find a conformal map from the strip $S = \{z \in \mathbb{C} : 0 < \mathrm{Im}\, z < 1\}$ onto B.

21. Construct a conformal map of the half-strip $\Omega = \{z \in \mathbb{C} : \mathrm{Im}\, z > 0 \text{ and } 0 < \mathrm{Re}\, z < 1\}$ onto B.

22. Let $f \in \mathscr{H}(B(0, r))$ be such that

$$|f'(z) - f'(0)| < |f'(0)|$$

for every $z \in B(0, r)$. Show that f is injective.

§4. Isolated Singularities and Residues

2.4.1. Definition. Let f be a holomorphic function defined in an open set Ω of $\mathbb{C}$. A point $a \in \mathbb{C}$ will be called an ***isolated singularity*** of f if a is an isolated point of $\partial\Omega$.

It is the same as saying that there is $r > 0$ such that $B(a, r) \setminus \{a\} \subseteq \Omega$ and $a \notin \Omega$. Note that this concept is very dependent upon the set Ω where we consider the function f, for example, the function

$$f(z) = \frac{1}{z} \quad \text{in} \quad \mathbb{C}\backslash\{0\}$$

has an isolated singularity at $z = 0$. But the function

$$g(z) = \frac{1}{z} \quad \text{in} \quad \{\operatorname{Re} z > 0\}$$

does not have an isolated singularity at $z = 0$. This ambiguity will be removed later when we discuss analytic continuation.

Examples with the origin as an Isolated Singularity.

(1) $f_1(z) = \dfrac{1 - \cos z}{z}$ in $\mathbb{C}\backslash\{0\}$.

(2) $f_2(z) = \dfrac{1}{z}$ in $\mathbb{C}\backslash\{0\}$.

(3) $f_3(z) = e^{1/z}$ in $\mathbb{C}\backslash\{0\}$.

The reader should note that f_1 is bounded in a neighborhood of 0 and can be defined in the whole $\mathbb{C}$; $\lim_{z\to 0} |f_2(z)| = \infty$ and $\lim_{z\to\infty} |f_3(z)|$ does not exist. Moreover, for every $\rho > 0$ one can verify $f_3(B(0,\rho)\backslash\{0\}) = \mathbb{C}\backslash\{0\}$. We will see that the distinction between these three types is a general phenomenon.

Let us remark also that if f is not identically zero in a connected open set Ω and $f \in \mathscr{H}(\Omega)$, then the function $1/f$ is holomorphic in $\Omega\backslash Z(f)$, where $Z(f) = \{z \in \Omega : f(z) = 0\}$, and every point of $Z(f)$ is an isolated singularity of the second type for $1/f$ $\left(\text{i.e., } \lim_{z\to z_0} |1/f(z)| = \infty \text{ if } z_0 \in Z(f)\right)$.

2.4.2. Proposition. *A function f that is holomorphic and bounded in a punctured disk $B(a,r)\backslash\{a\}$ is the restriction of a unique function $\tilde{f}$, holomorphic in the disk $B(a,r)$.*

Proof. Since f is bounded, $f \in L^1_{\text{loc}}(\Omega)$, $\Omega = B(a,r)$. From §2.2.9(5) we conclude it suffices to prove that for every $\varphi \in \mathscr{D}(\Omega)$ one has

$$\int_\Omega f \frac{\partial \varphi}{\partial \bar{z}} \, dz \wedge d\bar{z} = 0,$$

because, in that case, there is $\tilde{f} \in \mathscr{H}(\Omega)$ such that $\tilde{f} = f$ a.e., hence $\tilde{f} \equiv f$ in $B(a,r)\backslash\{a\}$.

For $\varphi \in \mathscr{D}(\Omega)$ we have

$$\int_\Omega f \frac{\partial \varphi}{\partial \bar{z}} \, dz \wedge d\bar{z} = \lim_{\varepsilon\to 0} \int_{\varepsilon < |z-a| < r} f \frac{\partial \varphi}{\partial \bar{z}} \, dz \wedge d\bar{z}$$

$$= \lim_{\varepsilon\to 0} \int_{|z|=\varepsilon} f\varphi \, dz,$$

since $d(f\varphi\,dz) = \bar{\partial}(f\varphi\,dz) = -f\dfrac{\partial\varphi}{\partial\bar{z}}dz \wedge d\bar{z}$ and we can apply Stokes' formula. The functions $|f|$, $|\varphi|$ being bounded, say by M, we have

$$\left|\int_{|z|=\varepsilon} f\varphi\,dz\right| \le 2\pi\varepsilon M^2,$$

which ends the proof. □

A different way of proving the same statement is the following: The function $g(z) = (z-a)f(z)$ for $z \ne a$, $g(a) = 0$ is continuous in $B(a,r)$ and holomorphic in $B(a,r)\backslash\{a\}$. By one of the corollaries of Morera's theorem, g is holomorphic in $B(a,r)$. Since it vanishes at $z = a$ we know also that $g(z) = (z-a)\tilde{f}(z)$ for some $\tilde{f}$ holomorphic in the whole disk. Evidently $f = \tilde{f}$ in $B(a,r)\backslash\{a\}$.

2.4.3. Definition. An isolated singularity of $f \in \mathscr{H}(\Omega)$ such that f is the restriction of a function $\tilde{f}$ holomorphic in $\Omega \cup \{a\}$ is called a ***removable singularity***.

Therefore, if f is holomorphic and bounded in $B(a,r)\backslash\{a\}$, the singularity a is removable.

2.4.4. Proposition (Classification of Isolated Singularities). *Let f be holomorphic in $B(a,r)\backslash\{a\}$. Only the following three cases are possible*:

(1) *The singularity is removable and this holds if and only if*

$$\limsup_{z\to a} |f(z)| < \infty.$$

(2) *If* $\limsup_{z\to a}|f(z)| = +\infty$,

(i) *either* $\lim_{z\to a}|f(z)|$ *exists in* $\bar{\mathbb{R}}$ *and equals* $+\infty$, *or*

(ii) $\lim_{z\to a}|f(z)|$ *does not exist in* $\bar{\mathbb{R}}$ *and this holds if and only if* $f(B(a,\rho)\backslash\{a\})$ *is dense in* $\mathbb{C}$ *for every* ρ, $0 < \rho < r$.

This theorem is often referred to as the Casorati-Weierstrass theorem. A stronger version, Picard's theorem, will be proved in §2.7.

Proof. It is enough to prove the equivalence of the two conditions stated in (2), item (ii).

It is clear that if, for every $\rho \in]0,r[$, $f(B(a,\rho)\backslash\{a\})$ is dense in $\mathbb{C}$, then $\limsup_{z\to a}|f(z)| = +\infty$ and that $\lim_{z\to a}|f(z)|$ does not exist since, for instance $\liminf_{z\to a}|f(z)| = 0$.

Let us show that if the $\lim_{z\to a}|f(z)|$ does not exist in $\bar{\mathbb{R}}$ then the density condition is satisfied. Suppose there are ρ $(0 < \rho < r)$, $w \in \mathbb{C}$ and $\delta > 0$ such that $|f(z) - w| > \delta$ in $B(a,\rho)\backslash\{a\}$. Then, the function $g(z) = 1/(f(z) - w)$ is holomorphic and bounded in $B(a,\rho)\backslash\{a\}$. Its singularity at the point a will be removable and g extends to $\tilde{g}$, holomorphic and bounded in $B(a,\rho)$. It is not possible that $\tilde{g}(a) \neq 0$, otherwise f would be bounded in a neighborhood of a. On the other hand if $\tilde{g}(a) = 0$ then

$$\lim_{z\to a} \frac{1}{|g(z)|} = \lim_{z\to a} |f(z) - w| = +\infty$$

and, hence, $\lim_{z\to a} |f(z)|$ exists in $\overline{\mathbb{R}}$ and has the value $+\infty$, thereby contradicting the assumptions. □

2.4.5. Definition. Let a be an isolated singularity of f which is not removable, if $\lim_{z\to a} |f(z)| = +\infty$ we say that a is a ***pole*** of f. If the limit does not exist we say that a is an ***essential singularity***.

2.4.6. Remark. If a is a pole of $f \in \mathscr{H}(B(a,r)\setminus\{a\})$ there is $r_0 \in]0, r[$ such that $|f(z)| \geq 1$ if $0 < |z - a| < r_0$. Hence the function $g = 1/f$ has a removable singularity at a and it takes the value 0 there. If k is the multiplicity of this zero of g one can write $f(z) = h(z)/(z - a)^k$ with h holomorphic in a neighborhood of a and $h(a) \neq 0$. The number k will be called the ***order of the pole*** a of f.

Using the Taylor expansion of h about $z = a$, one can find complex numbers A_j, $-k \leq j \leq -1$, and a function φ holomorphic in a neighborhood of 0 such that $A_{-k} \neq 0$ and

$$f(z) = \frac{A_{-k}}{(z - a)^k} + \cdots + \frac{A_{-1}}{(z - a)} + \varphi(z - a).$$

Conversely, if f can be written this way, then $z = a$ is a pole of f of order k. Therefore, to say that a is a pole of f means that f can be extended to a continuous map $\tilde{f} : B(a,r) \to S^2$ with $\tilde{f}(a) = \infty$ and $1/\tilde{f}$ holomorphic in a neighborhood of a (we define for this purpose $0 = 1/\infty$).

2.4.7. Definition. Let Ω be an open subset of $\mathbb{C}$, $g : \Omega \to S^2$ a continuous function such that $g^{-1}(\infty)$ is a discrete subset of Ω. We say g is a ***meromorphic function*** in Ω if $g|(\Omega\setminus g^{-1}(\infty))$ is a holomorphic function.

In other words, if g is meromorphic in Ω, the set $P(g) := g^{-1}(\infty)$ is exactly the set of poles of the holomorphic function $g|\Omega'$, $\Omega' = \Omega\setminus P(g)$.

The collection of meromorphic functions in Ω will be denoted $\mathscr{M}(\Omega)$.

Let us also observe that if A is a discrete subset of an open set Ω in $\mathbb{C}$, $g \in \mathscr{H}(\Omega\setminus A)$ and every point of A is either a removable singularity or a pole of g, then there is a unique continuous $\tilde{g} : \Omega \to S^2$ which extends g. Clearly the function $\tilde{g}$ is meromorphic in Ω.

A function f which is holomorphic in the set $\mathbb{C}\setminus B(a,r)$, can be considered as a function defined in S^2 in a punctured neighborhood V of the point of ∞. In that sense ∞ is an isolated singularity for this function. By introducing the change of variables $z \mapsto 1/z$ one can classify this singularity the way we did earlier, namely, consider the function $g(z) := f(1/z)$ defined in a punctured neighborhood of $z = 0$.

For instance, ∞ is a removable singularity of the functions $1/z$, $1/z^n$, $1/p(z)$ ($p(z) = a_n z^n + \cdots + a_0, a_n \neq 0$). It is an essential singularity of every entire function which is not a polynomial. In fact, ∞ is a pole of the entire function f if and only if $g(z) = f(1/z)$ has a pole at 0. In that case, $g(z) = h(z)/z^k$ for some integer $k \geq 1$ and a function h holomorphic in a neighborhood of 0 with $h(0) \neq 0$. It follows that for $|z|$ large there is a constant $M > 0$ such that

$$|f(z)| \leq M|z|^k.$$

From the Cauchy inequalities it follows that f is a polynomial of degree less than or equal to k.

From now on we will also consider the spaces $\mathscr{H}(\Omega)$ and $\mathscr{M}(\Omega)$ when Ω is an open subset of S^2. Assume f is a holomorphic function in the annulus $0 \leq r < |z| < \infty$, with ∞ being a removable singularity or a pole of f. Then $g(w) = f(1/w)$ will be defined in $0 < |w| < 1/r$, with $w = 0$ being at most a pole and, hence, it can be written as

$$g(w) = \sum_{j \geq -k} A_j w^j.$$

As a consequence

$$f(z) = A_{-k} z^k + \cdots + A_{-1} z + \sum_{j \geq 0} \frac{A^j}{z^j}.$$

If the singularity at ∞ were essential, we are going to see that the polynomial part in the expansion will have to be replaced by an infinite series.

Note that the concept of spherical derivative $f^{\#}$ from Exercise 2.3.3 makes sense for functions meromorphic in an open set of S^2. It is enough to note that if ∞ is not a singularity of f, then $f'(\infty)$ can be defined as $\frac{d}{dz} f(1/z)$ evaluated at $z = 0$, and $f^{\#}(\infty) := \lim_{z \to 0} \frac{\sigma(f(1/z), f(\infty))}{|z|} = \frac{2|f'(\infty)|}{1 + |f(\infty)|^2}$ holds.

2.4.8. Proposition. *A holomorphic function f in an annulus $A(a; r_1, r_2) := \{z \in \mathbb{C} : 0 \leq r_1 < |z - a| < r_2 \leq \infty\}$ can be written in a unique way as*

$$f(z) = \sum_{n \in \mathbb{Z}} a_n (z - a)^n,$$

in such a way that:

(i) *The series $\sum_{n \geq 0} a_n w^n$ converges in $B(0, r_2)$, defining a holomorphic function $f_1(w)$.*
(ii) *The series $\sum_{n > 0} a_{-n} w^n$ converges in $B(0, 1/r_1)$ defining a holomorphic function $f_2(w)$*

$$f(z) = f_1(z - a) + f_2\left(\frac{1}{z - a}\right).$$

Furthermore, the coefficients a_n are given by the formula

$$a_n = \frac{1}{2\pi i}\int_{|t-a|=\rho} \frac{f(t)}{(t-a)^{n+1}}\,dt,$$

where ρ is an arbitrary value in the interval $r_1 < \rho < r_2$.

PROOF. For $z \in A(a; r_1, r_2)$ choose ρ_1, ρ_2 such that $r_1 < \rho_1 < \rho_2 < r_2$ and $z \in A(a; \rho_1, \rho_2)$. We have then

$$f(z) = \frac{1}{2\pi i}\int_{|t-a|=\rho_2} \frac{f(t)}{t-z}\,dt - \frac{1}{2\pi i}\int_{|t-a|=\rho_1} \frac{f(t)}{t-z}\,dt.$$

For $|t-a| = \rho_2$ let us write

$$\frac{1}{t-z} = \frac{1}{(t-a)-(z-a)} = \frac{1}{(t-a)\left(1-\dfrac{z-a}{t-a}\right)} = \sum_{n\geq 0}\frac{(z-a)^n}{(t-a)^{n+1}}.$$

The series converges uniformly for t in the circle $|t-a| = \rho_2$.

For $|t-a| = \rho_1$ we can write

$$\frac{1}{t-z} = \frac{1}{(t-a)-(z-a)} = \frac{-1}{(z-a)\left(1-\dfrac{t-a}{z-a}\right)} = -\sum_{n\geq 0}\frac{(t-a)^n}{(z-a)^{n+1}}$$

and the series converges uniformly for t in the circle $|t-a| = \rho_1$. Hence

$$\frac{1}{2\pi i}\int_{|t-a|=\rho_2} \frac{f(t)}{t-z}\,dt = \sum_{n\geq 0} a_n(z-a)^n = f_1(z-a),$$

$$a_n = \frac{1}{2\pi i}\int_{|t-a|=\rho_2} \frac{f(t)}{(t-a)^{n+1}}\,dt,$$

which shows that $|a_n| \leq C(\rho_2)\rho_2^{-n}$ for some constant $C(\rho_2) > 0$. Therefore the series $\sum\limits_{n\geq 0} a_n w^n$ converges for $|w| < r_2$, because ρ_2 could be chosen arbitrarily close to r_2. Similarly,

$$-\frac{1}{2\pi i}\int_{|t-a|=\rho_1} \frac{f(t)}{t-z}\,dt = \sum_{n\geq 1} a_{-n}\frac{1}{(z-a)^n} = f\left(\frac{1}{z-a}\right),$$

$$a_{-n} = \frac{1}{2\pi i}\int_{|t-a|=\rho_1} f(t)(t-a)^{n-1}\,dt,$$

which shows that $\sum\limits_{n\geq 1} a_{-n}w^n$ is convergent for $|w| < 1/r_1$ by the same reasoning as earlier.

Finally, we remark that by Cauchy's Theorem 1.1.4, the integrals $\dfrac{1}{2\pi i}\displaystyle\int_{|t-a|=\rho} f(t)/(t-a)^{n+1}\,dt$ are independent of $\rho \in \,]r_1, r_2[$.

The uniqueness of the expansion being evident (by integration), this concludes the proof of 2.4.8. □

2.4.9. Definitions

(1) The expansion of f obtained in §2.4.8 is called the ***Laurent series expansion*** of f in the annulus $A(a; r_1, r_2)$.

(2) In the case when a is an isolated singularity for $f \in \mathscr{H}(\Omega)$ the annulus becomes $A(a; 0, d(a, (\Omega \cup \{a\})^c)$ and the series $\sum_{n<0} a_n(z-a)^n$ is called the ***principal part*** of f at a.

(3) When $\Omega = A(0; r, \infty)$, the Laurent development of f is called the Laurent series about $z = \infty$, and the principal part is now $\sum_{n>0} a_n z^n$.

2.4.10. Proposition. *Let a be an isolated singularity of a function f holomorphic in $B(a,r)\setminus\{a\}$. We have the following equivalences*:

(i) *The singularity is removable if and only if the principal part of f at a is zero.*
(ii) *The singularity is a pole if and only if the principal part is a polynomial in $1/(z-a)$.*
(iii) *The singularity is essential otherwise.*

PROOF. The proof is left to the reader. □

EXERCISES 2.4

$B = B(0, 1)$.

1. Let Ω be an open subset of $\mathbb{C}$, $g : \Omega \to S^2$ continuous. Let $Z(g) = \{z \in \Omega : g(z) = 0\}$ and $P(g) = \{z \in \Omega : g(z) = \infty\}$. Assume that:
   (i) g is not identically equal to ∞ in any component of Ω;
   (ii) g is holomorphic in $\Omega \setminus P(g)$;
   (iii) $\dfrac{1}{g}$ is holomorphic in $\Omega \setminus Z(g)$. Show that $g \in \mathscr{M}(\Omega)$.

2. Is the function $z \mapsto \left(\sin\left(\dfrac{1}{z}\right)\right)^{-1}$ meromorphic in $\mathbb{C}$? In $\mathbb{C}^*$?

3. If Ω is a connected open set then $\mathscr{H}(\Omega)$ is an integral domain and $\mathscr{M}(\Omega)$ is a field. Moreover, the quotient field of $\mathscr{H}(\Omega)$ is a subfield of $\mathscr{M}(\Omega)$.

4. Show that $\mathscr{H}(S^2) = \mathbb{C}$.

5. Let $f \in \mathscr{M}(\Omega)$, Ω connected, and f not constant, then $f(\Omega)$ is an open subset of S^2.

6. Find the isolated singular points of the following functions and determine the type of singularity:

   (a) $\dfrac{z}{\sin z}$;

(b) $\cot z - \frac{1}{z}$;

(c) $\frac{1}{z^2 - 1} \cos \frac{\pi z}{z + 1}$;

(d) $z(e^{1/z} - 1)$;

(e) $\frac{1 - \cos z}{z^2}$.

7. Let f, g be holomorphic functions in a neighborhood of $z = 0$ such that $f(0) = g(0) = 0$. L'Hospital's rule $\frac{0}{0}$ is normally stated as

$$\lim_{z \to 0} \frac{f(z)}{g(z)} = \lim_{z \to 0} \frac{f'(z)}{g'(z)}$$

when $g'(0) \neq 0$. Why is it correct? Generalize to the case $g'(0) = 0$ and also study the indeterminacies $\frac{\infty}{\infty}$ and $0 \cdot \infty$. What can you do if the indeterminacy occurs at $z = \infty$?

8. Let f be a function holomorphic in $B(0, r) \setminus \{0\}$ such that $|f(z)| \leq c|z|^{-1/2}$. Show that 0 is a removable singularity of f.

9. Let $f \in \mathscr{H}(B \setminus \{0\})$, $M(r) = \max_{|z|=r} |f(z)|$, $0 < r < 1$. Show that for every $n \in \mathbb{N}$, $\lim_{r \to 0} r^n M(r) = \infty$ if and only if 0 is an essential singularity of f. Conclude that if $g \in \mathscr{H}(B) \setminus \{0\}$, $g(0) = 0$, $g(B) \subseteq B$, and 0 is an essential singularity for f, then $f \circ g$ has an essential singularity at $z = 0$.

10. Let $f \in \mathscr{M}(\mathbb{C})$ whose set of poles is $\mathbb{Z}$. Show that for every $r > 0$, $f(\bar{B}(0, r)^c)$ is dense in S^2. (Hint: The proof is similar to the Casorati-Weierstrass theorem.)

11. Let $f \in \mathscr{H}(B \setminus \{0\})$, $\operatorname{Re} f(z) \geq 0$. What kind of singularity can $z = 0$ be? What happens if we only know that $f(z) \notin [-1, 1] \subseteq \mathbb{R}$? $\left(\text{Hint: For the second part consider the function } \sqrt{\frac{f(z) - 1}{f(z) + 1}}.\right)$

12. Let $B_+ = \{z \in B : \operatorname{Im} z > 0\}$ and $f \in \mathscr{H}(B_+) \cap \mathscr{C}(\bar{B}_+)$ such that $|f(x)| = 1$ for $x \in]-1, 1[$. Show that f has a unique meromorphic extension to the whole disk B. Do poles actually occur?

13. For each example that follows define a holomorphic function in an annulus such that it has the following as its Laurent series development:

(a) $\sum_{n=-\infty}^{\infty} \frac{z^n}{|n|!}$;

(b) $\sum_{n=-\infty}^{\infty} 2^{-|n|}(z - 1)^n$.

14. If $f(z) = \sum_{n=-\infty}^{\infty} a_n z^n$ is the Laurent series of f in $A(0; r_1, r_2)$, which is the Laurent series of f' in the same annulus?

15. If f and g have Laurent series $\sum_{n=-\infty}^{\infty} a_n z^n$ and $\sum_{n=-\infty}^{\infty} b_n z^n$ in the annulus $A(0; r_1, r_2)$, find the Laurent series development of the product $f \cdot g$ in the same annulus.

16. Determine the Laurent series development of $f(z) = (z^2 - 1)^{-1}$ in the following regions:
(a) $B(0, 1)$;
(b) $A(1; 0, 2)$;
(c) $A(-1; 0, 2)$;
(d) $A(0; 1, \infty)$.

17. Let $f(z) = \sqrt{\dfrac{z-1}{z+1}}$ in $|z| > 1$, $f(2) > 0$. Find its Laurent series development in $1 < |z| < \infty$.

18. Let $f \in \mathscr{H}(\Omega \setminus \{a_1, a_2, \ldots, a_n\})$, where the a_j are distinct points of Ω. Let $f_1, \ldots, f_n$ be the corresponding principal parts of the Laurent series development of f about $z = a_1, \ldots, z = a_n$, respectively. Show that $f - (f_1 + \cdots + f_n)$ has removable singularities at every a_j.

19. Let f be an entire function such that $f(z + 1) \equiv f(z)$. Show that there is a holomorphic function g in $\mathbb{C} \setminus \{0\}$ such that $f(z) = g(\exp(2\pi i z))$. Conclude that f can be represented in the form

$$f(z) = \sum_{k=-\infty}^{\infty} a_k e^{2\pi i k z},$$

the series being convergent in $\mathscr{H}(\mathbb{C})$. It is called the ***Fourier series*** of f.

Show that the Fourier series converges uniformly in every strip $a < \operatorname{Im} z < b$. Furthermore, the coefficients can be computed from the formula

$$a_k = \int_0^1 f(x + ib) e^{-2\pi i k(x+ib)}\, dx, \qquad k \in \mathbb{Z},$$

for any fixed $b \in \mathbb{R}$.

Generalize this expansion to the case when the periodic function f is not entire, but only holomorphic in $b_1 < \operatorname{Im} z < b_2$.

20. Show that the Fourier series expansion of $f(z) = \operatorname{cotan} \pi z$ in $\operatorname{Im} z > 0$ is given by

$$f(z) = -i\left(1 + 2 \sum_{k=1}^{\infty} e^{2\pi i k z}\right).$$

*21. (Bessel functions of order, $n \in \mathbb{Z}$).
(a) Show that if the differential equation

$$z^2 f''(z) + z f'(z) + (z^2 - \lambda^2) f(z) = 0$$

admits an entire solution $f \not\equiv 0$, then the parameter λ must be an integer $n \in \mathbb{Z}$. Prove that the converse is also true.

For $\lambda = n \geq 0$, the solution of this equation that starts its Taylor series development about $z = 0$ with the term $\left(\frac{z}{2}\right)^n$ is called J_n, the ***Bessel function (of the first kind) of order n***. Show that

$$J_n(z) = \left(\frac{z}{2}\right)^n \sum_{k=0}^{\infty} \frac{(-1)^k (z/2)^{2k}}{k!(n+k)!}.$$

One defines J_n, $n = -1, -2, \ldots$, by $J_{-n}(z) = (-1)^n J_n(z)$.

(b) Let z be a fixed complex number. Show that the Laurent expansion in $\mathbb{C}^*$ of the function $t \mapsto \exp\left[\frac{z}{2}\left(t - \frac{1}{t}\right)\right]$ is given by

$$\exp\left[\frac{z}{2}\left(t - \frac{1}{t}\right)\right] = \sum_{n \in \mathbb{Z}} J_n(z) t^n.$$

(c) From (b) obtain the development in Fourier series of $\theta \mapsto e^{iz\sin\theta}$ and conclude that

$$J_n(z) = \frac{1}{\pi} \int_0^{\pi} \cos(z \sin\theta - n\theta)\, d\theta.$$

(d) Show that for $n \in \mathbb{N}$:

(i) $J_{n-1}(z) + J_{n+1}(z) = \dfrac{2n}{z} J_n(z)$. (Hint: Use Part (c).)

(ii) $\dfrac{d}{dz}(z^n J_n(z)) = z^n J_{n-1}(z)$. (Hint: Use Taylor series expansion.)

(iii) $\dfrac{d}{dz}\left(\dfrac{J_n(z)}{z^n}\right) = -\dfrac{J_{n+1}(z)}{z^n}$;

(iv) $J_{n-1}(z) - J_{n+1}(z) = 2J_n'(z)$.

*22. Assume f to be a holomorphic function in the strip $S = \{z \in \mathbb{C} : -\rho < \operatorname{Im} z < \rho\}$ $(\rho > 0)$. Suppose further that the series $\sum_{k=-\infty}^{\infty} f(z+k)$ converges in $\mathscr{H}(S)$. Let F be the holomorphic function thus defined.

(a) Show that one can expand F in a Fourier series:

$$F(z) = \sum_{n=-\infty}^{\infty} A_n e^{2\pi i n z},$$

$A_n = \int_{-\infty}^{\infty} f(x) e^{-2\pi i n x}\, dx$. (The integrals converge as $\sum_{k=-\infty}^{\infty} \int_k^{k+1} f(x) e^{-2\pi i n x}\, dx$ due to the hypothesis.)

(b) (Poisson summation formula). Show that

$$\sum_{k=-\infty}^{\infty} f(k) = \sum_{n=-\infty}^{\infty} \int_{-\infty}^{\infty} f(x) e^{-2\pi i n x}\, dx.$$

(c) Let $t > 0$. Apply the last two parts to the function $f(z) = e^{-\pi t z^2}$ to show that if $\theta(t) := \sum_{k=-\infty}^{\infty} e^{-\pi k^2 t}$, then

$$\theta(t) = \sum_{n=-\infty}^{\infty} \int_{-\infty}^{\infty} e^{-\pi t x^2 + 2\pi i n x}\, dx.$$

Use part (a) of Exercise 2.1.9 to compute the integrals and conclude that the following functional equation holds for $t > 0$:

$$\theta(t) = \frac{1}{\sqrt{t}}\,\theta\left(\frac{1}{t}\right).$$

23. Show that if the f_n are functions in $\mathscr{H}(B\setminus\{0\})$ with a pole of order exactly n at $z = 0$, then the function F_n,

$$F_n := f_1 + \cdots + f_n,$$

has also a pole of order exactly n at $z = 0$.

What happens with the F_n, when $f_n = \dfrac{1}{2^{n-1}}\dfrac{1}{z^{n-1}} - \dfrac{1}{2^n}\dfrac{1}{z^n}$ and we let $n \to \infty$?

24. Let f be a rational function and

$$f(z) = a_m z^m + \cdots + a_0 + \frac{a_{-1}}{z} + \frac{a_{-2}}{z^2} + \cdots$$

be its Laurent development about $z = \infty$. Show the a_j satisfy a recurrence relation. Is the converse true?

§5. Residues and the Computation of Definite Integrals

2.5.1. Definitions. Let Ω be an open subset of $\mathbb{C}$, $f \in \mathscr{H}(\Omega)$.

(1) Let $a \in \partial\Omega$ be an isolated singularity. We call ***residue of f at the point*** a, $\operatorname{Res}(f, a)$, the residue of the closed form $\omega = f(z)\,dz$ with respect to the hole $\{a\}$ of Ω.

(2) If $\Omega \supseteq \mathbb{C}\setminus B(0, R)$ we call ***residue of f at*** ∞, $\operatorname{Res}(f, \infty)$, the residue of the closed form $\omega = -f\left(\dfrac{1}{w}\right)\dfrac{dw}{w^2} = \varphi^*(f(z)\,dz)$ with respect to the hole $\{0\}$ of $\varphi^{-1}(\Omega)$, where $z = \varphi(w) = \dfrac{1}{w}$.

2.5.2. Remark. The reader should consult §1.10 for the background material on residues of a closed form. We recall here that in case (1), if γ is the loop $\gamma(t) = a + re^{2\pi i t}$ ($r > 0$ sufficiently small), then

$$\operatorname{Res}(f, a) = \operatorname{Res}_{f\,dz}(\{a\}) = \frac{1}{2\pi i}\int_\gamma f\,dz.$$

In case (2), denote γ the loop $\gamma(t) = re^{2\pi i t}$ ($0 < r < 1/R$), then

$$\operatorname{Res}(f,\infty) = \operatorname{Res}_{\varphi^*(f\,dz)}(\{0\}) = \frac{1}{2\pi i}\int_\gamma \varphi^*(f\,dz)$$

$$= -\frac{1}{2\pi i}\int_\gamma \frac{f\left(\frac{1}{w}\right)}{w^2}\,dw = \frac{1}{2\pi i}\int_\Gamma f(z)\,dz,$$

where Γ is the loop $\Gamma(t) = (\varphi \circ \gamma)(t) = \dfrac{e^{-2\pi i t}}{r}$.

The oddity of the definition of $\operatorname{Res}(f,\infty)$ disappears a bit when we imagine $\{\infty\}$ as a "hole" of Ω, the orientation of the loop Γ being such that the "index" of this loop with respect to the hole $\{\infty\}$ is $+1$. Since the index should be something like $\dfrac{1}{2\pi i}\displaystyle\int_\Gamma \frac{dw}{w}$, and this integral is exactly $\operatorname{Ind}_\Gamma(0)$, and can be seen to be -1 by direct computation, we find ourselves justified in the choice of definition.

2.5.3. Proposition. *Let Ω be an open set of $\mathbb{C}$, $f \in \mathscr{H}(\Omega)$:*

(1) *If $a \in \partial\Omega$ is an isolated singularity of f and the Laurent expansion of f about a is given by*

$$f(z) = \sum_{n\in\mathbb{Z}} a_n(z-a)^n,$$

then

$$\operatorname{Res}(f,a) = a_{-1}.$$

(2) *If $\Omega \supseteq \mathbb{C}\setminus B(0,R)$ and if the Laurent expansion of f about ∞ is given by*

$$f(z) = \sum_{n\in\mathbb{Z}} a_n z^n,$$

then

$$\operatorname{Res}(f,\infty) = -a_{-1}.$$

Proof. It is enough to compute $\dfrac{1}{2\pi i}\displaystyle\int_\gamma f\,dz$ in the first case, and $\dfrac{1}{2\pi i}\displaystyle\int_\Gamma f\,dz$ in the second, using the notation from §2.5.2. □

2.5.4. Proposition. *Let f be holomorphic in $B(a,r)\setminus\{a\}$ and have a pole at $z = a$ (i.e., f is meromorphic in $B(a,r)$). Then*

(1) *If a is a simple pole of f (i.e., the order of the pole is $k = 1$), then*

$$\operatorname{Res}(f,a) = \lim_{z\to a}(z-a)f(z).$$

In particular, if $f(z) = P(z)/Q(z)$, P and Q are holomorphic in a neighborhood of a, $P(a) \neq 0$, and Q has a simple zero at a, then a is a simple pole of f and

$$\operatorname{Res}(f, a) = \frac{P(a)}{Q'(a)}.$$

(2) *If a is a pole of order k of f, then the function $g(z) = (z - a)^k f(z)$ is holomorphic in a neighborhood of a. Let $\sum_{n \geq 0} b_n(z - a)^n$ be the Taylor expansion of g. Then we have*

$$\operatorname{Res}(f, a) = b_{k-1} = \lim_{z \to a} \frac{1}{(k-1)!}\left(\frac{d^{k-1}}{dz^{k-1}}\{(z - a)^k f(z)\}\right).$$

PROOF. (1) The first statement is clear. The second follows from the fact that $\dfrac{Q(z)}{z - a}$ can be extended to be a nonvanishing holomorphic function in a neighborhood of a, hence the order of the pole is 1, and

$$(z - a)f(z) = \frac{P(z)}{\dfrac{Q(z) - Q(a)}{z - a}} \qquad (\text{since } Q(a) = 0)$$

has $P(a)/Q'(a)$ as a limit when z tends to a.

(2) We have $g(z) = b_0 + b_1(z - a) + \cdots + b_{k-1}(z - a)^{k-1} + (z - a)^k h(z)$, with h holomorphic near a. Hence

$$f(z) = \frac{b_0}{(z - a)^k} + \frac{b_1}{(z - a)^{k-1}} + \cdots + \frac{b_{k-1}}{(z - a)} + h(z),$$

and the formula for $\operatorname{Res}(f, a)$ follows from §2.5.3, (1). □

2.5.5. Example. The function $f(z) = \dfrac{1}{(z^2 + 1)(z - 1)^3} = \dfrac{1}{(z + i)(z - i)(z - 1)^3}$ has simple poles at $\pm i$ and a triple pole at 1. We have

$$\operatorname{Res}(f, i) = \lim_{z \to i} (z - i)f(z) = \frac{1}{2i(i - 1)^3} = -\frac{1 + i}{8}$$

$$\operatorname{Res}(f, -i) = \lim_{z \to -i} (z + i)f(z) = \frac{1}{(-2i)(-i - 1)^3} = \overline{\operatorname{Res}(f, i)} = -\frac{1 - i}{8}.$$

For the computation of $\operatorname{Res}(f, 1)$, we need to find the expansion of $(z - 1)^3 f(z) = \dfrac{1}{1 + z^2}$ near $z = 1$. Let $z = 1 + h$, we obtain

$$\frac{1}{1 + z^2} = \frac{1}{2}\left(1 - h + \frac{h^2}{2} + 0(h^3)\right),$$

hence

$$\operatorname{Res}(f, 1) = \frac{1}{4}.$$

Before we draw some consequences of the Residue Formula we will apply it to the computation of definite integrals. We recall that the Residue Formula 1.11.4, (4) is the following.

Let Ω be an admissible open set and δ a 1-cycle of Ω, then

$$\frac{1}{2\pi i}\int_\delta f(z)\,dz = \sum_T \operatorname{Ind}_\delta(T)\operatorname{Res}_{f\,dz}(T),$$

where the sum takes place over all the holes of Ω.

The method of computation of integrals using residues uses quite frequently the idea of taking limits on the cycles, and it depends on the following classical lemmas (usually attributed to C. Jordan).

2.5.6. Lemma

(1) *Let f be a continuous function on the set*

$$S_1 = \{z \in \mathbb{C} : 0 < |z-a| < r, \alpha_1 \leq \operatorname{Arg}(z-a) \leq \alpha_2\},$$
$$a \in \mathbb{C}, \quad r > 0, \quad \alpha_1 < \alpha_2.$$

Let γ_ρ be the arc of the circle of center a and radius ρ contained in S_1.

The condition

$$\lim_{\substack{z\to a\\ z\in S_1}} (z-a)f(z) = 0$$

implies

$$\lim_{\rho\to 0^+}\int_{\gamma_\rho} f(z)\,dz = 0.$$

(2) *Let f be a continuous function in the set*

$$S_2 = \{z \in \mathbb{C} : |z-a| > R, \alpha_1 \leq \operatorname{Arg}(z-a) \leq \alpha_2\}, a \in \mathbb{C}, R \geq 0, \alpha_1 < \alpha_2.$$

As before, let γ_ρ be the arc of the circle of radius ρ and center a contained in S_2.

The condition

$$\lim_{\substack{|z|\to\infty\\ z\in S_2}} (z-a)f(z) = 0$$

implies

$$\lim_{\rho\to+\infty}\int_{\gamma_\rho} f(z)\,dz = 0.$$

Proof. In the two cases the length of γ_ρ is $(\alpha_2-\alpha_1)\rho$. Then

$$\left|\int_{\gamma_\rho} f(z)\,dz\right| \leq (\alpha_2-\alpha_1)\rho \sup_{z\in\gamma_\rho}|f(z)| = (\alpha_2-\alpha_1)\sup_{z\in\gamma_\rho}|(z-a)f(z)|,$$

and the result is clear. □

2.5.7. Lemma. *Let g be a continuous function in the half-plane* $\operatorname{Im} z \geq 0$ *tending to zero when $|z| \to \infty$ in that half-plane. Let C_ρ be the semicircle of center* 0 *and radius ρ in that half-plane. Then, for $\alpha > 0$ fixed, we have*

$$\lim_{\rho \to \infty} \int_{C_\rho} e^{i\alpha z} g(z)\, dz = 0.$$

PROOF. It depends on the inequality (see Figure 2.3)

$$\frac{2\theta}{\pi} \leq \sin\theta \leq \theta \qquad \text{for} \qquad 0 \leq \theta \leq \pi/2.$$

Let us write $z = \rho e^{i\theta}$, $0 \leq \theta \leq \pi$. The integral becomes

$$I_\rho = \int_{C_\rho} g(z) e^{i\alpha z}\, dz = \int_0^\pi g(\rho e^{i\theta}) e^{i\alpha\rho(\cos\theta + i\sin\theta)} i\rho e^{i\theta}\, d\theta.$$

Let M_ρ be the maximum value of $|g(z)|$ on C_ρ, then

$$|I_\rho| \leq \rho \int_0^\pi |g(\rho e^{i\theta})| e^{-\alpha\rho\sin\theta}\, d\theta \leq \rho M_\rho \int_0^\pi e^{-\alpha\rho\sin\theta}\, d\theta.$$

$$\leq 2\rho M_\rho \int_0^{\pi/2} e^{-\alpha\rho 2\theta/\pi}\, d\theta \leq 2\rho M_\rho \frac{1 - e^{-\alpha\rho}}{\alpha\rho 2/\pi} \leq \frac{\pi}{\alpha} \sup_{z \in C_\rho} |g(z)|.$$

The lemma follows since $\lim_{\rho \to \infty} M_\rho = 0$ by hypothesis. □

2.5.8. Lemma. *Let a be a simple pole of a function f, γ_ρ an arc of circle positively oriented of angular opening α, center a and radius ρ. Then*

$$\lim_{\rho \to 0^+} \int_{\gamma_\rho} f(z)\, dz = i\alpha \operatorname{Res}(f, a).$$

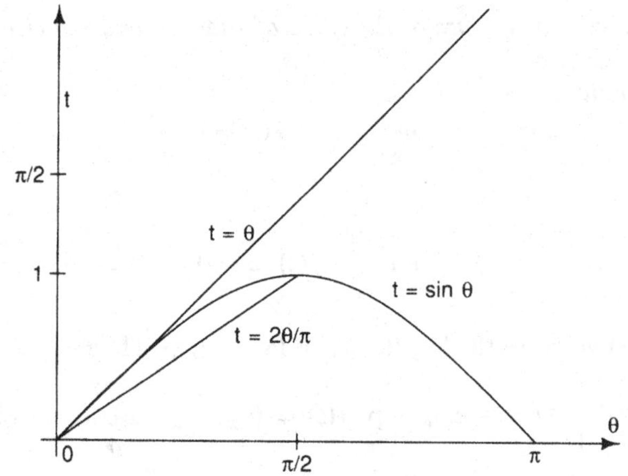

Figure 2.3

PROOF. Since a is a simple pole, the Laurent expansion of f about $z = a$ gives

$$f(z) = \frac{\operatorname{Res}(f, a)}{z - a} + \varepsilon(z),$$

where $|\varepsilon(z)| \leq M$, if z is sufficiently close to a. Therefore,

$$\int_{\gamma_\rho} f(z)\,dz = \operatorname{Res}(f, a) \int_{\gamma_\rho} \frac{dz}{z - a} + \int_{\gamma_\rho} \varepsilon(z)\,dz.$$

A direct computation yields

$$\int_{\gamma_\rho} \frac{dz}{z - a} = i\alpha.$$

On the other hand

$$\left| \int_{\gamma_\rho} \varepsilon(z)\,dz \right| \leq M\ell(\gamma_\rho) = M\alpha\rho \to 0 \quad \text{as} \quad \rho \to 0. \qquad \square$$

2.5.9. Examples. (1) *Integrals of rational functions.* Let $f(z) = P(z)/Q(z)$ be a rational function with no poles on the real axis and $\deg P \leq \deg Q - 2$. Therefore, $\lim_{|z| \to \infty} zf(z) = 0$ and hence

$$\lim_{R \to +\infty} \int_{C_R} f(z)\,dz = 0,$$

where C_R is the semicircle of center 0 and radius R in the half-plane $\operatorname{Im} z \geq 0$ (by §2.5.6).

Passing to the limit in the following identity, valid for R sufficiently large

$$\int_{C_R} f(z)\,dz + \int_{[-R, R]} f(z)\,dz = 2\pi i \sum_{\operatorname{Im} a > 0} \operatorname{Res}(f, a),$$

we find

$$\int_{-\infty}^{\infty} f(x)\,dx = 2\pi i \sum_{\operatorname{Im} a > 0} \operatorname{Res}(f, a),$$

where the sum is over all poles of f in the upper half-plane.

As an application we obtain

$$\int_0^{\infty} \frac{x^2\,dx}{(x^2 + 9)(x^2 + 4)^2} = \frac{1}{2} \int_{-\infty}^{\infty} \frac{x^2\,dx}{(x^2 + 9)(x^2 + 4)^2} = \frac{\pi}{200}.$$

$$\int_{-\infty}^{\infty} \frac{x^p}{1 + x^{2n}}\,dx = \begin{cases} 0 & \text{if } p \text{ is an odd integer} \\ \dfrac{\pi}{n \sin((p+1)\pi/2n)} & \text{if } p \text{ is an even integer} \end{cases} \qquad (0 \leq p \leq 2n - 2).$$

(2) *Trigonometric integrals.* Let $R(u, v)$ be a rational function of the two variables u and v which is continuous on $u^2 + v^2 = 1$. Consider the integral

$$I = \int_0^{2\pi} R(\cos\theta, \sin\theta)\, d\theta.$$

Introducing the variable $z = e^{i\theta}$, we can write $\cos\theta = \frac{1}{2}\left(z + \frac{1}{z}\right)$, $\sin\theta = \frac{1}{2i}\left(z - \frac{1}{z}\right)$, and the integral becomes

$$I = \int_C R\left(\frac{1}{2}\left(z + \frac{1}{z}\right), \frac{1}{2i}\left(z - \frac{1}{z}\right)\right)\frac{dz}{iz},$$

where C is the unit circle $t \mapsto e^{2\pi i t}$. If

$$f(z) = \frac{1}{iz} R\left(\frac{1}{2}\left(z + \frac{1}{z}\right), \frac{1}{2i}\left(z - \frac{1}{z}\right)\right)$$

then

$$I = 2\pi i \sum_{|a|<1} \operatorname{Res}(f, a).$$

As examples we obtain

$$\int_0^{2\pi} \frac{\cos m\theta}{5 + 3\cos\theta}\, d\theta = \frac{(-1)^m \pi}{2 \cdot 3^m}, \qquad m \in \mathbb{N};$$

$$\int_0^{2\pi} \frac{d\theta}{r^2 - 2r\cos\theta + 1} = \frac{2\pi}{1 - r^2}, \qquad 0 \le r < 1.$$

(3) *Fourier-type integrals.* By this we mean an integral of the form

$$I = \int_{-\infty}^{+\infty} f(x) e^{i\alpha x}\, dx,$$

where $\alpha > 0$, f holomorphic on $\operatorname{Im} z \ge 0$, with the possible exception of a discrete set A which does not intersect the real axis. We also assume

$$\lim_{\substack{|z| \to \infty \\ \operatorname{Im} z \ge 0}} |f(z)| = 0.$$

(We remind the reader that f holomorphic on $\operatorname{Im} z \ge 0$ means that $f \in \mathscr{H}(\Omega)$, Ω open, $\{z : \operatorname{Im} z \ge 0\} \subseteq \Omega$. Moreover, $A \subseteq \{z : \operatorname{Im} z > 0\}$.) It follows from the classification of singularities that A is finite and we have, as a consequence of §2.5.7,

$$I := \lim_{r \to +\infty} \int_{-r}^{r} f(x) e^{i\alpha x}\, dx = 2\pi i \sum_{a \in A} \operatorname{Res}(f(z) e^{i\alpha z}, a).$$

Note that the integral I is defined to be the limit on the right-hand side.

By way of example:

$$\int_0^{\infty} \frac{\cos x}{1 + x^2}\, dx = \frac{1}{2} \operatorname{Re}\left(\int_{-\infty}^{\infty} \frac{e^{ix}}{1 + x^2}\, dx\right) = \frac{\pi}{2e}.$$

(4) *Principal-value of Fourier-type integrals.* The same kind of integrals as in (3) except that we now allow for a finite number of simple poles at the points $\alpha_1 < \alpha_2 < \cdots < \alpha_p$ of the real axis. A will still denote the finite set of isolated singularities of f in $\{\operatorname{Im} z > 0\}$.

There is first the question of how we should define the integral I in (3). Let us set

$$I(r, \varepsilon_1, \ldots, \varepsilon_p) = \int_{-r}^{-\alpha_1-\varepsilon_1} + \int_{\alpha_1+\varepsilon_1}^{\alpha_2-\varepsilon_2} + \cdots + \int_{\alpha_{p-1}+\varepsilon_{p-1}}^{\alpha_p-\varepsilon_p} + \int_{\alpha_p+\varepsilon_p}^{r} (f(x)e^{i\alpha x})\,dx,$$

where $r > 0$ is sufficiently large, $\varepsilon_j > 0$ are sufficiently small. We set

$$I := \lim_{\substack{r\to\infty\\ \varepsilon_j\to 0\\ 1\le j\le p}} I(r, \varepsilon_1, \ldots, \varepsilon_p).$$

This definition determines the so-called ***principal value of the integral** I*.

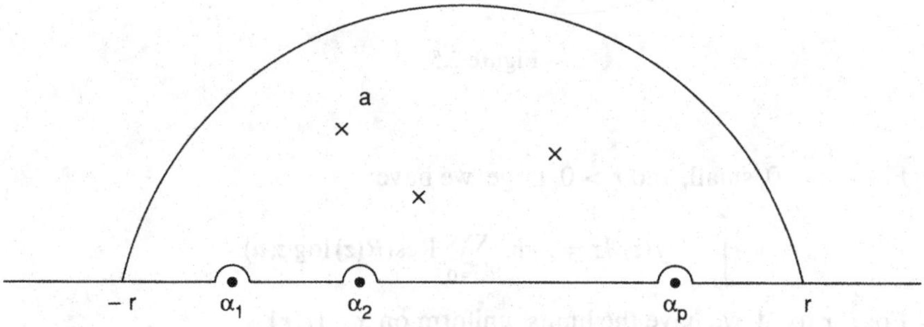

Figure 2.4

We apply the residue theorem to the contour and use Lemmas 2.5.7 and 2.5.8 to obtain

$$I = i\pi\left\{\sum_{1\le j\le p} \operatorname{Res}(f(z)e^{i\alpha z}, \alpha_j) + 2\sum_{a\in A} \operatorname{Res}(f(z)e^{i\alpha z}, a)\right\}.$$

For instance,

$$\int_0^\infty \frac{\sin x}{x}\,dx = \frac{1}{2i}\int_{-\infty}^\infty \frac{e^{ix}}{x}\,dx = \frac{\pi}{2}.$$

(5) *Integrals of the type* $\int_0^\infty R(x)\,dx$, *R rational function.* We assume $R(z) = P(z)/Q(z)$ is a rational function with no poles on the positive real axis $[0, \infty)$ and $\deg P \le \deg Q - 2$. We are going to apply the residue theorem to the contour $\gamma_{r,\varepsilon,h}$ shown in Figure 2.5.

The idea is to apply that theorem to an auxiliary function, meromorphic in $\mathbb{C}\setminus[0, \infty[$, $f(z) := R(z)\log z$, where $\operatorname{Im}\log z = \arg z \in\,]0, 2\pi[$. Its only poles, a finite number, are located at the zeros of $Q(z)$.

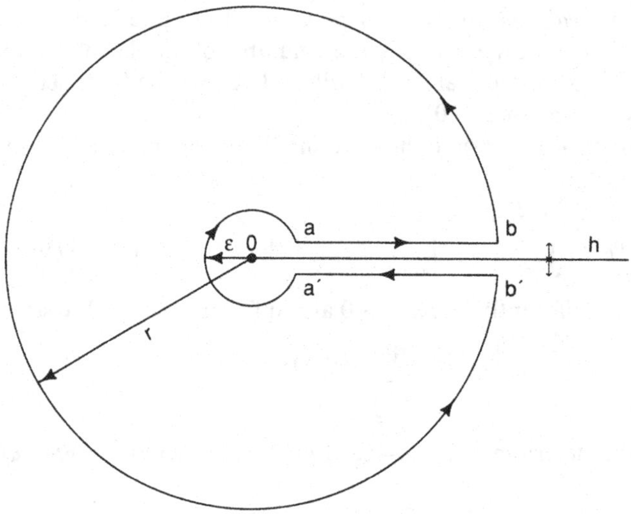

Figure 2.5

For ε, $h > 0$, small, and $r > 0$, large, we have

$$\int_{\gamma_{r,\varepsilon,h}} f(z)\,dz = 2\pi i \sum_{Q(a)=0} \operatorname{Res}(R(z)\log z, a).$$

For ε, r fixed, we have the limits, uniform on $x \in [\varepsilon, r]$,

$$\lim_{\substack{h\to 0+\\ \operatorname{Im} z>0}} R(z)\log z = R(x)\log x$$

$$\lim_{\substack{h\to 0+\\ \operatorname{Im} z<0}} R(z)\log z = R(x)(\log x + 2\pi i).$$

Therefore, if γ_ρ denotes the circle of center 0 and radius ρ, positively oriented, we have

$$\int_{\gamma_r} R(z)\log z\,dz - \int_{\gamma_\varepsilon} R(z)\log z\,dz - 2\pi i\int_\varepsilon^r R(x)\,dx = 2\pi i \sum_a \operatorname{Res}(R(z)\log z, a).$$

The first two integrals satisfy the hypothesis of §2.5.6, where now S denotes the sector $S = \{z \in \mathbb{C}^* : 0 \leq \arg z \leq 2\pi\}$, and we have

$$\lim_{z\to 0} zR(z)\log z = 0 \qquad \text{and} \qquad \lim_{|z|\to\infty} zR(z)\log z = 0.$$

As a corollary, one obtains

$$\int_0^\infty R(x)\,dx = -\sum_a \operatorname{Res}(R(z)\log z, a).$$

For example: if p and q are integers, $0 \leq p \leq q - 2$,

$$\int_0^\infty \frac{x^p}{1 + x^q} dx = \frac{\pi}{q \sin\left(\frac{p+1}{q}\pi\right)}.$$

(6) *Integrals of the type* $\int_0^\infty R(x) x^{-\alpha} dx$, $0 < \alpha < 1$. R is again a rational function without poles in the ray $x \geq 0$. We assume that $\lim_{|z| \to \infty} R(z) = 0$.

We note that the last hypothesis means that if $R = P/Q$ then $\deg P \leq \deg Q - 1$. We consider the function $f(z) = R(z)/z^\alpha$, $z^\alpha = e^{\alpha \log z}$, with the same choice of logarithm as in (5). Integrating on $\gamma_{r,\varepsilon,h}$ and letting first $h \to 0$, as was done earlier, we obtain

$$\int_{\gamma_r} \frac{R(z)}{z^\alpha} dz - \int_{\gamma_\varepsilon} \frac{R(z)}{z^\alpha} dz + (1 - e^{-2\pi i \alpha}) \int_\varepsilon^r \frac{R(x)}{x^\alpha} dx = 2\pi i \sum_{Q(a)=0} \operatorname{Res}\left(\frac{R(z)}{z^\alpha}, a\right).$$

We also have here that $\lim_{z \to \infty} z f(z) = 0$ and hence,

$$\int_0^\infty \frac{R(x)}{x^\alpha} dx = \frac{2\pi i}{(1 - e^{-2\pi i \alpha})} \sum_a \operatorname{Res}\left(\frac{R(z)}{z^\alpha}, a\right).$$

For instance,

$$\int_0^\infty \frac{dx}{(1 + x) x^\alpha} = \frac{\pi}{\sin(\pi\alpha)} \qquad (0 < \alpha < 1).$$

(7) *Integrals of the type* $\int_0^\infty R(x) \log x \, dx$. Here R is a rational function without poles in $x \geq 0$ such that

$$\lim_{x \to \infty} x R(x) = 0.$$

We proceed as in (5) and (6) with $f(z) = R(z)(\log z)^2$. This yields the relation

$$\int_0^\infty R(x) \log^2 x \, dx - \int_0^\infty R(x)(\log x + 2\pi i)^2 \, dx = 2\pi i \sum_a \operatorname{Res}(f(z), a).$$

Therefore

$$-2 \int_0^\infty R(x) \log x \, dx - 2\pi i \int_0^\infty R(x) \, dx = \sum_a \operatorname{Res}(R(z)(\log z)^2, a),$$

and one can use (5) to compute the second integral. If $R(x)$ is real-valued for $x \in \mathbb{R}$, then one can separate the real and imaginary parts in this relation and obtain

$$\int_0^\infty R(x) \log x \, dx = -\frac{1}{2} \operatorname{Re}\left(\sum_a \operatorname{Res}(R(z)(\log z)^2, a)\right)$$

$$\int_0^\infty R(x)\,dx = -\frac{1}{2\pi}\operatorname{Im}\left(\sum_a \operatorname{Res}(R(z)(\log z)^2, a)\right).$$

In this way we have obtained

$$\int_0^\infty \frac{\log x}{(1+x)^3}\,dx = -\frac{1}{2}.$$

(8) One can apply the preceding method to $\int_0^\infty R(x)\log x\,dx$ when R has a simple pole at the point $x = 1$. In this case the vanishing of $\log x$ at $x = 1$ keeps the integral convergent. One needs to modify the previous contour as shown in Figure 2.6.

If R is real-valued on the real axis one obtains the relation

$$\int_0^\infty R(x)\log x\,dx = \pi^2 \operatorname{Res}(R, 1) - \frac{1}{2}\operatorname{Re}\sum_{a\neq 1} \operatorname{Res}(R(z)(\log z)^2, a).$$

An example of an application of this identity is given by:

$$\int_0^\infty \frac{\log x}{x^2-1}\,dx = \frac{\pi^2}{4}.$$

One can compute integrals of the form $\int_0^\infty R(x)(\log x)^p\,dx$, $p \in \mathbb{N}$, by integrating $R(z)(\log z)^{p+1}$. A priori one assumes R has no poles in $x \geq 0$ and $xR(x) \to 0$ as $x \to \infty$, but one can accept a pole of order less than or equal to p at $x = 1$ as we have just done.

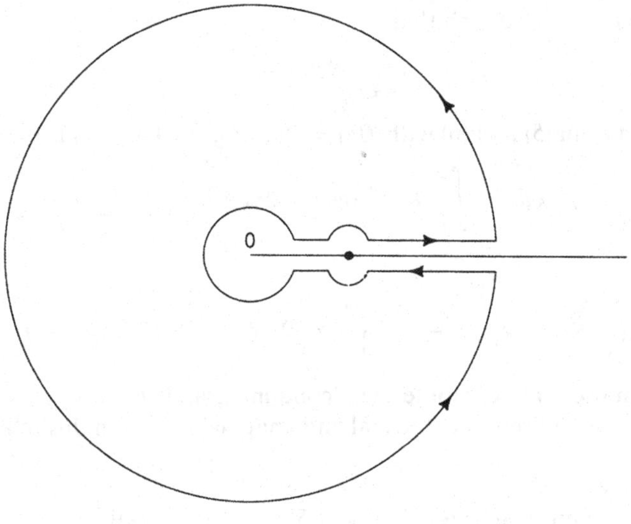

Figure 2.6

(9) *Integrals of the type* $\int_a^b f(x)\left(\frac{x-a}{b-x}\right)^\alpha dx$, $0 < \alpha < 1$. Here we assume that f is holomorphic in the whole complex plane with the exception of a finite number of points, none of them lying in $]a, b]$ and having at most a simple pole at a $(-\infty < a < b < \infty)$.

Let $\zeta = \frac{z-a}{z-b}$, this map is a biholomorphism from $\mathbb{C}\backslash[a,b]$ onto $\mathbb{C}\backslash(]-\infty, 0] \cup \{1\})$. Hence, if $\operatorname{Log}\zeta$ is the principal determination of the logarithm ($\operatorname{Arg}\zeta \in]-\pi, \pi[$), we have that the function $g(z) = f(z)e^{\alpha \operatorname{Log}\zeta(z)}$ has only the same singularities as f has in $\mathbb{C}\backslash[a,b]$.

Consider the 1-cycle $\gamma_{r,\varepsilon,h}$ in $\mathbb{C}\backslash[a,b]$ suggested by Figure 2.7.

For r, ε, h conveniently chosen we have

$$\int_{\gamma_{r,\varepsilon,h}} g(z)\,dz = 2\pi i \sum_{w \neq a} \operatorname{Res}(g, w).$$

Now, if we fix ε, r, and let $h \to 0+$, we obtain

$$\begin{aligned}\lim_{h\to 0+} \int_{[a',b']} g(z)\,dz &= \int_{a+\varepsilon}^{b-\varepsilon} f(x)e^{\alpha[\log((x-a)/(b-x))-i\pi]}\,dx\\ &= \int_{a+\varepsilon}^{b-\varepsilon} f(x)\left(\frac{x-a}{b-x}\right)^\alpha e^{-i\alpha\pi}\,dx,\end{aligned}$$

since $|\zeta| \to \frac{x-a}{b-x}$ and $\operatorname{Arg}\zeta \to -\pi$ when $z \in [a', b']$.

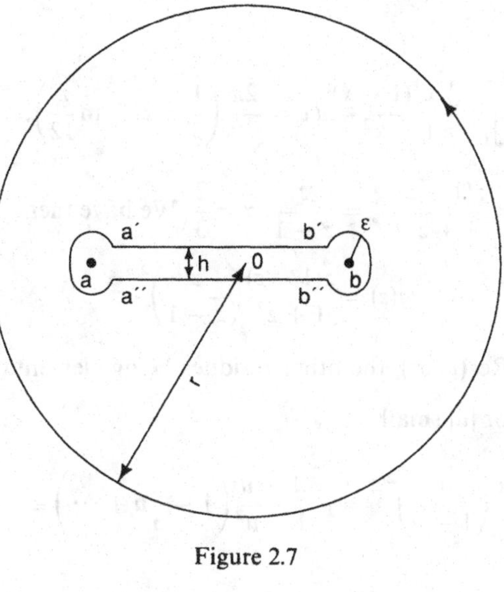

Figure 2.7

$$\lim_{h\to 0+}\int_{[a'',b'']} g(z)\,dz = -\int_{a+\varepsilon}^{b-\varepsilon} f(x)\left(\frac{x-a}{b-x}\right)^{\alpha} e^{i\alpha\pi}\,dx,$$

this time $\operatorname{Arg}\zeta \to \pi$.

Let $\gamma_{\varepsilon,a}$ and $\gamma_{\varepsilon,b}$ be the positively oriented circles of radius ε and centers a and b respectively. Then

$$\lim_{\varepsilon\to 0+}\int_{\gamma_{\varepsilon,a}} g(z)\,dz = \lim_{\varepsilon\to 0+}\int_{\gamma_{\varepsilon,b}} g(z)\,dz = 0,$$

since

$$\lim_{z\to a}(z-a)g(z) = \lim_{z\to b}(z-b)g(z) = 0.$$

(Note that the first identity is valid even if f has a simple pole at $z = a$.)

Therefore, if I denotes the integral we are trying to compute and C_r is the positively oriented circle of center 0 and radius r we have

$$\int_{C_r} g(z)\,dz + I\cdot(e^{-i\alpha\pi} - e^{i\alpha\pi}) = 2\pi i \sum_{w\neq a}\operatorname{Res}(g, w).$$

Our assumption indicates that g is holomorphic in $\mathbb{C}\setminus B(0,r)$, hence

$$\int_{C_r} g(z)\,dz = -2\pi i\operatorname{Res}(g,\infty).$$

Summarizing

$$\int_a^b f(x)\left(\frac{x-a}{b-x}\right)^{\alpha} dx = \frac{-\pi}{\sin\pi\alpha}\sum_{w\in S^2\setminus\{a\}}\operatorname{Res}(g, w).$$

Let us see that

$$\int_0^1 \frac{\sqrt[3]{x^5(1-x)}}{1+x^2}\,dx = \frac{2\pi}{\sqrt{3}}\left(\frac{1}{3} - 2^{1/6}\sin\frac{\pi}{12}\right).$$

We take $f(z) = \dfrac{z(1-z)}{1+z^2}$, $\zeta = \dfrac{z}{z-1}$, $\alpha = \dfrac{2}{3}$. We have then

$$g(z) = \frac{z(1-z)}{1+z^2}\left(\frac{z}{z-1}\right)^{2/3}.$$

We compute $\operatorname{Res}(g,\infty)$, the other residues being elementary to compute. Let $z = \dfrac{1}{u}$, then for $|u|$ small

$$g\left(\frac{1}{u}\right) = -\frac{1-u}{1+u^2}\left(\frac{1}{1-u}\right)^{2/3} = -\frac{1-u}{1+u^2}\left(1+\frac{2}{3}u+\cdots\right) \equiv -1+\frac{1}{3}u+\cdots.$$

Hence $g(z) = -1 + \frac{1}{3}\frac{1}{z} + \cdots$ for $|z|$ large and

$$\operatorname{Res}(g, \infty) = -\frac{1}{3}.$$

EXERCISES 2.5

1. Evaluate, using residue calculus,

$$\int_{-\pi}^{\pi} \frac{d\theta}{re^{i\theta} - \zeta}, \qquad |\zeta| \neq r.$$

Use this to compute the Cauchy transform of χ_B, $B = B(0, 1)$, namely,

$$\hat{\chi}_B(\zeta) = -\frac{1}{\pi}\int_B \frac{dx\,dy}{z - \zeta}.$$

2. Let $-1 < a < b < 1$, $n \in \mathbb{N}^*$, and $B = B(0, 1)$, evaluate

$$\int_{\partial B} z^n \operatorname{Log}\left(\frac{z - a}{z - b}\right) dz.$$

3. Evaluate

$$\int_0^{2\pi} \exp(\cos\theta)\cos(n\theta - \sin\theta)\,d\theta.$$

4. Compute $\displaystyle\int_0^{\infty} \left(\frac{\sin x}{x}\right)^2 dx.$

5. Compute $\displaystyle\int_{-\infty}^{\infty} \frac{e^{ax}}{1 + e^x}\,dx$ $(0 < a < 1)$. (Hint: Consider the function $f(z) = \dfrac{e^{az}}{1 + e^z}$ along the rectangle of vertices $-R$, R, $R + 2\pi i$, $-R + 2\pi i$.)

6. Let $\operatorname{Im} a > 0$, $\operatorname{Im} b > 0$, and $b \neq a$. Show that

$$\frac{\operatorname{Im} a}{\pi}\int_{-\infty}^{\infty} \frac{\log|x - b|}{|x - a|^2}\,dx = \log|b - a|.$$

7. Compute the integral $\displaystyle\int_0^{\infty} \frac{\sqrt{x}}{(x^4 + 1)^3}\,dx$ using as an auxiliary contour the boundary of the sector $0 \leq \operatorname{Arg} z \leq \dfrac{\pi}{4}$.

8. Alternative proof of the Jordan Lemma 2.5.7: Show that if $0 \leq \lambda < 1$ then

$$R^\lambda \int_0^{\pi/2} e^{-R\sin\theta}\,d\theta \to 0, \qquad \text{as } R \to \infty$$

by dividing the interval of integration into $\left[0, \dfrac{\pi}{3}\right]$ and $\left[\dfrac{\pi}{3}, \dfrac{\pi}{2}\right]$. In the first one use that $\cos\theta \geq \frac{1}{2}$ to conclude that

$$\int_0^{\pi/3} e^{-R\sin\theta}\,d\theta \leq 2\int_0^{\pi/2} e^{-R\sin\theta}\cos\theta\,d\theta.$$

9. Show that for $a > 0$, $b > 0$:

(a) $\displaystyle\int_0^\infty \frac{x\sin bx}{x^4+a^4}\,dx = \frac{\pi}{2a^2}e^{-ba/\sqrt{2}}\sin\frac{ba}{\sqrt{2}};$

(b) $\displaystyle\int_0^\infty \frac{\cos bx}{x^4+a^4}\,dx = \frac{\pi}{2a^3}e^{-ba/\sqrt{2}}\sin\left(\frac{ba}{\sqrt{2}}+\frac{\pi}{4}\right);$

(c) $\displaystyle\int_0^\infty \frac{x^3\sin bx}{x^4+a^4}\,dx = \frac{\pi}{2}e^{-ba/\sqrt{2}}\sin\frac{ba}{\sqrt{2}}.$

10. If $0 < a < 2$, then

$$\int_0^\infty \frac{x^{a-1}}{1+x+x^2}\,dx = \frac{2\pi}{\sqrt{3}}\frac{\cos\left(\frac{\pi}{3}a+\frac{\pi}{6}\right)}{\sin\pi a}.$$

11. Show that

$$\int_0^{2\pi}\cos^n\theta\,d\theta = \begin{cases} 0 & \text{if } n = 1, 3, 5, \ldots \\ 2\pi\dfrac{1\cdot 3\cdot 5\cdots(n-1)}{2\cdot 4\cdots n} & \text{if } n = 2, 4, 6, \ldots \end{cases}$$

12. For $r > 0$, $b > 0$, and $0 < a < 2$, prove that

$$\int_0^\infty x^{a-1}\sin\left(\frac{a\pi}{2}-bx\right)\frac{dx}{x^2-r^2} = \frac{\pi}{2}r^{a-2}\cos\left(\frac{a\pi}{2}-br\right).$$

13. Substituting $x = \tan\theta$, show that if $-1 < a < 1$ and $-1 < r < 1$,

$$\int_0^{\pi/2}\frac{1-r\cos 2\theta}{1-2r\cos 2\theta+r^2}(\tan\theta)^a\,d\theta = \frac{\pi}{4\cos(\pi a/2)}\left[1+\left(\frac{1-r}{1+r}\right)^a\right].$$

14. Prove that $\displaystyle\int_0^\infty \frac{\log^2 x}{1+x+x^2} = \frac{16}{81}\frac{\pi^3}{\sqrt{3}}.$

15. Show that

(a) $\displaystyle\int_0^\infty \frac{\sin(x^2)}{x}\,dx = \frac{\pi}{4}$

(b) $\displaystyle\int_0^\infty \frac{\cos x}{\sqrt{x}}\,dx = \sqrt{\frac{\pi}{2}}.$

16. Let $a > 0$, use the rectangle $[0, R] \times [0, 1]$, conveniently indented, to compute

$$\int_0^\infty \frac{\sin ax}{e^{2\pi x}-1}\,dx.$$

17. Find a function $f \in \mathscr{M}(\mathbb{C})$ such that

$$f(\alpha) = \int_0^\infty \frac{t^\alpha}{1+t^2}\,dt \qquad \text{when } 0 < \alpha < 1.$$

18. Let $f = P/Q$ be a rational function with $\deg P \leq \deg Q - 2$. Evaluate

$$\lim_{n\to\infty} \int_{\Gamma_n} f(z) \operatorname{cotan} \pi z \, dz$$

and

$$\lim_{n\to\infty} \int_{\Gamma_n} f(z) \operatorname{cosec} \pi z \, dz,$$

where $n \in \mathbb{N}$ and Γ_n is the square with vertices $\pm\left(n + \frac{1}{2}\right) \pm i\left(n + \frac{1}{2}\right)$. Evaluate

(a) $\displaystyle\sum_{n=1}^{\infty} \frac{1}{n^2}$;

(b) $\displaystyle\sum_{n=0}^{\infty} \frac{1}{n^2 + a^2}$ $\quad (0 < a < 1)$;

(c) $\displaystyle\sum_{-\infty}^{\infty} \frac{(-1)^n}{n^2 + a^2}$ $\quad (0 < a < 1)$;

(d) $\displaystyle\sum_{n=0}^{\infty} \frac{1}{n^2 - \omega^2}$ $\quad (w \in \mathbb{C}\setminus\mathbb{Z})$;

(e) $\displaystyle\sum_{n=1}^{\infty} \frac{1}{n^{2k}}$ $\quad (k \in \mathbb{N}^*)$.

19. Show that if $f \in \mathscr{H}(B(0,R))$ and f does not vanish, then for $0 < r < R$ we have

$$\log f(0) = \frac{1}{2\pi i} \int_{|z|=r} \frac{\log f(z)}{z}\,dz.$$

Use this formula to evaluate, for $a \in \mathbb{C}$ fixed,

$$\int_0^{2\pi} \log|e^{i\theta} - a|\,d\theta.$$

Deduce from this computation the Poisson-Jensen formula: Let $f \in \mathscr{H}(\bar{B}(0,R))$, $0 \leq m$, the multiplicity of the origin as zero of f, $a_1, \ldots, a_n$ (counted with multiplicities), the zeros of f in $\bar{B}(0,R)\setminus\{0\}$, then

$$\log\left(\frac{|f^{(m)}(0)|}{m!} R^m \prod_{j=1}^{n} \frac{R}{|a_j|}\right) = \frac{1}{2\pi} \int_0^{2\pi} \log|f(Re^{i\theta})|\,d\theta.$$

20. Let $f \in \mathscr{H}(\bar{B}(0,r))$, $f(z) \neq 0$ for $|z| \leq r$. Show that

$$\frac{1}{2\pi} \int_{-\pi}^{\pi} e^{-i\theta} \log|f(Re^{i\theta})|\,d\theta = \frac{1}{2} r \frac{f'(0)}{f(0)}.$$

Replace the hypothesis that f does not vanish by the hypothesis that $f(0) \neq 0$ and let $z_1, \ldots, z_n$ be the zeros of f (counted with multiplicity) in $\bar{B}(0, r)$. Show that

$$\frac{1}{2\pi}\int_{-\pi}^{\pi} e^{-i\theta}\log|f(Re^{i\theta})|\,d\theta = \frac{1}{2}r\frac{f'(0)}{f(0)} + \frac{1}{2}\sum_{k=1}^{n}\left(\frac{r}{z_k} - \frac{\bar{z}_k}{r}\right)$$

(cf. Exercise 2.5.19).

21. Use the loop in Figure 2.8 to integrate $\int_\gamma \frac{e^{2\pi i z^2/n}}{e^{2\pi i z} - 1}\,dz$. Use this computation to evaluate the Gauss sums $\sum_{k=0}^{n-1} e^{2\pi i k^2/n} = \sqrt{n}\,\frac{i + i^{1-n}}{i+1}$.

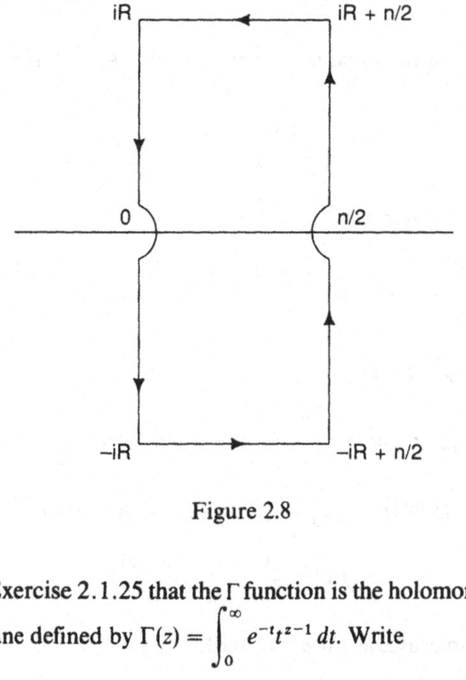

Figure 2.8

22. Recall from Exercise 2.1.25 that the Γ function is the holomorphic function in the right-hand plane defined by $\Gamma(z) = \int_0^\infty e^{-t}t^{z-1}\,dt$. Write

$$\Gamma(z) = \int_1^\infty e^{-t}t^{z-1}\,dt + \int_0^1 e^{-t}t^{z-1}\,dt,$$

expand the function e^{-t} in a Taylor series about $t = 0$, and evaluate the second integral term by term (why is it permissible?) to obtain

$$\Gamma(z) = \int_1^\infty e^{-t}t^{z-1}\,dt + \sum_{k=0}^{\infty}\frac{(-1)^k}{k!}\frac{1}{z+k}.$$

Conclude that Γ has an extension to the whole plane as a meromorphic function with simple poles at $z = -k$, $k \in \mathbb{N}$, with corresponding residues $\frac{(-1)^k}{k!}$.

§6. Other Applications of the Residue Theorem

We are now going to explore several other interesting consequences of the residue theorem and the argument principle (cf. §1.10.10 and §1.6.30).

2.6.1. Proposition. *Let Ω be an admissible open subset of $\mathbb{C}$, $\delta \in Z_1(\Omega; \mathbb{Z})$, f, g holomorphic in Ω, f taking values in $\mathbb{C}^*$. Then*

$$\frac{1}{2\pi i}\int_\delta g\frac{df}{f} = \sum_T \operatorname{Ind}_\delta(T)\operatorname{Res}_{(g/f)\,df}(T),$$

the sum takes place over all the holes of Ω.

PROOF. It is a direct application of the residue theorem to the closed form $g\dfrac{df}{f}$. □

2.6.2. Corollary. *Let Ω be an admissible open subset of $\mathbb{C}$, f a meromorphic function in Ω which is not $\equiv 0$ on any component of Ω. Let $Z(f)$, $P(f)$ be set of zeros and poles of f, respectively. Assume $\tilde{\Omega} = \Omega\backslash(Z(f)\cup P(f))$ is also admissible. Let $g \in \mathscr{H}(\Omega)$ and $\delta \in Z_1(\tilde{\Omega}; \mathbb{Z})$. The following relation holds:*

$$\frac{1}{2\pi i}\int_\delta g\frac{df}{f} = \sum_T \operatorname{Ind}_\delta(T)\operatorname{Res}_{(g/f)\,df}(T) + \sum_{a\in Z(f)} m(f,a)g(a)\operatorname{Ind}_\delta(a)$$
$$- \sum_{a\in P(f)} m(f,a)g(a)\operatorname{Ind}_\delta(a),$$

where $m(f, a)$ is the multiplicity of a as a zero of f (resp. pole of f).

PROOF. It is enough to observe that if $a \in Z(f)$,

$$\operatorname{Res}_{(g/f)\,df}(\{a\}) = m(f,a)g(a)$$

and if $a \in P(f)$,

$$\operatorname{Res}_{(g/f)\,df}(\{a\}) = -m(f,a)g(a).$$ □

2.6.3. Corollary. *Assume the same hypotheses as in §2.6.2, assume moreover that $\delta \in B_1(\Omega; \mathbb{Z})$ (for instance the 1-cycle σ_γ associated to a loop γ homotopic to a point), then*

$$\frac{1}{2\pi i}\int_\delta g\frac{df}{f} = \sum_{a\in Z(f)} m(f,a)g(a)\operatorname{Ind}_\delta(a) - \sum_{a\in P(f)} m(f,a)g(a)\operatorname{Ind}_\delta(a).$$

PROOF. $\operatorname{Ind}_\delta(T) = 0$ for every hole T of Ω since δ is a 1-boundary. □

2.6.4. Corollary (Argument Principle). *Let f be a meromorphic function in an open set U, α a Jordan curve in U whose interior* $\mathrm{Int}(\alpha)$ *is contained in U. Assume α does not intersect $Z(f) \cup P(f)$, then*

$$\frac{1}{2\pi i}\int_\alpha \frac{f'(z)}{f(z)}\,dz = n_Z - n_P = \frac{1}{2\pi i}\Delta_\alpha \log f = \frac{1}{2\pi}\Delta_\alpha \arg f,$$

where

$$n_Z = \#(\mathrm{Int}(\alpha) \cap Z(f)),$$
$$n_P = \#(\mathrm{Int}(\alpha) \cap P(f)),$$

the points being counted according to their multiplicity. Here $\Delta_\alpha \arg f$ (resp. $\Delta_\alpha \log f$) indicates the variation along α of a branch of the argument of f (resp. a branch of $\log f$*).*

Proof. The hypotheses imply that $f \not\equiv 0$ in the component of U which contains α ($f \not\equiv \infty$ by our definition of meromorphic function), hence $n_Z < \infty$ and $n_P < \infty$. We can construct Ω, admissible open set, with

$$\alpha \cup \mathrm{Int}(\alpha) \subseteq \Omega \subseteq U,$$

and apply §2.6.3. The last identities are a consequence of Remark 1.6.30. □

Note that when α is piecewise-C^1 one can consider $\dfrac{f'(z)}{f(z)}$ as $\dfrac{d}{dz}(\log f(z))$, where $\log f(z)$ is a continuous (but multiple-valued) determination of $\log f$. Locally, this derivative makes sense, so that

$$\frac{1}{2\pi i}\int_\alpha \frac{f'(z)}{f(z)}\,dz = \frac{1}{2\pi i}\int_\alpha d(\log f(z))\,dz.$$

2.6.5. Proposition (Rouché's Theorem for Holomorphic Functions—Strong Version). *Let Ω be an open subset of $\mathbb{C}$, α a Jordan curve in Ω with its interior contained in Ω, and f and g meromorphic functions in Ω without poles on α such that*

$$|f(z) - g(z)| < |f(z)| + |g(z)| \qquad (z \in \alpha).$$

Let $n_{Z(f)}, n_{Z(g)}, n_{P(f)}, n_{P(g)}$ be the number of zeros and poles of f and g in $\mathrm{Int}(\alpha)$*, counted according to their multiplicities. Then*

$$n_{Z(f)} - n_{P(f)} = n_{Z(g)} - n_{P(g)}.$$

Proof. In fact, from §1.8.6 the two loops in $\mathbb{C}^*$, $f \circ \alpha$ and $g \circ \alpha$, have the same degree, i.e.,

$$\mathrm{Ind}_{f\circ\alpha}(0) = \mathrm{Ind}_{g\circ\alpha}(0).$$

On the other hand

$$\operatorname{Ind}_{f\circ\alpha}(0) = \frac{1}{2\pi i}\int_\alpha \frac{f'(z)}{f(z)}\,dz,$$

$$\operatorname{Ind}_{g\circ\alpha}(0) = \frac{1}{2\pi i}\int_\alpha \frac{g'(z)}{g(z)}\,dz.$$

Now we can apply §2.6.4. □

2.6.6. Proposition (Rouché's Theorem for Meromorphic Functions—Usual Form). *Let Ω, α be as in §2.6.5 and f, g meromorphic functions in Ω without poles on α such that*

$$|f(z) - g(z)| < |g(z)| \qquad (z \in \alpha).$$

With the same notation as in §2.6.5 we have

$$n_{Z(f)} - n_{P(f)} = n_{Z(g)} - n_{P(g)}.$$

Proof. It can be obtained as a corollary of §2.6.5 or directly from §1.8.7 and §2.6.4. □

2.6.7. Example (D'Alembert-Gauss' Fundamental Theorem of Algebra). Let $f(z) = a_n z^n + \cdots + a_0$, $a_n \neq 0$, and $g(z) = a_n z^n$. Let R be sufficiently large so that

$$|f(z) - g(z)| = |a_{n-1}z^{n-1} + \cdots + a_0| < |g(z)| = |a_n| R^n$$

on $\alpha = \{z \in \mathbb{C} : |z| = R\}$. Proposition 2.6.6 tells us that the number of zeros of the polynomial f inside α equals that of g, that is, n.

2.6.8. Proposition (Hurwitz's Theorem). *Let Ω be an open subset of $\mathbb{C}$, Λ a topological space, $f : \Omega \times \Lambda \to \mathbb{C}$ a continuous map such that, for every $\lambda \in \Lambda$, the function $z \mapsto f_\lambda(z) := f(z, \lambda)$ is holomorphic in Ω. Let z_0 be a zero of multiplicity k $(1 \leq k < \infty)$ of f_{λ_0}. There is an $\varepsilon_0 > 0$ such that for every ε, $0 < \varepsilon \leq \varepsilon_0$, there is a neighborhood V_ε of λ_0 such that f_λ has exactly k zeros (counted with multiplicities) in the disk $B(z_0, \varepsilon)$ if $\lambda \in V_\varepsilon$.*

Proof. Since z_0 is an isolated zero of f_{λ_0}, there is $\varepsilon_0 > 0$ such that f_{λ_0} does not vanish on $\bar{B}(z_0, \varepsilon_0)\setminus\{z_0\}$. For $\varepsilon \in \,]0, \varepsilon_0]$ let

$$m(\varepsilon) = \inf_{|z - z_0| = \varepsilon} |f_{\lambda_0}(z)| > 0.$$

Since $f : \partial B(z_0, \varepsilon) \times \Lambda \to \mathbb{C}$ is continuous, the family (f_λ) converges uniformly to f_{λ_0} on $\partial B(z_0, \varepsilon)$ when $\lambda \to \lambda_0$. Therefore, there exists a neighborhood V_ε of λ_0 such that if $\lambda \in V_\varepsilon$ and $|z - z_0| = \varepsilon$ we have

$$|f_\lambda(z) - f_{\lambda_0}(z)| < m(\varepsilon) \leq |f_{\lambda_0}(z)|.$$

Hence, by §2.6.6, the functions f_λ and f_{λ_0} have the same number of zeros in $B(z_0, \varepsilon)$. □

2.6.9. Examples. (1) Let $P(z; a_0, \ldots, a_n) = a_n z^n + \cdots + a_0$ be a polynomial of degree n in the variable z. Assume that for some choice $a_0^0, \ldots, a_n^0$ of coefficients the value z_0 is a simple root of P. Hence, for coefficients $a_0, \ldots, a_n$ sufficiently close to $a_0^0, \ldots, a_n^0$ the corresponding polynomials have exactly one root in a neighborhood of z_0. This shows that, locally, the root $z = z(a_0, \ldots, a_n)$ is a continuous function of the coefficients. In fact, we can show that this function is differentiable (and even holomorphic in each a_j). Namely, we can write

$$z(a_0, \ldots, a_n) = \frac{1}{2\pi i} \int_{|z_0 - t| = \varepsilon} t \frac{\dfrac{\partial P}{\partial t}(t; a_0, \ldots, a_n)}{P(t; a_0, \ldots, a_n)} dt$$

for $\varepsilon > 0$ sufficiently small and one can differentiate under the integral sign.

(2) The function $f(z) = \lambda - z - e^{-z}$ has a single root (in fact, a real root) in $\operatorname{Re} z \geq 0$ for each λ real > 1. In fact, it is easy to see that $f(x)$ must vanish for some x real, $x \in]0, \lambda[$. If we set $g(z) = \lambda - z$ and consider the contour α_R suggested by Figure 2.9, one verifies that $|e^{-z}| = |f(z) - g(z)| \leq 1 < |\lambda - z|$ if $z \in \alpha_R$. Therefore, the number of roots of $f = 0$ in the interior of α_R does not change with R and remains equal to 1, the number of roots of g.

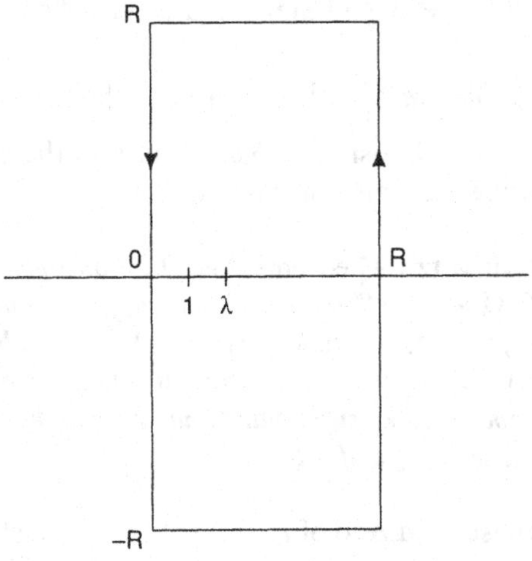

Figure 2.9

2.6.10. Proposition. *Let f be a holomorphic function in $|z| < R$ such that $f(0) = 0$, $f'(0) \neq 0$. Let $r \in]0, R]$ be such that $0 < |z| \leq r$ implies $f(z) \neq 0$. Then, for $0 < \rho < R$, the relation*

$$g(w) := \frac{1}{2\pi i} \int_{|z| = \rho} \frac{t f'(t)}{f(t) - w} dt$$

determines a holomorphic function defined in $|w| < m := \inf_{|t| = \rho} |f(t)|$. *For such* *values of w, $z = g(w)$ is the unique solution of the equation $f(z) = w$, which tends to zero when $w \to 0$ (in fact, $g(w)$ also lies in $|z| < \rho$). The coefficients of the Taylor expansion of g about 0, $g(w) = \sum_{n\geq 1} a_n w^n$ can be computed by*

$$a_n = \frac{1}{2\pi i n}\int_{|t|=\rho}\frac{dt}{(f(t))^n} = \frac{1}{n!}\left\{\frac{d^{n-1}}{dt^{n-1}}\left[\left(\frac{t}{f(t)}\right)^n\right]\right\}(0).$$

Proof. For $|w| < m$ and $|z| = \rho$ we have $|f(z)| \geq m > |w|$. Hence Rouche's theorem asserts that the two functions $f(z)$ and $f(z) - w$ have the same number of zeros in $|z| < \rho$, that is, exactly one root. Denote $g(w)$ the root of $f(z) - w = 0$ in $|z| < \rho$. It is given by the integral formula

$$g(w) = \frac{1}{2\pi i}\int_{|t|=\rho} t\frac{f'(t)}{f(t) - w}\,dt,$$

which shows that it is a holomorphic function of w (by differentiation under the integral sign). We remark that for $|t| = \rho$,

$$\frac{1}{f(t) - w} = \sum_{n\geq 0}\frac{w^n}{(f(t))^{n+1}},$$

which converges uniformly and we can integrate termwise. This implies $g(w) = \sum_{n\geq 0} a_n w^n$

$$a_n = \frac{1}{2\pi i}\int_{|t|=\rho}\frac{tf'(t)}{(f(t))^{n+1}}\,dt.$$

Note that $a_0 = 0$ (as it should) since the corresponding value $g(0)$ is the root of $f(z) = 0$ in $|z| < \rho$, which is exactly $z = 0$ by the hypotheses.

Since $(t/(f(t))^n)' = 1/(f(t))^n - ntf'(t)/(f(t))^{n+1}$, we find

$$a_n = \frac{1}{2\pi i n}\int_{|t|=\rho}\frac{dt}{(f(t))^n},$$

which is the residue at zero of the function $\frac{1}{n}(1/f(t))^n$. The origin is a pole of order n for this function, therefore

$$a_n = \frac{1}{(n-1)!}\left\{\frac{d^{n-1}}{dt^{n-1}}\left[\frac{1}{n}\frac{t^n}{(f(t))^n}\right]\right\}(0). \qquad \square$$

2.6.11. Corollary. *Let f be holomorphic in a neighborhood of $\bar{B}(a, R)$ and injective in $B(a, R)$. If $\Omega = f(B(a, R))$ and $\gamma = \partial B(a, R)$ then the inverse function f^{-1} is defined for $w \in \Omega$ by the relation*

$$f^{-1}(w) = \frac{1}{2\pi i}\int_\gamma \frac{tf'(t)}{f(t) - w}\,dt.$$

2.6.12. Remark. In many applications the equation one tries to solve has the form

$$z - w\varphi(z) = 0, \qquad \varphi(0) \neq 0.$$

This equation fits the Proposition 2.6.10 with $f(z) = z/\varphi(z)$. Therefore the solution $z = g(w)$ has the expansion $g(w) = \sum_{n\geq 1} a_n w^n$ with

$$a_n = \frac{1}{n!}\left\{\frac{d^{n-1}}{dt^{n-1}}[(\varphi(t))^n]\right\}(0).$$

Let us apply this to the equation

$$\lambda - z - e^{-z} = 0 \qquad (\operatorname{Re} z \geq 0, \lambda > 1)$$

from §2.6.9, (2). Intuitively the solution is close to λ when λ is large. We transform this equation to bring it to the preceding form by writing $z = \lambda + s$, then

$$s + e^{-\lambda}e^{-s} = 0.$$

Set $w = -e^{-\lambda}$, then we find $s - we^{-s} = 0$. The solution z of the original equation is now given by

$$z = \lambda + s = \lambda - \sum_{n\geq 1} \frac{n^{n-1}}{n!} e^{-n\lambda}.$$

Stirling's formula gives the asymptotic behavior of the factorial:

$$n! \sim \sqrt{2\pi}\, n^{n+1/2} e^{-n},$$

and therefore

$$\frac{n^{n-1}}{n!} e^{-n\lambda} \sim \frac{1}{\sqrt{2\pi}} \frac{e^{-n(\lambda-1)}}{n^{3/2}},$$

so that the series is convergent for $\lambda \geq 1$ (and it also allows us to estimate very well its value).

Let us recall that a meromorphic function in S^2 is a function which is meromorphic in $\mathbb{C}$ and has an isolated singularity at ∞ which is at most a pole. Clearly any rational function is meromorphic in S^2.

2.6.13. Proposition. *The class of meromorphic functions in S^2 coincides with the class of rational functions.*

PROOF. Let $m(z)$ be a meromorphic function in S^2. The total number of poles is finite. Let us denote $z_1, \ldots, z_n$ the poles in $\mathbb{C}$. For each j, let

$$m_j(z) = \frac{A_{k_j,j}}{(z - z_j)^{k_j}} + \cdots + \frac{A_{1,j}}{z - z_j}$$

be the principal part of the Laurent expansion of m about z_j. The function

$$P(z) := m(z) - \sum_j m_j(z)$$

is an entire function and at ∞ it has at most a pole, hence P is a polynomial. □

Note that we have simultaneously proven the validity of the well-known partial fraction expansion for rational functions.

2.6.14. Proposition. *Let m be a rational function ($m \not\equiv 0$). The sum of the residues over all poles of m and the point of ∞ is zero. In particular, the number of zeros and poles of m in S^2, counted with multiplicities, is the same.*

PROOF. Let $R > 0$ be sufficiently large so that all the poles of m in $\mathbb{C}$ lie in $B(0, R)$. Then the residue formula gives:

$$-\operatorname{Res}(m, \infty) = \frac{1}{2\pi i} \int_{\partial B(0,R)} m(z)\,dz = \sum_{\substack{a \in \mathbb{C} \\ \text{poles}}} \operatorname{Res}(m, a).$$

The second part of the proposition is a consequence of the first applied to m'/m. □

2.6.15. Example (Abel's Formula). Let P, Q be polynomials, $\deg P \leq \deg Q - 2$, every zero of Q is simple. Then we have

$$\sum_{\{a : Q(a) = 0\}} \frac{P(a)}{Q'(a)} = 0.$$

2.6.16. Proposition (Cauchy's Expansion in Partial Fractions). *Let f be a meromorphic function in $\mathbb{C}$ for which there is a family $(\gamma_n)_{n \geq 1}$ of rectifiable Jordan curves such that $\gamma_n \subseteq \operatorname{Int}(\gamma_{n+1})$ and $\mathbb{C} = \bigcup_{n \geq 1} \operatorname{Int}(\gamma_n)$. Assume further that*

$$\lim_{n \to \infty} \int_{\gamma_n} \frac{|f(z)|\,|dz|}{|z|} = 0.$$

Denote by $(a_k)_{k \geq 1}$ the poles of f, assume no a_k lies in any curve γ_n. Then we have the following expansion:

$$f(z) = \sum_{n \geq 1} \sum_{a_k \in D_n} P_k\left(\frac{1}{z - a_k}\right),$$

where $D_1 = \operatorname{Int}(\gamma_1)$, $D_n = \operatorname{Int}(\gamma_n) \setminus \overline{\operatorname{Int}(\gamma_{n-1})}$ ($n \geq 2$), and P_k is the principal part of f at a_k. The series whose general term is $\sum_{a_k \in D_n} P_k\left(\frac{1}{z - a_k}\right)$ converges uniformly on every compact subset of $\mathbb{C} \setminus \{a_k : k \geq 1\}$.

PROOF. For $z \in D_n \setminus \{a_k : k \geq 1\}$, we have

$$\frac{1}{2\pi i}\int_{\gamma_n} \frac{f(t)}{t-z} = \sum_{a_k \in D_n} \operatorname{Res}\left(\frac{f(t)}{t-z}, a_k\right) + \operatorname{Res}\left(\frac{f(t)}{t-z}, t = z\right).$$

Now, it is immediate that

$$\operatorname{Res}\left(\frac{f(t)}{t-z}, t = z\right) = f(z), \qquad \text{and}$$

$$\operatorname{Res}\left(\frac{f(t)}{t-z}, a_k\right) = -P_k\left(\frac{1}{z-a_k}\right).$$

On the other hand, for K compact in $\mathbb{C}\setminus\{a_k : k \geq 1\}$, $z \in K$, one has

$$\limsup_{n\to\infty}\left|\int_{\gamma_n} \frac{f(t)}{t-z}\,dt\right| \leq C(K)\limsup_{n\to\infty}\int_{\gamma_n} \frac{|f(t)|\,|dt|}{|t|} = 0.$$

The expansion is therefore valid. □

2.6.17. Example. Let $f(z) = \dfrac{\operatorname{cotan} z}{z}$ and γ_n be the circle of center 0 and radius $(2n+1)\dfrac{\pi}{2}$. The periodicity of $\operatorname{cotan} z$ implies that there is $M > 0$ such that if $|z - n\pi| \geq \pi/4$ for every $n \in \mathbb{Z}$, then $|\operatorname{cotan} z| \leq M$. Hence

$$\int_{\gamma_n} \left|\frac{\operatorname{cotan} z}{z}\right| \frac{|dz|}{|z|} \leq \frac{C}{n}.$$

The principal part at $z = 0$ of $\dfrac{\operatorname{cotan} z}{z}$ is $\dfrac{1}{z^2}$ and the principal part at $z = n\pi$ $(n \neq 0)$ is $\dfrac{1}{n\pi(z - n\pi)}$, since the points $z = n\pi$ are simple poles of residue $\dfrac{1}{n\pi}$. Therefore, for $z \notin \pi\mathbb{Z}$

$$\frac{\operatorname{cotan} z}{z} = \frac{1}{z^2} + \lim_{n\to\infty} \sum_{-n\leq k\leq n}{}' \frac{1}{k\pi(z - k\pi)},$$

where $\sum_k{}'$ means that $k \neq 0$ in the summation. It follows that

$$\operatorname{cotan} z = \frac{1}{z} + 2z \sum_{n\geq 1} \frac{1}{z^2 - n^2\pi^2}.$$

The last formula follows from the previous one by adding together the terms of index k and $-k$. Note that this series is absolutely and uniformly convergent on compact subsets of $\mathbb{C}\setminus\pi\mathbb{Z}$.

2.6.18. Proposition (Hurwitz's Theorem—Second Version). *Let Ω be an open subset of $\mathbb{C}$, Λ a topological space, $f : \Omega \times \Lambda \to \mathbb{C}$ a continuous function such* *that, for every* $\lambda \in \Lambda$,

$$f_\lambda : z \mapsto f(z, \lambda)$$

is holomorphic in Ω. *Let* $\lambda_0 \in \Lambda$ *and* γ *be a Jordan curve in* Ω *such that* $\overline{\operatorname{Int}(\gamma)} \subseteq \Omega$ *and* f_{λ_0} *does not vanish on* γ. *There is a neighborhood* V *of* λ_0 *in* Λ *such that for any* $\lambda \in V$, f_λ *and* f_{λ_0} *have the same number of zeros in* $\operatorname{Int}(\gamma)$.

PROOF. Let $m := \inf_{z \in \gamma} |f_{\lambda_0}(z)| > 0$. By the uniform continuity of $(f_\lambda)_{\lambda \in \Lambda}$ on γ we have a neighborhood of V of λ_0 in Λ such that if $\lambda \in V$ and $z \in \gamma$

$$|f_\lambda(z) - f_{\lambda_0}(z)| < m \leq |f_{\lambda_0}(z)| \qquad (z \in \gamma).$$

The proposition follows now from Rouché's theorem. □

2.6.19. Proposition. *Let* Ω *be a connected open subset of* $\mathbb{C}$, $(f_n)_{n \geq 1}$ *a sequence of injective holomorphic functions that converges in* $\mathscr{H}(\Omega)$ *to* f. *Then, either* f *is constant or* f *is injective.*

PROOF. Assume $f(z_1) = f(z_2) = a$ for some $z_1 \neq z_2$. If $f \not\equiv a$ then Hurwitz's Theorem 2.6.18 (with $\Lambda = \mathbb{N} \cup \{\infty\}$ and $f_\infty = f$) implies that $f_n - a$ has at least one zero in $B(z_1, \rho)$ and one in $B(z_2, \rho)$, for n sufficiently large and ρ conveniently chosen satisfying $0 < 2\rho < |z_1 - z_2|$. This contradicts the injectivity of f_n. □

2.6.20. Proposition. *Let* $B = \{z \in \mathbb{C} : |z| < 1\}$ *and let* Ω *be a connected open subset of* B *containing the origin. Let* $\mathscr{B} \subseteq \mathscr{A} \subseteq \mathscr{H}(\Omega)$ *be defined by*:

$$\mathscr{A} := \{f \in \mathscr{H}(\Omega) : f(\Omega) \subseteq B, f \text{ injective}, f(0) = 0\}$$

$$\mathscr{B} := \{f \in \mathscr{A} : |f'(0)| \geq 1\}.$$

Then the set $\mathscr{B}$ *is compact in* $\mathscr{H}(\Omega)$ *and there is* $f \in \mathscr{B}$ *such that*

$$|f'(0)| = \sup_{g \in \mathscr{A}} |g'(0)|.$$

If we assume further that Ω *is simply connected then this function* f *is a biholomorphism of* Ω *onto* B.

PROOF. Since $\Omega \subseteq B$, at least the identity function $z \mapsto z$ belongs to $\mathscr{B}$. Let us show that $\mathscr{B}$ is compact in $\mathscr{H}(\Omega)$; it will then follow that

$$\sup_{g \in \mathscr{A}} |g'(0)| = \sup_{g \in \mathscr{B}} |g'(0)|$$

is, in fact, achieved at some $f \in \mathscr{B}$, since $g \mapsto |g'(0)|$ is a continuous function on $\mathscr{B}$.

Since $\mathscr{B}$ is a bounded subset of $\mathscr{H}(\Omega)$, it is enough to show that it is closed. If $\varphi \in \overline{\mathscr{B}}$ there is a sequence $\varphi_n \to \varphi$ in $\mathscr{H}(\Omega)$, $\{\varphi_n\}_{n \geq 1} \subseteq \mathscr{B}$. Hence $\varphi(0) = 0$ and $|\varphi'(0)| \geq 1$. It follows that φ is not constant and hence, by §2.6.19, φ is injective. Since $\varphi_n(\Omega) \subseteq B$, it follows $\varphi(\Omega) \subseteq \bar{B}$. But the maximum principle implies now that $\varphi(\Omega) \subseteq B$. Therefore $\varphi \in \mathscr{B}$.

Let $f \in \mathscr{B}$ have the largest possible derivative at 0. To show that in the case when Ω is simply connected f is a biholomorphism onto B we only have to prove f is surjective. Assume there is $a \in B \setminus f(\Omega)$. We want to construct $g \in \mathscr{A}$ such that $|g'(0)| > |f'(0)|$. This will contradict the choice of f.

The simple connectivity of Ω assures us of the existence of the logarithm of any nonvanishing holomorphic function. Let

$$F(z) := \log\left(\frac{f(z) - a}{1 - \bar{a}f(z)}\right).$$

F is then holomorphic in Ω and since $\left|\dfrac{f(z) - a}{1 - \bar{a}f(z)}\right| < 1$ we also have $\operatorname{Re} F < 0$. Clearly, it is also injective.

To construct a function in $\mathscr{A}$ we use the inequality

$$\operatorname{Re} u < 0, \quad \operatorname{Re} v < 0 \quad \text{implies} \quad \left|\frac{v - u}{v + \bar{u}}\right| < 1.$$

(For u fixed, this is just a conformal map of $\{\operatorname{Re} v < 0\}$ onto B.) Therefore the function

$$g(z) := \frac{F(z) - F(0)}{F(z) + \overline{F(0)}} \in \mathscr{A}.$$

Let us now compute $g'(0)$. We note that

$$F'(0) = \left(\bar{a} - \frac{1}{a}\right)f'(0) \qquad \text{and} \qquad g'(0) = \frac{F'(0)}{F(0) + \bar{F}(0)}.$$

Hence,

$$\frac{|g'(0)|}{|f'(0)|} = \frac{1 - |a|^2}{-2|a|\log|a|}.$$

Consider the auxiliary function $\psi : t \mapsto 1 - t^2 + 2t\log t$ in $0 < t < 1$. It is strictly convex, $\psi'(1) = 0$, and $\psi(1) = 0$, therefore $\psi(t) > 0$. This implies that

$$|g'(0)| > |f'(0)|,$$

which is the contradiction we were looking for. Therefore f is a biholomorphism of Ω onto B. □

2.6.21. Riemann Mapping Theorem. *Every proper simply connected open subset of $\mathbb{C}$ is biholomorphic to the unit disk B.*

PROOF. Let us assume Ω_1 is a simply connected proper open subset of $\mathbb{C}$. To show it is biholomorphic to B it is enough to show that it is biholomorphic to a bounded open subset of $\mathbb{C}$. The rest follows by applying a similarity and the result obtained in Proposition 2.6.20.

If there is a closed ball $\bar{B}(a, \rho)$ disjoint from Ω_1, the map $\theta : z \mapsto \dfrac{1}{z - a}$ is a biholomorphism for Ω_1 onto $\theta(\Omega_1)$, which is simply connected and bounded.

In general, let $b \in \partial\Omega_1$ (it exists since $\Omega_1 \neq \mathbb{C}$), the holomorphic function $z \mapsto z - b$ does not vanish on Ω_1. Hence there are two determinations R_+, R_- of $\sqrt{z - b}$ in Ω_1. We have $R_- = -R_+$. The open sets $U_+ = R_+(\Omega_1)$ and $U_- = R_-(\Omega_1)$ are disjoint, otherwise if $R_+(z_1) = R_-(z_2)$ we would have $z_1 - b = z_2 - b$. Therefore $z_1 = z_2$ but $R_+(z_1) = R_-(z_1) = -R_+(z_1)$ implies $z_1 = b$, which is impossible. It follows that Ω_1 is biholomorphic to U_+, but U_+ is disjoint from the nonempty open set U_-. Hence U_+ is disjoint from some closed ball and we are back in the previous case. □

The ramifications of the Riemann mapping theorem will be explored in §2.8. Meanwhile, we conclude this section with another application of the argument principle.

Given a connected open set $\Omega \subseteq \mathbb{C}$, we have that $\mathcal{M}(\Omega) \subseteq \mathscr{C}(\Omega, S^2)$, the space of all continuous mappings from Ω into S^2. Since the Riemann sphere is a metric space with the chordal distance, we can consider in the space $\mathscr{C}(\Omega, S^2)$ the topology of uniform convergence on compact subsets of Ω. When restricted to the subspace $\mathscr{H}(\Omega)$ of $\mathscr{C}(\Omega, S^2)$, this topology actually coincides with the usual topology of $\mathscr{H}(\Omega)$.

2.6.24. Definition. A sequence $(f_n)_{n \geq 1} \subseteq \mathcal{M}(\Omega)$ ***converges normally*** to a function $f : \Omega \to S^2$ if $\sigma(f_n(z), f(z)) \to 0$ as $n \to \infty$ uniformly on compact subsets of Ω, where σ is the chordal distance in S^2 defined in Exercise 2.3.1.

The crucial property of this notion is the following proposition, whose proof is left to the reader.

2.6.25. Proposition. *Let $(f_n)_{n \geq 1} \subseteq \mathcal{M}(\Omega)$ converge normally to f, Ω connected open set in $\mathbb{C}$. Then, either $f \in \mathcal{M}(\Omega)$ or $f \equiv \infty$. Moreover, if all the f_n are holomorphic and $f \not\equiv \infty$, then $f \in \mathscr{H}(\Omega)$ and $f_n \to f$ in the topology of $\mathscr{H}(\Omega)$.*

The concept of normal convergence can be extended to sequences in $\mathscr{C}(\Omega, S^2)$ and, even further, assume that Ω is a connected open set in S^2. The previous proposition still makes sense and it can be translated into the following.

2.6.26. Proposition. *Let Ω be an open connected set in S^2, $\mathscr{C}(\Omega, S^2)$ is a Fréchet space when considered with the topology of normal convergence. The closure of $\mathcal{M}(\Omega)$ in this topology is $\mathcal{M}(\Omega) \cup \{\infty\}$, ∞ representing the constant function identically equal to ∞ in Ω.*

Finally, we have the following generalization of the concept of relatively compact families in $\mathscr{H}(\Omega)$.

2.6.27. Definition. A family $\mathscr{F} \subseteq \mathscr{M}(\Omega)$ (Ω open connected subset of S^2) is said to be a ***normal family*** if every sequence $(f_n)_{n\geq 1} \subseteq \mathscr{F}$ has a normally convergent subsequence (to an element in $\mathscr{M}(\Omega) \cup \{\infty\}$).

EXERCISES 2.6

Ω represents an open connected subset of $\mathbb{C}$, though some statements are clearly valid also when $\Omega \subseteq S^2$. $B = B(0, 1) \subseteq \mathbb{C}$.

1. Let Ω be an open connected subset of S^2, $f \in \mathscr{M}(\Omega)$. Then $f(\Omega)$ is open and connected. Moreover, if f is injective then $f^{-1} \in \mathscr{M}(f(\Omega))$.
2. Prove Propositions 2.6.25 and 2.6.26.
3. Show that any bounded family in $\mathscr{H}(\Omega)$ is a normal family in the sense of Definition 2.6.27. Is the converse true? Is it true that if $\mathscr{F} \subseteq \mathscr{H}(\Omega)$ and the family $\mathscr{F}' = \{f' : f \in \mathscr{F}\}$ is normal, then $\mathscr{F}$ is normal? Is the family $\mathscr{F} = \{f \equiv c, c \in \mathbb{C}\}$ normal?
4. Show that the family $\mathscr{F} = \{f \in \mathscr{H}(\Omega) : \operatorname{Re} f > 0\}$ is normal.
5. Show that if $f \in \mathscr{H}(S^2)$ then $f \equiv c \in \mathbb{C}$.

*6. Let $f \in \mathscr{H}(\mathbb{C})$ and $0 < r < R < \infty$ be two fixed numbers. Consider the family $\mathscr{F}$ of functions f_k, $f_k(z) = f(kz)$, $k \in \mathbb{N}^*$, for $z \in A(0; r, R)$. Show that $\mathscr{F}$ is normal if and only if f is a polynomial.

*7. (Marty's Theorem). A family $\mathscr{F} \subseteq \mathscr{M}(\Omega)$ is normal if and only if the family $\mathscr{F}^\# = \{f^\# : f \in \mathscr{F}\}$ of spherical derivatives is locally uniformly bounded. (Hint: See Exercise 2.3.3 for the definition of spherical derivatives. Use this exercise together with Arzelá-Ascoli's theorem to obtain the result.)

8. Show that all the fixed points of the meromorphic function $\tan z$ are real and simple.
9. How many roots does the polynomial $z^4 - 5z + 1$ have in $B(0, 1)$? Write down the first two terms of the power series expansion about $\lambda = 1$ of the root $z(\lambda) \in B(0, 1)$ of the equation
$$z^4 - 5z + \lambda = 0.$$
10. How many roots does $p(z) = z^5 + 12z^3 + 3z^2 + 20z + 3$ have in the annulus $\{z : 1 < |z| < 2\}$.
11. Show that all the roots of the equation $z^3 - z + \lambda = e^{-z}(z + 2)$ lie in the half-plane $\operatorname{Re} z < 0$, if $\lambda > 2$.
12. Let $f \in \mathscr{H}(B)$, $f(0) \neq 0$, $m \in \mathbb{N}^*$. Show that there exists $\rho > 0$ such that for every w, $0 < |w| < \rho$, the equation $z^m = wf(z)$ has m distinct roots in B.

*13. Let φ be a continuous real-valued nondecreasing (or nonincreasing) function in the interval $[0, 1]$. Show that the entire function

$$f(z) = \int_0^1 \varphi(t)\cos(tz)\,dt$$

has only real zeros. (Hint: Use Exercise 1.8.5.)

14. Let $\mathscr{F} = \{f \in \mathscr{H}(\Omega) : f(\Omega) \subseteq B, f \text{ injective}\}$. Determine the closure of the family $\mathscr{F}$ in $\mathscr{H}(\Omega)$.

15. Let $f \in \mathscr{H}(B)$, $f(0) = 0$, $f'(0) = 1$, f injective, $\|f\|_\infty \leq 1$. Consider the family $\{f^{[n]}\}_{n \geq 1}$ of iterates $f^{[n]} = f \circ \cdots \circ f$. Show that every limit function of this family is a conformal map of B onto an open set $\Omega \subseteq B$.

*16. Show that if Ω is a proper connected subset of $\mathbb{C}$ such that every nonvanishing holomorphic function admits a continuous square root, then Ω is simply connected. (Hint: This requires a modification of the construction of the auxiliary function F in the proof of Proposition 2.6.20.)

17. Let $S = \{f \in \mathscr{H}(B) : f(0) = 0, f'(0) = 1, f \text{ injective}\}$. Show that S is a closed subset of $\mathscr{H}(B)$.

18. Use Example 3.6.17 to compute $\sum_{n=1}^{\infty} \frac{1}{n^2}$.

19. Let $f : B \to B$ be holomorphic and have the property that there is a unique point $z_0 \in B$ such that $f(z_0) = z_0$. Show that if $\mathscr{F}$ is a family of holomorphic functions in B such that $g(B) \subseteq B$ and $f \circ g = g \circ f$ for every $g \in \mathscr{F}$, then $g(z_0) = z_0$ for every $g \in \mathscr{F}$.

20. Let $f \in \mathscr{H}(B)$ and $f(B) \subseteq B$. Assume there is some increasing sequence $n_k \to \infty$ of integers and some $g \in \operatorname{Aut}(B)$ such that $f^{[n_k]} \to g$ in $\mathscr{H}(B)$. Show that $f \in \operatorname{Aut}(B)$.

21. (a) Prove Abel's Formula 2.6.15.
    (b) Let $\zeta, a_1, \ldots, a_n$ be distinct complex numbers, P a polynomial of degree $n - 1$ such that

$$P(a_j) = b_j, \qquad j = 1, \ldots, n.$$

    Compute $P(\zeta)$ using Abel's formula and $Q(z) = (z - \zeta)(z - a_1)\ldots(z - a_n)$ (You will obtain Lagrange's interpolation formula).
    (c) Let $\zeta, a_1, \ldots, a_n$ as in (b), $\nu_1, \ldots, \nu_n \in \mathbb{N}^*$, $d = \nu_1 + \cdots + \nu_n - 1$, P a polynomial of degree d such that

$$\frac{P^{(k)}(a_j)}{k!} = b_{j,k}, \qquad 1 \leq j \leq n, \quad 0 \leq k \leq \nu_j - 1.$$

    Find a formula for $P(\zeta)$ using Abel's formula as in (b).

22. Let $f \in \mathscr{H}(\bar{B}(0, 3))$ and $f \neq 0$ on $|z| = 3$. Assume that

$$\frac{1}{2\pi i} \int_{|z|=3} \frac{f'(z)}{f(z)}\,dz = 2,$$

$$\frac{1}{2\pi i} \int_{|z|=3} z\frac{f'(z)}{f(z)}\,dz = 2,$$

and

$$\frac{1}{2\pi i}\int_{|z|=3} z^2\frac{f'(z)}{f(z)}\,dz = -4$$

Find the roots of f in $B(0,3)$.

§7. The Area Theorem

In this section we consider some simple properties of a particularly interesting class of conformal mappings of the disk, the class S. These properties are consequences of the area theorem (Lemma 2.7.3). We will use variants of these ideas to obtain Picard's theorem about the behavior of a holomorphic function near an essential singularity. The main theme is the interplay between covering properties of holomorphic maps and the concepts of bounded and normal families.

2.7.1. Definition. We denote by S the class of holomorphic functions in $B(0,1)$ which are injective and satisfy the normalization conditions $f(0) = 0$ and $f'(0) = 1$.

This notation comes from the German word *schlicht* which means precisely injective. We are going to show that S is a compact subset of $\mathscr{H}(B(0,1))$. We already know that as a consequence of Hurwitz's theorem, S is closed (cf. Exercise 2.6.17). To prove the compactness, all we need to do is to obtain a uniform estimate for each $r < 1$ of the values $|f(z)|$, where $|z| \le r < 1$ and $f \in S$. This will follow from an estimation of the second coefficient a_2 of the Taylor development of f at the origin

$$f(z) = z + a_2z^2 + a_3z^3 + \cdots.$$

2.7.2. Proposition. *For every* $f \in S$, $|a_2| \le 2$.

Proof. To every $f \in S$ we will associate a function g injective and holomorphic in the region $1 < |z| < \infty$; furthermore it will turn out that

$$g(z) = z + b_0 + \frac{b_1}{z} + \frac{b_2}{z^2} + \cdots.$$

Before making the b_j more precise, we consider the following useful lemma.

2.7.3. Lemma (Area Theorem). *Let* $g(z) = z + b_0 + \dfrac{b_1}{z} + \dfrac{b_2}{z^2} + \cdots$ *be holomorphic and injective in* $1 < |z| < \infty$, *then*

$$\sum_{n\ge 1} n|b_n|^2 \le 1.$$

Proof. Let E be the compact set $E := \mathbb{C}\backslash g(\{z : 1 < |z| < \infty\})$. Since g is injective then the image of the circle $|z| = r > 1$ is a Jordan curve whose interior E_r contains E (this follows from $g(\infty) = \infty$). Therefore the area $m(E_r) \geq m(E) \geq 0$ can be computed as follows:

$$m(E_r) = \frac{-1}{2i}\int_{E_r} dw \wedge d\bar{w} = \frac{1}{2i}\int_{\partial E_r} \bar{w}\,dw = \frac{1}{2i}\int_{|z|=r} g^*(\bar{w}\,dw)$$

$$= \frac{1}{2i}\int_{|z|=r} g'(z)\overline{g(z)}\,dz$$

$$= \frac{1}{2}\int_0^{2\pi}\left(re^{-i\theta} + \sum_{n\geq 0} \bar{b}_n \frac{e^{in\theta}}{r^n}\right)\left(1 - \sum_{n\geq 1} nb_n \frac{e^{-i(n+1)\theta}}{r^{n+1}}\right) re^{i\theta}\,d\theta$$

$$= \pi\left(r^2 - \sum_{n\geq 1} n\frac{|b_n|^2}{r^{2n}}\right).$$

Letting $r \to 1$ we obtain

$$0 \leq m(E) = \pi\left(1 - \sum_{n\geq 1} n|b_n|^2\right).$$

This proves the lemma. □

From the lemma it follows that $|b_1| \leq 1$. Returning to the proof of Proposition 2.7.2, to every $f \in S$ we associate the function

$$g(z) = \frac{1}{\sqrt{f(1/z^2)}} = z - \frac{a_2}{2}\frac{1}{z} + \cdots.$$

Once we show that g is well defined, satisfies the properties of Lemma 2.7.3, and has the coefficient $b_1 = \dfrac{-a_2}{2}$, then the proposition will be proven.

First we note that $\sqrt{f(w^2)}$ makes sense since

$$f(w^2) = w^2 + a_2 w^4 + a_3 w^6 + \cdots = w^2(1 + a_2 w^2 + a_3 w^4 + \cdots),$$

and the function $(1 + a_2 w^2 + \cdots)$ is holomorphic in $B(0, 1)$ and never vanishes since f is injective (hence $f(w^2)$ only vanishes at $w = 0$, where it has a double zero). Therefore we have a well-defined holomorphic square root $(1 + a_2 w^2 + \cdots)^{1/2}$ which takes the value 1 at $w = 0$,

$$(1 + a_2 w^2 + \cdots)^{1/2} = 1 + \frac{a_2}{2} w^2 + \cdots$$

and

$$h(w) := \sqrt{f(w^2)} = w\left(1 + \frac{a_2}{2} w^2 + \cdots\right),$$

is holomorphic in $B(0, 1)$, and vanishes only for $w = 0$. Let us check that this function is still injective (and hence it is in the class S). If $h(w_1) = h(w_2)$, w_1, $w_2 \in B(0, 1)$, $w_1 \neq w_2$, then $f(w_1^2) = f(w_2^2)$. Hence $w_1^2 = w_2^2$ and $w_1 = -w_2$. But h is an odd function, $h(w_2) = h(-w_1) = -h(w_1)$ implies $h(w_1) = 0$, hence

$w_1 = w_2 = 0$, which is impossible. (The reader should compare this argument with the proof of the Riemann mapping theorem.)

It follows that $g(z) := 1/h(1/z)$ is holomorphic and injective in the annulus $1 < |z| < \infty$. Furthermore,

$$g(z) = \frac{z}{1 + \frac{a_2}{2}\frac{1}{z^2} + \cdots} = z - \frac{a_2}{2}\frac{1}{z} + \cdots.$$

The proposition has therefore been proven. □

Recall that for $a \in B(0, 1)$, we denote by φ_a the Moebius transformation $\varphi_a(z) = (z - a)/(1 - \bar{a}z)$. For $f \in S$, consider the function $f \circ \varphi_{-a}$. The latter is injective in $B(0, 1)$ but it does not satisfy the normalization conditions to be in S. Its derivative at the origin is $(f \circ \varphi_{-a})'(0) = f'(a)\varphi'_{-a}(0) = (1 - |a|^2)f'(a)$. Hence, the function

$$F(z) := \frac{f \circ \varphi_{-a}(z) - f(a)}{(1 - |a|^2)f'(a)}$$

is in the class S. The Taylor development of F at the origin is $F(z) = z + A_2(a)z^2 + \cdots$. The chain rule gives

$$(f \circ \varphi_{-a})''(0) = f''(a)(1 - |a|^2)^2 - 2\bar{a}f'(a)(1 - |a|^2),$$

whence

$$A_2(a) = \frac{1}{2}\left((1 - |a|^2)\frac{f''(a)}{f'(a)} - 2\bar{a}\right).$$

From here we obtain the following.

2.7.4. Proposition. *For every $f \in S$ and $a \in B(0, 1)$ we have*

$$\left| a\frac{f''(a)}{f'(a)} - \frac{2r^2}{1 - r^2} \right| \le \frac{4r}{1 - r^2},$$

where $|a| = r$.

Proof. From the preceding computation we get

$$2a\frac{A_2(a)}{1 - |a|^2} = a\frac{f''(a)}{f'(a)} - \frac{2|a|^2}{1 - |a|^2}.$$

Using now that $|a| = r$ and that, by Proposition 2.7.3, $|A_2(z)| \le 2$, we obtain the desired inequality. □

For a function $f \in S$ we have that $f'(z)$ does not vanish and $f'(0) = 1$, therefore we can define the holomorphic function $g(z) = \operatorname{Log} f'(z)$. We would like to estimate its real part, $\operatorname{Re} g(z) = \log|f'(z)|$. For that purpose, we consider its derivative in the radial direction:

$$\frac{\partial}{\partial r}\operatorname{Re} g(re^{i\theta}) = \operatorname{Re}\left(\frac{\partial}{\partial r} g(re^{i\theta})\right) = \operatorname{Re}\left(\frac{f''(re^{i\theta})}{f'(re^{i\theta})} e^{i\theta}\right).$$

Hence, for $z = re^{i\theta}$,

$$r\frac{\partial}{\partial r}\log|f'(re^{i\theta})| = \operatorname{Re}\left(\frac{zf''(z)}{f'(z)}\right).$$

Note that the right-hand side contains one of the quantities that appears in Proposition 2.7.4. From that proposition we conclude that

$$\frac{2r^2 - 4r}{1 - r^2} \leq \operatorname{Re}\left(\frac{zf''(z)}{f'(z)}\right) \leq \frac{2r^2 + 4r}{1 - r^2}, \qquad |z| = r.$$

Divide by r and integrate in r between 0 and ρ, $0 \leq \rho < 1$, we obtain

$$\log\left(\frac{1-\rho}{(1+\rho)^3}\right) \leq \log|f'(\rho e^{i\theta})| \leq \log\left(\frac{1+\rho}{(1-\rho)^3}\right).$$

It follows that

$$\frac{1-\rho}{(1+\rho)^3} \leq |f'(\rho e^{i\theta})| \leq \frac{1+\rho}{(1-\rho)^3},$$

for any $f \in S$. Therefore if $|z| = r < 1$ we have

$$|f(z)| \leq \int_0^r |f'(\rho e^{i\theta})|\, d\rho \leq \int_0^r \frac{1+\rho}{(1-\rho)^3}\, d\rho = \frac{r}{(1-r)^2}.$$

Since this estimate is independent of f, it follows that S is a bounded family. We already know that S is a closed subset of $\mathscr{H}(B(0, 1))$. This proves the following.

2.7.6. Theorem. *The family S is compact in $\mathscr{H}(B(0, 1))$.*

We have not used the lower bound on $|f'(z)|$ obtained earlier. If we do, we obtain a more precise result.

2.7.7. Proposition. *If $f \in S$, then*

$$\frac{r}{(1+r)^2} \leq |f(z)| \leq \frac{r}{(1-r)^2}, \qquad |z| = r < 1.$$

Proof. We have already proven the upper bound. To prove the lower bound, note that $\dfrac{r}{(1+r)^2} < \dfrac{1}{4}$, hence the lower bound is automatically valid if $|f(z)| \geq \dfrac{1}{4}$. In the other case, let us assume for the moment that the segment

$L = [0, f(z)]$ lies entirely in the image of f. In that case, let γ be the Jordan arc in $B(0, 1)$ going from 0 to z, which is the inverse image of L. Over γ we have that $\operatorname{Arg}\left(f'(\zeta)\dfrac{d\zeta}{ds}\right) \equiv \operatorname{Arg} f(z) = \theta$, where s = arc length in γ. Therefore

$$f(z) = \int_\gamma f'(\zeta)\,d\zeta = e^{i\theta}\int_\gamma |f'(\zeta)|\,|d\zeta|.$$

If we consider $|\zeta|^2$ as a function of s, we see that there are only finitely many intervals where this function is not increasing (use that $|\zeta|^2$ is a real analytic function of s). Disregarding those intervals, we obtain a chain $\tilde{\gamma}$ where $|\zeta|$ takes values from 0 to r. Moreover, it is easy to see that $\left|\dfrac{d\zeta}{d\rho}\right| \geq 1$ in $\tilde{\gamma}$, where $\rho = |\zeta|$. These observations, plus the previous identity, show that

$$|f(z)| \geq \int_{\tilde{\gamma}} |f'(\zeta)|\,|d\zeta| \geq \int_0^r \frac{1-\rho}{(1+\rho)^3}\,d\rho = \frac{r}{(1+r)^2}.$$

We have to justify now our assumption. This is also a consequence of Proposition 2.7.2, as seen from the following.

2.7.8. Lemma (Koebe's One-Quarter Theorem). *If $f \in S$ then $f(B(0,1)) \supseteq B(0, 1/4)$.*

Proof. If $\zeta \notin f(B(0, 1))$, consider the auxiliary function

$$F(z) := \frac{\zeta f(z)}{\zeta - f(z)} = z + \left(a_2 + \frac{1}{\zeta}\right)z^2 + \cdots.$$

This function is in the class S, hence by Proposition 2.7.2,

$$\left|a_2 + \frac{1}{\zeta}\right| \leq 2.$$

Therefore,

$$\left|\frac{1}{\zeta}\right| \leq 4,$$

which shows that the lemma is correct. □

As we said earlier, the only case where we needed the segment $L = [0, f(z)]$ to be entirely contained in the image of f occurred when $|f(z)| < 1/4$. This is now assured by Lemma 2.7.8. □

We have seen the importance of the estimate $|a_2| \leq 2$. This estimate was originally obtained by L. Bieberbach. He also conjectured that if $(a_n)_{n \geq 0}$ are the Taylor coefficients of a function f in the class S, then $|a_n| \leq n$. This conjecture stood open for close to 70 years; its simplicity motivated a large amount of research on variational methods and other aspects of the theory of functions in the class S. In 1984, L. de Branges [de B] gave a beautiful and relatively simple proof of the correctness of this conjecture. The reader will find some references to these questions in the notes to this chapter. Let us also point out the fact that the only functions in the class S for which some coefficient a_n satisfies $|a_n| = n$, are the Koebe functions:

$$k(z) = \frac{z}{(1-z)^2} = z + 2z^2 + 3z^3 + \cdots$$

and

$$k_\theta(z) = e^{-i\theta} k(e^{i\theta} z), \qquad (\theta \in [0, 2\pi[).$$

These functions also have the remarkable property that they are the only functions in the class S for which $(f(B(0,1)))^c \cap \partial B(0, 1/4) \neq \varnothing$ (cf. [Du]).

We present here another application of the one-quarter theorem, which will be used in the proof of Lemma 3.7.2. The reader could very well bypass this statement until he arrives at that point in Chapter 3.

2.7.9. Lemma. *Let K be a compact connected set in $\mathbb{C}$ of diameter greater than or equal to $r > 0$. Assume further that K^c is connected. Then there is a unique biholomorphic map $F: B(0,1) \to S^2 \backslash K$ such that*

$$F(z) = \frac{a}{z} + \sum_{n \geq 0} b_n z^n, \qquad a > 0.$$

Moreover, $a \geq \dfrac{r}{4}$.

Proof. The hypothesis on K shows that $S^2 \backslash K$ is simply connected and biholomorphic to the unit disk by the Riemann mapping theorem. The uniqueness of F follows from the normalization conditions $F(0) = \infty$ and $a > 0$.

In order to estimate a using the one-quarter theorem, we construct an auxiliary function φ in the class S. Let $w_0 \in K$, then

$$\varphi(z) = \frac{a}{F(z) - w_0} = z - \frac{b_0 - w_0}{a} z^2 + \cdots.$$

This function is clearly holomorphic in $B(0,1)$ and a conformal map as a composition of F with a Moebius transformation. Hence $\varphi \in S$. It is clear that for any $w_1 \in K$ we have $\dfrac{a}{w_1 - w_0} \notin \varphi(B(0,1))$. Therefore §2.7.8 implies that

$$a \geq \frac{1}{4} |w_1 - w_0|.$$

Since w_0, w_1 are arbitrary, the condition diameter of $K \geq r$ immediately shows that

$$a \geq \frac{r}{4}. \qquad \square$$

This lemma is related to the concept of *analytic capacity* $\alpha(K)$ of a compact set K. The analytic capacity is defined by

$$\alpha(K) := \sup\{|\operatorname{Res}(f, \infty)| : f \in \mathscr{H}(S^2 \setminus K), \|f\|_\infty \leq 1\}.$$

Considering the function $f = F^{-1}$ in Lemma 2.7.9 we see that the lemma states that $\alpha(K) \geq \frac{r}{4}$, when K is a connected compact set of $\mathbb{C}$ with diameter greater than or equal to $r > 0$ and $S^2 \setminus K$ is simply connected.

In Chapter 4 we will also consider the logarithmic capacity $C(K)$ of a compact set. It will not be hard to see then that $C(K) \geq \alpha(K)$ (cf. Exercise 4.9.12). Otherwise, the properties of $\alpha(K)$ still are not well understood (see [Za1], [Vi]).

It is natural to ask whether the one-quarter theorem can be generalized to functions that are not injective. This is the content of the following theorem of Landau.

2.7.10. Proposition. *Let f be holomorphic in $\bar{B}(0, 1)$ and $f'(0) = 1$. Then $f(\bar{B}(0, 1))$ contains a disk of radius greater than or equal to $\frac{1}{16}$.*

This theorem has been further strengthened by Bloch, who showed that under the same hypotheses, there is an absolute constant β (Bloch's constant) and a disk Δ in $B(0, 1)$ such that f is injective in Δ and $f(\Delta)$ contains a disk of radius $\geq \beta$. The largest constant λ that can replace $\frac{1}{16}$ in Proposition 2.7.10 is called Landau's constant [Ah2].

Proof of Proposition 2.7.10. There is a largest value $0 \leq r < 1$ such that $\max_{|z|=r} |f'(z)|(1 - |z|) = 1$. In fact, for $r = 0$ we have equality, and $r = 1$ is not possible. Let z_0 be a point with $|z_0| = r$, $|f'(z_0)| = \frac{1}{1-r}$. Consider the auxiliary function $g(\zeta) := f(\zeta + z_0) - f(z_0)$. It is clearly enough to show that the image of g covers a disk of radius greater than or equal to $\frac{1}{16}$. The function g is holomorphic in $|\zeta| \leq 1 - r$, $g(0) = 0$ and $|g'(0)| = \frac{1}{1-r}$. Moreover, for $|\zeta| = \frac{1-r}{2}$, we have

$$|g'(\zeta)| = |f'(z_0 + \zeta)| \leq \max_{|z|=(1+r)/2} |f'(z)|$$
$$= \frac{2}{1-r} \max_{|z|=(1+r)/2} (1 - |z|)|f'(z)| < \frac{2}{1-r},$$

by the definition of r. It follows that for $|\zeta| \leq \frac{1-r}{2}$,

$$|g(\zeta)| = \left| \int_0^\zeta g'(w)\, dw \right| < 1.$$

Suppose that a point $w_0 \notin g\left(B\left(0, \frac{1-r}{2}\right)\right)$. Clearly $w_0 \neq 0$ and we can consider the holomorphic function h in $B\left(0, \frac{1-r}{2}\right)$

$$h(z) = \sqrt{1 - g(z)/w_0} = 1 + a_1 z + a_2 z^2 + \cdots.$$

We have $a_1 = -\frac{g'(0)}{2w_0}$, hence

$$|a_1| = \frac{1}{2|w_0|(1-r)}.$$

Moreover,

$$|h(z)| \leq 1 + \frac{|g(z)|}{|w_0|} \leq 1 + \frac{1}{|w_0|}.$$

From Exercise 2.1.31 one obtains

$$1 + |a_1|^2 \left(\frac{1-r}{2}\right)^2 \leq \sum_{n \geq 0} |a_n|^2 \left(\frac{1-r}{2}\right)^{2n}$$
$$= \frac{1}{2\pi} \int_{-\pi}^{\pi} \left| h\left(\frac{1-r}{2} e^{i\theta}\right) \right|^2 d\theta \leq 1 + \frac{1}{|w_0|}.$$

It is now immediate, from this inequality and the value of $|a_1|$, that

$$|w_0| \geq \frac{1}{16}. \qquad \square$$

This proposition and Lemma 1.11.15 allow us to prove the little Picard theorem. A different proof, depending on the uniformization theorem, can be found in Chapter 5.

2.7.11. Theorem (Little Picard Theorem). *Any entire function whose range omits at least two values is a constant.*

PROOF. Let $a, b \in \mathbb{C}$ be two distinct values not taken by the entire function F, then the function $f(z) = \dfrac{F(z) - a}{b - a}$ is an entire function that omits the values 0 and 1. We can therefore apply Lemma 1.11.5 and obtain an entire function h such that

$$f(z) = \exp(2\pi i \cosh(h(z))).$$

It is clear that h must omit the values in the set $\mathscr{L}$,

$$\mathscr{L} := \{\pm\cosh^{-1}(n+1) + 2m\pi i; n \in \mathbb{N}, m \in \mathbb{Z}\}$$

(cf. Exercise 1.11.12), otherwise f would take the value 1. The successive differences $\cosh^{-1}(n+2) - \cosh^{-1}(n+1)$ decrease with n, the largest value, for $n = 0$, is about 1.317. A quick glance at the set $\mathscr{L}$ will convince the reader that any point of the plane is at a distance at most $\frac{1}{2}(\pi^2 + 1.317^2)^{1/2} \cong 3.22$ from a point in $\mathscr{L}$. Therefore any disk of radius $R = 4$ contains a point omitted by h.

Assume that h is not constant. Let z_0 be such that $h'(z_0) \neq 0$. Consider the auxiliary function

$$H(z) := \frac{1}{64} h\left(\frac{64z}{h'(z_0)} + z_0\right).$$

Then $H'(0) = 1$ and we can apply Proposition 2.7.10 to it. That means the image of $\bar{B}(0, 1)$ by H contains a disk of radius at least $\frac{1}{16}$. Therefore the image of $\bar{B}\left(z_0, \dfrac{64}{|h'(z_0)|}\right)$ by h covers a disk of radius at least 4. This is clearly a contradiction to the preceding discussion. It follows that h, and hence f, is constant. □

The reader should remark that it easily follows from this that, whenever $f \in \mathscr{M}(\mathbb{C})$ and it omits three distinct values $a, b, c \in S^2$, then f is constant.

There are several ways of obtaining the big Picard theorem from here. It is more standard to again use this reasoning to prove a normality result due to Montel, and then obtain Picard's big theorem. We prefer to use a nice proof of Zalcman of a heuristic principle conjectured originally by A. Robinson, which leads directly from Picard's little theorem to Montel's normality criterion (we follow [Za2]).

We first need some precision on the concept of function, so that we indicate its domain of definition. For D a connected open set in $\mathbb{C}$, one denotes $\langle f, D\rangle$ a function $f \in \mathscr{H}(D)$ (or $f \in \mathscr{M}(D)$). For instance, $\langle e^z, \mathbb{C}\rangle \neq \langle e^z, B(0, 1)\rangle$.

A **property** P of holomorphic (resp. meromorphic) functions is just a set of elements $\langle f, D\rangle$, e.g., the property P of being a solution of the differential equation $f' - f = 0, f(0) = 1$, coincides with the set $P = \{\langle e^z, D\rangle : D \text{ domain in } \mathbb{C}\}$.

We say a property P is ***invariant under linear transformations*** if whenever $\langle f, D\rangle \in P$ and $\varphi(z) = az + b$, $a \neq 0$, then $\langle f \circ \varphi, \varphi^{-1}(D)\rangle \in P$ also.

Recall from §2.6.24 that the natural mode of convergence for a sequence of meromorphic functions is that of normal convergence, i.e., $\sigma(f_n(z), f(z)) \to 0$ uniformly on compact sets. We say that a property P is ***complete*** if:

(i) whenever $\langle f_n, D_n\rangle \in P$, $D_1 \subseteq D_2 \subseteq \cdots$, $D = \bigcup D_n$, and $f \in \mathscr{H}(D)$ (resp. $f \in \mathscr{M}(D)$) such that $f_n \to f$ normally in D, then $\langle f, D\rangle \in P$. (Why does it make sense to say that $f_n \to f$ normally, even though f_n is only defined in D_n?); and

(ii) if $\langle f, D\rangle \in P$ and $D' \subseteq D$, then $\langle f, D'\rangle \in P$.

2.7.12. Theorem. *Let P be a property of holomorphic (or meromorphic) functions which is invariant under linear transformations and complete. Assume further that whenever $\langle f, \mathbb{C}\rangle \in P$ it follows that f is a constant function. Then, for any fixed domain D, the family $\mathscr{F} = \{f : \langle f, D\rangle \in P\}$ is a normal family in $\mathscr{H}(D)$ (resp. $\mathscr{M}(D)$).*

We leave the exercises in this section to show that the completeness and the invariance under linear transformations are necessary conditions. Exercise 2.6.6 shows that under the assumption of completeness, the final condition is also necessary.

The proof depends on Marty's theorem, which was stated as Exercise 2.6.7. For the sake of completeness we sketch a proof of this result. Therefore, let $\mathscr{F} \subseteq \mathscr{M}(\Omega)$, Ω a domain in S^2, and we assume that the family of spherical derivatives $\mathscr{F}^{\#} = \{f^{\#} : f \in \mathscr{F}\}$ is locally uniformly bounded in Ω. Since S^2 is a compact space and $\mathscr{M}(\Omega) \subseteq \mathscr{C}(\Omega, S^2)$, then all we have to prove to conclude that $\mathscr{F}$ is a normal family is that it is locally equicontinuous. Let D be a closed disk in Ω, M an upper bound for all $f^{\#}(z)$, $z \in D$, $f \in \mathscr{F}$, then for any $z, w \in D$, we can apply Exercise 2.3.3, (d) and obtain

$$\sigma(f(z), f(w)) \leq \int_{[z,w]} f^{\#}(\zeta)|d\zeta| \leq M(z - w|.$$

This clearly shows the equicontinuity.

The converse is an elementary consequence of the fact that $\sigma(a, b) = \sigma\left(\dfrac{1}{a}, \dfrac{1}{b}\right)$ and $f^{\#}(z) = \left(\dfrac{1}{f}\right)^{\#}(z)$. Note also that $f^{\#}$ is a continuous function in Ω.

Proof of Theorem 2.7.12. We argue by contradiction. If there were a domain D for which the family $\mathscr{F} = \{f : \langle f, D\rangle \in P\}$ is not normal, by the definition of normality there would also be a disk $\Delta \subseteq D$ on which the family $\mathscr{F}$ is not normal. Hence we can assume $D = \Delta$; moreover, since P is invariant under linear transformations, one can further assume $\Delta = B(0, 1)$. By Marty's theorem, there is a value r_0, $0 \le r_0 < 1$, and a sequence $f_n \in \mathscr{F}_n$ such that $\max_{|z| \le r_0} |f_n^{\#}(z)| \to \infty$. The proof that follows imitates in part the proof of Landau's Proposition 2.7.10. Let $|w_n| \le r_0$ be such that $f_n^{\#}(w_n) \to \infty$. Choose any $r \in]r_0, 1[$ and consider

$$M_n = \max_{|z| \le r} \left(1 - \frac{|z|^2}{r^2}\right) f_n^{\#}(z).$$

We clearly have

$$\left(1 - \frac{r_0^2}{r^2}\right) f_n^{\#}(w_n) \le \left(1 - \frac{|w_n|^2}{r^2}\right) f_n^{\#}(w_n) \le M_n,$$

which shows that $M_n \to \infty$. Let z_n, $|z_n| \le r$, be such that

$$M_n = \left(1 - \frac{|z_n|^2}{r^2}\right) f_n^{\#}(z_n).$$

Let $\rho_n = 1/f_n^{\#}(z_n)$, then $0 < \rho_n < \infty$ and we can define functions g_n by

$$g_n(\zeta) = f_n(z_n + \rho_n \zeta).$$

The function g_n is defined at least for $|\zeta| < R_n$, $R_n = \dfrac{r - |z_n|}{\rho_n}$. Moreover, $\langle g_n, B(0, R_n) \rangle \in P$ and

$$R_n = \frac{r - |z_n|}{\rho_n} = (r - |z_n|) f_n^{\#}(z_n) = \frac{1}{r + |z_n|} \left(1 - \frac{|z_n|^2}{r^2}\right) f_n^{\#}(z_n) = \frac{M_n}{r + |z_n|} \ge \frac{M_n}{2r}.$$

We can therefore assume $B(0, R_1) \subseteq B(0, R_2) \subseteq \cdots$ and $\bigcup B(0, R_n) = \mathbb{C}$. In order to use the completeness of P we need to show that the g_n (or a subsequence) converges to a holomorphic (meromorphic) function in $\mathbb{C}$. For that purpose we fix $R > 0$ and we consider $|\zeta| \le R$ and n such that $R < R_n$.

Hence $z_n + \rho_n \zeta \in B(0, r)$ and

$$g_n^{\#}(\zeta) = \rho_n f_n^{\#}(z_n + \rho_n \zeta) \le \rho_n M_n \left(1 - \frac{|z_n + \rho_n \zeta|^2}{r^2}\right)^{-1}$$

$$= \left(1 - \frac{|z_n|^2}{r^2}\right)\left(1 - \frac{|z_n + \rho_n \zeta|^2}{r^2}\right)^{-1} \le \frac{r + |z_n|}{r + |z_n| + \rho_n R} \frac{r - |z_n|}{r - |z_n| - \rho_n R}$$

$$\le \frac{R_n}{R_n - R}.$$

Therefore we can extract a subsequence, still called g_n for simplicity, such that $g_n \to g$ normally to some function g defined everywhere. Moreover,

$$g_n^{\#}(0) = \rho_n f_n^{\#}(z_n) = 1.$$

This implies that g cannot be constantly equal to ∞ (or any other constant, for that matter), hence g is a holomorphic (or meromorphic) function defined everywhere. By the completeness of P, $\langle g, \mathbb{C}\rangle \in P$. This implies that g is constant, and we have a contradiction. The theorem is therefore correct. □

2.7.13. Corollaries

(a) *Montel's Normality Theorem: Let Ω be a domain in $\mathbb{C}$, $\mathscr{F} \subseteq \mathscr{M}(\Omega)$ be such that there are three distinct values $w_1, w_2, w_3 \in S^2$ such that $w_j \notin f(\Omega)$ for every $j = 1, 2, 3$, and every $f \in \mathscr{F}$. Then $\mathscr{F}$ is a normal family.*

(b) *Extended Montel's Theorem: Let Ω be a domain in $\mathbb{C}$, $\mathscr{F} \subseteq \mathscr{M}(\Omega)$ and $\varepsilon > 0$ be such that for every $f \in \mathscr{F}$ there are three values $w_1, w_2, w_3 \in S^2$ omitted by f and such that $\sigma(w_1, w_2)\sigma(w_2, w_3)\sigma(w_3, w_1) \geq \varepsilon$. Then $\mathscr{F}$ is a normal family.*

PROOF. It is clear that (b) is stronger than (a) since the values w_1, w_2, w_3 may depend on f. Let P be the property that a meromorphic function f omits at least three values w_1, w_2, w_3 (depending on f) such that $\sigma(w_1, w_2)\sigma(w_2, w_3)\sigma(w_3, w_1) \geq \varepsilon$. In order to apply Theorem 2.7.12 we only need to check what happens when a sequence of meromorphic functions in D satisfying the property P converges normally to a meromorphic function f on D. One can assume f is not constant, otherwise $\langle f, D\rangle \in P$ is clear.

Let a_n, b_n, c_n be the points omitted by f_n, $\sigma(a_n, b_n)\sigma(b_n, c_n)\sigma(c_n, a_n) \geq \varepsilon$. By the compactness of S^2 we can assume $a_n \to a$, $b_n \to b$, $c_n \to c$. Clearly $\sigma(a, b)\sigma(b, c)\sigma(c, a) \geq \varepsilon$, hence these points are distinct. We need to show that f does not take any of the values a, b, and c. Assume there is $z_0 \in D$ such that $f(z_0) = a$. Assume $a \neq \infty$ and let $\bar{B}(z_0, r) \subseteq D$ be so small that f is holomorphic in a neighborhood of $\bar{B}(z_0, r)$. Then the functions $f_n(z_n) - a_n$ converge uniformly to the function $f(z) - a$ in $\bar{B}(z_0, r)$. Since this last function is not constant and vanishes at least once, by Hurwitz's theorem every f_n takes the value a_n in $\bar{B}(z_0, r)$ (for n sufficiently large). This is clearly impossible. Hence f omits the value a if $a \neq \infty$. The same argument works with $a = \infty$, replacing f, f_n by $1/f$ and $1/f_n$, respectively.

Therefore, P is complete and we can apply Theorem 2.7.12 to conclude the proof. □

It is now easy to prove the following.

2.7.14. Theorem (Big Picard Theorem). *In a neighborhood of an isolated essential singularity a holomorphic function takes every value in $\mathbb{C}$ infinitely often with at most one exception.*

PROOF. Assume that $f \in \mathscr{H}(B(0, \varepsilon)\setminus\{0\})$, 0 is an essential singularity, and there are two values $a \neq b$ such that both equations $f(z) = a$ and $f(z) = b$ have only finitely many solutions z, $0 < |z| < \varepsilon$. Hence there is a δ, $0 < \delta \leq \varepsilon$ such that there are no solutions with $0 < |z| < \delta$. By a simple change of variable we can assume $\delta = 2$.

Consider now the domain $D = \{z \in \mathbb{C} : \frac{1}{2} < |z| < 2\}$ and the functions $f_n(z) = f\left(\frac{z}{2^n}\right)$, $n \in \mathbb{N}$, $z \in D$. Note that the function f_n takes in D the same values the function f takes in the annulus $2^{-n-1} < |z| < 2^{-n+1}$. In particular, none of the functions f_n takes the values a or b in D. It follows from Montel's Normality Theorem 2.7.13 (a) that $\mathscr{F} = \{f_n : n \in \mathbb{N}\}$ is a normal family in D. Therefore there is a subsequence $n_1 < n_2 < \cdots$ such that $f_{n_k} \to f$ normally in D. Either $f \equiv \infty$ or f is holomorphic. If $f \not\equiv \infty$, then we can assume that the f_{n_k} are uniformly bounded on the unit circle $\partial B(0, 1)$, which is a compact subset of D. If not $f_{n_k} \to \infty$ uniformly there and we can assume $1/f_{n_k}$ is bounded in $\partial B(0, 1)$. Let us say we are in the first case, and $|f_{n_k}(z)| \leq M < \infty$ for every $k \in \mathbb{N}^+$, $z \in \partial B(0, 1)$. Then we have

$$|f(z)| \leq M \quad \text{for} \quad |z| = 1 \quad \text{and} \quad |z| = 2^{-n_k}.$$

By the maximum principle

$$|f(z)| \leq M \quad \text{whenever} \quad z^{-n_k} \leq |z| \leq 1.$$

Hence, $|f(z)| \leq 1$ in $B(0, 1)\backslash\{0\}$, since $2^{-n_k} \to 0$ as $k \to \infty$. The singularity is therefore removable. The second case implies 0 is a removable singularity for $1/f$, also impossible. □

It is very easy to strengthen this theorem to the following result of G. Julia: If 0 is an essential singularity of a holomorphic function f in $B(0, 1)\backslash\{0\}$, then there is z_0, $0 < |z_0| < 1$, such that for every $\varepsilon > 0$ the function f takes every value, with at most one exception, infinitely often on the union of the disks $B\left(\frac{z_0}{2^n}, \frac{\varepsilon}{2^n}\right)$, $n \in \mathbb{N}$ (cf. [Za2], [SZ]).

EXERCISES 2.7.

1. Let f be a function of the class S such that all its Taylor coefficients about $z = 0$ are real, $f(z) = z + a_2 z^2 + a_3 z^3 + \cdots$. Fix r, $0 < r < 1$ and define the auxiliary function g on $[-\pi, \pi]$ by

$$g(\theta) := (f(re^{i\theta}) - f(re^{-i\theta})) \sin \theta.$$

   (a) Show that g is even and does not vanish except for $\theta = 0, \pm\pi$.

   (b) Show that $\int_{-\pi}^{\pi} g(\theta)\, d\theta = \pi r$. Conclude $g \geq 0$.

   (c) Let n be an integer greater than or equal to 2, show that

$$0 \leq \frac{2}{\pi r} \int_{-\pi}^{\pi} g(\theta)(1 \pm \cos n\theta)\, d\theta = 2 \pm \left(a_{n+1} - \frac{a_{n-1}}{r^2}\right) r^n.$$

   (d) Conclude from (c) that

$$|a_{n+1} - a_{n-1}| \leq 2 \qquad (n \geq 2).$$

   (e) Prove that $|a_n| \leq n$ for all $n \geq 1$. That is, the Bieberbach conjecture holds for the functions of class S that are real on the axis. (This result was proved in 1931 by J. Dieudonné.)

2. Let $f(z) = \dfrac{e^{5z} - 1}{5}$, show that f omits the value $-\frac{1}{5}$, $f(0) = 0$, $f'(0) = 1$. Why doesn't this function contradict the one-quarter theorem?

3. Use Picard's little theorem to show that if p is a nonzero polynomial, then the equation $e^z - p(z) = 0$ has infinitely many solutions.

4. State and prove the Picard little and big theorems for meromorphic functions.

5. Let $f \in \mathscr{H}(B(0, R))$ be a conformal map onto a region Ω of finite area. For $0 < r < R$, let $\Omega_r = f(B(0, r))$. Show that

$$\frac{m(\Omega)}{m(\Omega_r)} \geq \left(\frac{R}{r}\right)^2.$$

(Hint: Compare with the proof of Lemma 2.7.3 or with Exercise 2.1.31.)

6. Use the function $f(z) = \frac{1}{2}\operatorname{Log}\frac{1+z}{1-z}$ to show that Landau's constant λ is at most $\frac{\pi}{4}$.

7. Show that given any $a > 0$ there is a value $R = R(a) > 0$ such that if $|f(0)| = a$, $f(z) \neq 0, 1$, $f \in \mathscr{H}(B(0, R))$, then there is a singular point of f on $\partial B(0, R)$. (Hint: Compare with the proof of the little Picard theorem.)

8. Let $f \in S$ and assume $D = f(B(0, 1))$ is a convex set. The objective of this exercise is to show that $D \supseteq B\left(0, \frac{1}{2}\right)$.

(a) Let $0 < \rho < 1$, $t = e^{i\theta} \in \partial B(0, 1)$ show that

$$\frac{1}{2}\rho t = \frac{1}{2\pi i}\int_{|z|=\rho} f(z)\left(1 + \frac{z}{2\rho t} + \frac{\rho t}{2z}\right)\frac{dz}{z}.$$

Conclude that

$$\frac{1}{2}\rho e^{i\theta} = \frac{1}{\pi}\int_{-\pi}^{\pi} f(\rho e^{i\varphi})\cos^2\left(\frac{\theta - \varphi}{2}\right)d\varphi.$$

(b) Use the fact that $\frac{1}{\pi}\int_{-\pi}^{\pi}\cos^2\left(\frac{\theta}{2}\right)d\theta = 1$ and D is convex to show that $\frac{1}{2}\rho e^{i\theta} \in D$.

Conclude that $B\left(0, \frac{1}{2}\right) \subseteq D$.

(c) Use the function $f(z) = \dfrac{z}{1 - z}$ to show that $\frac{1}{2}$ is the best possible constant.

9. Let $f \in S$. Verify that for a fixed $a \in B(0, 1)$ the function

$$F(z) = \frac{f \circ \varphi_{-a}(z) - f(a)}{(1 - |a|^2)f'(a)}$$

belongs to the class S (cf. the proof of Theorem 2.7.6). Conclude that $\frac{1}{r}\frac{1-r}{1+r} \leq \left|\frac{f'(z)}{f(z)}\right| \leq \frac{1}{r}\frac{1+r}{1-r}$ for $0 < |z| = r < 1$.

10. Let $f \in \mathscr{H}(B(0, 1))$, $f(z) = z + a_2z^2 + a_3z^3 + \cdots$. Show that if

$$\sum_{n\geq 2} n|a_n| \leq 1$$

then $f \in S$.

11. Prove that the family of polynomials which belong to S is dense in S.

§8. Conformal Mappings

Let us recall that a differentiable homeomorphism between two planar domains Ω_1, Ω_2 (or even two open sets in S^2) with nonvanishing Jacobian, that preserves the orientation and the angles, is a conformal mapping, i.e., a biholomorphic mapping. We say that the domains are ***conformally equivalent***.

The Riemann mapping theorem can be restated as follows.

2.8.1. Theorem. *Let Ω be a simply connected open set in S^2, then Ω is conformally equivalent to exactly one of the following three domains:* (i) S^2, (ii) $\mathbb{C}$, *or* (iii) $B = B(0, 1)$.

The first case only occurs if $\Omega = S^2$, the second if $\Omega = S^2 \setminus \{a\}$, the third if $\Omega \subseteq S^2 \setminus \{a, b\}$, $a \neq b$.

PROOF. The proof is left to the reader as an exercise. □

There is a corollary of the proof of the Riemann mapping theorem which is worth mentioning here.

2.8.2. Proposition. *Let Ω be a domain $\mathbb{C}$, then Ω is simply connected if and only if every nonvanishing holomorphic function in Ω admits a continuous logarithm.*

PROOF. Clearly we can assume $\Omega \neq \mathbb{C}$. Now we only have to observe that in the proofs of §2.6.20 and §2.6.21 the only place where Ω being simply connected played a role was in the use of logarithms and square roots to construct auxiliary functions. Therefore, in this case we have a conformal map from Ω onto B and the domain must be simply connected. □

A theorem of Caratheodory asserts that for a Jordan domain, the conformal mapping onto B extends continuously to the boundary. Our objective is to explain the analytic ideas in the proof of this fact. For that reason we will state first a number of purely topological lemmas about simply connected domains in $\mathbb{C}$, whose proofs the reader can take for granted.

***2.8.3. Lemma.** *Given any two distinct points z_1, z_2 in a domain Ω, then there is a polygonal Jordan arc in Ω, with sides parallel to the axes, starting at z_1 and ending at z_2. (Jordan arc means that it is parameterized by an injective continuous map $\varphi : [0, 1] \to \mathbb{C}$.)*

PROOF. It is an elementary exercise. □

***2.8.4. Lemma.** *Let Ω be a domain and Γ a polygonal Jordan arc in Ω. Then $\Omega\backslash\Gamma$ is connected.*

PROOF. It an elementary exercise, draw a picture. □

***2.8.5. Lemma.** *Let Ω be a Jordan domain, i.e., $\Omega = \operatorname{Int}(\Gamma)$, Γ a Jordan curve in $\mathbb{C}$, and let $z_0 \in \Gamma$. Then, for every $\varepsilon > 0$ there is a $\delta > 0$ such that every pair of points $z_1, z_2 \in \Omega \cap B(z_0, \delta)$ can be joined by a Jordan polygonal arc $\gamma \subseteq \Omega \cap B(z_0, \varepsilon)$. In particular, any two points of $\Omega \cap B(z_0, \delta)$ are in the same component of $\Omega \cap B(z_0, \varepsilon)$.*

PROOF. This is the crucial and entirely topological property. A proof can be found in [Bu, 4.48]. It is also a consequence of Schoenflies' version of the Jordan curve theorem (see references in [Bu]). See Figure 2.10. □

***2.8.6. Lemma.** *Let Ω, z_0 be as in Lemma 2.8.5. Let $z_n \in \Omega$, $\lim_{n\to\infty} z_n = z_0$. Then, for every $\varepsilon > 0$, there is a subsequence $(z_{n_k})_k$ and an injective curve γ passing through all the z_{n_k}, contained in $B(z_0, \varepsilon) \cap \Omega$ except for its endpoint, which is z_0. (That is $\gamma : [0, 1] \to \mathbb{C}$ is a continuous injective path, $\gamma([0, 1[) \subseteq \Omega$, $\gamma(1) = z_0$, and there is a sequence $0 < t_{n_k} \nearrow 1$ such that $\gamma(t_{n_k}) = z_{n_k}$.)*

PROOF. Let $\varepsilon_1 = \varepsilon$ be given, let $\delta_1 > 0$ be the value obtained from Lemma 2.8.5 and let n_1 be chosen so that $z_n \in B(z_0, \delta_1)$ for every $n \geq n_1$. Let $n_2 = n_1 + 1$. Choose γ_1 polygonal Jordan arc in $\Omega \cap B(z_0, \varepsilon_1)$, starting at z_{n_1}, and ending at z_{n_2}. Let $\varepsilon_2 = \frac{1}{2}\operatorname{dist}(\gamma_1, z_0)$, and choose a corresponding δ_2 and $n_3, n_4 = n_3 + 1$ by the same procedure. There is then a Jordan polygonal arc γ_2' starting at z_{n_3}, ending at z_{n_4} and contained in $\Omega \cap B(z_0, \varepsilon_2)$. The points z_{n_2} and z_{n_3} both belong to $\Omega \cap B(z_0, \delta_1)$, hence to the same component C_1 of $\Omega \cap B(z_0, \varepsilon_1)$. Moreover, by the choice of ε_1, we have that γ_1 and γ_2' are disjoint. Applying twice Lemma 2.8.4 we conclude that $C_1 \backslash(\gamma_1 \cup \gamma_2')$ is connected. Choose two

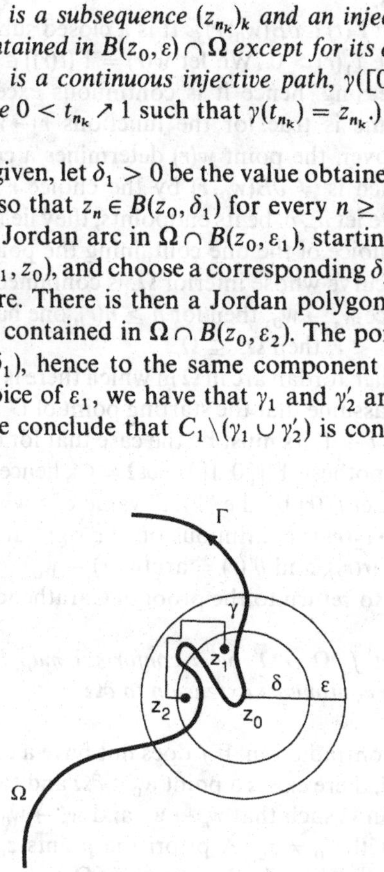

Figure 2.10

very small disks Δ_1, Δ_2 centered at z_{n_2}, z_{n_3}, respectively, contained in C_1 so that they are disjoint from γ_2' and γ_1, respectively, but γ_1 intersects the boundary of the first one and γ_2' the boundary of the second. Choose points z_{n_2}' in $\partial\Delta_1\backslash\gamma_1$, $z_{n_3}' \in \partial\Omega_2\backslash\gamma_2'$. We can connect z_{n_2}' and z_{n_3}' in $C_1\backslash(\gamma_1 \cup \gamma_2')$ by a polygonal Jordan arc, and adding Jordan polygonal arcs contained in $\bar{\Delta}_1$ and $\bar{\Delta}_2$, respectively, we construct a Jordan polygonal arc γ_2 entirely contained in C_1, starting at z_{n_1} passing through z_{n_2}, z_{n_3} and ending at z_{n_4}. We let now $\varepsilon_3 = \frac{1}{2}\mathrm{dist}(z_0, \gamma_2)$ and δ_3, n_4, $n_5 = n_4 + 1$ as earlier. The points z_{n_3} and z_{n_4} are in the same connected component C_2 of $B(z_0, \varepsilon_2) \cap \Omega$. We can repeat the procedure constructing first γ_3' in $B(z_0, \varepsilon_3) \cap \Omega$, and then joining it to γ_2 as done earlier. This procedure clearly answers the question posed. □

We need one more consequence of §2.8.5. Let Ω be a Jordan domain, $w_0 \in \partial\Omega$, and Γ a Jordan arc in Ω defined in $[0, 1)$ such that there is a sequence of values $t_k \to 1$ with $\Gamma(t_k) \to w_0$. Let $r_0 < \mathrm{dist}(w_0, \Gamma(0))$. Then, for any $0 < r \leq r_0$, there is a connected subdomain of Ω with w_0 in its boundary, constructed as follows.

Let $F_r = \{t \in [0, 1] : \Gamma(t) \in \partial B(w_0, r)\}$. It is a closed subset of $[0, 1[$. Hence, it has a smallest value $t_1(r) > 0$. We let $w(r) = \Gamma(t(r)) \in \partial B(w_0, r) \cap \Omega$. Note that $r \to t_1(r)$ is decreasing, hence it is continuous except for a countable set. Therefore the same is true for the functions $r \mapsto w(r)$ and $r \mapsto \theta(r) = \arg(w(r) - w_0)$. Moreover, the point $w(r)$ determines a component C_r of the set $\partial B(w_0, r) \cap \Omega$ (which is $\neq \partial B(w_0, r)$ by the choice $r \leq r_0$). This arc is a cross-cut of Ω, i.e., if we let a_r, b_r be its endpoints, they lie in $\partial\Omega$ and determine two arcs of $\partial\Omega$. The choice of the one containing the point w_0, say γ_r, allows us to define a Jordan curve whose interior Ω_r is contained in Ω. It is not hard to see that if a sequence $w_n \to w_0$, then for $n \geq n(r)$, one has $w_n \in \Omega_r$. It follows from this that if $0 < r' < r$, then $\Omega_{r'} \subseteq \Omega_r$.

Let Γ' now be another Jordan arc in Ω in which there is a sequence of points converging to w_0 and assume that the starting point of Γ' also lies outside Ω_{r_0}. Since $\Gamma'(t) \to w_0$ when $t \to 1$, its must be the case that for every r $\Gamma'(t) \in \Omega_r$ for some $0 < t < 1$. By hypothesis $\Gamma'([0, 1[) \cap \partial\Omega = \varnothing$, hence $\Gamma'([0, 1[) \cap C_r \neq \varnothing$ for every $0 < r < r_0$. Let $t_2(r)$ be the "first" value of t where Γ' crosses C_r. It is clear again that $r \mapsto t_2(r)$ is continuous on the right and hence the same is true for $r \mapsto w'(r) = \Gamma'(t(r))$, and $\theta'(r) = \arg(w'(r) - w_0)$.

We are now ready to return to the proof of Caratheodory's assertion.

2.8.7. Proposition. *Let $f : \Omega \to \Omega'$ be a conformal map between two Jordan domains. Then f has a continuous extension to $\partial\Omega$.*

Proof. We argue by contradiction. If f does not have a continuous extension to the boundary of $\partial\Omega$, there exists a point $w_0 \in \partial\Omega$ and two sequences $(w_n)_{n\geq 1}$ and $(w_n')_{n\geq 1}$ of points in Ω such that $w_n \to w_0$ and $w_n' \to w_0$ but $z_n = f(w_n) \to z_0$ and $z_n' = f(w_n') \to z_0'$ with $z_0 \neq z_0'$. A priori the points z_0, $z_0' \in \bar{\Omega}'$, but using Hurwitz's theorem one sees easily that z_0, $z_0' \in \partial\Omega$.

Let $\alpha = |z_0 - z_0'| > 0$; we can assume that for all n, $|z_n - z_0| < \frac{\alpha}{3}$ and $|z_n' - z_0'| < \frac{\alpha}{3}$. Using Lemma 2.8.6 we can construct two Jordan arcs γ, γ' such that γ lies in $B\left(z_0, \frac{\alpha}{3}\right) \cap \Omega'$ except for its endpoint z_0 and γ' lies in $B\left(z_0', \frac{\alpha}{3}\right) \cap \Omega'$ except for its endpoint z_0'. Furthermore, γ passes through a subsequence z_{n_k} of $(z_n)_{n\geq 1}$ and γ' passes through a subsequence of $(z_n')_{n\geq 1}$. Note that by construction we have $|\gamma(t) - \gamma'(t')| \geq \frac{\alpha}{3}$ for any $t', t \in [0, 1]$.

Let $\Gamma(t) = f^{-1}(\gamma(t))$ and $\Gamma'(t) = f^{-1}(\gamma'(t))$. These are two arcs in Ω to which we can apply the prior reasoning. Therefore we can choose for $0 < r \leq r_0$ a pair of values $\theta(r)$, $\theta'(r)$ in $[0, 2\pi[$ that is a continuous choice except at countably many values, such that the points $w(r) = z_0 + re^{i\theta(r)}$, $w'(r) = z_0 + re^{i\theta'(r)}$ belong to Γ and Γ', respectively, and the arc from $\theta(r)$ to $\theta'(r)$ is entirely contained in C_r. It follows that

$$\frac{\alpha}{3} \leq |f(w(r)) - f(w'(r))| \leq \int_{\theta(r)}^{\theta'(r)} \left|\frac{df}{dz}(w_0 + re^{it})\right| r\,dt,$$

since we can assume $\theta(r) \leq \theta'(r)$.

By Schwarz's inequality we obtain

$$\begin{aligned}\left(\frac{\alpha}{3}\right)^2 &\leq r^2\left(\int_{\theta(r)}^{\theta'(r)} |f'(w_0 + re^{it})|^2\,dt\right)(\theta'(r) - \theta(r)) \\ &\leq 2\pi r^2 \int_{\theta(r)}^{\theta'(r)} |f'(w_0 + re^{it})|^2\,dt.\end{aligned}$$

Divide by r and integrate in r between ε and r_0. Then

$$\begin{aligned}\left(\frac{\alpha}{3}\right)^2 \log\left(\frac{r_0}{\varepsilon}\right) &\leq \int_\varepsilon^{r_0} r\,dr \int_{\theta(r)}^{\theta'(r)} |f'(w_0 + re^{it})|^2\,dt \\ &\leq 2\pi \int_{\Omega_{r_0}} |f'(w)|^2\,dm(w) \leq 2\pi m(f(\Omega_{r_0})) \\ &\leq 2\pi m(f(\Omega)) \leq 2\pi m(\Omega') < \infty.\end{aligned}$$

This is clearly impossible when $\varepsilon \to 0$. This contradiction shows f extends continuously to the boundary of Ω. □

2.8.8. Theorem (Caratheodory). *Let $f : \Omega \to \Omega'$ be a biholomorphic mapping between two Jordan domains. Then there is a homeomorphism $\tilde{f} : \bar{\Omega} \to \bar{\Omega}'$ such that $\tilde{f}|\Omega = f$.*

PROOF. By the previous proposition f has a continuous extension $\tilde{f} : \bar{\Omega} \to \bar{\Omega}'$. Similarly, $g = f^{-1}$ has a continuous extension $\tilde{g} : \bar{\Omega}' \to \bar{\Omega}$. On the other hand,

$g \circ f = id_\Omega$, and by continuity we have $\tilde{g} \circ \tilde{f} = id_{\bar{\Omega}}$. Similarly $\tilde{f} \circ \tilde{g} = id_{\bar{\Omega}'}$. Therefore, $\tilde{f}$ is a homeomorphism. □

2.8.9. Corollary. *If Ω is a Jordan domain, then $\bar{\Omega}$ is homeomorphic to $\bar{B}(0, 1)$.*

In Chapter 4 we give two other independent proofs of the extension of the conformal maps to the boundary of Jordan domains, but under the assumption they have either C^∞ regular boundaries or real analytic boundaries. On the other hand, the extensions will be respectively C^∞ and real analytic. The reader will probably find those proofs in Chapter 4 much simpler to follow than the previous one. The reason is that the topological difficulties are eliminated due to the regularity of the boundaries.

We must also point out that the correct setting to study the question of extension to the boundary is the theory of prime ends. The books [CL] and [Pom] provide an excellent introduction to the difficult subject of the boundary behavior of conformal mappings.

The problem of explicitly finding the conformal equivalence between a given Jordan domain and $B(0, 1)$ is very difficult. Some examples will be given later. For many practical applications, e.g., in aerodynamics, this is done numerically. We recommend [Bie], [Dic], and [Hen] for explicit examples of conformal maps and numerical methods.

2.8.10. Examples of Conformal Maps. (1) The group $\operatorname{Aut}(\mathbb{C})$ of conformal automorphisms of $\mathbb{C}$: An entire function $f: \mathbb{C} \to \mathbb{C}$ which is injective cannot have the point $\infty \in S^2$ as an essential singularity. It is therefore a polynomial and the injectivity implies the degree is exactly one. Hence the group $\operatorname{Aut}(\mathbb{C})$ of holomorphic automorphisms of $\mathbb{C}$ consists of the affine mappings $T_{a,b}: z \mapsto az + b$, $a \neq 0$.

The transformation $T_{1,b}$ is a translation. If $a \neq 1$, $T_{a,b}$ admits a unique fixed point $z_0 = b/(1 - a)$. The group $\operatorname{Aut}(\mathbb{C})$ is clearly transitive. The stabilizer of a point $z_0 \in \mathbb{C}$ is the subgroup of the transformations of the form $z \mapsto a(z - z_0) + z_0$, $a \neq 0$.

(2) The group $\operatorname{Aut}(S^2)$ of conformal automorphisms of S^2: Consider the family Γ of Moebius transformations of the form $T_A: z \mapsto \dfrac{az + b}{cz + d}$, with $A = \begin{pmatrix} a & b \\ c & d \end{pmatrix} \in GL(2, \mathbb{C})$.

One has $T_A(\infty) = a/c$ and $T_A\left(-\dfrac{d}{c}\right) = \infty$ if $c \neq 0$. Otherwise $T_A(\infty) = \infty$. The map T_B is the inverse to T_A if $B = (\det A)A^{-1}$. Moreover, A and λA ($\lambda \neq 0$) define the same Moebius transformation. The family Γ forms a group of meromorphic maps of S^2 into S^2. This group Γ is precisely the image of the group homomorphism

$$T: GL(2, \mathbb{C}) \to \Gamma(S^2)$$
$$A \mapsto T_A,$$

where $\Gamma(S^2)$ is the group of all meromorphic automorphisms of S^2. The kernel of T is the subgroup D of diagonal matrices.

The Riemann sphere does not have any other meromorphic automorphisms, i.e., T is surjective. One way to see this surjectivity is the following. Let $S_\infty \subseteq \mathrm{Aut}(S^2)$ be the stabilizer group of $\infty \in S^2$. Clearly $S_\infty = \mathrm{Aut}(\mathbb{C})$. By the preceding characterization of $\mathrm{Aut}(\mathbb{C})$ we see that $S_\infty = T(N)$, N is the subgroup of $GL(2, \mathbb{C})$ of matrices with $c = 0$. On the other hand we have the following general fact.

Let Ω be an open set of S^2 and let G be a transitive subgroup of the group $\mathrm{Aut}(\Omega)$ of meromorphic automorphisms of Ω. Then, if there exists a point $z_0 \in \Omega$ whose stabilizer S_{z_0} is contained in G, it follows that $G = \mathrm{Aut}(\Omega)$.

In fact, for any $S \in \mathrm{Aut}(\Omega)$ there is a $T \in G$ such that $T(z_0) = S(z_0)$. Hence $R = T_0^{-1}S \in S_{z_0} \subseteq G$ and $S = T_0 R \in G$.

Applying this result to $\Omega = S^2$ and $G = T(GL(2, \mathbb{C}))$ we obtain the surjectivity of T. $\mathrm{Aut}(S^2)$ can now be identified to $GL(2, \mathbb{C})/D$.

(3) The group $\mathrm{Aut}(B(0, 1))$: The Moebius transformation $S: z \mapsto \dfrac{z - i}{z + i}$ transforms the upper half-plane $H = \{z \in \mathbb{C} : \mathrm{Im}\, z > 0\}$ onto the unit disk $B(0, 1)$. We conclude that the map

$$S^* : \mathrm{Aut}(B(0, 1)) \to \mathrm{Aut}(H)$$
$$\varphi \mapsto S^{-1} \circ \varphi \circ S$$

is a group isomorphism. We want to characterize $\mathrm{Aut}(B(0, 1))$ and $\mathrm{Aut}(H)$.

The subgroup G of $\mathrm{Aut}(S^2)$ of transformations which send the real axis $\mathbb{R}$ into itself is $G = T(GL(2, \mathbb{R}))$. One has $G \cap \mathrm{Aut}(H) = T(GL^+(2, \mathbb{R}))$, where

$$GL^+(2, \mathbb{R}) := \{A \in GL(2, \mathbb{R}) : \det A > 0\}.$$

This is a consequence of the identity

$$\mathrm{Im}(T_A(z)) = \det A \cdot \frac{\mathrm{Im}\, z}{|cz + d|^2} \quad \text{if} \quad A = \begin{pmatrix} a & b \\ c & d \end{pmatrix}.$$

It is easy to see that $G_+ = T(GL^+(2, \mathbb{R}))$ is transitive in H. Namely, $x + iy = T_A(i)$ with $A = \begin{pmatrix} y & x \\ 0 & 1 \end{pmatrix} \in GL^+(2, \mathbb{R})$ if $y > 0$. Moreover, the stabilizer S_i of the point i in $\mathrm{Aut}(H)$ is contained in G_+. To see this, it is the same to find the set of transformations in $\mathrm{Aut}(B(0, 1))$ leaving the origin fixed (using S^*). But Schwarz's lemma ensures that such a transformation φ satisfies $|\varphi(z)| \le |z|$ and $|\varphi^{-1}(w)| \le |w|$. The last inequality applied to $w = \varphi(z)$ yields $|\varphi(z)| = |z|$ and hence, $\varphi(z) = e^{i\theta} z$ for some $\theta \in \mathbb{R}$. Using the isomorphism S^* one finds that if $R \in S_i$ one has

$$R(z) = -\frac{\cos(\theta/2)z + \sin(\theta/2)}{\sin(\theta/2)z - \cos(\theta/2)}$$

for some $\theta \in \mathbb{R}$. This shows that $S_i \subseteq G_+$. By the general result proved earlier we obtain $\mathrm{Aut}(H) = G_+ = T(GL^+(2, \mathbb{R}))$. To determine $\mathrm{Aut}(B(0, 1))$ we use again the isomorphism S^*. One finds that $\operatorname{Aut}(B(0,1))$ consists of those Moebius transformations of the form

$$R(z) = e^{i\theta}\frac{z + z_0}{1 + \bar{z}_0 z} \qquad (\theta \in \mathbb{R}, |z_0| < 1).$$

This coincides with the result obtained in §2.3.12 by a different method, in that notation $R = e^{i\theta}\varphi_{-z_0}$.

(4) The transformation $z \mapsto w = \dfrac{(z+1)^2 - i(z-1)^2}{(z+1)^2 + i(z-1)^2}$ is the composition of $z \mapsto \dfrac{1+z}{1-z}$, $z \mapsto z^2$, $z \mapsto \dfrac{z-i}{z+i}$, and transforms $B^+(0,1) = B(0,1) \cap H$ conformally onto $B(0,1)$, leaving the points $1, -1, i$ fixed.

(5) Let $\Omega = \left\{z \in \mathbb{C} : \left|\dfrac{z}{a} - 1\right| > 1 > \left|\dfrac{z}{b} - 1\right|\right\}$ $(0 < a < b)$, it is transformed onto $B(0,1)$ by $z \mapsto i\tan\left\{\dfrac{ab}{b-a}\pi\left(\dfrac{1}{z} - \dfrac{1}{2b}\right) - \dfrac{\pi}{4}\right\}$.

One simply observes that $z \mapsto \dfrac{1}{z}$ transforms Ω into the strip $\dfrac{1}{2b} < \operatorname{Re} z < \dfrac{1}{2a}$ and that this strip is transformed into H by the function

$$z \mapsto \exp\left\{\frac{2a}{b-a}\pi i\left(z - \frac{1}{2b}\right)\right\}.$$

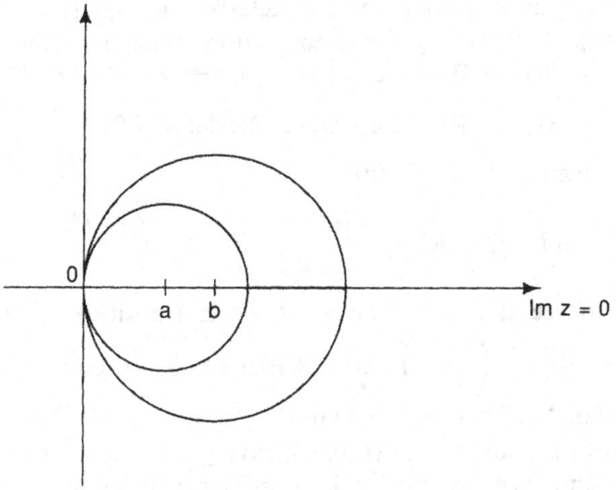

Figure 2.11

(6) The map $z \mapsto \left(\dfrac{z-1}{z+1}\right)^2$ transforms $H \setminus \bar{B}(0,1)$ into H.

(7) The mapping $z \mapsto i\sinh\dfrac{\pi z}{2h}$ transforms the region $R = \{z \in \mathbb{C} : \operatorname{Re} z > 0$ and $|\operatorname{Im} z| < h\}$ into H.

Let us see a few consequences of the argument principle that are useful to check whether a map is conformal.

2.8.11. Proposition. *Let γ be a Jordan curve. Assume that $f: \overline{\mathrm{Int}(\gamma)} \to \mathbb{C}$ is continuous, holomorphic in* $\mathrm{Int}(\gamma)$ *and injective on γ. Then f is a homeomorphism of $\overline{\mathrm{Int}(\gamma)}$ onto $\overline{\mathrm{Int}(\Gamma)}$ and a biholomorphism of* $\mathrm{Int}(\gamma)$ *onto* $\mathrm{Int}(\Gamma)$, *where Γ is the Jordan curve $f \circ \gamma$.*

PROOF. From the hypotheses we can conclude that $\Omega = f(\mathrm{Int}\,\gamma)$ is a connected open set. Let $w_1 \in \mathrm{Int}(\Gamma)$, the number n of roots of $f(z) - w_1 = 0$ with $z \in \mathrm{Int}(\gamma)$ is given by

$$n = \frac{1}{2\pi i} \int_{\gamma_t} \frac{f'(z)}{f(z) - w_1} dz,$$

where γ_t is the Jordan curve constructed as follows. Since $f^{-1}(\{w_1\})$ is a compact subset of $\mathrm{Int}(\gamma)$ and $\overline{\mathrm{Int}(\gamma)}$ is homeomorphic to $\bar{B}(0,1)$, there is a value $t_0, 0 < t_0 < 1$ such that for no $t \in]t_0, 1]$ the curve $\gamma_t(\theta) = \varphi(te^{i2\pi\theta})(0 \leq \theta \leq 1)$ intersects $f^{-1}(\{w_1\})$, where $\varphi : \bar{B}(0,1) \to \overline{\mathrm{Int}(\gamma)}$ is a homeomorphism such that $\varphi(e^{i2\pi\theta}) = \gamma(\theta)$. Therefore $n = \mathrm{Ind}_{\Gamma_t}(w_1)$, with $\Gamma_t = f \circ \gamma_t$ for every $t_0 < t < 1$. Since Γ and Γ_t are homotopic, we have $n = \mathrm{Ind}_\Gamma(w_1) = 1$.

Hence, for every $w_1 \in \mathrm{Int}(\Gamma)$ there is a unique $z_1 \in \mathrm{Int}(\gamma)$ such that $f(z_1) = w_1$. In particular, $\Omega \supseteq \mathrm{Int}(\Gamma)$.

Let now $w_2 \in \mathrm{Ext}(\Gamma)$. The preceding reasoning also shows that $\Omega \cap \mathrm{Ext}(\Gamma) = \varnothing$.

We have also $\Gamma \cap \Omega = \varnothing$. This is a consequence of the fact that f is an open mapping. If $w \in \Gamma$ could be written as $w = f(z)$ for some $z \in \mathrm{Int}(\gamma)$, then there would be a neighborhood V of z such that $V \subseteq \mathrm{Int}(\gamma)$ and $f(V)$ is a neighborhood of w. Therefore $f(V) \cap \mathrm{Ext}(\Gamma) \neq \varnothing$, which is impossible since $f(V) \subseteq \Omega$.

This shows that $f(\mathrm{Int}(\gamma)) = \Omega = \mathrm{Int}(\Gamma)$, f is a biholomorphic mapping and a homeomorphism from $\overline{\mathrm{Int}(\gamma)}$ onto $\overline{\mathrm{Int}(\Gamma)}$. □

There is an analogous result for the exterior of γ considered as a set in S^2, i.e., $\mathrm{Ext}_\infty(\gamma) = S^2 \backslash \overline{\mathrm{Int}(\gamma)}$.

2.8.12. Proposition. *If $f : \overline{\mathrm{Ext}_\infty(\gamma)} \to \mathbb{C}$ is a continuous function, holomorphic in* $\mathrm{Ext}_\infty(\gamma)$, *which is injective on γ, then f is a homeomorphic map from $\overline{\mathrm{Ext}_\infty(\gamma)}$ onto $\overline{\mathrm{Int}(\Gamma)}$ and biholomorphic from* $\mathrm{Ext}_\infty(\gamma)$ *onto* $\mathrm{Int}(\Gamma)$, *where $\Gamma = f \circ \gamma$.*

PROOF. Let $z_0 \in \mathrm{Int}(\gamma)$. The transformation $z \mapsto \zeta = \dfrac{1}{z - z_0}$ transforms γ into a Jordan curve γ^* in $\mathbb{C}$ and $\mathrm{Ext}_\infty(\gamma)$ into $\mathrm{Int}(\gamma^*)$. Consider $f^*(\zeta) = f\left(z_0 + \dfrac{1}{\zeta}\right)$. The preceding result can now be applied to γ^* and f^*, hence the conclusion of the proposition holds. □

2.8.13. Examples. (1) The transformation $z \mapsto w = z^2$, $(z = re^{i\varphi}, w = \rho e^{i\theta})$, injectively transforms the circle $r = \cos\varphi$ into the cardioid $\rho = \frac{1}{2}(1 + \cos 2\theta)$. By §2.8.11, $z \mapsto z^2$ is a conformal map of the disk $B(\frac{1}{2}, 1/2)$ onto the interior of the cardioid.

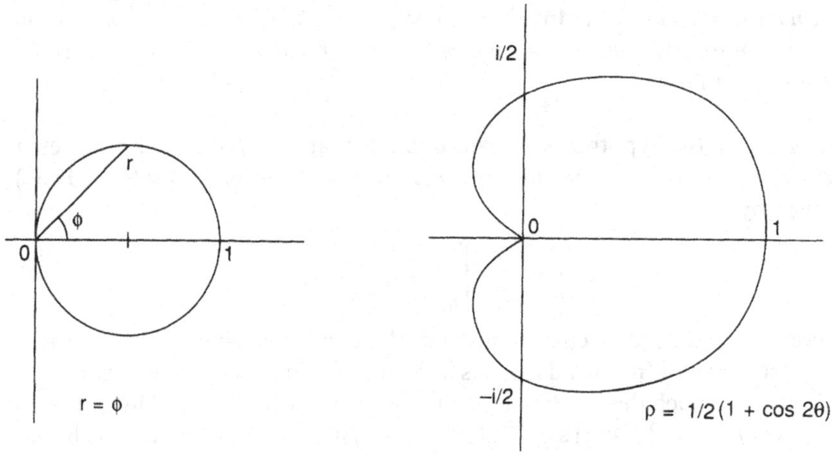

Figure 2.12

(2) The transformation $z \mapsto \sqrt{z} = e^{1/2 \operatorname{Log} z} = w$ transforms the disk $B(\frac{1}{2}, \frac{1}{2})$ into the right petal of the lemmiscate $\rho = \sqrt{\cos 2\theta}$.

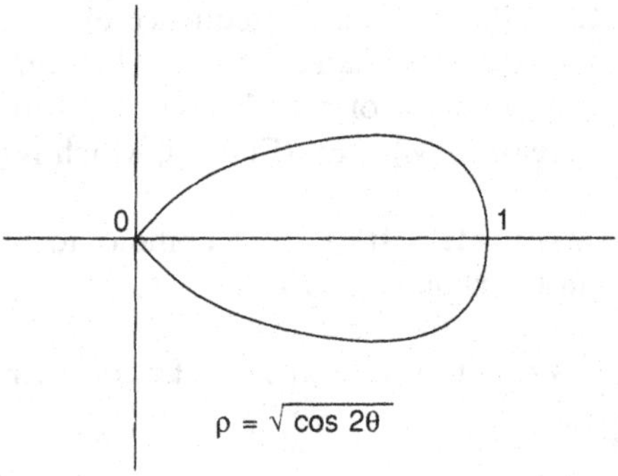

Figure 2.13

Let us now consider the situation analogous to §2.8.11 when the Jordan curve γ in S^2 is unbounded, i.e., $\infty \in \gamma$.

2.8.14. Proposition. *Let γ be a Jordan curve in S^2 with $\infty \in \gamma$. Let D_1 and D_2 be the two open components of $S^2 \backslash \gamma$. Assume f is a complex-valued continuous function on $\bar{D}_1$, holomorphic in D_1, injective on γ. Then f is a biholomorphism of D_1 onto* $\operatorname{Int}(\Gamma)$, $\Gamma = f \circ \gamma$, *which is a homeomorphism of $\bar{D}_1$ onto* $\overline{\operatorname{Int}(\Gamma)}$.

PROOF. Let $z_0 \in D_2$ and γ^* the Jordan curve in $\mathbb{C}$ image of γ by $z \mapsto \zeta = \dfrac{1}{z - z_0}$. We can apply §2.8.11 to γ^* and $f^*(\zeta) = f\left(z_0 + \dfrac{1}{\zeta}\right)$. Now f is injective from $\operatorname{Int}(\gamma^*)$ onto $\operatorname{Int}(\Gamma)$. □

An interesting particular case of §2.8.14 is the following.

If f is continuous on $\bar{H} \cup \{\infty\}$, H the open upper half-plane $\bar{H} = \{z \in \mathbb{C} : \operatorname{Im} z \geq 0\}$, f holomorphic in H, and injective on $\gamma = \mathbb{R} \cup \{\infty\}$, then f is a conformal map of H onto $\operatorname{Int}(\Gamma)$, Γ the Jordan curve $f \circ \gamma$.

To conclude this section let us consider a class of functions that falls within the last situation. They are the ***Schwarz-Christoffel transformations.***

Let $-\infty < a_1 < a_2 < \cdots < a_n < \infty$ be n real numbers, and $\alpha_1, \ldots, \alpha_n$ other n positive real numbers such that $\alpha_1 + \cdots + \alpha_n + 1 < n$. Let $\beta(t) := (t - a_1)^{\alpha_1 - 1} \ldots (t - a_n)^{\alpha_n - 1}$, then $\displaystyle\int_{-\infty}^{\infty} |\beta(t)|\, dt < \infty$. We choose the determination of β in such a way that the argument of $(t - a_k)^{\alpha_k - 1} = \exp((\alpha_k - 1)\log(t - a_k))$ is equal to $\pi(\alpha_k - 1)$ if $t < a_k$. The argument of $\beta(t)$ is therefore $\pi[(\alpha_1 + \cdots + \alpha_n) - \pi]$ for $t < a_1$. In $]a_{k-1}, a_k[$, we have $\arg \beta(t) = (\alpha_k - 1)\pi + \cdots + (\alpha_n - 1)\pi$, if $2 \leq k \leq n$, and $\arg \beta(t) = 0$ in $]a_n, \infty[$. Let us denote $a_0 = -\infty$, $a_{n+1} = +\infty$ and $c > 0$ a constant. Define $n + 2$ complex numbers w_k by

$$w_k := c \int_0^{a_k} (t - a_1)^{\alpha_1 - 1} \ldots (t - a_n)^{\alpha_n - 1}\, dt, \qquad 0 \leq k \leq n + 1.$$

The function f defined in $\bar{H} = \{z \in \mathbb{C} : \operatorname{Im} z \geq 0\}$ by

$$f(z) = c \int_0^z \beta(t)\, dt,$$

is holomorphic in H, and on the real axis it satisfies

$$f(x) = w_{k-1} + c \int_{a_{k-1}}^{x} \beta(t)\, dt = w_{k-1} + c e^{i[(\alpha_k - 1)\pi + \cdots + (\alpha_n - 1)\pi]} \int_{a_{k-1}}^{x} |\beta(t)|\, dt,$$

if $x \in\,]a_{k-1}, a_k[$ $(1 \leq k \leq n + 1)$. Therefore, $f(x) - w_{k-1}$ has always the same argument $[(\alpha_k - 1)\pi + \cdots + (\alpha_n - 1)\pi]$ in that interval, and its absolute value grows from 0 to ℓ_k,

$$\ell_k := c \int_{a_{k-1}}^{a_k} |\beta(t)|\, dt.$$

Hence, when x traverses the interval $[a_{k-1}, a_k]$ the function f traverses the straight line segment $[w_{k-1}, w_k] = \Delta_{k-1}$ of length ℓ_k and determines an angle of opening equal to $(\alpha_k - 1)\pi + \cdots + (\alpha_n - 1)\pi$ with the direction of the positive real axis.

Let us show that $w_0 = w_{n+1}$. It is enough to show that for every $\varepsilon > 0$ there is a number $R > 0$ such that if $z \in \bar{H}$, $|z| \geq R$ then

$$|w_0 - f(z)| \leq \varepsilon.$$

This will show that $\lim_{z \to \infty} f(z) = w_0 = w_{n+1}$. Since $\beta \in L^1(\mathbb{R})$, clearly there exists $R_1 > 0$ such that if $-\infty < x < -R_1$ then

$$\left| w_0 - c \int_0^x \beta(t)\,dt \right| \leq \int_{-\infty}^{-R_1} |\beta(t)|\,dt \leq \varepsilon/2.$$

We can certainly assume that $R_1 \geq \max\{|a_1|, \ldots, |a_n|\}$. For $z_0 = \rho_0 e^{i\theta_0}$, $\rho_0 \geq R_1$, $0 \leq \theta_0 \leq \pi$ we have

$$|f(z_0) - f(\rho_0)| = c \left| \int_0^{\theta_0} (\rho_0 e^{i\theta} - a_1)^{\alpha_1 - 1} \ldots (\rho_0 e^{i\theta} - a_n)^{\alpha_n - 1} \rho_0 e^{i\theta}\, d\theta \right|$$

$$\leq c\rho_0(\rho_0 - R_1)^{\alpha_1 + \cdots + \alpha_n - n}.$$

By the hypothesis on the α_i the last term goes to zero as $\rho_0 \to \infty$. Hence, there is an $R \geq R_1$ such that

$$|f(z_0) - f(\rho_0)| \leq \varepsilon/2$$

if $\rho_0 = |z_0| \geq R$. This proves that $|f(z) - w_0| \leq \varepsilon$ if $|z| \geq R$, $\operatorname{Im} z \geq 0$.

Therefore, f maps $\mathbb{R} \cup \{\infty\}$ onto the closed polygonal line Δ whose sides are the segments $\Delta_0, \Delta_1, \ldots, \Delta_{n+1}$ and whose vertices are the points w_0, w_1, $\ldots$, w_n, $w_{n+1} = w_0$. It is not possible to conclude that f is injective on $\mathbb{R} \cup \{\infty\}$ since Δ could have self-intersections. This function f will be injective on $\mathbb{R}$ (and on $\mathbb{R} \cup \{\infty\}$) if and only if Δ is a Jordan curve. In that case we can apply §2.8.14 and find that the Schwarz-Christoffel transformation f is a homeomorphism of $\bar{H} \cup \{\infty\}$ onto $\overline{\operatorname{Int}(\Delta)}$, which is a biholomorphism of H onto $\operatorname{Int}(\Delta)$.

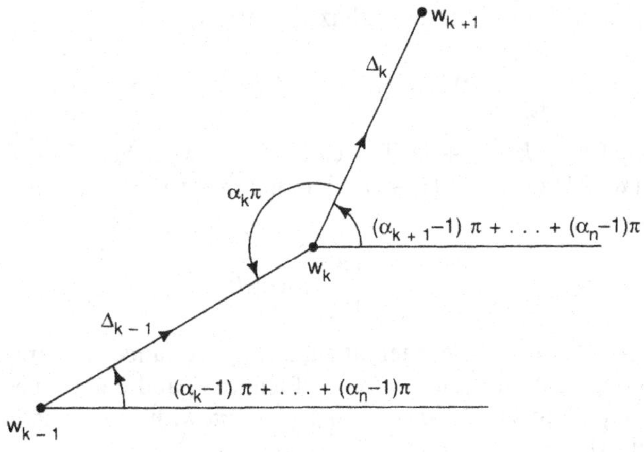

Figure 2.14

The interior angle of the polygon at the vertex w_k $(0 < k < n + 1)$ is $\alpha_k\pi (\text{mod } 2\pi)$. If one imposes the additional condition that $0 < \alpha_k < 2$, then this angle is exactly $\alpha_k\pi$.

In fact, one can write $\beta(t) = \beta_k(t)(t - a_k)^{\alpha_k - 1}$, with β_k holomorphic in a neighborhood V of a_k. In this neighborhood β_k has the series expansion

$$\beta_k(t) = a_{0,k} + a_{1,k}(t - a_k) + \cdots, \qquad a_{0,k} \neq 0.$$

Hence, for $z \in V \cap \bar{H}$ we have

$$\begin{aligned} f(z) &= w_k + c \int_{a_k}^{z} (a_{0,k} + a_{1,k}(t - a_k) + \cdots)(t - a_k)^{\alpha_k - 1}\, dt \\ &= w_k + c\frac{a_{0,k}}{\alpha_k}(z - a_k)^{\alpha_k}\left(1 + \frac{\alpha_k}{\alpha_{k+1}}\frac{a_{1,k}}{a_{0,k}}(z - a_k) + \cdots\right). \end{aligned}$$

If z tends towards a_k along a line making an angle θ with the positive real axis, it follows that $f(z)$ approaches w_k along a curve whose tangent at w_k has the direction $\arg(a_{0,k}) + \alpha_k\theta$, since $c > 0$ and $\alpha_k > 0$. When θ goes from 0 to π and z remains on a small circle centered at a_k, $f(z)$ traverses a Jordan arc joining a point in Δ_k to a point in Δ_{k-1}. This reasoning shows that while remaining always in the interior angle of vertex w_k, the point $f(z)$ has an argument that grows from $\arg(a_{0,k})$ to $\arg(a_{0,k}) + \alpha_k\pi$. Therefore, $\alpha_k\pi$ is the measure of this interior angle.

At the vertex $w_0 = w_{n+1}$ the interior angle of the polygon Δ is $((n - 1) - (\alpha_1 + \cdots + \alpha_n))\pi > 0$. The sum of the interior angles of Δ must be $(n + 1 - 2)\pi$. Therefore

$$((n - 1) - (\alpha_1 + \cdots + \alpha_n)) < 2\pi.$$

The condition

$$\alpha_1 + \cdots + \alpha_n > n - 1$$

is therefore necessary for the injectivity of f but not sufficient. (See Figure 2.16)

In the special case that $\alpha_1 + \cdots + \alpha_n = n - 2$ one finds that the angle at w_0 is exactly π, which means that $w_0 = w_{n+1}$ is only a fictitious vertex of the

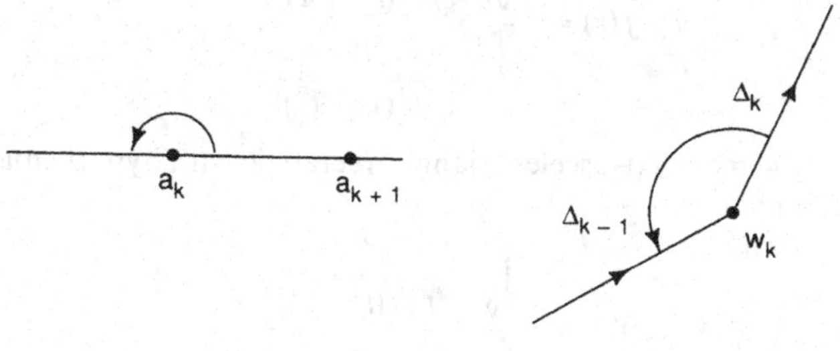

Figure 2.15

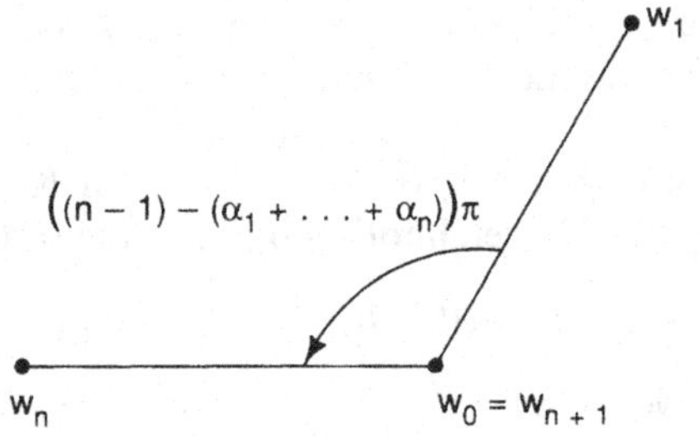

Figure 2.16

polygon since w_0 is in fact interior to the segment $[w_n, w_1]$. In this case f maps H onto a polygon with n sides.

As an example let us map H onto a triangle with angles $\alpha_1\pi$, $\alpha_2\pi$, $\alpha_3\pi$ $(0 < \alpha_k, \alpha_1 + \alpha_2 + \alpha_3 = 1)$ such that the length of the side opposite to the angle of $\alpha_1\pi$ is ℓ. We can do this in two ways:

(1) Take $n = 3$, $a_1 = -1$, $a_2 = 0$, $a_3 = 1$,

$$f(z) = c\int_0^z (t+1)^{\alpha_1-1}t^{\alpha_2-1}(t-1)^{\alpha_3-1}\,dt.$$

The constant c can be found from the condition

$$\ell = c\int_0^1 |(t+1)^{\alpha_1-1}t^{\alpha_2-1}(t-1)^{\alpha_3-1}|\,dt = c\int_0^1 (t+1)^{\alpha_1-1}t^{\alpha_2-1}(1-t)^{\alpha_3-1}\,dt.$$

Hence

$$f(z) = \frac{\displaystyle\int_0^z (t+1)^{\alpha_1-1}t^{\alpha_2-1}(t-1)^{\alpha_3-1}\,dt}{\displaystyle\int_0^1 (t+1)^{\alpha_1-1}t^{\alpha_2-1}(1-t)^{\alpha_3-1}\,dt}$$

maps conformally H onto such a triangle.

If $\alpha_1 = \alpha_2 = \alpha_3 = \frac{1}{3}$ (equilateral triangle), one obtains

$$f(z) = \ell\frac{\displaystyle\int_0^z \frac{dt}{\sqrt[3]{t^2(t^2-1)^2}}}{\displaystyle\int_0^1 \frac{dt}{\sqrt[3]{t^2(1-t^2)^2}}}.$$

If $\alpha_1 = \frac{1}{2}$, $\alpha_2 = \alpha_3 = \frac{1}{4}$ (isosceles triangle, rectangle with hypothenuse ℓ) the map f is given by

$$f(z) = \ell\frac{\displaystyle\int_0^z \frac{dt}{\sqrt{t}\sqrt[4]{(t^2-1)^3}}}{\displaystyle\int_0^1 \frac{dt}{\sqrt{t}\sqrt[4]{(1-t^2)^3}}}.$$

(2) If we take $n = 2$ and we pick a_1, a_2 arbitrarily that will be mapped onto the vertices of angles $\alpha_1\pi, \alpha_2\pi$ of a triangle, the third vertex must be the image of the point ∞. (Note that $\alpha_1 + \alpha_2 \neq n - 2 = 0$.) For instance, let us take $a_1 = 0$, $a_2 = 1$. Then

$$f(z) = c\int_0^z t^{\alpha_1-1}(t-1)^{\alpha_2-1}\,dt.$$

We determine c by

$$\ell = c\int_1^\infty t^{\alpha_1-1}(t-1)^{\alpha_2-1}\,dt,$$

then

$$f(z) = \ell\,\frac{\displaystyle\int_0^z t^{\alpha_1-1}(t-1)^{\alpha_2-1}\,dt}{\displaystyle\int_1^\infty t^{\alpha_1-1}(t-1)^{\alpha_2-1}\,dt}.$$

The case of an equilateral triangle corresponds to

$$f(z) = \ell\,\frac{\displaystyle\int_0^z \frac{dt}{\sqrt[3]{t^2(t-1)^2}}}{\displaystyle\int_1^\infty \frac{dt}{\sqrt[3]{t^2(t-1)^2}}}.$$

The Schwarz-Christoffel transformations will reappear when we study elliptic functions in the following volume. We suggest [Hen] for a thorough discussion of these transformations and their practical applications.

Some of the exercises that follow pertain not only to this section but to the preceding material.

Exercises 2.8.

As in previous sections, $B = B(0, 1)$; Ω is a domain in $\mathbb{C}$.

1. Let γ be a Jordan curve (with the positive orientation), f holomorphic in $\mathrm{Int}(\gamma)$ and continuous on $\overline{\mathrm{Int}(\gamma)}$ be such that $f(z) \neq 0$ on γ. Show that the number of zeros of f in $\mathrm{Int}(\gamma)$ is equal to $\mathrm{Ind}(f \circ \gamma)$. (Hint: Compare with the proof of Proposition 2.8.11.)
2. Apply the previous exercise to show that if γ is a Jordan curve, f is continuous in $\overline{\mathrm{Int}(\gamma)}$ and holomorphic in $\mathrm{Int}(\gamma)$, then for every $z_0 \in \mathrm{Int}(\gamma)$ there is a pair of distinct points z_1, z_2 on γ such that

$$\frac{f(z_2) - f(z_1)}{z_2 - z_1} = f'(z_0).$$

(Hint: Consider the function $g(z) = f(z) - f'(z_0)z$.)

Apply this result to the function $f(z) = z^2$ to obtain that every point $z_0 \in \mathrm{Int}(\gamma)$ lies on a chord $[z_1, z_2]$ with endpoints on γ.

3. (a) Let $\varphi : \partial B \to \partial B$ be a homeomorphism. Show that $\Phi(z) := |z|\varphi(z/|z|)$ if $z \neq 0$ and $\Phi(0) := 0$, is a homeomorphism of $\mathbb{C}$ onto itself.
   (b) Use part (a) and Caratheodory's theorem to show that if $h : \partial B \to \gamma$ is a parameterization of a Jordan curve and $\Omega = \operatorname{Int}(\gamma)$, then there is a homeomorphism $H : \bar{B} \to \bar{\Omega}$ such that $H|\partial B = h$. (This statement is usually called Schoenflies' theorem.)

4. Let Ω be a simply connected domain in S^2 such that $\#(S^2 \setminus \Omega) \geq 2$, $z_0 \in \Omega$, $\alpha \in [0, 2\pi[$. Show there is a unique biholomorphic map $f : \Omega \to B$ such that $f(z_0) = 0$, $\operatorname{Arg} f'(z_0) = \alpha$.

5. Let Ω be a simply connected proper open set in $\mathbb{C}$, symmetric with respect to the real axis. Let $x_0 \in \Omega \cap \mathbb{R}$, and $f : B(0, 1) \to \Omega$ be the biholomorphism such that $f(0) = x_0$ and $f'(0) > 0$. Show that $f^{(k)}(0) \in \mathbb{R}$ every $k \geq 1$.

6. Let $f \in \mathscr{H}(B)$. Assume there is an open arc $I \subseteq \partial B$ such that for any sequence $(z_n)_{n \geq 1} \subseteq B$ with limit $z_0 \in I$ we have

$$\lim_{n \to \infty} f(z_n) = 0.$$

   Show that $f \equiv 0$.

7. Let f be a conformal map of the half-disk D onto the triangle T shown in Figure 2.17 such that f is continuous in $\bar{D}$, $f(1) = 1$, $f(i) = i$, and $f(-1) = -1$. Prove that f can be extended to a conformal map F of B onto the square Q. Show that there is a function $G \in \mathscr{H}(B)$ such that $F(z) = zG(z^4)$. (The hypothesis of continuity of f up to the boundary of D is not really necessary.)

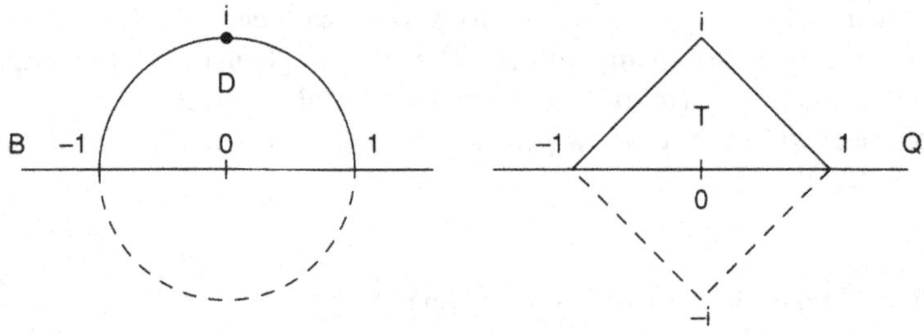

Figure 2.17

8. Generalize Schwarz's Reflection Principle 2.1.10 by showing that if $f \in \mathscr{H}(B)$, I is an open arc of ∂B, and f is continuous on $B \cup I$ and real valued in I, then the function F defined by

$$F(z) = \begin{cases} f(z) & \text{if } z \in B \cup I \\ \overline{f(1/\bar{z})} & \text{if } z \notin \bar{B} \end{cases}$$

   is holomorphic in the domain $B \cup I \cup \bar{B}^c$.

   How should one define F if we know that $|f(z)| = 1$ when $z \in I$, instead of assuming f was real-valued on I?

9. Let Ω be the region whose boundaries are the rays $\operatorname{Arg} z = \pm\frac{\pi}{4}$ and the branch

of the hyperbola $x^2 - y^2 = 1$ lying in $\operatorname{Re} z > 0$. Find a conformal map of Ω onto the unit disk.

10. Let $-\infty < a < b < \infty$. Show that the function $z \mapsto 1/\operatorname{Log}\left(\dfrac{z-a}{z-b}\right)$ is a conformal map in $\mathbb{C}\setminus[a,b]$. Find its image.

11. Let $f \in \mathscr{H}(\Omega)$ be locally injective, the ***Schwarzian derivative*** of f at the point z is defined by
$$\{f,z\} := \left(\frac{f''(z)}{f'(z)}\right)' - \frac{1}{2}\left(\frac{f''(z)}{f'(z)}\right)^2.$$
Assume that φ is a Moebius transformation:
(a) Show that $\{\varphi \circ f, z\} = \{f, z\}$
(b) Compute $\{f \circ \varphi, z\}$.
(c) What can you say about f if $\{f, z\} \equiv 0$?

12. Let $f \in \mathscr{H}(\bar{B})$ be a conformal map from B onto Ω. Give an example showing that f' can vanish on ∂B. Can it have a double zero?

13. Let $f \in \mathscr{H}(\mathbb{C})$ be real-valued on the real axis and satisfy $\operatorname{Im} f(z) > 0$ whenever $\operatorname{Im} z > 0$. Show that $f(z) = az + b$, $a > 0$, $b \in \mathbb{R}$.

*14. Let $f \in \mathscr{H}(B)$ and $c > 0$ be such that the connected component ω of $\{z \in B : |f(z)| < c\}$ is relatively compact in B and $f' \neq 0$ on $\partial\omega$. Let $N =$ the number of zeros (counted with multiplicities) of f in ω. Show that the number of zeros of f' inside ω is exactly $N - 1$ (counted with multiplicities). (Hint: Show first that $\partial\omega$ is a Jordan curve. Prove then that if s is the arc length parameter on $\partial\omega$ and $\theta = \arg f(z)$, then $\dfrac{d\theta}{ds}$ is real and does not vanish. Use $\dfrac{d\theta}{ds} = \dfrac{d\theta}{dz}\dfrac{dz}{ds}$ to compute $\Delta_{\partial\omega}\dfrac{d\theta}{dz}$. Conclude the proof writing $f = ce^{i\theta}$, $f' = cie^{i\theta}\dfrac{d\theta}{dz}$ on the curve $\partial\omega$.)

15. Let $f \in \mathscr{H}(B)$. Show that if f is not injective, there is r, $0 < r < 1$, such that $f|\partial B(0,r)$ is not injective.

16. Use the previous exercise to show that the polynomial $p(z) = z + \dfrac{z^2}{2} + \cdots + \dfrac{z^n}{n}$ belongs to the class S. (Hint: if $\operatorname{Re} p(re^{i\alpha}) = \operatorname{Re} p(re^{i\beta})$, $0 \le \alpha < \beta < 2\pi$, there is $\gamma \in\,]\alpha, \beta[$ such that $\left(\dfrac{d}{d\theta}\operatorname{Re} p\right)(re^{i\gamma}) = 0$.)

17. Let $A_1 = \{z \in \mathbb{C} : 0 < r_1 < |z| < R_1 < \infty\}$, $A_2 = \{z \in \mathbb{C} : 0 < r_2 < |z| < R_2 < \infty\}$, and $f : \bar{A}_1 \to \bar{A}_2$ a homeomorphism such that f is holomorphic in A_1. Apply the reflection principle of Exercise 2.8.8 to show that $\dfrac{R_2}{r_2} = \dfrac{R_1}{r_1}$ and determine f. (With the help of the more general reflection principle proved in Chapter 4 you will be able to reach the same conclusion if we only assume $f : A_1 \to A_2$ is a conformal map.)

18. (a) Let D be the square with vertices $0, \dfrac{1}{2}, \dfrac{1+i}{2}$, and $\dfrac{i}{2}$. Assume f is a conformal

map from D onto the upper half-plane H. Use Theorem 2.8.8 and Schwarz's reflection principle to show that f can be extended to a function $F \in \mathcal{M}(\mathbb{C})$, F which is *doubly periodic*, i.e.

$$F(z+1) = F(z), \qquad F(z+i) = F(z).$$

19. Let Ω be a bounded simply connected polygon with vertices at the points $w_0, w_1, \ldots, w_n$, with corresponding interior angles $\alpha_0\pi, \ldots, \alpha_n\pi$, $0 < \alpha_k < 2$. Let f be a conformal map of the upper half-plane $H = \{z \in \mathbb{C} : \operatorname{Im} z > 0\}$ onto Ω, let $z_0 = \infty$, $z_1, \ldots, z_n$ be the points corresponding to $w_0, \ldots, w_n$, respectively. (Why is f a homeomorphism of $\bar{H} \cup \{\infty\}$ onto $\bar{\Omega}$?) Define $F(z) := \log f'(z)$. Show that $F \in \mathcal{H}(H)$ and F' admits an analytic continuation to the whole plane with simple poles at $z_1, \ldots, z_n$, F' vanishes at ∞ and $\operatorname{Res}(F', z_k) = \alpha_k - 1$ for $k = 1, \ldots, n$. Conclude that f satisfies the differential equation

$$f''(z) - \sum_{k=1}^{n} \frac{\alpha_k - 1}{z - z_k} f'(z) = 0.$$

Show that f is of the form $\varphi \circ \psi$, where ψ is the Schwarz-Christoffel transformation

$$\psi(z) = c \int_0^z \prod_{k=1}^{n} (t - z_k)^{\alpha_k - 1}\, dt$$

and $\varphi(w) = aw + b$ is a linear map which sends Ω onto itself.

20. Find a conformal map of the upper half-plane H onto the regular n-gon with vertices $w_k = e^{2\pi i k/n}$, $k = 0, \ldots, n$.

21. Show that the function $f \in \mathcal{H}(B)$ given by

$$f(z) = \int_0^z \frac{dt}{(1 - t^n)^{2/n}},$$

where $f(x) > 0$ when $0 < x < 1$, defines a conformal map from B onto a regular n-gon. What is the relation between this function and that found in the previous exercise?

22. Let $0 < \kappa < 1$ and f be the Schwarz-Christoffel transformation

$$f(z) = \int_0^z \frac{dt}{\sqrt{(1 - t^2)(1 - \kappa^2 t^2)}}.$$

Show that f maps the upper half-plane onto the rectangle of vertices $-a, a, a + ib, -a + ib$, $a, b > 0$, given by

$$a = \int_0^1 [(1 - t^2)(1 - \kappa^2 t^2)]^{-1/2}\, dt$$

$$b = \int_\kappa^1 [(1 - t^2)(t^2 - \kappa^2)]^{-1/2}\, dt.$$

By a change of variables $t = \sin\theta$ show that a is given by the hypergeometric series

$$a = \frac{\pi}{2} \sum_{n=0}^{\infty} \left[\frac{(1/2)_n}{n!}\right]^2 \kappa^{2n},$$

where $(\frac{1}{2})_0 = 1$, $(\frac{1}{2})_1 = \frac{1}{2}$, $(\frac{1}{2})_2 = (\frac{1}{2})(\frac{1}{2} + 1), \ldots$.

23. Let Ω be the polygon from Exercise 2.8.19 and let $\beta_k = 2 - \alpha_k$. Consider the conformal map g of the upper half-plane H onto $\bar{\Omega}^c$ such that the points b_k on the real axis correspond to w_k, $b_0 = \infty$. Show that g satisfies the differential equation

$$\frac{g''(z)}{g'(z)} = \sum_{k=1}^{n} \frac{\beta_k - 1}{z - b_k} - \frac{2}{z - w_0} - \frac{2}{z - \overline{w_0}}.$$

24. Suppose $f \in \mathscr{M}(\mathbb{C})$ has only simple poles at $z_1, z_2, z_3, \ldots, 0 < |z_1| \le |z_2| \le \cdots$, $\operatorname{Res}(f, z_n) = A_n$. Assume there is a sequence of Jordan curves Γ_n such that no poles of f lie on any of them, $\Gamma_n \subseteq \operatorname{Int}(\Gamma_{n+1})$, $R_n := \min_{z \in \Gamma_n} |z| \to \infty$, $\ell(\Gamma_n) = O(R_n)$, and $\max_{z \in \Gamma_n} |f(z)| = O(R_n)$. Show that

$$f(z) = f(0) + \sum_{n \ge 0} A_n \left(\frac{1}{z - z_n} + \frac{1}{z_n} \right),$$

by considering $\displaystyle\int_{\Gamma_n} \frac{f(\zeta)}{\zeta} \frac{d\zeta}{\zeta - z}$. How should the series be understood in order that it converges in $\mathscr{H}(\mathbb{C} \backslash \{z_n\}_{n \ge 1})$?

25. Show that if f is an entire function with finite L^1 norm in $\mathbb{C}$, then $f \equiv 0$. (Is the same statement true for L^2, L^∞?)

26. Find all the meromorphic functions f in $\mathbb{C}$ such that $\|f\|_{L^2(\mathbb{C})} < \infty$. How about $\|f\|_{L^1(\mathbb{C})} < \infty$ or $\|f\|_{L^\infty(\mathbb{C})} < \infty$?

27. Show that if $f \in \mathscr{H}(B \backslash \{0\})$ and $f \in L^1(B)$, then the singularity is removable.

28. Let $f \in \mathscr{M}(\mathbb{C}) \backslash \mathbb{C}$. Show that $f(\mathbb{C})$ is dense in S^2.

29. Assume $f \in \mathscr{M}(\mathbb{C})$ and P is a polynomial such that

$$\operatorname{Re} f(z) \le \operatorname{Im} P(z)$$

at every point which is not a pole of f. Show that f is a polynomial and find it explicitly.

30. Let f, g be two holomorphic functions in a neighborhood of the disk $|w| \le r$, $f(0) \ne 0$. Show that for $\rho > 0$ sufficiently small, the equation

$$w = zf(w) + z^2 g(w)$$

has exactly one solution $w = \varphi(z)$ in $|w| < r$, and no solutions on $|w| = r$, as long as $|z| \le \rho$. The function φ is holomorphic and admits the Taylor series development $\varphi(z) = \sum_{n \ge 0} a_n z^n$,

$$a_n = \sum_{0 \le k \le n/2} \frac{1}{k!(n - 2k)!} \frac{d^{n-k-1}}{dw^{n-k-1}} [(f(w))^{n-2k} (g(w))^k]_{w=0}.$$

31. Let F be a C^∞ function of two complex variables z, w (i.e., as a function of four real variables) such that $\dfrac{\partial F}{\partial \bar{z}} = 0$, $\dfrac{\partial F}{\partial \bar{w}} = 0$ when $|z| \le R$, $|w| \le r$. Assume further that $F(0, w) = 0$ admits only the simple zero $w = 0$ in $|w| \le r$. Show that for some ρ, $0 < \rho < R$, there is a unique holomorphic function $\varphi(z)$ in $|z| < \rho$ such that $\varphi(0) = 0$ and $F(z, \varphi(z)) \equiv 0$.

32. Let $P(w) = w^m + A_1 w^{m-1} + \cdots + A_m$ be a polynomial and $w_1, \ldots, w_m$ its roots. (Each distinct root appearing as often as its multiplicity indicates). We know that the A_k can be determined from the identity

$$P(w) = \prod_{k=1}^{m} (w - w_k).$$

We want to show in this exercise that they can also be obtained from the Newton sums

$$B_k := w_1^k + \cdots + w_m^k, \qquad k = 0, \ldots, m.$$

For each k, write $B_{0,k} = 1$ and

$$\frac{P(w)}{w - w_k} = \sum_{j=0}^{m} B_{j,k} w^{m-j}.$$

Obtain the relation

$$A_j = B_{j,k} - B_{j-1,k} w_k.$$

Write $A_0 = 1$ and conclude that

$$\sum_{k=1}^{m} \frac{P(w)}{w - w_k} = A_0 B_0 w^{m-1} + (A_1 B_0 + A_0 B_1) w^{m-2} + \cdots.$$

Using that the left-hand side equals $P'(w)$ obtain expressions for the A_k in terms of the B_k.

*33. Let F be as in Exercise 2.8.31, except that we assume only that $F(0, w)$ vanishes at $w = 0$, but it is not identically zero in $|w| \le r$. Show there is an integer m, values $0 < r_1 \le r$, $0 < R_1 \le R$ and functions $A_1(z), \ldots, A_m(z)$ in $|z| \le R_1$, such that the functions

$$P(z, w) := w^n + A_1(z) w^{m-1} + \cdots + A_m(z)$$

and

$$G(z, w) := F(z, w)/P(z, w)$$

are C^∞ and holomorphic in each variable separately in the bidisk $|z| \le R_1$, $|w| \le r_1$. Moreover, $G(z, w)$ does not vanish anywhere in the bidisk. (Hint: Let m be the multiplicity of $w = 0$ as a zero of $F(0, w) = 0$ and choose r_1 so that $w = 0$ is the only zero of this equation in $|w| \le r_1$. Evaluate the functions

$$B_k(z) = \frac{1}{2\pi i} \int_{|w|=r_1} w^k \frac{\dfrac{\partial F}{\partial w}(z, w)}{F(z, w)} dw$$

in terms of the roots of the equation $F(z, w) = 0$. Use Exercise 2.8.32 to define A_k and P.)

34. Let Ω be a Jordan domain and f a conformal map of Ω onto B. Show that if $g \in \mathscr{H}(\Omega) \cap \mathscr{C}(\bar{\Omega})$, and $\varepsilon > 0$, there is a polynomial P such that $\|g - P \circ f\|_{L^\infty(\Omega)} \le \varepsilon$.

35. Let $\mathscr{F} \subseteq \mathscr{H}(B) \cap \mathscr{C}(\bar{B})$, $f \in \mathscr{F}$, and $n_k \nearrow \infty$ such that $f^{[n_k]} \to z_0$ in $\mathscr{H}(B)$ (z_0 denotes

the constant function). Assume $g \circ f = f \circ g$ for every $g \in \mathscr{F}$. Show that $g(z_0) = z_0$ for every $g \in \mathscr{F}$ which is not a constant function.

36. Use Exercises 2.3.16, 2.6.19, and 2.6.21 to show that, if a family $\mathscr{F} \subseteq \mathscr{H}(B) \cap \mathscr{C}(\bar{B})$ of commuting maps into $\bar{B}$ contains an element $f \in \mathrm{Aut}(B)\setminus\{id_B\}$, then there is a common fixed point for every $g \in \mathscr{F}$ which is not the constant map.

37. Let $f \in \mathscr{H}(B)$, $\|f\|_\infty \leq 1$ and $n_k \nearrow \infty$ be such that $f^{[n_k]} \to h$ in $\mathscr{H}(B)$ and $h \neq$ constant.
 (a) Assume $m_k = n_k - n_{k-1} \nearrow \infty$, show there is a subsequence m_{k_j} such that $f^{[m_{k_j}]}$ is convergent in $\mathscr{H}(B)$ to a function g.
 (b) Writing $f^{[n_{k_j}]} = f^{[m_{k_j}]} \circ f^{[n_{k_j-1}]}$, conclude with the help of Exercise 2.2.12 that $g \circ h = h$.
 (c) Show that $g = id_B$. Conclude that $f, h \in \mathrm{Aut}(B)$.

38. Use Exercises 2.6.20, 2.6.21, and 2.6.35–37 to prove the following theorem of A. Shields [Shi]: Let $\mathscr{F} \subseteq \mathscr{H}(B) \cap \mathscr{C}(\bar{B})$ be a commuting family of maps into $\bar{B}$. Then there is a common fixed point $z_0 \in \bar{B}$ for all the elements of $\mathscr{F}$. (Hint: Show first that one can assume that no $g \in \mathscr{F}$ is a constant function. Consider later separately the cases $\mathscr{F} \subseteq \mathrm{Aut}(B)$ or not.)

Notes to Chapter 2

1. We have assumed a certain familiarity with the elementary properties of power series and numerical series. They can be found in [Ah1] and [Mar]. For more about them one can consider, e.g., [Kn]. The books [Ah1], [Mar], and [JS] also provide a very clear introduction to the hyperbolic and spherical geometries which we introduced in the exercises.

2. There are many sufficient conditions for holomorphicity. The Cauchy-Goursat theorem treats one of them. Example 2.2.9, (5) treats another, that locally integrable functions satisfying the homogeneous Cauchy-Riemann equation "are" holomorphic. A third one is Morera's theorem. These last two examples can be greatly generalized; some of these generalizations appear in the next volume. Even though these kinds of conditions appear on first sight to be really weaker than Cauchy-Goursat's, they are not, since they involve a certain kind of "uniformity" and are amenable to systematic study. The Cauchy-Goursat theorem and its analogs are "pointwise" results and tend to require ad hoc and very delicate arguments to prove them. The interested reader is referred to [Za5] for a lively discussion of this point.

3. A number of other results appear natural in the context of distributions. For instance, the Cauchy transform $\hat{\mu}$ of a measure μ of compact support will be later shown to satisfy the equation $\dfrac{\partial \hat{\mu}}{\partial \bar{z}} = \mu$ in the sense of distributions. In particular, $\hat{\mu} = 0$ implies $\mu = 0$. Similarly, the Schwarz Reflection Principle 2.1.11 has a distributional version, the Edge-of-the-Wedge Theorem 3.6.23, which we will later use as a motivation to introduce the concepts of distributions and hyperfunctions.

On the other hand, the Schwarz reflection principle also shows clearly the dichotomy between "uniform" and "pointwise" results. As we shall see in Chapter 4, the boundary values of a holomorphic function can exist in a very weak sense, as long as they are "real," and still guarantee that functions will have a holomorphic extension across a real analytic curve. This will prove helpful in extending the conformal mappings obtained with the help of the Riemann mapping theorem across the boundary of the unit disk.

4. For detailed historical remarks and techniques of the calculus of residues, see [MK]. At the end of Chapter 3, we relate the concept of residues to the theory of distributions, in a way that leads naturally to the study of residues in several complex variables.

5. The calculus of residues is related to interpolation theory. A nice treatment with special emphasis on numerical procedures can be found in [Hen]. Interpolation is one of the crucial tools in our study of convolution equations in the next volume, and therefore the reader should not be surprised to see it appear many times throughout this text. We also recommend [Ge], which was written with applications to transcendental number theory and analytic number theory in mind.

6. For applications of the Lagrange Formula 1.5.19 and other related questions, such as the location of zeros of polynomials, we refer the reader to [Hen], and the bibliography there. This is a subject of great usefulness in control theory and other engineering areas.

7. One interesting application of the concept of normal families lies in questions about iteration of holomorphic maps: Julia, Fatou, and Mandelbrojt sets. Several of the exercises provide a test of the complicated behavior of functions under iteration. A very good introduction into this subject is [Ab], see also [Bl]. The present interest in chaos, fractals, and iteration theory exemplifies the ubiquitous role of complex analysis in the applications of mathematics, see, e.g., [Dev].

8. The interesting story of de Branges' proof of the Bierbach conjecture can be found in [BH] and [Ko]. A great deal of mathematics related to this conjecture appears in [Ba].

9. The properties of $\mathscr{H}(\Omega)$ and $\mathscr{M}(\Omega)$ as topological vector spaces play an important part in this book, e.g., in Chapter 3, and are central to the next volume. In particular, the concept of bounded and normal families which are, one could even say unexpectedly, related to the question of omitted values as shown, e.g., by the Heuristic Principle 2.7.11. A very sharp version of this principle can be found in [Hem]. The reader will find in [Schwick] a substantial number of normality criteria which can be obtained with the help of §2.7.12. It is also important to point out there are limitations to §2.7.12, see [Rub1] for examples in this direction.

10. The natural context for the Picard theorems and their generalizations lies in Nevanlinna theory; see [Ha]. For a very nice introduction to Nevanlinna theory in several variables, where the geometric point of view of Ahlfors appears very clearly, we suggest [Gri1].

CHAPTER 3

The $\bar{\partial}$-Equation

§1. Runge's Theorem

It is in this chapter that the difference between our textbook and more classical ones appears markedly. As stated in the preface, we have attempted to use, as systematically as possible, the inhomogeneous Cauchy-Riemann equation $\frac{\partial f}{\partial \bar{z}} = g$ to study holomorphic functions (also called the $\bar{\partial}$-equation). The reader should note the irony here. To better comprehend the solutions of the homogeneous equation $\frac{\partial f}{\partial \bar{z}} = 0$ one is forced to study a more complex object! Our presentation owes much to Hörmander's beautiful treatise on several complex variables [Hol].

Let us recall that if K is an arbitrary set in $\mathbb{C}$, we denote $\mathscr{H}(K)$ the family of functions which are holomorphic in some neighborhood of K. Note that the neighborhood may depend on the function.

It is evident that $\mathscr{H}(K) \subseteq \mathscr{C}(K)$, the space of continuous functions on K. Recall that if K is a compact set, the space $\mathscr{C}(K)$ is a Banach space with the norm

$$\|f\|_K = \max_{z \in K} |f(z)|.$$

The topological dual vector space $\mathscr{C}'(K)$ can be identified to the space of complex valued measures supported by K (***Radon measures***). In this context, the Hahn-Banach theorem implies that if $f \in \mathscr{C}(K)$ and

$$\int_K f \, d\mu = 0$$

for every Radon measure μ with support in K, then $f \equiv 0$. Conversely, if H is a subspace of $\mathscr{C}(K)$, and $H^\perp = \{\mu \in \mathscr{C}'(K) : \int f\, d\mu = 0 \text{ for every } f \in H\} = \{0\}$, it follows that H is dense in $\mathscr{C}(K)$. We write $\mu \perp H$ to indicate $\mu \in H^\perp$; we say μ is orthogonal to H. (See [Ru] and [HS] for the Hahn-Banach theorem and the few elements of functional analysis used here.)

Finally, we remind the reader that the total variation $\|\mu\|_K$ of a measure $\mu \in \mathscr{C}'(K)$ is given by

$$\|\mu\|_K = \sup\left\{\left|\int f\, d\mu\right| : f \in \mathscr{C}(K), \|f\|_K \le 1\right\}.$$

For Ω open in $\mathbb{C}$, K compact, $K \subseteq \Omega$, we have another subspace of $\mathscr{C}(K)$, $\mathscr{H}(\Omega)|K$, the space of restrictions to K of functions holomorphic in Ω. Clearly $\mathscr{H}(\Omega)|K \subseteq \mathscr{H}(K)$. The next theorem allows us to decide whether $\mathscr{H}(\Omega)|K$ is dense in $\mathscr{H}(K)$ (as subspaces of the Banach space $\mathscr{C}(K)$) in purely geometrical terms.

3.1.1. Theorem (Runge). *Let Ω be an open subset of $\mathbb{C}$ and K a compact subset of Ω. The following statements about the pair (Ω, K) are equivalent*:

(a) *Every function holomorphic in a neighborhood of K is the uniform limit over K of functions in $\mathscr{H}(\Omega)$.*
(b) *None of the connected components of $\Omega \backslash K$ is relatively compact in Ω.*
(c) *For every $z \in \Omega \backslash K$ there is a function $f \in \mathscr{H}(\Omega)$ such that*

$$|f(z)| > \sup_{\zeta \in K} |f(\zeta)| = \|f\|_K.$$

PROOF. We are going to show $a \Leftrightarrow b \Leftrightarrow c$.

(1) (c $\Rightarrow$ b): We want to show that condition (c) and the negation of (b) cannot hold simultaneously.

Let us assume that $\Omega \backslash K$ has a component $\mathscr{O}$ which is relatively compact in Ω. Hence $\partial \mathscr{O} = \bar{\mathscr{O}} \backslash \mathscr{O} \subseteq K$. Namely, $\partial \mathscr{O} = \bar{\mathscr{O}} \backslash \mathscr{O} \subseteq \Omega$ and at the same time, $\partial \mathscr{O} \subseteq (\Omega \backslash K)^c \subseteq \Omega^c \cup K$, so $\partial \mathscr{O} \subseteq K$. Therefore, for every $f \in \mathscr{H}(\Omega)$ we have

$$\sup_{z \in \bar{\mathscr{O}}} |f(z)| = \sup_{\partial \mathscr{O}} |f(z)| \le \sup_{z \in K} |f(z)|,$$

which contradicts (c).

(2) (a $\Rightarrow$ b): If (b) does not hold, let $\mathscr{O}$ be as in (1). For $\zeta \in \mathscr{O}$ the function $f : z \mapsto 1/(z - \zeta)$ is holomorphic in a neighborhood of K. By (a) there is a sequence $(h_n)_{n \ge 1}$ of functions in $\mathscr{H}(\Omega)$ converging uniformly to f on K.

By the same argument as in (1) we have

$$\sup_{z \in \bar{\mathscr{O}}} |h_p(z) - h_q(z)| \le \sup_{z \in K} |h_p(z) - h_q(z)|.$$

Therefore the sequence $(h_n)_{n \ge 1}$ converges uniformly on $\bar{\mathscr{O}} \cup K$ to a function F. This function will be holomorphic in $\mathscr{O}$ and will coincide with f on K. It follows that the function $z \mapsto (z - \zeta)F(z) - 1$ vanishes identically on $\partial\mathcal{O}$ and, by the maximum principle, it vanishes identically in $\mathcal{O}$. Evaluating it at $z = \zeta$ we obtain a contradiction.

(3) (b $\Rightarrow$ a): If we show that every Radon measure μ in K which vanishes on $\mathscr{H}(\Omega)$, also vanishes on $\mathscr{H}(K)$, then from the Hahn-Banach theorem we conclude that

$$\mathscr{H}(K) \subseteq \overline{\mathscr{H}(\Omega)|K},$$

which is what we need to prove (the closure taken in $\mathscr{C}(K)$, of course).

Let $\mu \in \mathscr{C}'(K)$ be a Radon measure orthogonal to $\mathscr{H}(\Omega)$. Let

$$\hat{\mu}(\zeta) = -\frac{1}{\pi}\int_K \frac{d\mu(z)}{z - \zeta}$$

be its Cauchy transform (cf. §2.1.5). We are going to show first that $\hat{\mu} \equiv 0$ in $K^c = \mathbb{C}\backslash K$.

1°. In the unbounded connected component of K^c: For $|\zeta| > r$, r sufficiently large, we have that the series

$$\frac{1}{z - \zeta} = -\sum_{n \geq 0} \frac{z^n}{\zeta^{n+1}}$$

converges uniformly for $z \in K$. Hence

$$\hat{\mu}(\zeta) = \frac{1}{\pi}\sum_{n \geq 0}\left(\int_K z^n \, d\mu\right)\Big/\zeta^{n+1}.$$

Since μ is orthogonal to $\mathscr{H}(\Omega)$, we have $\displaystyle\int_K z^n \, d\mu = 0$ for $n \geq 0$. Therefore $\hat{\mu} \equiv 0$ for $|\zeta| > r$ and hence, in the unbounded connected component of K^c.

2°. In any bounded connected component U of K^c: We have $U \cap \Omega^c \neq \varnothing$, otherwise, since $\partial U \subseteq K \subseteq \Omega$, we would have $\bar{U} \subseteq \Omega$ and hence, U would be a component of $\Omega\backslash K$, relatively compact in Ω, which contradicts the hypothesis.

Let $\zeta \in U \cap \Omega^c$. For every $k \geq 0$, the function

$$z \mapsto \frac{1}{(z - \zeta)^{k+1}}$$

is holomorphic in Ω, hence

$$\hat{\mu}^{(k)}(\zeta) = -\frac{k!}{\pi}\int_K \frac{d\mu(z)}{(z - \zeta)^{k+1}} = 0.$$

Therefore $\hat{\mu} \equiv 0$ in U.

We want to show now that $\hat{\mu} \equiv 0$ in K^c implies that μ vanishes on $\mathscr{H}(K)$. Let $f \in \mathscr{H}(K)$, then for some open set ω, $K \subseteq \omega \subseteq \Omega$, we have $f \in \mathscr{H}(\omega)$. Let $\psi \in \mathscr{D}(\omega)$ be such that $\psi \equiv 1$ in a neighborhood of K. Let $K_1 := \operatorname{supp}\left(\dfrac{\partial \psi}{\partial \bar{z}}\right)$.

For $z \in K$ we have, by Pompeiu's formula 2.1.2,

$$f(z) = \psi(z)f(z) = \frac{1}{2\pi i}\int_{K_1} f(\zeta)\frac{\partial\psi(\zeta)}{\partial\bar{\zeta}}\frac{1}{\zeta - z}\,d\zeta \wedge d\bar{\zeta}.$$

Since $K \cap K_1 = \varnothing$, the function $(z, \zeta) \mapsto f(\zeta)\dfrac{\partial\psi}{\partial\bar{\zeta}}(\zeta)\dfrac{1}{\zeta - z}$ is continuous in $K \times K_1$. Applying Fubini's theorem we obtain

$$\int_K f\,d\mu = \int_K \left\{\frac{1}{2\pi i}\int_{K_1} f(\zeta)\frac{\partial\psi}{\partial\bar{\zeta}}(\zeta)\frac{1}{\zeta - z}\,d\zeta \wedge d\bar{\zeta}\right\} d\mu(z)$$
$$= -\frac{1}{2i}\int_{K_1} f(\zeta)\frac{\partial\psi}{\partial\bar{\zeta}}(\zeta)\hat{\mu}(\zeta)\,d\zeta \wedge d\bar{\zeta} = 0.$$

The last identity follows from the fact that $K_1 \subseteq K^c$ and $\hat{\mu} \equiv 0$ on K^c.

(4) (b $\Rightarrow$ c): Let $z \in \Omega \setminus K$, and let $L = \bar{B}(z, \rho) \subseteq \Omega \setminus K$. The connected components of $\Omega \setminus (K \cup L)$ are exactly the same as those of $\Omega \setminus K$ except for that one from which we have taken away L (see Figure 3.1).

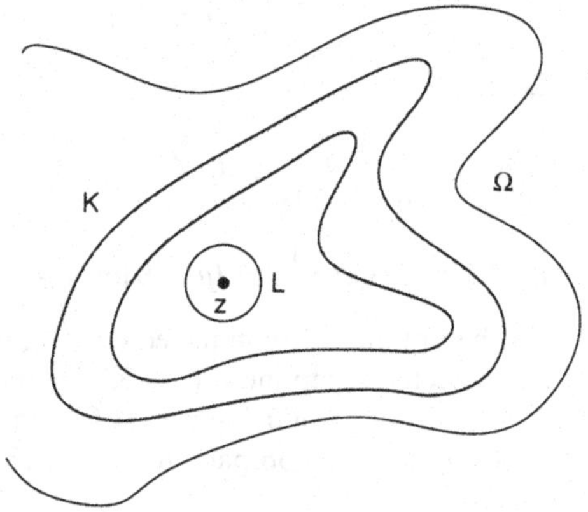

Figure 3.1

It follows that property (b) still holds for the pair $(\Omega, K \cup L)$. Since we have already shown that b $\Rightarrow$ a, we have that property (a) holds for $(\Omega, K \cup L)$. Let $f \equiv 0$ in a neighborhood of K and $f \equiv 1$ in a neighborhood of L. This function belongs to $\mathscr{H}(K \cup L)$ and hence, we have a sequence $(f_n)_{n \geq 1}$ in $\mathscr{H}(\Omega)$ converging uniformly to f on $K \cup L$. For n sufficiently large we have

$$|f_n| \leq 1/3 \quad \text{on } K, \qquad |f_n| \geq 2/3 \quad \text{on } L.$$

This shows that (c) holds and concludes the proof of Runge's theorem. □

In particular, when $\Omega = \mathbb{C}$ we have the following corollary.

3.1.2. Corollary. *Every function holomorphic in a neighborhood of the compact set K can be approximated uniformly on K by polynomials if and only if K^c is connected.*

PROOF. Apply §3.1.1 and the fact that the polynomials are dense in $\mathscr{H}(\mathbb{C})$. □

3.1.3. Definition. Let Ω be an open subset of the complex plane and K a compact subset of Ω. We call ***holomorphically convex hull*** of K in Ω the set

$$\hat{K}_\Omega := \{z \in \Omega : \forall f \in \mathscr{H}(\Omega)\ |f(z)| \le \|f\|_K\},$$

where, as before, $\|f\|_K = \max_{z \in K} |f(z)|$. If there is no possibility of confusion about Ω, we shall write $\hat{K}$ instead of $\hat{K}_\Omega$.

3.1.4. Proposition. *Let Ω be an open subset of $\mathbb{C}$, K compact in Ω. The following two properties hold:*

(a) *$K \subseteq \hat{K}$ and $\hat{K}$ is a compact subset of Ω.*
(b) *$d(K,\Omega^c) = d(\hat{K},\Omega^c)$.*

PROOF. $\hat{K}$ is relatively closed in Ω since it is an intersection of relatively closed sets,

$$\hat{K} = \bigcap_{f \in \mathscr{H}(\Omega)} \{z \in \Omega : |f(z)| \le \|f\|_K\}.$$

From the definition, one easily sees that if $r = \max_{z \in K} |z|$, then $\hat{K} \subseteq \bar{B}(0,r)$, hence $\hat{K}$ is bounded and (a) holds if $\Omega = \mathbb{C}$.

If $\Omega \ne \mathbb{C}$, let us show first (b).

The inclusion $K \subseteq \hat{K}$ implies that

$$d(\hat{K},\Omega^c) \le d(K,\Omega^c).$$

On the other hand, for $\zeta \in \Omega^c$, the function $z \mapsto 1/(z - \zeta)$ is holomorphic in Ω and hence, for $z \in \hat{K}$ we have

$$\frac{1}{|z-\zeta|} \le \sup_{w \in K} \frac{1}{|w-\zeta|} = \frac{1}{\inf_{w \in K} |w-\zeta|},$$

therefore $\inf_{w \in K} |w - \zeta| \le |z - \zeta|$ and

$$d(K,\Omega^c) = \inf_{\substack{w \in K \\ \zeta \in \Omega^c}} |w-\zeta| \le \inf_{\substack{z \in \hat{K} \\ \zeta \in \Omega^c}} |z-\zeta| = d(\hat{K},\Omega^c).$$

In conclusion, $\hat{K}$ is a relatively closed subset of Ω, bounded in $\mathbb{C}$ and $d(\hat{K},\Omega^c) = d(K,\Omega^c) > 0$. It is then clear that K is a compact subset of Ω. □

Note that if K is compact, $K \subseteq \Omega_1 \subseteq \Omega_2$, Ω_j open in $\mathbb{C}$, then

$$\hat{K}_{\Omega_1} \subseteq \Omega_1 \cap \hat{K}_{\Omega_2}.$$

If K is a compact set, then its convex hull $cv(K)$ is always compact. We have that if $\hat{K} = \hat{K}_{\mathbb{C}}$, $\hat{K}$ is always contained in $cv(K)$. This follows from the fact that

$$cv(K) = \bigcap_{\varphi \in \mathscr{L}_{\mathbb{R}}(\mathbb{C})} \left\{\zeta \in \mathbb{C} : \varphi(\zeta) \leq \sup_K \varphi\right\}.$$

In fact, every $\varphi \in \mathscr{L}_{\mathbb{R}}(\mathbb{C})$ is of the form $\varphi = \operatorname{Re}\theta$, $\theta \in \mathscr{L}_{\mathbb{C}}(\mathbb{C})$, and $\theta(z) = az$ for some $a \in \mathbb{C}$. If $z \in \hat{K}$ we have

$$e^{\operatorname{Re}(az)} = |e^{az}| \leq \sup_{\zeta \in K} |e^{a\zeta}| = \exp\left[\sup_{\zeta \in K} \operatorname{Re}(a\zeta)\right]$$

and, by the monotonicity of the exponential, we obtain $\varphi(z) \leq \sup_{\zeta \in K} \varphi(\zeta)$. Hence $\hat{K} \subseteq cv(K)$. Therefore, for any $\Omega \supseteq K$ we have

$$\hat{K}_\Omega \subseteq cv(K) \cap \Omega.$$

It is also clear that if $K \subseteq K_1$ and both are compact subsets of Ω, then $\hat{K} \subseteq \hat{K}_1$.

3.1.5. Definition. A compact subset K of an open set Ω is called ***holomorphically convex*** in Ω if $K = \hat{K}_\Omega$.

Evidently $(\hat{K}_\Omega)\hat{}_\Omega = \hat{K}_\Omega$ for every K and therefore, $\hat{K}_\Omega$ is holomorphically convex on Ω. Furthermore, by the preceding remark, if K is a compact convex set then it is holomorphically convex (in any Ω which contains K).

3.1.6. Proposition. *The set $\hat{K}_\Omega$ is the union of K and the connected components of $\Omega \setminus K$ which are relatively compact in Ω.*

PROOF. Let $\mathcal{O}$ be a component of $\Omega \setminus K$, relatively compact in Ω. We have $\bar{\mathcal{O}} \subseteq \hat{K}$ since $\partial\mathcal{O} \subseteq K$ and hence, by the maximum principle,

$$\sup_{\zeta \in \bar{\mathcal{O}}} |f(\zeta)| \leq \sup_{\zeta \in K} |f(\zeta)|,$$

for every $f \in \mathscr{H}(\Omega)$.

Let K_1 be the union of K and the relatively compact components of $\Omega \setminus K$. Then $K_1 \subseteq \hat{K}$. Moreover, the set $\Omega \setminus K_1$ is also the union of components of $\Omega \setminus K$, hence $\Omega \setminus K_1$ is open. It follows that K_1 is a compact subset of $\hat{K}$. By the definition of K_1, none of the components of $\Omega \setminus K_1$ is relatively compact in Ω. By part (c) of Runge's theorem we have $K_1 = \hat{K}_1$. Since $K \subseteq K_1$ we also have $\hat{K} \subseteq \hat{K}_1 = K_1$. □

3.1.7. Proposition. *For every open set $\Omega \subseteq \mathbb{C}$ there is an exhaustion of Ω by compact subsets which are holomorphically convex in Ω.*

PROOF. Let $(L_j)_{j \geq 1}$, be an exhaustion of Ω by compact subsets. Let $K_1 = \hat{L}_1$. There is an index $j_2 > 2$ such that $\mathring{L}_{j_2} \supseteq K_1$. Therefore $\mathring{L}_{j_2} \supseteq K_1 \cup L_2$. Let

$K_2 = \hat{L}_{j_2}$. There is $j_3 > 3$ such that $\mathring{L}_{j_3} \supseteq K_2$, hence $\mathring{L}_{j_3} \supseteq K_2 \cup L_3$. Let $K_3 = \hat{L}_{j_3}$, etc. □

EXERCISES 3.1

1. Give an example of a compact set K and two planar open connected sets Ω_1, Ω_2 such that $K \subseteq \Omega_1 \cap \Omega_2$ and $\hat{K}_{\Omega_1} \neq \hat{K}_{\Omega_2}$.

2. Let K be a compact subset of $\mathbb{C}$ and $f \in \mathscr{H}(K)$. Using Cauchy's formula, show that f can be uniformly approximated in K by rational functions whose poles are simple and lie in K^c. (Hint: Show first that given Ω open, $K \subseteq \Omega$, there is a finite number of polygonal Jordan curves $\gamma_1, \ldots, \gamma_n$ such that $K \subseteq \bigcup_{j=1}^{n} \operatorname{Int}(\gamma_j) \subseteq \Omega$ (cf. Chapter 1).) It is instructive to find another proof using Theorem 3.1.1.

3. Using Theorem 3.1.1, show that if K is a compact subset of $\mathbb{C}$ and $A = \{a_j\}$ is a collection of points in $S^2 \setminus K$, one in each connected component, then any function in $\mathscr{H}(K)$ can be uniformly approximated in K by rational functions with poles in A.

*4. Let $\Omega_1 \subseteq \Omega_2$ be open subsets of $\mathbb{C}$. Prove that the following five conditions are equivalent:
   (i) $\mathscr{H}(\Omega_2)$ is dense in $\mathscr{H}(\Omega_1)$.
   (ii) If $\Omega_2 \setminus \Omega_1 = K \cup F$, $K \cap F = \varnothing$, K compact and F closed in Ω_2, then $K = \varnothing$.
   (iii) For every compact subset K of Ω_1, $\hat{K}_{\Omega_2} = \hat{K}_{\Omega_1}$.
   (iv) For every compact subset K of Ω_1, $\hat{K}_{\Omega_2} \cap \Omega_1 = \hat{K}_{\Omega_1}$.
   (v) For every compact subset K of Ω_1, $\hat{K}_{\Omega_2} \cap \Omega_1$ is compact.

5. Let Ω be an open subset of $\mathbb{C}$ such that Ω^c has no bounded components. Show that the polynomials are dense in $\mathscr{H}(\Omega)$.

6. Show directly that the function $f(z) = \dfrac{1}{z}$ cannot be the uniform limit of polynomials in the closed annulus $1 \leq |z| \leq 2$.

7. (i) Let $n \in \mathbb{N}^*$, $0 < b_n < a_n < n$. Show that there is a polynomial p_n such that $|p_n(z)| \geq n$ if $\operatorname{Im} z = b_n$ and $z \in \bar{B}(0, n)$, and $|p_n(z)| \leq \dfrac{1}{n}$ whenever $z \in \bar{B}(0, n)$ and either $\operatorname{Im} z \leq 0$ or $\operatorname{Im} z \geq a_n$.
   (ii) Use the preceding construction to show the existence of a sequence of polynomials p_n such that $\lim_{n\to\infty} p_n(z) = 0$ everywhere, the limit is uniform in every compact subset of $\mathbb{C} \setminus \mathbb{R}$, but it is not uniform on any neighborhood of a real point.
   (iii) Modify the preceding procedure to construct a sequence of polynomials q_n such that $\lim_{n\to\infty} q_n(z) = 0$ if $z \in \mathbb{R}$ and $\lim_{n\to\infty} q_n(z) = 1$ if $z \in \mathbb{C} \setminus \mathbb{R}$.

*8. (Construction of a function in $\mathscr{H}(B)$ that has no radial limits at any point $\zeta \in \partial B$, $B = B(0, 1)$). Define a collection of subsets B_n of B as follows:

$$B_0 = \left\{|z| \leq \frac{1}{2}\right\},$$

$$B_1 = \left\{\frac{3}{4} \le |z| \le \frac{7}{8}, |\mathrm{Arg}\, z| \le \frac{3\pi}{4}\right\} \cup \left\{\frac{15}{16} \le |z| \le \frac{31}{32}, |\mathrm{Arg}\, z| \ge \frac{\pi}{4}\right\},$$

$$B_2 = \left\{\frac{63}{64} \le |z| \le \frac{127}{128}, |\mathrm{Arg}\, z| \le \frac{3\pi}{4}\right\} \cup \left\{\frac{255}{256} \le |z| \le \frac{511}{512}, |\mathrm{Arg}\, z| \ge \frac{\pi}{4}\right\},$$

and so on (Draw a picture.)

Show there is a sequence $\{P_n\}_{n\ge 0}$ of polynomials with the following properties: $P_0 \equiv 0$, assuming that $P_1, \ldots, P_{2n-2}$ have been found, then

$$|P_{2n-1}(z) - P_{2n-2}(z)| \le \frac{1}{2^{2n-1}} \quad \text{if} \quad |z| \le 1 - 2^{-8n+8},$$

$$|P_{2n-2}(z) - 1| \le \frac{1}{2^{2n}} \quad \text{if} \quad z \in B_{2n-1},$$

and

$$|P_{2n}(z) - P_{2n-1}(z)| \le \frac{1}{2^{2n}} \quad \text{if} \quad |z| \le 1 - 2^{-8n+4},$$

$$|P_{2n}(z)| \le \frac{1}{2^{2n}} \quad \text{if} \quad z \in B_{2n}.$$

Prove that there is a function f such that

$$P_n \to f \quad \text{in } \mathscr{H}(B).$$

Show that this function satisfies

$$\|f - 1\|_{B_{2n-1}} \le \frac{1}{2^{2n-2}}, \quad \|f\|_{B_{2n}} \le \frac{1}{2^{2n-1}},$$

and hence, it cannot have any radial limits.

*9. (Construction of a *universal* entire function). Let $\{Q_n\}_{n\ge 0}$ be an ordering of the collection of all polynomials with coefficients in $\mathbb{Q} + i\mathbb{Q}$. Let $K_n = \bar{B}(n^3, n)$. Let $P_1 = 0$, choose P_n, $n > 1$, by induction as follows:

$$|P_n(z) - P_{n-1}(z)| \le \frac{1}{2^n} \quad \text{in } \bar{B}(0, (n-1)^3)$$

$$|P_n(z) - Q_n(z - n^3)| \le \frac{1}{2^n} \quad \text{in } K_n.$$

(Why is this possible?) Show that the function $f := \sum_{n\ge 2} (P_n - P_{n-1})$ is entire. Moreover, on $\bar{B}(0, n)$ it satisfies

$$|f(z + n^3) - Q_n(z)| \le \frac{1}{2^{n-1}}.$$

Let now Ω be a bounded simply connected domain in $\mathbb{C}$, $\varphi \in \mathscr{H}(\Omega)$. Show that there is a sequence of integers $n_k \to \infty$, $Q_{n_k} \to \varphi$ in $\mathscr{H}(\Omega)$. Conclude that the translates of f, $f(z + n_k^3)$, converge to φ in $\mathscr{H}(\Omega)$ (cf. [BR1], [BR2]).

10. The object of this exercise is to prove the following extension of Runge's theorem. Let K be a compact subset of $\mathbb{C}$ such that $\overline{\mathring{K}} = K$. Let f be a function which is of class C^1 in a neighborhood of K and $f \in \mathscr{H}(\mathring{K})$, then $f \in \overline{\mathscr{H}(K)}$ (it is understood that the closure takes place in $\mathscr{C}(K)$ and we really mean $f|K$ is in the closure of $\mathscr{H}(K)$).
   (a) Let μ be any Radon measure with $\operatorname{supp}(\mu) \subseteq K$. Show that $\hat{\mu} \in L^1_{\text{loc}}(\mathbb{C})$. $\left(\text{Hint: Just show that for any } K_1 \subset\subset \mathbb{C}, \text{ one has } \int_{K_1} dm(z) \int \frac{d|\mu(\zeta)|}{|z-\zeta|} < \infty \text{ by Fubini's theorem.}\right)$
   (b) Use the proof of Theorem 3.1.1 to show that if μ is a Radon measure with $\operatorname{supp}(\mu) \subseteq K$, then $\mu \perp \mathscr{H}(K)$ if and only if $\hat{\mu} = 0$ in K^c.
   (c) Show we can assume $f \in \mathscr{D}_1(\mathbb{C})$ and represent it as

$$f(z) = -\frac{1}{\pi} \int_{\mathbb{C}} \frac{\partial f}{\partial \bar{\zeta}}(\zeta) \frac{dm(\zeta)}{\zeta - z}.$$

   (d) One also has $\frac{\partial f}{\partial \bar{\zeta}}(\zeta) = 0$ for $\zeta \in K$. Why?
   (e) Show that $\int f \, d\mu = 0$.

   Conclude that $(f|K) \in \overline{\mathscr{H}(K)}$.

§2. Mittag-Leffler's Theorem

3.2.1. Theorem. *The operator* $\bar{\partial} : \mathscr{E}^0(\Omega) \to \mathscr{E}^{0,1}(\Omega)$ *is surjective. In other words, for every* $f \in \mathscr{E}^{0,1}(\Omega)$ *there is* $u \in \mathscr{E}^0(\Omega)$ *such that* $\frac{\partial u}{\partial \bar{z}} dz = f$.

Proof. Let $(K_j)_{j \geq 0}$ be an exhaustion of Ω by holomorphically convex compact subsets. Let $\psi_j \in \mathscr{D}(\Omega)$, $0 \leq \psi_j \leq 1$, $\psi_j \equiv 1$ on a neighborhood of K_j ($j \geq 1$) and $\psi_0 = 0$. Set $\varphi_j = \psi_j - \psi_{j-1}$ ($j \geq 1$). Hence $\varphi_j \in \mathscr{D}(\Omega)$, $\varphi_j \equiv 0$ in a neighborhood of K_{j-1} and $\sum_{j=1}^{\infty} \varphi_j = 1$.

Since $\varphi_j f \in \mathscr{D}(\Omega)$, we can appeal to §2.1.5 to show there exists $u_j \in \mathscr{E}(\Omega)$ such that $\frac{\partial u_j}{\partial \bar{z}} = \varphi_j f$. The function u_j is holomorphic in a neighborhood of the holomorphically convex compact set K_{j-1}. By Runge's theorem, there is a function $v_j \in \mathscr{H}(\Omega)$ such that

$$\sup_{K_{j-1}} |u_j - v_j| < \frac{1}{2^j}.$$

Therefore, the series $u = \sum_{j \geq 1} (u_j - v_j)$ is uniformly convergent over every compact subset of Ω. Furthermore, for any $l \geq 1$, the functions $(u_j - v_j)$,

$j \geq l+1$, are holomorphic in $\mathring{K}_l$, hence their sum $\sum_{j\geq l+1} (u_j - v_j)$ is also a holomorphic function in $\mathring{K}_l$. A posteriori, it is of class C^∞ in $\mathring{K}_l$. Since u is the sum of the tail end of the series and a finite number of other C^∞ functions, then u is also C^∞ in $\mathring{K}_l$. This shows $u \in \mathscr{E}(\Omega)$.

Moreover, in $\mathring{K}_l$ we have

$$\frac{\partial u}{\partial \bar{z}} = \sum_{j=1}^{l} \frac{\partial}{\partial \bar{z}}(u_j - v_j) + \frac{\partial}{\partial \bar{z}} \sum_{j\geq l+1} (u_j - v_j) = \sum_{j=1}^{l} \frac{\partial u}{\partial \bar{z}} u_j = \left(\sum_{j=1}^{l} \varphi_j\right) f$$

$$= \left(\sum_{j\geq 1} \varphi_j\right) f = f,$$

since $\varphi_j \equiv 0$ on K_l if $j \geq l+1$. The index l being arbitrary, we conclude that in fact $\frac{\partial u}{\partial \bar{z}} = f$ everywhere in Ω. □

Let Ω be open in $\mathbb{C}$ and $(\Omega_i)_{i\in I}$ an open covering of Ω. With the convention that $\mathscr{H}(\varnothing) = \{0\}$ we have a sequence of complex vector spaces and of $\mathbb{C}$-linear maps:

$$0 \to \mathscr{H}(\Omega) \xrightarrow{a} \prod_j \mathscr{H}(\Omega_j) \xrightarrow{b} \prod_{j,k} \mathscr{H}(\Omega_j \cap \Omega_k) \xrightarrow{c} \prod_{j,k,l} \mathscr{H}(\Omega_j \cap \Omega_k \cap \Omega_l)$$

given by:

$$a(f) = (f|\Omega_j)_{j\in I};$$

$$b((f_j)_{j\in I}) = (g_{jk})_{(j,k)\in I\times I},$$

where

$$g_{jk} = (f_k|(\Omega_j \cap \Omega_k)) - (f_j|(\Omega_j \cap \Omega_k));$$

$$c((g_{jk})_{(j,k)\in I\times I}) = (c_{jkl})_{(j,k,l)\in I\times I\times I},$$

with

$$c_{jkl} = (g_{jk}|(\Omega_j \cap \Omega_k \cap \Omega_l)) + (g_{kl}|(\Omega_j \cap \Omega_k \cap \Omega_l)) + (g_{lj}|(\Omega_j \cap \Omega_k \cap \Omega_l)).$$

(Compare with §1.7.2.)

It is easy to see that the map a is injective, $\operatorname{Im} a = \operatorname{Ker} b$, and $\operatorname{Im} b \subseteq \operatorname{Ker} c$. We are going to show that it is also true that $\operatorname{Im} b = \operatorname{Ker} c$.

Let us note that if $(g_{jk})_{j,k} \in \operatorname{Ker} c$ then $g_{jj} = 0$ for every $j \in I$ and more generally $g_{jk} = -g_{kj}$ for every pair $(j,k) \in I \times I$.

Now, the corresponding sequence is exact (i.e., $\operatorname{Im} b = \operatorname{Ker} c$) when we replace holomorphic functions by C^∞ functions. In fact, let (g_{ij}) be an element of $\operatorname{Ker} c$ in the C^∞ case. Let

$$f_k = \sum_{\nu\geq 0} \varphi_\nu g_{r(\nu),k} \qquad \text{in } \Omega_k,$$

where $(\varphi_\nu)_{\nu\geq 0}$ is a C^∞ partition of unity subordinate to the covering $(\Omega_j)_{j\in I}$ by C^∞ functions of compact support, $r: \mathbb{N} \to I$ is a function such that

$\varphi_\nu \in \mathscr{D}(\Omega_{r(\nu)})$, and the family $(\operatorname{supp}\varphi_\nu)_{\nu\geq 0}$ is locally finite. Therefore, to see that $f_k \in \mathscr{E}(\Omega_k)$ it is enough to show that $\varphi_\nu g_{r(\nu),k} \in \mathscr{E}(\Omega_k)$. But this function is given by

$$\begin{cases} \varphi_\nu g_{r(\nu),k} & \text{in } \Omega_k \cap \Omega_{r(\nu)} \\ 0 & \text{in } \Omega_k \setminus \operatorname{supp}\varphi_\nu \end{cases},$$

which shows that it is C^∞ in all of Ω_k. Finally, in $\Omega_j \cap \Omega_k$,

$$f_k - f_j = \sum_{\nu\geq 0} \varphi_\nu(g_{r(\nu),k} - g_{r(\nu),j}) = \left(\sum_{\nu\geq 0} \varphi_\nu\right) g_{jk} = g_{jk}.$$

We prove now the exactness in the holomorphic case.

3.2.2. Theorem. *Let $(\Omega_j)_{j\in I}$ be an open covering of Ω. Let $g_{jk} \in \mathscr{H}(\Omega_j \cap \Omega_k)$ be a collection of functions indexed by $(j, k) \in I \times I$ such that*

(i) $g_{jk} = -g_{kj}$ *in* $\Omega_j \cap \Omega_k$ *for every* $(j, k) \in I \times I$

(ii) $g_{jk} + g_{kl} + g_{lj} = 0$ *in* $\Omega_j \cap \Omega_k \cap \Omega_l$ *for every triple* $(j, k, l) \in I \times I \times I$.

Then, there are functions $g_j \in \mathscr{H}(\Omega_j)$, $j \in I$, such that

$$g_{jk} = g_k - g_j \quad \text{in } \Omega_j \cap \Omega_k$$

for every $(j, k) \in I \times I$.

PROOF. Consider the commutative diagram

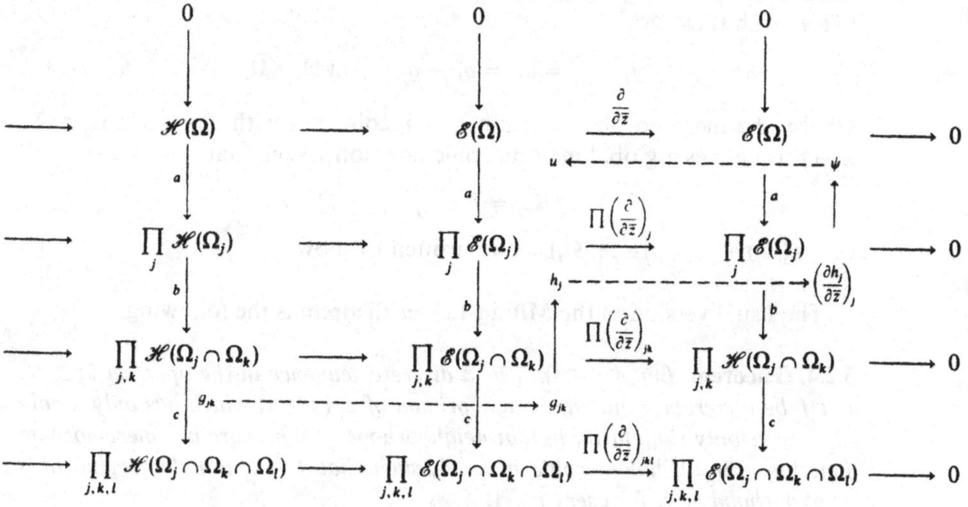

Every row and every column, except for the first column, is exact.

Let $(g_{jk}) \in \operatorname{Ker} c$ with $g_{jk} \in \mathscr{H}(\Omega_j \cap \Omega_k)$. One can consider (g_{jk}) in $\operatorname{Ker} c$ in the second column. Since this column is exact we can find $(h_j)_j \in \prod_j \mathscr{E}(\Omega_j)$ such that $g_{jk} = h_k - h_j$. Therefore $\left(\frac{\partial h_j}{\partial \bar{z}}\right)_j \in \operatorname{Ker} b$, since $\frac{\partial h_k}{\partial \bar{z}} - \frac{\partial h_j}{\partial \bar{z}} = \frac{\partial}{\partial \bar{z}} g_{jk} = 0$ in $\Omega_j \cap \Omega_k$. The exactness of the third column says that we have a global C^∞ function ψ whose restriction to Ω_j is $\frac{\partial h_j}{\partial \bar{z}}$. The exactness of the first row produces $u \in \mathscr{E}(\Omega)$ such that $\frac{\partial u}{\partial \bar{z}} = \psi$. Consider now the "corrected" functions:

$$g_j = h_j - (u|\Omega_j).$$

We also have $g_{jk} = g_k - g_j$. Moreover, this time we have $\frac{\partial g_j}{\partial \bar{z}} = \frac{\partial h_j}{\partial \bar{z}} - (\psi|\Omega_j) = 0$. This concludes the proof. □

We are now going to prove the Mittag-Leffler theorem for meromorphic functions.

3.2.3. Theorem. *Let $(\Omega_j)_{j\in I}$ be an open covering of an open set $\Omega \subseteq \mathbb{C}$. For every $j \in I$, let f_j be a meromorphic function in Ω_j. We assume $f_j - f_k \in \mathscr{H}(\Omega_j \cap \Omega_k)$ for every pair $(j, k) \in I \times I$ (i.e., f_j and f_k have the same principal parts in $\Omega_j \cap \Omega_k$). One can find a meromorphic function f in Ω such that $(f|\Omega_j) - f_j \in \mathscr{H}(\Omega_j)$ for every $j \in I$ (i.e., one can find a meromorphic function with given principal parts).*

PROOF. Let us define (g_{jk}) by $g_{jk} = f_j - f_k$ if $\Omega_j \cap \Omega_k \neq \varnothing$ and zero otherwise. These functions verify the conditions in §3.2.2. Therefore, there are $g_j \in \mathscr{H}(\Omega_j)$ $(j \in I)$ such that

$$f_j - f_k = g_{jk} = g_k - g_j \qquad \text{in } \Omega_j \cap \Omega_k.$$

Hence, the meromorphic function $f_j + g_j$ coincides with $f_k + g_k$ in $\Omega_j \cap \Omega_k$, and this defines a global meromorphic function f such that

$$f|\Omega_j = f_j + g_j \qquad \text{in } \Omega_j,$$

i.e., $(f|\Omega_j) - f_j = g_j \in \mathscr{H}(\Omega_j)$, as we wanted to show. □

The usual version of the Mittag-Leffler theorem is the following.

3.2.4. Theorem. *Let $A = (z_j)_{j\geq 1}$ be a discrete sequence in the open set $\Omega \subseteq \mathbb{C}$. Let f_j be a meromorphic in a neighborhood of z_j $(j \geq 1)$ which has only a pole at z_j as its only singularity in that neighborhood. Then, there is a meromorphic function f in Ω, holomorphic in $\Omega \setminus A$, such that $f - f_j$ is holomorphic in a neighborhood of z_j for every $j \geq 1$.*

PROOF. Let $\Omega_0 = \Omega \setminus A$ and $f_0 = 0$. For every $j \geq 1$, let Ω_j be disks centered at the point z_j, mutually disjoint, and such that f_j is meromorphic in Ω_j whose only singularity is a pole at z_j. By §3.2.3 there is a meromorphic function f in Ω such that

$$f - f_0 \text{ is holomorphic in } \Omega_0$$
$$f - f_j \text{ is holomorphic in } \Omega_j \ (j \geq 1).$$

This is the statement of the theorem. □

3.2.5. Example. Let Ω_1, Ω_2 be two open subsets of $\mathbb{C}$, $f \in \mathscr{H}(\Omega_1 \cap \Omega_2)$. There are $f_1 \in \mathscr{H}(\Omega_1)$, $f_2 \in \mathscr{H}(\Omega_2)$ such that

$$f = (f_1|(\Omega_1 \cap \Omega_2)) - (f_2|(\Omega_1 \cap \Omega_2)).$$

In fact, let $g_{11} = g_{22} = 0$, $g_{12} = -g_{21} = f$. This choice verifies the conditions of §3.2.2 with $\Omega = \Omega_1 \cup \Omega_2$ and the covering (Ω_1, Ω_2).

In the exercises we shall see how this theorem relates to Cauchy's theorem on partial fraction expansion of the previous chapter.

As will be seen in Exercise 3.2.2, the classical proof of Theorem 3.2.4 in the case $\Omega = \mathbb{C}$ is the following. Let $z_0 = 0$, $0 < |z_1| \leq |z_2| \leq \cdots \nearrow \infty$ and G_n, $n \in \mathbb{N}$, polynomials so that the functions $G_n\left(\dfrac{1}{z - z_n}\right)$ represent the desired principal parts. One finds polynomials P_n, with $P_0 = 0$, such that the series

$$f(z) = \sum_{n \geq 0} \left(G_n\left(\frac{1}{z - z_n}\right) - P_n(z)\right)$$

is uniformly convergent in compact subsets of $\mathbb{C} \setminus \{z_j : j \geq 0\}$. This series is usually called a ***Mittag-Leffler expansion*** of the function f. Note that Proposition 2.6.16 represented a partial solution to the converse problem: given a meromorphic f, find, if possible, a Mittag-Leffler expansion of f. In this way we found in Example 2.6.17 the expansion of cotan z:

$$\operatorname{cotan} z = \frac{1}{z} + \sum_{n \geq 1} \frac{2z}{z^2 - (n\pi)^2} = \frac{1}{z} + \sum_{n \neq 0} \left(\frac{1}{z - n\pi} + \frac{1}{n\pi}\right).$$

More generally, a Mittag-Leffler expansion for $f \in \mathscr{M}(\mathbb{C})$ is a representation of the form

$$f(z) = \sum_{n \geq 0} \left(G_n\left(\frac{1}{z - z_n}\right) - P_n(z)\right) + F(z),$$

where $F \in \mathscr{H}(\mathbb{C})$.

Exercises 3.2

1. Use the same proof of Theorem 3.2.4 to show that if $A = (z_j)_{j \geq 1}$ is a discrete sequence of points in an open set Ω, and if $(P_j(z))_j$ is a corresponding collection of principal parts $\left(\text{i.e., } P_j(z) = f_j\left(\dfrac{1}{z - z_j}\right),\ f_j \in \mathscr{H}(\{0\})\right)$ then there is a function $f \in \mathscr{H}(\Omega \setminus A)$ such that for every z_j the principal part of the Laurent development of f about $z = z_j$ is exactly P_j.

2. (Mittag-Leffler's expansion). A constructive proof of Theorem 3.2.4 for meromorphic functions in $\mathbb{C}$ is the following. Assume we have a sequence $\{z_n\}_{n\geq 1}$, $0 < |z_1| \leq |z_2| \leq \cdots$, $\lim_{n\to\infty} z_n = \infty$, and finite principal parts $G_n\left(\frac{1}{z - z_n}\right)$. Choose $P_n(z)$ to be the initial Taylor polynomial of $G_n\left(\frac{1}{z - z_n}\right)$ in $B(0, |z_n|)$, of sufficiently high degree so that

$$\left|G_n\left(\frac{1}{z - z_n}\right) - P_n(z)\right| \leq \varepsilon_n \qquad \text{if } |z| \leq \frac{1}{2}|z_n|,$$

for some sequence $\varepsilon_n > 0$ such that $\sum_{n\geq 1} \varepsilon_n < \infty$. Show that

$$F(z) = \sum_{n\geq 1} \left(G_n\left(\frac{1}{z - z_n}\right) - P_n(z)\right),$$

represents a meromorphic function in $\mathbb{C}$ with the desired poles and principal parts. What should you do if we also want a pole at $z = 0$? (Compare with §2.6.16).

3. Apply the previous exercise to the construction of meromorphic functions F in $\mathbb{C}$ with the following properties:
(a) simple poles at $z_n = n$, $n \in \mathbb{Z}^+$, $\operatorname{Res}(F, z_n) = n$;
(b) simple poles at $z_n = \sqrt{n}$, $n \in \mathbb{N}$, residues $= 1$;
(c) simple poles at $n \in \mathbb{Z}$, residues $= 1$;
(d) simple poles at $(\mathbb{Z} + i\mathbb{Z})\backslash\{0\}$, residues $= 1$;
(e) poles at $\mathbb{Z} + i\mathbb{Z}$ with principal parts $(z - (m + ni))^{-3}$.

4. The function $z \mapsto \int_0^1 e^{-t} t^{z-1}\, dt$ has an analytic continuation to the whole complex plane as a meromorphic function: by expanding e^{-t} in a power series, show that its Mittag-Leffler expansion is

$$\sum_{n\geq 0} \frac{(-1)^n}{n!} \frac{1}{z + n}.$$

Use this expansion to show that the Γ function introduced in §2.5.22 has a Mittag-Leffler expansion

$$\Gamma(z) = f(z) + \sum_{n\geq 0} \frac{(-1)^n}{n!} \frac{1}{z + n},$$

$f \in \mathscr{H}(\mathbb{C})$.

5. Find all the solutions $f \in \mathscr{E}(\mathbb{C})$ of the equation $\frac{\partial f}{\partial \bar{z}} = g$ for: (a) $g(z) = z$; (b) $g(z) = \bar{z}$; (c) $g(z) = z^m \bar{z}^n$, $m, n \in \mathbb{N}$; (d) $g(z) = e^{\bar{z}}$; (e) $g \in \mathscr{H}(\mathbb{C})$.

6. We say that a function f belongs to $C^\infty(\bar{B})$, $B = B(0, 1)$, if f is the restriction to $\bar{B}$ of a function which is defined and C^∞ in a neighborhood of $\bar{B}$. Let $g(z) = \exp\left(-\frac{1}{1 - |z|^2}\right)$ for $z \in B$, $g|\partial B = 0$. Show there is $f \in C^\infty(\bar{B})$ such that $\frac{\partial f}{\partial \bar{z}} = g$ in $\bar{B}$.

*7. Let $K \subset\subset \Omega$ open in $\mathbb{C}$. Show there is a constant $C > 0$ such that for every $u \in \mathscr{E}(\Omega)$ one has

$$\|u\|_K \leq C\left(\left\|\frac{\partial u}{\partial \bar{z}}\right\|_\Omega + \int_\Omega |u|\,dm\right),$$

where dm is the Lebesgue measure.

8. Let $f \in \mathscr{C}(\mathbb{C})$ be such that $\displaystyle\int_{|z|\geq 1} (|f(z)|/|z|)\,dm(z) < \infty$. Find a solution of the equation $\dfrac{\partial u}{\partial \bar{z}} = f$.

*9. (a) Let $\mu(k) := \#\{m + in \in \mathbb{Z} + i\mathbb{Z} : \max(|n|, |m|) = k \geq 0\}$. Show that $\sum_{k\geq 1} \dfrac{\mu(k)}{k^3} < \infty$.
(b) Show that the function $f(z) := \sum_{\omega \in \mathbb{Z}+i\mathbb{Z}} (z - \omega)^{-3}$ is meromorphic in $\mathbb{C}$ and doubly periodic, i.e., $f(z + 1) = f(z + i) = f(z)$ for every $z \notin \mathbb{Z} + i\mathbb{Z}$.
(c) Use the previous part to show that

$$g(z) := \frac{1}{z^2} + \sum_{0 \neq \omega \in \mathbb{Z}+i\mathbb{Z}} \left[\frac{1}{(z - \omega)^2} - \frac{1}{\omega^2}\right]$$

is a meromorphic function with double poles at every point in $\mathbb{Z} + i\mathbb{Z}$ and residue zero.
(d) Verify that $g'(z) = -2f(z)$. Conclude that $g(z + 1) - g(z) = \text{constant}$, $g(z + i) - g(z) = \text{constant}$.
(e) Use that g is an even function to conclude that g is doubly periodic.

10. The Bernoulli numbers B_{2m} are defined by

$$2^{2m-1} B_{2m} = \frac{(-1)^{m-1}(2m)!}{\pi^{2m}} \sum_{k=1}^{\infty} \frac{1}{k^{2m}}, \qquad m = 1, 2, \ldots.$$

Use the Mittag-Leffler expansion of cotan z obtained in this section to show that

$$\operatorname{cotan} z - \frac{1}{z} = \sum_{m=1}^{\infty} (-1)^m \frac{2^{2m} B_{2m} z^{2m-1}}{(2m)!}$$

for z near zero. What is the radius of convergence of this series?

11. Verify that, if the numbers B_n are defined by the identity

$$\frac{z}{2} + \sum_{n=0}^{\infty} \frac{B_n}{n!} z^n = \frac{z}{2}\frac{e^z + 1}{e^z - 1},$$

then $B_0 = 1$, $B_1 = -\frac{1}{2}$, $B_{2n+1} = 0$ for $n > 0$, and the B_{2n} coincide with those defined in the previous exercise. They also satisfy the recurrence relation

$$B_0 = 1, \quad \sum_{k=0}^{n} \binom{n}{k} B_k = B_n \qquad (n \geq 2).$$

The Bernoulli polynomials are defined by

$$B_n(z) := \sum_{k=0}^{n} \binom{n}{k} B_k z^{n-k};$$

verify that

$$\frac{e^{zt}}{e^t - 1} = \sum_{k=0}^{\infty} B_n(z) \frac{t^{n-1}}{n!}, \qquad 0 < |t| < 2\pi.$$

12. The following is another way to show that the identity

$$\pi \operatorname{cotan} \pi z = \frac{1}{z} + \sum_{n \in \mathbb{Z}^*} \left(\frac{1}{z+n} - \frac{1}{n} \right) \tag{$*$}$$

holds. Define

$$g(z) = \pi \operatorname{cotan} \pi z - \frac{1}{z} - \sum_{n \in \mathbb{Z}^*} \left(\frac{1}{z+n} - \frac{1}{n} \right).$$

(a) Show that g is entire and $g(0) = 0$.
(b) Show that

$$g'(z) = \frac{-\pi^2}{(\sin \pi z)^2} + \sum_{n \in \mathbb{Z}} \frac{1}{(z+n)^2}.$$

(c) Show that g' satisfies the functional relation

$$4g'(z) = g'\left(\frac{z}{2}\right) + g'\left(\frac{z+1}{2}\right)$$

(d) Conclude that if $M = \max_{|z| \leq 2} |g'(z)|$ then

$$4M \leq M + M.$$

This implies the identity ($*$) holds. Why?

13. (a) Let f be holomorphic in the annulus $0 \leq r < |z| < R \leq 0$. Show explicitly a way to decompose $f = f_1 - f_2$, $f_1 \in \mathscr{H}(B(0, R))$ and $f_2 \in \mathscr{H}(\bar{B}(0, r)^c)$ (it can be chosen holomorphic at $z = \infty$).
(b) Let f be holomorphic in a neighborhood of the closed strip $|\operatorname{Im} z| \leq 1$, assume $|f(z)| \leq \dfrac{C}{1 + |z|}$. Use Cauchy's formula to find an explicit decomposition $f = f_1 - f_2$ where f_1 is holomorphic for $\operatorname{Im} z < 1$ and f_2 is holomorphic for $\operatorname{Im} z > -1$.
(c) Let f be holomorphic in the open strip $|\operatorname{Im} z| < 1$; show there are functions f_1, f_2, f_1 holomorphic in the half-plane $\operatorname{Im} z < 1$, f_2 holomorphic for $\operatorname{Im} z > -1$, and $f = f_1 - f_2$ in $|\operatorname{Im} z| < 1$.

§3. Weierstrass' Theorem

3.3.1. Theorem. *Let $(z_j)_{j \geq 1}$ be a discrete sequence of distinct points in an open subset Ω of $\mathbb{C}$. Let $(m_j)_{j \geq 1}$ be a sequence of nonzero integers. There is a meromorphic function f in Ω which is holomorphic and never zero in $\Omega \backslash \{z_j : j \geq 1\}$, such that for every j:*

$$f(z) = (z - z_j)^{m_j} f_j(z)$$

in a neighborhood V_j of z_j, where f_j is a nonvanishing holomorphic function in V_j.

PROOF. We can assume all the $m_j \geq 1$. The general case is obtained by taking the quotient of two such holomorphic functions.

The idea of the proof is the following: Find $\varphi \in \mathscr{E}(\Omega)$, $\varphi(z) \neq 0$ for $z \notin \{z_j : j \geq 1\}$ and so that for every k, there is a disk Δ_k of center z_k such that $\varphi | \Delta_k = (z - z_k)^{m_k} h_k(z)$, $h_k \in \mathscr{H}(\Delta_k)$, h_k without zeros. (The disks can be chosen pairwise disjoint). Once φ has been found we choose f of the form $f = \varphi e^{\psi}$, $\psi \in \mathscr{E}(\Omega)$.

Let us first explain the last step. That is, we assume we have a φ and we want to find $\psi \in \mathscr{E}(\Omega)$ such that $f = \varphi e^{\psi}$ satisfies all the conditions of the theorem. Clearly we need $\dfrac{\partial f}{\partial \bar{z}} = 0$, that is,

$$e^{\psi}\left(\frac{\partial \varphi}{\partial \bar{z}} + \varphi \frac{\partial \psi}{\partial \bar{z}}\right) = 0,$$

so that ψ is a solution of the equation

$$\frac{\partial \psi}{\partial \bar{z}} = -\frac{1}{\varphi} \frac{\partial \varphi}{\partial \bar{z}}.$$

Since φ is holomorphic in any Δ_k, $\dfrac{\partial \varphi}{\partial \bar{z}} = 0$ in that disk and hence, we can also take the second member to be identically zero in every Δ_k. This is why the right-hand side is a C^∞ function in Ω. Therefore, the inhomogeneous Cauchy-Riemann equation has at least one solution $\psi \in \mathscr{E}(\Omega)$. Furthermore, ψ is holomorphic in each disk Δ_k. The function e^{ψ} is never zero, hence the holomorphic function $f = \varphi e^{\psi}$ only vanishes when φ does, that is, only in the set $\{z_j : j \geq 1\}$. In any Δ_k we have

$$f(z) = (z - z_k)^{m_k} h_k(z) e^{\psi(z)},$$

as we have already pointed out, the function $h_k(z)e^{\psi(z)}$ is holomorphic, and never zero in Δ_k. This shows that the problem reduces to that of finding φ.

Translating Ω, if necessary, we may assume that $0 \in \Omega$ and $z_j \neq 0$ for every j. We denote Ω^* the open subset of S^2 obtained from Ω by means of the map $z \mapsto w = \dfrac{1}{z}$. The point $\infty \in \Omega^*$, hence $\partial\Omega^*$ is a compact subset of $\mathbb{C}$.

We can reorder the points z_j so that the points $w_j = 1/z_j$ satisfy $d(w_j, \partial\Omega^*) \geq d(w_{j+1}, \partial\Omega^*)$. The discreteness assures that $d(w_j, \partial\Omega^*) \to 0$ unless the set $\{z_j : j \geq 1\}$ is finite. Let w_k' be a point in $\partial\Omega^*$ so that $|w_j - w_j'| = d(w_j, \partial\Omega^*)$.

Let $r_j = 2d(w_j, \partial\Omega^*)$ and $U_j = B(w_j', r_j)$. This disk contains the segment $[w_j, w_j']$. The family of open sets $(U_j \cap \Omega^*)_{j \geq 1}$ is locally finite even when $\{z_j : j \geq 1\}$ is infinite, since $w_j' \in \partial\Omega^*$ and $r_j \to 0$.

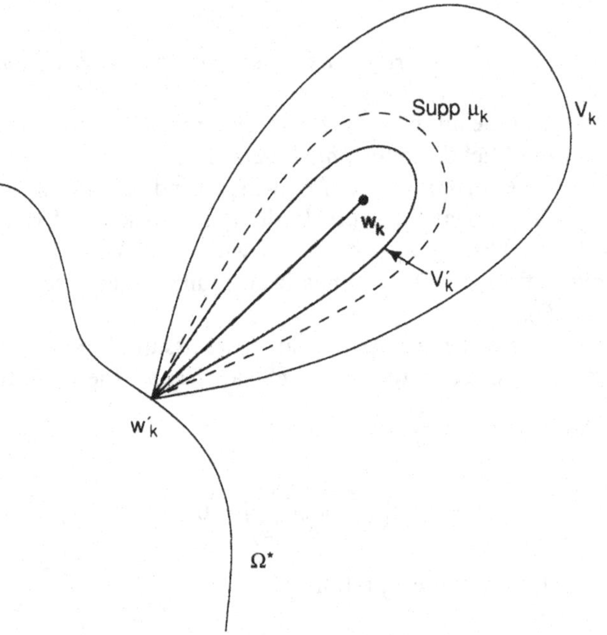

Figure 3.2

For every $k \geq 1$, let $I_k = \{j \geq 1 : w_j \in [w_k, w'_k[\}$ and $J_k = \mathbb{N}^* \setminus I_k$.

There are open neighborhoods V_k of $[w_k, w'_k[$, $V_k \subseteq U_k \cap \Omega^*$, such that for every $j \in J_k$, $w_j \notin \bar{V}_k$. In fact, this follows from $U_k \cap (\Omega^* \setminus \{w_j : j \in J_k\})$ being an open set in Ω^*. (Recall the sequence is discrete in Ω^*.) There exists a function $\mu_k \in \mathscr{E}(\Omega^*)$ with support (relative to Ω^*) contained in V_k and μ_k takes identically the value m_k on a neighborhood V'_k of $[w_k, w'_k[$, $V'_k \subseteq V_k$ (see Figure 3.2).

Let us recall that there is a determination of $\log \dfrac{w - w_k}{w - w'_k}$ outside the segment $[w_k, w'_k]$. Therefore, the functions

$$g_k(w) := \begin{cases} \exp\left(\mu_k(w) \log \dfrac{w - w_k}{w - w'_k}\right) & \text{in } \Omega^* \setminus [w_k, w'_k[\\ \left(\dfrac{w - w_k}{w - w'_k}\right)^{m_k} & \text{in } V'_k \end{cases}$$

are smooth in Ω^*, since the two formulas coincide in $V'_k \setminus [w_k, w'_k[$. They have the value 1 outside V_k and equal $((w - w_k)/(w - w'_k))^{m_k}$ in V'_k.

For every index $l \geq 1$ there is only a finite number of g_k which are $\not\equiv 1$ in a neighborhood of w_l. This follows from the fact that $V_k \subseteq U_k \cap \Omega^*$ and the family $(U_k \cap \Omega^*)_{k \geq 1}$ is locally finite. Therefore, the function

$$g(w) = \prod_{k\geq 1} g_k(w)$$

is a locally finite product of functions in $\mathscr{E}(\Omega^*)$, hence $g \in \mathscr{E}(\Omega^*)$. This function vanishes only on the set $\{w_k : k \geq 1\}$.

Consider a function g_k which is not identically equal to one in a neighborhood of w_l. If $l \in J_k$, this is impossible because $w_l \notin \bar{V}_k$. Therefore $l \in I_k$, that is, $w_l \in [w_k, w_k'[$, hence $g_k(w) = \left(\dfrac{w - w_k}{w - w_k'}\right)^{m_k}$ in a neighborhood of w_l and does not vanish at w_l unless $l = k$. Hence, there is a neighborhood N_l of w_l where g is holomorphic and clearly it vanishes at w_l with multiplicity exactly m_l. To conclude the proof, it is enough to define $\varphi(z) = g(1/z)$ and take the disks $\Delta_k \subseteq \{z : 1/z \in N_k\}$. □

The preceding proof was taken from [BT]. A different proof of the Weierstrass' theorem for entire functions in terms of infinite products appears in the next chapter and in the exercises of this one.

3.3.2. Corollary. *Every meromorphic function F in an open set Ω of $\mathbb{C}$ can be written in the form $F = f/g$ with $f, g \in \mathscr{H}(\Omega)$. In particular, if Ω is connected, then the ring $\mathscr{H}(\Omega)$ is an integral domain whose quotient field is the field $\mathscr{M}(\Omega)$ of meromorphic functions in Ω.*

Proof. If F has poles at the points z_j of order n_j, one can find a holomorphic function g in Ω, vanishing only at those points and with multiplicity exactly n_j. The function $f = Fg$ is then holomorphic in Ω.

If Ω is connected, $f, g \in \mathscr{H}(\Omega)$ and $fg \equiv 0$, then one of these two functions must be identically zero since it cannot have isolated zeros. The fact that $\mathscr{M}(\Omega)$ is the quotient field of $\mathscr{H}(\Omega)$ follows from the first part of the corollary. □

3.3.3. Corollary. *Let Ω be an open subset of $\mathbb{C}$. There is a function $f \in \mathscr{H}(\Omega)$ which cannot be analytically continued to any open set larger than Ω, even as a meromorphic function: there is no meromorphic function g on a disk $B(a,r)$ centered at a point $a \in \partial\Omega$, such that g coincides with f on $B(a,r) \cap \Omega$.*

Proof. We only have to construct a discrete sequence $\{z_j\}_{j\geq 1}$ (of different points) in Ω accumulating at every point in $\partial\Omega$. In fact, in such a case there exists a function $f \in \mathscr{H}(\Omega)$ which vanishes exactly at the points z_j and nowhere else. Suppose that g is a meromorphic function on a disk $B(a,r)$, $a \in \partial\Omega$, and $g = f$ on $B(a,r) \cap \Omega$. Then g vanishes at all the $z_j \in B(a,r)$, and these points accumulate at a. Therefore g must be identically zero, which contradicts the fact that f only vanishes at the points z_j.

To construct the sequence $\{z_j\}_{j\geq 1}$, write the points in $\Omega \cap (\mathbb{Q} + i\mathbb{Q})$ in a sequence $\{w_j\}_{j\geq 1}$. Let $r_j = d(w_j, \Omega^c)$ and $\{K_j\}_{j\geq 1}$ be a sequence of compact sets exhausting Ω. For every j choose $z_j \in K_j^c$ such that $|z_j - w_j| < r_j$ and, moreover, so that $z_j \neq z_k$ for $j > k$. Then it is clear that $\{z_j\}_{j\geq 1}$ is a discrete sequence (of different points) in Ω. If $a \in \partial\Omega$ and $\varepsilon > 0$, there is j such that $|a - w_j| \leq \varepsilon/2$, so $r_j \leq \varepsilon/2$ and hence $|a - z_j| \leq |a - w_j| + |w_j - z_j| \leq \frac{1}{2}\varepsilon + r_j \leq \varepsilon$. Thus $\{z_j\}_{j\geq 1}$ accumulates at every point in $\partial\Omega$, and the proof is complete. □

In the situation of this corollary we say that Ω is the ***domain of holomorphy*** of f.

EXERCISES 3.3

1. Derive the Weierstrass Theorem 3.3.1 from the Mittag-Leffler Theorem 3.2.4.

2. (Classical proof of Weierstrass' theorem for entire functions).
   (a) Given a sequence $\{z_n\}_{n\geq 1}$, $0 < |z_1| \leq |z_2| \leq \cdots$, $\lim_{n\to\infty} z_n = \infty$, show there is a sequence of integers $m_n \in \mathbb{N}$ such that
   $$\sum_{n\geq 1} \left|\frac{z}{z_n}\right|^{m_n} < \infty \qquad (\forall z \in \mathbb{C}).$$
   (b) Let $E(z,m) = (1 - z)\exp\left(z + \frac{z^2}{2} + \cdots + \frac{z^m}{m}\right)$ if $m \geq 1$, $E(z,0) = 1 - z$, be the ***Weierstrass primary factors***. Show that the ***Weierstrass product***
   $$f(z) = \prod_{n\geq 1} E\left(\frac{z}{z_n}, m_n\right)$$
   is an entire function vanishing exactly at z_n. Moreover, the multiplicity of a zero α of f is exactly the number of times it appears in the sequence $\{z_n\}_{n\geq 1}$. What should you do to account for a zero at the origin? (For the theory of infinite products see [Ahl1], [Ma], or Chapter 4.)

3. Find the Weierstrass product expansion corresponding to the following sequences: (a) $\mathbb{Z}$; (b) $z_n = \log n$, $n \geq 1$; (c) $\mathbb{Z} + i\mathbb{Z}$. (In parts (a) and (c) order the sequence with increasing absolute values as in §3.3.2.)

4. Let Ω be simply connected in $\mathbb{C}$. How should one choose the integers $m_j \in \mathbb{N}^*$ in Weierstrass' Theorem 3.3.1 in order that f has a holomorphic square root? A holomorphic cubic root?

5. Let Ω be a simply connected open set in $\mathbb{C}$, $f, g \in \mathscr{H}(\Omega)$. What are the necessary and sufficient conditions on f and g in order that the equation $X^2 + fX + g = 0$ admits a solution $X \in \mathscr{M}(\Omega)$?

*6. Show that the family of functions whose domain of holomorphy is Ω is dense in $\mathscr{H}(\Omega)$. (Hint: Consider $g + tf$, with f given by Corollary 3.3.3, $t > 0$.)

7. Let Ω_1 be a connected open set, $\Omega_1 \subsetneq \Omega_2$, Ω_2 a simply connected open set in $\mathbb{C}$, and $f \in \mathscr{H}(\Omega_2)$, all whose zeros are of even order. Suppose $g \in \mathscr{H}(\Omega_1)$ is such that $g^2 = f$ in Ω_1. Show Ω_1 is not the domain of holomorphy of g.

8. Integrating the expression obtained for $\operatorname{cotan} z$ at the end of §3.2, show that

$$\sin z = z \prod_{n=1}^{\infty}\left(1 - \frac{z}{n\pi}\right)e^{z/n\pi} \prod_{n=1}^{\infty}\left(1 + \frac{z}{n\pi}\right)e^{-z/n\pi}$$
$$= z \prod_{n=1}^{\infty}\left(1 - \frac{z^2}{n^2\pi^2}\right).$$

9. Derive the formulas:

(a) $\sinh \pi z = \pi z \prod_{n=1}^{\infty}\left(1 + \frac{z^2}{n^2}\right)$;

(b) $\cos \pi z = \prod_{n=0}^{\infty}\left[1 - \left(\frac{z}{n + 1/2}\right)^2\right]$;

(c) $e^z - 1 = ze^{z/2} \prod_{n=1}^{\infty}\left(1 + \frac{z^2}{4\pi^2 n^2}\right)$.

10. Assume f is an entire function with the property that it has simple zeros at the points a_j, $0 < |a_1| \le |a_2| \le \cdots$, and there is a sequence of piecewise-C^1 Jordan curves Γ_n such that $\Gamma_n \subseteq \operatorname{Int}(\Gamma_{n+1})$, f does not vanish on any Γ_n, $\operatorname{dist}(\Gamma_n, 0) = R_n \to \infty$ as $n \to \infty$, $\ell(\Gamma_n) = O(R_n)$ and

$$\max_{\Gamma_n}\left|\frac{f'(z)}{f(z)}\right| = O(R_n).$$

Show that

$$f(z) = f(0)e^{\alpha z} \prod_{k=1}^{\infty}\left(1 - \frac{z}{a_k}\right)e^{z/a_k},$$

$\alpha = f'(0)/f(0)$, and the factors are grouped together for those zeros a_k lying between Γ_n and Γ_{n+1} in order to make the infinite product convergent.

11. Use the previous exercise to derive the expansions

(a) $\sin z - z\cos z = \frac{1}{3}z^3 \prod_{n=1}^{\infty}\left(1 - \frac{z^2}{\lambda_n^2}\right)$, where λ_n is the root of the equation $\tan z = z$ lying in the interval $(n\pi, (n + \frac{1}{2})\pi)$, $n \in \mathbb{N}^*$.

(b) $\cos\frac{\pi z}{4} - \sin\frac{\pi z}{4} = \prod_{n=0}^{\infty}\left(1 - \frac{(-1)^n z}{2n + 1}\right)$.

*12. We return to the Γ function. By Exercise 3.2.4 we know it is a meromorphic function in $\mathbb{C}$ with simple poles at $z = 0, -1, -2, \ldots$. One of the aims of this exercise is to show that Γ does not have any zeros. Instead of the power series expansion of e^{-t} used in Exercise 3.2.4, we use here a different approximation of e^{-t}.

(a) Show that for $0 \le t \le n$, $n \in \mathbb{N}^*$, one has

$$0 \le e^{-t} - \left(1 - \frac{t}{n}\right)^n \le \frac{t^2 e^{-t}}{n}.$$

(b) Let $\varphi_n(z) = \int_0^n \left(1 - \frac{t}{n}\right)^n t^{z-1}\,dt$. Show that

$$\varphi_n \to \Gamma \qquad \text{in } \mathscr{H}(\{\operatorname{Re} z > 0\}).$$

(c) Use the change of variables $t = n\tau$ in the integral defining φ_n to show that

$$\varphi_n(z) = \frac{n!n^z}{z(z+1)\dots(z+n)}.$$

(d) Let γ be the Euler-Mascheroni constant, $\gamma = \lim_{n\to\infty}\left(\sum_{j=1}^{n}\frac{1}{j} - \log n\right)$, and $\gamma_n = \sum_{j=1}^{n}\frac{1}{j} - \log n - \gamma$. Rewriting n^z as

$$n^z = \exp\left(\left[\sum_{j=1}^{n}\frac{1}{j} - \gamma - \gamma_n\right]z\right),$$

show that

$$\Gamma(z) = \lim_{n\to\infty} e^{-\gamma z}\Big/\left(z\prod_{j=1}^{n}\left(1+\frac{z}{j}\right)e^{-z/j}\right),$$

a priori only when $\operatorname{Re} z > 0$.

(e) Use Exercise 3.3.2 to show that

$$ze^{\gamma z}\sum_{j=1}^{\infty}\left(1+\frac{z}{j}\right)e^{-z/j}$$

is an entire function. Conclude that this function coincides with $\frac{1}{\Gamma}$ everywhere.

(f) Does Γ have any zeros in $\mathbb{C}$?

(g) Show that $\Gamma(z)\Gamma(1-z) = \dfrac{\pi}{\sin \pi z}$ and $\Gamma(z+1) = z\Gamma(z)$.

(h) Use (g) to show that $\Gamma\left(\frac{1}{2}\right) = \sqrt{\pi}$, and a posteriori

$$\Gamma\left(n+\frac{1}{2}\right) = \frac{1\times 3\times\cdots\times(2n-1)}{2^n}\sqrt{\pi}, \qquad n \in \mathbb{N}^*.$$

13. (a) Show that any meromorphic solution of the equation

$$zf(z) = f(z+1)$$

satisfying the condition $f(1) = 1$ must have simple poles at $z = -n$, $n \in \mathbb{N}^*$, with residues $\dfrac{(-1)^n}{n!}$.

(b) Show that if f is a solution of (a) then

$$f(z) = \varphi(z)\Gamma(z),$$

where φ is a meromorphic function of period 1, i.e., $\varphi(z+1) = \varphi(z)$.

*14. Show that given any entire function f with real coefficients there is an entire function g with rational coefficients with exactly the same set of zeros (including multiplicities). Conclude that given any real number a there is an equation of the form

$$r_0 + r_1 z + r_2 z^2 + \cdots = 0,$$

with $r_j \in \mathbb{Q}$, and whose only root is a simple root $z = a$. Similarly, show that given any sequence, finite or infinite, of real numbers $a_1, a_2, \ldots$, such that $\lim_{n\to\infty} a_n = \infty$ if the sequence is infinite, then there is an entire function g with rational coefficients having exactly those values as zero.

The above statements hold for any entire function f if one replaces $\mathbb{Q}$ by $\mathbb{Q} + i\mathbb{Q}$.

§4. Interpolation Theorem

Let us recall that given m distinct points $z_1, \ldots, z_m$ and m values $a_1, \ldots, a_m$ we can find polynomials p that interpolate these values, that is,

$$p(z_j) = a_j, \qquad j = 1, \ldots, m.$$

In fact this polynomial will be unique if we require that $\deg p \leq m - 1$. In the same way we could prescribe a certain number of derivatives at each point z_j, or what amounts to the same, prescribe the first terms of the Taylor development of p at the point z_j. This amounts to requiring

$$\frac{p^{(k)}(z_j)}{k!} = a_{j,k}, \qquad j = 1, \ldots, m, \qquad 0 \leq k \leq n_j - 1.$$

Both problems can be solved using explicit formulas or the calculus of residues (see Exercises 2.6.21 and 3.4.1). We want to show that this interpolation problem has a solution even if we consider it for an infinite discrete sequence of points. This is the meaning of the following theorem.

3.4.1. Theorem. *Let $(z_j)_{j\geq 1}$ be a discrete sequence of distinct points in an open subset Ω of $\mathbb{C}$ and $(n_j)_{j\geq 1}$ be a sequence of integers greater than or equal to 1. Let $(a_{j,k})_{j,k}$ be a sequence of complex numbers indexed by $j \geq 1$ and $0 \leq k \leq n_j - 1$. There exists $g \in \mathscr{H}(\mathbb{C})$ such that*

$$g^{(k)}(z_j) = k!\,a_{j,k} \qquad \text{for all } j \in \mathbb{N}^*, 0 \leq k \leq n_j - 1.$$

Proof. From Weierstrass' Theorem 3.3.1 we can conclude that there is a function $f \in \mathscr{H}(\Omega)$ which vanishes exactly at the points z_j with multiplicity n_j. Let $(\varepsilon_j)_{j\geq 1}$ be a sequence of positive numbers such that the disks $\Delta_j = B(z_j, 2\varepsilon_j)$ are disjoint. Consider the polynomials

$$P_j(z) = \sum_{0\leq k\leq n_j-1} a_{j,k}(z - z_j)^k \qquad (j \geq 1),$$

and for each j, let $\varphi_j \in \mathscr{D}(\Delta_j)$ be such that $0 \leq \varphi_j \leq 1$ and $\varphi_j \equiv 1$ on $\bar{B}(z_j, \varepsilon_j)$. For $\psi \in \mathscr{E}(\Omega)$, let g be given by

$$g(z) = \sum_{j\geq 1} P_j(z)\varphi_j(z) - f(z)\psi(z).$$

Since the supports of the functions φ_j are disjoint, the series contains at most one nonzero term in the neighborhood of any point of Ω. Therefore $g \in \mathscr{E}(\Omega)$. We need to choose ψ so that g is holomorphic. As should be familiar by now, this means that the equation $\dfrac{\partial g}{\partial \bar{z}} = 0$ leads to

$$\sum_{j\geq 1} P_j(z)\frac{\partial \varphi_j}{\partial \bar{z}}(z) = f(z)\frac{\partial \psi}{\partial \bar{z}}.$$

Let us call $h(z)$ the left-hand side of this equation. In $\bigcup_j \bar{B}(z_j, \varepsilon_j)$ we have that h is identically zero, so that the function $h(z)/f(z)$ (taken to be zero at all the z_j) is well defined and C^∞ in Ω. Choosing a C^∞ solution ψ of the inhomogeneous Cauchy-Riemann equation $\partial\psi/\partial\bar{z} = h/f$, we obtain that g is holomorphic in Ω. Furthermore, ψ is holomorphic in the neighborhood of every z_j and, since f vanishes with multiplicity n_j at z_j, we have

$$g^{(k)}(z_j) = P_j^{(k)}(z_j) = k!\,a_{j,k} \qquad \text{if } 0 \leq k \leq n_j - 1,$$

as we wanted to show. □

EXERCISES 3.4

1. Let γ be a piecewise-C^1 Jordan curve, $\Omega \supseteq \overline{\text{Int}(\gamma)}$, $f, g \in \mathscr{H}(\Omega)$, and g does not vanish on γ. Define a function F by

$$F(z) := \frac{1}{2\pi i}\int_\gamma \frac{f(\zeta)}{g(\zeta)}\,\frac{g(z) - g(\zeta)}{z - \zeta}\,d\zeta.$$

   Show that $F \in \mathscr{H}(\text{Int}(\gamma))$. (In fact, you can prove $F \in \mathscr{H}(\omega)$ for some ω open, $\omega \supseteq \overline{\text{Int}(\gamma)}$.) Moreover, if z_0 is a zero of g of multiplicity m, $z_0 \in \text{Int}(\gamma)$, then

$$F^{(k)}(z_0) = f^{(k)}(z_0), \qquad 0 \leq k \leq m - 1.$$

   Show that there is a function $h \in \mathscr{H}(\text{Int}(\gamma))$ (in fact, holomorphic in some neighborhood of $\overline{\text{Int}(\gamma)}$) such that

$$F = f + hg.$$

   Show that if g is a polynomial, then F is a polynomial.

2. Why do we need the sequence $(z_j)_{j\geq 1}$ to be discrete in Theorem 3.4.1?

3. Let f be an entire function with simple zeros $(z_n)_{n\geq 1}$, $|z_n| \leq |z_{n+1}|$. Given any sequence $(a_n)_{n\geq 1}$ of complex numbers, show there is a sequence of nonnegative integers $(q_n)_{n\geq 1}$ such that

$$g(z) = \sum_{n\geq 1} \frac{a_n}{f'(z_n)}\,\frac{f(z)}{z - z_n}\left(\frac{z}{z_n}\right)^{q_n}$$

   represents an entire function such that $g(z_n) = a_n$.

*4. Show that if $(a_n)_{n\in\mathbb{Z}}$ satisfies the condition $\sum_n |a_n|^2 < \infty$, then the function

$$g(z) := \sum_n (-1)^n a_n \frac{\sin z}{z - n\pi}$$

is entire, takes the values $g(n\pi) = a_n$ and $\int_{-\infty}^{\infty} |g(x)|^2\,dx < \infty$.

5. (Newton's divided differences). Let $z_1, \ldots, z_n$ be distinct points in Ω, $g \in \mathscr{H}(\Omega)$. Let h be the polynomial of degree $n-1$ interpolating the values $h(z_j) = g(z_j)$, written as

$$h(z) = b_0 + b_1(z - z_1) + b_2(z - z_1)(z - z_2) + \cdots + b_{n-1}(z - z_1)\ldots(z - z_{n-1}).$$

Define the divided differences $\Delta^{(k)}[g; z_1, \ldots, z_{k+1}]$ by

$$\Delta^{(0)}[g; z_1] = g(z_1)$$

$$\Delta^{(k)}[g; z_1, \ldots, z_{k+1}] := \frac{\Delta^{(k-1)}[g; z_2, \ldots, z_{k+1}] - \Delta^{(k-1)}[g, z_1, \ldots, z_k]}{z_{k+1} - z_1}$$

Show that $\Delta^{(k)}[g; z_1, \ldots, z_{k+1}]$ is a symmetric function of the points $z_1, \ldots, z_{k+1}$. Compute $b_0, b_1, \ldots, b_{n-1}$ in terms of the divided differences of g at the points $z_1, \ldots, z_n$. Write a formula similar to that in Exercise 3.4.1 to represent h. What happens when two points coincide? (We recommend [Ge] for an extensive and clear study of these operators $\Delta^{(k)}$ and their applications.)

6. Let P, Q be two polynomials without common zeros. Show there exist polynomials A, B such that

$$PA + QB = 1.$$

Is the pair A, B unique? Show it is uniquely determined by the extra conditions $\deg A < \deg Q$ and $\deg B < \deg P$. Show the problem can be reduced to an interpolation problem and can be solved using Exercise 3.4.1. The formulas can be made explicit using residue calculus.

§5. Closed Ideals in $\mathscr{H}(\Omega)$

Let Ω be an open subset of $\mathbb{C}$, the family $\mathscr{H}(\Omega)$ of all holomorphic functions in Ω is a topological algebra, that is, it is an algebra over $\mathbb{C}$ and the topology of uniform convergence on compact subsets of Ω is compatible with the algebraic operations. This fact will prove to be important in the applications to harmonic analysis given in the second volume. The main objects of interest in an algebra are its ideals. Weierstrass' theorem will provide the main tool to study them. As a first step, we are going to show here that every closed ideal in $\mathscr{H}(\Omega)$ is a principal ideal. One should not assume that every ideal in $\mathscr{H}(\Omega)$ is closed (they are not!). We follow [LR] and [BT].

3.5.1. Definition. We call ***multiplicity variety*** in Ω, a sequence of pairs $(z_n, n_k)_{k \geq 1}$ with $Z = \{z_k : k \geq 1\}$ a discrete subset of Ω and n_k integers ≥ 1. Whenever $f \in \mathscr{H}(\Omega)$ satisfies $f^{(j)}(z_k) = 0$ for $0 \leq j \leq n_k - 1$ and every k, we say that f **vanishes on** V.

3.5.2. Proposition. *Let $V = (z_k, n_k)_{k\geq 1}$ be a multiplicity variety in Ω. The set*

$$I(V) = \{f \in \mathscr{H}(\Omega) : f(z_k) = \cdots = f^{(n_k-1)}(z_k) = 0, k \geq 1\}$$

is a closed ideal in $\mathscr{H}(\Omega)$. Moreover, $I(V)$ is a principal ideal.

$I(V)$ is called the ***ideal of the variety*** *V.*

PROOF. It is easy to verify that $I(V)$ is an ideal. It is closed because the convergence of a sequence in $\mathscr{H}(\Omega)$ implies the convergence of the derivatives of any order. Finally, Weierstrass' theorem provides us with a function $f \in \mathscr{H}(\Omega)$ which vanishes only at the points z_k and exactly with multiplicity n_k. Hence $f \in I(V)$.

Let g be another function in the same ideal. We claim that $g/f \in \mathscr{H}(\Omega)$. In fact, g/f is holomorphic in $\Omega \setminus Z$ since f does not vanish there. (Recall $Z = \{z_k : k \geq 1\}$.) In a neighborhood of a point $z_k \in Z$ we can write $g(z) = (z - z_k)^{m_k} h(z)$ with $m_k \geq n_k$ and $f(z) = (z - z_k)^{n_k} \varphi(z)$, $\varphi(z_k) \neq 0$, h, φ holomorphic. Therefore, in a small neighborhood of z_k we have $g(z)/f(z) = (z - z_k)^{m_k - n_k} h(z)/\varphi(z)$ $(z \neq z_k)$ and the right-hand side is holomorphic at $z = z_k$. It follows that f generates the ideal $I(V)$. □

3.5.3. Definition. Let I be an ideal of $\mathscr{H}(\Omega)$. We denote by $Z(I)$ ***the set of zeros of*** I, that is, the set

$$Z(I) := \bigcap_{f \in I} Z(f),$$

where $Z(f) := \{z \in \Omega : f(z) = 0\}$ is the zero set of f.

3.5.4. Remarks. (1) We have $Z(\mathscr{H}(\Omega)) = \varnothing$ (since $1 \in \mathscr{H}(\Omega)$ and $Z(1) = \varnothing$), but there exist ideals $I \neq \mathscr{H}(\Omega)$ such that $Z(I) = \varnothing$. For instance, let $(z_j)_{j\geq 1}$ be an infinite discrete sequence in Ω. Let I be the set of $f \in \mathscr{H}(\Omega)$ for which there is an integer $j_f \geq 0$ such that for every $j \geq j_f + 1$ we have $f(z_j) = 0$. Weierstrass' theorem shows that $I \neq \{0\}$. Since $1 \notin I$ we have $I \neq \mathscr{H}(\Omega)$. Clearly $Z(I) = \varnothing$. We leave as an exercise to the reader to verify that I is an ideal.

(2) If Ω is connected then $Z(I)$ is a discrete subset of Ω unless $I = \{0\}$.

(3) The proof of Proposition 3.5.2 also shows that any principal ideal is closed.

3.5.5. Definition. Let Ω be a connected open subset of $\mathbb{C}$ and I a proper ideal in $\mathscr{H}(\Omega)$ (that is, $\{0\} \neq I \neq \mathscr{H}(\Omega)$). We denote by $V = V(I)$ the ***multiplicity variety of the ideal*** I, where the sequence V of pairs $(z_k, n_k)_{k\geq 1}$ is defined by enumerating the discrete set $Z(I)$ as $(z_k)_{k\geq 1}$ and for each k we let n_k be the integer greater than or equal to 1 given by

$$n_k := \inf\{m(f, z_k) : f \in I\},$$

where $m(f, z_k)$ is the multiplicity of z_k as a zero of f.

We can assume without loss of generality that, when $Z(I)$ is infinite, $d(z_k, \Omega^c) \searrow 0$. (If $\Omega = \mathbb{C}$ we take this to mean $|z_k| \nearrow \infty$.)

Note that $V(f)$ = multiplicity variety of the principal ideal generated by f, consists precisely of the pairs given by the zeros of f and their respective multiplicities. If $V(f) = (z_k, n_k)_{k \geq 1}$ we have $Z(f) = \{z_k : n_k \geq 1\}$. Furthermore, it is clear that $I(V(f))$ is generated by f. Therefore, the statement of Proposition 3.5.2 already includes the observation made in Remark 3.5.4, (3).

We consider now a few properties of finitely generated ideals in $\mathscr{H}(\Omega)$.

3.5.6. Lemma. *Let $f_1, \ldots, f_n \in \mathscr{H}(\Omega)$ be such that $\bigcap_{1 \leq j \leq n} Z(f_j) = \varnothing$. There are $g_1, \ldots, g_n \in \mathscr{H}(\Omega)$ such that $\sum_{1 \leq j \leq n} f_j g_j = 1$.*

Proof. We shall proceed by induction on the number n of functions. We can assume Ω is connected (otherwise we work in each connected component separately). The case $n = 1$ is evident, but we need the case $n = 2$ to start the induction.

Let $f_1, f_2 \in \mathscr{H}(\Omega)$ such that $Z(f_1) \cap Z(f_2) = \varnothing$. Hence the sets $\Omega_1 := \Omega \backslash Z(f_1)$ and $\Omega_2 := \Omega \backslash Z(f_2)$ form an open covering of Ω.

In Ω_1 we can find $\alpha_1, \alpha_2 \in \mathscr{H}(\Omega_1)$ such that $1 = \alpha_1 f_1 + \alpha_2 f_2$, e.g., $\alpha_2 = 0$, $\alpha_1 = 1/f_1$. Similarly, in Ω_2 we can find β_1, β_2 such that $\beta_1 f_1 + \beta_2 f_2 = 1$.

In $\Omega_1 \cap \Omega_2 = \Omega \backslash (Z(f_1) \cup Z(f_2))$ we have

$$1 = \alpha_1 f_1 + \alpha_2 f_2 = \beta_1 f_1 + \beta_2 f_2,$$

hence

$$(\alpha_1 - \beta_1) f_1 = (\beta_2 - \alpha_2) f_2.$$

Therefore, we can define the function

$$g_{1,2} = \frac{\alpha_1 - \beta_1}{f_2} = \frac{\beta_2 - \alpha_2}{f_1} \in \mathscr{H}(\Omega_1 \cap \Omega_2).$$

Let $g_{2,1} = -g_{1,2}, g_{1,1} = g_{2,2} = 0$. These g_{jk} verify the conditions of §3.2.2. It follows that there are $g_1 \in \mathscr{H}(\Omega_1)$, $g_2 \in \mathscr{H}(\Omega_2)$ such that

$$g_{1,2} = g_2 - g_1 \qquad \text{in } \Omega_1 \cap \Omega_2.$$

This implies that in $\Omega_1 \cap \Omega_2$ we have the two identities

$$\frac{\alpha_1 - \beta_1}{f_2} = g_2 - g_1 \qquad \text{and} \qquad \frac{\beta_2 - \alpha_2}{f_1} = g_2 - g_1.$$

In other words,

$$\alpha_1 + f_2 g_1 = \beta_1 + g_2 f_2,$$

and

$$\alpha_2 - g_1 f_1 = \beta_2 - f_1 g_2.$$

Note that in the last identity we have $\alpha_2 - g_1 f_1 \in \mathscr{H}(\Omega_1)$ and $\beta_2 - f_1 g_2 \in \mathscr{H}(\Omega_2)$ and, since they coincide in $\Omega_1 \cap \Omega_2$, they define a single holomorphic function v in Ω. Similarly, the previous identity has the left-hand side $\alpha_1 + f_2 g_1 \in \mathscr{H}(\Omega_1)$ and the right-hand side $\beta_1 + g_2 f_2 \in \mathscr{H}(\Omega_2)$ and jointly, they define a function $u \in \mathscr{H}(\Omega)$.

We claim that $uf_1 + vf_2 = 1$ in Ω. Let us verify this in Ω_1, for instance. We have

$$(uf_1 + vf_2)|\Omega_1 = (\alpha_1 + f_2 g_1)f_1 + (\alpha_2 - g_1 f_1)f_2 = \alpha_1 f_1 + \alpha_2 f_2 = 1.$$

The same way we verify $(uf_1 + vf_2)|\Omega_2 = 1$. Hence the case $n = 2$ has been completely proven.

Let us now assume the result is valid for any set of $n - 1$ functions without any common zero.

Let $f_1, \ldots, f_n \in \mathscr{H}(\Omega)$ be given functions without any common zeros. It is possible that $f_1, \ldots, f_{n-1}$ have common zeros, otherwise we are done. Let the multiplicity variety of the common zeros of these first $n - 1$ functions be $V = (z_k, n_k)_{k \geq 1}$. By Weierstrass' theorem we have an $f \in \mathscr{H}(\Omega)$ with $V = V(f)$. Therefore, the functions $\varphi_1 := f_1/f, \ldots, \varphi_{n-1} := f_{n-1}/f$ are holomorphic in Ω and have no common zeros. By the induction hypothesis we know there are $u_1, \ldots, u_{n-1}$ in $\mathscr{H}(\Omega)$ such that

$$\sum_{1 \leq j \leq n-1} u_j \varphi_j = 1,$$

that is,

$$\sum_{1 \leq j \leq n-1} u_j f_j = f.$$

Now we are in the situation where f and f_n have no common zeros. By the case $n = 2$, which we proved earlier, there are functions $u, v \in \mathscr{H}(\Omega)$ such that $uf + vf_n = 1$. Hence

$$(uu_1)f_1 + \cdots + (uu_{n-1})f_{n-1} + vf_n = 1.$$

This proves the lemma. □

3.5.7. Remarks. In the case $n = 2$ of Lemma 3.5.6, we can construct the functions u, v directly, using the $\bar{\partial}$-equation. Let φ_1, φ_2 be a C^∞ partition of unity subordinate to the covering Ω_1, Ω_2 of Ω given in the proof of the lemma ($\Omega_j = \Omega \backslash Z(f_j)$). We look for u and v of the form

$$u = \frac{\varphi_1}{f_1} + \omega f_2,$$

$$v = \frac{\varphi_2}{f_2} - \omega f_1,$$

where $\omega \in \mathscr{E}(\Omega)$ is an unknown function. Note that $uf_1 + vf_2 = \varphi_1 + \varphi_2 = 1$, and that φ_1/f_1 and $\varphi_2/f_2 \in \mathscr{E}(\Omega)$. Writing down the conditions that assure that u and v are holomorphic, we arrive at the following pair of equations for

$\dfrac{\partial\omega}{\partial\bar{z}}$:

$$\frac{1}{f_1}\frac{\partial\varphi_1}{\partial\bar{z}} + f_2\frac{\partial\omega}{\partial\bar{z}} = 0$$

$$\frac{1}{f_2}\frac{\partial\varphi_2}{\partial\bar{z}} - f_1\frac{\partial\omega}{\partial\bar{z}} = 0.$$

These two equations are the same since $\dfrac{\partial\varphi_1}{\partial\bar{z}} = -\dfrac{\partial\varphi_2}{\partial\bar{z}}$. Therefore,

$$\frac{\partial\omega}{\partial\bar{z}} = -\frac{1}{f_2}\left(\frac{1}{f_1}\frac{\partial\varphi_1}{\partial\bar{z}}\right) = \frac{1}{f_1}\left(\frac{1}{f_2}\frac{\partial\varphi_2}{\partial\bar{z}}\right),$$

where, if we use the middle term in Ω_2 and the last one in Ω_1, we see that $\dfrac{\partial\omega}{\partial\bar{z}}$ must equal a well-defined C^∞ function on Ω. The surjectivity of $\dfrac{\partial}{\partial\bar{z}}$ assures the existence of a solution ω and, a posteriori, the existence of u and v.

A small variation on the same theme is achieved by defining $\psi_1 := \bar{f}_1/(|f_1|^2 + |f_2|^2)$, $\psi_2 := \bar{f}_2/(|f_1|^2 + |f_2|^2)$. These two functions are in $\mathscr{E}(\Omega)$ and $f_1\psi_1 + f_2\psi_2 = 1$. Let us now consider

$$u = \psi_1 + \omega f_2,$$

$$v = \psi_2 - \omega f_1.$$

The same reasoning as before leads to an inhomogeneous Cauchy-Riemann equation for ω. This time we did not need to use a partition of unity.

In any case, the main point to remember is that "to reduce the problem to a $\bar{\partial}$-problem" is an almost universal recipe.

3.5.8. Corollary. *Every finitely generated ideal in $\mathscr{H}(\Omega)$ is principal (and hence closed).*

Proof. We can assume Ω is connected. Let $I = I(f_1, \ldots, f_n)$ be the ideal generated by $f_1, \ldots, f_n$. If $Z(I) = \varnothing$, then $I = \mathscr{H}(\Omega)$ by §3.5.6. If $Z(I) \neq \varnothing$ then we construct a "greatest common divisor" $f \in \mathscr{H}(\Omega)$ such that $\bigcap_j Z(f_j/f) = \varnothing$ as was done in the proof of Lemma 3.5.6. Hence $f = \sum_{1 \le j \le n} u_j f_j \in I$ for some coefficients $u_j \in \mathscr{H}(\Omega)$ and it is also clear that every element of I is divisible by f. I is closed as pointed out in Remark 3.5.4, (3). □

3.5.9. Proposition. *Every ideal I of $\mathscr{H}(\Omega)$ such that $Z(I) = \varnothing$ is dense in $\mathscr{H}(\Omega)$.*

Proof. Let $T \in \mathscr{H}'(\Omega)$ (i.e., an analytic functional) orthogonal to I. If we can show that $T = 0$ then the Hahn-Banach theorem allows us to conclude that $\bar{I} = \mathscr{H}(\Omega)$.

Recall that there is a compact subset L of Ω and a constant $C \geq 0$ such that

$$|\langle T, h \rangle| \leq C \|h\|_L,$$

that is, T is already continuous for the "uniform convergence on L." Let $K = \hat{L}_\Omega$ (or any other holomorphically compact subset of Ω containing L). Since $Z(I) = \varnothing$ we have

$$Z(I) \cap K = \bigcap_{f \in I} (Z(f) \cap K) = \varnothing.$$

The compactness of K lets us conclude that there is a finite number of functions $f_1, \ldots, f_n \in I$ such that $Z(f_1) \cap \cdots \cap Z(f_n)$ is disjoint from K. Hence there is an open set ω, $K \subseteq \omega \subseteq \Omega$, such that $f_1, \ldots, f_n$ have no common zeros in ω. By §3.5.6 we can find $g_1, \ldots, g_n \in \mathscr{H}(\omega)$ such that $\sum_{1 \leq j \leq n} f_j g_j = 1$ in ω.

Since $K = \hat{K}_\Omega$, Runge's theorem provides sequences $(g_{j,k})_{k \geq 1}$, $1 \leq j \leq n$, of elements of $\mathscr{H}(\Omega)$ such that $\lim_{k \to \infty} g_{j,k} = g_j$ uniformly on K.

If f is any function in $\mathscr{H}(\Omega)$ we have

$$f = \lim_{k \to \infty} \sum_{1 \leq j \leq n} g_{j,k} f_j f,$$

uniformly on K, hence on L. Therefore,

$$\langle T, f \rangle = \lim_{k \to \infty} \left\langle T, \sum_{1 \leq j \leq n} g_{j,k} f_j f \right\rangle = 0,$$

since $\sum_{1 \leq j \leq n} g_{j,k} f_j f \in I$ and T is orthogonal to I. Since f was arbitrary, $T = 0$, and the proposition has been proved. □

3.5.10. Corollary. *Every closed proper ideal I of $\mathscr{H}(\Omega)$ satisfies $Z(I) \neq \varnothing$.*

3.5.11. Theorem. *Let I be a closed ideal of $\mathscr{H}(\Omega)$. Then $I = I(V(I))$. In particular:*

(a) *every closed ideal is principal;*
(b) *every closed maximal ideal has the form $(z - z_0)\mathscr{H}(\Omega)$ for some $z_0 \in \Omega$.*

PROOF. The inclusion $I \subseteq I(V(I))$ is evident. Let f be a generator of the principal ideal $I(V(I))$ (use §3.5.2). Let

$$I_1 := \{u \in \mathscr{H}(\Omega) : fu \in I\}.$$

It is easy to see that I_1 is an ideal of $\mathscr{H}(\Omega)$; it is usually called the **conductor** of I in $I(V(I))$. Clearly $I \subseteq I_1$, hence $Z(I_1) \subseteq Z(I)$. We claim that $Z(I_1) = \varnothing$. In fact, if $z_0 \in Z(I)$, then there is $f_0 \in I$ such that $m(f_0, z_0) = m(f, z_0)$ by definition of $V(I)$ and f. Therefore z_0 is not a zero of the holomorphic function $u = f_0/f$. But u belongs to I_1, hence $z_0 \notin Z(I_1)$.

From §3.5.9 we conclude that I_1 is a dense ideal in $\mathscr{H}(\Omega)$. Let θ be the continuous self-mapping of $\mathscr{H}(\Omega)$ given by $\theta(g) = fg$. We have

$$I(V(I)) = \theta(\mathscr{H}(\Omega)) = \theta(\bar{I}_1) \subseteq \overline{\theta(I_1)} \subseteq \bar{I} = I.$$

Hence $I = I(V(I))$ and it coincides with the principal ideal $I(f)$.

Finally, if I is a closed maximal ideal, then $Z(I) \neq \varnothing$ and $Z(I)$ cannot contain more than a single point, and this point has to be of multiplicity one, otherwise I would not be maximal. □

3.5.12. Remarks

(1) There are maximal ideals which are not closed, for example, every maximal ideal containing the ideal from Remark 3.5.4.

(2) If I is a proper dense ideal, then for any $f_1, \ldots, f_n \in I$ we have $Z(f_1) \cap \cdots \cap Z(f_n) \neq \varnothing$. Otherwise $1 \in I$ and $I = \mathscr{H}(\Omega)$. In particular, I cannot contain any non-zero polynomial P. If $P \in I$ and $z_1, \ldots, z_n$ are its roots we can find $f_1, \ldots, f_n \in I$ with $f_j(z_j) \neq 0$ (since $Z(I) = \varnothing$). Therefore $Z(P) \cap Z(f_1) \cap \cdots \cap Z(f_n) = \varnothing$ and that would be a contradiction.

EXERCISES 3.5

Ω represents an open connected set in $\mathbb{C}$.

1. Let I be a closed maximal ideal of $\mathscr{H}(\Omega)$. Show that $\mathscr{H}(\Omega)/I \cong \mathbb{C}$.

2. Recall that $\mathbb{C}[z]$ is the space of polynomials in the variable z, and $\mathbb{C}(z)$ is the field of rational functions. Let I be a maximal dense proper ideal in $\mathscr{H}(\Omega)$. Show that the map

$$\mathbb{C}[z] \to \mathscr{H}(\Omega)/I$$
$$p \mapsto p(\text{mod } I)$$

is injective. Conclude that $\mathscr{H}(\Omega)/I$ contains a subfield isomorphic to $\mathbb{C}(z)$.

3. Let $I =$ ideal generated by $\left\{\sin\left(\frac{z}{n}\right), n \in \mathbb{N}^*\right\}$. Show that I is not a closed ideal in $\mathscr{H}(\mathbb{C})$.

4. Let $f_1, \ldots, f_n \in \mathscr{H}(\Omega)$, show there is a $gcd(f_1, \ldots, f_n) = f$ in $\mathscr{H}(\Omega)$, and that there are functions $g_1, \ldots, g_n \in \mathscr{H}(\Omega)$ such that

$$g_1 f_1 + \cdots + g_n f_n = f.$$

*5. Let I be a maximal ideal in $\mathscr{H}(\Omega)$. The object of this exercise is to show that the field $\mathscr{H}(\Omega)/I$ is algebraically closed.
   (i) Show that we can reduce ourselves to the case $Z(I) = \varnothing$.
   (ii) Let $f_0 X^n + f_1 X^{n-1} + \cdots + f_n = 0$, $f_0, \ldots, f_n \in \mathscr{H}(\Omega)$, $f_0 \notin I$, be the equation we want to solve in $\mathscr{H}(\Omega)/I$. Show that we can also assume $f_n \notin I$.
   (iii) Prove that there are functions $\alpha \in \mathscr{H}(\Omega)$ and $g \in I\setminus\{0\}$ such that $\alpha f_0 f_n + g = 1$.
   (iv) Let $h_k = \alpha f_n f_k$. Show that it is enough to find functions $\beta, \gamma \in \mathscr{H}(\Omega)$ such that $(h_0 \beta^n + h_1 \beta^{n-1} + \cdots + h_n) = g\gamma$. (Then we can take $X = \beta$.)

(v) Let $V(g) = (z_j, m_j)_j$. For each j choose an arbitrary root $x = b_j \in \mathbb{C}$ of the equation

$$h_0(z_j)x^n + \cdots + h_n(z_j) = 0.$$

Show that $b_j \neq 0$.

(vi) We are going to choose $\beta \in \mathscr{H}(\Omega)$ so that $\beta(z_j) = b_j$ for each j. Differentiate the equation $h_0\beta^n + \cdots + h_n = g\gamma$, ν times, where $1 \leq \nu \leq m_j - 1$, and set $z = z_j$. Show that the values $\beta^{(\nu)}(z_j)$ are uniquely determined.

(vii) Interpolate to obtain β and conclude the proof.

6. Let $H = \{z \in \mathbb{C} : \operatorname{Re} z > 0\}$ and $A := \{f \in \mathscr{H}(H) : f \in \mathscr{C}(\bar{H})$ and $f(z) \to 0$ when $|z| \to \infty\}$.
   (i) Show A is a Banach algebra with the norm $\|f\| = \sup\limits_{z \in H} |f(z)|$.
   (ii) Show $e^{-z} \notin A$.
   (iii) Prove that $I := \{fe^{-z} : f \in A\}$ is a closed ideal in A, $Z(I) = \varnothing$, and $I \neq A$.

7. The object of this exercise is to show that there is a natural structure associated to the family of all ideals in the algebra $\mathscr{H}(\Omega)$.
   (i) Let I be a proper ideal in $\mathscr{H}(\Omega)$, associate to it a family $\mathscr{B}(I)$ of sets in Ω defined as follows:

$$\mathscr{B}(I) := \{Z(f) : f \in I\}.$$

   Show that $\mathscr{B}(I)$ is a prefilter in Ω, i.e., (a) if $A, B \in \mathscr{B}(I)$ then $A \cap B \neq \varnothing$; and (b) for any $A, B \in \mathscr{B}(I)$ there is $C \in \mathscr{B}(I)$ such that $C \subseteq A \cap B$.
   (ii) Let $\mathscr{F}(I)$ be the filter generated by $\mathscr{B}(I)$ as follows

$$\mathscr{F}(I) := \{A \text{ subset of } \Omega : \exists f \in I \text{ with } Z(f) \subseteq A\}.$$

   Show that $\mathscr{F}(I)$ satisfies the three axioms of filters: (a) $\varnothing \notin \mathscr{F}(I)$; (b) $A, B \in \mathscr{F}(I)$ implies that $A \cap B \in \mathscr{F}(I)$; and (c) if B is a subset of Ω and there is $A \in \mathscr{F}(I)$ with $A \subseteq B$, then $B \in \mathscr{F}(I)$.
   (iii) Given a filter $\mathscr{F}$ in Ω, define $I(\mathscr{F}) := \{f \in \mathscr{H}(\Omega) : Z(f) \in \mathscr{F}\}$. Show that $I(\mathscr{F})$ is an ideal in $\mathscr{H}(\Omega)$, $I(\mathscr{F}) \neq \mathscr{H}(\Omega)$. Moreover, $I(\mathscr{F}) \neq \{0\}$ if and only if $\mathscr{F}$ contains a discrete subset of Ω.

   We denote by D this property for a filter, i.e., $\mathscr{F} \in D$ means that $I(\mathscr{F})$ is a proper ideal of $\mathscr{H}(\Omega)$.
   (iv) Show that one always has the following relations:
   (a) I, J proper ideals in $\mathscr{H}(\Omega)$, $I \subseteq J \Rightarrow \mathscr{F}(I) \subseteq \mathscr{F}(J)$.
   (b) $\mathscr{F}_1 \subseteq \mathscr{F}_2 \Rightarrow I(\mathscr{F}_1) \subseteq I(\mathscr{F}_2)$.
   (c) $\mathscr{F} \in D \Rightarrow \mathscr{F}(I(\mathscr{F})) = \mathscr{F}$.
   (d) I proper ideal in $\mathscr{H}(\Omega) \Rightarrow I \subseteq I(\mathscr{F}(I))$.
   In general there is no equality in (d). Is it true that the radical $\sqrt{I} = \sqrt{I(\mathscr{F}(I))}$?
   (v) Show that if $\mathscr{F}, \mathscr{G} \in D$, $\mathscr{F} \neq \mathscr{G}$, then $I(\mathscr{F}) \neq I(\mathscr{G})$.

   Recall that an ultrafilter is a maximal element for the family of filters partially ordered by inclusion.
   (vi) Show that $\mathscr{U} \mapsto I(\mathscr{U})$ and $M \mapsto \mathscr{F}(M)$ are inverse mappings between the family of ultrafilters satisfying the property D and the family of maximal ideals in $\mathscr{H}(\Omega)$. To the trivial ultrafilter $\mathscr{U}_a$ of all sets containing the point $a \in \Omega$, corresponds the maximal ideal generated by the function $f(z) = z - a$.

§6. The Operator $\frac{\partial}{\partial \bar{z}}$ Acting on Distributions

In this section we assume the reader is acquainted with the theory of distributions and of topological vector spaces [S], [Sch]. The reader will find an elementary introduction to the subject in the last section of this chapter. The portions that depend heavily on these concepts can be skipped in a first reading.

We are going to show first that a meromorphic function h in an open set Ω determines a distribution. The image of this distribution by $\frac{\partial}{\partial \bar{z}}$ is a distribution whose support is exactly the set of poles of h. The contribution of each pole to the latter distribution is computed using the principal part of h at that pole.

Later on, we identify $\mathscr{H}'(\Omega)$ to $\mathscr{E}'(\Omega) \Big/ \left(\frac{\partial}{\partial \bar{z}} \mathscr{E}'(\Omega)\right)$ and use this identification to characterize the orthogonal of a closed ideal and to reformulate the interpolation theorem proved earlier.

Finally, in this section we shall also discuss the relation between boundary values of holomorphic functions and distributions.

As a corollary to Pompeiu's formula we have shown that, for $\varphi \in \mathscr{D}(\mathbb{C})$,

$$\varphi(z) = \frac{1}{2\pi i} \int_{\mathbb{C}} \frac{\partial \varphi}{\partial \bar{\zeta}}(\zeta) \frac{\partial \zeta \wedge \partial \bar{\zeta}}{\zeta - z} = -\frac{1}{\pi} \int_{\mathbb{C}} \frac{\partial \varphi}{\partial \bar{\zeta}}(\zeta) \frac{dm(\zeta)}{\zeta - z},$$

where, we remind the reader, $dm(\zeta)$ denotes the Lebesgue measure of $\mathbb{R}^2$, $dm(\zeta) = d\xi\, d\eta$ if $\zeta = \xi + i\eta$. For $z = 0$ this formula becomes

$$\varphi(0) = -\frac{1}{\pi} \int_{\mathbb{C}} \frac{\partial \varphi}{\partial \bar{\zeta}}(\zeta) \frac{dm(\zeta)}{\zeta}.$$

This means that, in the sense of distributions, this formula is

$$-\frac{1}{\pi} \left\langle z, \frac{\partial \varphi}{\partial \bar{z}} \right\rangle = \varphi(0),$$

hence

$$\frac{\partial}{\partial \bar{z}} \frac{1}{\pi z} = \delta,$$

since $\left\langle \frac{\partial}{\partial \bar{z}} \frac{1}{\pi z}, \varphi \right\rangle = -\frac{1}{\pi} \left\langle \frac{1}{z}, \frac{\partial}{\partial \bar{z}} \varphi \right\rangle = \varphi(0) = \langle \delta, \varphi \rangle.$

Note that since $\frac{1}{z}$ is a locally integrable function in $\mathbb{C}$, then $\frac{1}{\pi z}$ defines a distribution and the preceding manipulations are just the way the distributional derivative $\dfrac{\partial}{\partial \bar{z}} \dfrac{1}{\pi z}$ is defined. We are now going to introduce other distributions generalizing $\dfrac{1}{\pi z}$.

3.6.1. Definition. For n integer greater than or equal to 1 we denote $pv\left(\dfrac{1}{\pi z^n}\right)$, the ***principal value of*** $\dfrac{1}{\pi z^n}$, the distribution defined by

$$\left\langle pv\left(\frac{1}{\pi z^n}\right), \varphi \right\rangle := \lim_{\varepsilon \to 0^+} \frac{1}{\pi} \int_{|\zeta| \geq \varepsilon} \frac{\varphi(\zeta)\, dm(\zeta)}{\zeta^n}.$$

When $n = 1$ we have $pv\left(\dfrac{1}{\pi z}\right) = \dfrac{1}{\pi z}$ since $\dfrac{1}{z}$ is locally integrable. This fails for $n > 1$ and even the existence of the limit is not a priori clear.

For an integer $n \geq 1$, we can use Taylor's formula about $\zeta = 0$ for $\varphi \in \mathscr{D}(\mathbb{C})$, and obtain

$$\varphi(\zeta) = \sum_{0 \leq p+q \leq n} \frac{1}{p!\,q!} \frac{\partial^{p+q}\varphi}{\partial \zeta^p \partial \bar{\zeta}^q}(0) \zeta^p \bar{\zeta}^q + \Phi_n(\zeta),$$

with $\Phi_n(\zeta) = O(|\zeta|^{n+1})$. To see this, one applies Taylor's formula with integral remainder to the functions of one real variable, $g_\zeta(t) = \varphi(t\zeta)$, and shows, by induction, that

$$\frac{d^k}{dt^k} g_\zeta(t) = \sum_{p+q=k} \frac{k!}{p!\,q!} \zeta^p \bar{\zeta}^q \frac{\partial^k \varphi}{\partial \zeta^p \partial \bar{\zeta}^q}(t\zeta).$$

Therefore, if $\operatorname{supp} \varphi \subseteq B(0, A)$, we have

$$\int_{|\zeta| \geq \varepsilon} \frac{\varphi(\zeta)\, dm(\zeta)}{\zeta^n} = \sum_{0 \leq p+q \leq n} \frac{1}{p!\,q!} \frac{\partial^{p+q}\varphi}{\partial \zeta^p \partial \bar{\zeta}^q}(0) \int_\varepsilon^A \int_0^{2\pi} \rho^{p+q-n+1} e^{i(p-q-n)\theta}\, d\rho\, d\theta$$
$$+ \int_{\varepsilon \leq |\zeta| \leq A} \frac{\Phi_n(\zeta)}{\zeta^n}\, dm(\zeta).$$

Now,

$$\int_0^{2\pi} e^{i(p-q-n)\theta}\, d\theta = \begin{cases} 0 & \text{if } p - q - n \neq 0 \\ 2\pi & \text{if } p - q - n = 0. \end{cases}$$

Since $0 \leq p + q \leq n$ (and they are both nonnegative) the only possibility for the integral to be different from zero occurs when $p = n$ and $q = 0$. The remainder term involves $\Phi_n(\zeta)/\zeta^n$, which is integrable in $0 \leq |\zeta| \leq A$, hence

$$\left\langle pv\left(\frac{1}{\pi z^n}\right), \varphi \right\rangle = \lim_{\varepsilon \to 0} \frac{1}{\pi} \int_{|\zeta| \geq \varepsilon} \frac{\varphi(\zeta)}{\zeta^n}\, dm(\zeta) = \frac{A^2}{n!} \frac{\partial^n \varphi}{\partial z^n}(0) + \frac{1}{\pi} \int_{|\zeta| \leq A} \frac{\Phi_n(\zeta)}{\zeta^n}\, dm(\zeta).$$

Furthermore, the integral expression

$$\Phi_n(\zeta) = \frac{1}{n!}\int_0^1 (1-t)^n \frac{d^{n+1}}{dt^{n+1}} g_\zeta(t)\,dt$$
$$= \sum_{p+q=n+1} \frac{n+1}{p!q!}\zeta^p\bar{\zeta}^q \int_0^1 (1-t)^n \frac{\partial^{n+1}\varphi(t\zeta)}{\partial\zeta^p\partial\bar{\zeta}^q}\,dt,$$

ensures that $\varphi \mapsto \frac{1}{\pi}\int_{|\zeta|\le A} (\Phi_n(\zeta)/\zeta^n)\,dm(\zeta)$ is a distribution. It follows that $pv\left(\frac{1}{\pi z^n}\right)$ is well defined as a distribution.

3.6.2. Proposition. *The relation*

$$\frac{\partial}{\partial z} pv\left(\frac{1}{\pi z^n}\right) = -n\, pv\left(\frac{1}{\pi z^{n+1}}\right)$$

holds for any $n \ge 1$.

PROOF. We have

$$\left\langle \frac{\partial}{\partial z} pv\left(\frac{1}{\pi z^n}\right), \varphi \right\rangle = -\left\langle pv\left(\frac{1}{\pi z^n}\right), \frac{\partial\varphi}{\partial z}\right\rangle = -\lim_{\varepsilon\to 0}\frac{1}{\pi}\int_{|z|\ge\varepsilon} \frac{1}{z^n}\frac{\partial\varphi}{\partial z}(z)\,dm(z).$$

On the other hand, for $z \ne 0$,

$$\frac{\partial}{\partial z}\left(\frac{\varphi(z)}{z^n}\right) = \frac{1}{z^n}\frac{\partial\varphi}{\partial z}(z) - n\frac{\varphi(z)}{z^{n+1}},$$

hence

$$\left\langle \frac{\partial}{\partial z} pv\left(\frac{1}{\pi z^n}\right), \varphi \right\rangle = -\lim_{\varepsilon\to 0}\frac{1}{\pi}\int_{|z|\ge\varepsilon} \frac{\partial}{\partial z}\left(\frac{\varphi(z)}{z^n}\right) dm(z) - n\left\langle pv\left(\frac{1}{\pi z^{n+1}}\right), \varphi\right\rangle.$$

But, using Stokes' formula and that supp φ is compact,

$$-\frac{1}{\pi}\int_{|z|\ge\varepsilon} \frac{\partial}{\partial z}\left(\frac{\varphi(z)}{z^n}\right) dm(z) = \frac{1}{2\pi i}\int_{|z|\ge\varepsilon} \frac{\partial}{\partial z}\left(\frac{\varphi(z)}{z^n}\right) dz \wedge d\bar{z}$$
$$= \frac{1}{2\pi i}\int_{|z|\ge\varepsilon} \partial\left(\frac{\varphi(z)}{z^n}\,d\bar{z}\right) = -\frac{1}{2\pi i}\int_{|z|=\varepsilon} \frac{\varphi(z)}{z^n}\,d\bar{z}.$$

Developing φ according to Taylor's formula, we have

$$-\frac{1}{2\pi i}\int_{|z|=\varepsilon} \frac{\varphi(z)}{z^n}\,d\bar{z} = \frac{1}{2\pi}\sum_{0\le p+q\le n-1} \frac{1}{p!q!}\frac{\partial^{p+q}\varphi}{\partial z^p\partial\bar{z}^q}(0)\varepsilon^{p+q-n+1}\int_0^{2\pi} e^{i(p-q-n-1)\theta}\,d\theta$$
$$+ \frac{1}{2\pi}\int_0^{2\pi} e^{-i(n+1)\theta}\frac{\Phi_{n-1}(\varepsilon e^{i\theta})}{\varepsilon^{n-1}}\,d\theta.$$

Since $p + q \le n - 1$, all the trigonometric integrals in the sum are zero. Furthermore, $|\Phi_{n-1}(\varepsilon e^{i\theta})| = O(\varepsilon^n)$, hence the limiting value of the last integral is zero as $\varepsilon \to 0$. It follows that

$$\left\langle \frac{\partial}{\partial z} pv\left(\frac{1}{\pi z^n}\right), \varphi \right\rangle = -n \left\langle pv\left(\frac{1}{\pi z^{n+1}}\right), \varphi \right\rangle,$$

as we wanted to show. □

3.6.3. Proposition. *For $n \geq 1$ we have the identities*

$$\frac{\partial}{\partial \bar z} pv\left(\frac{1}{\pi z^n}\right) = \frac{(-1)^{n-1}}{(n-1)!} \frac{\partial^{n-1}}{\partial z^{n-1}} \delta,$$

where $\delta = \delta_0$ is the Dirac mass at 0.

PROOF. We have already shown that

$$\frac{\partial}{\partial \bar z}\left(\frac{1}{\pi z}\right) = \delta,$$

which is the statement of this proposition when $n = 1$. Let us proceed by induction and assume the statement to be correct for n, therefore

$$\frac{\partial}{\partial \bar z} pv\left(\frac{1}{\pi z^{n+1}}\right) = -\frac{1}{n}\frac{\partial}{\partial \bar z}\left(\frac{\partial}{\partial z} pv\left(\frac{1}{\pi z^n}\right)\right) = -\frac{1}{n}\frac{\partial}{\partial z}\left(\frac{(-1)^{n-1}}{(n-1)!}\frac{\partial^{n-1}}{\partial z^{n-1}}\delta\right)$$

$$= \frac{(-1)^n}{n!}\frac{\partial^n}{\partial z^n}\delta.$$ □

Let f be a holomorphic function in an open set Ω which contains the origin. Define $pv\left(\frac{f}{\pi z^n}\right)$ as the product of f and $pv\left(\frac{1}{\pi z^n}\right)$ (restricted to Ω),

$$pv\left(\frac{f}{\pi z^n}\right) := f \cdot \left(pv\left(\frac{1}{\pi z^n}\right)\Big|\Omega\right).$$

For $\varphi \in \mathscr{D}(\Omega)$,

$$\left\langle pv\left(\frac{f}{\pi z^n}\right), \varphi \right\rangle = \left\langle pv\left(\frac{1}{\pi z^n}\right), f\varphi \right\rangle = \lim_{\varepsilon \to 0} \int_{|z| \geq \varepsilon} \frac{f(z)\varphi(z)}{\pi z^n}\, dm(z),$$

which shows that we could have used the same definition as in §3.6.1. We have then

$$\left\langle \frac{\partial}{\partial \bar z} pv\left(\frac{f}{\pi z^n}\right), \varphi \right\rangle = -\left\langle pv\left(\frac{f}{\pi z^n}\right), \frac{\partial \varphi}{\partial \bar z} \right\rangle = -\left\langle pv\left(\frac{1}{\pi z^n}\right), f\frac{\partial \varphi}{\partial \bar z} \right\rangle$$

$$= -\left\langle pv\left(\frac{1}{\pi z^n}\right), \frac{\partial}{\partial \bar z}(f\varphi) \right\rangle = \left\langle \frac{\partial}{\partial \bar z} pv\left(\frac{1}{\pi z^n}\right), f\varphi \right\rangle$$

$$= \left\langle f\frac{\partial}{\partial \bar z} pv\left(\frac{1}{\pi z^n}\right), \varphi \right\rangle,$$

hence

$$\frac{\partial}{\partial \bar{z}} pv\left(\frac{f}{\pi z^n}\right) = f \frac{\partial}{\partial \bar{z}} pv\left(\frac{1}{\pi z^n}\right).$$

Let us make more explicit a few cases. For instance, $n = 1$ becomes

$$\frac{\partial}{\partial \bar{z}} pv\left(\frac{f}{\pi z}\right) = f(0)\delta,$$

and $n = 2$ gives

$$\frac{\partial}{\partial \bar{z}} pv\left(\frac{f}{\pi z^2}\right) = f'(0)\delta - f(0)\frac{\partial \delta}{\partial z}.$$

In fact,

$$\left\langle \frac{\partial}{\partial \bar{z}} pv\left(\frac{f}{\pi z^2}\right), \varphi \right\rangle = -\left\langle f\frac{\partial \delta}{\partial z}, \varphi \right\rangle = -\left\langle \frac{\partial \delta}{\partial z}, f\varphi \right\rangle = \left\langle \delta, \frac{\partial}{\partial z}(f\varphi) \right\rangle$$

$$= f'(0)\varphi(0) + f(0)\frac{\partial \varphi(0)}{\partial z},$$

which is exactly the preceding statement.

This example shows also a general fact, multiplication by holomorphic functions commutes with the operator $\dfrac{\partial}{\partial \bar{z}}$ acting on distributions, that is, $\dfrac{\partial}{\partial \bar{z}}$ is $\mathscr{H}(\Omega)$-linear acting on the $\mathscr{H}(\Omega)$-module $\mathscr{D}'(\Omega)$. In effect, $\dfrac{\partial}{\partial \bar{z}} : \mathscr{D}'(\Omega) \to \mathscr{D}'(\Omega)$ is the transpose of $-\dfrac{\partial}{\partial \bar{z}} : \mathscr{D}(\Omega) \to \mathscr{D}(\Omega)$ and the $\mathscr{H}(\Omega)$-linearity is then evident.

The computations we have just done allow us to embed the space $\mathscr{M}(\Omega)$ of meromorphic functions in Ω as a subspace of $\mathscr{D}'(\Omega)$. This map is called the ***Cauchy principal value***

$$pv : \mathscr{M}(\Omega) \to \mathscr{D}'(\Omega)$$

and it is defined as follows: Let $(z_k)_{k \geq 1}$ be the indexing of the poles of f (this sequence could be finite or even empty; if not, we can assume $\lim\limits_{k \to \infty} d(z_k, \Omega^c) = 0$). Let $\Omega_\varepsilon = \{z \in \Omega : |z - z_k| \geq \varepsilon \text{ for every } k\}$, define for $\varphi \in \mathscr{D}(\Omega)$

$$\langle pv(f), \varphi \rangle := \lim_{\varepsilon \to 0} \int_{\Omega_\varepsilon} f(z)\varphi(z)\, dm(z).$$

Note that if $\operatorname{supp} \varphi$ contains no z_k, then for all ε sufficiently small but positive, $\displaystyle\int_{\Omega_\varepsilon} f\varphi\, dm(z) = \int_{\Omega} f\varphi\, dm(z)$, which is well defined. If z_k is in $\operatorname{supp} \varphi$, let

$$f(z) = \sum_{-\infty < \nu \leq \nu_k} \frac{A_{k,\nu}}{(z - z_k)^\nu}$$

be the Laurent development of f in the neighborhood of z_k, where ν_k is the order of the pole z_k of f. Then we see that

$$\langle pv(f), \varphi \rangle = \sum_{k \geq 0} \left(\sum_{1 \leq \nu \leq \nu_k} A_{k,\nu} \left\langle pv\left(\frac{1}{(z - z_k)^\nu}\right), \varphi \right\rangle \right) + \langle h, \varphi \rangle,$$

where the first sum is finite and the function h is holomorphic in a neighborhood of supp φ. This shows that $pv(f)$ defines a distribution in $\mathscr{D}'(\Omega)$. We have

$$\frac{\partial}{\partial \bar{z}} pv(f) = \pi \sum_{k \geq 0} \left(\sum_{1 \leq \nu \leq \nu_k} \frac{(-1)^{\nu-1}}{(\nu - 1)!} A_{k,\nu} \frac{\partial^{\nu-1}}{\partial z^{\nu-1}} \delta_{z_k} \right),$$

where δ_{z_k} is the Dirac mass at z_k and the sum is meaningful as a distribution in Ω since it is locally finite in Ω. In the literature this distribution is sometimes called the ***residue distribution of*** f, i.e., $\mathrm{Res}(f) := \dfrac{\partial}{\partial \bar{z}} pv(f)$.

This should not be confused with the residues $\mathrm{Res}(f, z_k)$ of the differential form $f(z)\,dz$ at the point z_k. The relation between these two notions can be summarized as follows: Let V_k be a neighborhood of z_k not containing any other pole of f, then

$$\mathrm{Res}(f) | V_k = \pi \,\mathrm{Res}(f, z_k) \delta_{z_k} + T_k,$$

where T_k is a distribution with supp $T_k = \{z_k\}$ involving derivatives of δ_{z_k} and T_k is not zero unless $\nu_k = 1$. Note that only derivatives of the form $\dfrac{\partial^\nu}{\partial z^\nu}$ appear, i.e., there are not derivatives with respect to $\bar{z}$.

It is the fact that a meromorphic function is not locally integrable, unless all the poles are of order 1, which requires we introduce the map pv. On the other hand, if $f \in L^1_{\mathrm{loc}}(\Omega)$ and f is holomorphic outside a relatively closed subset F of Ω, we can define the residue distribution of f, $\mathrm{Res}(f)$, by

$$\mathrm{Res}(f) := \frac{\partial}{\partial \bar{z}} T_f,$$

where, as usual, $\langle T_f, \varphi \rangle = \displaystyle\int f\varphi \, dm(z)$ for every $\varphi \in \mathscr{D}(\Omega)$. It is clear that $\mathrm{supp}(\mathrm{Res}(f)) \subseteq F$.

Recall that if ω is a relatively compact open subset of Ω with $\partial\omega$ piecewise regular of class C^1, then the connected components $C_1, \ldots, C_n$ of $\partial\omega$ are piecewise-C^1 Jordan curves that are oriented in such a way that the Stokes formula is valid. We can define distributions T, T_j of "integration over $\partial\omega$," resp. "over C_j," given by

$$\langle T, \varphi \rangle := \int_{\partial\omega} \varphi(z)\, dz$$

and

$$\langle T_j, \varphi \rangle := \int_{C_j} \varphi(z)\, dz.$$

We have therefore $T = \sum_{1 \le j \le n} T_j$. If χ_ω is the characteristic function of ω, the following proposition holds.

3.6.4. Proposition. *Let ω be a relatively compact open subset of Ω with piecewise regular boundary $\partial\omega$ of components $C_1, \ldots, C_n$, then*

$$\operatorname{Res}(T_{\chi_\omega}) = \frac{\partial}{\partial \bar{z}} \chi_\omega = -\frac{1}{2i} T = -\frac{1}{2i} \sum_{1 \le j \le n} T_j.$$

Furthermore, let $\zeta \in \omega$ and let S be the distribution defined by the locally integrable function of z, $z \mapsto \frac{1}{\pi} \frac{\chi_\omega(z)}{z - \zeta}$. This function is holomorphic outside $\partial\omega \cup \{\zeta\}$, and

$$\operatorname{Res}(S) = \frac{\partial S}{\partial \bar{z}} = \delta_\zeta - \frac{1}{2\pi i} \frac{1}{z - \zeta} T.$$

PROOF. It is just a translation into the language of distributions of the formulas of Stokes and Pompeiu. Note that the function $z \mapsto \frac{1}{z - \zeta}$ is smooth (even holomorphic) in a neighborhood of $\operatorname{supp}(T)$ $(= \partial\omega)$. □

We are now going to return to the study of analytic functionals, that is, elements of $\mathscr{H}'(\Omega)$. We have the exact sequence of Fréchet-Schwartz spaces

$$0 \longrightarrow \mathscr{H}(\Omega) \xrightarrow{\ i\ } \mathscr{E}(\Omega) \xrightarrow{\frac{\partial}{\partial \bar{z}}} \mathscr{E}(\Omega) \longrightarrow 0,$$

where i is the inclusion map. The transposes of these maps are ${}^t\!\left(\frac{\partial}{\partial \bar{z}}\right) = -\frac{\partial}{\partial \bar{z}}$ and ${}^t i = r =$ restriction of the action of distributions of compact support to the space of holomorphic functions (which is surjective by the Hahn-Banach theorem, cf. [Sch]). We obtain the exact sequence:

$$0 \longrightarrow \mathscr{E}'(\Omega) \xrightarrow{\frac{\partial}{\partial \bar{z}}} \mathscr{E}'(\Omega) \xrightarrow{\ r\ } \mathscr{H}'(\Omega) \longrightarrow 0.$$

We can conclude that $\mathscr{H}'(\Omega)$ can be identified to $\mathscr{E}'(\Omega) \Big/ \left(\frac{\partial}{\partial \bar{z}}(\mathscr{E}'(\Omega))\right)$.

We are going to show now, using the same type of ideas, that the operator

$$\frac{\partial}{\partial \bar{z}} : \mathscr{D}'(\Omega) \to \mathscr{D}'(\Omega)$$

is surjective. With that purpose in mind, we prove first a regularity result for the Cauchy-Riemann operator $\frac{\partial}{\partial \bar{z}}$ and the Laplace operator Δ. Recall that $\Delta = 4\frac{\partial^2}{\partial z \partial \bar{z}}$ and that a function h is called harmonic if it is C^2 and $\Delta h = 0$. Proposition 3.6.5 shows h is automatically C^∞. (Example 2.2.9, (3) is similar in spirit to the following proposition.)

3.6.5. Proposition. *Let Ω be an open subset of $\mathbb{C}$ and $T \in \mathscr{D}'(\Omega)$ satisfying the equation $\Delta T = 0$. Then there is a harmonic function h in Ω such that $T_h = T$. Similarly, if $R \in \mathscr{D}'(\Omega)$ satisfies $\frac{\partial R}{\partial \bar{z}} = 0$, there is a holomorphic function f in Ω such that $T_f = R$.*

PROOF. It is enough to prove the statement for the Laplace operator. A fundamental solution E for this operator is $E = \frac{1}{2\pi}\log|z|$, that is, $\Delta E = \delta$ in the sense of distributions. Let now Ω' be open and $\Omega' \subset\subset \Omega$. We shall show "$T$ is C^∞ in Ω'," that is, there is a C^∞ function in Ω' (necessarily harmonic since its Laplacian will be zero) which defines T in Ω'. Let $g \in \mathscr{D}(\Omega)$ be such that $g \equiv 1$ in a neighborhood of $\bar{\Omega}'$, then

$$\Delta(gT) = g\Delta T + S = S$$

with $S \in \mathscr{E}'(\Omega)$ and $\operatorname{supp} S \subseteq \bar{\Omega}'^c$. Therefore

$$gT = \Delta(E * (gT)) = E * \Delta(gT) = E * S.$$

We want to show that this implies that T is smooth in Ω'. Let $z_0 \in \Omega'$ and $\varepsilon > 0$ sufficiently small so that $V_\varepsilon = \{z \in \mathbb{C} : d(z, \Omega'^c) > \varepsilon\}$ is a neighborhood of z_0. Let now $\theta_\varepsilon \in \mathscr{D}(\mathbb{C})$ be equal to 1 for $|z| \le \varepsilon/2$ and zero for $|z| \ge \varepsilon$. We can write

$$E * S = ((\theta_\varepsilon E) * S) + ((1 - \theta_\varepsilon)E) * S.$$

Remark that

(i) $((1 - \theta_\varepsilon)E) * S$ is C^∞ since $(1 - \theta_\varepsilon)E$ is C^∞ and S has compact support,
(ii) $\operatorname{supp}((\theta_\varepsilon E) * S) \subseteq \operatorname{supp}(\theta_\varepsilon E) + \operatorname{supp} S \subseteq \{z : d(z, \operatorname{supp}(S)) \le \varepsilon\}$.

The second condition ensures that $(\theta_\varepsilon E) * S$ is zero in V_ε, hence $T|V_\varepsilon = (gT)|V_\varepsilon = (((1 - \theta_\varepsilon)E) * S)|V_\varepsilon$ which is C^∞ by (i). Therefore, T is smooth in the neighborhood of every point of $z_0 \in \Omega'$. Since Ω' was an arbitrary open relatively compact subset of Ω, we have T is C^∞ in Ω. □

3.6.6. Theorem. *The operator $\frac{\partial}{\partial \bar{z}} : \mathscr{D}'(\Omega) \to \mathscr{D}'(\Omega)$ is surjective.*

PROOF. First, let us point out that for every $S \in \mathscr{E}'(\Omega)$ there is $T \in \mathscr{D}'(\mathbb{C})$ such that $\left(\frac{\partial T}{\partial \bar{z}}\right)\Big|\Omega = S$. In fact, the distribution $T := \frac{1}{\pi z} * S \in \mathscr{D}'(\mathbb{C})$ and $\frac{\partial T}{\partial \bar{z}} = S$ in Ω (or everywhere if we extend S by zero outside Ω, this is meaningful since $\operatorname{supp}(S)$ is a compact subset of Ω).

Let now $S \in \mathscr{D}'(\Omega)$ be given. Let $(K_j)_{j\geq 1}$ be an exhaustion of Ω by a sequence of compact holomorphically convex subsets of Ω. Let $\psi_j \in \mathscr{D}(\mathbb{C})$ be such that $0 \leq \psi_j \leq 1$, $\psi_j \equiv 1$ in a neighborhood of K_j, $\text{supp}(\psi_j) \subseteq K_{j+1}$. For every j let $T_j \in \mathscr{D}'(\mathbb{C})$ be such that $\frac{\partial T_j}{\partial \bar{z}} = \psi_j S$. For $j \geq 2$, the distribution $T_j - T_{j-1}$ satisfies $\frac{\partial}{\partial \bar{z}}(T_j - T_{j-1}) = 0$ in a neighborhood of K_{j-1}. Since K_{j-1} is holomorphically convex in Ω, for $j \geq 2$ we can find $h_j \in \mathscr{H}(\Omega)$ such that

$$\sup_{z \in K_{j-1}} |T_j - T_{j-1} - h_j| \leq 2^{-j}.$$

Therefore, the series

$$T_1 + \sum_{j\geq 2} (T_j - T_{j-1} - h_j)$$

converges in $\mathscr{D}'(\Omega)$. Let T be its sum, then

$$\frac{\partial T}{\partial \bar{z}} = \frac{\partial T_1}{\partial \bar{z}} + \sum_{j\geq 2} \left(\frac{\partial T_j}{\partial \bar{z}} - \frac{\partial T_{j-1}}{\partial \bar{z}}\right) = \lim_{n\to\infty} \frac{\partial T_n}{\partial \bar{z}} = S. \qquad \square$$

3.6.7. Proposition. *Let Ω be an open connected subset of $\mathbb{C}$, I a closed proper ideal of $\mathscr{H}(\Omega)$, f a generator of I, $V = (z_k, m_k)_{k\geq 1}$ the multiplicity variety of I. The orthogonal $I^\perp$ in $\mathscr{H}'(\Omega)$ can be identified to the quotient space $\mathscr{E}'(V(I))\Big/\left(\frac{\partial}{\partial \bar{z}}\mathscr{E}'(V(I))\right)$, where $\mathscr{E}'(V(I))$ is the set of distributions $R \in \mathscr{E}'(\Omega)$ such that fR is the zero distribution. Furthermore, $I^\perp$ can also be identified to the space of all distributions of the form*

$$\sum_{k\geq 1} \sum_{0\leq \nu \leq m_k - 1} A_{k,\nu} \frac{\partial^\nu}{\partial z^\nu} \delta_{z_k}$$

with only finitely many nonzero coefficients $A_{k,\nu}$.

Proof. Let us point out that the last statement of the theorem can be obtained easily from the Interpolation Theorem 3.4.1. We do not want to do that here since we would like to show that Theorem 3.4.1 can be obtained as a corollary of this proposition.

We first observe that if $g \in \mathscr{H}(\Omega) \cap f\mathscr{E}(\Omega)$, i.e., there is a function $h \in \mathscr{E}(\Omega)$ such that $g = fh$, then $g \in I$. In fact, g will vanish on $V(I)$.

We already know that the following sequence is exact:

$$0 \longrightarrow \mathscr{H}(\Omega) \xrightarrow{\ i\ } \mathscr{E}(\Omega) \xrightarrow{\frac{\partial}{\partial \bar{z}}} \mathscr{E}(\Omega) \longrightarrow 0,$$

From here it is easy to conclude the exactness of the following one:

$$0 \longrightarrow f\mathscr{H}(\Omega) \xrightarrow{\ i\ } f\mathscr{E}(\Omega) \xrightarrow{\frac{\partial}{\partial \bar{z}}} f\mathscr{E}(\Omega) \longrightarrow 0.$$

Passing to the quotient we obtain the commutative diagram

$$
\begin{array}{ccccccccc}
 & & 0 & & 0 & & 0 & & \\
 & & \downarrow & & \downarrow & & \downarrow & & \\
0 & \longrightarrow & I & \xrightarrow{i} & f\mathscr{E}(\Omega) & \xrightarrow{\frac{\partial}{\partial \bar{z}}} & f\mathscr{E}(\Omega) & \longrightarrow & 0 \\
 & & \downarrow & & \downarrow & & \downarrow & & \\
0 & \longrightarrow & \mathscr{H}(\Omega) & \xrightarrow{i} & \mathscr{E}(\Omega) & \xrightarrow{\frac{\partial}{\partial \bar{z}}} & \mathscr{E}(\Omega) & \longrightarrow & 0 \\
 & & \downarrow & & \downarrow & & \downarrow & & \\
0 & \longrightarrow & \mathscr{H}(\Omega)/I & \xrightarrow{\alpha} & \mathscr{E}(\Omega)/f\mathscr{E}(\Omega) & \xrightarrow{\beta} & \mathscr{E}(\Omega)/f\mathscr{E}(\Omega) & \longrightarrow & 0 \\
 & & \downarrow & & \downarrow & & \downarrow & & \\
 & & 0 & & 0 & & 0 & &
\end{array}
$$

where α (resp. β) is $i \left(\text{resp. } \frac{\partial}{\partial \bar{z}}\right)$ passed to the quotient. Every column is exact and we want to show the third row is also exact. First, β is surjective since $\frac{\partial}{\partial \bar{z}}$ is known to be surjective. The map α is injective because, if $\alpha(\tilde{g}) = 0$, where $\tilde{g}$ is the class of $g \in \mathscr{H}(\Omega)$ modulo I, then we have $g \in f\mathscr{E}(\Omega) \cap \mathscr{H}(\Omega) = I$, hence $\tilde{g} = 0$. Finally, if $\beta(\tilde{g}) = 0$ for $g \in \mathscr{E}(\Omega)$, this means that $\frac{\partial}{\partial \bar{z}} g = fk$, for some $k \in \mathscr{E}(\Omega)$. Since $k = \frac{\partial}{\partial \bar{z}} k_1$ for some $k_1 \in \mathscr{E}(\Omega)$, we obtain $\frac{\partial}{\partial \bar{z}}(g - fk_1) = 0$. Hence $h = g - fk_1 \in \mathscr{H}(\Omega)$ and $\alpha(\tilde{h}) = [g - fk_1]^{\sim} = \tilde{g}$. This ends the proof of the exactnesses.

Let us note that the subspace $f\mathscr{E}(\Omega)$ is closed in $\mathscr{E}(\Omega)$. In fact, let $\varphi_n = fu_n$ be a sequence that converges in $\mathscr{E}(\Omega)$ to φ. We want to show that φ is divisible by f in $\mathscr{E}(\Omega)$. It is enough to prove this locally, and this is shown to be true by the following lemma.

3.6.8. Lemma. *Let g be a C^∞ function in a neighborhood of* 0 *such that for some fixed integer $k \geq 1$ satisfies*

$$\frac{\partial^{i+j}}{\partial z^i \partial \bar{z}^j} g(0) = 0$$

for any (i,j) such that $0 \leq i \leq k - 1$. Then the function $z \mapsto g(z)/z^k$ is C^∞ in the same neighborhood of zero.

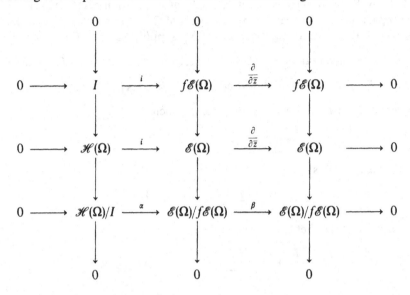

PROOF. For any $n > k$ we can rewrite the Taylor formula for g about $z = 0$ as follows

$$g(z) = z^k\left(\sum_{\substack{0 \le p+q \le n,\\ k \le p}} \frac{1}{p!\,q!}\frac{\partial^{p+q}}{\partial z^p \partial \bar{z}^q}(0) z^{p-k}\bar{z}^q + \frac{\theta_n(z)}{z^k}\right),$$

where $\theta_n(z) = O(|z|^{n+1})$. It follows that $g(z)/z^k$ has derivatives of every order which extend continuously to $z = 0$. This proves the lemma. □

Since the converse of this lemma is obviously true (e.g., by considering the Taylor formula of $z^k\psi(z)$ at $z = 0$ when ψ is C^∞ near 0), then $\varphi = \lim\limits_n fu_n$ will satisfy the conditions of the lemma near each zero of f. Therefore φ is divisible by f and $f\mathscr{E}(\Omega)$ is closed.

It follows that all the quotient spaces in the last row of the diagram are Hausdorff, and hence Fréchet-Schwartz spaces. Identify, as usual, $(f\mathscr{E}(\Omega))^\perp$ to $(\mathscr{E}(\Omega)/f\mathscr{E}(\Omega))'$ and $I^\perp$ to $(\mathscr{H}(\Omega)/I)'$. Taking transposes in the last row of that sequence (here we use the fact they are Fréchet-Schwartz) we obtain

$$I^\perp \cong (f\mathscr{E}(\Omega))^\perp \Big/ \frac{\partial}{\partial \bar{z}}((f\mathscr{E}(\Omega))^\perp).$$

(See [Sch] for a proof of these statements.)

To conclude the proof of the proposition we have to analyze a bit more in detail the meaning of $T \in (f\mathscr{E}(\Omega))^\perp$. First we have $T \in \mathscr{E}'(\Omega)$ and $\langle T, f\varphi\rangle = 0$ for every $\varphi \in \mathscr{D}(\Omega)$. This means that $fT = 0$ as a distribution and one can conclude that $\operatorname{supp}(T) \subseteq Z(f)$. The compactness of $\operatorname{supp}(T)$ now implies that T is a finite sum of derivatives of δ_{z_k} with $z_k \in Z(f)$. Furthermore, Lemma 3.6.8 shows that T is a finite sum of the form

$$T = \sum_{\substack{0 \le \nu \le m_k - 1\\ k \ge 1, \mu \ge 0}} A_{k,\nu,\mu}\frac{\partial^{\nu+\mu}}{\partial z^\nu \partial \bar{z}^\mu}\delta_{z_k}.$$

Therefore, if $\mathscr{E}'(V(I))$ denotes this space of finite sums we have

$$(f\mathscr{E}(\Omega))^\perp = \mathscr{E}'(V(I)).$$

Evidently, any such sum is congruent to

$$\sum_{\substack{0 \le \nu \le m_k - 1\\ k \ge 1}} A_{k,\nu,0}\frac{\partial^\nu}{\partial z^\nu}\delta_{z_k}$$

modulo $\frac{\partial}{\partial \bar{z}}(\mathscr{E}'(V(I)))$. This ends the proof of Proposition 3.6.7. □

We denote by $\mathscr{H}(V(I))$, the ***space of holomorphic functions on the multiplicity variety*** $V(I)$, the space of sequences $(a_{k,l})_{\substack{k \ge 1\\ 0 \le l \le m_k - 1}}$. There is a natural ***restriction*** map ρ,

$$\rho : \mathscr{H}(\Omega) \to \mathscr{H}(V(I))$$

$$\rho : \varphi \mapsto \left(\frac{\varphi^{(l)}(z_k)}{l!} \right)_{\substack{0 \le l \le m_k - 1 \\ k \ge 1}}.$$

The interpolation Theorem 3.4.1 can now be obtained as a corollary of Proposition 3.6.7.

3.6.9. Corollary. *The map $\rho : \mathscr{H}(\Omega) \to \mathscr{H}(V(I))$ is surjective.*

Proof. We only need to show that $\mathscr{H}(\Omega)/I$ can be naturally identified to $\mathscr{H}(V(I))$. By duality, from the proof of 3.6.7, we have that $\mathscr{E}(\Omega)/f\mathscr{E}(\Omega)$ can be identified to the space of sequences of the form

$$(a_{k,p,q})_{\substack{1 \le p \le m_k - 1, \\ q \ge 0, k \ge 0}}$$

with the projection map identified to the map

$$\varphi \mapsto \left(\frac{1}{p!q!} \frac{\partial^{p+q}}{\partial z^p \partial \bar{z}^q} \varphi(z_k) \right),$$

(where the indices run over the set $0 \le p \le m_k - 1, q \ge 0, k \ge 1$). The operator $\beta \left(\text{the passage of } \frac{\partial}{\partial \bar{z}} \text{ to the quotient} \right)$ is then identified to

$$(a_{k,p,q}) \mapsto ((q+1)a_{k,p,q+1})$$

and its kernel is then isomorphic to $\{(a_{k,p,0})\}$. This is precisely the identification of $\mathscr{H}(\Omega)/I$ with $\mathscr{H}(V(I))$. □

We end this section with a discussion of the notion of the boundary values of a holomorphic function in the sense of distributions. The aim is to prove an important analytic continuation theorem called the edge-of-the-wedge theorem. This theorem is a generalization of the Schwarz reflection principle of Chapter 2.

3.6.10. Definition. Let Ω be the union of the rectangles Ω_+ and Ω_-, with $\Omega_+ =]a, b[+ i]0, c[$ (resp. $\Omega_- =]a, b[+ i]-c, 0[$), where a, $b \in \bar{\mathbb{R}}$ $(= R \cup \{\pm\infty\})$ and $0 < c \le \infty$. We say that a function f, holomorphic in Ω_+ (resp. Ω_-), ***admits a boundary value in the sense of distributions*** if the limit

$$\lim_{\varepsilon \to 0+} \int_a^b f(x + i\varepsilon)\varphi(x)\, dx$$

$$\left(\text{resp. } \lim_{\varepsilon \to 0+} \int_a^b f(x - i\varepsilon)\varphi(x)\, dx \right)$$

exists for every $\varphi \in \mathscr{D}(]a, b[)$.

If f is holomorphic in $\Omega = \Omega_+ \cup \Omega_- =]a, b[+ i(]-c, c[\setminus\{0\})$ we write it as (f_+, f_-) where $f_+ = f|\Omega_+$ and $f_- = f|\Omega_-$. We say that f admits a boundary value in the sense of distributions if both f_+ and f_- admit boundary values in the sense of distributions.

It follows from ([S], Theorem XIII, page 74) that the mapping which to each $\varphi \in \mathscr{D}(]a, b[)$ assigns $\lim_{\varepsilon \to 0+} \int_a^b f(x + i\varepsilon)\varphi(x)\,dx \left(\text{resp. } \lim_{\varepsilon \to 0+} \int_a^b f(x - i\varepsilon)\varphi(x)\,dx\right)$ is a distribution which we denote $b_+(f)$ (resp. $b_-(f)$). If $f = (f_+, f_-)$ admits boundary values, we denote by $b(f)$ the distribution

$$b(f) := b_+(f) - b_-(f).$$

The distributions $b(f)$, $b_+(f)$, and $b_-(f)$ are called the ***boundary values of*** f, f_+, and f_-, respectively.

We keep the sets Ω_+, Ω_-, and Ω, fixed throughout the remainder of this section.

We are going to show that a necessary and sufficient condition for a holomorphic function to admit boundary values in the sense of distributions is that it is of "slow growth" in the following sense.

3.6.11. Definition. A function f holomorphic in Ω_+ (resp. Ω_-, or Ω) is said to be of ***slow growth*** if, for every compact subset K of $]a, b[$, there exist an integer k and two positive constants ε, C such that

$$|f(z)| \le \frac{C}{|\operatorname{Im} z|^k} \quad \text{for} \quad \operatorname{Re} z \in K,\ 0 < |\operatorname{Im} z| \le \varepsilon.$$

Let us denote by $\mathscr{H}_b(\Omega_+)$, $\mathscr{H}_b(\Omega_-)$, and $\mathscr{H}_b(\Omega)$ the respective spaces of functions.

3.6.12. Proposition. *If f is a holomorphic function of slow growth in Ω_+, it admits a boundary value in the sense of distributions.*

PROOF. Let $K = [\alpha, \beta]$ be a compact subinterval of $]a, b[$, k, $c > 0$ and $\varepsilon > 0$ as in §3.6.11, and $z_0 = x_0 + iy_0$ a fixed point in Ω_+. Let us denote by $\gamma_0, \gamma_1, \gamma_2 \cdots$ the successive primitives of f in Ω_+ vanishing at z_0.

$$\gamma_0(z) := \int_{z_0}^{z} f(w)\,dw, \qquad \gamma_1(z) := \int_{z_0}^{z} \gamma_0(w)\,dw, \ldots.$$

One can show by recurrence that there are positive constants $C_0, \ldots, C_k$ and C_k' such that, for $z = x + iy$, $x \in K$, $0 < y \le \varepsilon$,

$$|\gamma_0(z)| \le C_0/y^{k-1}, \ldots, |\gamma_{k-2}(z)| \le C_{k-2}/y$$

$$|\gamma_{k-1}(z)| \le C_{k-1}|\log y|, \qquad |\gamma_k(z)| \le C_k y |\log y| + C_k'.$$

Since γ_k is bounded, it follows that γ_{k+1} can be extended as a continuous function to $[\alpha, \beta] + i[0, c[$. Therefore the family of distributions $T_y \in \mathscr{D}'(]\alpha, \beta[)$

defined by

$$\langle T_y, \varphi \rangle := \int_\alpha^\beta \gamma_{k+1}(x+iy)\varphi(x)\,dx, \qquad 0 < y \le \varepsilon$$

admits a limit T_0 in the sense of distributions when $y \to 0+$. From

$$\frac{d^{k+2}}{dx^{k+2}}\gamma_{k+1} = \frac{d^{k+2}}{dz^{k+2}}\gamma_{k+1} = f$$

one obtains

$$\left\langle \frac{d^{k+2}}{dx^{k+2}} T_y, \varphi \right\rangle = \int_\alpha^\beta f(x+iy)\varphi(x)\,dx$$

which has the distribution $\frac{d^{k+2}}{dx^{k+2}} T_0$ as a limit, when $y \to 0+$. Hence f admits this last distribution as boundary value in the sense of $\mathscr{D}'(]\alpha,\beta[)$. Since the subinterval $K = [\alpha,\beta]$ was arbitrary and a distribution is determined by its values locally, then f admits a boundary value $b_+(f)$ in the sense of distributions. □

Under the conditions of Proposition 3.6.12, let us denote by $\tilde{\gamma}_{k+1}$ the function (and its associated distribution) defined in $[\alpha,\beta] + i[-\infty, c[$ as follows:

$$\tilde{\gamma}_{k+1}(x+iy) = \begin{cases} \gamma_{k+1}(x+iy) & \text{if } 0 < y < c \\ \gamma_{k+1}(x+i0) = \lim\limits_{y\to 0+} \gamma_{k+1}(x+iy) & \text{if } y = 0 \\ 0 & \text{if } y < 0. \end{cases}$$

3.6.13. Proposition. *Let f satisfy the hypotheses of the previous proposition. For a compact subinterval $[\alpha,\beta]$ of $]a,b[$ define a distribution $f_{\alpha,\beta}$ in $\mathscr{D}'(]\alpha,\beta[+ i]-\infty,c[)$ by*

$$f_{\alpha,\beta} := \frac{\partial^{k+2}}{\partial x^{k+2}}\tilde{\gamma}_{k+1},$$

with $\tilde{\gamma}_{k+1}$ as defined earlier. Then we have the following identity:

$$\frac{\partial}{\partial \bar{z}} f_{\alpha,\beta} = -\frac{1}{2i}(b_+(f)|]\alpha,\beta[) \otimes \delta_0(y),$$

in the sense of distributions.

Proof. We keep the notation from the previous proposition and let $\varphi \in \mathscr{D}(]\alpha,\beta[+ i]-\infty,c[)$. Denote $\tilde{f} := f_{\alpha,\beta}$ and $\tilde{\gamma} := \tilde{\gamma}_{k+1}$ for simplicity. We have

$$\left\langle \frac{\partial \tilde{f}}{\partial \bar{z}}, \varphi \right\rangle = -\left\langle \tilde{f}, \frac{\partial \varphi}{\partial \bar{z}} \right\rangle = (-1)^{k+3} \left\langle \tilde{\gamma}, \frac{\partial^{k+3}}{\partial \bar{z} \partial x^{k+2}} \varphi \right\rangle$$

$$= \frac{(-1)^{k+3}}{2i} \int_{\alpha}^{\beta} \int_{0}^{c} \tilde{\gamma} \frac{\partial^{k+3}}{\partial x^{k+2} \partial \bar{z}} \varphi \, d\bar{z} \wedge dz$$

$$= \lim_{\varepsilon \to 0+} \frac{(-1)^{k+3}}{2i} \int_{\alpha}^{\beta} \int_{\varepsilon}^{c} \gamma_{k+1} \frac{\partial^{k+3}}{\partial x^{k+2} \partial \bar{z}} \varphi \, d\bar{z} \wedge dz,$$

where the last step used the continuity of $\tilde{\gamma}_{k+1}$ up to the real axis. Note that the support of φ does not intersect three of the sides of the rectangle where the integration is taking place. Hence, using that

$$\gamma_{k+1} \frac{\partial^{k+3}}{\partial \bar{z} \partial x^{k+2}} \varphi \, d\bar{z} \wedge dz = d\left(\gamma_{k+1} \frac{\partial^{k+2}}{\partial x^{k+2}} \varphi \, dz \right),$$

and applying Stokes' formula, we are led to

$$\left\langle \frac{\partial \tilde{f}}{\partial \bar{z}}, \varphi \right\rangle = \lim_{\varepsilon \to 0+} \frac{(-1)^{k+3}}{2i} \int_{\alpha}^{\beta} \gamma_{k+1}(x + i\varepsilon) \frac{\partial^{k+2}}{\partial x^{k+2}} (x + i\varepsilon) \, dx$$

$$= \frac{(-1)^{k+3}}{2i} \int_{\alpha}^{\beta} \gamma_{k+1}(x + i0) \frac{\partial^{k+2} \varphi}{\partial x^{k+2}} (x + i0) \, dx$$

$$= \frac{(-1)^{k+3}}{2i} \left\langle T_0(x), \frac{\partial^{k+2} \varphi}{\partial x^{k+2}} (x + i0) \right\rangle$$

$$= -\frac{1}{2i} \langle b_+(f), \varphi(x + i0) \rangle = -\left\langle \frac{1}{2i} b_+(f) \otimes \delta_0(y), \varphi \right\rangle,$$

which is the desired formula. □

3.6.14. Remarks. (1) If f^* is another extension of f to a distribution in $]\alpha, \beta[+ i]-\infty, c]$ with $\operatorname{supp}(f^*) \subseteq]\alpha, \beta[+ i[0, c[$, which satisfies

$$\frac{\partial f^*}{\partial \bar{z}} = -\frac{1}{2i} (b_+(f) \otimes \delta_0(y)),$$

then $\frac{\partial}{\partial \bar{z}} (\tilde{f} - f^*) = 0$. Therefore, there is a holomorphic function h in $]\alpha, \beta[+ i]-\infty, c[$ such that $\tilde{f} - f^* = h$ in the sense of distributions, but this implies $h(z) = 0$ when $\operatorname{Im} z < 0$, therefore $h \equiv 0$ and $f^* = \tilde{f}$.

Note also that if $b_+(f) |]\alpha, \beta[= 0$, then $\tilde{f}$ is holomorphic in $]\alpha, \beta[+ i]-\infty, c[$ and zero in the lower half-plane, hence zero throughout. This implies $f \equiv 0$ in Ω_+.

(2) If we start with $f \in \mathscr{H}(\Omega_-)$ of slow growth, we obtain

$$\frac{\partial \tilde{f}}{\partial \bar{z}} = \frac{1}{2i} b_-(f_-) \otimes \delta_0(y).$$

3.6.15. Proposition. *Let $\mathfrak{E}$ be the space of those distributions in $]a, b[+ i]-c, c[$ whose support lies in $]a, b[+ i\{0\}$. The quotient space $\mathfrak{E}\Big/\left(\frac{\partial}{\partial \bar{z}}\mathfrak{E}\right)$ is isomorphic to $\mathscr{D}'(]a, b[)$. Moreover, for any $T \in \mathfrak{E}$ there is a unique $R \in \mathscr{D}'(]a, b[)$ such that*

$$T \equiv R \otimes \delta_0(y) \bmod \frac{\partial}{\partial \bar{z}}\mathfrak{E}.$$

Proof. Let $a < x_0 < b$. For a sufficiently small open interval U_{x_0} about the point x_0, an element $T \in \mathfrak{E}$ can be written as a finite sum of the form

$$T = \sum_j T_j(x) \otimes \delta_0^{(j)}(y) \qquad \text{in } U_{x_0} + i]-c, c[,$$

where T_j are distributions acting on the variable x, and $\delta_0^{(j)}(y) = \frac{d^j}{dy^j}\delta_0(y)$ which are distributions acting on the y variable (see [S], Theorem XXXVI, p. 101). On the other hand, for $j \geq 1$,

$$\frac{\partial}{\partial \bar{z}}(T_j(x) \otimes \delta_0^{(j-1)}(y)) = \frac{1}{2}\frac{d}{dx}T_j(x) \otimes \delta_0^{(j-1)}(y) + \frac{i}{2}T_j(x) \otimes \delta_0^{(j)}(y).$$

Therefore,

$$T_j(x) \otimes \delta_0^{(j)}(y) \equiv i\frac{d}{dx}T_j(x) \otimes \delta_0^{(j-1)}(y) \bmod \frac{\partial}{\partial \bar{z}}\mathfrak{E}.$$

It follows that, at least in $U_{x_0} + i]-c, c[$,

$$T \equiv \left(\sum_j i^j T_j^{(j)}(x)\right) \otimes \delta_0(y) \bmod \frac{\partial}{\partial \bar{z}}\mathfrak{E}.$$

To obtain a global result we need the following lemma.

3.6.16. Lemma. *If a distribution of the form $T(x) \otimes \delta_0(y)$ in $\mathscr{D}'(]\alpha, \beta[+ i]-\gamma, \gamma[)$ can be written in the form $\frac{\partial U}{\partial \bar{z}}$, with $\operatorname{supp}(U)$ contained in the real axis, then $T = 0$.*

Proof. Locally U can be written in a unique way as $\sum_{i \leq j \leq n} U_j(x) \otimes \delta_0^{(j)}(y)$. Hence

$$\frac{\partial}{\partial \bar{z}}U = \frac{1}{2}\sum_{0 \leq j \leq n}\frac{d}{dx}U_j(x) \otimes \delta_0^{(j)}(y) + \frac{i}{2}\sum_{0 \leq j \leq n}U_j(x) \otimes \delta_0^{(j+1)}(y) = T(x) \otimes \delta_0(y).$$

From the uniqueness of the representation we conclude that

$$U_n = 0, \frac{dU_n}{dx} + iU_{n-1} = 0, \ldots, \frac{dU_1}{dx} + iU_0 = 0, T = \frac{1}{2}\frac{dU_0}{dx}.$$

Therefore, $T = 0$. □

Let us return to the proof of Proposition 3.6.15. We have found so far a covering $(U_k)_{k\geq 1}$ of $]a,b[$ by open intervals and distributions $R_k \in \mathscr{D}'(U_k)$, $S_k \in \mathscr{D}'(V_k)$ ($V_k = U_k + i]-c,c[$), $\operatorname{supp}(S_k) \subseteq U_k + i\{0\}$ such that

$$T = R_k \otimes \delta_0(y) + \frac{\partial}{\partial \bar{z}} S_k \quad \text{in} \quad V_k$$

$$(R_k - R_j) \otimes \delta_0(y) = \frac{\partial}{\partial \bar{z}}(S_j - S_k) \quad \text{in} \quad V_k \cap V_j.$$

By Lemma 3.6.16 we have $R_k = R_j$ in $U_k \cap U_j$, hence they define a distribution R in $]a,b[$. Introducing R into the preceding equations we obtain

$$T = R \otimes \delta_0(y) + \frac{\partial}{\partial \bar{z}} S_k = R \otimes \delta_0(y) + \frac{\partial}{\partial \bar{z}} S_j \quad \text{in} \quad V_k \cap V_j.$$

Hence $S_k - S_j$ is holomorphic in $V_k \cap V_j$, but it has support in the real axis. It follows that $S_k - S_j = 0$ in $V_k \cap V_j$ and the family $\{S_k\}$ defines a distribution $S \in \mathfrak{E}$ such that

$$T = R(x) \otimes \delta_0(y) + \frac{\partial}{\partial \bar{z}} S.$$

Moreover, by Lemma 3.6.16, we have that this equation determines R uniquely, hence the map $T \mapsto R$ is well defined from $\mathfrak{E}$ into $\mathscr{D}'(]a,b[)$. Its kernel is $\frac{\partial \mathfrak{E}}{\partial \bar{z}}$, also by Lemma 3.6.16. This concludes the proof of the proposition. □

We would like to show how every distribution in $\mathscr{D}'(]a,b[)$ can be obtained as boundary value of a function holomorphic in $\Omega = \Omega_+ \cup \Omega_-$. We start by showing this is the case for distributions in $\mathscr{E}'(]a,b[)$.

3.6.17. Definition. For a distribution $S \in \mathscr{E}'(\mathbb{C})$ we define its Cauchy transform $\hat{S}$ by

$$\hat{S} := \frac{1}{\pi z} * S.$$

This definition coincides with that from §2.1.5 in case S in a Radon measure.

For a distribution $T \in \mathscr{E}'(\mathbb{R})$, its Cauchy transform $\hat{T}$ is the distribution in $\mathbb{C}$ given by

$$\hat{T} := (T(x) \otimes \delta_0(y))^\wedge.$$

Note that these distributions are holomorphic in $\mathbb{C} \backslash \operatorname{supp} S$ and $\mathbb{C} \backslash \operatorname{supp} T$, respectively. Moreover they verify

$$\frac{\partial \hat{S}}{\partial \bar{z}} = S \qquad \text{and} \qquad \frac{\partial \hat{T}}{\partial \bar{z}} = T(x) \otimes \delta_0(y),$$

since $\dfrac{1}{\pi z}$ is a fundamental solution for $\dfrac{\partial}{\partial \bar{z}}$, and S (resp. $T \otimes \delta_0(y)$) has compact support.

3.6.18. Lemma. *For every $T \in \mathscr{E}'(\mathbb{R})$, the Cauchy transform $\hat{T}$ has $-2iT$ as boundary value in the sense of distributions.*

Proof. Let us observe first that, T being of finite order, say n, $\hat{T}$ is of slow growth. In fact, there are positive constants C and R such that

$$|\langle T, \varphi \rangle| \le C \sup_{x \in [-R,R]} \sup_{0 \le j \le n} \left| \frac{d^j \varphi}{dx^j}(x) \right|.$$

Therefore, if $z \notin [-R, R]$ we have

$$|\hat{T}(z)| = \left| \left\langle T(x), \frac{1}{\pi} \frac{1}{(z-x)} \right\rangle \right| \le \frac{C}{\pi} \sup_{x \in [-R,R]} \sup_{0 \le j \le n} \left| \frac{j!}{(z-x)^{j+1}} \right|.$$

It follows that, for a conveniently chosen constant $C_1 > 0$, we have

$$|\hat{T}(z)| \le \frac{C_1}{|\operatorname{Im} z|^{n+1}} \qquad \text{if } \operatorname{Im} z \ne 0.$$

Let us denote by f_+ and f_-, respectively, the restrictions $\hat{T}|\{\operatorname{Im} z > 0\}$ and $\hat{T}|\{\operatorname{Im} z < 0\}$, and by $\tilde{f}_+, \tilde{f}_-$ the respective extensions to $\mathscr{D}'(\mathbb{C})$ with support in $\{\operatorname{Im} z \ge 0\}$ and $\{\operatorname{Im} z \le 0\}$, respectively. Their existence is a consequence of §3.6.12. We know that

$$\frac{\partial \tilde{f}_+}{\partial \bar{z}} = -\frac{1}{2i} b_+(f_+) \otimes \delta_0(y), \qquad \frac{\partial \tilde{f}_-}{\partial \bar{z}} = \frac{1}{2i} b_-(f_-) \otimes \delta_0(y)$$

and

$$\frac{\partial \hat{T}}{\partial \bar{z}} = T \otimes \delta_0(y).$$

Therefore, $\hat{T} - (\tilde{f}_+ + \tilde{f}_-)$ is a distribution in $\mathbb{C}$ with support in the real axis which satisfies

$$\frac{\partial}{\partial \bar{z}}(\hat{T} - (\tilde{f}_+ + \tilde{f}_-)) = \left(T - \frac{1}{2i}(-b_+(f_+) + b_-(f_-)) \right) \otimes \delta_0(y).$$

By Lemma 3.6.16 we have

$$T = -\frac{1}{2i}(b_+(f_+) - b_-(f_-)),$$

This is precisely the statement $T = -\dfrac{1}{2i} b(\hat{T})$. □

3.6.19. Theorem. *Every distribution in $\mathscr{D}'(]a,b[)$ is a boundary value of a holomorphic function of slow growth in Ω.*

PROOF. Recall that $\Omega = (]a,b[+i]-c,0[) \cup (]a,b[+i]0,c[) = \Omega_+ \cup \Omega_-$. Let $\tilde{\Omega} =]a,b[+i]-c,c[$. Let now $K_n = [a_n, b_n]$ be a nested sequence of intervals with $a_{n+1} < a_n < b_n < b_{n+1}$, $a_n \to a$, $b_n \to b$. Choose a sequence $\varphi_n \in \mathscr{D}(\mathbb{R})$, $0 \leq \varphi_n \leq 1$, $\varphi_n \equiv 1$ in a neighborhood of K_n and $\operatorname{supp} \varphi_n \subseteq]a_{n+1}, b_{n+1}[$. Finally, choose a sequence of positive numbers c_n, $c_n \nearrow c$.

Define $T_1 := \varphi_1 T$, $T_n := (\varphi_n - \varphi_{n-1})T$ for $n \geq 2$. We have $T = \sum_{n \geq 1} T_n$. For every $n \geq 1$ the function

$$\hat{T}_n := (T_n(x) \otimes \delta_0(y))^\wedge$$

is holomorphic outside $\operatorname{supp}(T_n) + i\{0\}$. In particular, for $n \geq 2$, $\hat{T}_n$ is holomorphic in a neighborhood of the closed rectangle

$$L_{n-1} := \{z \in \mathbb{C} : a_{n-1} \leq \operatorname{Re} z \leq b_{n-1}, -c_{n-1} \leq \operatorname{Im} z \leq c_{n-1}\}.$$

Since this set is convex, it is holomorphically convex in $\mathbb{C}$, therefore we can find polynomials h_n which satisfy

$$\sup_{z \in L_{n-1}} |\hat{T}_n(z) - h_n(z)| \leq \frac{1}{2^n} \qquad (n \geq 2).$$

Therefore, the series

$$S := \hat{T}_1 + \sum_{n \geq 2} (\hat{T}_n - h_n)$$

converges in $\mathscr{D}'(\tilde{\Omega})$ and its sum S defines a holomorphic function of slow growth in Ω. Furthermore,

$$\frac{\partial S}{\partial \bar{z}} = \frac{\partial \hat{T}_1}{\partial \bar{z}} + \sum_{n \geq 2} \frac{\partial \hat{T}_n}{\partial \bar{z}} = \left(\sum_{n \geq 1} T_n\right) \otimes \delta_0(y) = T \otimes \delta_0(y).$$

For every relatively compact subinterval $]\alpha, \beta[$ of $[a,b]$ we have that if $\tilde{S}_+, \tilde{S}_-$ denote the extensions of $S|(]\alpha,\beta[\times \{\operatorname{Im} z > 0\})$ and $S|(]\alpha,\beta[\times \{\operatorname{Im} z < 0\})$ with respective supports in $\{\operatorname{Im} z \geq 0\}$ and $\{\operatorname{Im} z \leq 0\}$, then

$$\frac{\partial \tilde{S}_+}{\partial \bar{z}} = -\frac{1}{2i}(b_+(S_+) \otimes \delta_0(y)), \qquad \frac{\partial \tilde{S}_-}{\partial \bar{z}} = \frac{1}{2i}(b_-(S_-) \otimes \delta_0(y)).$$

Hence

$$\frac{\partial}{\partial \bar{z}}((S|(]\alpha,\beta[+i]-c,c[)) - (\tilde{S}_+ + \tilde{S}_-))$$

$$= \left((T|]\alpha,\beta[) + \frac{1}{2i}(b_+(S_+) - b_-(S_-)\right) \otimes \delta_0(y)$$

and one concludes, as before, that

$$T = -\frac{1}{2i} b(S). \qquad \square$$

We would like now to prove the converse of Proposition 3.6.12, that is, every holomorphic function in Ω_+ which admits boundary values in the sense of distributions must be of slow growth.

3.6.20. Lemma. *Let $f \in \mathscr{H}(\Omega_+)$ having $T \in \mathscr{D}'(]a, b[)$ as a boundary value. Then, for every $\varphi \in \mathscr{D}(]a, b[+ i]-\infty, c[)$, the function $g : [0, c[\to \mathbb{C}$ defined by*

$$g(\varepsilon) := \begin{cases} \displaystyle\int_a^b f(x + i\varepsilon)\varphi(x + i\varepsilon)\, dx & \text{for } 0 < \varepsilon < c \\ \langle T, \varphi(x + i0)\rangle & \text{for } \varepsilon = 0 \end{cases}$$

is continuous.

PROOF. The proof is left to the reader. $\square$

3.6.21. Lemma. *Under the same hypotheses as in Lemma 3.6.20, the integral*

$$S(\varphi) := \int_0^c \left(\int_a^b f(x + i\varepsilon)\varphi(x + i\varepsilon)\, dx \right) d\varepsilon$$

exists and $S : \varphi \mapsto S(\varphi)$ defines a distribution in the half-strip $]a, b[+ i]-\infty, c[$, with support in $]a, b[+ i[0, c[$, which extends f to the half-strip.

PROOF. The existence of $S(\varphi)$ follows from Lemma 3.6.20. From [S], (Theorem XIII, p. 74), we conclude that S is a distribution. $\square$

3.6.22. Theorem. *Under the same hypotheses of Lemma 3.6.20 we conclude that f has slow growth and*

$$\frac{\partial S}{\partial \bar{z}} = -\frac{1}{2i} b_+(f) \otimes \delta_0(y) = -\frac{1}{2i}(T \otimes \delta_0(y)),$$

with S defined by Lemma 3.6.21.

PROOF. Let $K = [\alpha, \beta] \subset\subset]a, b[\subseteq \mathbb{R}$ and $0 < \varepsilon < \inf(b - \beta, \alpha - a)$. Let ψ be a standard radial function in $\mathbb{C}$ with support in $B(0, 1)$, $\psi_\delta(z) = \delta^{-2}\psi(z/\delta)$ ($\delta > 0$). For $x_0 \in K$ fixed, denote by $\varphi_\delta(\zeta)$ the function $\psi_\delta(\zeta - (x_0 + i\delta))$. Since S has a finite order n in the compact set $F := \{z \in \mathbb{C} : d(z, K) \leq \varepsilon\}$, there is a positive constant C such that if $\varphi \in \mathscr{D}(\mathbb{C})$ has support in $\mathring{F} \cap \{\operatorname{Im} z > 0\}$ then

$$|\langle S, \varphi\rangle| = \left| \int \varphi f\, dm \right| \leq C \sup_{\zeta \in F} \sup_{0 \leq p+q \leq n} \left| \frac{\partial^{p+q}\varphi}{\partial z^p \partial \bar{z}^q}(\zeta) \right|.$$

On the other hand

$$\frac{\partial^{p+q}\varphi_\delta}{\partial z^p \partial \bar{z}^q}(\zeta) = \frac{1}{\delta^{p+q+2}} \frac{\partial^{p+q}\psi}{\partial z^p \partial \bar{z}^q}\left(\frac{\zeta - (x_0 + i\delta)}{\delta}\right).$$

Therefore, there is a constant $C_K > 0$ such that if $0 < \delta \leq \varepsilon$,

$$\left| \int \varphi_\delta(\zeta) f(\zeta)\, dm(\zeta) \right| \leq C_K \delta^{-n-2}.$$

By the mean value property of holomorphic functions the left-hand side is precisely $|f(x_0 + i\delta)|$. In other words,

$$|f(x_0 + i\delta)| \leq C_K \delta^{-n-2}, \quad x_0 \in K, \quad 0 < \delta \leq \varepsilon. \qquad \square$$

To close this section we obtain a generalization of the Schwarz Reflection Principle 2.1.11 known as the edge-of-the-wedge theorem.

3.6.23. Theorem (Edge-of-the-Wedge). *Let $f = (\tilde{f}_+, \tilde{f}_-)$ be a holomorphic function in $\Omega_+ \cup \Omega_-$ admitting boundary values $b_+(f_+)$, $b_-(f_-)$ which coincide, i.e., $b(f) = 0$. Then f has a holomorphic extension $\tilde{f}$ to the connected set $\tilde{\Omega}$.*

Proof. Let $S = \tilde{f}_+ + \tilde{f}_-$ defined as earlier in $\tilde{\Omega}$. We have

$$\frac{\partial}{\partial \bar{z}} S = -\frac{1}{2i} b(f) \otimes \delta_0(y) = 0.$$

Therefore S is a holomorphic function in $\tilde{\Omega}$ which coincides with f_+ in Ω_+ and f_- in Ω_-. $\square$

3.6.24. Corollary. *Let Ω and $\tilde{\Omega}$ be as earlier, then the map b induces an isomorphism*

$$\tilde{b} : \mathscr{H}_b(\Omega)/\mathscr{H}(\Omega) \to \mathscr{D}'(]a, b[).$$

Exercises 3.6

1. Let $T \in \mathscr{E}'(\mathbb{R})$. Show that for $|z|$ sufficiently large, we have

$$\hat{T}(z) = \frac{a_1}{z} + \frac{a_2}{z^2} + \cdots.$$

Identify the coefficients. Is this expansion still true when $T \in \mathscr{H}'(\mathbb{C})$?

2. Show that if $f \in \mathscr{H}_b(\mathbb{C})$, then $f' \in \mathscr{H}_b(\mathbb{C})$. What can you say about $f \in \mathscr{H}_b(\mathbb{C})$ if $b(f^{(n)}) = 0$ for some $n \in \mathbb{N}$? What about the case $f = \hat{T}$, $T \in \mathscr{D}'(\mathbb{R})$?

3. Let φ be a continuous function on the real axis such that for some $\alpha > 0$ it satisfies the estimate $|\varphi(x)| = O(|x|^{-\alpha})$ when $x \to \pm\infty$. Show directly that $\phi(z) = \frac{1}{\pi z} * (\varphi(x) \otimes \delta_0(y))$ is well defined and $b(\phi) = -2i\varphi$. In fact

$$\lim_{\varepsilon \to 0+} (\phi(x + i\varepsilon) - \phi(x + i\varepsilon)) = -2i\varphi(x)$$

locally uniformly. (Remark: Essentially the same statement is valid if we replace the continuity and decay conditions on φ by $\varphi \in L^1(\mathbb{R}, dx)$ or $\varphi \in L^2(\mathbb{R}, dx)$.)

4. Let $\varphi \in L^2(\mathbb{R}, dx)$ and $g(\xi) = \mathscr{F}\varphi(\xi) = \int_{-\infty}^{\infty} \varphi(x)e^{-ix\xi}\,d\xi$ its Fourier transform. Show that the Cauchy transform $\hat{g}$ of g is given by

$$\hat{g}(z) = \begin{cases} -2i\displaystyle\int_{-\infty}^{0} e^{-itz}\varphi(t)\,dt & \text{if } \operatorname{Im} z > 0 \\ 2i\displaystyle\int_{0}^{\infty} e^{-itz}\varphi(t)\,dt & \text{if } \operatorname{Im} z < 0. \end{cases}$$

The same result holds if we assume φ, g are both L^1 functions.

5. The Heaviside function H is given by $H := \chi_{[0,\infty)}$. It satisfies $\frac{d}{dx}H = \delta_0 = \delta$ in the sense of distributions on $\mathbb{R}$. We also know that $\hat{\delta}(z) = \frac{1}{\pi z}$. Find $f \in \mathscr{H}_b(\mathbb{C})$ such that $b(f) = H$.

6. Let $f(z) = (\operatorname{Log} z)^2$ (as a holomorphic function in $\mathbb{C}\setminus\mathbb{R}$). Find $b(f)$.

7. The principal value of $\frac{1}{x}$, $pv\left(\frac{1}{x}\right)$ is the distribution in $\mathscr{D}'(\mathbb{R})$ defined by

$$\left\langle pv\left(\frac{1}{x}\right), \varphi \right\rangle = \lim_{\varepsilon \to 0+} \int_{|x|>\varepsilon} \frac{\varphi(x)}{x}\,dx = \int_0^{\infty} \frac{\varphi(x) - \varphi(-x)}{x}\,dx.$$

Show that if

$$f(z) = \begin{cases} \dfrac{1}{2z} & \text{for } \operatorname{Im} z > 0 \\ -\dfrac{1}{2z} & \text{for } \operatorname{Im} z < 0 \end{cases}$$

then $b(f) = pv\left(\frac{1}{x}\right)$. Use this to show that $\left(pv\left(\frac{1}{x}\right)\right)^{\wedge}(z) = -2if(z)$.

8. The distribution $(x + i0)^{-1}$ is defined as $b(g)$, where g is the function given by

$$g(z) = \begin{cases} \dfrac{1}{z} & \text{if } \operatorname{Im} z > 0 \\ 0 & \text{if } \operatorname{Im} z < 0. \end{cases}$$

Show that

$$(x + i0)^{-1} = pv\left(\frac{1}{x}\right) - i\pi\delta.$$

9. Let Γ be a C^1-regular Jordan arc with endpoints a and b, and let f be a function on Γ that satisfies a Hölder condition of order μ, $0 < \mu \leq 1$ (i.e.,

$|f(z) - f(\zeta)| \le M|z - \zeta|^\mu$ for any pair $z, \zeta \in \Gamma$). For any $z_0 \in \Gamma\setminus\{a, b\}$ define the value $F(z_0)$ as follows:

$$F(z_0) := pv\frac{1}{2\pi i}\int_\Gamma \frac{f(z)}{z - z_0}\,dz := \lim_{\varepsilon\to 0+}\frac{1}{2\pi i}\int_{\Gamma_\varepsilon}\frac{f(z)}{z - z_0}\,dz,$$

where $\Gamma_\varepsilon = \Gamma\setminus B(z_0, \varepsilon)$. We want to show this value $F(z_0)$ is well defined.

(a) Show that

$$\int_{\Gamma_\varepsilon}\frac{f(z)}{z - z_0}\,dz = \int_\Gamma \frac{f(z) - f(z_0)}{z - z_0}\,dz + f(z_0)\,\mathrm{Log}\frac{b - z_0}{a - z_0} + i\pi f(z_0) + O(\varepsilon),$$

so that

$$F(z_0) = \frac{1}{2\pi i}\int_\Gamma \frac{f(z) - f(z_0)}{z - z_0}\,dz + \frac{1}{2\pi i}f(z_0)\,\mathrm{Log}\frac{b - z_0}{a - z_0} + \frac{f(z_0)}{2}.$$

If Γ is closed we can drop the logarithmic term.

(b) For Γ a C^1 Jordan curve, f as earlier, let Φ be $\left(\frac{i}{2}\right)$ times the Cauchy transform of the Radon measure $\mu = f(z)\,dz$, with support in Γ, that is,

$$\Phi(z) = \frac{1}{2\pi i}\int_\Gamma \frac{f(\zeta)}{\zeta - z}\,d\zeta, \qquad z \notin \Gamma.$$

For $z_0 \in \Gamma$, let

$$\Phi^+(z_0) := \lim_{\substack{z\to z_0 \\ z\in \mathrm{Int}(\Gamma)}} \Phi(z),$$

$$\Phi^-(z_0) := \lim_{\substack{z\to z_0 \\ z\in \mathrm{Ext}(\Gamma)}} \Phi(z),$$

where the limits are understood to be along the normal direction to Γ at z_0. Show that these limits exist and

$$\Phi^+(z_0) = F(z_0) + \tfrac{1}{2}f(z_0)$$

$$\Phi^-(z_0) = F(z_0) - \tfrac{1}{2}f(z_0).$$

Finally,

$$\Phi^+(z_0) - \Phi^-(z_0) = f(z_0).$$

These formulas are usually called the Plemelj-Sokhotski formulas.

10. (a) Let f be a holomorphic function in the upper half-plane $H = \{\mathrm{Im}\, z > 0\}$ satisfying the inequality $|f(z)| \le \dfrac{1}{\mathrm{Im}\, z}$. Show that $g(z) = \displaystyle\int_i^z (z - w)f(w)\,dw$ satisfies the upper bound $|g(z)| \le C(1 + |z|)^2$ for some constant C ($z \in H$).

(b) State and prove a similar result when $|f(z)| \le 1/(\mathrm{Im}\, z)^\alpha$ for some $\alpha > 0$.

(c) Use part (a) to show that an entire function satisfying $|f(z)| \le \dfrac{1}{|\mathrm{Im}\, z|}$ everywhere must be identically zero. Is the result true when $1/|\mathrm{Im}\, z|$ is replaced by $1/|\mathrm{Im}\, z|^\alpha$, $\alpha > 0$?

§7. Mergelyan's Theorem

Runge's theorem shows that if K is a compact subset of the open set $\Omega \subseteq \mathbb{C}$, and $K = \hat{K}_\Omega$ then $\mathscr{H}(K) \subseteq \overline{\mathscr{H}(\Omega)|K}$ (the closure in $\mathscr{C}(K)$). It is a natural question to try to figure out exactly which continuous functions on K belong to $\overline{\mathscr{H}(\Omega)|K}$. The following subalgebra of $\mathscr{C}(K)$ is worth considering:

$$A(K) := \{f \in \mathscr{C}(K) : f \text{ is holomorphic in } \mathring{K}\}.$$

$A(K)$ can be defined even if K is not holomorphically convex in Ω or compact in $\mathbb{C}$. Furthermore, note that if $\mathring{K} = \varnothing$ then $A(K) = \mathscr{C}(K)$. It is clear that

$$\overline{\mathscr{H}(\Omega)|K} \subseteq A(K).$$

It is not always true that both coincide; Mergelyan's theorem concerns a particularly interesting case, $\Omega = \mathbb{C}$.

3.7.1. Theorem (Mergelyan). *Let K be a compact set which is holomorphically convex in $\mathbb{C}$, then*

$$A(K) = \overline{\mathscr{H}(\mathbb{C})|K}.$$

In other words, if K is a compact set such that K^c is connected, then any function in $A(K)$ can be uniformly approximated by polynomials.

The original proof of Mergelyan, which we follow, is another nice example of the applications of Pompeiu's Formula 2.1.2. Let us first sketch the idea in a particular case. Assume that $K = \mathring{K}$, $m(\partial K) = 0$, and $f \in A(K) \cap \mathscr{D}_1(\mathbb{C})$. Then we have

$$f(z) = \frac{1}{2\pi i} \int_{\mathbb{C}} \frac{\partial f}{\partial \bar{\zeta}}(\zeta) \frac{d\zeta \wedge d\bar{\zeta}}{\zeta - z} = \frac{1}{2\pi i} \int_{K_1} \frac{\partial f}{\partial \bar{\zeta}}(\zeta) \frac{d\zeta \wedge d\bar{\zeta}}{\zeta - z},$$

where $K_1 = (\operatorname{supp} f) \backslash \mathring{K}$. For each z fixed, the function $\dfrac{\partial f}{\partial \bar{\zeta}} \dfrac{1}{\zeta - z}$ is integrable over the compact set K_1. Moreover, it is easy to convince ourselves that given $\varepsilon > 0$ there is a $\delta > 0$ such that for any $z \in K$ and any Borel set E Lebesgue of measure $m(E) \le \delta$ one has

$$\left| \frac{1}{2\pi i} \int_{E \cap K_1} \frac{\partial f}{\partial \bar{\zeta}}(\zeta) \frac{d\zeta \wedge d\bar{\zeta}}{\zeta - z} \right| \le \varepsilon.$$

Now, cover ∂K by a finite collection of small open disks B_j so that if $E = \bigcup_j B_j$ then $m(E) \le \delta$ and $\operatorname{dist}(K, K_1 \backslash E) > 0$. The function g defined as

$$g(z) = \frac{1}{2\pi i} \int_{K_1 \backslash E} \frac{\partial f}{\partial \bar{\zeta}}(\zeta) \frac{d\zeta \wedge d\bar{\zeta}}{\zeta - z}$$

is then holomorphic in a neighborhood of K and

$$\|f - g\|_K \le \varepsilon.$$

By Runge's theorem, there is a polynomial h such that $\|g - h\|_K \leq \varepsilon$, hence

$$\|f - h\|_K \leq \varepsilon + \|g - h\|_K \leq 2\varepsilon.$$

Of course, if one wants to extend this proof, there are two problems. We have to approximate functions in $A(K)$. They are only continuous and, furthermore, it is not necessarily true that $K = \mathring{K}$ and that $m(\partial K) = 0$ in general. The second condition has been dispensed with in Exercise 3.1.10. Theorem 3.7.1 does not require $\bar{\mathring{K}} = K$ either.

Before proceeding to the proof of the general case, let us make a few observations. First, given a function $f \in \mathscr{C}(K)$ we can always assume $f \in \mathscr{D}_0(\mathbb{C})$ (i.e., continuous and with compact support). The reason is that the Tietze-Urisohn theorem [Arm] guarantees the existence of a continuous extension to the whole plane. Then we multiply this extended function by a cutoff $\chi \in \mathscr{D}(\mathbb{C})$, $\chi = 1$ on a neighborhood of K.

Second, a function $f \in \mathscr{D}_0(\mathbb{C})$ is uniformly continuous, hence its modulus of continuity ω,

$$\omega(\delta) := \max\{|f(z) - f(w)| : |z - w| \leq \delta\} \qquad (\delta > 0),$$

has the property that $\omega(\delta) \to 0$ as $\delta \to 0$.

Thirdly, assume $f \in A(\bar{B}(z_0, \delta))$ and that $k(z) = k(|z|) \in \mathscr{D}_0(\bar{B}(0, \delta))$ satisfies

$$\int_{\mathbb{C}} k(z)\, dm(z) = 2\pi \int_0^\delta rk(r)\, dr = 1.$$

Then, as an easy consequence of Cauchy's formula, we see that

$$f(z_0) = \int_{\mathbb{C}} f(z_0 - z)k(z)\, dm(z).$$

We have already used this kind of argument in §2.2.9. This mean value property will be exploited in full in the next chapter.

We are now ready to start the proof.

Proof of Theorem 3.7.1. We want to approximate a function $f \in A(K)$ uniformly on K by polynomials. We know that we can assume that $f \in \mathscr{H}(\mathring{K}) \cap \mathscr{D}_0(\mathbb{C})$. Let ω be its modulus of continuity. We will first try to approximate f by a function φ_δ in $\mathscr{D}_1(\mathbb{C})$ which is holomorphic in a "large" part of $\mathring{K}$. For that purpose, consider the auxiliary radial function $k \in \mathscr{D}_1(\mathbb{C})$ given by $k(z) = k(|z|)$, where

$$k(r) = \begin{cases} \dfrac{3}{\pi}(1 - r^2)^2 & \text{if } 0 \leq r \leq 1 \\ 0 & \text{if } r > 1. \end{cases}$$

It is immediate that $\int_{\mathbb{C}} k(z)\, dm(z) = 1$. For any fixed δ, $0 < \delta < 1$, let

$k_\delta(z) = \delta^{-2}k(z/\delta)$. Then $\operatorname{supp} k_\delta = \bar{B}(0,\delta)$, $\displaystyle\int_{\mathbb{C}} k_\delta\, dm = 1$. We define now the approximation φ_δ by

$$\varphi_\delta(z) = \int_{\mathbb{C}} f(z-\zeta)k_\delta(\zeta)\, dm(\zeta),$$

as we did in §2.2.9. It follows that $\varphi_\delta \in \mathscr{D}_1(\mathbb{C})$ and

$$\frac{\partial}{\partial \bar{z}}\varphi_\delta(z) = -\int_{\mathbb{C}} f(z-\zeta)\frac{\partial}{\partial \bar{\zeta}}k_\delta(\zeta)\, dm(\zeta).$$

Let $\Omega_\delta = \{z \in K : d(z, K^c) > \delta\}$. Then $\Omega_\delta \subseteq \mathring{K}$ and $\varphi_\delta = f$ in Ω_δ by the previous remarks. Hence $\operatorname{supp}\left(\dfrac{\partial}{\partial \bar{z}}\varphi_\delta\right) \cap \Omega_\delta = \emptyset$. Moreover, we have

$$|\varphi_\delta(z) - f(z)| = \left|\int_{\mathbb{C}} (f(z-\zeta) - f(z))k_\delta(\zeta)\, dm(\zeta)\right| \le \omega(\delta)\int_{\mathbb{C}} k_\delta(\zeta)\, dm(\zeta) = \omega(\delta),$$

so that φ_δ is in fact an approximation of f.

To continue the proof we need to obtain an upper bound for $\dfrac{\partial}{\partial \bar{z}}\varphi_\delta$. Since k_δ has compact support, it follows from Stokes' theorem that

$$\int_{\mathbb{C}} \frac{\partial}{\partial \bar{\zeta}}k_\delta(\zeta)\, dm(\zeta) = 0.$$

Therefore, the formula for $\dfrac{\partial}{\partial \bar{z}}\varphi_\delta$ can be rewritten as follows:

$$\begin{aligned}\frac{\partial}{\partial \bar{z}}\varphi_\delta(z) &= -\int_{\mathbb{C}} f(z-\zeta)\frac{\partial}{\partial \bar{\zeta}}k_\delta(\zeta)\, dm(\zeta)\\ &= -\int_{\mathbb{C}} (f(z-\zeta) - f(z))\frac{\partial}{\partial \bar{\zeta}}k_\delta(\zeta)\, dm(\zeta).\end{aligned}$$

A direct computation shows that if $r = |\zeta|$ then

$$\left|\frac{\partial}{\partial \bar{\zeta}}k_\delta(\zeta)\right| = \frac{1}{2}\frac{1}{\delta^3}\left|\frac{dk}{dr}\left(\frac{r}{\delta}\right)\right| = \frac{6}{\pi\delta^3}\left(1 - \frac{r^2}{\delta^2}\right)\frac{r}{\delta}$$

when $0 \le r \le \delta$ and zero otherwise. Hence

$$\left|\frac{\partial}{\partial \bar{z}}\varphi_\delta(z)\right| \le \frac{8}{5}\frac{\omega(\delta)}{\delta} \le 2\frac{\omega(\delta)}{\delta}.$$

We will use this estimate and Pompeiu's formula to approximate φ_δ by functions in $\mathscr{H}(K)$. The crucial step is the following lemma, which in turn depends on Proposition 2.7.9, that is, on Koebe's one-quarter theorem.

3.7.2. Lemma. *Let E be a compact connected set of diameter greater than or equal to $r > 0$, E^c connected, and B an open disk of radius r such that $E \subseteq B$. Then there is a function $Q \in \mathscr{H}(S^2 \backslash E)$ and $\beta \in \mathbb{C}$ such that*

$$R(\zeta, z) = Q(z) + (\zeta - \beta)(Q(z))^2$$

satisfies the inequalities

$$|R(\zeta, z)| \leq \frac{c_1}{r}$$

and

$$\left| R(\zeta, z) - \frac{1}{z - \zeta} \right| \leq \frac{c_2 r^2}{|\zeta - z|^3},$$

for all $z \in E^c$ and $\zeta \in B$. Here c_1, c_2 are two absolute constants, i.e., positive numbers independent of r, E, etc.

Proof of Lemma 3.7.2. It is clear that if we translate the z and ζ variables by the same quantity the inequalities do not change and only the value β is affected. Therefore we can assume that $B = B(0, r)$. Recall that from Lemma 2.7.9 we have a biholomorphic mapping

$$F : B(0, 1) \to S^2 \backslash E$$

$$z = F(w) = \frac{a}{w} + b_0 + b_1 w + \cdots, \quad a \geq \frac{r}{4}.$$

Let $Q : S^2 \backslash E \to B\left(0, \frac{1}{a}\right)$, also a biholomorphic mapping, given by

$$w = Q(z) := \frac{1}{a} F^{-1}(z).$$

We have $Q(\infty) = 0$. Furthermore, from the definition of Q we have for $z = F(w)$

$$aQ(z) = aQ(F(w)) = F^{-1}(F(w)) = w.$$

Hence

$$zQ(z) = \frac{wF(w)}{a}.$$

It follows that

$$\lim_{z \to \infty} zQ(z) = \lim_{w \to 0} \frac{wF(w)}{a} = 1.$$

Note that, for any ζ fixed, we also obtain

$$\lim_{z \to \infty} (z - \zeta)Q(z) = 1.$$

From the hypotheses on E we have that $\{|z| \geq r\}$ is entirely contained in the simply connected region $S^2 \setminus E$ of the Riemann sphere. Therefore, the following expansion is valid

$$Q(z) = \frac{1}{z} + \frac{\beta}{z^2} + \cdots$$

in $\{|z| \geq r\}$, and

$$\beta = \frac{1}{2\pi i} \int_{|z|=\rho} zQ(z)\,dz, \qquad r \leq \rho < \infty.$$

We have $|Q(z)| \leq \frac{1}{a} \leq \frac{4}{r}$, hence $|\beta| \leq 4r$ (just take $\rho = r$).

We are now ready to prove the first estimate for $R(\zeta, z)$, $|\zeta| < r$, $z \in E^c$

$$|R(\zeta, z)| \leq |Q(z)| + (|\zeta| + |\beta|)|Q(z)|^2 \leq \frac{84}{r}.$$

In order to obtain the other estimate, fix $\zeta \in B$ and consider the Laurent expansion in $|z - \zeta| > 2r$

$$Q(z) = \frac{1}{z - \zeta} + \frac{c_{-2}}{(z - \zeta)^2} + \cdots,$$

$$c_{-2} = \frac{1}{2\pi i} \int_{|z|=3r} (z - \zeta)Q(z)\,dz = \frac{1}{2\pi i} \int_{|z|=3r} zQ(z)\,dz - \frac{\zeta}{2\pi i} \int_{|z|=3r} Q(z)\,dz$$
$$= \beta - \zeta,$$

using the expansion in powers of z^{-1}. Summarizing, in $|z - \zeta| > 2r$ we have

$$Q(z) = \frac{1}{z - \zeta} + \frac{\beta - \zeta}{(z - \zeta)^2} + O((z - \zeta)^{-3}),$$

and

$$Q^2(z) = \frac{1}{(z - \zeta)^2} + O((z - \zeta)^{-3}).$$

Hence for ζ fixed in B,

$$z \mapsto (z - \zeta)^3 \left[R(\zeta, z) - \frac{1}{z - \zeta} \right]$$

is certainly holomorphic in $|z - \zeta| > 2r$, including ∞, and the function between brackets is holomorphic in $E^c \setminus \{\zeta\}$. The possible pole at $z = \zeta$ is killed by the factor $(z - \zeta)^3$ and therefore we have constructed a function holomorphic in $S^2 \setminus E$. If we find an upper bound in $(S^2 \setminus E) \cap B$, the maximum principle will guarantee this is an upper bound everywhere in $S^2 \setminus E$. When $z \in B$ we have

$|z - \zeta| \le 2r$, then

$$\left|(z-\zeta)^3\left[R(\zeta,z) - \frac{1}{z-\zeta}\right]\right| \le |z-\zeta|^3|R(\zeta,r)| + |z-\zeta|^2$$

$$\le 8r^3 \times \frac{84}{r} + 4r^2 \le 676r^2.$$

This concludes the proof of the lemma; we can take $c_1 = 84$, $c_2 = 676$. □

We go back to the proof of Theorem 3.7.1. From Pompeiu's formula and previous analysis we have

$$\varphi_\delta(z) = \frac{1}{2\pi i}\int_{\Omega_\delta^c} \frac{\partial \varphi_\delta}{\partial \bar\zeta}(\zeta)\frac{d\zeta \wedge d\bar\zeta}{z-\zeta}.$$

From the definition of Ω_δ we can see immediately that there is a finite covering of $S := \operatorname{supp}\left(\frac{\partial \varphi_\delta}{\partial \bar\zeta}\right)$ by open disks $B_1, \ldots, B_m$ of radius 2δ and centers lying in K^c. Let $S_1, \ldots, S_m$ be disjoint Borel sets such that $S = \bigcup_j S_j$, $S_j \subseteq B_j$. Then

$$\varphi_\delta(z) = \sum_{j=1}^m \frac{1}{2\pi i}\int_{S_j} \frac{\partial \varphi_\delta}{\partial \bar\zeta}(\zeta)\frac{d\zeta \wedge d\bar\zeta}{z-\zeta}.$$

Since $S^2\backslash K$ is connected, we can find a path Γ_j joining the center of B_j with ∞ and such that $S^2\backslash\Gamma_j$ is also connected. Extending it a little bit beyond the center of B_j and keeping a portion inside B_j which starts very near ∂B_j, we can define a compact connected set $E_j \subseteq B_j$, diameter $(E_j) \ge 2\delta$, $S^2\backslash E_j$ connected and $E_j \cap K = \varnothing$. We can now apply Lemma 3.7.2 to each E_j, $r = 2\delta$, giving approximations $R_j(\zeta, z) = Q_j(z) + (\zeta - \beta_j)Q_j^2(z)$ to the Cauchy kernel $1/(z-\zeta)$ in B_j. Consider the function

$$\begin{aligned}\Phi_\delta(z) &= \sum_{j=1}^m \frac{1}{2\pi i}\int_{S_j} \frac{\partial \varphi_\delta}{\partial \bar\zeta}(\zeta)R_j(\zeta,z)\,d\zeta \wedge d\bar\zeta\\ &= \sum_{j=1}^m Q_j(z)\frac{1}{2\pi i}\int_{S_j} \frac{\partial \varphi_\delta}{\partial \bar\zeta}(\zeta)\,d\zeta \wedge d\bar\zeta\\ &\quad + \sum_{j=1}^m Q_j^2(z)\frac{1}{2\pi i}\int_{S_j} \frac{\partial \varphi_\delta}{\partial \bar\zeta}(\zeta)(\zeta - \beta_j)\,d\zeta \wedge d\bar\zeta.\end{aligned}$$

This is a linear combination of Q_j and Q_j^2. They are holomorphic in $S^2\backslash E_j$; in particular they are all holomorphic in $\Omega = S^2\backslash(E_1 \cup \cdots \cup E_m)$, which is an open neighborhood of K. By Runge's theorem, we will be able to approximate Φ_δ uniformly on K by polynomials. Therefore, to end the proof of the theorem we have to show that Φ_δ approximates φ_δ uniformly on K when $\delta \to 0$. We have

$$|\Phi_\delta(z) - \varphi_\delta(z)| \le \sum_{j=1}^m \frac{1}{2\pi} \int_{S_j} \left|\frac{\partial \varphi_\delta}{\partial \bar{\zeta}}(\zeta)\right| \left|R_j(\zeta, z) - \frac{1}{z-\zeta}\right| dm(\zeta)$$

$$\le \frac{\omega(\delta)}{\pi\delta} \sum_{j=1}^m \int_{S_j} \left|R_j(\zeta, z) - \frac{1}{z-\zeta}\right| dm(\zeta).$$

For a fixed $z \in \Omega$, introduce polar coordinates $\zeta = z + \rho e^{i\theta}$. Then the sets S_j can be divided into two disjoint pieces, $S_j' = S_j \cap \{\zeta : |z - \zeta| \le 3\delta\}$, $S_j'' = S_j \backslash S_j'$. For $\zeta \in S_j'$ we have

$$\left|R_j(\zeta, z) - \frac{1}{z-\zeta}\right| \le \frac{c_1}{2\delta} + \frac{1}{\rho}.$$

Hence

$$\sum_{j=1}^m \int_{S_j'} \left|R_j(\zeta, z) - \frac{1}{z-\zeta}\right| dm(\zeta) \le \int_{\bar{B}(z, 3\delta)} \left(\frac{c_1}{2\delta} + \frac{1}{\rho}\right) dm(\zeta)$$

$$= 2\pi\left(\frac{c_1}{2\delta} \cdot \frac{(3\delta)^2}{2} + 3\delta\right) = c_1'\delta.$$

In S_j'' we have

$$\left|R_j(\zeta, z) - \frac{1}{z-\zeta}\right| \le c_2 \frac{4\delta^2}{\rho^3}.$$

Hence

$$\sum_{j=1}^m \int_{S_j''} \left|R_j(\zeta, z) - \frac{1}{z-\zeta}\right| dm(\zeta) \le \int_{3\delta}^\infty 8\pi c_2 \delta^2 \frac{d\rho}{\rho^2} = c_2'\delta.$$

In conclusion, for $z \in \Omega$,

$$|\Phi_\delta(z) - \varphi_\delta(z)| \le \frac{\omega(\delta)}{\pi\delta}(c_1' + c_2')\delta = c\omega(\delta).$$

Summarizing, we have found a constant $c > 0$ (independent of K, δ, f, etc.) and a function $\Phi_\delta \in \mathscr{H}(\Omega)$, Ω open neighborhood of K (Ω depends on δ) so that

$$\|f - \Phi_\delta\|_K \le (1 + c)\omega(\delta).$$

Therefore, given $\varepsilon > 0$ we first choose $\delta > 0$ so that $2(1 + c)\omega(\delta) \le \varepsilon$. Then we approximate Φ_δ in K by a polynomial P so that $\|P - \Phi_\delta\|_K \le \varepsilon/2$ and we have

$$\|f - P\|_K \le \varepsilon. \qquad \square$$

There are several generalizations and different proofs of Mergelyan's theorem. The study of $A(K)$ per se leads to the subject of uniform algebras. We refer the reader to the exercises and notes at the end of this chapter for some references to these questions.

It is also often useful to approximate uniformly by entire functions, functions that are continuous in a unbounded closed subset E of $\mathbb{C}$ and holomorphic in $\mathring{E}$. In fact, in this case one might want to do better: given $\omega : E \to \mathbb{R}_+$ continuous and $f \in \mathscr{H}(\mathring{E}) \cap \mathscr{C}(E)$ (i.e., $f \in A(E)$), find $h \in \mathscr{H}(\mathbb{C})$ such that

$$|f(z) - h(z)| < \omega(z), \qquad \text{for all } z \in E.$$

This problem was considered by Arakelian who found necessary and sufficient conditions for this type of approximation to be possible. The conditions on E are similar to that in the theorem of Mergelyan. Recall that a hole is a bounded component of E^c. Then, one needs

(i) E has no holes, and
(ii) for every closed disk B, the holes of $B \cup E$ lie in a bounded set.

For an arbitrary closed set E satisfying (i) and (ii) the essentially best conditions on ω, when $\omega(z) = \varepsilon(|z|)$ and $\varepsilon(t) \to 0$ as $t \to +\infty$ are that

$$\int_1^\infty \frac{\log \varepsilon(t)}{t^{3/2}}\, dt < \infty,$$

but when $\mathring{E} = \emptyset$, the last condition can be dispensed with. We refer to [Ful] and [Ga] for details. As an interesting application of Arakelian's theorem we mention an elementary construction in [Za6] of a nonzero entire function f with the following properties:

(a) on every line ℓ in $\mathbb{C}$, $f(z) \to 0$ as $z \to \infty$, $z \in \ell$;
(b) f is integrable on every line ℓ, i.e., $\int_\ell |f(z)|\, ds(z) < \infty$ (ds is the element of length in ℓ);
(c) for every line ℓ, $\int_\ell f(z)\, ds(z) = 0$.

In a recent article, J. P. Rosay and W. Rudin [RR] have given a very elementary derivation of Arakelian's theorem from Mergelyan's theorem and a Mittag-Leffler type of argument in the particular cases were either $\omega \equiv$ constant or $\mathring{E} = \emptyset$. We proceed to reproduce it here.

3.7.3. Theorem. *Let E be a closed set in $\mathbb{C}$ satisfying conditions* (i) *and* (ii). *Given $f \in A(E)$ and $\varepsilon > 0$, there is an entire function h such that*

$$|f(z) - h(z)| \le \varepsilon \quad \text{for all} \quad z \in E.$$

PROOF. For $n \in \mathbb{N}^*$, let $B_n = \bar{B}(0, R_n)$, $R_n < R_{n+1} \to \infty$ be chosen so that $\mathring{B}_{n+1} \supseteq B_n \cup \bar{H}_n$, where $H_n =$ union of all the holes of $E \cup B_n$. This is possible by property (ii). By property (i) it follows that if $E_0 = E$ and $E_n = E \cup B_n \cup \bar{H}_n$, then the compact sets $E_{n-1} \cap B_{n+1}$ have no holes and hence Mergelyan's theorem holds for them. Note that $E_n \subseteq E_{n+1}$ and $\bigcup_n E_n = \mathbb{C}$.

We define now a sequence of holomorphic functions h_n, which for $n \geq 1$ are going to be continuous in E_n and holomorphic in $\mathring{E}_n$, and the $\lim\limits_{n\to\infty} h_n = h$ will be uniform over compact sets, whence h will be entire.

Let $h_0 = f$ and assume h_{n-1} has already been chosen. Let $\psi_n \in C_0^\infty(\mathbb{C})$ be such that $0 \leq \psi_n \leq 1$, $\psi_n = 1$ on a neighborhood U_n of $B_n \cup \bar{H}_n$ and $\operatorname{supp}(\psi_n) \subseteq B_{n+1}$. We know that $z \mapsto \frac{1}{\pi}\int\left|\frac{\partial\psi_n}{\partial\bar{w}}(w)\right|\frac{dm(w)}{|z-w|}$ is a continuous function on $\mathbb{C}$ which tends to zero at infinity. Let $M_n \geq 1$ be an upper bound for this function. As we pointed out, we can apply Mergelyan's theorem to the function h_{n-1} on $E_{n-1} \cap B_{n+1}$ and find a polynomial P_n such that

$$|h_{n-1}(z) - P_n(z)| < \frac{\varepsilon}{2^{n+1}M_n} \qquad \text{on } E_{n-1} \cap B_{n+1}. \qquad (*)$$

Let now

$$r_n(z) := \frac{1}{\pi}\int_{E_{n-1}} [h_{n-1}(w) - P_n(w)]\frac{\partial\psi_n(w)}{\partial\bar{w}}\frac{dm(w)}{z-w} \qquad (z \in \mathbb{C})$$

and define

$$h_n := \psi_n P_n + (1-\psi_n)h_{n-1} + r_n \qquad \text{in } E_n.$$

We have that $\operatorname{supp}\left(\frac{\partial\psi_n}{\partial\bar{w}}\right) \subseteq B_{n+1}\backslash U_n$, hence $r_n \in \mathscr{H}(U_n)$. Moreover, when $z \in U_n$ we have $\psi_n(z) = 1$, hence $h_n(z) = P_n(z) + r_n(z)$. Therefore, h_n is well defined and holomorphic in U_n. Hence, h_n is continuous in $E_{n-1} \cup U_n$, in particular, in E_n.

We also have in $\mathring{E}_{n-1}$

$$\frac{\partial h_n}{\partial\bar{z}} = P_n\frac{\partial\psi_n}{\partial\bar{z}} - h_{n-1}\frac{\partial\psi_n}{\partial\bar{z}} + \frac{\partial r_n}{\partial\bar{z}} = 0$$

by Pompeiu's formula. In other words, h_n is continuous in E_n and holomorphic in $\mathring{E}_n$.

The condition $(*)$ implies that

$$|r_n(z)| < \frac{\varepsilon}{2^{n+1}}, \qquad \text{for all } z \in \mathbb{C}.$$

Hence, for $z \in E_{n-1}$ we have

$$|h_n(z) - h_{n-1}(z)| \leq |r_n(z)| + \psi_n(z)|P_n(z) - h_{n-1}(z)| < \frac{\varepsilon}{2^n}.$$

This implies immediately that the sequence $(h_n)_{n\geq m}$ converges uniformly on E_m. Namely, for $n \geq m$ and $p > 0$,

$$\max_{z\in E_m}|h_{n+p}(z) - h_n(z)| < \frac{\varepsilon}{2^n},$$

so it defines a Cauchy sequence and therefore an entire function h. Clearly $h = h_0 + \sum_{n\geq 1} (h_n - h_{n-1})$. It follows that on E we have

$$|h - f| \leq \sum_{n\geq 1} |h_n - h_{n-1}| \leq \varepsilon. \qquad \square$$

3.7.4. Corollary. *Let E be as in the previous theorem and assume further that $\mathring{E} = \emptyset$. Given a continuous function $\omega : E \to \mathbb{R}_+$ and f any continuous function on E, there is an entire function h such that*

$$|h(z) - f(z)| < \omega(z), \qquad z \in E.$$

PROOF. From the previous theorem there is an entire function g_1 such that

$$|g_1(z) - \log \omega(z)| < 1 \qquad (z \in E).$$

Let $g_2(z) = g_1(z) - 1$, we have

$$\operatorname{Re} g_2(z) = \operatorname{Re} g_1(z) - 1 < \log \omega(z) \qquad (z \in E).$$

By the same theorem we can find $g_3 \in \mathscr{H}(\mathbb{C})$ such that

$$|g_3(z) - f(z)e^{-g_2(z)}| < 1 \qquad (z \in E).$$

Hence

$$|g_3(z)e^{g_2(z)} - f(z)| < |e^{g_2(z)}| < \omega(z) \qquad (z \in E),$$

which concludes the proof of the corollary. $\square$

A simple but very interesting case of this corollary occurs when $E = \mathbb{R}$. We shall have occasion to use this remark in the following chapter. See also, Exercise 3.7.6.

EXERCISES 3.7

1. Use Theorem 3.7.1 to prove the classical Weierstrass approximation theorem in $\mathbb{R}$: Any function continuous in an interval $[a, b]$ can be uniformly approximated there by polynomials of a single real variable.

2. Use Mergelyan's theorem to prove the following version of Cauchy's theorem. Let Γ be a rectifiable Jordan curve, $K = \Gamma \cup \operatorname{Int}(\Gamma)$, $f \in A(K)$. Then

$$\int_\Gamma f(z)\, dz = 0.$$

*3. Assume K is a compact subset of the plane such that K^c has finitely many components. State and prove an appropriate version of Mergelyan's theorem.

4. (Alice Roth's Swiss cheese). Let $\{a_j\}_{j\in\mathbb{N}}$, $a_0 = 0$, $0 < r_j < 1$ be chosen in such a way such that the disks $B_j = B(a_j, r_j)$ satisfy
   (1) $\bar{B}_j \subseteq B(0, 1) = B$.
   (2) $\sum_{j=0}^{\infty} r_j < 1$.

(3) $\bar{B}_j \cap \bar{B}_k = \varnothing$ if $j \neq k$.

(4) $K = \bar{B} \setminus \left(\bigcup_{j \geq 0} \bar{B}_j \right)$ has empty interior.

(Can you find such a_j, r_j?)

(a) Show that if f is a rational function with poles in K^c then

$$\int_{\partial B} f(z)\, dz = \sum_{j \geq 0} \int_{\partial B_j} f(z)\, dz.$$

(b) Show that the same is true if $f \in \overline{\mathscr{H}(K)}$.

(c) Is the function $f(z) = \frac{|z|}{z}$ in $A(K)$? In $\overline{\mathscr{H}(K)}$?

5. Let $B = B(0, 1)$, μ a Radon measure in $\bar{B}$ such that for any $\alpha \in B$ it satisfies

$$\int \frac{z - \bar{\alpha}}{1 - \alpha z}\, d\mu(z) = 0.$$

Show that this implies that the functions f, g defined in B by

$$f(\alpha) := \int \frac{z}{1 - \alpha z}\, d\mu(z), \qquad g(\alpha) = \int \frac{d\mu(z)}{1 - \alpha z}, \qquad \alpha \in B,$$

vanish identically. Conclude from this that if $|\beta| > 1$ then

$$\int \frac{d\mu(z)}{\beta - z} = 0.$$

Prove now that $\int h\, d\mu = 0$ for every $h \in A(\bar{B})$.

Find the closed linear span of the Moebius functions $\frac{z - \alpha}{1 - \bar{\alpha} z}$ $(\alpha \in B)$ in $\mathscr{C}(\bar{B})$.

*6. Using Corollary 3.7.4, show directly that any continuous function f on $\mathbb{R}$ can be obtained as boundary values $b(g)$, of a function $g \in \mathscr{H}(\mathbb{C} \setminus \mathbb{R})$. (Hint: Reduce the problem to the case where $|f(t)| \leq \frac{c}{1 + |t|}$. Now the Cauchy transform $\hat{f}$ makes sense. Show $f = -\frac{1}{2i} b(\hat{f})$, cf. Exercise 3.6.9.)

§8. Short Survey of the Theory of Distributions: Their Relation to the Theory of Residues

In this section we give a very quick introduction to the fundamental properties of distributions, only insofar as letting the reader operate with them confidently. We hope that this section will be enough to be able to read §3.6 for those who have not seen distributions before. For the general theory we recommend [S], [GS], and [Ho2, Vol. 1]. We conclude this section with a recent approach to the theory of residues.

3.8.1. Definition. A ***distribution*** T on an open subset Ω of $\mathbb{R}^n$ is a $\mathbb{C}$-linear map $T: \mathscr{D}(\Omega) \to \mathbb{C}$ such that, for every compact subset K of Ω, if $(\varphi_j)_{j\geq 1}$ is a sequence of elements of $\mathscr{D}(\Omega)$ such that

(i) $\operatorname{supp}(\varphi_j) \subseteq K$ $(j \geq 1)$

(ii) for every $\alpha \in \mathbb{N}^n$, $\lim_{j\to\infty} \sup_{x\in K} |D^\alpha \varphi_j(x)| = 0$, (i.e., $\varphi_j \to 0$ in $\mathscr{E}(\Omega)$), we have

$$\lim_{j\to\infty} \langle T, \varphi_j \rangle = 0.$$

(It is standard to denote $T(\varphi)$ by $\langle T, \varphi\rangle$.) The space of all distributions in Ω is denoted by $\mathscr{D}'(\Omega)$.

One can show from this definition that if $T \in \mathscr{D}'(\Omega)$ and $K \subset\subset \Omega$, then there is $N \in \mathbb{N}$ and $C > 0$ such that

$$|\langle T, \varphi\rangle| \leq C \sup_{\substack{|\alpha|\leq N \\ x\in K}} |D^\alpha \varphi(x)|$$

for every $\varphi \in \mathscr{D}(\Omega)$ with $\operatorname{supp}(\varphi) \subseteq K$.

The simplest example of a distribution is obtained starting from a function $f \in L^1_{\text{loc}}(\Omega)$. One defines $T = T_f \in \mathscr{D}'(\Omega)$ by

$$\langle T_f, \varphi\rangle := \int_\Omega f\varphi \, dx.$$

If $f, g \in L^1_{\text{loc}}(\Omega)$ and $T_f = T_g$ then $f = g$ a.e. in Ω. By abuse of language, one often writes f instead of T_f when $f \in L^1_{\text{loc}}(\Omega)$.

If $f \in C^1(\Omega)$, then one can associate not only to f, but also to every partial derivative $f_{x_j} = \dfrac{\partial f}{\partial x_j}$, a distribution T_f, $T_{f_{x_j}}$, respectively. The relation between them is obtained by integration by parts:

$$\langle T_{f_{x_j}}, \varphi\rangle = \int_\Omega \frac{\partial f}{\partial x_j} \varphi \, dx = -\int_\Omega f \frac{\partial \varphi}{\partial x_j} dx = -\left\langle T_f, \frac{\partial \varphi}{\partial x_j}\right\rangle.$$

For this reason we define the derivative $\dfrac{\partial T}{\partial x_j}$ of a distribution $T \in \mathscr{D}'(\Omega)$ by

$$\left\langle \frac{\partial T}{\partial x_j}, \varphi \right\rangle := -\left\langle T, \frac{\partial \varphi}{\partial x_j}\right\rangle.$$

It is easy to see that $\dfrac{\partial T}{\partial x_j}$ is again a distribution in Ω. It follows that for $\alpha \in \mathbb{N}^n$,

$$D^\alpha = \frac{\partial^{|\alpha|}}{\partial x^\alpha},$$

$$\langle D^\alpha T, \varphi\rangle = (-1)^{|\alpha|} \langle T, D^\alpha \varphi\rangle.$$

It is clear that

$$D^\alpha(D^\beta T) = D^{\alpha+\beta}(T).$$

As an example, let δ denote as always the Dirac mass at 0, i.e., the distribution $\delta : \varphi \mapsto \langle \delta, \varphi \rangle = \varphi(0)$ for $\varphi \in \mathscr{D}(\mathbb{R})$. If $H = \chi_{[0,\infty[}$ (the Heaviside function) and $T = T_H$, then

$$\frac{dT_H}{dx}\left(= \frac{dH}{dx}\right) = \delta.$$

3.8.2. Definition. If $\Omega_1 \subseteq \Omega_2$ are two open sets in $\mathbb{R}^n$ and $T \in \mathscr{D}'(\Omega_2)$ one can define the ***restriction*** $T|\Omega_1$ of T to Ω_1:

$$\langle T|\Omega_1, \varphi\rangle = \langle T, \varphi\rangle \quad \text{for} \quad \varphi \in \mathscr{D}(\Omega_1).$$

This makes sense, since $\mathscr{D}(\Omega_1) \subseteq \mathscr{D}(\Omega_2)$.

With the help of this definition one can define the support of a distribution.

3.8.3. Definition. Let $T \in \mathscr{D}'(\Omega)$. One denotes $\operatorname{supp}(T)$, ***support of*** T, the complement in Ω of the union of all open sets $\omega \subseteq \Omega$ such that $T|\omega = 0$.

It follows that $\Omega \setminus \operatorname{supp}(T)$ is the largest open subset of Ω on which T is zero (i.e., it restricts to the zero distribution). For a continuous function in Ω, $\operatorname{supp}(T_f) = \operatorname{supp}(f)$.

Sometimes there is some doubt about the variable on which a distribution acts. For instance, in the case $\varphi \in \mathscr{D}(\mathbb{R}^n \times \mathbb{R}^n)$ and $T \in \mathscr{D}'(\mathbb{R}^n)$, then we can write T_x to indicate that for each $y \in \mathbb{R}^n$ one has a function

$$y \mapsto \langle T_x, \varphi(x, y)\rangle,$$

which can be seen to belong to $\mathscr{D}(\mathbb{R}^n)$. In this way one can define the ***tensor product*** $T \otimes S$ of two distributions in $\mathscr{D}'(\mathbb{R}^n)$ as a distribution in $\mathscr{D}'(\mathbb{R}^{2n})$ given by the formula

$$\langle T \otimes S, \varphi\rangle := \langle T_x, \langle S_y, \varphi(x, y)\rangle\rangle = \langle S_y, \langle T_x, \varphi(x, y)\rangle\rangle.$$

The last identity is a generalization of the theorem of Fubini, which one proves first for $\varphi \in \mathscr{D}(\mathbb{R}^{2n})$ of the form $\varphi(x, y) = \psi(x)\theta(y)$, $\psi, \theta \in \mathscr{D}(\mathbb{R}^n)$. Then one uses the fact that linear combinations of such products are dense in $\mathscr{D}(\mathbb{R}^{2n})$. One finds that

$$\operatorname{supp}(T \otimes S) = \operatorname{supp}(T) \times \operatorname{supp}(S).$$

Starting from this concept of tensor product one can introduce the extremely important concept of convolution that generalizes the usual convolution of functions. One observes that, if for a certain compact $K \subseteq \mathbb{R}^n$ and distributions $T, S \in \mathscr{D}'(\mathbb{R}^n)$, the set

$$\tilde{K} := \{(x, y) \in \mathbb{R}^n \times \mathbb{R}^n : x + y \in K\} \cap (\operatorname{supp}(T) \times \operatorname{supp}(S))$$

is compact, then one can give a meaning to the expression

$$\langle T_x \otimes S_y, \varphi(x+y)\rangle$$

when $\varphi \in \mathscr{D}(K)$ (i.e., $\varphi \in \mathscr{D}(\mathbb{R}^n)$, $\operatorname{supp}(\varphi) \subseteq K$). Namely, choose $\chi \in \mathscr{D}(\mathbb{R}^{2n})$ which is identically 1 in a neighborhood of $\tilde{K}$, then the value

$$\langle T \otimes S, \chi(x, y)\varphi(x+y)\rangle$$

is independent of the choice of χ. Hence, one declares it to be $\langle T_x \otimes S_y, \varphi(x+y)\rangle$.

3.8.4. Definition. Given $S, T \in \mathscr{D}'(\mathbb{R}^n)$ one says they can be convolved if for any $K \subset\subset \mathbb{R}^n$, the set $\tilde{K}$ is compact in $\mathbb{R}^{2n}$. In that case the ***convolution*** $T * S$ is the distribution in $\mathbb{R}^n$ given by

$$\langle T * S, \varphi\rangle = \langle T_x \otimes S_y, \varphi(x+y)\rangle.$$

3.8.5. Remarks. (1) A sufficient condition that S, T can be convolved is that one of them has compact support.

(2) Convolution is commutative but, in general, not associative. On the other hand, if $\operatorname{supp} S_1 \subset\subset \mathbb{R}^n$, $\operatorname{supp} S_2 \subset\subset \mathbb{R}^n$ then

$$(T * S_1) * S_2 = T * (S_1 * S_2) = (T * S_2) * S_1.$$

(3) The Dirac mass is the identity for the convolution product. For any $T \in \mathscr{D}'(\mathbb{R}^n)$

$$T * \delta = \delta.$$

(4) If f, g are two functions in $L^1_{\mathrm{loc}}(\mathbb{R}^n)$ and one of them has compact support, then

$$T_f * T_g = T_{f*g},$$

where $f * g(x) = \displaystyle\int_{\mathbb{R}^n} f(x-y)g(y)\,dy$ is the usual convolution of functions. For instance, if $\varphi \in \mathscr{D}(\mathbb{R}^n)$ and $T \in \mathscr{D}'(\mathbb{R}^n)$, then $T * T_\varphi$ coincides with T_f where f is the C^∞ function

$$f(x) = \langle T_y, \varphi(x-y)\rangle.$$

One writes $T * \varphi$ for this expression in order to simplify the notation.

3.8.6. Definition. One denotes $\mathscr{E}'(\mathbb{R}^n)$ the family of those $T \in \mathscr{D}'(\mathbb{R}^n)$ such that $\operatorname{supp}(T) \subset\subset \mathbb{R}^n$.

If $T \in \mathscr{E}'(\mathbb{R}^n)$ and $\chi \in \mathscr{D}(\mathbb{R}^n)$ is such that $\chi \equiv 1$ in a neighborhood of $\operatorname{supp}(T)$, then for any $\varphi \in \mathscr{E}(\mathbb{R}^n)$ we have $\chi\varphi \in \mathscr{D}(\mathbb{R}^n)$ and it makes sense to compute $\langle T, \chi\varphi\rangle$, which also turns out to be independent of χ. One concludes that T extends to a $\mathbb{C}$-linear map $\mathscr{E}(\mathbb{R}^n) \to \mathbb{C}$ defined by $\langle T, \varphi\rangle := \langle T, \chi\varphi\rangle$. It can be proved that $\varphi \mapsto \langle T, \varphi\rangle$ is a continuous map in the Frechet space $\mathscr{E}(\mathbb{R}^n)$. Moreover, every element of the topological dual of $\mathscr{E}(\mathbb{R}^n)$ corresponds to a unique distribution $T \in \mathscr{E}'(\mathbb{R}^n)$.

One of the fundamental theorems in the theory of distributions is the following.

3.8.7. Proposition (Theorem of Titchmarsh-Lions-Schwartz). *The space $\mathscr{E}'(\mathbb{R}^n)$ is a commutative algebra under convolution with unit δ. Moreover, it is an integral domain as a consequence of the identity*

$$cv(\text{supp}(T * S)) = cv(\text{supp}(T)) + cv(\text{supp}(S)),$$

valid for any $T, S \in \mathscr{E}'(\mathbb{R}^n)$. (We denote $cv(\text{supp}(T))$ = convex hull of the support of T and, for subsets A, B of $\mathbb{R}^n$, $A + B = \{a + b : a \in A, b \in B\}$.)

There is a further property of the convolution worth mentioning. If $\delta^{(\alpha)}$ represents the derivative $D^\alpha\delta$ of the Dirac distribution, then

$$D^\alpha T = \delta^{(\alpha)} * T.$$

Therefore, for a linear partial differential operator with constant coefficients $P(D) = \sum_{|\alpha| \leq N} a_\alpha D^\alpha$, we have

$$P(D)T = P(\delta) * T,$$

where $P(\delta) = \sum_{|\alpha| \leq N} a_\alpha \delta^{(\alpha)}$. This formula can be generalized as follows: let $T \in \mathscr{D}'(\mathbb{R}^n)$, $S \in \mathscr{E}'(\mathbb{R}^n)$, then

$$D^\alpha(T * S) = (D^\alpha T) * S = T * (D^\alpha S).$$

3.8.8. Definition. A ***fundamental solution*** E of a differential operator $P(D)$ is a distribution $E \in \mathscr{D}'(\mathbb{R}^n)$ such that $P(D)E = \delta$.

If $S \in \mathscr{E}'(\mathbb{R}^n)$ (or if $E * S$ makes sense) we can solve the distribution equation

$$P(D)T = S,$$

by taking $T = E * S$. In fact,

$$P(D)T = P(D)(E * S) = (P(D)E) * S = \delta * S = S.$$

For instance, for $n = 2$, $\mathbb{R}^2 = \mathbb{C}$, we have seen that the distribution associated to the $L^1_{\text{loc}}(\mathbb{C})$ function $z \mapsto \dfrac{1}{\pi z}$ is a fundamental solution for the differential operator $\dfrac{\partial}{\partial \bar{z}}$.

Another operation on distributions is the following. Let $T \in \mathscr{D}'(\mathbb{R}^n)$ and $f \in C^\infty(\mathbb{R}^n)$, then one can define a new distribution $fT \in \mathscr{D}'(\mathbb{R}^n)$ by

$$\langle fT, \varphi \rangle := \langle T, f\varphi \rangle.$$

Clearly $\text{supp}(fT) \subseteq \text{supp}(f) \cap \text{supp}(T)$. Note that the inclusion can be strict, e.g., $x\delta_0(x) = 0$. Furthermore, Leibnitz' rule for the derivatives is also valid for products of functions and distributions. For instance, $(xT)' = T + xT'$.

Using a partition of unity $(\varphi_j)_{j \geq 1}$ on a set Ω one can write any distribution

$T \in \mathscr{D}'(\Omega)$ as a locally finite sum (in the sense of supports)

$$T = \sum_j (\varphi_j \cdot T) = \sum_j T_j,$$

$T_j \in \mathscr{D}'(\mathbb{R}^n)$, $\operatorname{supp}(T_j) \subseteq \operatorname{supp}(\varphi_j)$. This remark was used in §3.6.

This last remark, together with the following "representation" theorem, allows us to operate with distributions as if they were functions.

3.8.9. Proposition. *Let $T \in \mathscr{E}'(\mathbb{R}^n)$, $\operatorname{supp}(T) \subseteq \bar{B}(0, R)$. There exist an $N \in \mathbb{N}$ and a finite number of continuous functions f_α, $\alpha \in \mathbb{N}^n$, $|\alpha| \leq N$ such that $\operatorname{supp}(f_\alpha) \subseteq \bar{B}(0, R)$ and*

$$T = \sum_{|\alpha| \leq N} D^\alpha T_{f_\alpha}.$$

In the exercises following this section, the reader will see the family of distributions $|x|^\lambda$ in $\mathbb{R}$ (more precisely, $T_{|x|^\lambda}$) indexed by a complex parameter λ, $\operatorname{Re}\lambda > 0$, given by

$$\langle |x|^\lambda, \varphi \rangle = \int_{\mathbb{R}} |x|^\lambda \varphi(x)\, dx \qquad (\varphi \in \mathscr{D}(\mathbb{R})).$$

The transformation $\lambda \mapsto \langle |x|^\lambda, \varphi \rangle$ is holomorphic in the half-plane $\operatorname{Re}\lambda > 0$ and its analytic continuation, as a function of λ, is related to the distribution $pv\left(\frac{1}{x}\right)$ defined by

$$\left\langle pv\left(\frac{1}{x}\right), \varphi \right\rangle = \lim_{\varepsilon \to 0} \int_{|x| \geq \varepsilon} \frac{\varphi(x)}{x}\, dx;$$

cf. Exercise 3.6.7.

In §3.6, we also considered the distributions in $\mathscr{D}'(\mathbb{C})$ defined by $\frac{1}{z}$, $pv\left(\frac{1}{z^n}\right)$, $n \geq 2$, and, more generally, $pv(f)$, for $f \in \mathscr{M}(\Omega)$. We saw there that they are related to the concept of residue in the classical sense, as it was introduced in Chapters 1 and 2. In fact, we have three slightly different notions of residue so far. In the remainder of this section we will introduce a less known systematic way of considering the residue, which can be traced back to Poincaré, and more recently to Leray, Gelfand, Dolbeault, Grothendieck, Herrera, Lieberman, Passare, Yger, etc., and that promises to be very useful in several applications of complex analysis.

In §3.6 we have already shown that for Ω open subset of $\mathbb{C}$, the space $\mathscr{M}(\Omega)$ can be embedded as a subspace of $\mathscr{D}'(\Omega)$ via the Cauchy principal value map

$$pv : \mathscr{M}(\Omega) \to \mathscr{D}'(\Omega)$$

given by

$$\langle pv(f), \varphi \rangle = \lim_{\varepsilon \to 0} \int_{\Omega_\varepsilon} f(z)\varphi(z)\, dm(z),$$

where $\Omega_\varepsilon = \{z \in \Omega : |z - a| \geq \varepsilon > 0 \text{ for every } a \in P(f)\}$. We have also obtained that

$$\frac{\partial}{\partial \bar{z}} pv(f) = \pi \sum_{a \in P(f)} \sum_{v} \frac{(-1)^{v-1}}{(v-1)!} A_{a,v} \frac{\partial^{v-1}}{\partial z^{v-1}} \delta_a,$$

where the inner sum in v runs over the index set $1 \leq v \leq m = m_a =$ order of a as a pole of f, and the principal part P_a of f at $z = a$ is given by

$$P_a(z) = \sum_{1 \leq v \leq m} \frac{A_{a,v}}{(z-a)^v}.$$

We pointed out that it is standard to call residue distribution of f the distribution $\operatorname{Res}(f) = \dfrac{\partial}{\partial \bar{z}} pv(f)$, so that if V is an open neighborhood of a such that $V \cap P(f) = \{a\}$ we have

$$\operatorname{Res}(f)|V = \pi \operatorname{Res}(f, a)\delta_a + T_a,$$

where T_a is a distribution with $\operatorname{supp}(T_a) = \{a\}$, involving only $\dfrac{\partial^v}{\partial z^v}\delta_a$, $1 \leq v \leq m - 1$, which is not zero unless $m = 1$.

In what follows, given $f \in \mathscr{H}(\Omega)$ we will recover $\operatorname{Res}(1/f)$ in terms of the family of distributions defined by $|f|^\lambda$.

We start by observing that for a fixed value $\mu \in \mathbb{C}$, if $\operatorname{Re}\mu > 0$, the function

$$z \mapsto \frac{1}{\pi}|z|^{2(\mu-1)}$$

is locally integrable in $\mathbb{C}$, and as such, it defines a distribution, still denoted $\dfrac{1}{\pi}|z|^{2(\mu-1)}$,

$$\left\langle \frac{1}{\pi}|z|^{2(\mu-1)}, \varphi \right\rangle = \frac{1}{2\pi i}\int_{\mathbb{C}} |z|^{2(\mu-1)}\varphi(z)\, d\bar{z} \wedge dz \qquad (\varphi \in \mathscr{D}(\mathbb{C})).$$

Moreover, for each fixed $\varphi \in \mathscr{D}(\Omega)$, we can consider the map

$$\mu \mapsto \left\langle \frac{1}{\pi}|z|^{2(\mu-1)}, \varphi \right\rangle$$

and prove without difficulty that it is also holomorphic in the same half-plane $\operatorname{Re}\mu > 0$. It is customary to express this fact by saying that the map

$$h : \mu \mapsto \frac{1}{\pi}|z|^{2(\mu-1)}, \qquad \{\operatorname{Re}\mu > 0\} \to \mathscr{D}'(\mathbb{C})$$

is holomorphic. Note that this means that for each λ_0, $\operatorname{Re}\lambda_0 > 0$, we have a Taylor series expansion about λ_0,

$$h(\lambda) = \sum_{k \geq 0} T_k(\lambda - \lambda_0)^k,$$

with $T_k \in \mathscr{D}'(\mathbb{C})$. The meaning of this expansion is that for every $\varphi \in \mathscr{D}(\mathbb{C})$ one has

$$\langle h(\lambda), \varphi \rangle = \sum_{k \geq 0} \langle T_k, \varphi \rangle (\lambda - \lambda_0)^k,$$

which is convergent for all λ close to λ_0. Similarly for a meromorphic function g with values in $\mathscr{D}'(\Omega)$, the Laurent development near a point λ_0 is formally given by

$$g(\lambda) = \sum_{k \geq -m} T_k(\lambda - \lambda_0)^k = \frac{T_{-m}}{(\lambda - \lambda_0)^m} + \cdots + \frac{T_{-1}}{\lambda - \lambda_0} + T_0 + T_1(\lambda - \lambda_0) + \cdots,$$

with $T_k \in \mathscr{D}'(\Omega)$. It means that for $\varphi \in \mathscr{D}(\Omega)$, and $0 < |\lambda - \lambda_0|$ small (smallness depending on φ),

$$\langle g(\lambda), \varphi \rangle = \sum_{k \geq -m} (\lambda - \lambda_0)^k \langle T_k, \varphi \rangle.$$

It is then justified to say that the distribution T_{-1} is the residue $\operatorname{Res}(g(\lambda), \lambda = \lambda_0)$.

3.8.10. Proposition. *The map* $h: \mu \mapsto \frac{1}{\pi}|z|^{2(\mu-1)}$, *with values in* $\mathscr{D}'(\mathbb{C})$, *admits an analytic continuation to the whole complex plane as a distribution valued meromorphic function with poles at* $\mu = 0, -1, -2, \ldots$. *All the poles are simple and their residues are*

$$\operatorname{Res}(h(\mu), \mu = -k) = \frac{1}{(k!)^2} \frac{\partial^{2k}\delta}{\partial z^k \partial \bar{z}^k}, \qquad k \in \mathbb{N}.$$

PROOF. For $z \neq 0$, $p \in \mathbb{N}$, $\mu \in \mathbb{C}$, and $\varphi \in \mathscr{D}(\mathbb{C})$, we have the identity

$$\bar{\partial}\left(\frac{|z|^{2\mu+2p}}{z^{p+1}} \frac{\partial^p \varphi(z)}{\partial \bar{z}^p} dz\right) = (\mu + p) \frac{|z|^{2\mu+2p-2}}{z^p} \frac{\partial^p \varphi(z)}{\partial \bar{z}^p} d\bar{z} \wedge dz$$

$$+ \frac{|z|^{2\mu+2p}}{z^{p+1}} \frac{\partial^{p+1}\varphi(z)}{\partial \bar{z}^{p+1}} d\bar{z} \wedge dz.$$

Therefore, whenever $\operatorname{Re} 2\mu + p > 2$, we can apply Stokes' formula and obtain

$$\frac{(-1)}{\mu + p} \int \frac{|z|^{2\mu+2p}}{z^{p+1}} \frac{\partial^{p+1}\varphi(z)}{\partial \bar{z}^{p+1}} d\bar{z} \wedge dz = \int \frac{|z|^{2\mu+2p-2}}{z^p} \frac{\partial^p \varphi(z)}{\partial \bar{z}^p} d\bar{z} \wedge dz.$$

In terms of our distribution valued function $h(\mu) = \frac{1}{\pi}|z|^{2(\mu-1)}$, it means that for $\operatorname{Re}\mu$ sufficiently large we have

$$\langle h(\mu), \varphi \rangle = \frac{1}{2\pi i} \int |z|^{2\mu-2} \varphi(z)\, d\bar{z} \wedge dz = \frac{(-1)}{2\pi i \mu} \int \frac{|z|^{2\mu}}{z} \frac{\partial \varphi(z)}{\partial \bar{z}} d\bar{z} \wedge dz$$

$$= \frac{(-1)^2}{2\pi i \mu(\mu + 1)} \int \frac{|z|^{2\mu+2}}{z^2} \frac{\partial^2 \varphi(z)}{\partial \bar{z}^2} d\bar{z} \wedge dz = \cdots$$

$$= \frac{(-1)^{p+1}}{2\pi i\mu(\mu+1)\dots(\mu+p)}\int \frac{|z|^{2\mu+2p}}{z^{p+1}}\frac{\partial^{p+1}\varphi(z)}{\partial\bar{z}^{p+1}}\,d\bar{z}\wedge dz.$$

The function $z\mapsto |z|^{2\mu+2p}/z^{p+1}$ and its derivative with respect to μ are clearly integrable as long as $\operatorname{Re} 2\mu + p - 1 > -2$, that is, for $\operatorname{Re}\mu > -\frac{p+1}{2}$. By Morera's theorem, the last integral is a holomorphic function in the half-plane $\operatorname{Re}\mu > -\frac{p+1}{2}$. This shows that $\langle h(\mu), \varphi\rangle$ is a holomorphic function of μ whenever $\operatorname{Re}\mu > -\frac{p+1}{2}$ except at the points $\mu = -k$, $k \in \mathbb{N}$, $k < \frac{p+1}{2}$, where it has a simple pole. (Remark we only move to the left by $\frac{1}{2}$ at each stage of this procedure.) Since p is an arbitrary natural integer, this proves the proposition except for the computation of the residue of $h(\mu)$ at $\mu = -k, k \in \mathbb{N}$. We treat here the cases $k = 0$ and $k = 1$, and leave the other ones to the reader as an exercise.

For $k = 0$, we use the previous identity with $p = 0$:

$$\lim_{\mu\to 0}\mu\langle h(\mu),\varphi\rangle = \lim_{\mu\to 0}\left(-\frac{1}{2\pi i}\int |z|^{2\mu}\frac{\partial\varphi}{\partial\bar{z}}\frac{d\bar{z}\wedge dz}{z}\right) = -\frac{1}{2\pi i}\int\frac{\partial\varphi}{\partial\bar{z}}\frac{d\bar{z}\wedge dz}{z}$$
$$= \varphi(0)$$

by Corollary 2.1.4 of Pompeiu's formula.

For $k = 1$, we need $p = 2$ in the identity, hence

$$(\mu+1)\langle h(\mu),\varphi\rangle = \frac{(-1)^3}{2\pi i\mu(\mu+2)}\int\frac{|z|^{2\mu+4}}{z^3}\frac{\partial^3\varphi(z)}{\partial\bar{z}^3}\,d\bar{z}\wedge dz.$$

Let $\mu = -1$ to compute the residue. Then we have

$$\operatorname{Res}(\langle h(\mu),\varphi\rangle, \mu = -1) = \frac{1}{2\pi i}\int\frac{\bar{z}}{z^2}\frac{\partial^3\varphi(z)}{\partial\bar{z}^3}\,d\bar{z}\wedge dz = \left\langle pv\left(\frac{1}{\pi z^2}\right), \bar{z}\frac{\partial^3\varphi(z)}{\partial\bar{z}^3}\right\rangle.$$

From Proposition 3.6.2, $pv\left(\frac{1}{\pi z^2}\right) = -\frac{\partial}{\partial z}pv\left(\frac{1}{\pi z}\right)$, hence

$$\left\langle pv\left(\frac{1}{\pi z^2}\right),\bar{z}\frac{\partial^3\varphi(z)}{\partial\bar{z}^3}\right\rangle = -\left\langle\frac{\partial}{\partial z}pv\left(\frac{1}{\pi z}\right),\bar{z}\frac{\partial^3\varphi(z)}{\partial\bar{z}^3}\right\rangle$$
$$= \left\langle pv\left(\frac{1}{\pi z}\right),\bar{z}\frac{\partial^4\varphi(z)}{\partial z\partial\bar{z}^3}\right\rangle = \left\langle\bar{z}\,pv\left(\frac{1}{\pi z}\right),\frac{\partial}{\partial\bar{z}}\frac{\partial^3\varphi(z)}{\partial z\partial\bar{z}^2}\right\rangle$$
$$= -\left\langle\frac{\partial}{\partial\bar{z}}\left(\bar{z}\,pv\left(\frac{1}{\pi z}\right)\right),\frac{\partial^3\varphi(z)}{\partial z\partial\bar{z}^2}\right\rangle.$$

We know that $\frac{\partial}{\partial\bar{z}}\left(\bar{z}\,pv\left(\frac{1}{\pi z}\right)\right) = pv\left(\frac{1}{\pi z}\right) + \bar{z}\delta = pv\left(\frac{1}{\pi z}\right)$, since $\bar{z}$ is zero for $z = 0$. Hence

$$\left\langle pv\left(\frac{1}{\pi z^2}\right), \bar{z}\frac{\partial^3\varphi(z)}{\partial\bar{z}^3}\right\rangle = -\left\langle pv\left(\frac{1}{\pi z}\right), \frac{\partial}{\partial\bar{z}}\frac{\partial^2\varphi(z)}{\partial\bar{z}\partial z}\right\rangle = \left\langle \frac{\partial}{\partial\bar{z}}pv\left(\frac{1}{\pi z}\right), \frac{\partial^2\varphi(z)}{\partial\bar{z}\partial z}\right\rangle$$
$$= \frac{\partial^2\varphi(0)}{\partial\bar{z}\partial z},$$

as we wanted to show. □

The last part of the computation used the relation between $pv\left(\frac{1}{\pi z^n}\right)$ and its derivatives. A similar result is the following.

3.8.11. Lemma. *Let $\varphi \in \mathcal{D}(\mathbb{C})$, $m \in \mathbb{N}$. Then*

$$\frac{1}{2\pi i}\int \frac{\bar{z}^m}{z^m}\frac{\partial^m\varphi(z)}{\partial\bar{z}^m}\, d\bar{z}\wedge dz = (-1)^m m! \left\langle pv\left(\frac{1}{\pi z^m}\right), \varphi\right\rangle.$$

Proof. The proof is left to the reader. $\Bigg($It is enough to use that the left-hand side is $\left\langle pv\left(\frac{1}{\pi z^m}\right)\varphi, \bar{z}^m\frac{\partial^m\varphi(z)}{\partial\bar{z}^m}\right\rangle$ and apply Propositions 3.6.2 and 3.6.3.$\Bigg)$ □

Proposition 3.8.10 can be interpreted as saying that

$$\operatorname{Res}\left(\frac{1}{\pi}|z|^{2(\mu-1)}, \mu = 0\right) = \frac{1}{\pi}\operatorname{Res}\left(\frac{1}{z}\right) = \frac{1}{\pi}\frac{\partial}{\partial\bar{z}}T_{1/z},$$

where both sides are interpreted in the sense of distributions. Its natural generalization is the following proposition.

3.8.12. Proposition. *Let $f \in \mathcal{H}(\Omega)$, then the $\mathcal{D}'(\Omega)$-valued holomorphic function $\mu \mapsto \frac{1}{\pi}|f|^{2(\mu-1)}\overline{f'}$, defined for $\operatorname{Re}\mu > 1$, has an analytic continuation to the whole plane as a meromorphic function. It has a simple pole at $\mu = 0$ and its residue coincides with the distribution $\operatorname{Res}\left(\frac{1}{\pi f}\right)$. That is,*

$$\operatorname{Res}\left(\left\langle\frac{1}{\pi}|f(z)|^{2(\mu-1)}\overline{f'(z)}, \varphi(z)\right\rangle, \mu = 0\right) = \left\langle\frac{1}{\pi}\frac{\partial}{\partial\bar{z}}pv\left(\frac{1}{f}\right), \varphi\right\rangle.$$

Proof. Let $(a_j)_{j\geq 1}$ be the collection of zeros of f in Ω, Δ_j pairwise disjoint disks centered at a_j, $\Delta_j \subset\subset \Omega$, $\alpha_j \in \mathcal{D}(\Delta_j)$, $0 \leq \alpha_j \leq 1$, $\alpha_j = 1$ on a neighborhood of a_j, $\beta = 1 - \sum_j \alpha_j$. Let $\varphi \in \mathcal{D}(\Omega)$. When $\operatorname{Re}\mu > 1$ we can use Stokes' formula to verify that

$$\mu h(\mu) = \mu\left\langle \frac{1}{\pi}|f|^{2(\mu-1)}\bar{f}', \varphi \right\rangle = \frac{\mu}{2\pi i}\int_\Omega |f|^{2(\mu-1)}\bar{\partial}\bar{f} \wedge \varphi\, dz$$

$$= -\frac{1}{2\pi i}\int_\Omega \frac{|f|^{2\mu}}{f}\frac{\partial\varphi}{\partial\bar{z}}\, d\bar{z}\wedge dz$$

$$= -\frac{1}{2\pi i}\int_\Omega \frac{|f|^{2\mu}}{f}\beta\frac{\partial\varphi}{\partial\bar{z}}\, d\bar{z}\wedge dz + \sum_j\left(-\frac{1}{2\pi i}\int_\Omega \frac{|f|^{2\mu}}{f}\alpha_j\frac{\partial\varphi}{\partial\bar{z}}\, d\bar{z}\wedge dz\right).$$

The last sum is finite since $\operatorname{supp}(\varphi)$ is compact.

The first term is clearly an entire function of μ since f never vanishes on $\operatorname{supp}(\beta)$. Its value for $\mu = 0$ is obtained by direct evaluation. We obtain

$$-\frac{1}{2\pi i}\int_\Omega \frac{1}{f}\beta\frac{\partial\varphi}{\partial\bar{z}}\, d\bar{z}\wedge dz = \left\langle -pv\left(\frac{1}{\pi f}\right), \beta\frac{\partial\varphi}{\partial\bar{z}}\right\rangle = -\left\langle \beta pv\left(\frac{1}{\pi f}\right), \frac{\partial\varphi}{\partial\bar{z}}\right\rangle.$$

Let us deal with one of the terms of the sum. In Δ_j we have $f(z) = (z - a_j)^{m_j} g_j(z)$, g_j holomorphic and never vanishing in Δ_j. Define $\theta_j(\mu, z) = \dfrac{|g_j(z)|^{2\mu}}{g_j(z)}\alpha_j(z)\dfrac{\partial\varphi(z)}{\partial\bar{z}}$.

We then have (suppressing the index j to simplify the notation)

$$-\frac{1}{2\pi i}\int_\Delta \frac{|f|^{2\mu}}{f}\alpha\frac{\partial\varphi}{\partial\bar{z}}\, d\bar{z}\wedge dz = -\frac{1}{2\pi i}\int_\Delta \frac{|z-a|^{2m\mu}}{(z-a)^m}\theta(\mu, z)\, d\bar{z}\wedge dz.$$

It is clear that $\theta(\mu, z)$ is an entire function of μ, but the integral converges a priori only for $\operatorname{Re}\mu > \dfrac{1}{2} - \dfrac{1}{m}$, which is not enough for us unless $m = 1$. To get around this difficulty we transform the integral as in Proposition 3.8.10 by means of the following lemma.

3.8.13. Lemma. *One has the following identity*

$$-\frac{1}{2\pi i}\int_\Delta \frac{|z-a|^{2m\mu}}{(z-a)^m}\theta(\mu, z)\, d\bar{z}\wedge dz$$

$$= \frac{(-1)^{m+1}}{\prod_{1\le k\le m}(m\mu + k)}\frac{1}{2\pi i}\int_\Delta |z-a|^{2m\mu}\left(\frac{\bar{z}-\bar{a}}{z-a}\right)^m\frac{\partial^m}{\partial\bar{z}^m}\theta(\mu, z)\, d\bar{z}\wedge dz.$$

PROOF. It is easy to verify that for $z \neq a$

$$\frac{\partial^m}{\partial\bar{z}^m}\left(|z-a|^{2m\mu}\left(\frac{\bar{z}-\bar{a}}{z-a}\right)^m\right) = \left(\prod_{1\le k\le m}(m\mu + k)\right)\frac{|z-a|^{2m\mu}}{(z-a)^m}.$$

Then one uses Stokes's formula to verify the identity. □

As a first consequence of this lemma we see that the integral we are considering has an analytic continuation which is holomorphic for $\operatorname{Re}\mu > -\dfrac{1}{2m}$.

It follows that if $M = \max\{m_j : a_j \in \operatorname{supp}(\varphi)\}$, then $\mu h(\mu)$ has an analytic continuation to a function holomorphic for $\operatorname{Re}\mu > -\dfrac{1}{2M}$. Therefore, $\mu = 0$ is a simple pole of h, and to evaluate the residue we only need to let $\mu = 0$ in the integrals transformed using Lemma 3.8.13. By Lemma 3.8.11 we have ($m = m_j$)

$$\frac{(-1)^{m+1}}{m!}\frac{1}{2\pi i}\int_{\Delta}\left(\frac{\bar z - \bar a}{z - a}\right)^m \frac{\partial^m}{\partial \bar z^m}\theta(0, z)\, d\bar z \wedge dz$$

$$= -\left\langle pv\left(\frac{1}{\pi(z-a)^m}\right), \theta(0,z)\right\rangle = -\left\langle pv\left(\frac{1}{\pi(z-a)^m}\right), \frac{\alpha_j}{g_j}\frac{\partial\varphi}{\partial\bar z}\right\rangle$$

$$= -\left\langle \alpha_j pv\left(\frac{1}{\pi(z-a)^m g_j}\right), \frac{\partial\varphi}{\partial\bar z}\right\rangle = -\left\langle \alpha_j pv\left(\frac{1}{\pi f}\right), \frac{\partial\varphi}{\partial\bar z}\right\rangle.$$

(The next to last identity is due to the fact that g_j does not vanish on $\operatorname{supp}(\alpha_j)$.) As a consequence we obtain

$$\operatorname{Res}(h(\mu), \mu = 0) = \lim_{\mu\to 0}\mu h(\mu) = -\left\langle \beta pv\left(\frac{1}{\pi f}\right), \frac{\partial\varphi}{\partial\bar z}\right\rangle - \sum_j\left\langle \alpha_j pv\left(\frac{1}{\pi f}\right), \frac{\partial\varphi}{\partial\bar z}\right\rangle$$

$$= -\left\langle pv\left(\frac{1}{\pi f}\right), \frac{\partial\varphi}{\partial\bar z}\right\rangle = \left\langle \frac{\partial}{\partial\bar z}pv\left(\frac{1}{\pi f}\right), \varphi\right\rangle = \left\langle \operatorname{Res}\left(\frac{1}{\pi f}\right), \varphi\right\rangle$$

This is the main identity we wanted to prove.

With respect to the analytic continuation of h to the whole plane, one can easily reduce it to Proposition 3.8.10. We note that the poles will appear in the sequence $0, -\frac{1}{M}, -\frac{2}{M}, \ldots$, but the value M will be dependent on the function φ; more precisely, it is determined as in the proof in terms of the multiplicities of the zeros a_j of f in $\operatorname{supp}(\varphi)$. □

In order to give some applications of the last proposition, we state first a consequence of Pompeiu's formula.

3.8.14. Lemma (Pompeiu's Formula with Weights). *Let Ω be an open set with a C^1 regular boundary, $f \in C^1(\bar\Omega)$, and $\Phi \in C^1(\bar\Omega \times \bar\Omega)$ such that $\Phi(z, z) = 1$ at the point $z \in \Omega$. Then*

$$f(z) = \frac{1}{2\pi i}\int_{\partial\Omega}\frac{\Phi(z,\zeta)f(\zeta)}{\zeta - z}d\zeta - \frac{1}{2\pi i}\int_{\Omega}\frac{\Phi(z,\zeta)\bar\partial f(\zeta)\wedge d\zeta}{\zeta - z}$$
$$- \frac{1}{2\pi i}\int_{\Omega} f(\zeta)\frac{\bar\partial_\zeta\Phi(z,\zeta)\wedge d\zeta}{\zeta - z}.$$

PROOF. The proof is immediate from §2.1.2 applied to $\zeta \mapsto f(\zeta)\Phi(z,\zeta)$. □

A particular example occurs when G is an entire function, $G(1) = 1$, and

$\Phi(z,\zeta) = G(1 + (z - \zeta)q(z,\zeta))$, $q \in C^1(\bar{\Omega} \times \bar{\Omega})$. Then

$$\bar{\partial}_\zeta \Phi(z,\zeta) = G'(1 + (z - \zeta)q(z,\zeta))(z - \zeta)\frac{\partial q}{\partial \bar{\zeta}}(z,\zeta)\, d\bar{\zeta}.$$

3.8.15. Proposition. *Let $P_1, \ldots, P_m$ be non-constant polynomials without common zeros in $\mathbb{C}$. Let $g_j(z,\zeta) = \dfrac{P_j(z) - P_j(\zeta)}{z - \zeta}$ if $z \neq \zeta$, $g_j(z,z) = P_j'(z)$, $\|P(\zeta)\|^2 = \sum_{j=1}^{m} |P_j(\zeta)|^2$, $q(z,\zeta) = \dfrac{1}{\|P(\zeta)\|^2} \sum_{j=1}^{m} \overline{P_j(\zeta)} g_j(z,\zeta)$ and $Q(z,\zeta) = q(z,\zeta)\, d\zeta$. We have the identities*

(i) $$1 = \frac{1}{i\pi} \int_{\mathbb{C}} \sum_{1 \leq j \leq m} \frac{\overline{P_j(\zeta)} P_j(z)}{\|P(\zeta)\|^2} \bar{\partial}_\zeta Q(z,\zeta) \quad \textit{for every } z \in \mathbb{C};$$

(ii) $$1 = \left\langle \frac{\partial}{\partial \bar{\zeta}} pv\left(\frac{1}{\pi P_1(\zeta)}\right), \sum_{1 \leq j \leq m} \frac{\overline{P_j(\zeta)}}{\|P(\zeta)\|^2} \det \begin{bmatrix} g_1(z,\zeta) & g_j(z,\zeta) \\ P_1(z) & P_j(z) \end{bmatrix} \right\rangle.$$

Note that the g_j are polynomials in the two variables z, ζ. Also $\bar{\partial}_\zeta Q(z,\zeta) = \dfrac{\partial}{\partial \bar{\zeta}} q(z,\zeta)\, d\bar{\zeta} \wedge d\zeta$. It follows that both formulas produce polynomials A_j (resp. B_j) such that $\sum_j A_j P_j \equiv 1 \left(\text{resp. } \sum_j B_j P_j \equiv 1\right)$. This is usually called the algebraic Bezout identity, and though it can easily be solved by the Euclidean division algorithm for polynomials, formulas (i) and (ii) can be generalized to the case of several variables and they permit sometimes an a priori analysis of the A_j (resp. B_j) before finding them. Finally, this kind of formula is also valid for entire functions. We will see an application to deconvolution problems in the second volume.

PROOF. To prove formula (i), we let $G(t) = t^2$ as the auxiliary function used in the preceding example of Φ, $\Omega = B(0,R)$, and apply §3.8.14. Then we have for $f(z) \equiv 1$, that for any $z \in B(0,R)$

$$1 = \frac{1}{2\pi i} \int_{\partial B(0,R)} G(1 + (z - \zeta)q(z,\zeta)) \frac{d\zeta}{\zeta - z}$$
$$+ \frac{1}{2\pi i} \int_{B(0,R)} G'(1 + (z - \zeta)q(z,\zeta)) \frac{\partial q}{\partial \bar{\zeta}}(z,\zeta)\, d\bar{\zeta} \wedge d\zeta.$$

Since $1 + (z - \zeta)q(z,\zeta) = \sum_{j=1}^{m} \dfrac{\overline{P_j(\zeta)} P_j(z)}{\|P(\zeta)\|^2}$ and $\bar{\partial}_\zeta Q(z,\zeta) = \dfrac{\partial q}{\partial \bar{\zeta}}(z,\zeta)\, d\bar{\zeta} \wedge d\zeta$, we obtain

$$1 = \frac{1}{2\pi i} \int_{\partial B(0,R)} \left(\sum_{1 \leq j \leq m} \frac{\overline{P_j(\zeta)} P_j(z)}{\|P(\zeta)\|^2} \right)^2 \frac{d\zeta}{\zeta - z} + \frac{1}{i\pi} \int_{B(0,R)} \sum_{1 \leq j \leq m} \frac{\overline{P_j(\zeta)} P_j(z)}{\|P(\zeta)\|^2} \bar{\partial}_\zeta Q(z,\zeta).$$

Since none of the polynomials is constant we have $\|P(\zeta)\| \geq c(1 + |\zeta|)$ and $|P_j(\zeta)| \|P(\zeta)\|^{-2} \leq \dfrac{c}{(1 + |\zeta|)}$. It follows that if we let $R \to \infty$ the first integral tends to zero and the second one converges. It is clear that as functions of z the integrals

$$A_j(z) = \frac{1}{i\pi} \int_{\mathbf{C}} \frac{\overline{P_j(\zeta)}}{\|P(\zeta)\|^2} \sum_{k=1}^{m} \frac{\partial}{\partial \bar{\zeta}} \left[\frac{\overline{P_k(\zeta)}}{\|P(\zeta)\|^2} \right] g_k(z, \zeta)\, d\bar{\zeta} \wedge d\zeta$$

are polynomials and

$$\sum_{j=1}^{m} P_j A_j \equiv 1.$$

In order to prove (ii) we let $\varphi(z, \zeta) = 1 + (z - \zeta) q(z, \zeta) = \sum_{j=1}^{m} \dfrac{P_j(z)\overline{P_j(\zeta)}}{\|P(\zeta)\|^2}$ and $\theta_\lambda(z, \zeta) = 1 + \overline{P_1(\zeta)} |P_1(\zeta)|^{2(\lambda-1)} g_1(z, \zeta)(z - \zeta)$ for $\operatorname{Re} \lambda$ sufficiently large. Let χ be a radial function in $\mathscr{D}(B(0, 2))$, $\chi = 1$ in a neighborhood of $\bar{B}(0, 1)$. We let $f(z) \equiv 1$, $\Phi(z, \zeta) = \varphi(z, \zeta)\theta_\lambda(z, \zeta)\chi\left(\dfrac{\zeta}{R}\right)$. Then $\Phi(z, z) = 1$ for $|z| < R$. Let $\Omega = B(0, 2R)$ and apply §3.8.14. Since $\Phi(z, \zeta) = 0$ when $\zeta \in \partial\Omega$, for $z \in B(0, R)$ we have

$$\begin{aligned} 1 = &-\frac{1}{2\pi i R} \int_{\mathbf{C}} \varphi(z, \zeta)\theta_\lambda(z, \zeta) \frac{\bar{\partial}\chi(\zeta/R) \wedge d\zeta}{\zeta - z} \\ &+ \frac{\lambda}{2\pi i} \int_{\mathbf{C}} \varphi(z, \zeta) |P_1(\zeta)|^{2(\lambda-1)} \overline{P_1'(\zeta)} g_1(z, \zeta) \chi\left(\frac{\zeta}{R}\right) d\bar{\zeta} \wedge d\zeta \\ &+ \frac{1}{2\pi i} \int_{\mathbf{C}} \theta_\lambda(z, \zeta) \chi\left(\frac{\zeta}{R}\right) \frac{\partial q}{\partial \bar{\zeta}}(z, \zeta)\, d\bar{\zeta} \wedge d\zeta. \end{aligned}$$

The idea now is to apply Stokes' formula to the last integral. Observe that

$$\begin{aligned} \frac{\partial}{\partial \bar{\zeta}} \left(\theta_\lambda(z, \zeta) q(z, \zeta) \chi\left(\frac{\zeta}{R}\right) \right) &= \frac{1}{R} \frac{\partial \chi}{\partial \bar{\zeta}} \left(\frac{\zeta}{R}\right) q(z, \zeta)\theta_\lambda(z, \zeta) + \chi\left(\frac{\zeta}{R}\right) \frac{\partial q}{\partial \bar{\zeta}}(z, \zeta)\theta_\lambda(z, \zeta) \\ &\quad + \lambda\chi\left(\frac{\zeta}{R}\right) q(z, \zeta) g_1(z, \zeta)(z - \zeta) |P_1(\zeta)|^{2(\lambda-1)} \overline{P_1'(\zeta)}. \end{aligned}$$

This allows us to replace the last integral by a sum of two integrals. One obtains

$$\begin{aligned} 1 = &-\frac{1}{2\pi i R} \int_{\mathbf{C}} \frac{\partial \chi}{\partial \bar{\zeta}} \left(\frac{\zeta}{R}\right) \theta_\lambda(z, \zeta) \left(\frac{\varphi(z, \zeta)}{\zeta - z} + q(z, \zeta) \right) d\bar{\zeta} \wedge d\zeta \\ &+ \frac{\lambda}{2\pi i} \int_{\mathbf{C}} \varphi(z, \zeta) \chi\left(\frac{\zeta}{R}\right) g_1(z, \zeta) |P_1(\zeta)|^{2(\lambda-1)} \overline{P_1'(\zeta)}\, d\bar{\zeta} \wedge d\zeta \\ &- \frac{\lambda}{2\pi i} \int_{\mathbf{C}} \chi\left(\frac{\zeta}{R}\right) q(z, \zeta)(z - \zeta) g_1(z, \zeta) |P_1(\zeta)|^{2(\lambda-1)} \overline{P_1'(\zeta)}\, d\bar{\zeta} \wedge d\zeta. \end{aligned}$$

Let R be sufficiently large so that all zeros of P_1 lie in $|\zeta| < R$. We have then the right to let $\lambda \to 0$ and obtain that the second integral becomes

$$\left\langle \operatorname{Res}\left(\frac{1}{\pi P_1}\right), \chi\left(\frac{\zeta}{R}\right) g_1(z,\zeta)\varphi(z,\zeta) \right\rangle = \sum_{j=1}^{m} \left\langle \operatorname{Res}\left(\frac{1}{\pi P_1}\right), g_1(z,\zeta)\frac{\overline{P_j(\zeta)}}{\|P(\zeta)\|^2} \right\rangle P_j(z).$$

Under the same conditions, the last integral becomes

$$\begin{aligned} -&\left\langle \operatorname{Res}\left(\frac{1}{\pi P_1}\right), \chi\left(\frac{\zeta}{R}\right) q(z,\zeta)(P_1(z) - P_1(\zeta)) \right\rangle \\ &= -\left\langle \operatorname{Res}\left(\frac{1}{\pi P_1}\right), q(z,\zeta) \right\rangle P_1(z) \\ &= -P_1(z) \sum_{j=1}^{m} \left\langle \operatorname{Res}\left(\frac{1}{\pi P_1}\right), \frac{\overline{P_j(\zeta)} g_j(z,\zeta)}{\|P(\zeta)\|^2} \right\rangle, \end{aligned}$$

since $P_1(\zeta)q(z,\zeta)$ is killed by $\operatorname{Res}\left(\frac{1}{\pi P_1}\right)$.

Finally, we consider the first integral. We have that

$$\frac{\varphi(z,\zeta)}{\zeta - z} + q(z,\zeta) = \frac{1}{\zeta - z}.$$

Since the integrand is only different from zero when $R < |\zeta| < 2R$, we are allowed to let $\lambda = 0$ in θ_λ. We then have

$$\theta_0(z,\zeta) = 1 + \frac{\overline{P_1(\zeta)}}{|P_1(\zeta)|^2}(P_1(z) - P_1(\zeta)) = \frac{P_1(z)}{P_1(\zeta)}.$$

Therefore,

$$-\frac{1}{2\pi i R}\int_{\mathbb{C}} \frac{\partial \chi}{\partial \bar{\zeta}}\left(\frac{\zeta}{R}\right)\theta_0(z,\zeta)\frac{d\bar{\zeta} \wedge d\zeta}{\zeta - z} = -\frac{P_1(z)}{2\pi i R}\int_{\mathbb{C}} \frac{\partial \chi}{\partial \bar{\zeta}}\left(\frac{\zeta}{R}\right)\frac{d\bar{\zeta} \wedge d\zeta}{P_1(\zeta)(\zeta - z)} \to 0$$

as $R \to \infty$. We conclude that

$$1 = \left\langle \operatorname{Res}\left(\frac{1}{\pi P_1(\zeta)}\right), \sum_{j=1}^{m} \frac{\overline{P_j(\zeta)}}{\|P(\zeta)\|^2} \begin{vmatrix} g_1(z,\zeta) & g_j(z,\zeta) \\ P_1(z) & P_j(z) \end{vmatrix} \right\rangle. \qquad \square$$

We conclude this section with another application of the Pompeiu formula with weights and residues. The object is to find an explicit formula to write a division theorem with remainder for holomorphic functions. That is, given f, $h \in \mathscr{H}(\bar{B})$, $B = B(0,1)$, we want to find a systematic way to obtain α, $\beta \in \mathscr{H}(B)$ such that

$$h = \alpha f + \beta$$

and $\beta = 0$ if and only if h is a multiple of f.

In order to do so we let $g(z,\zeta) = \dfrac{f(z) - f(\zeta)}{z - \zeta}$, and consider two auxiliary functions $q(\zeta) = \dfrac{\bar{\zeta}}{|\zeta|^2 - 1}$, $p_\lambda(z, \zeta) = |f(\zeta)|^{2(\lambda-1)}\overline{f(\zeta)}g(z, \zeta)$. We have

$$(z - \zeta)q(\zeta) = \frac{z\bar{\zeta} - |\zeta|^2}{|\zeta|^2 - 1} \quad \text{and} \quad 1 + (z - \zeta)q(\zeta) = \frac{z\bar{\zeta} - 1}{|\zeta|^2 - 1}.$$

Moreover,

$$(z - \zeta)p_\lambda(\zeta) = |f(\zeta)|^{2(\lambda-1)}\overline{f(\zeta)}(f(z) - f(\zeta)),$$

and

$$1 + (z - \zeta)p_\lambda(\zeta) = |f(\zeta)|^{2(\lambda-1)}\overline{f(\zeta)}f(z) + (1 - |f(\zeta)|^{2\lambda}).$$

For a positive integer N, arbitrary for the moment, we let

$$\begin{aligned}\Phi(z, \zeta) &= (1 + (z - \zeta)q(\zeta))^{-N}(1 + (z - \zeta)p_\lambda(z, \zeta)) \\ &= \left(\frac{|\zeta|^2 - 1}{z\bar{\zeta} - 1}\right)^N (|f(\zeta)|^{2(\lambda-1)}\overline{f(\zeta)}f(z) + (1 - |f(\zeta)|^{2\lambda})).\end{aligned}$$

We fix a point $z \in B$ and apply Lemma 3.8.14 to the functions h and Φ. We have

$$\begin{aligned}h(z) &= \frac{1}{2\pi i}\int_{\partial B} \Phi(z, \zeta)h(\zeta)\frac{d\zeta}{\zeta - z} - \frac{1}{2\pi i}\int_B h(\zeta)\frac{\partial\Phi}{\partial\bar{\zeta}}(z, \zeta)\frac{d\bar{\zeta} \wedge d\zeta}{\zeta - z} \\ &= -\frac{1}{2\pi i}\int_B h(\zeta)\frac{\partial\Phi}{\partial\bar{\zeta}}(z, \zeta)\frac{d\bar{\zeta} \wedge d\zeta}{\zeta - z},\end{aligned}$$

since Φ vanishes on ∂B. For $\operatorname{Re}\lambda$ sufficiently large, we can easily compute

$$\begin{aligned}-\frac{h(\zeta)}{\zeta - z}\frac{\partial\Phi}{\partial\bar{\zeta}}(z, \zeta) = h(\zeta)\Bigg[&\frac{N(|\zeta|^2 - 1)^{N-1}}{(z\bar{\zeta} - 1)^{N+1}}(|f(\zeta)|^{2(\lambda-1)}\overline{f(\zeta)}f(z) + (1 - |f(\zeta)|^{2\lambda})) \\ &+ \lambda\left(\frac{|\zeta|^2 - 1}{z\bar{\zeta} - 1}\right)^N |f(\zeta)|^{2(\lambda-1)}\overline{f'(\zeta)}g(z, \zeta)\Bigg].\end{aligned}$$

Therefore, for $\operatorname{Re}\lambda$ sufficiently large, we have

$$\begin{aligned}h(z) = {}&\frac{N}{2\pi i}\left[\int_B h(\zeta)\frac{(|\zeta|^2 - 1)^{N-1}}{(z\bar{\zeta} - 1)^{N+1}}|f(\zeta)|^{2(\lambda-1)}\overline{f(\zeta)}\, d\bar{\zeta} \wedge d\zeta\right]f(z) \\ &+ \frac{N}{2\pi i}\int_B h(\zeta)\frac{(|\zeta|^2 - 1)^{N-1}}{(z\bar{\zeta} - 1)^{N+1}}(1 - |f(\zeta)|^{2\lambda})\, d\bar{\zeta} \wedge d\zeta \\ &+ \frac{\lambda}{2\pi i}\int_B h(\zeta)\left(\frac{|\zeta|^2 - 1}{z\bar{\zeta} - 1}\right)^N g(z, \zeta)|f(\zeta)|^{2(\lambda-1)}\overline{f'(\zeta)}\, d\bar{\zeta} \wedge d\zeta.\end{aligned}$$

We know that $\lambda \mapsto |f|^{2(\lambda-1)}\overline{f'}$ has an analytic continuation as a meromorphic function with values in $\mathscr{D}'(B(0, 1 + \varepsilon))$ for some $\varepsilon > 0$. Furthermore, to obtain this analytic continuation for the half-plane $c < \operatorname{Re}\lambda$, we only need to do a finite number of integrations by parts. Therefore this distribution will act on functions which are only required to have L continuous derivatives (L depends on c). Choosing N a sufficiently large integer, the function $\left(\frac{|\zeta|^2 - 1}{z\bar{\zeta} - 1}\right)^N$, extended as zero for ζ in $\bar{B}^c$, will be of class C^L in a neighborhood of $\bar{B}$, and will have support in $\bar{B}$. Therefore we can let $\lambda \to 0$ in the third integral and obtain

$$\beta(z) := \left\langle \operatorname{Res}\left(\frac{1}{\pi f}\right), h(\zeta)g(z,\zeta)\left(\frac{|\zeta|^2 - 1}{z\bar{\zeta} - 1}\right)^N \right\rangle.$$

This function $\beta \in \mathscr{H}(B)$ and, furthermore, if h is a multiple of f, then automatically $\beta \equiv 0$.

The second integral in the representation of h tends to zero as $\lambda \to 0$ since $(1 - |f(\zeta)|^{2\lambda}) \to 0$ almost everywhere in B.

Finally, when $\lambda \to 0$ the distributions $|f|^{2(\lambda-1)}\bar{f} = \dfrac{|f|^{2\lambda}}{f}$ tend to $pv\left(\dfrac{1}{f}\right)$ in $\mathscr{D}'(B(0, 1 + \varepsilon))$, acting on functions with only finitely many continuous derivatives by the same argument of analytic continuation. Letting N be sufficiently large for all these conditions to be satisfied, we can define

$$\alpha(z) := N\left\langle pv\left(\frac{1}{\pi f}\right), h(\zeta)\frac{(|\zeta|^2 - 1)^{N-1}}{(z\bar{\zeta} - 1)^{N+1}} \right\rangle$$

which is also in $\mathscr{H}(B)$. We have therefore

$$h = \alpha f + \beta,$$

with the properties we claimed.

A corollary of this decomposition is the following interesting result. Let Ω be a connected open set and $f, h \in \mathscr{H}(\Omega)$, then $h \in f\mathscr{H}(\Omega)$ if and only if the distribution $h \operatorname{Res}\left(\dfrac{1}{f}\right)$ is identically zero in $\mathscr{D}'(\Omega)$.

Another corollary of the explicit formula for α is that when $h = gf$ then $h\, pv\left(\dfrac{1}{\pi f}\right) = g$ and the decomposition of h becomes

$$g(z) = \frac{N}{2\pi i}\int_B g(\zeta)\frac{(|\zeta|^2 - 1)^{N-1}}{(z\bar{\zeta} - 1)^{N+1}}\, d\bar{\zeta} \wedge d\zeta.$$

This kernel, $\dfrac{N}{2\pi i}\dfrac{(|\zeta|^2 - 1)^{N-1}}{(z\bar{\zeta} - 1)^{N+1}}$ plays a crucial role in the work of P. Charpentier to obtain minimal solutions of the equation $\bar{\partial}u = f$ in the ball and the polydisk in $\mathbb{C}^n$. It is the kernel of the orthogonal projection of the Hilbert space $L^2\left(B, (1 - |\zeta|^2)^{N-1}\dfrac{d\bar{\zeta} \wedge d\zeta}{2i}\right)$ onto the closed subspace of holomorphic functions (see [Ch]).

Finally, note that the formula for the remainder term β depends only on the values of h near the zeros of f.

EXERCISES 3.8

1. Let $x_+ := \max\{0, x\}$, then x_+^λ is well defined for any $\lambda \in \mathbb{C}$ ($0^\lambda = 0$). Moreover, when $\operatorname{Re}\lambda > -1$ the function $x \mapsto x_+^\lambda$ is locally integrable in $\mathbb{R}$. It defines a distribution which we still denote x_+^λ, by

$$\langle x_+^\lambda, \varphi\rangle := \int_{-\infty}^{\infty} x_+^\lambda \varphi(x)\,dx = \int_0^{\infty} x^\lambda \varphi(x)\,dx, \qquad (\varphi \in \mathscr{D}(\mathbb{R})).$$

   (a) Show that for any $\varphi \in \mathscr{D}(\mathbb{R})$, the function

$$\lambda \mapsto \langle x_+^\lambda, \varphi\rangle$$

   is a holomorphic function in the half-plane $\{\lambda \in \mathbb{C} : \operatorname{Re}\lambda > -1\}$.

   (b) Show that, for $\operatorname{Re}\lambda > -1$,

$$\langle x_+^\lambda, \varphi\rangle = \int_0^1 x^\lambda[\varphi(x) - \varphi(0)]\,dx + \frac{\varphi(0)}{\lambda + 1} + \int_1^\infty x^\lambda \varphi(x)\,dx.$$

   Explain why the third term is an entire function of λ and the first one defines a holomorphic function for $\operatorname{Re}\lambda > -2$. Note that $\operatorname{Res}(x_+^\lambda, \lambda = -1) = \delta$, with the obvious meaning for the notation.

   (c) Show that $\lambda \mapsto x_+^\lambda$ has an analytic continuation to a meromorphic function in the whole plane such that

$$f_+^\lambda : \lambda \mapsto \frac{x_+^\lambda}{\Gamma(\lambda + 1)}$$

   is an entire function with values in $\mathscr{D}'(\mathbb{R})$. Find its values for $\lambda = -1, -2, \ldots$. Show that

$$\frac{d}{dx} f_+^\lambda = f_+^{\lambda - 1}.$$

   (d) Defining $x_- = \max\{-x, 0\}$, repeat the procedure to study $\lambda \mapsto x_-^\lambda$.

   (e) Define z^λ in $\mathbb{C}\setminus\mathbb{R}$ by using the principal branch of the argument, $-\pi < \operatorname{Arg} z < \pi$. Show that for $\operatorname{Re}\lambda > -1$, we have the identities

$$(x + i0)^\lambda = b_+(z^\lambda) = x_+^\lambda + e^{i\lambda\pi}x_-^\lambda$$
$$(x - i0)^\lambda = b_-(z^\lambda) = x_+^\lambda + e^{-i\lambda\pi}x_-^\lambda.$$

   These identities allow us to obtain the analytic continuations of $(x \pm i0)^\lambda$ as (possibly) meromorphic functions in $\mathbb{C}$ with values in $\mathscr{D}'(\mathbb{R})$. Show they are entire functions.

2. Let $f \in \mathscr{H}(\Omega)$, $f \not\equiv 0$. Show that $f \operatorname{Res}\left(\dfrac{1}{f}\right) = 0$ as a distribution.

3. Prove $\left\langle pv\left(\dfrac{1}{\pi z^m}\right), \bar{z}^m \dfrac{\partial^m \varphi}{\partial \bar{z}^m}\right\rangle = (-1)^m m! \left\langle pv\left(\dfrac{1}{\pi z^m}\right), \varphi\right\rangle.$

4. Show that for $k \in \mathbb{N}$, $\operatorname{Res}\left(\dfrac{1}{\pi}|z|^{2(\mu-1)}, \mu = -k\right) = \dfrac{1}{(k!)^2}\dfrac{\partial^{2k}\delta}{\partial z^k \partial \bar{z}^k}$.

5. Let $g \in \mathscr{H}(B(0, r))$, $g(z) \neq 0$ in $B(0, r)$, and let $f(z) = z^m g(z)$. Use the analytic continuation formula for $|z|^\lambda$ to obtain the result from Proposition 3.8.12. (Hint: write

$\langle |f|^{2(\mu-1)}\bar{f}', \varphi\rangle = \langle |z^m|^{2(\mu-1)}, \bar{f}'|g|^{2(\mu-1)}\varphi\rangle = \langle |z|^{2(\lambda-1)}, \bar{f}'|g|^{2(\lambda-1)/m}\varphi\rangle$, where $\lambda = 1 + m(\mu - 1)$. Hence $\mu = 0$ corresponds to $\lambda = -m + 1$. Use that the function $\lambda \mapsto |g|^{2(\lambda-1)/m}$, is holomorphic in $\mathscr{D}'(B(0,r))$ and the Exercise 3.8.4 to complete the proof.)

6. In the case of two polynomials P_1, P_2 without common zeros, this exercise provides a very easy proof of Proposition 3.8.15.
   (a) Let $R > 0$ be so large that $Z(P_1) \subseteq B(0,R)$. Show that for $z \in B(0,R)$ one has

$$\frac{1}{2\pi i}\sum_{m\geq 0}(P_1(z))^m\int_{|\zeta|=R}\frac{1}{(P_1(\zeta))^{m+1}}\frac{P_1(\zeta)-P_1(z)}{\zeta - z}\,d\zeta = 1.$$

$$\left(\text{Hint: Write } 1 = \frac{1}{2\pi i}\int_{|\zeta|=R}\frac{P_1(\zeta)-P_1(z)}{P_1(\zeta)-P_1(z)}\frac{d\zeta}{\zeta - z}.\right)$$

   (b) Show that if $\deg P_1 \geq 1$, then

$$\frac{1}{2\pi i}\int_{|\zeta|=R}\frac{1}{(P_1(\zeta))^{m+1}}\frac{P_1(\zeta)-P_1(z)}{\zeta - z}\,d\zeta = 0$$

   for all $m \geq 1$ and $z \in B(0,R)$. Conclude that for every $z \in B(0,R)$

$$\left\langle \operatorname{Res}\left(\frac{1}{\pi P_1(\zeta)}\right), \frac{P_1(\zeta)-P_1(z)}{\zeta - z}\right\rangle = 1.$$

   $\left(\text{It is understood the distribution } \operatorname{Res}\left(\frac{1}{\pi P_1(\zeta)}\right) \text{ acts on the variable } \zeta.\right)$

   (c) Using that $Z(P_1) \cap Z(P_2) = \varnothing$, show that (b) becomes

$$\left\langle \operatorname{Res}\left(\frac{1}{\pi P_1(\zeta)}\right), \frac{1}{P_2(\zeta)}\left[P_2(\zeta)\frac{P_1(\zeta)-P_1(z)}{\zeta - z}\right]\right\rangle.$$

   (d) Write $P_2(\zeta)\dfrac{P_1(\zeta)-P_1(z)}{\zeta - z} = \begin{vmatrix} \dfrac{P_1(z)-P_1(\zeta)}{\zeta - z} & 0 \\ 0 & P_2(\zeta)\end{vmatrix}$ and use properties of the determinant to conclude

$$P_2(\zeta)\frac{P_1(\zeta)-P_1(z)}{\zeta - z} = \begin{vmatrix} \dfrac{P_1(z)-P_1(\zeta)}{\zeta - z} & \dfrac{P_2(z)-P_2(\zeta)}{\zeta - z} \\ P_1(z) & P_2(z)\end{vmatrix} + P_1(\zeta)g(\zeta, z),$$

   g holomorphic in ζ.
   (e) With the help of Exercise 3.8.2, conclude that for $z \in B(0,R)$,

$$1 = \left\langle \operatorname{Res}\left(\frac{1}{\pi P_1(\zeta)}\right), \frac{1}{P_2(\zeta)}\begin{vmatrix} \dfrac{P_1(z)-P_1(\zeta)}{\zeta - z} & \dfrac{P_2(z)-P_2(\zeta)}{\zeta - z} \\ P_1(z) & P_2(z)\end{vmatrix}\right\rangle.$$

   Since R is arbitrary this identity holds for all $z \in \mathbb{C}$.

7. Let $C_\infty(\mathbb{R})$ denote the space of continuous function in $\mathbb{R}$ such that $\lim_{|x|\to\infty} f(x) = 0$. It is a Banach space with the norm $\|f\| := \max_{x\in\mathbb{R}} |f(x)|$. Any continuous linear

functional in $C_\infty(\mathbb{R})$ is given by integration against a Radon measure $d\mu$ such that $\int_{\mathbb{R}} d|\mu| < \infty$ (see [HS] or [Ru]).

(a) Show that if $\varphi \in \mathscr{D}(\mathbb{R})$ and $f \in C_\infty(\mathbb{R})$, then $\varphi * f \in C_\infty(\mathbb{R})$ and is a C^∞ function. Prove also that $\mathscr{D}(\mathbb{R})$ is dense in $C_\infty(\mathbb{R})$.

*(b) Let $(z_k)_{k\geq 1}$ be a sequence of complex numbers with $\operatorname{Im} z_k > 0$ and $z_k \to z_0$ for some z_0, with $\operatorname{Im} z_0 > 0$. Prove that the family of linear combinations of the functions $\dfrac{1}{x - z_k}, \dfrac{1}{x - \bar{z}_k}$ is dense in $C_\infty(\mathbb{R})$.

Notes to Chapter 3

1. The treatment of Runge's theorem, its relation to the Mittag-Leffler theorems, and the solvability of the inhomogeneous Cauchy-Riemann equation follow basically [Ho1], where presented as an introduction to the theory of several complex variables. The fact that solvablity results and approximation results are related extends to all linear partial differential operators and convolution operators; we refer the reader to Volume 2 of the treatise [Ho2].

2. As we mentioned in §3.7, approximation problems can be considered in many different lights. We refer to three excellent introductions to the subject, [Za1], [Ga], and [Fu1], for a rather complete treatment of Runge's and Mergelyan's theorems and their generalizations, as well as to some of the literature on uniform algebras (see also [Gam] and [Wer]). The proof of Theorem 3.7.1 is essentially from [Mer]. The role of analytic capacity, Proposition 2.7.9, is very crucial. There are many questions open about this capacity and we refer to [Vi] for details.

3. The proof of Weierstrass' theorem given in the text follows [BT]. The proofs in the exercises are classical and can be found in [Ah1] and [Mar]. We will return to these Weierstrass expansions in Chapter 4.

4. Mittag-Leffler Theorem 3.2.2 is clearly a theorem about vanishing of cohomology. In this form it generalizes to several complex variables.

5. We gave two proofs to the interpolation theorem, namely, §3.4.1 and §3.6.9, because they introduce ideas that are central to the relations between complex analysis and harmonic analysis to be explored in the second volume.

6. The theory of ideals in $\mathscr{H}(\Omega)$ follows the work of Helmer and Henriksen. The reader will find a similar treatment in [LR], where a functional analysis approach to complex analysis is also emphasized.

7. The "jump formulas" of Exercise 3.6.9 were first found by Sokhotski in 1873. They appear naturally in many problems of integral and differential equations [Gak]. These formulas lead naturally to consideration of more general boundary values of holomorphic functions, in particular in the sense of distributions. Heaviside and others, in the first decades of this century, were led to consider these "generalized functions" while developing systematic methods, known as "operational calculus," to solve partial differential equations. We recommend [Bre] and the first volume of the treatise [G-S] for a modern introduction to the systematic use of boundary values of holomorphic functions in the theory of distributions. In the rather recent past, M. Sato recognized the usefulness of assigning a boundary value to *every* holomorphic function in $\mathbb{C}\setminus\mathbb{R}$. This leads to the theory of hyperfunctions and the very powerful algebraic analysis

methods of Sato and his school. The proof of the Edge-of-the-Wedge Theorem 3.6.23 follows Martineau [Mart], who also recognized the link between this theorem and the theory of hyperfunctions. We will discuss hyperfunctions in the second volume. There are many new books on this emerging subject, none very elementary. Perhaps the best introduction is [KKK].

8. Given a Jordan domain Ω, clearly the function $f(z) = \bar{z}$ is far from being approximated uniformly in $\bar{\Omega}$ by polynomials. Nevertheless, the distance $d(\bar{z}, \mathbb{C}[z])$ ($\mathbb{C}[z]$ = space of all polynomials) in the metric of $\mathscr{C}(\bar{\Omega})$ appears already in the work of Ahlfors-Beurling. It is very curious that this quantity can be related to the isoperimetric inequality, as discovered by D. Khavinson [GK].

9. This chapter has clearly shown the important role of the inhomogeneous Cauchy-Riemann equation in complex analysis. There is a cognate equation that plays a fundamental role in the study of quasiconformal maps and Teichmüller theory. It is the Beltrami equation $\frac{\partial f}{\partial \bar{z}} = \mu \frac{\partial f}{\partial z}$, where $\mu = L^\infty$. We refer the reader to [Ah3] for an introduction to this subject.

CHAPTER 4

Harmonic and Subharmonic Functions

§1. Introduction

A large number of properties of holomorphic functions (maximum principle, Schwarz's lemma, convexity properties, etc.) still hold for a much larger class of functions. It is the class of subharmonic functions (see Definition 4.4.1). The relation between these two classes of functions is given by the fact that if f is a holomorphic function, then $\log|f|$ is a subharmonic function.

If f is a holomorphic function without zeros, then $V = \log|f|$ is a C^∞ function, which verifies the Laplace equation

$$\Delta V = 0,$$

where Δ is the second-order differential operator, ***Laplace operator***, given by

$$\Delta = \frac{\partial^2}{\partial x^2} + \frac{\partial^2}{\partial y^2} = 4\frac{\partial^2}{\partial z \partial \bar{z}}.$$

One says that V is a ***harmonic function***.

On the other hand, in a neighborhood of a zero z_0 of order k of f, the function $V = \log|f|$ behaves like $k\log|z - z_0|$, hence it tends to $-\infty$ as z approaches z_0. Its Laplacian in the sense of distributions exists since V is locally integrable. Moreover, as will be seen later, in a neighborhood of z_0 we have

$$\Delta V = 2\pi k \delta_{z_0}.$$

The properties of these functions, the subharmonic and harmonic functions, are essentially properties of their average values over circles and disks. In order to simplify the hypotheses necessary to compute these averages we are obliged first to extend the notion of integration to the case of measurable functions, bounded above and with values in $[-\infty, \infty[$. We refer to the excellent books [Ru] and [HS] for the elements of the classical theory of integration.

§2. A Remark on the Theory of Integration

4.2.1. Proposition. *Let $(X, \mathcal{T}, \mu)$ be a complete measure space with a positive finite measure μ ($\mu(X) < \infty$). Let $f: X \to [-\infty, \infty[$ be a measurable function defined a.e. on X and bounded above a.e. For every $M \in \mathbb{R}$, which is an upper bound of f a.e., the value $I(M) \in [-\infty, \infty[$ given by*

$$I(M) = M\mu(X) - \int_X (M - f)\,d\mu$$

*is well defined and independent of the upper bound (**majorant**) M.*

PROOF. (i) If $I(M_0) \in \mathbb{R}$ for some M_0 upper bound of f a.e. and if M is another a.e. upper bound, then $M_0\mu(X) \in \mathbb{R}$ and $0 \le \int_X (M_0 - f)\,d\mu < \infty$. Hence, from $M - f = M - M_0 + M_0 - f$, one concludes that $I(M) \in \mathbb{R}$ also. Furthermore

$$I(M) - I(M_0) = (M - M_0)\mu(X) - \int_X (M - f)\,d\mu + \int_X (M_0 - f)\,d\mu,$$

but $\int_X (M - f)\,d\mu = (M - M_0)\mu(X) + \int_X (M_0 - f)\,d\mu$, hence $I(M) = I(M_0)$.

(ii) This argument also shows that if $I(M_0) = -\infty$ for some a.e. upper bound M_0 of f, then it must be the case that $I(M) = -\infty$ for every other majorant. □

We define the generalized integral of f over X as

$$\int_X f\,d\mu := I(M),$$

for an arbitrary majorant M of f. This value coincides with the usual one if f is μ-integrable. It also preserves the properties of linearity and monotonicity of the usual definition.

4.2.2. Examples. (1) Let X be a compact metric space and μ a positive Radon measure on X. We can now integrate, in the generalized sense, any upper semicontinuous function in X with values in $[-\infty, \infty[$, which is not identically equal to $-\infty$ (we will abbreviate this to u.s.c.). This is precisely the case of interest in what follows.

(2) $\int_{-1}^{1} \left(-\frac{1}{x^2}\right) dx = -\infty.$

4.2.3. Proposition. *Let X be a compact metric space and μ a positive Radon measure on X. For every upper semicontinuous function f on X taking values in $[-\infty, \infty[$ we have*

$$\int_X f\,d\mu = \inf\left\{\int_X \varphi\,d\mu : \varphi \geq f,\ \varphi \text{ continuous}\right\}.$$

PROOF. We need first the following.

4.2.4. Lemma. *Let X be a compact metric space and f u.s.c. on X. There exists a decreasing sequence of real valued continuous functions on X whose pointwise limit is f.*

PROOF. Let d be a distance function on X. Consider the sequence of functions

$$f_n(z) := \sup_{\zeta \in X} (f(\zeta) - nd(z, \zeta)), \qquad z \in X.$$

It is easy to see that $f_1 \geq f_2 \geq \cdots \geq f_n \geq \cdots \geq f$ and the f_n are real valued. (It is here where one uses that f does not take the value $+\infty$ and, hence, $\sup_\zeta f(\zeta) < \infty$ by the upper semicontinuity.) To show that for every z, $\lim_{n\to\infty} f_n(z) = f(z)$, it is enough to show that for a given $\varepsilon > 0$ there is an n large enough such that for all $\zeta \in X$,

$$f(\zeta) - nd(z, \zeta) \leq f(z) + \varepsilon.$$

If $f(z) > -\infty$, there is a $\delta > 0$ such that $d(z, \zeta) < \delta$ implies that $f(\zeta) \leq f(z) + \varepsilon$ by the semicontinuity. Therefore, we also have $f(\zeta) - nd(z, \zeta) \leq f(z) + \varepsilon$ if $n \geq 1$. On the other hand, if $d(z, \zeta) \geq \delta$ then

$$f(\zeta) - nd(z, \zeta) \leq \left(\sup_{w \in X} f(w)\right) - n\delta.$$

It is clear that there is an $n_z \geq 1$ such that the right-hand side is less than $f(z) + \varepsilon$. Therefore we have

$$f(\zeta) - nd(z, \zeta) \leq f(z) + \varepsilon, \qquad n \geq n_z,$$

in $X \setminus B(z, \delta)$. In other words,

$$f(z) \leq f_n(z) \leq f(z) + \varepsilon \qquad \text{if } n \geq n_z,$$

which shows that $f_n(z) \searrow f(z)$ when $f(z) \neq -\infty$.

If $f(z) = -\infty$, for any $A \in \mathbb{R}$, there is $\delta > 0$ such that $d(z, \zeta) < \delta$ implies that $f(\zeta) - nd(z, \zeta) < A$ for every $n \geq 1$. If $d(z, \zeta) \geq \delta$, then we have $f(\zeta) - nd(z, \zeta) \leq \left(\sup_{w \in X} f(w)\right) - n\delta < A$ once

$$n \geq n_A = \left(1 + \sup_{w \in X} f(w) - A\right)\delta^{-1}.$$

Hence $n \geq n_A$ implies

$$f(z) = -\infty < f_n(z) \leq A.$$

Finally, the continuity of the f_n can be seen using the inequality

$$|(f(\zeta) - nd(z,\zeta)) - (f(\zeta) - nd(z',\zeta))| \leq nd(z,z'),$$

which implies

$$f(\zeta) - nd(z,\zeta) \leq f(\zeta) - nd(z',\zeta) + nd(z,z'),$$

and, passing to the upper bounds,

$$f_n(z) \leq f_n(z') + nd(z,z').$$

Interchanging the role of z and z' we obtain

$$|f_n(z) - f_n(z')| \leq nd(z,z').$$

This concludes the proof of the lemma. □

PROOF of 4.2.3. We have first

$$\inf\left\{\int_X \varphi\, d\mu : \varphi \geq f,\ \varphi \text{ continuous}\right\} \geq \int_X f\, d\mu.$$

This can be seen easily considering separately the cases $\int_X f\, d\mu = -\infty$ and $\int_X f\, d\mu$ finite.

Conversely, let $(f_n)_{n\geq 1}$ be a decreasing sequence of continuous functions coverging pointwise to f. Let M be a strict majorant of f. The compactness of X implies there is a positive integer n_0 such that if $n \geq n_0$ then $f_n < M$. We can therefore write

$$\int_X f_n\, d\mu = M\mu(X) - \int_X (M - f_n)\, d\mu$$

$$\int_X f\, d\mu = M\mu(X) - \int_X (M - f)\, d\mu.$$

The Beppo Levi theorem ensures now that the sequence $\int_X (M - f_n)\, d\mu$ converges in a monotonously increasing way to $\int_X (M - f)\, d\mu$. Therefore, the sequence $\int_X f_n\, d\mu$ converges decreasingly to $\int_X f\, d\mu$. □

4.2.5. Remark. Proposition 4.2.3. shows that the generalized integral we have defined coincides with the Daniell integral for u.s.c. functions (cf. [HS]).

4.2.6. Proposition (Dini-Cartan). *Let X be a compact metric space, μ a positive Radon measure on X, and $(f_i)_{i\in I}$ be a cofinal decreasing family of u.s.c. functions on X. Then*

$$\inf_{i\in I}\int_X f_i\,d\mu = \int_X \left(\inf_{i\in I} f_i\right)d\mu.$$

PROOF. Recall that the hypotheses that the family $(f_i)_{i\in I}$ is decreasing and cofinal means that for $i, j \in I$ there is $k \in I$ such that

$$f_k \le f_i \qquad \text{and} \qquad f_k \le f_j$$

everywhere in X.

First, consider the case $\inf_I f_i \equiv 0$. We claim that given $\varepsilon > 0$ there is $i_0 \in I$ such that $f_{i_0} < \varepsilon$. In fact, for every $z \in X$ there is i_z such that $f_{i_z}(z) < \varepsilon$. By the upper semicontinuity there is $\delta_z > 0$ such that, if $d(\zeta, z) < \delta_z$, then we still have $f_{i_z}(\zeta) < \varepsilon$. The compactness of X allows us to find $i_1, \ldots, i_n \in I$ such that $\inf(f_{i_1}, \ldots, f_{i_n}) < \varepsilon$ everywhere. Since the family is cofinal, there is an $i_0 \in I$ such that $f_{i_0} < \varepsilon$ as claimed.

Hence, $\int_X f_{i_0}\,d\mu \le \varepsilon\mu(X)$ and therefore, $\inf_I \int_X f_i\,d\mu \le 0 = \int_X \left(\inf_I f_i\right)d\mu$ in this particular case.

If $f = \inf_{i\in I} f_i \not\equiv 0$, and if $\lambda \in \mathbb{R}$ is such that $\int_X f\,d\mu < \lambda$, then by Proposition 4.2.3 there is a continuous function $\varphi \ge f$ such that

$$\int_X f\,d\mu \le \int_X \varphi\,d\mu < \lambda.$$

Let $g_i = \sup(f_i - \varphi, 0)$. The family $(g_i)_{i\in I}$ is now a decreasing, cofinal family of u.s.c. functions such that $\inf_I g_i = 0$. Therefore, $\inf_I \int_X g_i\,d\mu \le 0$. This implies that $\inf_I \int_X (f_i - \varphi)\,d\mu \le 0$, and hence

$$\inf_I \int_X f_i\,d\mu \le \int_X \varphi\,d\mu < \lambda.$$

A posteriori,

$$\inf_I \int_X f_i\,d\mu \le \int_X f\,d\mu = \int \left(\inf_I f_i\right)d\mu.$$

Since the reverse inequality is evident, the proposition holds. □

4.2.7. Proposition (Fubini-Tonelli). *Let $(X, \mathcal{T}, \mu)$ and $(Y, \mathcal{S}, \nu)$ be two measure spaces with μ, ν bounded positive Radon measures and $\mu \times \nu$ the product mea-*

sure. Let $f: X \times Y \to [-\infty, \infty[$ be an everywhere-defined measurable function for the product tribe $\mathscr{T} \times \mathscr{S}$, and let f be bounded above everywhere. Then

(1) *For every $y \in Y$ the function $x \mapsto f(x, y)$ is $\mathscr{T}$-measurable and for every $x \in X$ the function $x \mapsto f(x, y)$ is $\mathscr{S}$-measurable.*

(2) *The function $y \mapsto \int_X f(x, y)\, d\mu(x)$ takes values in $[-\infty, \infty[$ and is bounded above and $\mathscr{S}$-measurable. The function $x \mapsto \int_X f(x, y)\, dv(y)$ takes values in $[-\infty, \infty[$, is bounded,* and *is $\mathscr{T}$-measurable.*

(3) $\int_{X \times Y} fd(\mu \times v) = \int_X \left[\int_Y f\, dv\right] d\mu = \int_Y \left[\int_X f\, d\mu\right] dv.$

PROOF. Apply the usual theorem of Fubini-Tonelli to the function $M - f$ for a majorant M of f. □

4.2.8. Remark. In the case of compact subsets of $\mathbb{R}^n$, the usual theorem of change of variables is always valid for the integral we have just defined when the compacts are related by a C^1-diffeomorphism defined in their neighborhood.

§3. Harmonic Functions

4.3.1. Definitions. Let Ω be a nonempty open subset of $\mathbb{C}$.

(1) Let $f \in L^1_{\text{loc}}(\Omega)$ (resp. $f: \Omega \to [-\infty, \infty[$ measurable and bounded above on every compact subset of Ω). For every closed disk $\bar{B}(z, r) \subseteq \Omega$, we call ***area average*** of f over $\bar{B}(z, r)$, and denote it $A(f, z, r)$, the complex number (resp. the element of $[-\infty, \infty[$) given by

$$A(f, z, r) = \frac{1}{\pi r^2} \int_{\bar{B}(z,r)} f\, dm,$$

where dm represents the Lebesgue measure in $\mathbb{C}$, as usual.

(2) Let $f: \Omega \to \mathbb{C}$, $z \in \Omega$, $r > 0$ be such that $\bar{B}(z, r) \subseteq \Omega$ and

$$f \mid \partial B(z, r) \in L^1(\partial B(z, r), d\sigma),$$

where $d\sigma = r\, d\theta$ is the Lebesgue measure on $\partial B(z, r)$ (resp. $f: \Omega \to [-\infty, \infty[$, measurable and bounded above on every compact subset of Ω). We denote $\lambda(f, z, r)$ the ***circular average*** of f over $\partial B(z, r)$, i.e. the complex number (resp. the element in $[-\infty, \infty[$) given by

$$\lambda(f, z, r) = \frac{1}{2\pi r} \int_{\partial B(z,r)} f\, d\sigma.$$

4.3.2. Remarks

(1) If $f \in L^1_{\text{loc}}(\Omega)$, $z \mapsto A(f, z, r)$ is continuous in $\Omega_r := \{z \in \Omega : d(z, \Omega^c) > r\}$ ($r > 0$ fixed). For $z \in \Omega$, the function $r \mapsto A(f, z, r)$ is continuous in $]0, d(z, \Omega^c)[$.

(2) Integrating in polar coordinates one finds the relation

$$A(f, z, r) = \frac{2}{r^2} \int_0^r \lambda(f, z, \rho)\rho\, d\rho,$$

since, in the case $f \in L^1_{\text{loc}}(\Omega)$, the function $\theta \mapsto f(z + \rho e^{i\theta})$ belongs to $L^1(\partial B(z, \rho), d\theta)$ for almost every $\rho \in [0, r]$.

4.3.3. Proposition. *Let Ω be a nonempty open subset of $\mathbb{C}$. The following statements are equivalent for any $f : \Omega \to \mathbb{C}$.*

(1) *$f \in \mathscr{E}(\Omega)$ and $\Delta f = 0$.*
(2) *$f \in \mathscr{E}_0(\Omega)$ and $f(z) = \lambda(f, z, r)$ for every $\bar{B}(z, r) \subseteq \Omega$.*
(3) *$f \in \mathscr{E}_0(\Omega)$ and $f(z) = A(f, z, r)$ for every $\bar{B}(z, r) \subseteq \Omega$.*
(4) *$f \in L^1_{\text{loc}}(\Omega)$ and $f(z) = A(f, z, r)$ for every $\bar{B}(z, r) \subseteq \Omega$.*
(5) *$f \in L^1_{\text{loc}}(\Omega)$ and $f(z) = (f * \varphi_r)(z)$ in $\Omega_r = \{z \in \Omega : d(z, \Omega^c) > r\}$ if $r > 0$ is sufficiently small and φ is a standard function.*

In the case that f satisfies any of these properties, we say that it is a ***harmonic function*** in Ω.

PROOF. The plan of the proof is:

$$(1) \Rightarrow (2) \Rightarrow (3) \Rightarrow (2) \Rightarrow (5) \Rightarrow (3); \quad (3) \Leftrightarrow (4); \quad \text{and } (2) \Rightarrow (1);$$

although it is clear that some of the arrows are redundant.

(a) $(1) \Rightarrow (2)$. We have $\lambda(f, z, r) = \dfrac{1}{2\pi} \displaystyle\int_0^{2\pi} f(z + re^{i\theta})\, d\theta$. Since f is C^∞, it follows from this formula that $r \mapsto \lambda(f, z, r)$ is differentiable in $[0, d(z, \Omega^c)[$ and

$$\frac{d}{dr}\lambda(f, z, r) = \frac{1}{2\pi}\int_0^{2\pi} \frac{\partial}{\partial r} f(z + re^{i\theta})\, d\theta.$$

Now

$$\frac{\partial}{\partial r} f(z + re^{i\theta}) = \frac{\partial f}{\partial x}(z + re^{i\theta})\cos\theta + \frac{\partial f}{\partial y}(z + re^{i\theta})\sin\theta = \frac{\partial f}{\partial n}(z + re^{i\theta}).$$

Recall that Green's formula states that

$$\int_{\bar{B}(z,r)} \Delta f(\zeta)\, d\xi\, d\eta = r\int_0^{2\pi} \frac{\partial f}{\partial n}(z + re^{i\theta})\, d\theta \qquad (\zeta = \xi + i\eta).$$

Altogether this gives

$$\frac{d}{dr}\lambda(f,z,r) = \frac{1}{2\pi r}\int_{\bar{B}(z,r)} \Delta f(\zeta)\,d\xi\,d\eta.$$

Since $\Delta f = 0$ we get $\frac{d}{dr}\lambda(f,z,r) = 0$. Therefore $\lambda(f,z,r)$ is constant in $[0, d(z,\Omega^c)[$ and the value $\lambda(f,z,0) = \lim_{r\to 0}\lambda(f,z,r) = f(z)$, allows us to conclude that

$$\lambda(f,z,r) = f(z) \quad \text{for } r \in [0, d(z,\Omega^c)[.$$

(b) (2) $\Rightarrow$ (3). It follows from $A(f,z,r) = \frac{2}{r^2}\int_0^r \lambda(f,z,\rho)\rho\,d\rho$.

(c) (3) $\Rightarrow$ (2). If $f(z) = A(f,z,r)$ for every $r \in [0, d(z,\Omega^c)[$, we have

$$r^2 f(z) = r^2 A(f,z,r) = 2\int_0^r \lambda(f,z,\rho)\rho\,d\rho.$$

Since f is continuous, $\rho \mapsto \lambda(f,z,\rho)$ is also continuous, and we can differentiate the last identity with respect to r obtaining

$$2rf(z) = 2\lambda(f,z,r)r.$$

That is, $f(z) = \lambda(f,z,r)$ for $r > 0$, and for $r = 0$ by continuity.

(d) (3) $\Leftrightarrow$ (4). This follows from the fact that $z \mapsto A(f,z,r)$ is continuous in Ω_r if $f \in L^1_{\text{loc}}(\Omega)$.

(e) (2) $\Rightarrow$ (5). We have $(f * \varphi_r)(z) = \int_{|\zeta|\le r} f(z+\zeta)\varphi_r(-\zeta)\,d\zeta\,d\eta$ (where we write $\zeta = \xi + i\eta$). Hence

$$\begin{aligned}(f*\varphi_r)(z) &= \int_0^r \varphi_r(\rho)\left(\int_0^{2\pi} f(z+\rho e^{i\theta})\,d\theta\right)\rho\,d\rho \\ &= 2\pi\int_0^r \varphi_r(\rho)\lambda(f,z,\rho)\rho\,d\rho \\ &= f(z)\left[2\pi\int_0^r \varphi_r(\rho)\rho\,d\rho\right] = f(z)\left[\int_{\mathbb{C}} \varphi_r(\zeta)\,d\xi\,d\eta\right] \\ &= f(z).\end{aligned}$$

(f) (5) $\Rightarrow$ (3). Let $(\varphi_n)_{n\ge 2}$ be a sequence of standard functions, uniformly bounded, all of them with support in $\bar{B}(0,1)$, and depending only on $|z|$. Assume, moreover, they converge pointwise to $\frac{1}{\pi}\chi_{\bar{B}(0,1)}$. To construct the φ_n, choose $\alpha_n \in \mathcal{D}(\bar{B}(0,1))$, radially symmetric, equal to $\frac{1}{\pi}$ in a neighborhood of $\bar{B}\left(0, 1-\frac{1}{n}\right)$, $0 \le \alpha_n \le \frac{1}{\pi}$. It follows that

$$\left(1 - \frac{1}{n}\right)^2 \le a_n := \int_{\mathbb{C}} \alpha_n(\zeta)\, d\xi\, d\eta \le 1.$$

Set $\varphi_n = \alpha_n/a_n$. We have $\displaystyle\int_{\mathbb{C}} \varphi_n\, dm = 1, 0 \le \varphi_n \le \frac{1}{\pi(1 - 1/n)^2} < 4/\pi$, and

$$\lim_{n\to\infty} \varphi_n(z) = \frac{1}{\pi}\chi_{\bar{B}(0,1)}$$

for every $z \in \mathbb{C}$. We can conclude that $\displaystyle\lim_{n\to\infty} \varphi_{n,r} = \frac{1}{\pi r^2}\chi_{\bar{B}(0,r)}$.

If (5) holds, then $f(z) = (f * \varphi_{n,r})(z)$, which shows that f is a C^∞ function. Hence

$$f(z) = \lim_{n\to\infty} (f * \varphi_{n,r})(z) = \lim_{n\to\infty} \int_{|\zeta|\le r} f(z + \zeta)\varphi_{n,r}(\zeta)\, d\xi\, d\eta$$

$$= \frac{1}{\pi r^2}\int_{|\zeta|\le r} f(z + \zeta)\, d\xi\, d\eta = A(f, z, r).$$

Note we have only proved (3) for r sufficiently small. The next step will show that this is enough to show (1) and hence the implication will be valid for every disk $\bar{B}(z, r) \subseteq \Omega$.

(g) (2) ⇒ (1). Since (2) ⇒ (5) has already been proved, and the proof of (5) ⇒ (3) shows that $f \in C^\infty$, now $f = f * \varphi_r$ shows $D^\alpha f = D^\alpha f * \varphi_r$ for every $\alpha \in \mathbb{N}^2$. It follows again from (f) that $D^\alpha f = A(D^\alpha f, z, r)$ for r sufficiently small. Therefore, taking $\alpha = (2, 0)$ and $(0, 2)$ we obtain $\Delta f(z) = A(\Delta f, z, r)$. On the other hand, we have already shown in (a) that the first identity here holds,

$$\frac{d}{dr}\lambda(f, z, r) = \frac{1}{2\pi}\int_{\bar{B}(z,r)} \Delta f\, d\xi\, d\eta = \frac{r^2}{2}A(\Delta f, z, r) = \frac{r^2}{2}\Delta f(z).$$

Since $\lambda(f, z, r) = f(z)$, it follows that $\Delta f(z) = 0$. □

4.3.4. Examples. (1) Let Ω be an open subset of $\mathbb{C}$, $f \in \mathscr{H}(\Omega)$. Then f, $\operatorname{Re} f$, and $\operatorname{Im} f$ and harmonic functions.

(2) If Ω is a simply connected open set in $\mathbb{C}$, every real-valued harmonic function in Ω is the real part of a holomorphic function in Ω.

In fact, if u is a harmonic function, $\dfrac{\partial^2 u}{\partial \bar{z}\partial z} = 0$, which shows that $\dfrac{\partial u}{\partial z}$ is a holomorphic function in Ω. The differential form $\omega := 2\dfrac{\partial u}{\partial z}dz$ is closed since

$$d\omega = \bar{\partial}\left(2\frac{\partial u}{\partial z}dz\right) = 2\frac{\partial^2 u}{\partial \bar{z}\partial z}d\bar{z} \wedge dz = 0.$$

There is $h \in \mathscr{E}(\Omega)$ such that $\omega = dh$. We have, therefore,

$$\frac{\partial h}{\partial z} = 2\frac{\partial u}{\partial z} \quad \text{and} \quad \frac{\partial h}{\partial \bar{z}} = 0.$$

Hence h is a holomorphic function. Furthermore, since u is real valued we have

$$d\bar{h} = 2\frac{\partial u}{\partial \bar{z}}d\bar{z}$$

and

$$d\left(\frac{h+\bar{h}}{2}\right) = \frac{\partial u}{\partial z}dz + \frac{\partial u}{\partial \bar{z}}d\bar{z} = du.$$

It follows that for some real constant k we have $u = \operatorname{Re} h + k$. Letting $f = h + k$ we have obtained a holomorphic function in Ω such that $u = \operatorname{Re} f$. The function f is uniquely determined up to the addition of a purely imaginary constant.

A function v such that $u = \operatorname{Re} f$ and $v = \operatorname{Im} f$ for the same holomorphic function f is called a ***harmonic conjugate*** of u. This concept is invariant under holomorphic transformations.

We are now going to introduce Poisson's representation formula for harmonic functions. If h is a function holomorphic in a neighborhood of the unit disk $\bar{B}(0, 1)$ (or holomorphic in $B(0, 1)$ and continuous in $\bar{B}(0, 1)$), we have for $|z| < 1$,

$$h(z) = \frac{1}{2\pi i}\int_{|\zeta|=1}\frac{h(\zeta)}{\zeta - z}d\zeta = \int_{|\zeta|=1}\frac{\zeta h(\zeta)}{\zeta - z}\frac{d\zeta}{2\pi i\zeta}$$

and

$$0 = \frac{1}{2\pi i}\int_{|\zeta|=1}\frac{h(\zeta)}{\zeta - 1/\bar{z}}d\zeta = \int_{|\zeta|=1}\frac{\bar{z}h(\zeta)}{\bar{z} - \bar{\zeta}}\frac{d\zeta}{2\pi i\zeta},$$

since $\zeta\bar{\zeta} = 1$. By subtraction it follows that

$$h(z) = \int_{|\zeta|=1}h(\zeta)\left[\frac{\zeta}{\zeta - z} + \frac{\bar{z}}{\bar{\zeta} - \bar{z}}\right]\frac{d\zeta}{2\pi i\zeta}.$$

Note that the measure $\dfrac{d\zeta}{2\pi i\zeta}$ is actually $\dfrac{d\theta}{2\pi}$ for $\zeta = e^{i\theta}$, hence $\dfrac{1 - |z|^2}{|\zeta - z|^2}\dfrac{d\zeta}{2\pi i\zeta}$ is a positive measure on $|\zeta| = 1$. Conjugating the previous identity we obtain

$$\bar{h}(z) = \int_{|\zeta|=1}\bar{h}(\zeta)\frac{1 - |z|^2}{|\zeta - z|^2}\frac{d\zeta}{2\pi i\zeta}.$$

Therefore the real-valued harmonic function $u = \dfrac{h + \bar{h}}{2} = \operatorname{Re} h$ verifies also

$$\begin{aligned}u(z) &= \int_{|\zeta|=1}u(\zeta)\frac{1 - |z|^2}{|\zeta - z|^2}\frac{d\zeta}{2\pi i\zeta} = \int_{-\pi}^{\pi}u(e^{i\theta})\frac{1 - |z|^2}{|e^{i\theta} - z|^2}\frac{d\theta}{2\pi}\\ &= \int_{-\pi}^{\pi}u(e^{i\theta})\frac{1 - r^2}{1 - 2r\cos(\alpha - \theta) + r^2}\frac{d\theta}{2\pi} \qquad (z = re^{i\alpha}).\end{aligned}$$

If $\bar{B}(0,1)$ is replaced by $\bar{B}(z_0,r)$, we obtain for $z \in B(z_0,r)$, by a simple change of variables,

$$u(z) = \frac{1}{2\pi}\int_{-\pi}^{\pi} \frac{r^2 - |z - z_0|^2}{|z_0 + re^{i\theta} - z|^2} u(z_0 + re^{i\theta})\,d\theta$$

$$= \frac{1}{2\pi}\int_{-\pi}^{\pi} \frac{r^2 - \rho^2}{r^2 - 2\rho r\cos(\alpha - \theta) + \rho^2} u(z_0 + re^{i\theta})\,d\theta \qquad (z = z_0 + \rho e^{i\alpha}).$$

Since we have already shown that every real-valued harmonic function in a simply connected open set is the real part of a holomorphic function, we see that setting

$$P_r(\alpha) = \frac{1 - r^2}{1 - 2r\cos\alpha + r^2},$$

we have the ***Poisson integral representation***

$$u(re^{i\theta}) = \frac{1}{2\pi}\int_{-\pi}^{\pi} P_r(\alpha - \theta)u(e^{i\theta})\,d\theta$$

for every real-valued harmonic function u in $B(0,1)$ that is continuous in $\bar{B}(0,1)$. (Note that, strictly speaking, we do not know a priori that a holomorphic function h in $B(0,1)$ such that $\operatorname{Re} h = u$ is also continuous in $\bar{B}(0,1)$; a little limiting argument is necessary to bypass this point, we leave it as an exercise to the reader). This representation is a convolution on the group $\mathbb{T} = \{w \in \mathbb{C} : |w| = 1\}$. By linearity, it is also valid for complex-valued harmonic functions.

Moreover, if u is harmonic in $B(0,1)$, continuous in $\bar{B}(0,1)$, and real valued, we can use the relation

$$\frac{|\zeta|^2 - |z|^2}{|\zeta - z|^2} = \operatorname{Re}\frac{\zeta + z}{\zeta - z}$$

to obtain a representation of the unique holomorphic function h in $B(0,1)$ with $\operatorname{Im} h(0) = 0$ such that $\operatorname{Re} h = u$. Namely,

$$h(z) = \frac{1}{2\pi i}\int_{|\zeta|=1} \frac{\zeta + z}{\zeta - z} u(\zeta)\frac{d\zeta}{\zeta}, \qquad z \in B(0,1).$$

We have $h(0) = \dfrac{1}{2\pi i}\displaystyle\int_{|\zeta|=1} u(\zeta)\frac{d\zeta}{\zeta} = u(0) \in \mathbb{R}$.

The positive function $P_r(\theta)$ is called the ***Poisson kernel of the unit disk***. The function

$$z = re^{i\theta} \mapsto P_r(\theta)$$

is harmonic in the unit disk, hence it is C^∞ in θ for every fixed r, $0 \le r < 1$. The integral $\dfrac{1}{2\pi}\displaystyle\int_{-\pi}^{\pi} P_r(\alpha)\,d\alpha = 1$, as can be seen when representing the function $u \equiv 1$. Furthermore,

(i) $P_r(\alpha) = P_r(-\alpha)$,
(ii) $\lim_{r \nearrow 1} P_r(\alpha) = 0$ uniformly in $\delta \leq |\alpha| \leq \pi$, for each fixed $\delta \in]0, \pi[$, and
(iii) $P_r(\alpha) = \sum_{n \in \mathbb{Z}} r^{|n|} e^{in\alpha}$.

The Poisson representation of harmonic functions implies that if u is harmonic in $B(0, 1)$ and continuous up to the boundary of the unit disk, then its values in $B(0, 1)$ are completely determined by its values on $\partial B(0, 1)$. One could naturally ask which continuous functions on $B(0, 1)$ can be obtained as restrictions to $\partial B(0, 1)$ of functions continuous in $\bar{B}(0, 1)$ and harmonic in the interior. This questions is known as the ***Dirichlet problem***:

Given a continuous function f on $\partial B(0, 1)$ find a continuous function u in $\bar{B}(0, 1)$ such that

$$\begin{cases} \Delta u = 0 & \text{in } B(0, 1) \\ u|_{\partial B(0,1)} = f. \end{cases}$$

We will solve this problem with the help of the Poisson representation formula. Later we will consider the same problem in arbitrary open subsets Ω of $\mathbb{C}$.

4.3.5. Definition. Let f be an integrable function on $B(0, 1)$ (resp. $f: \partial B(0, 1) \to [-\infty, \infty[$ measurable and bounded above). The ***Poisson integral*** of f is the function $P(f)$ defined in $B(0, 1)$ by

$$P(f)(z) = \frac{1}{2\pi} \int_{-\pi}^{\pi} P_r(\alpha - \theta) f(e^{i\theta})\, d\theta,$$

for $z = re^{i\alpha}$, $0 \leq r < 1$.

More generally, replacing $B(0, 1)$ by $B(z_0, r)$, we call the Poisson integral of f over $\partial B(z_0, r)$ the function defined in $B(z_0, r)$ by

$$P(f)(z) = \frac{1}{2\pi} \int_{-\pi}^{\pi} \frac{r^2 - |z - z_0|^2}{|z_0 + re^{i\theta} - z|^2} f(z_0 + re^{i\theta})\, d\theta.$$

Observe that the notation is a bit ambiguous and attention should be paid as to which disk one is working with. Modulo these considerations, we remark that

$$P(f)(z_0) = \lambda(f, z_0, r),$$

where z_0 is the center of the disk $B(z_0, r)$, and

$$P(f)(z_0 + \rho e^{i\alpha}) = \sum_{n \in \mathbb{Z}} a_n \rho^{|n|} e^{in\alpha},$$

where

$$a_n = \frac{1}{2\pi r^n} \int_{-\pi}^{\pi} f(z_0 + re^{i\theta}) e^{-in\theta}\, d\theta,$$

which coincides with the nth Fourier coefficient of f when $z_0 = 0, r = 1$.

4.3.6. Proposition. *Let $f : \partial B(z_0, r) \to \mathbb{C}$ be an integrable function. Then we have*

(1) *$P(f)$ is harmonic in $B(z_0, r)$*
(2) *If $f \equiv \lambda_0 \in \mathbb{C}$ then $P(f) \equiv \lambda_0$.*
(3) *If f is continuous at a point $z_1 \in \partial B(z_0, r)$ then*

$$\lim_{\substack{z \to z_1 \\ z \in B(z_0, r)}} P(f)(z) = f(z_1)$$

(4) *If f is a continuous function in $\bar{B}(z_0, r)$, which is harmonic in $B(z_0, r)$ then*

$$f = P(f \mid \partial B(z_0, r)).$$

PROOF. (1) This is proved either by differentiation under the integral sign or by considering the averages of $P(f)$ over $\partial B(z, \rho)$ for disks $\bar{B}(z, \rho) \subseteq B(z_0, r)$ and using Fubini's theorem. In either case, one has to use the harmonicity of the function $z \mapsto \dfrac{r^2 - |z - z_0|^2}{|\zeta - z|^2}$ in $B(z_0, r)$ for a fixed $\zeta \in \partial B(z_0, r)$. We leave the details to the reader.

(2) By direct evaluation we see that

$$P(\lambda_0)(z_0 + \rho e^{i\alpha}) = \left[\frac{1}{2\pi} \int^{2\pi} P_{\rho/r}(\theta - \alpha)\, d\theta \right] \lambda_0 = \lambda_0.$$

(3) Using the previous result, $P(f(z_1)) \equiv f(z_1)$, we have

$$P(f)(z) = f(z_1) + \frac{1}{2\pi} \int^{2\pi} \frac{r^2 - |z - z_0|^2}{|z_0 + re^{i\theta} - z|^2} (f(z_0 + re^{i\theta}) - f(z_1))\, d\theta.$$

Let $z_1 = z_0 + re^{i\theta_1}$. Then for every $\varepsilon > 0$ there is $0 < \delta < 2\pi$ such that if $\zeta = z_0 + re^{i\theta}$, $|\theta - \theta_1| < \delta$ then $|f(\zeta) - f(z_1)| < \varepsilon$. Therefore, we can split the integral into two parts, one over $|\theta - \theta_1| < \delta$, the other over the complementary arc.

Recall that by property (ii) mentioned immediately after the definition of the Poisson kernel, $\lim_{r \nearrow 1} P_r(\alpha) = 0$ uniformly in α, $\delta \le |\alpha| \le \pi$. Using this fact and the Lebesgue-dominated convergence theorem, we see that the integral over the arc $\delta \le |\theta - \theta_1| \le \pi$ defines a function of z whose limit at z_1 exists and equals zero. On the other hand we have

$$\left| \frac{1}{2\pi} \int_{|\theta - \theta_1| < \delta} \frac{r^2 - |z - z_0|^2}{|z_0 + re^{i\theta} - z|^2} (f(z_0 + re^{i\theta}) - f(z_1))\, d\theta \right|$$

$$\le \left(\max_{|\theta - \theta_1| < \delta} |f(z_0 + re^{i\theta}) - f(z_1)| \right) \frac{1}{2\pi} \int_0^{2\pi} \frac{r^2 - |z - z_0|^2}{|z_0 + re^{i\theta} - z|^2}\, d\theta$$

$$= \max_{|\theta - \theta_1| < \delta} |f(z_0 + re^{i\theta}) - f(z_1)| < \varepsilon.$$

Finally, property (4) will follow from Corollary 4.3.8, applied to the function $\varphi = f - P(f)$, which is harmonic in $B(z_0, r)$, continuous in $\bar{B}(z_0, r)$, and zero on $\partial B(z_0, r)$ by (3). □

4.3.7. Lemma (Maximum Principle for Harmonic Functions). *Let Ω be an open connected subset of $\mathbb{C}$, $u : \Omega \to \mathbb{R}$ a harmonic function. If u is bounded above and attains its least upper bound in Ω, then u is constant.*

PROOF. Let $E := \left\{ z \in \Omega : u(z) = \sup_{\zeta \in \Omega} u(\zeta) = M < \infty \right\}$. This set E is closed in Ω and not empty by hypothesis. Let $a \in E$ and $r > 0$ such that $\bar{B}(a,r) \subseteq \Omega$. We have

$$M = u(a) = A(u,a,r) = A(u(a),a,r).$$

Hence $A(u(a) - u, a, r) = 0$. Since $u(a) - u \geq 0$ and it is continuous, it follows that $u(a) \equiv u$ in $B(a,r)$. Therefore E is also open. It follows that $E = \Omega$.

□

4.3.8. Corollary. *Let Ω be a connected open subset of $\mathbb{C}$ and $\varphi : \Omega \to \mathbb{C}$ a bounded harmonic function such that $|\varphi|$ takes its least upper bound in Ω. The function φ is then a constant.*

PROOF. Let $M = \sup_{\zeta \in \Omega} |\varphi(\zeta)|$ and $a \in \Omega$ such that $\varphi(a) = Me^{i\alpha}$ for some real number α. Let $g = e^{-i\alpha}\varphi = u + iv$. Then both u and v are real-valued harmonic functions, $g(a) = u(a) = M$. Clearly $u \leq |g| \leq M$. By the previous lemma, u is the constant function M. It also follows that $v \equiv 0$ since $|g| = \sqrt{M^2 + v^2} \leq M$. This proves the corollary. □

4.3.9. Proposition (Harnack's Inequality). *Let u be a nonnegative continuous function in $\bar{B}(z_0, r)$ and u harmonic in $B(z_0, r)$. Then the following inequality holds for $z \in B(z_0, r)$:*

$$\frac{r - |z - z_0|}{r + |z - z_0|} u(z_0) \leq u(z) \leq \frac{r + |z - z_0|}{r - |z - z_0|} u(z_0).$$

PROOF. It is easy to see that

$$\frac{r - |z - z_0|}{r + |z - z_0|} \leq \frac{r^2 - |z - z_0|}{|z_0 + re^{i\theta} - z|^2} \leq \frac{r + |z - z_0|}{r - |z - z_0|}.$$

From the Poisson representation and $u \geq 0$ it follows that

$$\frac{1}{2\pi} \int_0^{2\pi} \frac{r - |z - z_0|}{r + |z - z_0|} u(z_0 + re^{i\theta})\, d\theta \leq P(u)(z)$$

$$\leq \frac{1}{2\pi} \int_0^{2\pi} \frac{r + |z - z_0|}{r - |z - z_0|} u(z_0 + re^{i\theta})\, d\theta.$$

The desired inequality now follows from the identity

$$u(z_0) = P(u)(z_0) = \lambda(u, z_0, r) = \frac{1}{2\pi} \int_0^{2\pi} u(z_0 + re^{i\theta})\, d\theta.$$ □

4.3.10. Corollary. *A bounded harmonic function u in $\mathbb{C}$ is necessarily constant.*

PROOF. It is clearly sufficient to prove it for real-valued functions. By addition of a constant we can assume $u \geq 0$. Therefore for any r such that $0 \leq |z| < r$ we have

$$\frac{r - |z|}{r + |z|} u(0) \leq u(z) \leq \frac{r + |z|}{r - |z|} u(0).$$

For z fixed let $r \to +\infty$. It follows that $u(z) = u(0)$. □

4.3.11. Proposition (Harnack's Theorem). *Let Ω be a connected open subset of $\mathbb{C}$ and $\{u_n\}_{n \geq 0}$ be an increasing sequence of (real-valued) harmonic functions in Ω. Either the sequence $u_n(z)$ tends to $+\infty$ for every $z \in \Omega$ (and uniformly over any compact subset of Ω) or the sequence converges uniformly over every compact subset of Ω to a harmonic function u.*

PROOF. Since the limit of $u_n(z)$ exists (possibly $+\infty$) for every $z \in \Omega$, the function u can be defined by this limit as a map $u : \Omega \to]-\infty, \infty]$. Let us assume that there is some $z_0 \in \Omega$ such that $u(z_0) < \infty$. Given $\varepsilon > 0$ there is some n_ε such that if $n \geq m > n_\varepsilon$ then

$$u_n(z_0) - u_m(z_0) < \varepsilon.$$

If $\bar{B}(z_0, r) \subseteq \Omega$ and $|z - z_0| = \rho < r$ we have by one part of Harnack's inequality

$$0 \leq u_n(z) - u_m(z) < \varepsilon \frac{r + \rho}{r - \rho}.$$

Therefore, by the maximum principle, the sequence $\{u_n\}_{n \geq 0}$ coverges uniformly in $\bar{B}(z_0, \rho)$, and u is defined and continuous in $\bar{B}(z_0, \rho)$ and hence in $B(z_0, r)$. If, on the other hand, $u(z_0) = +\infty$, we can apply the other part of Harnack's inequality to $u_n(z) - u_0(z)$, see that $u(z) \equiv +\infty$ in $\bar{B}(z_0, \rho)$, and that the convergence is uniform in this disk. It follows that the sets Ω_1 and Ω_2 where $u < \infty$ and $u = \infty$, respectively, are both open subsets of Ω. One of the two must be empty. In either case the argument shows that the convergence is locally uniform and hence uniform over compact sets.

If $\Omega_2 = \varnothing$, then for any $\bar{B}(z_0, r) \subseteq \Omega$ we conclude that $P(u) = u$ in $B(z_0, r)$ since $P(u_n) = u_n$ and the convergence is uniform on $\bar{B}(z_0, r)$. Therefore u is harmonic. □

4.3.12. Remark. If D is a Jordan domain, by Theorem 2.8.8 there is a homeomorphism φ of $\bar{D}$ to $\bar{B}(0, 1)$, such that φ is holomorphic in D. Then we can use the preceding considerations to solve the Dirichlet problem in D by reducing it to $B(0, 1)$. If f is a continuous function on ∂D, the harmonic extension u of f to D is given by

$$u(z) = P(f \circ \varphi^{-1})(\varphi(z)).$$

4.3.13. Examples. (1) $D = B(z_0, r)$, $\varphi(z) = \dfrac{z - z_0}{r}$. One obtains the Poisson formula in $B(z_0, r)$ from the Poisson formula in $B(0, 1)$.

(2) $D = B(0,1) \cap \{\operatorname{Im} z > 0\}$, $w = \varphi(z) = \dfrac{(z+1)^2 - i(z-1)^2}{(z+1)^2 + i(z-1)^2}$ is a homeomorphism of $\bar{D}$ onto $\bar{B}(0,1)$, holomorphic in D, as one verifies from the fact that φ is the composition of the following three maps, $z \mapsto w_1 = \dfrac{1+z}{1-z}$, $w_1 \mapsto w_2 = w_1^2$, and $w_2 \mapsto w = \dfrac{w_2 - i}{w_2 + i}$.

(3) If f is continuous on $\mathbb{R}$ and bounded, then one finds in the same way a function u, harmonic in $\{\operatorname{Im} z > 0\}$, continuous in $\{\operatorname{Im} z \geq 0\}$, bounded and equal to f on $\mathbb{R}$, by taking

$$u = P(f \circ \varphi^{-1}) \circ \varphi$$

with $\varphi(z) = \dfrac{z - i}{z + i}$.

To conclude this section let us mention that under rather general conditions, a harmonic function in the unit disk has a Poisson representation $u = P(f)$. For instance, the Riesz-Herglotz theorem states that if u is harmonic in $B(0,1)$, $u \geq 0$, then there is a nonnegative Radon measure $d\mu$ in $\partial B(0,1)$ such that $|z| < 1$

$$u(z) = \frac{1}{2\pi} \int_{\partial B} \frac{1 - |z|^2}{|\zeta - z|^2} \, d\mu(\zeta),$$

and $\int_{\partial B} d\mu = u(0)$. (See, e.g., [Ru], [He2].)

EXERCISES 4.3

Here $B = B(0, 1)$; Ω is a domain in $\mathbb{C}$.

1. Use the solution of the Dirichlet problem in B to show that every continuous 2π-periodic function f in $\mathbb{R}$ (i.e., $f(x + 2\pi) = f(x)$) can be uniformly approximated by linear combinations of the trigonometric functions $1, \sin x, \cos x, \sin 2x, \cos 2x, \ldots$ (that is, by ***trigonometric polynomials***).
2. Use Exercise 4.3.1 to show that every continuous function on a closed interval $[a, b]$ of the real line can be uniformly approximated on $[a, b]$ by polynomials.
3. Let $f : \Omega \to \mathbb{R}$ be such that both f and f^2 are harmonic. Show that f is constant.
4. Let $f : \Omega \to \mathbb{R}$ be a harmonic function such that fg is harmonic for every other harmonic function g in Ω. Prove that f is a constant.
5. Introduce polar coordinates in B and show that

$$\Delta u = \frac{1}{4} \frac{\partial}{\partial r}\left(r \frac{\partial u}{\partial r}\right) + \frac{1}{r^2} \frac{\partial^2 u}{\partial \theta^2},$$

with the usual identification $x = r\cos\theta$, $y = r\sin\theta$.

Using this formula, determine the radial harmonic functions in $B\setminus\{0\}$. Are there any nonconstant radial harmonic functions in B?

6. Let h be harmonic in the annulus $0 \leq R_1 < |z| < R_2 \leq \infty$, let $f(r) = \lambda(h, r, 0)$. Show $f(r) = a \log r + b$ for two constants a, b.

7. Let $f \in \mathscr{H}(\Omega)$ be such that $g(z) = \bar{z}f(z)$ is harmonic. Show that f is a constant.

8. In this exercise we want to show that the Fourier series of a 2π-periodic function of class C^2 on $\mathbb{R}$ converges uniformly to the function, just using the solvability of the Dirichlet problem in B. With a little bit more effort and using Exercise 4.3.1, one can obtain the usual theory of Fourier series in $L^2([-\pi, \pi], dx)$.
   (a) If $f \in C^2(\mathbb{R})$ and f is 2π-periodic, show that f', f'' are 2π-periodic.
   (b) Define $a_n = \dfrac{1}{2\pi}\displaystyle\int_{-\pi}^{\pi} f(x)e^{-inx}\,dx$, and show that $\sum_{n \in Z} |a_n| < \infty$.
   (c) Recall from the text that
$$(Pf)(re^{i\theta}) = \sum_{n \in Z} a_n r^{|n|}$$
   to show that
$$\sum_{n \in Z} a_n e^{inx} = f(x),$$
   and the convergence is uniform.
   *(d) Since the previous result only depends on the fact that $\sum_{n \in Z} |a_n| < \infty$, show that the uniform convergence still holds under the weaker hypotheses that $f \in C^0(\mathbb{R})$ be 2π-periodic and piecewise C^1. (Hint: First show that if b_n is the nth Fourier coefficient of f', then $b_n = ina_n$. Second, show that $\sum_{n \in Z} |b_n|^2 \leq \dfrac{1}{2\pi}\displaystyle\int_{-\pi}^{\pi} |f'(x)|^2\,dx$.)

9. (a) Show that all the harmonic functions in an annulus $R_1 < |z| < R_2$, which are of the form $u(z) = v(r)w(\theta)$, v, w of class C^2, and w 2π-periodic, are solutions of a pair of ordinary differential equations
$$\begin{cases} r\dfrac{d}{dr}\left(r\dfrac{dv}{dr}\right) - \lambda v = 0 \\ w'' + \lambda w = 0 \end{cases}$$
   for $\lambda = n^2$, $n \in \mathbb{N}$. Conclude that
$$\begin{aligned} v(r) &= ar^n + br^{-n} \\ w(\theta) &= A\cos n\theta + B\sin n\theta \end{aligned}$$
   for some constants a, b, A, B.
   (b) Use this observation to look for solutions
$$u(re^{i\theta}) = a_0 + \sum_{n=1}^{\infty} (a_n \cos n\theta + b_n \sin n\theta)r^n,$$
   of the Dirichlet problem in $B(0, 1)$ with boundary values $f(\theta)$ given by, respectively,
   (i) $f(\theta) = \cos^2\theta$;

(ii) $f(\theta) = \sin^3\theta$;
(iii) $f(\theta) = \sin^4\theta + \cos^4\theta$.

(c) The Dirichlet problem in the annulus $1 < |z| < 2$ with the boundary data $f_1(\theta)$ on $|z| = 1$ and $f_2(\theta)$ on $|z| = 2$, consists in finding a continuous function u in $1 \le |z| \le 2$, C^2 in $1 < |z| < 2$, such that

$$\begin{cases} \Delta u = 0 & \text{in } 1 < |z| < 2 \\ u(e^{i\theta}) = f_1(\theta), & u(2e^{i\theta}) = f_2(\theta). \end{cases}$$

Use item (a) to solve this problem in the cases
(i) $f_1(\theta) = a$, $f_2(\theta) = b$, $a, b \in \mathbb{R}$; and
(ii) $f_1(\theta) = 1 + \cos^2\theta$, $f_2(\theta) = \sin^2\theta$.

10. Solve the problem

$$\begin{cases} \Delta u = -xy & \text{in } B \\ u|\partial B = 0 \end{cases}$$

(Hint: use that the function $v(z) = -xy(x^2 + y^2)/12$ has the property $\Delta v = -xy$, to reduce this problem to the Dirichlet problem in B.)

11. This exercise gives two ways of showing that if $h: B\backslash\{0\} \to \mathbb{R}$ is a harmonic function, then

$$h(re^{i\theta}) = a\log r + b + \sum_{n \in \mathbb{Z}^*} (a_n \cos n\theta + b_n \sin n\theta)r^n,$$

with the convergence being uniform in $0 < r_1 \le r \le r_2 < 1$, for any pair r_1, r_2.

(a) First method: For r fixed, develop the function $\theta \mapsto h(re^{i\theta})$ in a Fourier series of the form

$$h(re^{i\theta}) = A_0(r) + \sum_{n \ge 1} (A_n(r)\cos n\theta + B_n(r)\sin n\theta).$$

Show that $A_0(r)$, $A_n(r)\cos n\theta$, and $B_n(r)\sin n\theta$ are harmonic in $B\backslash\{0\}$.

(b) Second method: Use the fact that $\dfrac{\partial h}{\partial z} \in \mathscr{H}(B\backslash\{0\})$ to obtain the desired representation.

12. Let $h: B\backslash\{0\} \to \mathbb{R}$ be a harmonic function such that $h \ge 0$ everywhere. Show that $h(re^{i\theta}) = a\log r + b + \sum_{n\ge1} (a_n \cos n\theta + b_n \sin n\theta)r^n$. What can you say about a and b? (Hint: Use $h \ge 0$ to show $|\lambda(h(z)\cos n\theta, 0, r)| \le \lambda(h, 0, r)$.)

13. Let us recall that a ***real analytic function*** f in an open subset U of $\mathbb{R}$ is a C^∞ function such that for each $x_0 \in U$, the Taylor series $\sum_{n\ge0} \dfrac{f^{(n)}(x_0)}{n!}(x - x_0)^n$ converges to $f(x)$ for $|x - x_0|$ sufficiently small.
(i) Show that every loop in an open set $D \subseteq \mathbb{C}$ is homotopic to a real analytic loop.
(ii) Show every path in D is homotopic, with a fixed-point homotopy, to a real analytic path in D.

14. Let u_n $(n \ge 1)$ be a family of nonnegative harmonic functions in Ω such that for some $z_0 \in \Omega$, numerical series $\sum_{n\ge1} u_n(z_0)$ is convergent. Show that $u(z) = \sum_{n\ge1} u_n(z)$ is a harmonic function in Ω.

15. Let $(u_n)_{n\geq 1}$ be a sequence of harmonic functions in Ω that converges locally uniformly to a function u. Show that for any $\alpha \in \mathbb{N}^2$, the sequence

$$u_n^{(\alpha)} = \frac{\partial^{|\alpha|}}{\partial x^{\alpha_1} \partial y^{\alpha_2}} u_n \to u^{(\alpha)}$$

locally uniformly.

16. For a point $z \in B$, let us define a function $\varphi_z = \varphi : [0, 2\pi] \to \mathbb{R}$ as follows. The point $e^{i\varphi(\theta)}$ is the intersection of ∂B with the ray starting at z with direction $e^{i\theta}$ Moreover, we impose the restriction that $0 \leq \varphi(\theta) - \theta < 2\pi$.

(i) Show that φ is differentiable and

$$\frac{d\varphi}{d\theta} = \left| \frac{e^{i\varphi(\theta)} - z}{e^{i\theta} - z} \right|.$$

(ii) Show that if f is a continuous function on ∂B, then

$$Pf(z) = \frac{1}{2\pi} \int_0^{2\pi} f(e^{i\varphi_z(\theta)})\, d\theta.$$

17. Let $f \in \mathscr{H}(B)$, $f(0) = 0$, and $|\operatorname{Re} f| \leq A < \infty$ in B. Show that

$$|\operatorname{Im} f(z)| \leq \frac{2A}{\pi} \log \frac{1+r}{1-r} \qquad (|z| \leq r < 1).$$

(Hint: Compare the derivation of the formula for the Poisson kernel to represent $\operatorname{Im} f$ in terms of $\operatorname{Re} f$.)

18. Show that if a real-valued harmonic function u in $\mathbb{C}$ is bounded above, then u is constant.

19. Let γ be a C^1 regular Jordan curve, $\overline{\operatorname{Int}(\gamma)} \subseteq \Omega$, h a harmonic function in Ω, and $z \in \operatorname{Int}(\gamma)$. Show that

$$h(z) = -\frac{1}{2\pi} \int_\gamma \frac{\partial h}{\partial n}(\zeta) \log|\zeta - z|\,|d\zeta| - \frac{1}{2\pi} \int_\gamma h(\zeta) \frac{\partial}{\partial n} \log|\zeta - z|\,|d\zeta|.$$

20. Let Ω be a domain with piecewise regular boundary, u, $v \in C^2(\overline{\Omega})$. Define the Dirichlet scalar product

$$D(u, v) := \int_\Omega \left(\frac{\partial u}{\partial x} \frac{\partial v}{\partial x} + \frac{\partial u}{\partial y} \frac{\partial v}{\partial y} \right) dx\, dy,$$

and the Dirichlet norm $D(u) := D(u, u)$.

(i) Show that $D(u, v) = \displaystyle\int_{\partial\Omega} u \frac{\partial v}{\partial u} |dz| - \int_\Omega u\, \Delta v\, dx\, dy$.

(ii) Let $u|_{\partial\Omega} = 0$ and v harmonic in Ω. Show that

$$D(v) \leq D(u + v).$$

(iii) (Dirichlet principle) If f is a function in $C^2(\overline{\Omega})$, $f|\partial\Omega = g$ and u solves the Dirichlet problem with data g on $\partial\Omega$, then $D(u) \leq D(f)$.

21. (i) Prove the statement of Example 4.3.13, (3). Beware, the function $f \circ \varphi^{-1}$ might not be continuous on ∂B. (Why?)

(ii) Show that the Poisson representation formula for the upper half-plane H, i.e., $F = P(f \circ \varphi^{-1}) \circ \varphi$, becomes

$$F(z) = \frac{1}{\pi}\int_{-\infty}^{\infty} \frac{y}{(x-t)^2 + y^2} f(t)\,dt \qquad (\operatorname{Im} z > 0).$$

(iii) Why is this formula valid for any function $f \in L^1\left(\mathbb{R}, \frac{dt}{1+t^2}\right)$, in particular, for $f \in L^\infty(\mathbb{R})$? Show that, if $f \in L^1\left(\mathbb{R}, \frac{dt}{1+t^2}\right)$ and t_0 is a continuity point of f, then

$$\lim_{z \to t_0} F(z) = f(t_0).$$

22. Let f be a harmonic function in a neighborhood of $z_0 \neq 0$. Show that $g(z) = f(1/\bar{z})$ is harmonic in a neighborhood of $1/\bar{z}_0$. Can you generalize this to $f \circ h$, f harmonic on the image of h, and $\bar{h}$ holomorphic?

23. Let $f: \mathbb{R} \to \mathbb{R}$ be a C^2 function such that whenever $g: \Omega \to \mathbb{R}$ is harmonic, it follows that $f \circ g$ is harmonic. What can you say about f?

24. Let $f: B\backslash\{0\} \to \mathbb{R}$ be harmonic and satisfy $|f(z)| \leq A|\log|z||^{1/2} + B$. Show that f has a harmonic extension to B.

§4. Subharmonic Functions

4.4.1. Definition. Let Ω be an open subset of $\mathbb{C}$. A function $u: \Omega \to [-\infty, \infty[$ is called ***subharmonic in the wide sense*** in Ω if it verifies the following two conditions:

(i) u is upper semicontinuous (u.s.c.);
(ii) for every disk $\bar{B}(z, r) \subseteq \Omega$, $u(z) \leq \lambda(u, z, r)$.

4.4.2. Remarks

(1) Here the circular average λ is given by a generalized integral in the sense of §4.1.
(2) We will see later that one can replace the circular average by the area average $A(u, z, r)$ in condition (ii). From the identity

$$A(u, z, r) = \frac{2}{r^2}\int_0^r \lambda(u, z, \rho)\rho\,d\rho$$

we see that if u is subharmonic in the wide sense

$$u(z) \leq A(u, z, r).$$

4.4.3. Proposition. *Let Ω be an open connected subset of $\mathbb{C}$ and u function subharmonic in the wide sense in Ω. We have then either*

(1) $u \equiv -\infty$, *or*
(2) $u \in L^1_{\text{loc}}(\Omega)$.

In the second case we say that u is ***subharmonic***.

PROOF. Let $G = \{z \in \Omega : \exists V_z,\ V_z \text{ relatively compact neighborhood of } z \text{ in } \Omega, \text{ such that } u|V_z \in L^1(V_z)\}$. The set G is open by its very definition. Let $B = \Omega \setminus G$. If $B \neq \varnothing$ then case (2) holds and there is nothing to prove. Assume $B \neq \varnothing$ and let $z_0 \in B$. For every V, relatively compact neighborhood of z_0 in Ω, we have

$$\int_V u\,dx\,dy = -\infty.$$

Let $r > 0$ be such that $\bar{B}(z_0, r) \subseteq \Omega$. We are going to show that $u(z) = -\infty$ for every $z \in B(z_0, r/2)$. In fact, for such a z, $B(z, r/2)$ is a relatively compact neighborhood of z_0, hence $A(u, z, r/2) = -\infty$. By Remark 4.4.2(2) we have $u(z) = -\infty$. Therefore $u \equiv -\infty$ in $B(z_0, r/2)$. We claim that it follows that $B(z_0, r/2) \subseteq B$. If fact, if $z \in B(z_0, r/2)$ and V is a neighborhood of z, then $V \cap B(z_0, r/2)$ contains a set of positive measure where $u \equiv -\infty$; therefore it is not possible that $u \in L^1(V)$. Hence B is both open and closed in Ω and not empty since $z_0 \in B$. We conclude that $B = \Omega$ and the preceding proof also shows that $u \equiv -\infty$ in this case. □

4.4.4. Definitions

(1) A subharmonic function in the wide sense in an open set Ω is said to be ***subharmonic*** in Ω if it is not identically equal to $-\infty$ in any component of Ω. We denote $\text{SH}(\Omega)$ the set of all subharmonic functions in Ω.
(2) A subset P of Ω is said to be a ***polar subset*** of Ω if there is a subharmonic function u in Ω such that $P \subseteq \{z \in \Omega : u(z) = -\infty\}$.
(3) If u is subharmonic in Ω, the (relatively) closed subset

$$E := \{z \in \Omega : u(z) = -\infty\}$$

is called the ***polar set*** of u.

4.4.5. Definition (Maximum Principle). Let X be a topological space. A function $f : X \to [-\infty, \infty[$ verifies the maximum principle in X if the existence of a relative maximum $x_0 \in X$ implies that f is constant in a neighborhood of x_0.

4.4.6. Proposition (Maximum Property). *If a u.s.c. function f in a connected topological space X verifies the maximum principle and has an absolute maximum in X, then it is constant.*

PROOF. The set $Y = \{y \in X : f(y) = \sup_{x \in X} f(x)\}$ is both open and closed. □

4.4.7. Proposition. *A subharmonic function in an open subset of* $\mathbb{C}$ *verifies the maximum principle.*

PROOF. Let $z_0 \in \Omega$ be a relative maximum of f. Let $r > 0$ be such that $\bar{B}(z_0, r) \subseteq \Omega$ and such that if $z \in \bar{B}(z_0, r)$, then $f(z) \leq f(z_0)$. Hence

$$f(z_0) \leq A(f, z_0, r) \leq f(z_0).$$

If there is $w_0 \in B(z_0, r)$ such that $f(w_0) < f(z_0)$, there would be $\rho > 0$ such that $\bar{B}(w_0, \rho) \subseteq B(z_0, r)$ and $f(w) < f(z_0)$ for every $w \in \bar{B}(w_0, \rho)$. One would then have $A(f, z_0, r) < f(z_0)$. □

4.4.8. Corollary. *Let* Ω *be a connected open subset of* $\mathbb{C}$, f *subharmonic function bounded above in* Ω *such that* f *takes its least upper bound somewhere in* Ω. *Then* f *is constant.*

4.4.9. Proposition. *Let* u *be a real-valued function of class* C^2 *in an open set* Ω *in* $\mathbb{C}$. *The following two conditions are equivalent*:

(1) u *is subharmonic in* Ω
(2) $\Delta u \geq 0$ *in* Ω.

PROOF. (1) implies (2) by the following lemma.

4.4.10. Lemma. *For a function* f *of class* C^2 *in a neighborhood of* z *we have*

$$\Delta f(z) = \lim_{r \to 0} \frac{2}{r^2} [\lambda(f, z, r) - f(z)].$$

PROOF OF LEMMA 4.4.10. Integrate in the variable h the Taylor development

$$\begin{aligned} f(z+h) - f(z) &= h_1 \frac{\partial f}{\partial x}(z) + h_2 \frac{\partial f}{\partial y}(z) \\ &\quad + \frac{1}{2}\left(\frac{\partial^2 f}{\partial x^2}(z) h_1^2 + 2\frac{\partial^2 f}{\partial x \partial y}(z) h_1 h_2 + \frac{\partial^2 f}{\partial y^2}(z) h_2^2\right) + |h|^2 \varepsilon(h) \end{aligned}$$

where $\varepsilon(h) \to 0$ as $|h| \to 0$. By the symmetry of $\partial B(0, r)$ in the h_1 and h_2 directions we obtain for $r > 0$ sufficiently small

$$\begin{aligned} \lambda(f, z, r) - f(z) &= \frac{1}{2\pi r}\left[\frac{1}{2}\frac{\partial^2 f}{\partial x^2}(z) \int_{|h|=r} h_1^2 \, d\sigma(h) + \frac{1}{2}\frac{\partial^2 f}{\partial y^2}(z) \int_{|h|=r} h_2^2 \, d\sigma(h) \right. \\ &\quad \left. + r^2 \int_{|h|=r} \varepsilon(h) \, d\sigma(h)\right] \\ &= \frac{r^2}{2}[\Delta f(z) + o(r)]. \end{aligned}$$

□

(2) implies (1) by an application of Green's formula: for $r \in \,]0, d(z, \Omega^c)[$ we have

$$\frac{d}{dr}\lambda(u,z,r) = \frac{1}{2\pi}\int_0^{2\pi}\frac{\partial u}{\partial r}(z+re^{i\theta})\,d\theta = \frac{1}{2\pi r}\int_{\partial B(z,r)}\frac{\partial u}{\partial n}(\zeta)|d\zeta|$$
$$= \frac{1}{2\pi r}\int_{\partial B(z,r)}\Delta u(\zeta)\,d\xi\,d\eta \geq 0.$$

Hence $r \mapsto \lambda(u,z,r)$ is increasing and, since $\lambda(u,z,0) = u(z)$, we have $u(z) \leq \lambda(u,z,r)$. □

4.4.11. Proposition. *Let Ω be an open subset of $\mathbb{C}$. The following properties hold:*

(1) *If u is subharmonic in Ω and v is harmonic in Ω, then both $u+v$ and $u-v$ are subharmonic in Ω.*
(2) *Let $\{u_i\}_{i\in I}$ be a family of subharmonic functions in Ω. The function $u = \sup_{i\in I} u_i$ is subharmonic if it is $< +\infty$ at every point and u.s.c. In particular, u is subharmonic if I is finite.*
(3) *If $x_1,\ldots,x_n$ are real numbers ≥ 0 and $u_1,\ldots,u_n$ are subharmonic functions, then $\sum_{1\leq i\leq n} x_i u_i$ is a subharmonic function.*
(4) *The infimum of a decreasing net of subharmonic functions, $\inf_I u_i$, is subharmonic in the wide sense. In particular, the limit of a decreasing sequence of subharmonic functions is subharmonic in the wide sense.*
(5) *If u is harmonic (even if it is complex-valued) then $|u|$ is subharmonic.*
(6) *$\log|z-a|$ is subharmonic.*
(7) *If f is holomorphic then $\operatorname{Re} f$, $\operatorname{Im} f$, $|f|$, and $\log|f|$ are subharmonic functions. (The last one only holds if $f \not\equiv 0$ in each connected component of Ω.)*
(8) *If Ω is an open subset of $\mathbb{C}$ such that $\Omega^c \neq \varnothing$, then $-\log d(z,\Omega^c)$ is subharmonic in Ω.*

PROOF. (1) $(u+v)(z) = u(z) + v(z) \leq \lambda(u,z,r) + \lambda(v,z,r) = \lambda(u+v,z,r)$ if $B(z,r) \subseteq \Omega$.

(2) For every $i \in I$ we have $u_i(z) \leq \lambda(u_i,z,r)$ if $\bar{B}(z,r) \subseteq \Omega$. Hence $u(z) = \sup_{i\in I} u_i(z) \leq \lambda\left(\sup_{i\in I} u_i, z, r\right)$ if we assume that $\sup_I u_i$ is u.s.c. Therefore $u = \sup u_i$ is subharmonic in that case. If I is finite, then $\sup_{i\in I} u_i$ is always u.s.c.

(3) $\sum_{1\leq i\leq n} \alpha_i u_i(z) \leq \sum_i \alpha_i \lambda(u_i,z,r) = \lambda\left(\sum_i \alpha_i u_i, z, r\right).$

(4) The infimum $u = \inf_i u_i$ is u.s.c. and for every $j \in I$, $u(z) \leq \lambda(u_j,z,r)$ if $\bar{B}(z,r) \subseteq \Omega$. By the Dini-Cartan Proposition 4.2.6 we have

$$\inf_{i\in I} \lambda(u_i,z,r) = \lambda\left(\inf_{i\in I} u_i, z, r\right) = \lambda(u,z,r),$$

hence u is subharmonic.

(5) If u is harmonic in Ω and $\bar{B}(z,r) \subseteq \Omega$, then $u(z) = \lambda(u,z,r)$, it follows then that $|u(z)| \leq \lambda(|u|,z,r)$ even if u is complex-valued.

(6) For every $\varepsilon > 0$, the function $u_\varepsilon(z) = \frac{1}{2}\log(|z-a|^2+\varepsilon)$ is C^∞ and its Laplacian is $4\varepsilon/(|z-a|^2+\varepsilon)^2$. Therefore $\log|z-a|$ is subharmonic as a decreasing limit of a family of subharmonic functions (see (4)).

(7) It is clear that $\operatorname{Re} f$ and $\operatorname{Im} f$ are subharmonic. Moreover, by (5), $|f|$ is also subharmonic. If $f \not\equiv 0$ in every open set relatively compact and simply connected $\Omega_1 \subset\subset \Omega$, then we can write $f|\Omega_1 = h \prod_{1\le j\le q} (z-a_j)^{m_j}$, where the a_j, $1 \le j \le q < \infty$, are all the zeros of f in Ω_1, m_j their multiplicities, and $h = e^g$ for some holomorphic function g in Ω_1. Therefore, in Ω_1 we have

$$\log|f| = \sum_{1\le j\le q} m_j \log|z-a_j| + \operatorname{Re} g,$$

which shows that $\log|f|$ is subharmonic in Ω_1. This ensures that $\log|f|$ is subharmonic in disks whose closure is contained in Ω. This obviously implies $\log|f|$ is subharmonic in Ω.

(8) The function $-\log d(z,\Omega^c)$ is continuous in Ω. Moreover, for every $\zeta \in \Omega^c$ the function $-\log|\zeta - z|$ is harmonic in Ω, hence

$$-\log d(z,\Omega^c) = \sup_{\zeta\in\Omega^c} (-\log|z-\zeta|)$$

ensures that $-\log d(z,\Omega^c)$ is subharmonic by part (2). □

For the proof of item (7) of the preceding proposition we proved that $\log|f|$ was subharmonic essentially by a localization argument. In fact, subharmonicity is a local property as it is shown by:

4.4.12. Proposition. *Let Ω be an open subset of $\mathbb{C}$, $u : \Omega \to [-\infty, \infty[$ a u.s.c. function. The following properties are equivalent:*

(1) *For every compact $K \subseteq \Omega$ and every real-valued continuous function f in K that is harmonic in $\mathring{K}$, if $f \ge u$ on ∂K, then $f \ge u$ in K.*
(2) *If $D = \bar{B}(z_0, r) \subseteq \Omega$ and if $f \in \mathbb{C}[z]$ is a polynomial in z with complex coefficients such that $u \le \operatorname{Re} f$ on ∂D, then $u \le \operatorname{Re} f$ in D.*
(3) *For every $r > 0$, for every measure $\mu \ge 0$ on $[0,r]$, and for every $z \in \Omega_r = \{\zeta \in \Omega : d(\zeta, \Omega^c) > r\}$, we have*

$$\mu([0,r])u(z) \le \int_{[0,r]} \lambda(u,z,\rho)\,d\mu(\rho) \qquad (*)$$

(4) *For every $\delta > 0$ and $z \in \Omega$ there is a positive measure in $[0,\delta]$, with nonzero mass in $]0,\delta]$ such that $(*)$ is valid.*

PROOF. (1) $\Rightarrow$ (2) and (3) $\Rightarrow$ (4) are evident.

(2) $\Rightarrow$ (3). Let us recall that if a trigonometric polynomial $\varphi(\theta) = \sum_{-n}^{n} a_k e^{ik\theta}$ is real-valued, then we have $a_{-k} = \bar{a}_k$, $0 \le k \le n$.

If $z \in \Omega_r$, $0 < \rho \le r$, then $D = \bar{B}(z,\rho) \subseteq \Omega$. Suppose a real-valued trigonometric polynomial $\varphi(\theta)$ satisfies $u(z + \rho e^{i\theta}) \le \varphi(\theta)$ for every $\theta \in [0, 2\pi]$.

Then the polynomial $f = u_0 + \sum_{k=0}^{n} a_k(\zeta - z)^k/\rho^k$ will satisfy $u \leq \operatorname{Re} f$ on ∂D. Therefore $u \leq \operatorname{Re} f$ in D. In particular, at $\zeta = z$ we have

$$u(z) \leq \operatorname{Re} f(z) = a_0 = \frac{1}{2\pi}\int_0^{2\pi} \varphi(\theta)\, d\theta.$$

If ψ is an arbitrary real-valued continuous function in $[0, 2\pi]$ such that $u(z + \rho e^{i\theta}) \leq \psi(\theta)$, and if $\varepsilon > 0$ is given, we can find a trigonometric polynomial φ such that $\psi \leq \varphi \leq \psi + \varepsilon$. It follows that

$$u(z) \leq \varepsilon + \frac{1}{2\pi}\int_0^{2\pi} \psi(\theta)\, d\theta.$$

Since $\varepsilon > 0$ was arbitrary we can also let $\varepsilon = 0$ in this inequality. On the other hand

$$\int_0^{2\pi} u(z + \rho e^{i\theta})\, d\theta = \inf_{\psi \geq u} \int_0^{2\pi} \psi(\theta)\, d\theta, \qquad \psi \text{ continuous},$$

which shows that $u(z) \leq \lambda(u, z, \rho)$. Integrating this inequality with respect to the positive measure μ we obtain (3).

(4) $\Rightarrow$ (1). Let K be a compact subset of Ω and h a real-valued function that is continuous in K, harmonic in $\mathring{K}$, and $h \geq u$ on ∂K. If $M = \sup_K (u - h) > 0$, then the upper semicontinuity of $u - h$ implies that $u - h = M$ on a nonempty compact subset F of $\mathring{K}$. Let $z_0 \in F$ such that $d(z_0, \partial K) = d(F, \partial K)$. Let $0 < \delta < d(F, \partial K)$ and μ be a measure in $[0, \delta]$ with the property (4). By the very definition of z_0, for every r, $0 < r \leq \delta$, the circle $\partial B(z_0, r)$ contains at least a point in $\mathring{K} \setminus F$, hence, by the upper semicontinuity, it contains a whole arc of points such that $u(z) - h(z) < M$. Therefore,

$$\int_0^{2\pi} (u - h)(z_0 + re^{i\theta})\, d\theta < 2\pi M,$$

whence

$$\int_0^{\delta}\int_0^{2\pi} (u - h)(z_0 + re^{i\theta})\, d\theta\, d\mu(r) < 2\pi M \int_0^{\delta} d\mu(r) = (u - h)(z_0) 2\pi\mu([0, \delta]).$$

For the harmonic function h we have equality of the corresponding terms in this inequality, it follows that

$$\frac{1}{\mu([0, \delta])}\int_{[0,\delta]} \lambda(u, z_0, \rho)\, d\mu(\rho) < u(z_0),$$

which contradicts (4). □

4.4.13. Remarks. (1) To show that (1) $\Rightarrow$ (3) we could also reason as follows: if φ is continuous on $\partial B(z_0, \rho)$ $(0 < \rho \leq r)$ then $\varphi = P(\varphi)|\partial B(z_0, \rho)$, where $P(\varphi)$ is the Poisson integral of φ. If $u \leq \varphi$ on $\partial B(z_0, \rho)$ then it follows by (1) that

$u \le P(\varphi)$ in $\bar{B}(z_0, \rho)$. Hence, $u(z_0) \le \lambda(u, z_0, \rho)$ and, by integrating with respect to μ, the proof can be concluded.

(2) The same argument shows that $u \le P(u)$ in $\bar{B}(z_0, \rho)$, since

$$P(u) = P\Bigg(\inf_{\substack{\varphi \ge u \\ \varphi \text{ continuous}}} \varphi \Bigg) = \inf_{\varphi \ge u} P(\varphi) \ge u.$$

(3) Clearly condition (4) is satisfied if u is subharmonic in the wide sense. It follows that we can define subharmonicity in the wide sense using area averages instead of circular averages.

It is clear what we mean by the phrase: u is locally subharmonic (or locally subharmonic in the wide sense). We have the following corollary to Proposition 4.4.12.

4.4.14. Corollary. *Let Ω be an open subset of $\mathbb{C}$ and $u : \Omega \to [-\infty, \infty[$ a u.s.c. function.*

(a) *u is subharmonic in the wide sense if and only if u is locally subharmonic in the wide sense.*

(b) *If $u \in L^1_{\mathrm{loc}}(\Omega)$, u is subharmonic if and only if u is locally subharmonic.*

If $u \in \mathrm{SH}(\Omega)$ and $\bar{B}(z_0, r) \subseteq \Omega$, then we can define

$$M(u, z_0, r) := \max_{|z - z_0| = r} u(z).$$

By Propositions 4.4.6 and 4.4.7 we have that

(i) $M(u, z_0, r) = \max_{|z - z_0| \le r} u(z)$, and

(ii) $r \mapsto M(u, z_0, r)$ is increasing for $r \in]0, d(z_0, \Omega^c)[$.

The following proposition extends these properties to the circular and area averages of u. If $z_0 = 0$ then we often write $M(u, r)$ instead of $M(u, 0, r)$, especially in the case $\Omega = \mathbb{C}$.

4.4.15. Proposition. *Let Ω be an open subset of $\mathbb{C}$, u a subharmonic function in Ω, $z_0 \in \Omega$. In the interval $]0, d(z_0, \Omega^c)[$ the two following functions are increasing:*

$$\lambda(u, z_0) : r \mapsto \lambda(u, z_0, r) \in [-\infty, \infty[$$

$$A(u, z_0) : r \mapsto A(u, z_0, r) \in \mathbb{R}.$$

Furthermore,

(1) $A(u, z_0, r) \le \lambda(u, z_0, r)$ $(0 < r < d(z_0, \Omega^c))$

(2) $u(z_0) = \lim_{r \to 0} \lambda(u, z_0, r) = \lim_{r \to 0} A(u, z_0, r)$.

PROOF. Let $0 < r < R < d(z_0, \Omega^c)$.

(1) $\lambda(u, z_0)$ is increasing:

In $B(z_0, R)$ we have $u \leq P(u)$, therefore

$$\lambda(u, z_0, r) \leq \lambda(P(u), z_0, r) = P(u)(z_0) = \lambda(u, z_0, R).$$

(2) Inequality $A(u, z_0, r) \leq \lambda(u, z_0, r)$:

$$A(u, z_0, r) = \frac{2}{r^2}\int_0^r \lambda(u, z_0, t)t\,dt \leq \frac{2}{r^2}\int_0^r \lambda(u, z_0, r)t\,dt = \lambda(u, z_0, r).$$

(3) $A(u, z_0)$ is increasing:
Assume $0 < r < R$. Then

$$\begin{aligned}
A(u, z_0, R) &= \frac{2}{R^2}\int_0^R \lambda(u, z_0, \rho)\rho\,d\rho \\
&= \frac{2}{R^2}\int_r^R \lambda(u, z_0, \rho)\rho\,d\rho + \frac{r^2}{R^2}\frac{2}{r^2}\int_0^r \lambda(u, z_0, \rho)\rho\,d\rho \\
&\geq A(u, z_0, r)\frac{(R^2 - r^2)}{R^2} + \frac{r^2}{R^2}A(u, z_0, r) = A(u, z_0, r).
\end{aligned}$$

To conclude the proof it suffices to show that $\lim_{r \to 0} \lambda(u, z_0, r) = u(z_0)$. Let $\alpha \in \mathbb{R}$ such that $u(z_0) < \alpha$. There exists $\varepsilon > 0$ such that if $z \in B(z_0, \varepsilon)$, then $u(z) < \alpha$. Therefore $u(z_0) \leq \lambda(u, z_0, \rho) < \alpha$ for $0 < \rho < \varepsilon$. □

4.4.16. Proposition. *Let u be a subharmonic function in an open subset Ω of $\mathbb{C}$ and α a standard function in $\mathbb{C}$. For ρ sufficiently small, $u_\rho = u * \alpha_\rho$ is subharmonic in $\Omega_\rho = \{z \in \Omega : d(z, \Omega^c) > \rho\}$ and the sequence $\{u_{1/n}\}_{n \geq n_0}$ is decreasing and converges to u in Ω_{1/n_0} (n_0 integer sufficiently large).*

Proof. We have $u_\rho(z) = \int_{B(0,\rho)} u(z - \zeta)\alpha_\rho(\zeta)\,d\xi\,d\eta = \int_{B(z,\rho)} u(\zeta)\alpha_\rho(z - \zeta)\,d\xi\,d\eta$ for $z \in \Omega_\rho$. If $\bar{B}(z, r) \subseteq \Omega_\rho$ we can apply Fubini's Theorem 4.2.7 and obtain

$$\begin{aligned}
\lambda(u_\rho, z, r) &= \int_{B(0,\rho)} \left(\frac{1}{2\pi}\int_0^{2\pi} u(z + re^{i\theta} - \zeta)\,d\theta\right)\alpha_\rho(\zeta)\,d\xi\,d\eta \\
&\geq \int_{B(0,\rho)} u(z - \zeta)\alpha_\rho(\zeta)\,d\xi\,d\eta = u_\rho(z),
\end{aligned}$$

hence the C^∞ function u_ρ is subharmonic in Ω_ρ. To show that the family u_ρ decreases in ρ when $\rho \searrow 0$ and converges toward u, we just have to rewrite the definition of u_ρ for $z \in \Omega_\rho$:

$$\begin{aligned}
u_\rho(z) &= \int_{B(0,\rho)} u(z - \zeta)\alpha_\rho(\zeta)\,d\xi\,d\eta = \int_0^\rho \int_0^{2\pi} u(z - re^{i\theta})\alpha_\rho(r)\,dr\,d\theta \\
&= 2\pi \int_0^\rho \lambda(u, z, \rho)\alpha_\rho(r)\,dr = 2\pi \int_0^1 \lambda(u, z, \rho s)\alpha(s)\,ds.
\end{aligned}$$

From this identity we see that $u_\rho(z) \to u(z)$ if $\rho \to 0$, and that if $\rho_1 < \rho_2$ and $z \in \Omega_{\rho_2}$, then

$$u_{\rho_2}(z) - u_{\rho_1}(z) = 2\pi \int_0^1 (\lambda(u, z, \rho_2 s) - \lambda(u, z, \rho_1 s))\alpha(s)\, ds \geq 0. \qquad \square$$

4.4.17. Remark. There is an analogous statement for a convex function f on an open interval $]a, b[$ of the real axis. For any $[a', b'] \subseteq]a, b[$ we can define for $x \in [a', b']$

$$f_k(x) = \int_{-1/k}^{1/k} f(x - t)\alpha_{1/k}(t)\, dt,$$

if k is a sufficiently large integer and α is a standard function in $\mathbb{R}$. Set $\lambda(f, x, s) := (f(x + s) + f(x - s))/2$, then we have

$$\begin{aligned} f_k(x) &= \int_0^{1/k} (f(x + t) + f(x - t))\alpha_{1/k}(t)\, dt \\ &= \int_0^1 \left(f\left(x + \frac{s}{k}\right) + f\left(x - \frac{s}{k}\right)\right)\alpha(s)\, ds \\ &= 2\int_0^1 \lambda\left(f, x, \frac{s}{k}\right)\alpha(s)\, ds. \end{aligned}$$

The convexity of f ensures that if $s_1 \leq s_2$, then $\lambda(f, x, s_2) \geq \lambda(f, x, s_1)$. Therefore the sequence f_k converges decreasingly to f. One verifies as in §4.4.16 that f_k is convex on $]a', b'[$. It is also clear from the definition that if f is increasing, then f_k is also increasing.

4.4.18. Proposition. *Let $[a, b[\subseteq [-\infty, \infty[$, $f : [a, b[\to \mathbb{R}$ convex, increasing and such that $f(a) = \lim_{t \to a} f(t)$. Let $u : \Omega \to [a, b[$ be a subharmonic function in the open set $\Omega \subseteq \mathbb{C}$. The function $f \circ u$ is then subharmonic in Ω.*

Proof. (1) If f and u are both of class C^2, then an easy computation shows that

$$\Delta(f \circ u) = (f'' \circ u)\|\text{grad}\, u\|^2 + (f' \circ u)\Delta u \geq 0,$$

hence $f \circ u$ is subharmonic.

(2) If f is C^2 and u is subharmonic then (with the notations of §4.4.16) we have that $f \circ u_\rho$ is also subharmonic in Ω_ρ. These functions tend decreasingly towards $f \circ u$, which is therefore subharmonic.

(3) If $a > -\infty$, extend the definition of f to $]-\infty, b[$ by setting it to be equal to $f(a)$ in $]-\infty, a]$. This function is still convex and increasing, and we can approximate it in $[a, b']$ for any $b' < b$ by a decreasing sequence of C^∞ increasing convex functions f_k, as was done in §4.4.17. Therefore, if u is subharmonic, we have that the decreasing sequence of subharmonic functions $f_k \circ u$ converges toward $f \circ u$. This ends the proof of the proposition. $\square$

4.4.19. Examples. The last proposition allows us to construct numerous examples of subharmonic functions.

(1) If $u \geq 0$ is subharmonic and $\alpha \geq 1$, then u^α is subharmonic.
(2) If u is subharmonic, then e^u is subharmonic.
(3) If $\log u$ is subharmonic ($u \geq 0$), then u is subharmonic and u^p is also subharmonic for any $p > 0$.
(4) If f is holomorphic, then $|f|^p$ is subharmonic for any $p > 0$.
(5) If φ is a convex increasing function of a real variable, then $\varphi(|z|)$ is subharmonic.
(6) If φ, ψ are two convex functions in $\mathbb{R}$, then for any $\lambda, \mu \geq 0$ we have that $u(z) = \lambda\varphi(\operatorname{Re} z) + \mu\psi(\operatorname{Im} z)$ is a subharmonic function in $\mathbb{C}$.

4.4.20. Proposition. *Let Ω be an open set in $\mathbb{C}$ and $u : \Omega \to \mathbb{R}^+$ a u.s.c. function. Then $\log u$ is subharmonic in Ω if and only if for every $\alpha \in \mathbb{R}^2$, the function $v(z) := u(z)\exp(\alpha_1 x + \alpha_2 y)$ is subharmonic in Ω.*

PROOF. 1. The condition is necessary. If $\log u$ is subharmonic then

$$\log u + \alpha_1 x + \alpha_2 y$$

is also subharmonic as well as its exponential.

2. The condition is sufficient. Let us assume $u \in C^2$ with values in $]0, \infty[$. We then have

$$\Delta v = e^{\alpha_1 x + \alpha_2 y}\left[\Delta u + 2\left(\alpha_1 \frac{\partial u}{\partial x} + \alpha_2 \frac{\partial u}{\partial y}\right) + (\alpha_1^2 + \alpha_2^2)u\right] \geq 0$$

for every $\alpha \in \mathbb{R}^2$. Therefore, eliminating the exponential factor, multiplying by u, and completing squares, we have

$$\left[\left(\alpha_1 u + \frac{\partial u}{\partial x}\right)^2 + \left(\alpha_2 u + \frac{\partial u}{\partial y}\right)^2\right] + u\Delta u - \left[\left(\frac{\partial u}{\partial x}\right)^2 + \left(\frac{\partial u}{\partial y}\right)^2\right] \geq 0.$$

Since $\alpha \in \mathbb{R}^2$ is arbitrary, for each $z \in \Omega$ we can choose it so that the first term vanishes. Hence

$$u^2\Delta(\log u) = u\Delta u - \|\operatorname{grad} u\|^2 \geq 0.$$

It follows that $\log u$ is subharmonic. If u is not of class C^2 we regularize as in 4.4.16 and 4.4.18. If u is not strictly positive, consider $u + \varepsilon$, $\varepsilon > 0$. Then let $\varepsilon \to 0$. □

4.4.21. Example. If $u_j \geq 0$ and $\log u_j$ is subharmonic ($1 \leq j \leq q$) then

$$\log(u_1 + \cdots + u_q)$$

is subharmonic ($\log 0 = -\infty$).

It is enough to show this for $q = 2$. For $\alpha \in \mathbb{R}^2$, we have $u_j(z)e^{\alpha_1 x + \alpha_2 y}$ is subharmonic since $\log u_j$ is subharmonic. Therefore $(u_1(z) + u_2(z))e^{\alpha_1 x + \alpha_2 y}$ is also subharmonic for every $\alpha \in \mathbb{R}^2$, hence $\log(u_1 + u_2)$ is subharmonic.

Let $u \in L^1_{\text{loc}}(\Omega)$. We have seen in Chapter 3 that u can be considered as the element of the space of distributions $\mathscr{D}'(\Omega)$ given by

$$\varphi \mapsto \int_\Omega \varphi\,dx\,dy \qquad (\varphi \in \mathscr{D}(\Omega)).$$

As such we can define $D^\alpha u = \dfrac{\partial^{|\alpha|}u}{\partial x^{\alpha_1}\partial y^{\alpha_2}}$, for $\alpha = (\alpha_1, \alpha_2) \in \mathbb{N}^2$, $|\alpha| = \alpha_1 + \alpha_2$. In particular, the Laplacian of u in the sense of distributions is the linear functional

$$\varphi \mapsto \langle \Delta u, \varphi\rangle = \int_\Omega u\Delta\varphi\,d\xi\,d\eta, \qquad \varphi \in \mathscr{D}(\Omega).$$

Example 5 of §2.2.9 can be restated as saying that if $\dfrac{\partial u}{\partial \bar{z}} = 0$ in the sense of distributions then u coincides almost everywhere with a holomorphic function in Ω.

We recall that the Dirac measure at the point a is the linear functional $\delta_a : \varphi \mapsto \varphi(a)$. If $a = 0$ then we occasionally suppress the subscript.

4.4.22. Proposition. (1) *The Laplacian of* $\dfrac{1}{2\pi}\log|z|$ *in the sense of distributions is the Dirac measure* δ *(at the origin).*

(2) *If* $f \in \mathscr{H}(\Omega)$, $f \not\equiv 0$ *in any connected component of* Ω, *and if* $\{a_i\}_{i\in I}$ *is the discrete set of zeros of and* $m_i = m(f, a_i)$ *the multiplicity of those zeros, then the Laplacian of* $\log|f|$ *in the sense of distributions is a nonnegative measure:*

$$\Delta \log|f| = 2\pi \sum_{i\in I} m_i \delta_{a_i},$$

that is,

$$\langle \Delta \log|f|, \varphi\rangle = 2\pi \sum_{i\in I} m_i \varphi(a_i) \qquad \textit{for } \varphi \in \mathscr{D}(\Omega).$$

PROOF. Let us compute $\left\langle \dfrac{\partial^2}{\partial z \partial \bar{z}} \log|z|, \varphi \right\rangle$ for $\varphi \in \mathscr{D}(\mathbb{C})$. We have by definition

$$\left\langle \frac{\partial^2}{\partial z \partial \bar{z}} \log|z|, \varphi \right\rangle = \int_{\mathbb{C}} \log|z| \frac{\partial^2 \varphi}{\partial z \partial \bar{z}}\,dx\,dy = \frac{i}{4}\int_{\mathbb{C}} (\log|z|^2) \frac{\partial^2 \varphi}{\partial z \partial \bar{z}}\,dz \wedge d\bar{z}$$
$$= \frac{i}{4} \lim_{\varepsilon \to 0} \int_{|z|\geq\varepsilon} (\log|z|^2) \frac{\partial^2 \varphi}{\partial z \partial \bar{z}}\,dz \wedge d\bar{z}.$$

The last identity is valid because the function $\log|z|^2$ is locally integrable. On the other hand we have that for $z \neq 0$ the following computation is valid:

$$d\left(\frac{\partial\varphi}{\partial\bar{z}}(\log|z|^2)\,d\bar{z}\right) = \partial\left(\frac{\partial\varphi}{\partial\bar{z}}(\log|z|^2)\,d\bar{z}\right)$$
$$= \frac{\partial^2\varphi}{\partial z\partial\bar{z}}(\log|z|^2)\,dz \wedge d\bar{z} + \frac{\partial\varphi}{\partial\bar{z}}\frac{1}{z}\,dz \wedge d\bar{z}.$$

Therefore, for $\varepsilon > 0$ we have

$$\int_{|z|\geq\varepsilon}(\log|z|^2)\frac{\partial^2\varphi}{\partial z\partial\bar{z}}\,dz \wedge d\bar{z} = \int_{|z|\geq\varepsilon} d\left(\frac{\partial\varphi}{\partial\bar{z}}(\log|z|^2)\,d\bar{z}\right) - \int_{|z|\geq\varepsilon}\frac{\partial\varphi}{\partial\bar{z}}\frac{1}{z}\,dz \wedge d\bar{z}$$
$$= -\int_{\partial B(0,\varepsilon)}\frac{\partial\varphi}{\partial\bar{z}}(\log|z|^2)\,d\bar{z} - \int_{|z|\geq\varepsilon}\frac{\partial\varphi}{\partial\bar{z}}\frac{dz \wedge d\bar{z}}{z}.$$

It is easy to see that $\lim_{\varepsilon\to 0}\int_{\partial B(0,\varepsilon)}\frac{\partial\varphi}{\partial\bar{z}}\log|z|^2\,d\bar{z} = 0$, therefore

$$\left\langle\frac{\partial^2\log|z|}{\partial z\partial\bar{z}}, \varphi\right\rangle = -\frac{i}{4}\int_{\mathbb{C}}\frac{\partial\varphi}{\partial\bar{z}}\frac{1}{z}\,dz \wedge d\bar{z} = \frac{\pi}{2}\varphi(0) \quad \left(= \left\langle\frac{\pi}{2}\delta, \varphi\right\rangle\right).$$

The last identity is a consequence of Pompeiu's formula. This proves the first part of the proposition,

$$\Delta\left(\frac{1}{2\pi}\log|z|\right) = \delta.$$

(2) Let $\{\varepsilon_i\}_{i\in I}$ be a sequence of positive real numbers such that the closed disks $\bar{B}(a_i, \varepsilon_i) \subseteq \Omega$ are pairwise disjoint, $f(z) = (z - a_i)^{m_i}h_i(z)$ for some h_i holomorphic in Ω without any zeros in $\bar{B}(a_i, \varepsilon_i)$. Let $\Omega_0 = \Omega\backslash\{a_i : i \in I\}$, $\Omega_i = B(a_i, \varepsilon_i)$, $i \in I$. Consider α_0, α_i $(i \in I)$ a C^∞ partition of the unity subordinate to this open covering of Ω. For $\varphi \in \mathscr{D}(\Omega)$ we have

$$\langle\Delta\log|f|, \varphi\rangle = \int_\Omega(\log|f|)\Delta\varphi\,dx\,dy$$
$$= \int_\Omega(\log|f|)\Delta(\alpha_0\varphi)\,dx\,dy + \sum_i\int_{B(a_i,\varepsilon_i)}(\log|f|)\Delta(\alpha_i\varphi)\,dx\,dy.$$

Let Ω_0' be a regular open set with boundary piecewise C^1 such that $\text{supp}(\alpha_0\varphi) \subseteq \Omega_0' \subset\subset \Omega_0$. We can apply to the first term Green's formula and obtain

$$\int_\Omega(\log|f|)\Delta(\alpha_0\varphi)\,dx\,dy = \int_{\Omega_0'}(\log|f|)\Delta(\alpha_0\varphi)\,dx\,dy$$
$$= \int_{\Omega_0'}(\Delta\log|f|)\alpha_0\varphi\,dx\,dy = 0,$$

where the final identity follows from the fact that $\log|f|$ is harmonic in Ω_0.

We also have

$$\int_{B(a_i,\varepsilon_i)} (\log|f|)\Delta(\alpha_i\varphi)\,dx\,dy = m_i \int_{B(a_i,\varepsilon_i)} (\log|z-\alpha_i|)\Delta(\alpha_i\varphi)\,dx\,dy$$
$$+ \int_{B(a_i,\varepsilon_i)} (\log|h_i|)\Delta(\alpha_i\varphi)\,dx\,dy.$$

The second term on the right-hand side is zero, because h_i does not vanish on $B(a_i, \varepsilon_i)$. By (1) we can compute

$$\int_{B(a_i,\varepsilon_i)} \log|z-\alpha_i|\Delta(\alpha_i\varphi)\,dx\,dy = \int_{\mathbb{C}} \log|z-\alpha_i|\Delta(\alpha_i\varphi)\,dx\,dy$$
$$= 2\pi(\alpha_i\varphi)(a_i) = 2\pi\varphi(a_i),$$

since by the choice of the partition of unity, we have $\alpha_i(a_i) = 1$. Altogether we have obtained

$$\langle \Delta \log|f|, \varphi\rangle = 2\pi \sum_{i\in I} m_i \langle \delta_{a_i}, \varphi\rangle,$$

i.e.,

$$\Delta \log|f| = 2\pi \sum_{i\in I} m_i \delta_{a_i}. \qquad \square$$

4.4.23. Proposition. *Let $u \in L^1_{\mathrm{loc}}(\Omega)$. The function u is subharmonic (i.e., equal a.e. to a true subharmonic function in Ω) if and only if Δu is a nonnegative measure.*

PROOF. Let us recall that for $\varphi \in \mathscr{D}(\Omega)$ we have

$$\lim_{r\to 0} \frac{2}{r^2}[\lambda(\varphi, z, r) - \varphi(z)] = \Delta\varphi(z),$$

and the convergence is uniform (cf. §4.4.11). Let u be a subharmonic function. We claim that the following identity holds:

$$\int_\Omega [\lambda(u,z,r) - u(z)]\varphi(z)\,dx\,dy = \int_\Omega u(z)[\lambda(\varphi,z,r) - \varphi(z)]\,dx\,dy.$$

Assuming this claim for the moment, we have for $\varphi \geq 0$

$$\langle \Delta u, \varphi\rangle = \int_\Omega u(z)\Delta\varphi(z)\,dx\,dy = \lim_{r\to 0}\frac{2}{r^2}\int_\Omega u(z)[\lambda(\varphi,z,r) - \varphi(z)]\,dx\,dy$$
$$= \lim_{r\to 0}\frac{2}{r^2}\int_\Omega [\lambda(u,z,r) - u(z)]\varphi(z)\,dx\,dy \geq 0$$

since u is subharmonic. It is a standard fact in the theory of distributions that if $\langle \Delta u, \varphi\rangle \geq 0$ for $\varphi \in \mathscr{D}(\Omega)$, $\varphi \geq 0$, then Δu is a nonnegative measure (i.e., it makes sense to compute $\int_\Omega \psi \Delta u$ for any ψ continuous function with compact support), see [L. Schwartz, Ch. 1, §4, Theorem V].

To prove the claim, we observe that

$$\begin{aligned}\int_\Omega \lambda(u,z,r)\varphi(z)\,dx\,dy &= \int_\Omega \left(\frac{1}{2\pi}\int_0^{2\pi} u(z+re^{i\theta})\,d\theta\right)\varphi(z)\,dx\,dy\\ &= \frac{1}{2\pi}\int_0^{2\pi}\left(\int_\Omega u(z+re^{i\theta})\varphi(z)\,dx\,dy\right)d\theta\\ &= \frac{1}{2\pi}\int_0^{2\pi}\left(\int_\Omega u(z)\varphi(z-re^{i\theta})\,dx\,dy\right)d\theta\\ &= \int_\Omega\left(\frac{1}{2\pi}\int_0^{2\pi}\varphi(z-re^{i\theta})\,d\theta\right)u(z)\,dx\,dy\\ &= \int_\Omega \lambda(\varphi,z,r)u(z)\,dx\,dy,\end{aligned}$$

where we have used several times that $\operatorname{supp}\varphi \subseteq \Omega_r$.

This proves half of the statement of the proposition. To prove the other half we need two lemmas.

4.4.24. Lemma. (1) *If μ is a complex Radon measure with compact support K, then the function*

$$U^\mu(z) := \frac{1}{2\pi}\int_{\mathbb{C}} \log|z-\zeta|\,d\mu(\zeta)$$

is harmonic in $\mathbb{C}\setminus K$ and, for $|z|$ large, equals $\frac{1}{2\pi}\log|z| + h(z)$, where h is a harmonic function tending to zero at infinity.

(2) *If μ is also nonnegative then U^μ is subharmonic in $\mathbb{C}$ and $U^\mu(a) = -\infty$ for every atom a of μ.*

(3) (Riesz' Decomposition Theorem): *Let Ω_1 be a relatively compact open subset of an open set $\Omega \subseteq \mathbb{C}$, $u \in L^1_{\text{loc}}(\Omega)$ such that its Laplacian Δu (in the sense of distributions) is a nonnegative measure μ. Then, if we define a function u_0 in Ω_1 by means of the decomposition,*

$$u(z) = u_0(z) + \frac{1}{2\pi}\int_{\Omega_1} \log|z-\zeta|\,d\mu(\zeta),$$

we have that u_0 is a harmonic function in Ω_1.

The function U^μ is called the ***logarithmic potential of the measure*** μ.

PROOF. (1) The proof of (1) is left to the reader.

(2) Let $r > 0$. For $t \in K$ we have

$$\int_{|z| \le r} |\log|t - z|| \, dx\, dy \le \int_{|w| \le r+d} |\log|w|| \, dx\, dy = M = M(r, d) < \infty,$$

where d is chosen so that $K \subseteq \bar{B}(0, d)$. Therefore, the function

$$t \mapsto \int_{|z| \le r} |\log|t - z|| \, dx\, dy \in L^1(\mu)$$

since

$$\int \left[\int_{|z| \le r} |\log|t - z|| \, dx\, dy \right] d\mu(t) \le M\mu(K) < \infty.$$

By Fubini's theorem this implies that $U^\mu \in L^1_{\text{loc}}(\mathbb{C})$. Note that U^μ is only defined (up to now) outside a set of Lebesgue measure zero.

Let $h_\varepsilon(z) := \frac{1}{4\pi}(\log|z|^2 + \varepsilon)$ and $g_\varepsilon(z) := \int h_\varepsilon(z - \zeta)\, d\mu(\zeta)$. We verify immediately that $\Delta g_\varepsilon(z) = \frac{1}{\pi} \int \frac{\varepsilon}{(|\zeta - z|^2 + \varepsilon)^2} d\mu(\zeta) \ge 0$. Therefore the family $(g_\varepsilon)_{\varepsilon > 0}$ of subharmonic functions decreases towards U^μ, which is hence subharmonic in the wide sense. Since $U^\mu \in L^1_{\text{loc}}(\mathbb{C})$ then U^μ is subharmonic and defined everywhere.

The last statement of (2) follows from the observation that

$$U^{\delta_a}(z) = \frac{1}{2\pi} \log|z - a| \qquad \text{if } z \ne a.$$

(3) Let us show now that in Ω_1 we have the following property:

$$\int_{\Omega_1} u_0 \Delta\varphi \, dx\, dy = 0 \qquad \text{for every } \varphi \in \mathscr{D}(\Omega_1).$$

In fact, by the definition of u_0 we have

$$\int_{\Omega_1} u(z)\Delta\varphi(z)\, dx\, dy = \frac{1}{2\pi} \int_{\Omega_1} \left(\int_{\Omega_1} \log|z - \zeta| d\mu(\zeta) \right) \Delta\varphi(z)\, dx\, dy$$
$$+ \int_{\Omega_1} u_0(z)\Delta\varphi(z)\, dx\, dy.$$

We have already proved that $\frac{1}{2\pi} \int_{\Omega_1} \log|z - \zeta| \Delta\varphi(z)\, dx\, dy = \varphi(\zeta)$, since $\varphi \in \mathscr{D}(\Omega_1)$. By Fubini, the last identity becomes

$$\int_{\Omega_1} u(z)\Delta\varphi(z)\, dx\, dy = \int_{\Omega_1} \varphi(z)\, d\mu(z) + \int_{\Omega_1} u_0(z)\Delta\varphi(z)\, dx\, dy.$$

But $\int_{\Omega_1} u(z)\Delta\varphi(z)\,dx\,dy = \int_{\Omega_1} \varphi(z)\,d\mu(z)$ by the very definition of μ! Therefore we have shown that the claim holds.

We need an additional lemma.

4.4.25. Lemma (Weyl's Lemma). *Let $u_0 \in L^1_{\text{loc}}(\Omega_1)$ such that its Laplacian in the sense of distribution vanishes identically in Ω_1. Then u_0 is harmonic (i.e., it coincides a.e. with a harmonic function in Ω_1).*

PROOF. This lemma is a particular case of Proposition 3.6.5, nevertheless it is instructive to give a different proof here. Let α be a standard function. The family of C^∞ functions $U_\rho = u_0 * \alpha_\rho$, $\rho > 0$, converges to u in $L^1_{\text{loc}}(\Omega_1)$ when $\rho \to 0$. The function u_ρ is only defined in $\Omega_{1,\rho}$, but it is harmonic there. In fact, if $z \in \Omega_{1,\rho}$

$$\Delta u_\rho(z) = \int_{\Omega_1} u_0(\zeta)\Delta_z(\alpha_\rho(z-\zeta))\,d\xi\,d\eta = \int_{\Omega_1} u_0(\zeta)\Delta_\zeta(\alpha_\rho(z-\zeta))\,d\xi\,d\eta = 0,$$

since, for z fixed in $\Omega_{1,\rho}$ the function $\zeta \mapsto \alpha_\rho(z-\zeta)$ is in $\mathscr{D}(\Omega_1)$, and $\Delta u_0 = 0$ in the sense of distributions in Ω_1.

Let now $\rho, \sigma > 0$ be very small, we have in $\Omega_{1,\rho+\sigma}$,

$$u_0 * \alpha_\sigma * \alpha_\rho = u_0 * \alpha_\rho * \alpha_\sigma = u_0 * \alpha_\rho,$$

by §4.3.3, (5), since we have just shown that $u_0 * u_\rho$ is harmonic in $\Omega_{1,\rho}$. Therefore letting $\rho \to 0$ we obtain $u_0 * \alpha_\sigma = u_0$ a.e. in $\Omega_{1,\sigma}$. That is, u_0 coincides a.e. in $\Omega_{1,\sigma}$ with a function C^∞ and harmonic. A posteriori u_0 can be modified in a set of measure zero of Ω_1 and taken to be harmonic itself. This proves Lemma 4.4.25. □

This also concludes the last part of the proof of the Riesz's decomposition theorem. Lemma 4.4.24 is completely proven. □

We can now go back to the proof of Proposition 4.4.23. If $u \in L^1_{\text{loc}}(\Omega)$ and $\Delta u \geq 0$ in the sense of distributions, then for every open relatively compact set we have (up to a set of Lebesgue measure zero)

$$u|\Omega_1 = U^\mu + u_0, \qquad \mu = \Delta u.$$

Since U^μ is subharmonic in Ω_1 and u_0 is harmonic there, we can modify $u|\Omega_1$ so that it is defined everywhere in Ω_1 and subharmonic. Since the subharmonicity is a local property, u can be made subharmonic in Ω (after modifying it in a set of Lebesgue measure zero). □

4.2.26. Proposition (Riesz's Convexity Theorem). *Let u be a subharmonic function in the annulus $C = \{\zeta \in \mathbb{C} : 0 \leq R < |\zeta| < R' \leq \infty\}$. Then the function $\lambda : r \mapsto \lambda(u, 0, r)$, is a convex function of $\log r$ for $r \in]R, R'[$.*

PROOF. Let us first assume u of class C^2, let $t = \log r$, and compute $\frac{d^2\lambda}{dt^2}$. We have

$$\frac{d\lambda}{dt} = \frac{d\lambda}{dr}\frac{dr}{dt} = r\frac{d\lambda}{dr}$$

and

$$\frac{d^2\lambda}{dt^2} = r\frac{d}{dr}\left(r\frac{d\lambda}{dr}\right) = r^2\left(\frac{d^2\lambda}{dr^2} + \frac{1}{r}\frac{d\lambda}{dr}\right) = r^2\Delta\lambda.$$

The last identity is valid since λ is a radial function. Therefore, to show the convexity of λ with respect to $\log r$, we have to show that λ is subharmonic as a function of z in the annulus C. Let us introduce the auxiliary function $g(z) := \lambda(u, 0, |z|)$. Then

$$g(z) = \frac{1}{2\pi}\int_0^{2\pi} u(|z|e^{i\theta})\,d\theta = \frac{1}{2\pi}\int_0^{2\pi} u(ze^{i\alpha})\,d\alpha.$$

If $\bar{B}(z_0, \rho) \subseteq C$ then

$$\begin{aligned} A(g, z_0, \rho) &= \frac{1}{\pi\rho^2}\int_{\bar{B}(z_0,\rho)} g(z)\,dx\,dy = \frac{1}{2\pi}\int_0^{2\pi}\left(\frac{1}{\pi\rho^2}\int_{\bar{B}(z_0,\rho)} u(ze^{i\alpha})\,dx\,dy\right)d\alpha \\ &\geq \frac{1}{2\pi}\int_0^{2\pi} u(z_0e^{i\alpha})\,d\alpha = g(z_0). \end{aligned}$$

If u is not of class C^2 we can regularize and obtain $\lambda(u, 0, r)$ as a decreasing limit of $\lambda(u * \alpha_\rho, 0, r)$, which are convex functions of $\log r$. Therefore λ is also a convex function of $\log r$. □

4.4.27. Remarks. (1) It follows from §4.4.26 that λ is differentiable as a function of $\log r$, except possibly at a countable collection of points. The right and left derivatives always exist.

(2) The proof of §4.4.26 shows that if u is subharmonic in C and radial, and we write $u(z) = \varphi(|z|)$, then φ is a convex function of $\log r$. If u is subharmonic in $B(0, R)$ and radial, then φ is also increasing since $\varphi(r) = \lambda(u, 0, r)$.

(3) As it could already have been remarked after §4.4.15, let us note that if u is subharmonic in Ω and if E is the polar set of u, then not only E has Lebesgue measure zero but for any $\bar{B}(z_0, r) \subseteq \Omega$ the linear measure of $E \cap \partial B(z_0, r)$ is also zero. If not, $\lambda(u, z_0, r) = -\infty$, and we would have $\lambda(u, z_0, \rho) = -\infty$ for $0 \leq \rho \leq r$, contradicting the fact that $u \in L^1_{\text{loc}}(\Omega)$.

4.4.28. Proposition (Gauss' Theorem). *Let u be subharmonic in an open set Ω of $\mathbb{C}$. For every $z_0 \in \Omega$, $r \in \,]0, d(z_0, \Omega^c)[$ the function $\lambda : r \mapsto \lambda(u, z_0, r)$ admits right and left derivatives with respect to* $\log r$ *and we have*:

$$\left[\frac{\partial}{\partial \log r}\lambda(u, z_0 r)\right]_-(r) = \frac{1}{2\pi}\int_{B(z_0,r)} \Delta u = \frac{1}{2\pi}\Delta u(B(z_0,r)),$$

$$\left[\frac{\partial}{\partial \log r}\lambda(u, z_0 r)\right]_+(r) = \frac{1}{2\pi}\int_{\bar{B}(z_0,r)} \Delta u = \frac{1}{2\pi}\Delta u(\bar{B}(z_0,r))$$

where Δu denotes the measure ≥ 0 induced by the Laplacian of u in the sense of distributions. Since λ is a convex of $\log r$, these derivatives are equal except at most at a countable collection of points and we have the integral relation:

$$\lambda(u, z_0, R) - \lambda(u, z_0, r) = \frac{1}{2\pi}\int_r^R \frac{1}{t}\left(\int_{B(z_0,r)} \Delta u\right) dt = \frac{1}{2\pi}\int_r^R \frac{1}{t}\left(\int_{\bar{B}(z_0,r)} \Delta u\right) dt,$$

for $0 < r < R < d(z_0, \Omega^c)$.

PROOF. Let us suppose first that u is of class C^2, let $\sigma = \log r$. Green's formula gives

$$r\frac{d\lambda}{dr}(r) = \frac{d\lambda}{d\sigma}(r) = \frac{1}{2\pi}\int_{B(z_0,r)} \Delta u \, dm,$$

(dm = Lebesgue measure) and, by integration, if $0 < r < R < d(z_0, \Omega^c)$

$$\lambda(u, z_0, R) - \lambda(u, z_0, r) = \frac{1}{2\pi}\int_r^R \frac{1}{t}\left(\int_{B(z_0,r)} \Delta u \, dm\right) dt.$$

For u an arbitrary subharmonic function we regularize as usual, $u_\rho = u * \alpha_\rho$. We claim that

(1) $\lim_{\rho\to 0} \lambda(u_\rho, z_0, t) = \lambda(u, z_0, t)$, and it is decreasing.

(2) $\displaystyle\int_{B(z_0,t)} \Delta u \leq \varliminf_{\rho\to 0} \int_{B(z_0,t)} \Delta u_\rho \, dm \leq \varlimsup_{\rho\to 0} \int_{B(z_0,t)} \Delta u_\rho \, dm \leq \int_{\bar{B}(z_0,t)} \Delta u.$

We have observed (1) several times before. Property (2) follows from general properties of nonnegative measures. In fact, if $\varphi, \psi \in \mathscr{D}(\Omega)$ and

$$\varphi \leq \chi_{B(z_0,t)} \leq \chi_{\bar{B}(z_0,t)} \leq \psi,$$

then

$$\int_\Omega \varphi\Delta u = \int_\Omega u\Delta\varphi = \lim_{\rho\to 0}\int_\Omega u_\rho \Delta\varphi = \lim_{\rho\to 0}\int_\Omega \Delta u_\rho \cdot \varphi \leq \varliminf_{\rho\to 0}\int_{B(z_0,t)} \Delta u_\rho \, dm.$$

Hence

$$\int_{B(z_0,t)} \Delta u = \sup_{\substack{0\leq\varphi\leq 1 \\ \varphi\in\mathscr{D}(B(z_0,t))}} \int_\Omega \varphi\Delta u \leq \varliminf_{\rho\to 0}\int_{B(z_0,t)} \Delta u_\rho \, dm.$$

The same way we obtain

$$\int_\Omega \psi\Delta u = \int_\Omega u\Delta\psi = \lim_{\rho\to 0}\int_\Omega u_\rho\Delta\psi = \lim_{\rho\to 0}\int_\Omega \psi\Delta u_\rho \geq \overline{\lim_{\rho\to 0}}\int_{\bar{B}(z_0,t)} \Delta u_\rho\,dm,$$

hence

$$\int_{\bar{B}(z_0,t)} \Delta u = \inf_{\substack{\psi\in\mathscr{D}(\Omega)\\ \psi\geq\chi_{\bar{B}(z_0,t)}}}\int_\Omega \psi\Delta u \geq \overline{\lim_{\rho\to 0}}\int_{B(z_0,t)} \Delta u_\rho\,dm.$$

This leads to the chain of inequalities (2). We can now integrate them against the measure dt/t. One obtains, by Fatou's inequality,

$$\begin{aligned}\int_r^R \frac{1}{t}\left(\int_{B(z_0,t)} \Delta u\right)dt &\leq \int_r^R \frac{1}{t}\left(\underline{\lim}_{\rho\to 0}\int_{B(z_0,t)} \Delta u_\rho\,dm\right)dt\\ &\leq \underline{\lim}_{\rho\to 0}\int_r^R \frac{1}{t}\left(\int_{B(z_0,t)} \Delta u_\rho\,dm\right)dt\\ &= \underline{\lim}_{\rho\to 0}\,[\lambda(u_\rho, z_0, R) - \lambda(u_\rho, z_0, r)]\\ &= \lambda(u, z_0, R) - \lambda(u, z_0, r),\end{aligned}$$

since the limits exist by (1). Similarly we can use the other part of (2). This leads to

$$\int_r^R \frac{1}{t}\left(\int_{B(z_0,t)} \Delta u\right)dt \leq \lambda(u, z_0, R) - \lambda(u, z_0, r) \leq \int_r^R \frac{1}{t}\left(\int_{\bar{B}(z_0,r)} \Delta u\right)dt.$$

To compute the left derivative at R we need to divide by $R - r$ and let $r \nearrow R$. Now we have from the continuity properties of measures

$$\lim_{t\to R^+} \Delta u(B(z_0, t)) = \lim_{t\to R^-} \Delta u(\bar{B}(z_0, t)) = \Delta u(B(z_0, R)).$$

Hence the limit of the difference quotient is

$$\left[\frac{d}{ds}\lambda(u, z_0, s)\right]_-(R) = \frac{1}{2\pi R}\Delta u(B(z_0, R)).$$

The expression for the logarithmic derivative is obtained from this one by the chain rule. The right derivative is computed the same way. □

4.4.29. Corollary (Jensen's Formula I). *Let f be a holomorphic function in $B(0, R)$, $f(0) \neq 0$, $0 < r < R$ and $a_1, \ldots, a_N$ the zeros of f in $\bar{B}(0, r)$, counted according to multiplicity, $0 < |a_1| \leq |a_2| \leq \cdots \leq |a_N|$. We then have the expression*

$$\begin{aligned}\lambda(\log|f|, 0, r) &= \frac{1}{2\pi}\int_0^{2\pi} \log|f(re^{i\theta})|\,d\theta = \log|f(0)| + N\log r - \sum_{1\leq j\leq N}\log|a_j|\\ &= \log\left(|f(0)|\prod_{1\leq j\leq n}\frac{r}{|a_j|}\right).\end{aligned}$$

PROOF. Let us assume first that $|a_N| < r$. Denote $\alpha_1 < \cdots < \alpha_m$ the different values of $|a_j|$ and n_k the number of zeros (counted with multiplicities) of f in $\bar{B}(0,\alpha_k)$. Finally, let α_0 be an arbitrary value in $]0,\alpha_1[$ and $n_0 = 0$. We are going to obtain Jensen's formula from the integral version of Gauss' theorem. Recall that

$$\Delta \log|f| = 2\pi \sum_{1 \le j \le N} \delta_{a_j},$$

since this sum includes the multiplicities. For $0 < t \le r$ we have

$$v(t) := \frac{1}{2\pi}(\Delta \log|f|)(B(0,t)) = \begin{cases} n_{j-1} & \text{if } t \in [\alpha_{j-1}, \alpha_j[, j \le m \\ n_m = N & \text{if } t \in [\alpha_m, r]. \end{cases}$$

(Recall we are assuming no zeros of f lie in $\partial B(0,r)$.) Gauss' theorem shows that

$$\begin{aligned} &\lambda(\log|f|, 0, r) - \lambda(\log|f|, 0, \alpha_0) \\ &= \int_{\alpha_0}^{r} \frac{v(t)}{t}\,dt \\ &= \int_{\alpha_0}^{\alpha_1} n_0 \frac{dt}{t} + \int_{\alpha_1}^{\alpha_2} n_1 \frac{dt}{t} + \cdots + \int_{\alpha_{m-1}}^{\alpha_m} n_{m-1}\frac{dt}{t} + \int_{\alpha_m}^{r} n_m \frac{dt}{t} \\ &= -n_1 \log\alpha_1 + (n_1 - n_2)\log\alpha_2 + \cdots + (n_{m-1} - n_m)\log\alpha_{m-1} + n_m \log r \\ &= N \log r - \sum_{1 \le j \le N} \log|a_j|. \end{aligned}$$

On the other hand the function $\log|f|$ is harmonic in $B(0,\alpha_1)$, hence $\lambda(\log|f|, 0, \alpha_0) = \log|f(0)|$. This proves Jensen's formula when there are no zeros a_j, with $|a_j| = r$.

If now $\alpha_m = r$, we can write the preceding formula for $r' > r$, and let r' tend to r since every term is continuous on r'. Note the terms with $|a_j| = r$ do not really count at the end. □

4.4.30. Corollary (Jensen's Formula II). *The same conditions as in Corollary 4.4.29 except we do not assume $f(0) \neq 0$ but let k be the order of multiplicity of the origin as zero of f. Then one has*

$$\frac{1}{2\pi}\int_0^{2\pi} \log|f(re^{i\theta})|\,d\theta = \log\left(\frac{1}{k!}|f^{(k)}(0)|r^k \prod_{1 \le j \le N} \frac{r}{|a_j|}\right),$$

where $a_1, \ldots, a_N$ denote the zeros of f in $\bar{B}(0,r)\setminus\{0\}$, counted according to multiplicity.

PROOF. Apply Corollary 4.4.29 to the function $g(z) = f(z)/z^k$. □

4.4.31. Remark. (1) If u is a subharmonic function in Ω, $z_0 \in \Omega$ and $0 < r < er < d(z_0, \Omega^c)$, then

$$\Delta u(\bar{B}(z_0, r)) \le \lambda(u, z_0, er) - \lambda(u, z_0, r).$$

In fact, $\int_r^{er} \Delta u(B(z_0, r))\frac{dt}{t} = \lambda(u, z_0, er) - \lambda(u, z_0, r)$, and for $t \in]r, er[$ we have $\Delta(B(z_0, t)) \geq \Delta u(\bar{B}(z_0, r))$.

(2) In particular, if f is holomorphic in Ω, $f(z_0) = 1$ and $v(r)$ is defined by $v(r) := \frac{1}{2\pi}(\Delta \log|f|)(B(z_0, r))$, then

$$v(r) \leq \log M(|f|, z_0, er),$$

where $M(|f|, z_0, R) = \sup_{|\zeta - z_0| = R} |f(\zeta)|$.

In fact, $\lambda(\log|f|, z_0, er) \leq \log M(|f|, z_0, er)$ and

$$\lambda(\log|f|, z_0, r) \geq \log|f(z_0)| = 0.$$

Note that $v(r)$ coincides with the number of zeros of f in $\bar{B}(0, r)$.

4.4.32. Proposition. (Hadamard's Three Circles Theorem). *Let u be a subharmonic function in the annulus $C = \{\zeta \in \mathbb{C} : 0 \leq R < |\zeta| < R' \leq \infty\}$. The function $r \mapsto M(u, r) = \sup_{|\zeta| = r} u(\zeta)$, defined in $]R, R'[$, is an increasing convex function of* $\log r$.

Proof. For $\theta \in [0, 2\pi]$ fixed, the function $u_\theta(\zeta) := u(\zeta e^{i\theta})$ is subharmonic in C. Let $\mathcal{M}(\zeta) := \sup_{0 \leq \theta \leq 2\pi} u_\theta(\zeta)$. It will be subharmonic if one could prove it is u.s.c. If that were done, then $\mathcal{M}(\zeta) = M(u, r)$ when $|\zeta| = r$, hence $\mathcal{M}$ is radial and therefore equal to its circular average. We already know that the circular average of a subharmonic function is an increasing convex function of $\log r$.

So all we have to do is prove that $\mathcal{M}$ is u.s.c. Let $z_0 \in C$ and $\alpha > \mathcal{M}(z_0)$. Let ζ be any point with $|\zeta| = |z_0|$. Since u is u.s.c. there is a neighborhood V_ζ of ζ such that $u(z) < \alpha$ if $z \in V_\zeta$. There is $\varepsilon > 0$ such that

$$\bigcup_{|\zeta| = |z_0|} V_\zeta \supseteq \{z : R < |z_0| - \varepsilon < |z| < |z_0| + \varepsilon < R'\} = C'.$$

For $z \in C'$ we have $M(\zeta) \leq \alpha$. This inequality proves the proposition. □

4.4.33. Corollary. *There is no subharmonic function which is bounded above in $\mathbb{C}$, except for the constants.*

Note that this corollary includes Liouville's Theorem 2.2.20 as a particular case.

If u is subharmonic function in Ω and E is its polar set, then for $z \notin E$ we have

$$\chi(u, z) := \lim_{r \to 0} \frac{\lambda(u, z, r)}{\log r} = 0,$$

since $\lambda(u, z, r) \to u(z)$ as $r \to 0$ when $z \notin E$. More generally we have the following.

4.4.34. Proposition. *Let u be a subharmonic function in a neighborhood of $z \in \mathbb{C}$. The two limits*

$$\lim_{r\to 0} \frac{\lambda(u,z,r)}{\log r} \quad \textit{and} \quad \lim_{r\to 0} \frac{M(u,z,r)}{\log r}$$

exist, are finite, nonnegative, and they coincide.

The common value limit $\chi(u,z)$ is called the **Lelong number** of u at z.

PROOF. The existence of the two limits and that their values lie in $[0, \infty[$ are consequences of the fact that $\lambda(u,z,r)$ and $M(u,z,r)$ are increasing and convex functions of $\log r$ (cf. §4.4.15 and §4.4.32). From the preceding observation it is enough to prove their coincidence at points where $u(z) = -\infty$, otherwise they are both zero.

If $u(z) = -\infty$ there is R_0, $0 < R_0 < 1$ such that $u(\zeta) < 0$ when $\zeta \in B(z, R_0)$. Let $|\zeta - z| = r < \rho < R < R_0$. We have

$$u(\zeta) \le M(u,z,\rho) \le \sup_{|w-z|=\rho} \frac{1}{2\pi}\int_0^{2\pi} \frac{R^2-\rho^2}{|z+\mathrm{Re}^{i\theta}-w|^2} u(z+\mathrm{Re}^{i\theta})\,d\theta.$$

Now we have $\dfrac{R^2-\rho^2}{|z+\mathrm{Re}^{i\theta}-w|^2} \ge \dfrac{R^2-\rho^2}{(R+\rho)^2} = \dfrac{R-\rho}{R+\rho}$ and $u(z+\mathrm{Re}^{i\theta}) < 0$, therefore

$$u(\zeta) \le M(u,z,\rho) \le \frac{R-\rho}{R+\rho}\lambda(u,z,R).$$

Hence

$$\lambda(u,z,r) \le M(u,z,\rho) \le \frac{R-\rho}{R+\rho}\lambda(u,z,R).$$

By continuity we can replace ρ by r. Dividing by $\log r$ reverses the inequalities, so

$$\frac{\lambda(u,z,r)}{\log r} \ge \frac{M(u,z,r)}{\log r} \ge \frac{1-r/R}{1+r/R}\frac{\lambda(u,z,r)}{\log R}\frac{\log R}{\log r}.$$

To obtain the desired conclusion it suffices to let $R, r \to 0$ in such a way that $\log R \sim \log r$ and $r/R \sim 0$. For instance $R = -r\log r$. □

4.4.35. Corollary. *If f is a holomorphic function in a neighborhood of z, then*

$$\chi(\log|f|, z) = \lim_{r\to 0}\frac{\lambda(\log|f|,z,r)}{\log r} = \lim_{r\to 0}\frac{M(\log|f|,z,r)}{\log r} = m(f,z),$$

where $m(f,z)$ is the multiplicity of z as a zero of f.

PROOF. Use Jensen's formula 4.4.30 to evaluate $\chi(\log|f|, z)$. □

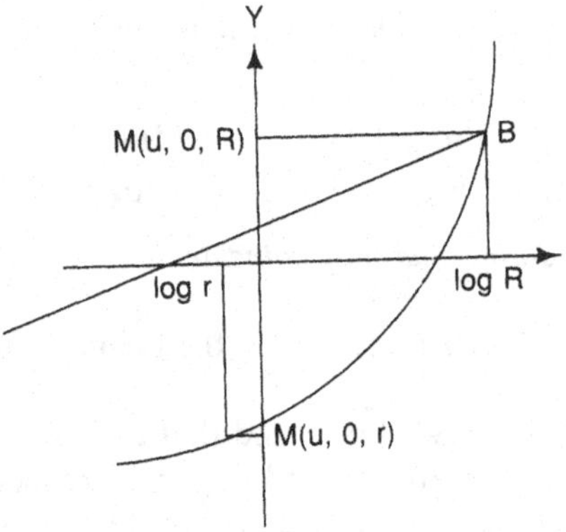

Figure 4.1

4.4.36. Proposition (Schwarz's Lemma for Subharmonic Functions). *Let u be subharmonic in $B(0, R')$ and such that $u(0) = -\infty$, $\chi(u, 0) > 0$. Let R be such that $0 < R < R'$. For every r with $0 < r \le R$ we have*

$$M(u,0,r) \le M(u,0,R) + \chi(u,0)\log\frac{r}{R}.$$

Proof. Consider the convex increasing function F in $]-\infty, \log R'[$ such that $M(u,0,r) = F(\log r)$ (see Figure 4.1). The corresponding graph in $]-\infty, \log R]$ lies below the straight line passing through the point $B = (\log R, M(u,0,R))$ with slope $\chi(u,0)$. The equation of this line is

$$Y = M(u,0,R) + \chi(u,0)[\log r - \log R] = M(u,0,R) + \chi(u,0)\log\frac{r}{R}.$$

The inequality follows. □

4.4.37. Corollary (Schwarz's Lemma for Holomorphic Functions). *Let f be a holomorphic function in a neighborhood of $\bar{B}(0, R)$ with a zero at the origin of multiplicity $m = m(f, 0) > 0$. For $0 \le |z| = r \le R$ we have*

$$|f(z)| \le M(|f|,0,R)\left(\frac{|z|}{R}\right)^m.$$

Proof. Let $u = \log|f|$ in the previous proposition. □

Let X be a Hausdorff topological space and $f: X \to \bar{\mathbb{R}} = [-\infty, \infty]$. We introduce the ***upper regularized function*** f^* of f and the ***lower regularized function*** f_* of f by

$$f^*(x) := \limsup_{y\to x} f(x), \qquad f_*(x) := \liminf_{y\to x} f(y).$$

The function f^* is upper semicontinuous (u.s.c.) and f_* is lower semicontinuos (l.s.c.). Furthermore, f^* is the smallest u.s.c. majorant of f and f_* the largest l.s.c. minorant of f. Therefore, f is u.s.c. if and only if $f = f^*$, and f is l.s.c. if and only if $f = f_*$.

4.4.38. Proposition. *Let Ω be an open set in $\mathbb{C}$ and $u : \Omega \to [-\infty, \infty[$ be such that $u \in L^1_{\mathrm{loc}}(\Omega)$ and $u(z) \leq A(u, z, r)$ for every $\bar{B}(z, r) \subseteq \Omega$. Then u^* is subharmonic in Ω.*

PROOF. We have $u(z) \leq u^*(z) \leq A(u, z, r)$ since $A(u, z, r)$ is continuous in Ω_r. Therefore $u^*(z) \leq A(u^*, z, r)$ and, since u^* is u.s.c., u^* is subharmonic. □

4.4.39. Remark. We have in §4.4.38 that $u = u^*$ a.e. Namely,

$$\lim_{r \to 0} A(u, z, r) = u(z)$$

at every Lebesgue point of u (cf. [HS]), that is, a.e. From the chain of inequalities

$$u(z) \leq u^*(z) \leq A(u, z, r) \leq A(u^*, z, r)$$

and u^* subharmonic, we also conclude that everywhere

$$u^*(z) = \lim_{r \to 0} A(u^*, z, r) = \lim_{r \to 0} A(u, z, r).$$

Therefore $u = u^*$ a.e.

4.4.40. Proposition (Hartogs's Lemma). *Let $\{u_n\}_{n \geq 1}$ be a sequence of subharmonic functions in an open set Ω of $\mathbb{C}$. Assume they are uniformly bounded above on every compact subset of Ω. Let $u := \limsup_{n \to \infty} u_n$ and u^* its upper regularization. Then*

(i) *the function u^* is subharmonic;*
(ii) *assume there is a continuous function g in Ω such that $u \leq g$ in Ω. Then, for every compact $K \subseteq \Omega$ and every $\varepsilon > 0$ there is an integer N such that*

$$u_n(z) \leq g(z) + \varepsilon \qquad \text{for } z \in K, \quad n \geq N.$$

PROOF. Let $\varphi_n := \sup_{k \geq n} u_k$. The functions φ_n and $u = \lim_{n \to \infty} \varphi_n$, satisfy the hypotheses of Proposition 4.4.38. Therefore φ_n^* and u^* are subharmonic. Furthermore

$$u^* = \lim_{n \to \infty} \varphi_n^*.$$

In fact, $\varphi_{n+1} \leq \varphi_n$ implies $\varphi_{n+1}^* \leq \varphi_n^*$. Therefore the limit of φ_n^* exist, is subharmonic, and clearly $u \leq \lim_{n \to \infty} \varphi_n^*$. It follows that $u^* \leq \lim_{n \to \infty} \varphi_n^*$. Moreover, both sides coincide except possibly in a set of Lebesgue measure zero (just use the previous remark). The identity follows now from the following lemma.

4.4.41. Lemma. *Let u_1, u_2 be two subharmonic functions in Ω such that $u_1 = u_2$ a.e., then $u_1 \equiv u_2$. In particular, $u = u^*$ whenever $u \in SH(\Omega)$.*

PROOF. It follows from the following more general statement.

4.4.42. Lemma. *Let u_1 be subharmonic and u_2 be u.s.c. in Ω. If $u_1 = u_2$ a.e. then $u_1 \le u_2$ everywhere.*

PROOF. Let us assume there is $z_0 \in \Omega$ such that $u_2(z_0) < u_1(z_0)$. Then there is $\delta > 0$ such that $u_2(z) < u_1(z_0)$ in $\bar{B}(z_0, \delta)$. For $0 \le r \le \delta$ we then have $\lambda(u_2, z_0, r) < u_1(z_0) \le \lambda(u_1, z_0, r)$, therefore $A(u_2, z_0, \delta) < A(u_1, z_0, \delta)$, which contradicts the hypothesis. □

Returning to the proof of Proposition 4.4.40, let now $K \subset\subset \Omega$, $\varepsilon > 0$ be given. Set $E_n := \{z \in K : \varphi_n^*(z) - g(z) \ge \varepsilon\}$. E_n is then a compact set and

$$\bigcap_{n \ge 1} E_n = \varnothing$$

since $u^* \le g$. There is therefore $n(\varepsilon, K)$ such that $n \ge n(\varepsilon, K)$ implies $E_n = \varnothing$. (The E_n are nested.) Therefore $\varphi_n^*(z) \le g(z) + \varepsilon$ for $z \in K$. Since $u_n \le u_n^* \le \varphi_n^*$, the proposition follows. □

4.4.43. Remarks. (1). Let $\{u_n\}_{n \ge 1}$ be an increasing sequence of functions in $\mathrm{SH}(\Omega)$ which is locally bounded above and $u(z) := \lim_{n \to \infty} u_n(z)$. Then $u^* \in \mathrm{SH}(\Omega)$ and the set $\{z \in \Omega\colon u(z) < u^*(z)\}$ has Lebesgue measure zero.

(2) The statement of §4.4.40 holds when we have a family u_t depending on the parameter $t > 0$, $(u_t)_{t>0}$ uniformly bounded above on compact sets and $u = \limsup_{t \to \infty} u_t$.

Some results comparing the averages of a subharmonic function in different points are given later. They will be applied to a division problem.

4.4.44. Proposition. *Let Ω be an open subset of $\mathbb{C}$, u subharmonic in Ω, and K a compact subset of Ω. There are two positive constants A, B depending only on u and K such that if z_0, z_1 are any two points in Ω, $0 < r_0$, r_1 such that $\bar{B}(z_0, r_0) \subseteq K$, $\bar{B}(z_1, r_1) \subseteq K$, and the segment $[z_0, z_1] \subseteq K$, then*

$$|\lambda(u, z_0, r_0) - \lambda(u, z_1, r_1)| \le A|z_1 - z_0| + B[\log^+(1/r_1) + \log^+(1/r_0) + 1].$$

PROOF. Let Ω_1 be a relatively compact subset of Ω such that $K \subseteq \Omega_1$. We can apply the Riesz's decomposition theorem to u and obtain

$$u = U^\mu + h \qquad \text{in } \Omega_1,$$

with $\mu = \Delta u$, h harmonic in Ω_1. We have evidently (since $[z_0, z_1] \subseteq K$)

$$|\lambda(h, z_0, r_0) - \lambda(h, z_1, r_1)| = |h(z_0) - h(z_1)| \le k_1|z_1 - z_0|,$$

for some constant k_1 that depends only on h and K. Our problem reduces to study only the case $u = U^\mu$. Let $0 < R < \operatorname{dist}(K, \Omega_1^c)$. We will start by estimating $\lambda(U^\mu, z_1, R) - \lambda(U^\mu, z_0, R)$. For that purpose recall that

$$\lambda(U^\mu, \zeta, R) = \int_{\Omega_1} \lambda(\log|\cdot - t|, \zeta, R)\, d\mu(t),$$

where the dot indicates that the average is taken with respect to that variable. Let $N_R(\zeta, t)$ represent the integrand, there are many ways to compute it explicitly, for instance, see §4.6.2,

$$N_R(\zeta, t) = \frac{1}{2\pi}\int_0^{2\pi} \log|t - \zeta - Re^{i\theta}|\, d\theta = \begin{cases} \log|t - \zeta| & \text{if } |t - \zeta| \ge R. \\ \log R & \text{if } |t - \zeta| < R. \end{cases}$$

Let us now estimate $|N_R(z_1, t) - N_R(z_0, t)|$. Let $z(s) = z_1 + s(z_0 - z_1)$, $0 \le s \le 1$, and $g(s) = N_R(z(s), t)$. We can compute $\dfrac{dg}{ds}(s)$ for the values of s such that $|z(s) - t| \ne R$. This excludes at most two values. We have

$$\frac{dg}{ds} = 0 \qquad \text{if} \qquad |z(s) - t| < R$$

and

$$\frac{dg}{ds} = -\frac{1}{2}\frac{(t - z)(\bar{z}_0 - \bar{z}_1) + (\bar{t} - \bar{z})(\bar{z}_0 - \bar{z}_1)}{|z(s) - t|^2} \qquad \text{if } |z(s) - t| > R.$$

Therefore, $\left|\dfrac{dg}{ds}\right| \le \dfrac{1}{2}\dfrac{|t - z||z_0 - z_1|}{R^2}$ in the second case. Clearly, for $t \in \Omega_1$, $z \in K$, we have $\dfrac{1}{2}|t - z|/R^2 \le C$. It follows that

$$|g(1) - g(0)| \le C|z_0 - z_1|.$$

This implies that

$$|\lambda(U^\mu, z_1, R) - \lambda(U^\mu, z_0, R)| \le C\mu(\Omega_1)|z_1 - z_0|$$

and

$$|\lambda(u, z_1, R) - \lambda(u, z_0, R)| \le A|z_1 - z_0|.$$

Let us now estimate $|\lambda(u, z_1, R) - \lambda(u, z_1, r_1)|$. Assume first $r_1 < R$. Then by Gauss' theorem

$$|\lambda(u, z_1, R) - \lambda(u, z_1, r_1)| = \left|\frac{1}{2\pi}\int_{r_1}^{R} \frac{1}{t}\left(\int_{B(z_1, t)} d\mu\right) dt\right|$$
$$\le \frac{\mu(\Omega_1)}{2\pi}\left[\log^+\frac{1}{r_1} + \log R\right].$$

The other case $r_1 \geq R$ is easier since $r_1 < R_0 < \infty$, with R_0 a constant that depends only on K. Since the point z_0 can be treated the same way we obtain

$$|\lambda(u, z_1, r_1) - \lambda(u, z_0, r_0)| \leq A|z_1 - z_0| + B\left[\log^+ \frac{1}{r_0} + \log^+ \frac{1}{r_1} + 1\right],$$

as we wanted to prove. □

4.4.45. Theorem. *Let u be a subharmonic function in an open subset Ω of $\mathbb{C}$, K a connected compact subset of Ω. There are two real constants a, b, $b > 0$, such that for every disk $\bar{B}(z, r) \subseteq K$ one has*

$$\lambda(u, z, r) \geq a + b \log r.$$

PROOF. Let $0 < \alpha < \operatorname{dist}(K, \Omega^c)$, $K_\alpha = \{z \in \mathbb{C} : \operatorname{dist}(z, K) \leq \alpha\}$. There is a finite number N of disks $B(z_i, \alpha)$ such that $z_i \in K$ and

$$K \subseteq \bigcup_{1 \leq i \leq N} B(z_i, \alpha) \subseteq K_\alpha.$$

Let ρ, $0 < \rho < \alpha/2$, be the Lebesgue number of this covering of K and $z_0 \in K$ be such that $u(z_0) > -\infty$. (We can suppose z_0 exists since we are only interested in the case K contains some disk.)

Let us show that there is a constant $C > 0$ such that

$$|\lambda(u, z, \rho) - \lambda(u, z_0, \rho)| \leq C|z - z_0| \quad (z \in K).$$

We are going to use that the proof of the preceding proposition shows that we can take $B = 0$ when $r_0 = r_1 = \rho$, independent of the points considered. The only problem to surmount is that the segment $[z, z_0]$ does not belong necessarily to K_α. The problem only arises if $\bar{B}(z, \rho)$ and $\bar{B}(z_0, \rho)$ are not both contained in the same disk $B(z_i, \alpha)$. In that case $|z - z_0| \geq \rho$. Since K is connected there is a chain of disks $B(z_{i_k}, \alpha)$, $1 \leq k \leq n$, such that $B(z_0, \rho) \subseteq B(z_{i_n}, \alpha)$, $B(z_{i_k}, \alpha) \cap B(z_{i_{k+1}}, \alpha) \neq \varnothing$ if $1 \leq k \leq n - 1$ and $B(z, \rho) \subseteq B(z_{i_1}, \alpha)$. We can also assume that in the intersection we can find x_k such that $\bar{B}(x_k, \rho) \subseteq B(z_{i_k}, \alpha) \cap B(z_{i_{k+1}}, \alpha)$, by the definition of Lebesgue number. Now we have that $z, x_1, x_2, \ldots, z_0$ are such that the successive segments lie in K_α, therefore

$$\begin{aligned} |\lambda(u, z, \rho) - \lambda(u, z_0, \rho)| &\leq |\lambda(u, z, \rho) - \lambda(u, x_1, \rho)| + \cdots \\ &\quad + |\lambda(u, x_n, \rho) - \lambda(u, z_0, \rho)| \\ &\leq AN\rho \leq C|z - z_0|. \end{aligned}$$

Let us consider now separately the cases $r \geq \rho$ and $0 < r < \rho$ for a disk $\bar{B}(z, r) \subseteq K$.

If $r \geq \rho$ then:

$$\begin{aligned} \lambda(u, z, r) &\geq \lambda(u, z, \rho) \geq \lambda(u, z_0, \rho) - A|z - z_0| \\ &\geq \inf_{z \in K} (\lambda(u, z_0, \rho) - A|z - z_0|) = \beta > -\infty. \end{aligned}$$

If $0 < r \leq \rho$ we have, since $B(z,\rho) \subseteq K_\alpha$,

$$\lambda(u,z,\rho) - \lambda(u,z,r) = \frac{1}{2\pi}\int_r^\rho \frac{1}{t}\left(\int_{B(z,t)} d\mu\right)dt \leq \frac{\mu(K_\alpha)}{2\pi}\log\frac{\rho}{r}.$$

Therefore

$$\lambda(u,z,r) \geq \lambda(u,z,\rho) - \frac{\mu(K_\alpha)}{2\pi}\log\frac{\rho}{r} \geq \lambda(u,z_0,\rho) - A|z-z_0| - \frac{\mu(K_\alpha)}{2\pi}\log\frac{\rho}{r}.$$

Let now $r_1 := \sup\{r > 0 : \exists z \in K \text{ such that } \bar{B}(z,r) \subseteq K\}$ and let $b := \dfrac{\mu(K_\alpha)}{2\pi}$.
Choose now a_1 so that $a_1 + b\log r_1 \leq \beta$.

If $0 < r < \rho$ then we have

$$\lambda(u,z,r) \geq \beta - b\log\rho + b\log r \geq a + b\log r.$$

If $\rho \leq r \leq r_1$ then we have

$$\lambda(u,z,r) \geq \beta \geq a_1 + b\log r_1 \geq a + b\log r,$$

for a convenient choice of a, and the theorem has been proved. □

4.4.46. Definition. Let Ω be a bounded open subset of $\mathbb{C}$, f holomorphic in Ω and u subharmonic in Ω. We say that f (resp. u) has ***polynomial growth*** in Ω if there is $M > 0$ and $\rho > 0$ such that

$$|f(z)| \leq \frac{M}{d(z,\Omega^c)^\rho} \qquad (z \in \Omega)$$

(resp. $u(z) \leq M - \rho\log d(z,\Omega^c)$, $z \in \Omega$).

Therefore f has polynomial growth if and only if $\log|f|$ has polynomial growth.

We have now the following "division" theorem.

4.4.47. Theorem. *Let Ω be an open connected set. Let u be a subharmonic function of polynomial growth in Ω and w a function subharmonic in a neighborhood W of $\bar{\Omega}$ such that $h = u - w$ is subharmonic in Ω. Then h is also of polynomial growth in Ω.*

Proof. We have $u(z) \leq M - \rho\log d(z,\Omega^c)$, $M \in \mathbb{R}$, $\rho > 0$. We also have

$$h(z) \leq \lambda\left(h, z, \frac{d(z,\Omega^c)}{2}\right).$$

For $\zeta \in \partial B\left(z, \dfrac{d(z,\Omega^c)}{2}\right)$ we have $d(\zeta,\Omega^c) \geq \dfrac{d(z,\Omega^c)}{2}$, therefore

$$\begin{aligned} h(\zeta) = u(\zeta) - w(\zeta) &\leq M - \rho\log d(\zeta,\Omega^c) - w(\zeta) \\ &\leq (M + \rho\log 2) - \rho\log d(z,\Omega^c) - w(\zeta). \end{aligned}$$

From the previous theorem we can conclude that there are $a, b \in \mathbb{R}$, $b > 0$, such that if $\bar{B}(z,r) \subseteq \Omega$, then

$$\lambda(w, z, r) \geq a + b \log r.$$

In particular, for $r = \dfrac{d(z, \Omega^c)}{2}$ we obtain

$$\lambda\left(w, z, \frac{d(z,\Omega^c)}{2}\right) \geq a - b\log 2 + b \log d(z, \Omega^c).$$

Therefore,

$$h(z) \leq \lambda\left(h, z, \frac{d(z,\Omega^c)}{2}\right) \leq M + \rho \log 2 - \rho \log d(z, \Omega^c) - \lambda\left(w, z, \frac{d(z,\Omega^c)}{2}\right)$$

$$\leq M_1 - \rho_1 \log d(z, \Omega^c).$$

with $\rho_1 = \rho + b$, $M_1 = M - a + (\rho + b)\log 2$. □

4.4.48. Corollary. *Let Ω be a connected bounded open subset of $\mathbb{C}$, f holomorphic in Ω of polynomial growth, g holomorphic in a neighborhood of $\bar{\Omega}$. If $h = f/g$ is holomorphic in Ω then h has polynomial growth in Ω.*

This corollary is related to very precise lower bounds on the absolute value of holomorphic functions, which go by the generic name of "minimum modulus" theorems. We will see other ones in this book. The quintessential one is the Cartan-Boutroux Lemma 4.5.13 (cf. [Le], see also [Momm] and references therein, [Val1]).

Exercises 4.4

Ω is a domain of $\mathbb{C}$ and $B = B(0, 1)$. We write $z = x + iy$.

1. Show that the subset of $\mathrm{SH}(\Omega)$, consisting of the continuous subharmonic functions, can be characterized as the set of real-valued functions u in Ω such that for every compact $K \subseteq \Omega$ there is an open neighborhood V of K, a decreasing sequence $u_n \in C^2(V)$, u_n subharmonic and u is the uniform limit of the u_n on K.

2. Let $u : \Omega \to [-\infty, \infty[$ be u.s.c., $u \not\equiv -\infty$. Assume that for every a, $b \in \mathbb{R}$, the function $z \mapsto e^{ax+by+u(z)}$ is subharmonic in Ω. Show that u is subharmonic. Is the converse true?

3. (i) Let $f \in C^2_{\mathbb{R}}(\mathbb{R})$, $f'(t) > 0$ for every t, and $v \in C^2_{\mathbb{R}}(\Omega)$. Assume that for every a, $b \in \mathbb{R}$, the function $z \mapsto f(ax + by + v(z))$ is subharmonic. Show that v is subharmonic.
   (ii) Let $f \in C^2_{\mathbb{R}}(\mathbb{R})$. Assume that for every a, $b \in \mathbb{R}$ and every $v \in \mathrm{SH}(\Omega) \cap C^2(\Omega)$, the function $z \mapsto f(ax + by + v(z))$ is in $\mathrm{SH}(\Omega)$. Show that f is a convex function.

4. Give an example of a u.s.c. function $f : \Omega \to [-\infty, \infty[$ such that for every compact $K \subseteq \Omega$, $f|K$ takes its maximum value on ∂K, but $f \notin \mathrm{SH}(\Omega)$.

5. Let $f \in \mathrm{SH}(\Omega)$ and $\varphi \in C_{\mathbb{R}}(\Omega)$ such that φ is subharmonic in $\Omega \setminus \{z \in \Omega : f(z) = -\infty\}$. Show that φ is subharmonic in Ω. (Hint: For $\varepsilon > 0$, show that $\varphi + \varepsilon f \in \mathrm{SH}(\Omega)$.)

6. Assume that $(u_n)_{n \geq 1} \subseteq \mathrm{SH}(\Omega)$ and $u_n \to u$ uniformly on Ω when $n \to \infty$. Show that $u \in \mathrm{SH}(\Omega)$. Show that it is enough to assume the convergence is uniform over compact subsets of Ω.

7. Consider the space $\mathrm{SH}(\Omega)$ with the topology of locally uniform convergence, suppose $\varnothing \neq \mathscr{F} \subseteq \mathrm{SH}(\Omega)$ is a relatively compact family. Show that the function $g(z) = \sup\{f(z) : f \in \mathscr{F}\}$ is subharmonic in Ω.

8. Let $H = \{z \in \mathbb{C} : \operatorname{Im} z = y > 0\}$. Let $v \in \mathrm{SH}(H)$ be such that $v(z) \leq A + By$ for some constants A, B. Show that

$$\varphi(y) = \sup_{x \in \mathbb{R}} v(x + iy)$$

is a convex function of y. (Hint: let $0 < a < b$, $L(y) = \alpha y + \beta$ be such that $\varphi(a) \leq L(a)$, $\varphi(b) \leq L(b)$. Let $\varepsilon > 0$ and consider the auxiliary function $v_\varepsilon(z) = v(z) - L(y) - \varepsilon(x^2 - y^2 + b^2)$. Show that $v_\varepsilon \leq 0$ in the strip $a \leq y \leq b$. Conclude that φ is convex.)

*9. Let $a_0, a_1, \ldots, a_n \in \mathscr{H}(\mathbb{C})$, $a_0(z) \neq 0$ for every $z \in \mathbb{C}$. Consider $P(z, w) = a_n(z)w^n + a_{n-1}(z)w^{n-1} + \cdots + a_1(w)z + a_0(z)$. Let $Z(P) = \{(x, w) \in \mathbb{C}^2 : P(z, w) = 0\}$.
   (i) Show that $Z(P)$ is a closed subset of $\mathbb{C}^2$ and $Z(P) \cap (\mathbb{C} \times \{0\}) = \varnothing$.
   (ii) Define $\varphi(z) = d(0, \{w : P(z, w) = 0\})$. Show φ is a continuous function with values in $]0, \infty[$. (Hint: use Hurwitz's theorem.)
   (iii) For every fixed z, consider the function $w \mapsto (P(z, w))^{-1}$. It is holomorphic in $B(0, \varphi(z))$. Let

$$\frac{1}{P(z, w)} = \sum_{k \geq 0} \alpha_k(z) w^k.$$

   Show that the α_k are entire functions.
   (iv) Show that $\log\left(\dfrac{1}{\varphi(z)}\right) = \limsup_{k \to \infty} \dfrac{1}{k} \log|\alpha_k(z)|$. Conclude that φ is subharmonic in $\mathbb{C}$.
   (v) Assume that there is $\rho > 0$ such that $(\mathbb{C} \times B(0, \rho)) \cap Z(P) = \varnothing$. Show that there is $R > 0$ such that for every z, the polynomial equation $P(z, w) = 0$ has a root of absolute value exactly R. (Hint: What can you say about φ under this assumption?)

*10. Let $0 \leq a_{ij} \in \mathrm{SH}(\Omega)$, $1 \leq i, j \leq n$. Let $A(z)$ be the $n \times n$ matrix whose entries are the functions a_{ij}. Suppose $A(z)$ is an orthogonal matrix for every $z \in \Omega$, show all the a_{ij} are constant functions.

11. Let $D = \{z \in \mathbb{C} : R_1 < |z| < R_2\}$, $0 \leq u_1 \in \mathrm{SH}(D)$. The aim of this exercise is to generalize §4.4.26 and show that

$$r \mapsto m(r) := \left(\frac{1}{2\pi} \int_0^{2\pi} (u(re^{i\theta}))^2 \, d\theta\right)^{1/2}$$

is a convex function of $\log r$ when $R_1 < r < R_2$.
   (i) Show that it is enough to prove it for $0 < u$ and $u \in C^\infty$.

(ii) Show that if one can prove that the function m is subharmonic we are done.
(iii) Show that if one could prove that, for every $n \in \mathbb{N}^*$, the function

$$u_n(z) = \left(\frac{1}{n}\sum_{k=0}^{n-1} (u(e^{2\pi ik/n}z))^2\right)^{1/2}$$

is subharmonic, the statement would be correct.
(iv) Let $v_1, \ldots, v_n \in C^\infty$, $0 < v_j \in \mathrm{SH}(D)$. Prove that $z \mapsto (v_1^2(z) + \cdots + v_n^2(z))^{1/2}$ is subharmonic. Note that for a fixed z_0 there are real nonnegative numbers $x_1, \ldots, x_n$ such that $x_1^2 + \cdots + x_n^2 = 1$ and $(v_1^2(z_0) + \cdots + v_n^2(z_0))^{1/2} = x_1 v_1(z_0) + \cdots + x_n v_n(z_0)$. Conclude the proof of (iv) and of the original claim.

*12. (Hopf's Lemma). Assume that Ω has a C^2 regular boundary. Let $f \in C^1_{\mathbb{R}}(\bar{\Omega})$, f harmonic in Ω and not constant. Let $z_0 \in \partial\Omega$ be such that z_0 is a strict local maximum for f (that is, $f(z_0) \geq f(z)$ for every $z \in \bar{B}(z_0, r) \cap \bar{\Omega}$, for some r small.) We want to conclude that $\dfrac{\partial f}{\partial n}(z_0) > 0$, $n =$ outer normal.
(i) Show we can find a disk $B(z_1, R)$ such that $z_0 \in \bar{B}(z_1, R) \subseteq \Omega \cap \{z_0\}$.
(ii) Consider the region $B' = B(z_1, R) \cap B(z_0, r)$, where r, $0 < r < R$, is chosen so that the hypothesis of the local maximum is satisfied. Show that for $\alpha > 0$ sufficiently large, the function $h(z) = e^{-\alpha|z - z_1|^2} - e^{-\alpha R^2}$ satisfies $\Delta h > 0$ in B'.
(iii) For $\varepsilon > 0$ small, consider the function $\varphi = f + \varepsilon h$. Show that for $z \in B(z_0, r) \cap \Omega$, $\varphi(z) < f(z_0)$. Conclude that $\varphi(z) < f(z_0)$ in $\partial B' \setminus \{z_0\}$, while $\varphi(z_0) = f(z_0)$. Deduce that

$$\frac{\partial \varphi}{\partial n}(z_0) \geq 0.$$

Show that this implies that $\dfrac{\partial f}{\partial n}(z_0) > 0$.

13. Let $u \geq 0$ in Ω, show that $\log u$ is subharmonic in Ω if and only if u is u.s.c., $u \not\equiv 0$, and for every open set U in Ω and every h harmonic in U, the function ue^h is subharmonic in U.

14. Let $u \not\equiv 0$, u.s.c. in Ω, $u \geq 0$. Show that $\log u$ is subharmonic in Ω if and only if u^α is subharmonic in Ω for every $\alpha > 0$.

15. (i) Let $0 < u$ be a continuous function in the annulus $D = \{z \in \mathbb{C} : 0 \leq R_1 < |z| < R_2 \leq \infty\}$. Show that

$$r \mapsto \log \lambda(u, 0, r)$$

is a convex function of $\log r$ for $r \in]R_1, R_2[$ if for every real α the function

$$r \mapsto r^\alpha \lambda(u, 0, r)$$

is a convex function of $\log r$.
(ii) With u as in (i), assume further that $\log u$ is subharmonic in D. Show that for real α, $z \mapsto \log u(z) + \alpha \log|z|$ is subharmonic and the same is true for $z \mapsto |z|^\alpha u(z)$.
(iii) Under the conditions of (ii), compute $\lambda(|z|^\alpha u(z), 0, r)$ in terms of $\lambda(u, 0, r)$.
(iv) Conclude that if $0 \leq u$ in D and $\log u \in \mathrm{SH}(D)$, then

$$r \mapsto \log \lambda(u, 0, r)$$

is a convex function of $\log r$, $r \in]R_1, R_2[$.

(v) Let $u \geq 0$ for which $\log u \in SH(B)$. Show that

$$r \mapsto \log A(u, 0, r)$$

is a convex function of $\log r$ in $]0, 1[$.

16. Let $u : B \to [-\infty, \infty[$ be u.s.c. and $\not\equiv -\infty$. Show that if for every $y \in]-1, 1[$ the function

$$x \mapsto u(x, y) \qquad (|x| < \sqrt{1 - y^2})$$

is a convex function of x, and for every $x \in]-1, 1[$, the function

$$y \mapsto u(x, y) \qquad (|y| < \sqrt{1 - x^2})$$

is a convex function of y, then $u \in SH(B)$. (Hints: (i) Show that if $(h, k) \in \mathbb{R}^2$ are such that $0 < h^2 + k^2 = r^2$ sufficiently small, then $u(x_0, y_0) \leq \frac{1}{4}[u(x_0 + h, y_0 + k) + u(x_0 + h, y_0 - k) + u(x_0 - h, y_0 + k) + u(x_0 - h, y_0 - k)]$. (ii) Let φ_n be a decreasing sequence of continuous functions converging to u in $\bar{B}(z_0, r)$ for some $z_0 \in B$, $0 < r$ small. Use (i) to conclude that $u(z_0) \leq \lambda(\varphi_n, z_0, r)$.)

17. Recall from the last chapter that $A(B) = \mathscr{H}(B) \cap \mathscr{C}(\bar{B})$.

(i) Let $f \in A(B)$. Let γ be an arc of ∂B of length $\alpha > 0$ and $\lambda(z) = \dfrac{\alpha}{2\pi} \dfrac{1 - |z|}{1 + |z|}$, $z \in B$. Show that

$$|f(z)| \leq \left(\max_{\zeta \in \gamma} |f(\zeta)|\right)^{\lambda(z)} \left(\max_{\zeta \in \bar{B}} |f(\zeta)|\right)^{1 - \lambda(z)}. \qquad (*)$$

(Hint: Let $M \geq m$ be such that $M > \max_{\zeta \in \bar{B}} |f(\zeta)|$, $m = \max_{\zeta \in \gamma} |f(\zeta)|$. Show that $\varphi(z) = \log \max(|f(z)|, m) = \log m + \log^+ \left(\dfrac{|f(z)|}{m}\right)$ is subharmonic in B and

$$\varphi(z) \leq h(z) := \frac{1}{2\pi} \int_0^{2\pi} \frac{1 - |z|^2}{|e^{i\theta} - z|^2} \varphi(e^{i\theta})\, d\theta.$$

Show that $h(z) \leq \lambda(z) \log m + (1 - \lambda(z)) \log M$ in B. Conclude that $(*)$ holds.)

(ii) Use (i) to show that if $(f_n)_{n \geq 1} \subseteq A(B)$, $\|f_n\|_{\bar{B}} \leq M$, and the sequence $(f_n)_n$ converges uniformly in some arc γ of length $\alpha > 0$, then $(f_n)_n$ converges in $\mathscr{H}(B)$ to a bounded holomorphic function f.

18. Let $f \in \mathscr{H}(\mathbb{C})$. Assume $|f(z)| \leq Ae^{B|\zeta|^\rho}$, $A, B, \rho > 0$. Estimate the function $v(r) = \dfrac{1}{2\pi}(\Delta \log|f|)(B(0, r))$, for large values of r. Apply to the case $f(z) = g(z) - a$, for $a \in \mathbb{C}$ fixed, $|g(z)| \leq Ae^{B|z|^\rho}$.

19. Let $f \in \mathscr{M}(\mathbb{C})$, compute $\Delta \log|f|$. Is $\log|f|$ subharmonic? Can you derive Jensen's formula in this case from the Riesz's decomposition theorem anyway?

20. Let $D = \{z \in \mathbb{C} : 0 \leq R_1 < r < R_2 < \infty\}$. Let $f \in A(D)$, $f(z) \neq 0$ for every point in $\bar{D}$. Let $m_1 = \min_{|z| = R_1} |f(z)|$, $m_2 = \min_{|z| = R_2} |f(z)|$. Estimate

$$\min_{|z| = r} |f(z)| \qquad (R_1 < r < R_2).$$

*21. Let Γ be a Jordan arc contained, except for its endpoints, in the open angle $|\operatorname{Arg} z| < \alpha < \pi$. Assume its endpoints a, b lie respectively on the rays $\operatorname{Arg} z = \alpha$,

$\operatorname{Arg} z = -\alpha$. Define a Jordan domain Ω as the interior of the Jordan curve formed by $[0,a[$, Γ and $]b,0[$. Let $f \in \mathscr{H}(\Omega)$ be such that

$$\overline{\lim_{z\to\zeta}}\,|f(z)| \le \begin{cases} m & \text{if } \zeta \in \Gamma \\ M & \zeta \in [0,a] \cup [b,0] \end{cases}$$

and let $R = \max\{|\zeta| : \zeta \in \Gamma\}$. Show that for any $r \in \Omega \cap \mathbb{R}$, one has the inequality

$$|f(r)| \le M^{1-(r/R)^{\pi/\alpha}} m^{(r/R)^{\pi/\alpha}}.$$

Note this estimate is only interesting if $m < M$. (Hint: for any $\varepsilon > 0$ consider the auxiliary function $\varphi_\varepsilon(z) := f(z)\exp(\varepsilon(z/r)^{\pi/\alpha})$, $z \in \Omega$. Obtain an estimate for $\sup_\Omega |\varphi_\varepsilon(z)|$, a corresponding estimate for $|f(r)|$, and minimize the latter over all $\varepsilon > 0$.)

Conclude that if a sequence $(f_n)_{n\ge1} \subseteq \mathscr{H}(\Omega)$ is uniformly bounded in Ω and converges uniformly to zero on Γ, then it converges to zero in $\mathscr{H}(\Omega)$.

22. Let $f \in \mathscr{H}(\bar{B})$, $p > 0$. Show that

$$\frac{1}{\pi}\int_B |f(z)|^p\,dm(z) \le \frac{1}{2\pi}\int_{-\pi}^{\pi} |f(e^{i\theta})|^p\,d\theta.$$

In particular,

$$\left(\int_B |f(z)|^2 \frac{dm(z)}{\pi}\right)^{1/2} \le \left(\int_{\partial B} |f(z)|^2 \frac{|dz|}{2\pi}\right)^{1/2}.$$

(The functions $f \in \mathscr{H}(B)$ for which $\lim_{r\to1^-}\int_{-\pi}^{\pi} |f(re^{i\theta})|^p \frac{d\theta}{2\pi} < \infty$ constitute the natural domain for the inequality. One says $f \in H^p(B)$. There is a very rich theory about the space $H^p(B)$, see [Du2], [Koo].)

*23. (a) Let $\alpha_0, \alpha_1, \ldots, \ge 0$ and $\beta_n := \sum_{j=0}^{n} \alpha_j\alpha_{n-j}$. Show that

$$\beta_n^2 \le (n+1)\sum_{j=0}^{n} \alpha_j^2\alpha_{n-j}^2.$$

(b) Let $f \in \mathscr{H}(\bar{B})$. Show that

$$\int_B |f(z)|^4 \frac{dm(z)}{\pi} \le \left(\int_{\partial B} |f(z)|^2 \frac{|dz|}{2\pi}\right)^2.$$

Why is this not an obvious consequence of Exercise 4.4.22? (Hint: introduce the function $\varphi(z) = (f(z))^2$, compute both sides in terms of their Taylor expansions and use part (a).)

(c) Let $f \in \mathscr{H}(\bar{B})$ show that

$$\int_B |f(z)|^2 \frac{dm(z)}{\pi} \le \left(\int_{\partial B} |f(z)| \frac{|dz|}{2\pi}\right)^2.$$

(Hint: Assume first f does not vanish on B.)

(d) Let $\psi \in \mathscr{H}(\bar{B})$ which is injective in ∂B, and let $\Omega = \psi(B)$, $\Gamma = \partial\Omega$. Use (c) to show that the isoparametric inequality holds:

$$m(\Omega) \le \frac{1}{4\pi}(\ell(\Gamma))^2.$$

(Note that one can refine the previous parts to show also that equality holds in the isoparametric inequality if and only if Ω is a disk. For related inequalities, see [Car]. Moreover, part (c) can also be used to show that if $u \in \mathscr{C}(D)$, $u > 0$ throughout, then $\log u \in \mathrm{SH}(D)$ is subharmonic if and only if for every disk $\bar{B}(z_0, r) \subseteq D$ one has

$$(A(u^2; z_0, r))^{1/2} \le \lambda(u; z_0, r),$$

see [Ra, 3.26] and references therein.)

24. Let u be a nonnegative continuous and subharmonic function in $\bar{B}(0, R)$. Let $z_0 \in B(0, R)$, C be a circle of center z_0 and contained in $B(0, R)$. Show that

$$\int_C u(z)|dz| \le \left(1 + \frac{|z_0|}{R}\right) \int_{\partial B(0,R)} u(z)|dz|.$$

(Hint: let v be the harmonic function in $B(0, R)$ that coincides with u on $\partial B(0, R)$. Compare $\int_C u(z)|dz|$ and $v(z_0)$. Then use the Poisson formula.) In the case where $u(z) = |f(z)|^\lambda$, $f \in A(B(0, R))$, $\lambda > 0$, and C is a convex curve inside $B(0, r)$, it is possible to prove the more general inequality

$$\int_C |f(z)|^\lambda |dz| \le 2 \int_{\partial B(0,R)} |f(z)|^\lambda |dz|$$

(see [Gab1], [Gab2]).

25. Use Theorem 4.4.5 to prove that if $f \in A(B(0, 1))$, $f \not\equiv 0$, then

$$\int_{-\pi}^{\pi} \log|f(e^{i\theta})|\, d\theta > -\infty.$$

(In fact, weaker conditions that continuity up to $\partial B(0, 1)$ suffice, e.g., f bounded.)

Conclude that if g is a holomorphic function in $H = \{z : \operatorname{Im} z > 0\}$, g is continuous in $\bar{H} = \{z : \operatorname{Im} z \ge 0\}$ and bounded, then

$$\int_{-\infty}^{\infty} \frac{|\log|g(x)||}{1 + x^2}\, dx = \infty$$

implies that $g \equiv 0$.

26. Give a different proof of 4.4.22(1), using that $\log|z|$ and $\frac{1}{z}$ are locally integrable; hence the computation of the derivative $\frac{\partial}{\partial z}\log|z| = \frac{1}{2}\frac{\partial}{\partial z}\log|z|^2 = \frac{1}{2z}$, which is correct for $z \ne 0$, holds everywhere in the sense of distributions. Use the previously proven $\frac{\partial}{\partial \bar{z}}\frac{1}{z} = \pi\delta$ to conclude the proof.

*27. Let $\Omega = \left\{z : \frac{1}{r} < |z| < r\right\}$, $B = B(0, 1)$, $f : \Omega \to B$ holomorphic, f continuous in $\bar{\Omega}$, and $f(\partial B) \subseteq \partial B$, and let $h(z) = \frac{1}{2}\frac{\log|z|}{\log r} + \frac{1}{2}$.

(a) Let $u(z) := \sum_{\zeta} h(\zeta)$, where the sum takes place over the solutions $\zeta \in \bar{\Omega}$ of the equation $f(\zeta) = f(z)$. Show that $u \in \mathrm{SH}(\Omega)$.
(b) Assume $f|\{z : |z| = r\}$ is injective. Show that $u(z) \leq 1$ everywhere in Ω.
(c) Use (b) to conclude that $f|\{z : 1 < |z| < r\}$ is injective.
(d) Use the function $g(z) = z + \frac{1}{z}$ to construct an example of a function f that is not injective in $\{z : \alpha < |z| < r\}$ for any $\alpha < 1$ (cf. [BGHT]).

§5. Order and Type of Subharmonic Functions in $\mathbb{C}$

As we have already seen, a function u which is subharmonic in $\mathbb{C}$ cannot be bounded above unless u is constant. In fact, $M(u,r)$ grows at least as fast as a linear function of $\log r$. It is therefore natural to compare the order of growth of $M(u,r)$ with that of different functions of $\log r$, and classify, if possible, the subharmonic functions according to the different orders of growth of $M(u,r)$.

In what follows, an order of growth will be a positive, nonconstant convex function of $\log r$. The typical functions are r^α, $\exp(r^\alpha)$, $\exp(\exp(r^\alpha)) = \exp_2(r^\alpha)$, $\ldots, \exp_p(r^\alpha)$, with p integer ≥ 1, $\alpha > 0$, and $\exp_k = \exp(\exp_{k-1})$. Let us denote $\log_2 x = \log(\log x)$ (for $x \geq e$), $\log_k x = \log(\log_{k-1} x)$ (for $x \geq \exp_k 1$). We can now introduce the concept of class of growth of a subharmonic function.

4.5.1. Definition. For a subharmonic function u in $\mathbb{C}$ we say that

- It is of ***class zero*** if

$$\limsup_{r\to\infty} \frac{M(u,r)}{\log r} < \infty.$$

- It is of ***finite class*** $p \geq 1$ if p is the smallest integer $k \geq 1$ such that

$$\limsup_{r\to\infty} \frac{\log_k M(u,r)}{\log r} < \infty.$$

- It is of ***infinite class*** if no such integer exists.

If $p = 1$ one says also that u is of ***finite order***. In this case, the number $\rho \in [0, \infty[$ defined by

$$\rho = \limsup_{r\to\infty} \frac{\log M(u,r)}{\log r}$$

is called the ***order of*** u.

For functions of positive order $\rho > 0$, we have also the concept of the ***type*** τ:

$$\tau = \limsup_{r\to\infty} \frac{M(u,r)}{r^\rho}.$$

We say that u is of ***minimal*** (resp. ***normal, maximal***) type if $\tau = 0$ (resp. it is finite and nonzero, infinite). The function u is said to be of ***finite type*** if it is of either minimal or normal type.

Finally, if f is an entire function we use for f the terminology used for $u = \log|f|$.

4.5.2. Examples and Remarks. (1) If f is an entire function, then f is of class zero if and only if f is a polynomial.

(2) $u(z) = \dfrac{\log|z| + \sqrt{(\log|z|)^2 + 4}}{2}$ is a subharmonic function of class zero which is not the logarithm of $|f|$, for any f entire.

(3) Let P be the polynomial $P(z) = a_n z^n + \cdots + a_0$, $z_n \neq 0$. Then $\exp(P)$ is an entire function of order n and type $|a_n|$. The function $\exp_2(P)$ is of class 2, etc.

(4) More generally, if f is an entire function of class p then $\exp(f)$ is of class $p + 1$.

(5) An entire function which is not a polynomial is called ***transcendental***. We will see that there are transcendental functions of order zero. Such a function f verifies the condition:

$$\forall \varepsilon > 0\ \exists c_\varepsilon \geq 0 : |f(z)| \leq c_\varepsilon \exp(|z|^\varepsilon) \qquad (z \in \mathbb{C}).$$

In the same way a function f will be of order $\rho \geq 0$ if and only if:

$$\forall \varepsilon > 0\ \exists c_\varepsilon \geq 0 : |f(z)| \leq c_\varepsilon \exp(|z|^{\rho+\varepsilon}) \qquad (z \in \mathbb{C}),$$

and ρ is the smallest number having this property.

The entire functions most often considered are those of finite order since they arise in a natural way in harmonic analysis. Quite often, they are of ***exponential type***. These are entire functions for which there are constants A, $B > 0$ such that

$$|f(z)| \leq A \exp(B|z|) \qquad (z \in \mathbb{C}).$$

In other words, these functions are precisely the entire functions of order < 1 together with the functions of order 1 and finite type.

(6) If u_1, u_2 are two subharmonic functions of the same class, then $u_1 + u_2$ is at most of the same class. If the classes are different, then $u_1 + u_2$ is of the largest class. One can verify the same property holds for orders.

(7) The class, order, type of an entire function, and those of its derivative coincide.

(8) If f is an entire function without any zeros and of finite order ρ (resp. class p) then the function $1/f$ is of the same order ρ (resp. class p).

The proofs of the nonevident parts of these statements will constitute the rest of this section.

4.5.3. Proposition. *The entire function* $f(z) = \sum_{n \geq 0} a_n z^n$ *is of finite order* ρ *if and only if*

$$\mu = \limsup_{n\to\infty} \frac{n \log n}{-\log|a_n|}$$

is finite. In that case $\rho = \mu$.

PROOF. We show first that $\mu \le \rho$. We will use that

$$|a_n| \le \frac{M(|f|, r)}{r^n}, \qquad r > 0.$$

If $\mu = 0$ we have clearly that $\mu \le \rho$. Let us assume $0 < \mu \le \infty$. If $\mu < \infty$, let $R = \mu - \varepsilon$, with ε small and positive, so that $R > 0$, and let $R > 0$ arbitrary if $\mu = +\infty$. It follows that for infinitely many indices $n \ge 1$ we have $\log|a_n| < 0$ and

$$n \log n \ge R \log \frac{1}{|a_n|}.$$

Therefore for these indices

$$\log|a_n| \ge \frac{-n \log n}{R}$$

and

$$\log M(|f|, r) \ge \log|a_n| + n \log r \ge n\left(\log r - \frac{\log n}{R}\right).$$

Set $r_n = (en)^{1/R}$. Then

$$\log M(|f|, r_n) \ge \frac{n}{R}$$

and

$$\frac{\log\log M(|f|, r_n)}{\log r_n} \ge R\left(\frac{\log n - \log R}{\log n + 1}\right).$$

We conclude that

$$\rho = \limsup_{r\to\infty} \frac{\log\log M(|f|, r)}{\log r} \ge \limsup_{n\to\infty} \frac{\log\log M(|f|, r_n)}{\log r_n} \ge R.$$

Since R is an arbitrary real number smaller than μ it follows that $\rho \ge \mu$.

Let us show now that $\rho \le \mu$. If $\mu = +\infty$ there is nothing to prove. Assume $\mu < \infty$ and let $\varepsilon > 0$. For n sufficiently large we have

$$0 \le \frac{n \log n}{-\log|a_n|} \le \mu + \varepsilon,$$

that is,

$$|a_n| \le n^{-n/(\mu+\varepsilon)}.$$

Note this condition also implies that f is entire. Since the addition of a polynomial does not change the order of a function we can assume the last inequality holds for every $n \geq 1$ and that $a_0 = 0$. We have therefore for $r \geq 1$:

$$M(|f|, r) \leq \sum_{n\geq 0} |a_n| r^n \leq \sum_{n\geq 1} n^{-n/(\mu+\varepsilon)} r^n = S_1 + S_2,$$

where

$$S_1 = \sum_{1 \leq n < (2r)^{\mu+\varepsilon}} n^{-n/(\mu+\varepsilon)} r^n, \; S_2 = \sum_{n \geq (2r)^{\mu+\varepsilon}} n^{-n/(\mu+\varepsilon)} r^n.$$

In S_2 we have $rn^{-1/(\mu+\varepsilon)} \leq 1/2$. Hence $S_2 \leq 1$. On the other hand

$$S_1 \leq r^{(2r)^{\mu+\varepsilon}} \sum_{n\geq 1} n^{-n/(\mu+\varepsilon)} \leq M \exp((2r)^{\mu+\varepsilon} \log r) \leq M_1 \exp(r^{\mu+2\varepsilon}).$$

for some constants $M > 0$ and $M_1 > 0$. It follows that $\rho \leq \mu + 2\varepsilon$. Since $\varepsilon > 0$ is arbitrary, the proposition has been proved. □

4.5.4. Proposition. *Let* $f(z) = \sum_{n\geq 0} a_n z^n$ *an entire function of order* ρ, $0 < \rho < \infty$. *Let*

$$\nu = \limsup_{n\to\infty} n|a_n|^{\rho/n}.$$

Then f *is of type* τ, $\nu = \rho e \tau$ *(with the understanding that* f *is of maximal type if and only if* $\nu = +\infty$.*)*

PROOF. If $\nu < \infty$ then f is at most of order ρ and if $\nu > 0$ then f is at least of order ρ. Namely, let $\varepsilon > 0$, then if $\nu < \infty$ and n large

$$n|a_n|^{\rho/n} \leq \nu + \varepsilon,$$

hence

$$\frac{n \log n}{-\log|a_n|} \leq \frac{\rho}{(1 - \log((\nu + \varepsilon)/n)}.$$

it follows from the previous proposition that the order of f is at most ρ. Analogously one shows the other statement holds.

Assume now that $0 < \nu < \infty$ and we will show that $\tau \leq \nu/e\rho$. Let $\varepsilon > 0$, for n large one has

$$|a_n| \leq \left(\frac{\nu + \varepsilon}{n}\right)^{\rho/n}$$

and one can suppose this inequality holds for $n \geq 1$ and $a_0 = 0$ since adding a polynomial to f will not change its type. Therefore

$$|f(z)| \leq \sum_{n\geq 1} |a_n| r^n \leq \sum_{n\geq 1} \left(\frac{r^\rho(\nu + \varepsilon)}{n}\right)^{n/\rho}.$$

Consider the function $\varphi(t) = \left(\frac{r^\rho(\nu + \varepsilon)}{t}\right)^{t/\rho}$ for $t > 0$. Its maximum value is taken at $t = \dfrac{(\nu + \varepsilon)r^\rho}{e}$ and it is $\exp\left(\dfrac{(\nu + \varepsilon)r^\rho}{\rho e}\right)$. We have thus, for some convenient $M > 0$,

$$S_1 = \sum_{1 \le n \le (\nu+2\varepsilon)r^\rho} \left(\frac{(\nu + \varepsilon)r^\rho}{n}\right)^{n/\rho} \le (\nu + 2\varepsilon)r^\rho \exp\left(\frac{(\nu + \varepsilon)r^\rho}{\rho e}\right)$$

$$\le M \exp\left(\frac{(\nu + 2\varepsilon)r^\rho}{\rho e}\right)$$

$$S_2 = \sum_{n > (\nu+2\varepsilon)r^\rho} \left(\frac{(\nu + \varepsilon)r^\rho}{n}\right)^{n/\rho} \le \sum_{n \ge 1} \left(\frac{\nu + \varepsilon}{\nu + 2\varepsilon}\right)^{n/\rho} = M_1 < \infty.$$

It follows that $\tau \le \nu/e\rho$. To show the opposite inequality, note that if $0 < \varepsilon < \nu$ there are infinitely many n such that

$$|a_n| \ge \left(\frac{\nu - \varepsilon}{n}\right)^{n/\rho}.$$

In the Cauchy inequality $|a_n| \le r^{-n} M(|f|, r)$, let us take $r_n^\rho = \dfrac{ne}{\nu - \varepsilon}$. One obtains

$$M(|f|, r_n) \ge |a_n| r_n^n \ge \left(\frac{r_n^\rho(\nu - \varepsilon)}{n}\right)^{n/\rho} = e^{n/\rho} = \exp\left(\frac{(\nu - \varepsilon)}{e\rho} r_n^\rho\right).$$

Therefore $\tau \ge \dfrac{\nu}{e\rho}$. □

Examples

(1) $f(z) = \sum_{n \ge 1} \left(\dfrac{\tau e^\rho}{n}\right)^{n/\rho} z^n$ is an entire function of order ρ and type τ.

(2) The same holds for $f(z) = \sum_{n \ge 0} \dfrac{(\tau z)^n}{\Gamma(1 + n/\rho)}$.

4.5.5. Proposition. *Let v_1, v_2 be two subharmonic functions in $\mathbb{C}$ such that the function $v = v_1 - v_2$ is also subharmonic. Assume $v_2(0) > -\infty$. For every $s > 0$ and $\lambda > 0$ one has*

$$M(v, s) \le M(v_1, (1 + \lambda)s) + \left[\left(1 + \frac{1}{\lambda}\right)^2 - 1\right] M(v_2, (1 + \lambda)s) - \left(1 + \frac{1}{\lambda}\right)^2 v_2(0).$$

Proof. Let $R = (1 + \lambda)s$, $r = R - s = \lambda s$ and

$$\varphi = v_2 - M(v_2, R).$$

This function is then subharmonic and negative in $\bar{B}(0, R)$. This last property implies that for any z with $|z| = s$ we have

$$\int_{B(0,R)} \varphi \, dx \, dy \le \int_{B(z,r)} \varphi \, dx \, dy,$$

which can be rewritten as

$$r^2 A(\varphi, z, r) \ge R^2 A(\varphi, 0, R),$$

or

$$r^2 A(v_2, z, r) - r^2 M(v_2, R) \ge R^2 \varphi(0) = R^2 v_2(0) - R^2 M(v_2, R),$$

because φ is subharmonic. This last inequality is equivalent to

$$A(v_2, z, r) \ge M(v_2, R)\left(1 - \left(1 + \frac{1}{\lambda}\right)^2\right) + \left(1 + \frac{1}{\lambda}\right)^2 v_2(0).$$

We use now the hypothesis that v is also subharmonic,

$$\begin{aligned} v(z) &\le A(v, z, r) = A(v_1, z, r) - A(v_2, z, r) \le M(v_1, R) - A(v_2, z, r) \\ &\le M(v_1, R) + M(v_2, R)\left(\left(1 + \frac{1}{\lambda}\right)^2 - 1\right) - \left(1 + \frac{1}{\lambda}\right)^2 v_2(0). \end{aligned}$$

Since z was arbitrary we have

$$M(v, s) \le M(v_1, (1 + \lambda)s) + M(v_2, (1 + \lambda)s)\left(\left(1 + \frac{1}{\lambda}\right)^2 - 1\right) - \left(1 + \frac{1}{\lambda}\right)^2 v_2(0).$$

which is the inequality we were looking for. □

4.5.6. Corollary. *If v_1 and v_2 are two subharmonic functions of order at most ρ (resp. finite class at most p) such that $v = v_1 - v_2$ is also subharmonic, then v is of order at most ρ (resp. finite class at most p).*

If for every $\varepsilon > 0$ there is an $\eta > 0$ such that whenever $|z| > \eta$ one has

$$v_1(z) \le (\tau_1 + \varepsilon)|z| \qquad \textit{and} \qquad v_2(z) \le (\tau_2 + \varepsilon)|z|,$$

then

$$\tau = \limsup_{r \to \infty} \frac{M(v, r)}{r} \le \begin{cases} \tau_1 & \textit{if } \tau_2 = 0 \\ 2\tau_2 & \textit{if } \tau_1 = 0 \\ (1 + \lambda_0)\tau_1 + \dfrac{(1 + \lambda_0)(1 + 2\lambda_0)}{\lambda_0^2}\tau_2 & \textit{if } \tau_1\tau_2 \ne 0, \\ \textit{where } \lambda_0 \textit{ is the positive root of} \\ (\tau_1/\tau_2)\lambda^3 - 3\lambda - 2 = 0. \end{cases}$$

Proof. Let us prove only the last statement, the other ones being similar. We can assume $v_2(0) > -\infty$ (otherwise a small translation would achieve this). For $\varepsilon > 0$ and s large we have

$$M(v,s) \le (\tau_1 + \varepsilon)(1+\lambda)s + (\tau_2 + \varepsilon)(1+\lambda)s\left(\left(1+\frac{1}{\lambda}\right)^2 - 1\right) - \left(1+\frac{1}{\lambda}\right)^2 v_2(0).$$

Hence

$$\tau = \limsup_{s\to\infty} \frac{M(r,s)}{s} \le \inf_{\lambda\ge 0}\left\{\tau_1(1+\lambda) + \tau_2(1+\lambda)\left(\left(1+\frac{1}{\lambda}\right)^2 - 1\right)\right\}.$$

If $\tau_2 = 0$ one finds $\tau \le \tau_1$. If $\tau_2 > 0$ one looks for the minimum of

$$(1+\lambda)\left\{\frac{\tau_1}{\tau_2} + \left(1+\frac{1}{\lambda}\right)^2 - 1\right\}$$

which is exactly $(1+\lambda_0)\left\{\frac{\tau_1}{\tau_2} + \left(1+\frac{1}{\lambda}\right)^2 - 1\right\}$ with λ_0 the only positive root of $(\tau_1/\tau_2)\lambda^3 - 3\lambda - 2 = 0$. □

4.5.7. Corollary. *If f_1, f_2 are two entire functions of exponential type at most τ_1 and τ_2 respectively, and if $f = f_1/f_2$ is entire, then f is of exponential type τ with $\tau \le \tau_1$ if $\tau_2 = 0$, $\tau \le 2\tau_2$ if $\tau_1 = 0$, and*

$$\tau \le (1+\lambda_0)\tau_1 + \frac{(1+\lambda_0)(1+2\lambda_0)}{\lambda_0^2}\tau_2 \qquad \text{if } \tau_1\tau_2 \ne 0,$$

where λ_0 is the positive root of $(\tau_1/\tau_2)\lambda^3 - 3\lambda - 2 = 0$.

4.5.8. Remark. It follows from §4.5.6. that if f is an entire function without any zeros and of finite order ρ (resp. finite class p) then the function $1/f$ is of order ρ (resp. class p). More generally, if f and g are entire functions of finite order ρ (resp. class p) and f/g is entire, then the function f/g is of order at most ρ (resp. class at most p).

The statement and proof of §4.5.5 are due to Avanissian [Av].

4.5.9. Proposition. (Borel-Caratheodory Inequality). *Let f be a holomorphic function in $B(0,\rho)$. For $0 \le r < \rho$ let $\mathscr{A}(f,r) = \max_{|z|=r} \operatorname{Re} f(z)$. The following inequality holds for any $0 \le r < R < \rho$*

$$M(|f|,r) \le [\mathscr{A}(f,R) - \operatorname{Re} f(0)]\frac{2r}{R-r} + |f(0)|.$$

Proof. It is easy to see that we can assume $f(0) = 0$, f not constant. Hence $\operatorname{Re} f$ is not constant either and $\operatorname{Re} f(z) < \mathscr{A}(f,R)$ if $|z| < R$. Consider the auxiliary function

$$g(z) = \frac{f(z)}{2\mathscr{A}(f,R) - f(z)},$$

which is holomorphic in $B(0, R)$. Furthermore, $|g(z)| \le 1$. In fact,

$$|g(z)|^2 = \frac{|f(z)|^2}{|f(z)|^2 + 4\mathscr{A}(f, R)(\mathscr{A}(f, R) - \operatorname{Re} f(z))} \le 1.$$

By Schwarz's lemma we have $|g(z)| \le |z|/R$. Writing f in terms of g one obtains

$$f(z) = \frac{2\mathscr{A}(f, R)g(z)}{1 + g(z)},$$

whence

$$|f(z)| \le \frac{2\mathscr{A}(f, R)(|z|/R)}{1 - (|z|/R)} \le \frac{2\mathscr{A}(f, R)r}{R - r}$$

if $|z| \le r < R$. The proposition follows immediately from this. □

4.5.10. Corollary. *Let f be a holomorphic function in $B(0, \rho)$, $f(0) = 1$ and assume further that $f(z) \neq 0$ in $\bar{B}(0, R)$, $0 < R < \rho$. Then*

$$\log|f(z)| \ge \frac{2r}{R - r} \log M(|f|, R)$$

for $|z| \le r < R$.

Proof. The function $g = \operatorname{Log} f$ is holomorphic in $\bar{B}(0, R)$ and zero at the origin. Recalling that $|\operatorname{Log} f| \ge -\operatorname{Re} \operatorname{Log} f = -\log|f|$ and applying the Borel-Caratheodory inequality to g we obtain for $|z| \le r$

$$-\log|f(z)| \le M(-\log|f|, r) \le M(|g|, r)$$

$$\le \mathscr{A}(g, R)\frac{2r}{R - r} = \frac{2r}{R - r} M(\log|f|, R),$$

which proves the corollary. □

Without much difficulty one can obtain now the following results.

4.5.11. Corollary. *If f is an entire function without zeros of class $p + 1$, then $\log f$ is an entire function of class p. If f is an entire function without any zeros of finite order at most ρ then $\log f$ is a polynomial of degree $\le [\rho]$.*

We have already seen that every transcendental function has a dense range and even that the image of every set $\{|z| > R\}$ is dense (Casorati-Weierstrass). This statement was made more precise by showing that for every α the equation $f(z) = \alpha$ always has infinitely many solutions, with the exception of at most a single value α. This result is Picard's Little Theorem 2.7.10. Here we can give a very elementary proof for the functions of finite class. (This argument is inspired on a theorem of E. Borel.)

4.5.12. Proposition. *Let f be an entire function of finite class $p \geq 1$. We have*

$$\#\{a \in \mathbb{C} : \#(f^{-1}(\{a\})) < \infty\} \leq 1.$$

PROOF. Let $a \neq b$ be two exceptional values:

$$\#(f^{-1}(\{a\})) < \infty, \qquad \#(f^{-1}(\{b\})) < \infty.$$

Therefore, there are two nonzero polynomials P, Q and two entire functions g, h such that

$$f = a + P\exp(g) = b + Q\exp(h).$$

Differentiate the two sides of the identity

$$a - b + P\exp(g) = Q\exp(h)$$

to obtain

$$(P' + Pg')\exp(g) = (Q' + Qh')\exp(h).$$

Let us assume first that $P' + Pg' \not\equiv 0$, $Q' + Qh' \not\equiv 0$. The functions g, h are of class exactly $p - 1$, therefore $P' + Pg'$, $Q' + Qh'$ are also of class at most $p - 1$. The last identity can be rewritten as

$$(P' + Pg')\exp(g - h) = Q' + Qh',$$

hence $\exp(g - h)$ if also of class at most $p - 1$. We have to consider two separate cases:

If $p = 1$, then h is a polynomial and $\exp(g - h)$ is also a polynomial, hence it is a constant since it does not vanish. Since

$$e^{-h} = \frac{1}{a - b}[Q - P\exp(g - h)],$$

it follows that e^{-h} is of class 0. This is impossible because h is a polynomial which cannot be constant, otherwise f could not be of class 1.

If $p > 1$, then e^{-h} would be of class at most $p - 1$, which is again impossible. Otherwise h would be of class at most $p - 2$.

The only possibility left is that $P' + Pg' \equiv 0$ and $Q' + Qh' \equiv 0$. This implies that P cannot vanish, otherwise g' would have poles. Therefore P is a constant, $P' = 0$ and $g' = 0$. This is again impossible because f would be a constant. □

The minimum modulus theorem at the end of this section is a far-reaching generalization of Corollary 4.5.10.

4.5.13. Lemma. (Cartan-Boutroux). *Let $z_1, \ldots, z_n$ be n arbitrary complex numbers and $P(z) = \prod_{1 \leq j \leq n} (z - z_j)$.* For every $H > 0$ the inequality

$$|P(z)| > \left(\frac{H}{e}\right)^n$$

holds outside a set which is the union of p disks, $p \le n$, whose radii add up to at most $2H$.

PROOF. The change of variables $\zeta = \frac{H}{n} z$ reduces the problem to the case $H = n$. That is, we must show $|P(z)| \ge \left(\frac{n}{e}\right)^n$ outside a collection of no more than n circles, the sum of whose radii does not exceed $2n$. The set $K = cv\{z_1, \ldots, z_n\}$ is a polygon whose vertices are among the z_j. Let z_k be one of these vertices, m_k the multiplicity of this value among the z_j. One can easily see (by drawing a picture) that there is at least one closed disk D of radius m_k such that $D \cap K = \{z_k\}$. Let us call E the collection of closed disks D such that $\#\{j : z_j \in D\} =$ radius of D. As we said, $E \ne \emptyset$. Of course, the radii of any disk in E is an integer $\le n$. Let r_1 be the largest value for those radii.

For any integer $\rho \ge r_1$ and any $a \in \mathbb{C}$ it is not possible that $\bar{B}(a, \rho)$ contains strictly more z_j than ρ, say $\rho' > \rho$. Otherwise, either $\bar{B}(a, \rho') \in E$ and $\rho' > \rho \ge r_1$, which contradicts the maximality of r_1, or $\bar{B}(a, \rho')$ contains $\rho'' > \rho'$ of the z_j. Then we see again that $\bar{B}(a, \rho'') \notin E$, which means it contains $\rho''' > \rho''$ of the z_j. And so on, but the z_j are a finite collection and this would lead to an infinite set. Therefore, if $\rho \ge r_1$, $\rho \in \mathbb{N}$, we have $\#\{j : z_j \in \bar{B}(a, \rho)\} \le \rho$. Choose $a_1 \in \mathbb{C}$ such that $\bar{B}(a_1, r_1) \in E$. The points $z_j \in \bar{B}(a_1, r_1)$ are said to belong to the first generation, there are exactly r_1 of them (counted with multiplicities). Let us take them out. We are left with $n - r_1$ points. If $n - r_1 > 0$ we can continue and obtain a disk $\bar{B}(a_2, r_2)$ by the same procedure as earlier $r_2 \le n - r_1$. Let us see that $r_2 \le r_1$. If not $\bar{B}(a_2, r_2)$ contains $r_2' \ge r_2 > r_1$ points of the initial set and we know this is impossible. The points in $\bar{B}(a_2, r_2)$ are said to be of the second generation. Iterating this procedure we obtain integers $r_1 \ge r_2 \ge \cdots \ge r_p \ge 1$, $r_1 + \cdots + r_p = n$ (hence $p \le n$), and $\bigcup_{j=1}^{p} \bar{B}(a_j, r_j) \supseteq \{z_1, \ldots, z_n\}$.

Our exceptional disks are going to be the $\bar{B}(a_j, 2r_j)$, $1 \le j \le p$. Clearly the sum of the radii equals $2n$. Let now $z \notin \bigcup_j \bar{B}(a_j, 2r_j)$. If $\rho \in \mathbb{N}^*$ then we must have

$$\bar{B}(z, \rho) \cap \bar{B}(a_j, r_j) = \emptyset \qquad \text{if } r_j \ge \rho,$$

otherwise $|z - a_j| \le \rho + r_j \le 2r_j$, which is false. Let k be the smallest index such that $r_k < \rho$, $\bar{B}(z, \rho)$ contains no points of the first $k - 1$ generations. Therefore $\bar{B}(z, \rho)$ must contain at most $\rho - 1$ points of the sequence, otherwise it would have been a competitor for the kth generation of points. Let us now reorder the sequence by increasing distance to z. Since $\bar{B}(z, 1)$ contains no points of the sequence, we have $|z - z_1| > 1$. The disk $\bar{B}(z, 2)$ contains at most one point of the sequence, therefore in the worst case this is z_1. Hence $|z - z_2| > 2$. And we can go on to show $|z - z_k| > k$ (after reordering). It follows that

$$|P(z)| = \prod_{j=1}^{n} |z - z_j| > n! \geq \left(\frac{n}{e}\right)^n.$$

Since z was arbitrary we have proved the lemma. □

4.5.14. Minimum Modulus Theorem. *Let f be holomorphic in the disk $B(0, 2eR)$ and continuous in the closure of this disk. Assume $f(0) = 1$ and let $\varepsilon > 0$ be such that $0 < \varepsilon < \dfrac{3e}{2}$. Then in the disk $|z| \leq R$, and outside a collection of closed disks the sum of whose radii does not exceed $4\varepsilon R$, we have*

$$\log|f(z)| > -\left(2 + \log\frac{3e}{2\varepsilon}\right)\log M(|f|, 2eR).$$

PROOF. Let $z_1, \ldots, z_n$ be the zeros of f in $|z| \leq 2R$, always counted with multiplicity. We can assume that none of them lies in $|z| = 2R$. (If not, replace in what follows $2R$ by a slightly bigger quantity R' and verify that the argument remains true.) Let φ be the auxiliary function

$$\varphi(z) = \frac{(-2R)^n}{z_1 z_2 \ldots z_n} \prod_{1 \leq k \leq n} \frac{2R(z - z_k)}{(2R)^2 - \bar{z}_k z},$$

holomorphic in $|z| \leq 2R$, $\varphi(0) = 1$, $|\varphi(2Re^{i\theta})| = \dfrac{(2R)^n}{|z_1 \ldots z_n|} \geq 1$. The function $\psi(z) = f(z)/\varphi(z)$ is now holomorphic and without zeros in $|z| \leq 2R$. From Corollary 4.5.10 we conclude that if $|z| \leq R$, then

$$\begin{aligned}\log|\psi(z)| &\geq -2\log M(|f|/|\varphi|, 2R) = -2\log M(|f|, 2R) + 2\log|\varphi(2Re^{i\theta})| \\ &\geq -2\log M(|f|, 2R) \geq -2\log M(|f|, 2eR).\end{aligned}$$

We need to find a lower bound for $|\varphi|$. On one hand, for $|z| \leq R$ the denominator of φ can be easily estimated:

$$\prod_{k=1}^{n} |(2R)^2 - \bar{z}_k z| \leq (6R^2)^n.$$

To the numerator we can apply Cartan-Boutroux' Lemma 4.5.13 with $H = 2\varepsilon R$, hence for any $z \in \mathbb{C}$ outside a family of p disks ($p \leq n$) whose radii add up to at most $4\varepsilon R$ we have

$$\prod_{k=1}^{n} |z - z_k| \geq \left(\frac{2\varepsilon R}{n}\right)^n.$$

Therefore, for those points in $|z| \leq R$ that lie outside those exceptional disks, we have

$$|\varphi(z)| \geq \frac{(2R)^n}{|z_1 \ldots z_n|}\left(\frac{2\varepsilon R}{n}\right)^n \frac{1}{(6R^2)^n} \geq \left(\frac{2\varepsilon}{3e}\right)^n.$$

On the other hand, by §4.4.31 (2), we have

$$n = \nu_f(2R) \leq \log M(|f|, 2eR),$$

that is, outside the exceptional disks

$$\log|\varphi(z)| \geq \log\left(\frac{2\varepsilon}{3e}\right)\log M(|f|, 2eR)$$

and

$$\log|f(z)| = \log|\varphi(z)| + \log|\psi(z)| \geq -\left(2 + \log\left(\frac{3e}{2\varepsilon}\right)\right)\log M(|f|, 2eR).$$

This proves the theorem. □

EXERCISES 4.5

1. Find the order and type of the following functions:

(a) $E_\rho(z) = \sum_{n\geq 0} \frac{z^n}{\Gamma(1 + n\rho)}$ $(\rho > 0)$ (b) $\sin z$ (c) $\cos z$

(d) $\cosh z$ (e) $\sinh z$

(f) $J_p(z) = \sum_{n\geq 0} \frac{(-1)^n z^{p+2n}}{2^{p+2n} n!(n+p)!}$ $(p \in \mathbb{N})$ (g) $Ae^{\lambda z} + Be^{\mu z^3}$

(h) $\cos\sqrt{z}$ (i) $\frac{(\cos\sqrt[4]{z} + \cos i\sqrt[4]{z})}{2}$ (j) $\sum_{n\geq 1}\left(\frac{z}{n}\right)^n$

(k) $\sum_{n\geq 1} \frac{(\log n)^{\alpha/n} z^n}{n^{\beta/n}}$ $(\alpha, \beta > 0)$ (l) $\int_0^1 e^{z^p t^2}\,dt$, $p \in \mathbb{N}^*$.

2. Compare the order, type, class of f and f'.

3. Let $f \in \mathscr{H}(\mathbb{C})$, $f(z) = \sum_{n\geq 0} a_n z^n$. Show that the condition for every $\varepsilon > 0$ there is C_ε such that

$$|f(z)| \leq C_\varepsilon e^{\varepsilon|z|} \qquad (\forall z \in \mathbb{C})$$

is equivalent to

$$\lim_{n\to\infty} \sqrt[n]{n!|a_n|} = 0.$$

4. Why is the Example 4.5.2 (2) correct? Verify also items 1 and 3 of §4.5.2.

5. Write a formula for the class of an entire function f in terms of the Taylor coefficients a_n of f about $z = 0$.

6. Let f be an entire function of order ρ and finite type, show that there are two positive constants, ε, $C > 0$ such that for any $R \geq 1$ one can find $r \in [R, 2R]$ such that

$$\min_{|z|=r} |f(z)| \geq \varepsilon e^{-Cr^\rho}.$$

7. Verify statement 4.5.2 (6). What can you say about the class, order, and type of the sum and of the product of two entire functions?

8. Let f be an entire function of class $p \geq 1$. Show that there cannot be two distinct constants a, b such that

$$f = a + Pe^g = b + Qe^h,$$

for some P, Q, entire functions of class $\leq p - 1$, and g, h entire.

*9. Let $g(z) = \sum_{n\geq 1} b_n z^n$ be an entire function of order ρ, $0 < \rho < \infty$, and type τ, $0 < \tau < \infty$. Let $f(z) := e^{g(z)} = \sum_{n\geq 1} a_n z^n$.

(i) Show that $\rho = \limsup_{r\to\infty} \dfrac{\log_3 M(|f|, r)}{\log r}$

$$\tau = \limsup_{r\to\infty} \frac{\log_2 M(|f|, r)}{r^\rho}$$

(ii) Let $v := \limsup_{n\to\infty} (\log n |a_n|^{\rho/n})$. We want to show that $\tau \leq v$.

(a) We can assume $v < \infty$ (why?). Let $v_1 = v + \varepsilon$ and $n_\varepsilon \geq 2$ be such that $\log n |a_n|^{\rho/n} \leq v_1$ for $n \geq n_\varepsilon$. Show first that

$$M(|f|, r) \leq \sum_{n=0}^{n_\varepsilon} |a_n| r^n + \sum_{n\geq 2} \left(\frac{v_1 r^\rho}{\log n}\right)^{n/\rho}$$

(b) Show that for $r \gg 1$, $\max_{t\geq 2} \left(\dfrac{v_1 r^\rho}{\log t}\right)^{t/\rho} \leq \exp\left(\dfrac{e^{v_1 r^\rho}}{\rho}\right)$. The maximum is achieved at a value $t = t(r)$ satisfying

$$\exp\left(\frac{e^{v_1 r^\rho}}{e}\right) \leq t(r) \leq \exp(v_1 r^\rho).$$

(c) Use (a) and (b) to conclude that

$$\limsup_{r\to\infty} \frac{\log_2 M(|f|, r)}{r^\rho} \leq v_1.$$

(iii) Show that $v/e^\rho \leq \tau$.

(iv) Let $G(z) := \sum_{n\geq 0} |b_n| z^n$. Find the order and type of G.

(v) Let $s_n(z) := \sum_{0\leq k\leq n} a_k z^k$. Show that for $|z| < r$ one has

$$|e^{g(z)} - s_n(z)| \leq \left(\frac{|z|}{r}\right)^{n+1} e^{G(r)}.$$

Conclude that if $|z| < r\exp\left(-\dfrac{2G(r)}{n}\right)$, then

$$|1 - s_n(z)e^{-g(z)}| < 1.$$

(vi) Let δ_n be the modulus of the smaller zero of s_n. Why does $\delta_n \to \infty$ as $n \to \infty$? Show that for any $r > 0$ we have

$$\delta_n \geq r\exp\left(-\frac{2G(r)}{n}\right).$$

Let α_n be the unique positive root of $G(\alpha_n) = n$. Show that

$$\delta_n \geq e^{-2}\alpha_n.$$

(vii) Show that $\rho = \limsup_{n\to\infty} \dfrac{n\log_2 n}{-\log|a_n|}$.

(One can also show that when g is not a polynomial

$$\rho = \limsup_{n\to\infty} \frac{\log_2 n}{\log \delta_n},$$

$$\tau = \limsup_{n\to\infty} \frac{\log n}{\delta_n^\rho},$$

cf. [Buck]).

10. Let $f(z) = \sum_{n\geq 0} a_n z^n$ be an entire function of exponential type.

(i) Show that if for every $z \in \mathbb{C}$ we have

$$|f(z)| \leq Ae^{\alpha|z|},$$

then

$$|a_n| \leq \inf_{r>0} \frac{Ae^{\alpha r}}{r^n} = \frac{Ae^n\alpha^n}{n^n}.$$

Use Stirling's formula,

$$n! = \frac{n^n\sqrt{2\pi n}}{e^n}(1 + \varepsilon(n)), \qquad \varepsilon(n) \to 0 \text{ as } n \to \infty,$$

to show that the radius of convergence of

$$g(w) = \sum_{n=0}^{\infty} n!\,a_n w^n$$

is at least $\dfrac{1}{\alpha}$.

(ii) Let $B(w) := \dfrac{1}{w} g\left(\dfrac{1}{w}\right)$ defined in $|w| > \alpha$ and 0 at ∞, show that for $\rho > \alpha$ one has

$$f(z) = \frac{1}{2\pi i}\int_{|w|=\rho} e^{zw}B(w)\,dw.$$

As usual, the circle $|w| = \rho$ is traversed in the positive sense.

(iii) Show that f is of exponential type zero, i.e., for every $\varepsilon > 0$ there is an A_ε such that $|f(z)| \leq A_\varepsilon e^{\varepsilon|z|}$, if and only if g is an entire function.

11. Throughout this exercise f will be a function as in part (i) of Exercise 4.5.10 with $0 < \alpha < \log 2$. Let $\Delta^{(n)}(f) := (-1)^n \sum_{0\leq k\leq n} (-1)^k \binom{n}{k} f(k)$. (These are exactly the divided differences defined in Exercise 3.4.5.)

(i) Show that for every $z \in \mathbb{C}$ fixed, the series

$$\sum_{n=0}^{\infty} \frac{z(z-1)\ldots(z-n+1)}{n!}(e^w - 1)^n$$

converges uniformly in every compact subset of $|w| < \log 2$ to the value e^{zw}. The term with $n = 0$ has the value 1. (Hint: use the Taylor series development of $u \mapsto (1+u)^z$ for $|u| < 1$.)

(ii) Use Exercise 4.5.10, (ii) to show that

$$f(z) = \sum_{n=0}^{\infty} \Delta^{(n)}(f) \frac{z(z-1)\dots(z-n+1)}{n!}$$

holds for every fixed $z \in \mathbb{C}$.

(iii) Show that for $n \geq 1$

$$\Delta^{(n)}(f) = \frac{n!}{2\pi i} \int_{|z|=2n} \frac{f(z)}{z(z-1)\dots(z-n)} dz.$$

Parameterize the circle $|z| = 2n$ by $z = \varphi(t) = 2ne^{it/\sqrt{n}}$, $-\pi\sqrt{n} \leq t \leq \pi\sqrt{n}$ to conclude that

$$|\Delta^{(n)}(f)| \leq \frac{2A}{\pi} \frac{(n!)^2 2^{2n}}{(2n)!\sqrt{n}} \left(\frac{e^\alpha}{2}\right)^{2n} \int_0^{\pi/n} \left|\frac{2n-1}{\varphi(t)-1}\right| \cdots \left|\frac{n}{\varphi(t)-n}\right| dt$$

$$\leq \frac{4A}{\pi} \left(\frac{e^\alpha}{2}\right)^{2n} \int_0^{\pi/n} \left|\frac{2n-1}{\varphi(t)-1}\right| \cdots \left|\frac{n}{\varphi(t)-n}\right| dt.$$

(iv) Using that for $\lambda = \frac{1}{8}\log 9$, one has $\frac{1}{1+y} \leq e^{-\lambda y}$ for $0 \leq y \leq 8$, show that for $0 \leq t \leq \pi\sqrt{n}$

$$\left|\frac{2n-k}{\varphi(t)-k}\right|^2 = \frac{1}{1 + (8nk/(2n-k)^2)\sin^2(t/2\sqrt{n})} \leq \exp\left(-\frac{8\lambda k}{(2n-k)^2} \frac{t^2}{\pi^2}\right).$$

Conclude that

$$\left|\frac{2n-1}{\varphi(t)-1}\right| \cdots \left|\frac{n}{\varphi(t)-n}\right| \leq \left(-\frac{(1-\log 2)4\lambda}{\pi^2} t^2\right).$$

Hence, there is an absolute constant $\kappa > 0$ such that

$$|\Delta^{(n)}(f)| \leq \kappa A \left(\frac{e^\alpha}{2}\right)^{2n}.$$

(v) Show that if $f(\mathbb{N}) \subseteq \mathbb{Z}$, then $\Delta^{(n)}(f) \in \mathbb{Z}$ for every $n \in \mathbb{N}$.

(vi) Conclude that if $0 < \alpha < \log 2$ and $|f(z)| \leq Ae^{\alpha|z|}$ for every $z \in \mathbb{C}$, then the condition $f(\mathbb{N}) \subseteq \mathbb{Z}$ implies that f is a polynomial. Estimate the degree. What happens if f vanishes for all $n \in \mathbb{N}$? (For different generalizations of this expansion and applications, we refer to [Ge].)

§6. Integral Representations

Our objective in this section is to obtain integral formulas representing a subharmonic function u in terms of the measure $\mu = \Delta u \geq 0$. In case $u = \log|f|$, f an entire function, one obtains, as a corollary, the Weierstrass and Hadamard representations of entire functions as infinite products of elementary factors. These factors are defined in terms of the zeros of f and the order of growth of f.

We start by generalizing the Jensen and Poisson formulas.

4.6.1. Lemma. *For a fixed ζ, $0 \le |\zeta| < R$, the function* $z \mapsto \log\left|\dfrac{R(z-\zeta)}{R^2 - z\bar{\zeta}}\right|$ *has the following properties*:

(1) *It is harmonic in $B(0, R)\backslash\{\zeta\}$.*
(2) *Continuous in $\bar{B}(0, R)\backslash\{\zeta\}$ and identically zero when $|z| = R$.*
(3) *The function* $z \mapsto \log|z-\zeta| - \log\left|\dfrac{R(z-\zeta)}{R^2 - z\bar{\zeta}}\right|$ *is harmonic in $B(0, R)$ and continuous in $\bar{B}(0, R)$.*

Proof. Since $|\zeta| < R$, we have $R^2 - z\bar{\zeta} \neq 0$ for $z \in \bar{B}(0, R)$. We can write in $\bar{B}(0, R)\backslash\{\zeta\}$

$$\log\left|\frac{R(z-\zeta)}{R^2 - z\bar{\zeta}}\right| = \log|z-\zeta| + \log R - \log|R^2 - z\bar{\zeta}|.$$

which shows (1) and the continuity part of (2). It also shows (3) holds. The only statement to check is the last part of (2).

If $|z| = R$ then $z\bar{z} = R^2$ and

$$|R^2 - z\bar{\zeta}| = |z|\,|\bar{z} - \bar{\zeta}| = R|z-\zeta| = |R(z-\zeta)|,$$

hence $\log\left|\dfrac{R(z-\zeta)}{R^2 - z\bar{\zeta}}\right| = \log 1 = 0$ when $|z| = R$. □

4.6.2. Lemma. *The following identity holds for $|z| < R$:*

$$\frac{1}{2\pi}\int_{-\pi}^{\pi} \frac{R^2 - |z|^2}{|Re^{i\theta} - z|^2} \log|w - Re^{i\theta}|\, d\theta$$

$$= \begin{cases} \log|z-w| - \log\left|\dfrac{R(z-w)}{R^2 - z\bar{w}}\right| & \text{if } |w| < R \\[2ex] \log|z-w| & \text{if } |w| \ge R. \end{cases}$$

Proof. (1) If $|w| > R$ then $\log|z-w|$ is a harmonic function of z in a neighborhood of $\bar{B}(0, R)$, hence by Poisson's formula we obtain

$$\log|z-w| = \frac{1}{2\pi}\int_{-\pi}^{\pi} \frac{R^2 - |z|^2}{|Re^{i\theta} - z|^2} \log|Re^{i\theta} - w|\, d\theta.$$

(2) If $|w| < R$, the function

$$h(z) = \log|z-w| - \log\left|\frac{R(z-w)}{R^2 - z\bar{w}}\right|$$

is harmonic in $B(0, R)$ and continuous in $\bar{B}(0, R)$ by Lemma 4.6.1, (3). Therefore for $|z| < R$ we have

$$h(z) = \log|z - w| - \log\left|\frac{R(z - w)}{R^2 - z\bar{w}}\right| = \frac{1}{2\pi}\int_{-\pi}^{\pi}\frac{R^2 - |z|^2}{|Re^{i\theta} - z|^2}h(Re^{i\theta})\,d\theta$$

$$= \frac{1}{2\pi}\int_{-\pi}^{\pi}\frac{R^2 - |z|^2}{|Re^{i\theta} - z|^2}\log|Re^{i\theta} - w|\,d\theta$$

since the other term vanishes by part (2) of the previous lemma.

(3) If $|w| = R$, let $w = e^{i\alpha}R$, consider in $\frac{1}{2} < t < 1$ the function

$$\varphi(t) = \frac{1}{2\pi}\int_{-\pi}^{\pi}\frac{R^2 - |z|^2}{|Re^{i\theta} - z|^2}\log|tw - Re^{i\theta}|\,d\theta,$$

which can be evaluated by the previous part of this proof as

$$\varphi(t) = \log|z - tw| - \log\left|\frac{R(z - tw)}{R^2 - zt\bar{w}}\right| = \log|R - zte^{-i\alpha}|$$

$$= \log|Re^{i\alpha} - zt| \xrightarrow[t\to 1]{} \log|w - z|.$$

On the other hand, the function $|\log|tw - Re^{i\theta}||$ can be bounded as follows (draw a picture!)

$$|w - Re^{i\theta}| \le \left|w - \frac{R}{t}e^{i\theta}\right| \le |w - 2Re^{i\theta}|.$$

Therefore,

$$|\log|tw - Re^{i\theta}|| \le \log 2 + \max(|\log|w - Re^{i\theta}||, |\log|w - 2Re^{i\theta}||) = g(\theta),$$

and $g(\theta)$ is an integrable function of θ. The Lebesgue-dominated convergence theorem allows us to take the limit under the integral defining φ and obtain

$$\log|w - z| = \lim_{t\to 1}\varphi(t) = \frac{1}{2\pi}\int_{-\pi}^{\pi}\frac{R^2 - |z|^2}{|Re^{i\theta} - z|^2}\log|w - Re^{i\theta}|\,d\theta.$$

(Recall that all along z is fixed, $|z| < R$). □

4.6.3. Proposition. (Poisson-Jensen's Formula for Subharmonic Functions). *Let u be a subharmonic function in a neighborhood of $\bar{B}(0, R)$. Then for $z \in B(0, R)$ we have*

$$u(z) = \frac{1}{2\pi}\int_{-\pi}^{\pi}\frac{R^2 - |z|^2}{|Re^{i\theta} - z|^2}u(Re^{i\theta})\,d\theta + \frac{1}{2\pi}\int_{B(0,R)}\log\left|\frac{R(z - \zeta)}{R^2 - z\bar{\zeta}}\right|(\Delta u)(\zeta).$$

Proof. Let $R' > R$ be such that u is subharmonic in $B(0, R')$. Then for $|\zeta| < R'$ we have (cf. §4.4.24(3))

$$u(\zeta) = \frac{1}{2\pi}\int_{B(0,R')}\log|\zeta - w|(\Delta u)(w) + h(\zeta),$$

h harmonic. Therefore, for $|z| < R$,

$$\frac{1}{2\pi}\int_{-\pi}^{\pi}\frac{R^2-|z|^2}{|Re^{i\theta}-z|^2}u(Re^{i\theta})\,d\theta$$
$$=\frac{1}{2\pi}\int_{-\pi}^{\pi}\frac{R^2-|z|^2}{|Re^{i\theta}-z|^2}\left\{\frac{1}{2\pi}\int_{B(0,R')}\log|Re^{i\theta}-w|\Delta u(w)+h(Re^{i\theta})\right\}d\theta$$
$$=h(z)+\frac{1}{2\pi}\int_{B(0,R')}\left\{\frac{1}{2\pi}\int_{-\pi}^{\pi}\frac{R^2-|z|^2}{|Re^{i\theta}-z|^2}\log|Re^{i\theta}-w|\,d\theta\right\}\Delta u(w).$$

By Lemma 4.6.2, the integral has the value

$$\frac{1}{2\pi}\int_{B(0,R')}\log|z-w|\Delta u(w)-\frac{1}{2\pi}\int_{B(0,R)}\log\left|\frac{R(z-w)}{R^2-z\bar{w}}\right|(\Delta u)(w).$$

The first term is exactly $u(z)-h(z)$, hence

$$\frac{1}{2\pi}\int_{-\pi}^{\pi}\frac{R^2-|z|^2}{|Re^{i\theta}-z|^2}u(Re^{i\theta})\,d\theta=u(z)-\frac{1}{2\pi}\int_{B(0,R)}\log\left|\frac{R(z-w)}{R^2-z\bar{w}}\right|(\Delta u)(w).\quad\square$$

4.6.4. Corollary. (Poisson-Jensen's Formula for Holomorphic Functions). *Let f be a holomorphic function in the neighborhood of $\bar{B}(0,R)$, and $a_1,\ldots,a_n$ its zeros in $B(0,R)$, counted with multiplicities. For $z\in B(0,R)$, we have*

$$\log|f(z)|=\frac{1}{2\pi}\int_{-\pi}^{\pi}\frac{R^2-|z|^2}{|Re^{i\theta}-z|^2}\log|f(Re^{i\theta})|\,d\theta+\sum_{1\le j\le n}\frac{|R(z-a_j)|}{|R^2-\bar{a}_jz|}.$$

In fact, one can add the zeros on $\partial B(0,R)$ since they do not contribute to the last sum.

We want to introduce now the ***Nevanlinna growth function*** N of a subharmonic function. It will be apparent from the following definition that it is convenient to assume that u is harmonic in a neighborhood of the origin. Assume u is subharmonic in a neighborhood of $\bar{B}(0,R)$. For $\varepsilon>0$ very small and fixed, we replace u by the harmonic extension to $B(0,\varepsilon)$ of $u|\partial B(0,\varepsilon)$. This way, one obtains a new subharmonic function in a neighborhood of $\bar{B}(0,R)$, which is harmonic in a neighborhood of the origin and coincides with the original function outside $B(0,\varepsilon)$. By abuse of language we will call the new function u also. Applying the Poisson-Jensen formula to this new function we obtain

$$u(0)=\frac{1}{2\pi}\int_{-\pi}^{\pi}u(Re^{i\theta})\,d\theta+\int_{B(0,R)}\left(\log\left|\frac{\zeta}{R}\right|\right)\Delta u(\zeta)$$
$$=\lambda(u,0,R)+\int_{B(0,R)}\left(\log\left|\frac{\zeta}{R}\right|\right)\Delta u(\zeta).$$

Let

$$u^+=\max\{u,0\},\qquad u^-=-\min\{u,0\}=\max\{-u,0\},$$

and define the auxiliary functions T, m, and ν by

$$T(u,0,r) = T(u,r) = \lambda(u^+,0,r),$$

$$m(u,0,r) = m(u,r) = \lambda(u^-,0,r),$$

$$\nu(r) = (\Delta u)(B(0,r)).$$

Note that $\nu(r)$ is increasing and $\nu(r) = 0$ for $0 \le r < \varepsilon$. We claim that

$$-\int_{B(0,R)} \left(\log\left|\frac{\zeta}{R}\right|\right)(\Delta u)(\zeta) = \int_0^R \frac{\nu(t)}{t}\,dt.$$

In fact, we know already from §4.4.28 that

$$\lambda(u,0,R) - \lambda(u,0,r) = \frac{1}{2\pi}\int_r^R \frac{\nu(t)}{t}\,dt.$$

Since $u(0) = \lim_{r\to 0} \lambda(u,0,r)$ we have also

$$\lambda(u,0,R) = u(0) + \frac{1}{2\pi}\int_0^R \frac{\nu(t)}{t}\,dt.$$

By comparison with the previous application of the Poisson-Jensen formula, we see that the claim is correct.

Let us now define

$$N(u,r) = \frac{1}{2\pi}\int_0^r \frac{\nu(t)}{t}\,dt.$$

The Poisson-Jensen formula becomes

$$\begin{aligned} u(0) &= \lambda(u,0,R) - \int_{B(0,R)} \left(\log\left|\frac{R}{\zeta}\right|\right)(\Delta u)(\zeta) = \lambda(u,0,R) - N(u,R) \\ &= \lambda(u^+,0,R) - \lambda(u^-,0,R) - N(u,R). \end{aligned}$$

This can be rewritten as the ***first fundamental formula of R. Nevanlinna***, namely,

$$T(u,R) = N(u,R) + m(u,R) + u(0).$$

This formula is the first step of a very rich and deep theory developed originally by R. Nevanlinna. It has many applications to the study of holomorphic functions of one and several complex variables. Its pursuit will take us too far afield, the reader will profit from consulting the references in the bibliography, e.g., [Ha] and [Gril1]. We remark that the two functions

$$\rho \mapsto T(u,\rho), \quad \rho \mapsto N(u,\rho)$$

are increasing and convex functions of $\log \rho$.

Let us return now to the question of integral representations of subharmonic functions in $\mathbb{C}$.

We start by describing the orders of growth of functions of the form $v(t) = \mu(B(0,t))$, μ a positive Borel measure. By analogy with §4.5 we define the ***order of growth*** ρ of v to be

$$\rho := \limsup_{r\to\infty} \frac{\log v(r)}{\log r}$$

We use throughout the Stieltjes integral and the formula of integration by parts for this integral without any further ado (see [HS]).

4.6.5. Lemma. *For $a > 0$, $q > 0$, and v an increasing nonnegative function, the following conditions are equivalent*:

(1) $\displaystyle\int_a^\infty \frac{dv(t)}{t^q} < \infty,$

(2) $\displaystyle\int_a^\infty \frac{v(t)}{t^{q+1}}\,dt < \infty.$

Let $N(t) = \displaystyle\int_a^\infty \frac{v(s)}{s}\,ds$, *then the third equivalent condition is*

(3) $\displaystyle\int_a^\infty \frac{N(t)}{t^{q+1}}\,dt < \infty.$

Every one of these three conditions implies $\displaystyle\lim_{t\to\infty} \frac{v(t)}{t^q} = 0$. *We also have then* $\displaystyle\lim_{t\to\infty} \frac{N(t)}{t^q} = 0.$

Proof. Integrating by parts one finds

$$\int_a^r \frac{dv(t)}{t^q} = \frac{v(t)}{t^q}\bigg|_a^r + q\int_a^r \frac{v(t)}{t^{q+1}}\,dt.$$

(1) ⇒ (2): We have

$$\frac{v(r)}{r^q} + q\int_a^r \frac{v(t)}{t^{q+1}}\,dt = \frac{v(a)}{a^q} + \int_a^r \frac{dv(t)}{t^q} \le \frac{v(a)}{a^q} + \int_a^\infty \frac{dv(t)}{t^q} < \infty.$$

Since the two terms on the left are nonnegative, then (2) follows. Furthermore, the functions $r \mapsto q\displaystyle\int_a^r \frac{v(t)}{t^{q+1}}\,dt$ and $r \mapsto \displaystyle\int_a^r \frac{dv(t)}{t^q}$ are increasing, therefore both have finite limits when $r \to \infty$. Hence, the preceding identity implies that $L = \displaystyle\lim_{r\to\infty} \frac{v(r)}{r^q}$ exists, $0 \le L < \infty$. Given $\varepsilon > 0$ there is an $r_\varepsilon > 0$ such that for $r \ge r_\varepsilon$ we have

$$q\int_r^\infty \frac{v(t)}{t^{q+1}}\,dt \le \varepsilon.$$

Hence, since v is increasing

$$\frac{v(r)}{r^q} = qv(r)\int_r^\infty \frac{dt}{t^{q+1}} \le \varepsilon q \int_r^\infty \frac{v(t)}{t^{q+1}}\,dt \le \varepsilon$$

and $L \le \varepsilon$ for $r \ge r_\varepsilon$. So, $\lim\limits_{r\to\infty} \dfrac{v(r)}{r^q} = 0$.

(2) $\Rightarrow$ (1): Note that the previous reasoning shows that when (2) holds, $\limsup\limits_{r\to\infty} \dfrac{v(r)}{r^q} = 0$. Since $\dfrac{v(r)}{r^q} \ge 0$ it follows that $\lim\limits_{r\to\infty} \dfrac{v(r)}{r^q} = 0$. Hence

$$\int_a^\infty \frac{dv(t)}{t^q} = \lim_{r\to\infty} \int_a^r \frac{dv(t)}{t^q} = -\frac{v(a)}{a^q} + \lim_{r\to\infty} \frac{v(r)}{r^q} + q\int_a^\infty \frac{v(t)}{t^{q+1}}\,dt$$

$$= -\frac{v(a)}{a^q} + q\int_a^\infty \frac{v(t)}{t^{q+1}}\,dt < \infty.$$

(2) $\Rightarrow$ (3):

$$\int_a^r \frac{N(t)}{t^{q+1}}\,dt = \frac{1}{q}\left[-\frac{N(t)}{t^q}\right]_a^r + \frac{1}{q}\int_a^r \frac{dN(t)}{t^q} = \frac{1}{q}\left[\frac{N(a)}{a^q} - \frac{N(r)}{r^q} + \int_a^r \frac{v(t)}{t^{q+1}}\,dt\right]$$

$$\le \frac{1}{q}\left[\frac{N(a)}{a^q} + \int_a^r \frac{v(t)}{t^{q+1}}\,dt\right] < \infty.$$

(3) $\Rightarrow$ (2): Since the integral in (3) is finite, $\dfrac{N(r)}{r^q} \to 0$ when $r \to \infty$, by the reasoning used in the part (2) $\Rightarrow$ (1), which only uses the function being increasing and nonnegative. Therefore,

$$\int_a^\infty \frac{v(t)}{t^{q+1}}\,dt = -\frac{N(a)}{a^q} + q\int_a^\infty \frac{N(t)}{t^{q+1}}\,dt,$$

which concludes the proof. □

It is clear the conditions in Lemma 4.6.5 are independent of the choice of $a > 0$.

4.6.6. Definitions. Let μ be a positive Borel measure in $\mathbb{C}$, $v(t) = \mu(B(0,t))$, $a > 0$.

(1) The number (or more precisely, the element of $[0,\infty]$)

$$\rho_1 := \inf\left\{s > 0 : \int_a^\infty \frac{dv(t)}{t^s} < \infty\right\} = \inf\left\{s > 0 : \int_a^\infty \frac{v(t)}{t^{s+1}}\,dt < \infty\right\}$$

is called the ***exponent of convergence*** of the measure μ.

(2) When $\rho_1 > 0$, the smallest nonnegative integer q such that $\displaystyle\int_a^\infty \frac{dv(t)}{t^{q+1}} < \infty$

is called the ***genus*** of the measure μ. Note that the genus q is also the smallest nonnegative integer such that $\int_a^\infty \frac{\nu(t)}{t^{q+2}}\,dt < \infty$.

4.6.7. Proposition. *Let μ, ν, a be related as in §4.6.6. Then*

(1) *The order ρ of the function ν, $\rho = \limsup_{r\to\infty} \frac{\log \nu(r)}{\log r}$, coincides with the exponent of convergence of measure μ.*
(2) *If ρ is not an integer, the genus q of μ is the largest integer $\le \rho$, that is $q = [\rho]$.*
(3) *If ρ is an integer, then $q \le \rho \le q + 1$. When $\rho = q$ then ν is of minimal type for the order ρ.*

Proof. (1) If $s > \rho_1$, $\varepsilon > 0$, there is $r_\varepsilon > a$ such that $r \ge r_\varepsilon$ implies

$$\nu(r)\int_r^\infty \frac{dt}{t^{s+1}} \le \int_r^\infty \frac{\nu(t)}{t^{s+1}}\,dt \le \varepsilon.$$

Hence, $\frac{\nu(r)}{r^s} \le s\varepsilon$ and

$$\log \nu(r) - s\log r \le \log(s\varepsilon),$$

from this inequality is follows that

$$\rho = \limsup_{r\to\infty} \frac{\log \nu(r)}{\log r} \le s.$$

Therefore $\rho \le \rho_1$, since s was an arbitrary number bigger than ρ_1.

Conversely, if $s > \rho$, let $\rho < \sigma < s$, then there is $r_\sigma > a$ such that for $r \ge r_\sigma$ one has $\frac{\log \nu(r)}{\log r} \le \sigma$, that is, $\frac{\nu(r)}{r^\sigma} \le 1$. Hence

$$\frac{\nu(r)}{r^{s+1}} \le \frac{1}{r^{(s-\sigma)+1}} \quad \text{for } r \ge r_\sigma,$$

and

$$\int_a^\infty \frac{\nu(t)}{t^{s+1}}\,dt \le \int_a^{r_\sigma} \frac{\nu(t)}{t^{s+1}}\,dt + \int_{r_\sigma}^\infty \frac{dt}{t^{(s-\sigma)+1}} < \infty.$$

Therefore $\rho_1 \le s$ for every $s > \rho$, and hence $\rho_1 \le \rho$.

The proofs of (2) and (3) are obvious. □

Examples. If $\{a_n\}_{n\ge1}$ is a sequence of nonzero complex numbers such that $|a_n| \le |a_{n+1}|$ for every $n \ge 1$, we assume that either $\{a_n\}_n$ is a finite sequence or $|a_n| \to \infty$ as $n \to \infty$. Let us consider the measure

$$\mu = \sum_{n \geq 1} \delta_{a_n}.$$

We have here

$$\nu(t) = \mu(B(0, t)) = \#\{n : |a_n| < t\},$$

which is an increasing function, continuous on the left.

The exponent of convergence of μ is here the exponent of convergence of the sequence $\{a_n\}_{n \geq 1}$,

$$\rho_1 = \inf\left\{s > 0 : \sum_{n \geq 1} \frac{1}{|a_n|^s} < \infty\right\}.$$

The genus q is defined only when $\rho_1 > 0$, hence the sequence is necessarily infinite. In this case it is the smallest integer such that $\sum_n \frac{1}{|a_n|^{q+1}} < \infty$.

For instance,

(i) if $a_n = n^{1/\rho}$, then the exponent of convergence is ρ while $\sum_{n=1}^{\infty} \frac{1}{a_n^\rho} = \infty$;
(ii) if $a_n = (n \log^2 n)^{1/\rho}$ $(n \geq 2)$, the exponent of convergence is also ρ but this time $\sum_{n=2}^{\infty} \frac{1}{a_n^\rho} < \infty$;
(iii) if $a_n = e^n$, the exponent of convergence is zero;
(iv) if $a_n = \log n$ $(n \geq 2)$, the exponent of convergence is $+\infty$ (recall that the infimum of an empty set of real numbers is $+\infty$).

We are now going to define a ***canonical potential*** associated to a Borel measure $\mu \geq 0$, using ***canonical kernels*** adapted to the behavior of μ at ∞.

For $\zeta \neq 0$ the function $z \mapsto \log|z - \zeta|$ is harmonic in $\mathbb{C}\backslash\{\zeta\}$ and has a development in a Taylor series valid for $|z| < |\zeta|$:

$$\log|z - \zeta| = \operatorname{Re}\{\log|\zeta| + \operatorname{Log}(1 - z/\zeta)\} = \operatorname{Re}\left\{\log|\zeta| - \sum_{n \geq 1} \frac{1}{n}\left(\frac{z}{\zeta}\right)^n\right\}.$$

Let

$$a_n(z, \zeta) = -\operatorname{Re}\left\{\frac{1}{n}\left(\frac{z}{\zeta}\right)^n\right\} \quad (n \geq 1).$$

One has

$$|a_n(z, \zeta)| \leq \frac{1}{n}\left(\frac{\rho}{r}\right)^n \quad \text{when} \quad |z| = \rho, \quad |\zeta| = r.$$

If $0 \leq \rho < r$ then the series $\log|\zeta| + \sum_{n \geq 1} a_n(z, \zeta)$ converges absolutely and uniformly to $\log|z - \zeta|$ for $|z| \leq \rho$, $|\zeta| = r$.

For q an integer ≥ 0, let K_q be the ***canonical kernel of genus q***:

$$K_q(z, \zeta) := \log|z - \zeta| - \log|\zeta| - \sum_{1 \leq n \leq q} a_n(z, \zeta) = \sum_{n \geq q+1} a_n(z, \zeta).$$

Note the last identity only holds with $|z| = \rho < r = |\zeta|$. The function $K_q(z, \zeta) - \log|z - \zeta|$ is a harmonic function of z in the whole plane, hence K_q is subharmonic in $\mathbb{C}$.

We are now going to give some estimates for the kernels K_q.

4.6.8. Lemma. *Let $r = |\zeta| > 0$, $\rho = |z|$, then*

(1) $|K_q(z,\zeta)| \leq \dfrac{2}{q+1}\left(\dfrac{\rho}{r}\right)^{q+1} \leq \dfrac{1}{2^q(q+1)}$ *if* $\rho \leq \dfrac{r}{2}$ $\quad (q \geq 1)$

(2) $K_0(z,\zeta) \leq \log\left(1 + \dfrac{\rho}{r}\right)$

(3) $K_q(z,\zeta) \leq 2^{q+1}\dfrac{\rho^q}{r^q}$ *if* $\dfrac{r}{2} < \rho$ $\quad (q \geq 1)$

(4) *For $q \geq 1$, one always has*

$$K_q(z,\zeta) \leq 2^{q+1}\frac{\rho^q}{r^q}\inf\left\{1, \frac{\rho}{r}\right\}.$$

Proof. We have, first of all,

$$K_0(z,\zeta) = \log|z - \zeta| - \log|\zeta| = \log\left|1 - \frac{z}{\zeta}\right| \leq \log\left(1 + \frac{|z|}{|\zeta|}\right) = \log\left(1 + \frac{\rho}{r}\right).$$

For $q \geq 1$ and $0 \leq \rho < r$ we have

$$|K_q(z,\zeta)| \leq \sum_{n \geq q+1} |a_n(z,\zeta)| \leq \sum_{n \geq q+1} \frac{1}{n}\left(\frac{\rho}{n}\right)^n = \frac{1}{q+1}\left(\frac{\rho}{r}\right)^{q+1} \frac{1}{1 - (\rho/r)}.$$

If $0 \leq \rho \leq r/2$ one obtains

$$|K_q(z,\zeta)| \leq \frac{2}{q+1}\left(\frac{\rho}{r}\right)^{q+1} \leq \frac{1}{2^q(q+1)}.$$

Still, in the case $q \geq 1$, if $r > \rho > r/2$ one has

$$\begin{aligned} K_q(z,\zeta) &= \log\left|1 - \frac{z}{\zeta}\right| - \sum_{1 \leq n \leq q} a_n(z,\zeta) \leq \log\left(1 + \frac{\rho}{r}\right) + \sum_{1 \leq n \leq q} \frac{1}{n}\left(\frac{\rho}{r}\right)^n \\ &\leq \frac{\rho}{r} + \left(\frac{\rho}{r}\right)^q\left(1 + \left(\frac{r}{\rho}\right) + \cdots + \left(\frac{r}{\rho}\right)^{q-1}\right) \\ &\leq \left(\frac{\rho}{r}\right)^q (2^{q-1} + 2^q) \leq 2^{q+1}\left(\frac{\rho}{r}\right)^q. \end{aligned}$$

Finally, (4) is a consequence of (1) and (3). □

4.6.9. Theorem. *Let μ be a Borel measure ≥ 0 in $\mathbb{C}$, $\nu(t) = \mu(B(0,t))$, and $q(t)$ a function continuous on the left with values in $\mathbb{N}$, such that for some $r_0 > 0$ and every $t_0 > 0$ one has*

$$\int_{r_0}^{\infty} \left(\frac{t_0}{t}\right)^{q(t)+1} d\nu(t) < \infty.$$

Then, there are subharmonic functions u in $\mathbb{C}$ such that $\Delta u = \mu$. One of them can be explicitly written down as

$$u(z) = \frac{1}{2\pi} \int_{|\zeta| \le r_0} \log|z - \zeta|\, d\mu(\zeta) + \int_{|\zeta| \le r_0} K_{q(|\zeta|)}(z,\zeta)\, d\mu(\zeta).$$

Furthermore,

$$\int_{|\zeta| \ge r_0} |K_{q(|\zeta|)}(z,\zeta)|\, d\mu(\zeta) < \infty.$$

The second term in the decomposition of u is called the ***canonical potential for*** μ.

Before proceeding to prove Theorem 4.6.9 let us remark that given an increasing, continuous-on-the-left function $n(t)$, one can always find a function $q(t)$, increasing, continuous on the left, with values in $\mathbb{N}$ such that for some $r_0 > 0$ and every $t_0 > 0$ one has

$$\int_{r_0}^{\infty} \left(\frac{t_0}{t}\right)^{q(t)+1} dn(t) < \infty.$$

For instance, we can assume $n(t) \ge 1$ for $t \ge r_0$ (otherwise we add the value $-n(r_0) + 1$ to it) and take $q(t) = [\log(n(t) + 1)] =$ integral part of $\log(n(t) + 1)$. Then for $t \ge t_1 = t_0 e^2$ we have

$$\left(\frac{t_0}{t}\right)^{q(t)+1} \le \exp(-2\log(n(t) + 1)) = \frac{1}{(1 + n(t))^2}$$

and

$$\int_{r_0}^{\infty} \left(\frac{t_0}{t}\right)^{q(t)+1} dn(t) \le \text{constant} + \int_{t_1}^{\infty} \frac{dn(t)}{(1 + n(t))^2} < \infty.$$

PROOF OF THEOREM 4.6.9. Let us assume $q(t) = q$ is constant in the interval $]t_1, t_2]$, $r_0 \le t_1 < t_2 < \infty$. Let

$$v(z) := \frac{1}{2\pi} \int_{t_1 < |\zeta| \le t_2} K_q(z,\zeta)\, d\mu(\zeta).$$

Since K_q is subharmonic in $\mathbb{C}$ (as a function of z), v is a subharmonic function in $\mathbb{C}$. Furthermore, the measure Δv is given by

$$\Delta v = \begin{cases} d\mu & \text{in } t_1 < |z| \le t_2 \\ 0 & \text{outside.} \end{cases}$$

In fact,

$$p(z) = \frac{1}{2\pi} \int_{t_1 < |\zeta| \le t_2} \log|z - \zeta|\, d\mu(\zeta)$$

is a subharmonic function with that property and (by §4.4.24) $v - p$ is a harmonic polynomial of degree $\leq q$.

Let $q \geq 1$, if $|z| \leq \rho$, $|\zeta| = t$, $\rho \leq \frac{t_1}{2} < t_1 \leq t \leq t_2$, then $\rho \leq t/2$, hence $|K_1(z, \zeta)| \leq \frac{2}{q+1}\left(\frac{\rho}{t}\right)^{q+1}$ and, therefore,

$$|v(z)| \leq \frac{2}{q+1} \int_{t_1 < |\zeta| \leq t_2} \left(\frac{\rho}{|\zeta|}\right)^{q+1} d\mu(\zeta) = \frac{2}{q+1} \int_{t_1 < |t| \leq t_2} \left(\frac{\rho}{t}\right)^{q+1} d\nu(t).$$

Let us now consider separately the two cases:

(1) The function q is not bounded:
Let $q_1 = q(r_0)$ and, for $q > q_1$, let $r_q = \inf\{r \geq r_0 : q(r) \geq q\}$.

(2) The function is bounded:
Let q_1 as previously, $q_2 = \sup\{q(t) : t > 0\}$. For $q \leq q_2$ let $r_q = \inf\{r \geq r_0 : q(r) \geq q\}$, and for $q > q_2$, let $r_q = r_{q_2} + q - q_2$.

In either case we have that $r_{q_1} = r_0, q(t) = q$ in $r_q < t \leq r_{q+1}$, and $\lim_{q\to\infty} r_q = \infty$.

Let us define

$$u_q(z) = \int_{r_q < |\zeta| \leq r_{q+1}} K_q(z, \zeta)\, d\mu(\zeta).$$

If ρ satisfies $r_0 \leq \rho \leq \frac{r_q}{2}$, $|z| \leq \rho$, then from the preceding considerations we have

$$|u_q(z)| \leq 2 \int_{r_q < t \leq r_{q+1}} \left(\frac{\rho}{t}\right)^{q(t)+1} d\nu(t).$$

For a fixed ρ, this inequality is valid for q sufficiently large and therefore, for any $q_0 \geq q_1$, the series $\sum_{q \geq q_0} u_q(z)$ converges uniformly and absolutely in $|z| \leq \rho$. Moreover, the sum is harmonic in $|z| \leq \rho$ if $r_{q_0} > 2\rho$. In fact, to show the uniform convergence we can assume that $r_{q_0} \geq 2\rho$. For a fixed integer $n \geq q_0$ we have

$$\begin{aligned}\sum_{q_0 \leq q \leq n} |u_q(z)| &\leq 2 \sum_{q_0 \leq q \leq n} \int_{r_q < t \leq r_{q+1}} \left(\frac{\rho}{t}\right)^{q(t)+1} d\nu(t) \\ &\leq 2 \int_{r_0}^{\infty} \left(\frac{\rho}{t}\right)^{q(t)+1} d\nu(t) < \infty\end{aligned}$$

by hypothesis. If $r_{q_0} > 2\rho$ then z is not in the support of Δu_q, hence all the u_q, $q \geq q_0$ are harmonic in a neighborhood of z, and the same holds for its sum.

Therefore, the function

$$u(z) = \frac{1}{2\pi} \int_{|\zeta| \leq r_0} \log|z - \zeta|\, d\mu(\zeta) + \sum_{q \geq q_1} u_q(z)$$

is subharmonic in $|z| < \rho$ and satisfies $\Delta u = \mu$ for every $\rho > r_0$. This is exactly the function from the statement of the theorem. □

4.6.10. Lemma. *Let μ be a Borel measure ≥ 0 in $\mathbb{C}$, $v(t) = \mu(B(0,t))$, and $r_0 > 0$ be such that $\mu(\bar{B}(0,r_0)) = 0$. Assume that q is the genus of the μ and let*

$$u(z) = \int_{|\zeta|>r_0} K_q(z,\zeta)\, d\mu(\zeta).$$

The function u is subharmonic in $\mathbb{C}$ and $\Delta u = \mu$. Moreover, if $r > r_0$, $|z| \leq r$, the following inequalities hold:

$$u(z) \leq 2^{q+1}\left\{qr^q \int_{r_0}^{r} \frac{v(t)\,dt}{t^{q+1}} + (q+1)r^{q+1}\int_r \frac{v(t)\,dt}{t^{q+2}}\right\} \qquad (q \geq 1),$$

$$u(z) \leq \int_{r_0}^{r} \frac{v(t)\,dt}{t^{q+1}} + r\int_r^\infty \frac{v(t)}{t^2}\,dt \qquad (q = 0).$$

Proof. From §4.6.9 and the definition of genus we have that the function u is a well-defined subharmonic function with $\Delta u = \mu$. (Recall that $\mu = 0$ on $\bar{B}(0,r_0)$.) We only need to prove the inequalities.

If $q \geq 1$, we can use that $K_q(z,\zeta) \leq 2^{q+1}\left(\frac{|z|}{|\zeta|}\right)^q \inf\left\{1, \frac{|z|}{|\zeta|}\right\}$. Then, with $|z| = r$, we have

$$\begin{aligned} u(z) &= \int_{r_0<|\zeta|\leq r} K_q(z,\zeta)\,d\mu(\zeta) + \int_{r<|\zeta|<\infty} K_q(z,\zeta)\,d\mu(\zeta) \\ &\leq 2^{q+1}\left\{\int_{r_0<|\zeta|\leq r}\left(\frac{|z|}{|\zeta|}\right)^q d\mu(\zeta) + \int_{r<|\zeta|<\infty}\left(\frac{|z|}{|\zeta|}\right)^{q+1} d\mu(\zeta)\right\} \\ &= 2^{q+1}\left\{r^q\int_{r_0}^{r}\frac{dv(t)}{t^q} + r^{q+1}\int_r^\infty \frac{dv(t)}{t^{q+1}}\right\}, \end{aligned}$$

which is exactly the first inequality after integration by parts.

If $q = 0$ we have

$$u(z) \leq \int_{r_0}^\infty \log\left(1 + \frac{r}{t}\right) dv(t) = r\int_{r_0}^\infty \frac{v(t)}{t(t+r)}\,dt \leq \int_{r_0}^{r} \frac{v(t)}{t}\,dt + r\int_r^\infty \frac{v(t)}{t^2}\,dt.$$

□

4.6.11. Theorem. (1) *Let μ be a Borel measure in $\mathbb{C}$, $\mu \geq 0$ such that $\mu(\bar{B}(0,r_0)) = 0$ for some $r_0 > 0$, and assume that μ is of genus $q > 0$. Then*

$$u(z) = \int_{|\zeta|>r_0} K_q(z,\zeta)\,d\mu(\zeta)$$

is a subharmonic function in $\mathbb{C}$, $\Delta u = \mu$, and the order of u equals the order of v ($v(t) = \mu(B(0,t))$). If the order of v equals $q+1$ then the type of u is minimal.

(2) *Let u be a subharmonic function in* $\mathbb{C}$ *of order at most* ρ, *such that u is harmonic in a neighborhood of* $\bar{B}(0, r_0)$ *for some* $r_0 > 0$. *Let q be the genus of* $\mu = \Delta u$, *then* $q \leq [\rho]$ *and*

$$u(z) = \int_{|\zeta|>r_0} K_q(z, \zeta)\, d\mu(\zeta) + v(z),$$

where v is a harmonic polynomial of degree $\leq [\rho]$.

Proof. Let us first prove part (1). The preceding lemma shows the function u is subharmonic, $\Delta u = \mu$, and satisfies the inequality

$$u(z) \leq 2^{q+1}(q+1)r^q \left\{ \int_{r_0}^{r} \frac{\nu(t)}{t^{q+1}}\, dt + r \int_{r}^{\infty} \frac{\nu(t)}{t^{q+2}}\, dt \right\},$$

with $|z| = r \geq r_0$. (This inequality holds irrespective of whether $q = 0$ or $q \geq 1$.)

We show now that the order of u is at most the order ρ_1 of ν. Since q is the genus of μ, then the exponent of convergence ρ_1 of μ satisfies $q \leq \rho_1 \leq q + 1$. One also has $\rho_1 = \limsup\limits_{r\to\infty} \dfrac{\log \nu(r)}{\log r}$.

Let us consider first the case $\rho_1 < q + 1$. Let $\rho_1 < \lambda < q + 1$. Hence there is $c = c_\lambda > 0$ such that for every $t > 0$, $ct^\lambda \geq \nu(t)$. Therefore

$$u(z) \leq 2^{q+1}(q+1)cr^q \left\{ \int_{r_0}^{r} t^{\lambda-q-1}\, dt + r \int_{r}^{\infty} t^{\lambda-q-2}\, dt \right\} = c' r^\lambda,$$

with $c' = c2^{q+1}\left[\dfrac{1}{\lambda - q} + \dfrac{1}{q+1-\lambda}\right]$. This shows that the order of u is at most ρ_1 in this case.

Let us now assume $\rho_1 = q + 1$. By §4.6.7 we have $\lim\limits_{t\to\infty} \dfrac{\nu(t)}{t^{q+1}} = 0$ and $\displaystyle\int_{r_0}^{\infty} \frac{\nu(t)}{t^{q+2}}\, dt < \infty$. The first condition implies that

$$\lim_{r\to\infty} \frac{1}{r} \int_{r_0}^{r} \frac{\nu(t)}{t^{q+1}}\, dt = 0,$$

while the second one implies that

$$\lim_{r\to\infty} \int_{r}^{\infty} \frac{\nu(t)}{t^{q+2}}\, dt = 0.$$

On the other hand, the preceding inequality on u can be rewritten as follows:

$$u(z) \leq 2^{q+1}(q+1)r^{q+1} \left\{ \frac{1}{r} \int_{r_0}^{r} \frac{\nu(t)}{t^{q+1}}\, dt + \int_{r}^{\infty} \frac{\nu(t)}{t^{q+1}}\, dt \right\},$$

which shows that for $\varepsilon > 0$ there is r_ε sufficiently large such that if $|z| = r \geq r_\varepsilon$ then $u(z) \leq \varepsilon r^{q+1}$. That is, u is of order at most $q + 1$, and of minimal type.

It remains to show that the order of ν is exactly that of u. The first part of the proof of part (2) accomplishes this. To prove part (2) we proceed as follows: By Gauss' Theorem 4.4.28 one has for $r \geq r_0$

$$\lambda(u, 0, r) = u(0) + \frac{1}{2\pi} \int_{r_0}^{r} \frac{\nu(t)}{t} dt,$$

where, as earlier, $\nu(t) = \mu(B(0, t)) = (\Delta u)(B(0, t))$. If we let $N(t) = \displaystyle\int_{r_0}^{r} \frac{\nu(t)}{t} dt$, $(r \geq r_0)$, then the hypothesis on u says that the order of N is at most ρ. Hence this also holds for the order of ν and the exponent of convergence of μ. Therefore the genus q of μ satisfies $q \leq \rho \leq q + 1$, furthermore, $q = [\rho]$ if ρ is not an integer. Consider the auxiliary function

$$u_0(z) = \int_{|\zeta| > r_0} K_q(z, \zeta) \Delta u(\zeta).$$

It is subharmonic, of order at most ρ by part (1), and

$$u = u_0 + v, \qquad v \text{ harmonic in } \mathbb{C}.$$

We have to show that v is a harmonic polynomial of degree $\leq [\rho]$. This function is of order at most ρ by §4.5.6. Hence for every $\varepsilon > 0$ there is $c_\varepsilon > 0$ such that

$$v(z) \leq c_\varepsilon (1 + |z|)^{\rho + \varepsilon}.$$

Let f be the entire function such that $\operatorname{Re} f = v$, $\operatorname{Im} f(0) = 0$. By the Borel-Caratheodory inequality

$$|f(z)| \leq c'_\varepsilon (1 + |z|)^{\rho + \varepsilon}$$

for some $c'_\varepsilon > 0$. Therefore f is a polynomial of degree $\leq \rho$.

Let us remark that if u is of minimal type then the degree of v is going to be strictly less than ρ. □

We are now going to apply these considerations to a measure of the form $\mu = \sum_{n \geq 1} \delta_{a_n}$, where $\{a_n\}_n$ is a sequence ordered so that $|a_n| \leq |a_{n+1}|$, $|a_n| \to \infty$ as $n \to \infty$ (the following arguments will, of course, also apply if the sequence $\{a_n\}_n$ is finite, i.e., $1 \leq n < N < \infty$). Mittag-Leffler's theorem assures that there are entire functions vanishing exactly at the points a_n. (If the point a_n appears with multiplicity m_n in the sequence we mean here vanishing with multiplicity m_n.) We will see now how to give a representation to these functions in the form of an infinite product, and relate this infinite product representation to the one obtained for the canonical potential of μ.

Let us start by recalling some well-known properties of infinite products.

Let $\{u_n\}_{n \geq 1}$ be a sequence of complex numbers and set

$$p_n = (1 + u_1)(1 + u_2) \dots (1 + u_n)$$

$$p_n^* = (1 + |u_1|)(1 + |u_2|)\dots(1 + |u_n|).$$

The following inequalities hold:

(a) $p_n^* \leq \exp(|u_1| + \cdots + |u_n|) \leq 1 + (|u_1| + \cdots + |u_n|)\exp(|u_1| + \cdots + |u_n|)$.

(b) $|p_n - 1| \leq p_n^* - 1 \leq (|u_1| + |u_2 + \cdots + |u_n|)\exp\left(\sum_{k=1}^{n} |u_k|\right)$

In fact,

(a) One starts from the known inequality for $x \in \mathbb{R}$, $1 + x \leq e^x$, to obtain

$$p_n^* = (1 + |u_1|)\dots(1 + |u_n|) \leq \exp(|u_1| + \cdots + |u_n|).$$

(b) We have $p_n = 1 + \sum_{\substack{1 \leq i_j \leq n \\ 1 \leq k \leq n}} u_{i_1} \dots u_{i_k}$, hence

$$|p_n - 1| = \left| \sum_{\substack{1 \leq i_j \leq n \\ 1 \leq k \leq n}} u_{i_1} \dots u_{i_k} \right| \leq \sum_{\substack{1 \leq i_j \leq n \\ 1 \leq k \leq n}} |u_{i_1}| \dots |u_{i_k}| = p_n^* - 1.$$

The last inequality follows from $e^x - 1 \leq xe^x$.

4.6.12. Proposition. *Let $\{u_n(s)\}_{n \geq 1}$ be a sequence of functions bounded on a set S such that $\sum_{n \geq 1} |u_n(s)|$ converges uniformly on S to a bounded function. Then the functions*

$$p_n(s) = \prod_{1 \leq k \leq n} (1 + u_k(s))$$

converge uniformly to a bounded function $f(s)$. We denote it by

$$f(s) = \prod_{n \geq 1} (1 + u_n(s)).$$

The order of the factors does not alter the limit function f and $f(s) = 0$ for an $s \in S$ if and only if $u_n(s) = -1$ for some n.

Proof. We have

$$|p_n(s)| \leq 1 + |p_n(s) - 1| \leq p_n^*(s) - 1 + 1 = p_n^*(s) \leq \exp\left(\sum_{1 \leq k \leq n} |u_k(s)| \right) \leq M,$$

for some $M > 0$ since the function $\sum_{n \geq 1} |u_n(s)|$ is bounded in S. If $n \geq m$ then

$$|p_n(s) - p_m(s)| = |p_m(s)| \left| \prod_{m+1}^{n} (1 + u_k(s)) - 1 \right| \leq M\left(\prod_{m+1}^{n} (1 + |u_k(s)|) - 1 \right)$$

$$\leq M\left[\exp\left(\sum_{m+1}^{n} |u_k(s)| \right) - 1 \right]$$

$$\leq M\left(\sum_{m+1}^{n} |u_k(s)| \right) \exp\left(\sum_{m+1}^{n} |u_k(s)| \right) \leq \varepsilon M e^{\varepsilon} \leq 2\varepsilon M$$

if $0 < \varepsilon < \log 2$, $m \geq m_\varepsilon$ so that $\sum_{m+1}^{\infty} |u_k(s)| \leq \varepsilon$ for every $s \in S$, which proves the uniform convergence of the p_n.

Note we could have kept $|p_m(s)|$ instead of M in the last inequality throughout. Letting $n \to \infty$ we obtain

$$|f(s) - p_m(s)| \leq 2\varepsilon |p_m(s)|.$$

If $\varepsilon < 1/2$ then we have, for any $m \geq m_\varepsilon$, $s \in S$,

$$|f(s)| \geq |p_m(s)|(1 - 2\varepsilon).$$

Therefore $f(s) = 0$ only if $p_m(s) = 0$, that is, only if one of the u_k takes the value -1 when $1 \leq k \leq m$.

Finally, if σ is any permutation of $\mathbb{N}^*$ and we define $q_n(s) = \prod_{j=1}^{n} (1 + u_{\sigma(j)}(s))$, $m \geq m_\varepsilon$ and n is sufficiently large so that $\{1, 2, \ldots, m\} \subset \{\sigma(1), \ldots, \sigma(n)\}$ then

$$|q_n - p_m| = p_m \left| \prod_{\sigma(j) \geq m+1} (1 + u_{\sigma(j)}) - 1 \right| \leq M(e^\varepsilon - 1) \leq 2\varepsilon M$$

and it follows that $\lim_{n\to\infty} q_n = \lim_{m\to\infty} p_m = f$. □

We can now introduce the ***Weierstrass primary factors***:

$$G(z, 0) = 1 - z,$$

$$G(z, q) = (1 - z)\exp\left(z + \frac{z^2}{2} + \cdots + \frac{z^q}{q}\right) \qquad (q \in \mathbb{N}^*).$$

One sees immediately that

$$\log|G(z, 0)| = \log|1 - z|$$

and

$$\log|G(z, q)| = \log|1 - z| + \operatorname{Re}\left(z + \frac{z^2}{2} + \cdots + \frac{z^q}{q}\right).$$

Therefore, for $\zeta \neq 0$ we have

$$K_0(z, \zeta) = \log\left|G\left(\frac{z}{\zeta}, 0\right)\right|$$

and

$$K_q(z, \zeta) = \log\left|G\left(\frac{z}{\zeta}, q\right)\right| \qquad (q \geq 1).$$

Hence the §4.6.8 yields estimates for $\log\left|G\left(\frac{z}{\zeta}, q\right)\right|$ $(q \geq 0)$. We have also the following estimate.

4.6.13. Lemma. *If* $|z| < 1, q \geq 0$ *then*

$$|1 - G(z, q)| \leq |z|^{q+1}.$$

PROOF. For $q = 0$ it is obvious. For $q \geq 1$ we observe that

$$-G'(z, q) = z^q \exp\left[z + \frac{z^2}{2} + \cdots + \frac{z^q}{q}\right]$$

has a zero of order q at the origin, and its development in a Taylor series about $z = 0$ has all its coefficients ≥ 0. Since

$$1 - G(z, q) = \int_{[0,z]} G'(u, q)\, du,$$

one sees that $1 - G(z, q)$ has a zero of order $q + 1$ at $z = 0$ and if $\varphi(z) = \frac{1 - G(z,q)}{z^{q+1}}$, then $\varphi(z) = \sum_{n \geq 0} a_n z^n$ with $a_n \geq 0$. Therefore $|\varphi(w)| \leq \varphi(1)$ if $|w| \leq 1$. But $\varphi(1) = 1$ and we are done. □

If $(z_n)_{n \geq 1}$ is a sequence of nonzero complex numbers such that $|z_n| \to \infty$ as $n \to \infty$, then there always exist sequences of integers $q_n \geq 0$ such that

$$\sum_{n \geq 1} \left(\frac{r}{|z_n|}\right)^{q_n+1} < \infty \qquad \text{for every } r > 0.$$

For instance, $q_n = n - 1$ will always work; just observe that for $n \geq n_0 = n_0(r)$, one has $\frac{r}{|z_n|} \leq \frac{1}{2}$. Under this convergence condition on the q_n, we can form the infinite product $f(z) = \prod_{n \geq 1} G\left(\frac{z}{z_n}, q_n\right)$, which defines an entire function vanishing only at the points z_n and with the correct multiplicity. In fact, for $|z| \leq r$, we have

$$\left|1 - G\left(\frac{z}{z_n}, q_n\right)\right| \leq \left|\frac{z}{z_n}\right|^{q_n+1} \leq \left(\frac{r}{|z_n|}\right)^{q_n+1}$$

which assures that the series

$$\sum_{n \geq 1} \left|1 - G\left(\frac{z}{z_n}, q_n\right)\right|$$

converges uniformly over every compact subset of $\mathbb{C}$. This observation is essentially the proof of the following theorem.

4.6.14. Theorem (Weierstrass). *Every entire holomorphic function g has a representation of the form*

$$g(z) = z^m e^{h(z)} \prod_{1 \leq n \leq \omega} G\left(\frac{z}{z_n}, q_n\right),$$

where h is entire, the $(z_n)_{n=1}^{\infty}$ form the family of zeros of g distinct from $z = 0$, each of them repeated as many times as its multiplicity, $\omega \in \mathbb{N}$ if that sequence is finite, $\omega = \infty$ if not. The value $m \in \mathbb{N}$ is the multiplicity of the origin as a zero of g.

PROOF. Since $f(z) = z^m \prod_{1 \le n \le \omega} G\left(\frac{z}{z_n}, q_n\right)$ is an entire function and $g(z)/f(z)$ is an entire function without zeros, then this quotient is an exponential of an entire function. □

The representation in Weierstrass' theorem is not unique. It can be considerably simplified if the series $\sum_{n \ge 1} \frac{1}{|z_n|^\lambda} < \infty$ for some $\lambda > 0$. If q is the genus of the measure $\sum_n \delta_{z_n}$ one can take $q_n = q$ for every n and then

$$\prod_{n \ge 1} G\left(\frac{z}{z_n}, q\right)$$

converges uniformly over every compact. It is called the ***canonical product of genus*** q for the sequence $(z_n)_{n \ge 1}$. The integer q is also called the genus of the canonical product.

The subharmonic function

$$\begin{aligned} u(z) = \log\left|\prod_{n \ge 1} G\left(\frac{z}{z_n}, q\right)\right| &= \sum_{n \ge 1} \log\left|G\left(\frac{z}{z_n}, q\right)\right| \\ &= \sum_{n \ge 1} \left[\log\left|1 - \frac{z}{z_n}\right| + \operatorname{Re}\left(\frac{z}{z_n} + \frac{1}{2}\left(\frac{z}{z_n}\right)^q + \cdots + \frac{1}{q}\left(\frac{z}{z_n}\right)^q\right)\right] \end{aligned}$$

is precisely the subharmonic function from Theorem 4.6.11, (1) corresponding to the measure $2\pi \sum_{n \ge 1} \delta_{z_n}$. Hence its order is exactly the exponent of convergence of the sequence $(z_n)_{n \ge 1}$.

It follows from §4.6.11, (2) that if f is an entire function of order at most ρ, and m is the multiplicity of the origin as a zero of f, then

$$\log|f(z)| = m \log|z| + u(z) + v(z),$$

with v a harmonic polynomial of degree $\le [\rho]$, u the canonical potential. Proof 4.6.11 showed that $v = \operatorname{Re} h$, h polynomial in z, degree $h \le [\rho]$.

This argument proves the following.

4.6.15. Hadamard's Factorization Theorem. *Let f be an entire function of finite order ρ, then*

$$f(z) = z^m e^{h(z)} \prod_{n \ge 1} G\left(\frac{z}{z_n}, q\right),$$

where h is a polynomial of degree $\leq \rho$, m the multiplicity of the origin as a zero of f, and the infinite product is the canonical product of genus q, $q \leq \rho$.

4.6.16. Remark. If $(a_n)_{n\geq 1}$ is a discrete sequence of nonzero (but not necessarily distinct) complex numbers and if u is a subharmonic function in $\mathbb{C}$ such that $\Delta u = 2\pi \sum_{n\geq 1} \delta_{a_n}$, one can find, without passing through the intermediate step of infinite products, an entire function f such that $u = \log|f|$. That is, f is a function that vanishes with correct multiplicity at every a_n and nowhere else. To see how to do this, let us assume we already have a function f such that $\log|f| = u$. Let $B(0, R)$ be a disk not containing any of the a_n. Since f is never zero in $B(0, R)$ then $g = \log f$ holomorphic and $\operatorname{Re} g = \log|f|$. So that $g = u + iv$. Now $\partial\bar{g} = \partial u - i\partial v = 0$, hence

$$dg = \partial g = \partial u + i\partial v = 2\partial u.$$

If γ_z is any path from 0 to z in $B(0, R)$ we have

$$g(z) = g(0) + 2\int_{\gamma_z} \partial u.$$

We can profit from this insight to construct f. Let $\Omega = \mathbb{C}\setminus\bigcup_{n\geq 1}\{a_n\}$. Let γ_z be a path joining 0 to z in Ω and define

$$g(\gamma_z) = 2\int_{\gamma_z} \partial u.$$

If $\gamma_{1,z}$, $\gamma_{2,z}$ are two paths joining 0 to z in Ω we have

$$g(\gamma_{1,z}) - g(\gamma_{2,z}) = 2\int_{\gamma_{1,z}\bar{\gamma}_{2,z}} \partial u = 2\left(2\pi i \sum_n \operatorname{Ind}_{\gamma_{1,z}\bar{\gamma}_{2,z}}(a_n)\operatorname{Res}(\partial u, a_n)\right),$$

since ∂u is a closed 1-form of class C^∞ in Ω. Moreover,

$$\operatorname{Res}(\partial u, a_n) = \frac{1}{2\pi i}\int_{|z-a_n|=r} \partial u,$$

if $r > 0$ is so small that $\bar{B}(a_n, r)$ does not contain any a_k, with $a_k \neq a_n$. (Recall that there could be several indices k with $a_k = a_n$.) In this little disk we have

$$u(z) = m_n \log|z - a_n| + h(z),$$

where m_n = multiplicity of a_n in the sequence, and h harmonic in $B(a_n, r)$. (This is simply a consequence of the fact that $\Delta u = 2\pi \sum_{n\geq 1} \delta_{a_n}$.) Therefore, in $B(a_n, r)\setminus\{a_n\}$,

$$\partial u = \frac{m_n}{2}\frac{dz}{z - a_n} + \frac{\partial h}{\partial z}dz,$$

and

$$\operatorname{Res}(\partial u, a_n) = \frac{m_n}{2}.$$

This shows that

$$g(\gamma_{1,z}) - g(\gamma_{2,z}) = 2\pi i\, N(\gamma_{1,z}, \gamma_{2,z}),$$

for some $N(\gamma_{1,z}, \gamma_{2,z}) \in \mathbb{Z}$.

It follows that the function given by

$$f(z) = e^{g(\gamma_z)} = \exp\left(2 \int_{\gamma_z} \partial u\right)$$

is well defined for $z \in \Omega$, since it does not depend on the path γ_z which joins 0 to z in Ω.

If $z_0 \in \Omega$, $B(z_0, r) \subset \Omega$, then we can take for $z \in B(z_0, r)$ the path $\gamma_z = \gamma_{z_0} \cdot s_z$, where s_z is the straight-line segment joining z_0 to z. Hence we have

$$g(\gamma_z) = 2 \int_{\gamma_{z_0}} \partial u + 2 \int_{z_0}^{z} \partial u,$$

which is a C^∞ function of z in $B(z_0, r)$. Furthermore, $d_z g(\gamma_z) = 2\partial u$. Therefore f is C^∞ in Ω. Since $df = e^g\, dg = 2e^g \partial u$ is a $(1, 0)$ form, it follows that f is actually holomorphic in Ω. Moreover, f never vanishes in Ω.

Let us now study the behavior of f near one of the points a_n. We have seen that

$$\partial u(\zeta) = \frac{m_n}{2} \frac{d\zeta}{\zeta - a_n} + \partial h(\zeta).$$

in the punctured disk $B(z_n, r)\backslash\{a_n\}$. For $z \in B(a_n, r)\backslash\{a_n\}$ the path γ_z will be chosen as follows. Suppose first $z \notin [a_n - r, a_n]$. Let γ_0 be a fixed path in Ω joining 0 to $a_n + r$, followed by the segment $z = [a_n + r, z]$, (see Figure 4.2 in the following page) then

$$f(z) = \exp\left(2 \int_{\gamma_0} \partial u\right) \exp\left(\int_{s_z} \partial h\right) \exp\left(m_n \int_{s_z} \frac{d\zeta}{\zeta - a_n}\right).$$

The two first terms define a holomorphic nonvanishing function in $B(z_n, r)$. The last term is exactly $(z - a_n)^{m_n}$. If z lies in $[a_n - r, a_n[$ then we let $z_1 = z + i\varepsilon$, $0 < \varepsilon$ and small, and join 0 to z with the help of the obvious path passing through z_1. We obtain that for $z \in B(a_n, r)\backslash\{a_n\}$

$$f(z) = F(z)(z - a_n)^{m_n},$$

where F is a holomorphic nonvanishing function in $B(z_n, r)$. This identity shows that f extends to a holomorphic function in $B(a_n, r)$, vanishing at a_n with multiplicity exactly m_n.

It is clear then that we have constructured an entire function f such that in Ω, $\partial \log|f| = \partial u$, hence after multiplying f by a positive real number we will have $\log|f| = u$.

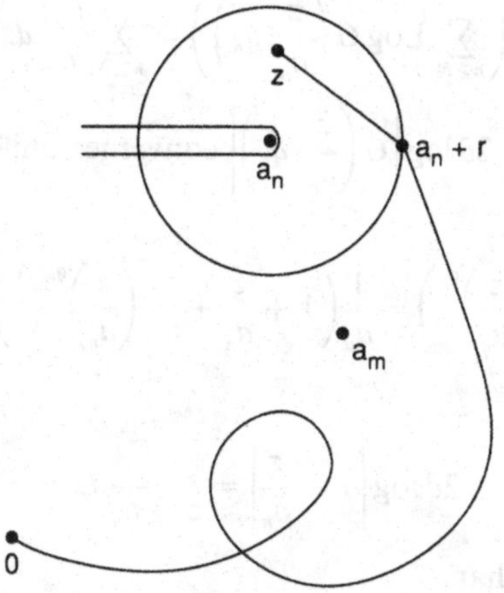

Figure 4.2

If we had known that u was given by a canonical potential,

$$u(z) = \sum_{n\geq 1} K_{q_n}(z, a_n) = \sum_{n\geq 1} \log\left|G\left(\frac{z}{a_n}, q_n\right)\right|,$$

then, given $r > 0$ let $N = N(r)$ be such that $n \geq N$ implies $|a_n| > r$. The function

$$g_n(z) := \operatorname{Log} G\left(\frac{z}{a_n}, q_n\right) = \log\left|G\left(\frac{z}{a_n}, q_n\right)\right| + iv_n(z)$$

is then holomorphic in $B(0, r)$ and

$$dg_n(z) = 2\partial \log\left|G\left(\frac{z}{a_n}, q_n\right)\right|.$$

Since $\left|\frac{z}{a_n}\right| < 1$, we can use the series representation of g_n and differentiate it term by term:

$$g_n(z) = \operatorname{Log}\left(1 - \frac{z}{a_n}\right) + \frac{z}{a_n} + \cdots + \frac{1}{q_n}\left(\frac{z}{a_n}\right)^{q_n} = -\sum_{j\geq q_n+1} \frac{1}{j}\left(\frac{z}{a_n}\right)^j,$$

$$\frac{\partial g_n}{\partial z}(z) = -\frac{1}{a_n}\cdot \sum_{j\geq q_n+1} \left(\frac{z}{a_n}\right)^{j-1},$$

hence

$$\left|\frac{\partial g_n}{\partial z}(z)\right| \leq \left|\frac{z}{a_n}\right|^{q_n} \frac{1}{|a_n| - |z|}.$$

This guarantees the uniform convergence on compact subsets of $B(0, r)$ of the series $\sum_{n\geq N}\left|\frac{\partial g_n}{\partial z}(z)\right|$ and, therefore, the identity

$$\partial\left(\sum_{n\geq N}\operatorname{Log}G\left(\frac{z}{a_n},q_n\right)\right)=\sum_{n\geq N}\frac{\partial g_n}{\partial z}dz$$

holds. Thus $2\partial u=\sum_{n\geq 1}2\partial\log\left|G\left(\frac{z}{a_n},q_n\right)\right|$ converges uniformly in every compact subset of Ω. Now

$$2\partial\operatorname{Re}\left(\frac{z}{a_n}+\cdots+\frac{1}{q_n}\left(\frac{z}{a_n}\right)^{q_n}\right)=\frac{1}{a_n}\left(1+\frac{z}{a_n}+\cdots\left(\frac{z}{a_n}\right)^{q_n-1}\right)dz=\frac{1-(z/a_n)^{q_n}}{a_n(1-z/a_n)}dz$$

and

$$2\partial\log\left|1-\frac{z}{a_n}\right|=\frac{1}{z-a_n}dz.$$

From here it follows that

$$2\partial\log\left|G\left(\frac{z}{a_n},q_n\right)\right|=\left(\frac{1}{z-a_n}+\frac{1}{a_n-z}-\frac{(z/a_n)^{q_n}}{a_n-z}\right)dz=-\left(\frac{z}{a_n}\right)^{q_n}\frac{1}{a_n-z}dz.$$

Hence

$$\begin{aligned}2\int_{\gamma_z}\partial u&=\sum_{n\geq 1}\int_{\gamma_z}\left(\frac{z}{a_n}\right)^{q_n}\frac{dz}{z-a_n}\\&=\sum_{1\leq n<N}\left[\log\left(1-\frac{z}{a_n}\right)+\left(\frac{z}{a_n}+\cdots+\frac{1}{q_n}\left(\frac{z}{a_n}\right)^{q_n}\right]+\sum_{n\geq N}g_n(z),\end{aligned}$$

where the first terms differ from g_n by $2\pi ik$, $k\in\mathbb{Z}$. It follows that

$$f(z)=e^{2\int_{\gamma_z}\partial u}=\prod_{n\geq 1}G\left(\frac{z}{a_n},q_n\right).$$

In this way one obtains Theorems 4.6.14 and 4.6.15 starting from the representation of subharmonic functions in terms of canonical potentials.

This procedure will be exploited in the second volume where we will need to construct functions satisfying more complicated growth conditions than just being of finite order.

4.6.17. Examples of Factorization. (1) The function $f(z)=\dfrac{\sin\pi\sqrt{z}}{\pi\sqrt{z}}$ is entire of order $\rho=1/2$ and has simple zeros at $z=n^2$, $n=1,2,3,\ldots$. By Theorem 4.6.14

$$f(z)=\prod_{n\geq 1}\left(1-\frac{z}{n^2}\right),$$

since $f(0)=1$. If z is replaced by z^2 one finds

$$\frac{\sin\pi z}{\pi z}=\prod_{n\geq 1}\left(1-\frac{z^2}{n^2}\right).$$

Note that the terms of this product are not primary factors, but the product of two of them, $G\left(\frac{z}{n},1\right)$ and $G\left(-\frac{z}{n},1\right)$.

(2) The function $\Gamma(z)$ can be defined by the formula

$$\frac{1}{\Gamma(z)} = ze^{\gamma z}\prod_{n\geq 1}\left(1+\frac{z}{n}\right)e^{-z/n},$$

where γ is the Euler-Mascheroni constant:

$$\gamma := \lim_{n\to\infty}\left(1+\frac{1}{2}+\cdots+\frac{1}{n}-\log n\right)$$

(see Exercise 3.5.2). We will recover some of its properties from this representation. Let us remark that the canonical product

$$P(z) = \prod_{n\geq 1}\left(1+\frac{z}{n}\right)e^{-z/n} = \prod_{n\geq 1} G\left(\frac{z}{n},1\right)$$

has genus 1 and P is entire of order 1. It is clear that Γ is a meromorphic function in $\mathbb{C}$ with simple poles at $z = 0, -1, -2, \ldots$.

The constant γ has been chosen so that $\Gamma(1) = 1$ since

$$\prod_{n\geq 1}\left(1+\frac{1}{n}\right)e^{-1/n} = \lim_{N\to\infty}\prod_{1}^{N}\left(1+\frac{1}{n}\right)e^{-1/n} = \lim_{N\to\infty}(N+1)e^{-\sum_1^N(1/n)} = e^{-\gamma}.$$

The function $P(z-1)$ is also entire of order 1, and has simple zeros at $z = 0, -1, -2, \ldots$. By §4.6.14 one can write

$$P(z-1) = ze^{Az+B}P(z),$$

for some constants A and B. These constants can be computed taking logarithmic derivatives of both sides. One has the identity

$$\sum_{n\geq 1}\left(\frac{1}{z-1+n}-\frac{1}{n}\right) = \frac{1}{z} + A + \sum_{n\geq 1}\left(\frac{1}{z+n}-\frac{1}{n}\right).$$

Both series are uniformly and absolutely convergent in compact subsets of $\mathbb{C}\backslash\mathbb{N}^*$. By rearranging the first one we conclude that $A = 0$. Taking now $z = 1$, we see that

$$1 = P(0) = e^B P(1) = e^B e^{-\gamma},$$

hence

$$P(z-1) = ze^{\gamma}P(z).$$

It follows that

$$\Gamma(z+1) = z\Gamma(z).$$

In particular, $\Gamma(n) = (n-1)!$ for every n integer ≥ 1.

Finally, from

$$P(z)P(-z) = \frac{\sin \pi z}{\pi z}$$

one obtains the important formula

$$\Gamma(z)\Gamma(1-z) = \frac{\pi}{\sin \pi z}.$$

Exercises 4.6

1. Find the exponent of convergence of the sequences $(e^n)_{n\in\mathbb{N}}$, $(n^a)_{n\in\mathbb{N}}$ $(a > 0)$, $(\log n)_{n\in\mathbb{N}^*}$.

2. Prove that the following infinite product expansions are correct:

(a) $\sin z = z \prod_{n=1}^{\infty}\left(1 - \frac{z^2}{n^2\pi^2}\right)$; (b) $\cos z = \prod_{n=1}^{\infty}\left(1 - \frac{z^2}{((n-1/2)\pi)^2}\right)$

(c) $\cosh z = \prod_{n=1}^{\infty}\left[1 + \left(\frac{2z}{(2n+1)\pi}\right)^2\right]$ (d) $e^z - 1 = ze^{z/2}\prod_{n=1}^{\infty}\left(1 + \frac{z^2}{4n^2\pi^2}\right)$

(e) $\frac{\sin(\lambda - z)}{\sin \lambda} = e^{-z\,\mathrm{cotan}\,\lambda}\prod_{n\in\mathbb{Z}}\left(1 - \frac{z}{\lambda + n\pi}\right)e^{z/(\lambda+n\pi)}$ $(\lambda \notin \pi\mathbb{Z})$

(f) $\cosh z - \cos z = z^2 \prod_{n=1}^{\infty}\left(1 + \frac{z^4}{4n^4\pi^4}\right)$

(g) $e^{z^2} + e^{2z-1} = 2e^{(z^2+2z-1)/2}\prod_{n=1}^{\infty}\left(1 + \frac{(z-1)^4}{\pi^2(2n-1)^2}\right).$

3. Let f be an entire function of order $\rho < \infty$, $\rho \notin \mathbb{N}$.
   (a) Show that f has infinitely many zeros.
   (b) More precisely, show that if $v_a(t) = \frac{1}{2\pi}\Delta(\log|f - a|)(B(0,t))$, then the order of growth of v_a equals ρ for every $a \in \mathbb{C}$.
   (c) Conclude that the function $f(z) = \frac{\sin\sqrt{z}}{\sqrt{z}}$ has infinitely many fixed points.
   (d) Give an example showing that (b) may not hold if $\rho \in \mathbb{N}^*$.

4. Let f be an entire function of finite order $\rho \in \mathbb{N}^*$ and ρ_a the order of growth of v_a defined in Exercise 4.6.3. Show that there could be at most one value $a \in \mathbb{C}$ such that $\rho_a < \rho$. (Hint: study the proof of Proposition 4.5.12.)

5. Let $f(z) = e^z + p(z)$, p polynomial $\neq 0$. Show f has infinitely many zeros, using Hadamard's factorization theorem.

6. Let f be a transcendental entire function of order 0. Show that for every $a \in \mathbb{C}$, the equation $f(z) = a$ has infinitely many solutions.

7. Given a sequence $(z_n)_{n\geq 1}$ which has a finite exponent of convergence, find explicitly a meromorphic function with simple poles exactly at the points of the sequence.

*8. Let f be an entire function which is real on the axis and has genus $q = 0$ or 1. Show that all the zeros of f' are real. (Hint: compute the imaginary part of f'/f for $z \in \mathbb{C}\setminus\mathbb{R}$.)

9. (i) From Exercise 2.5.20 we know that if f is entire and $f(0) \neq 0$, then for any

$r > 0$ we have

$$\frac{1}{2\pi}\int_{-\pi}^{\pi} e^{-i\theta}\log|f(re^{i\theta})|\,d\theta = \frac{1}{2}r\frac{f'(0)}{f(0)} + \frac{1}{2}\sum_{k=1}^{n}\left[\frac{r}{a_k} - \frac{\bar{a}_k}{r}\right],$$

where $a_1, \ldots, a_n$ are the zeros of f in $\bar{B}(0, r)$ counted with multiplicities. Use this identity to show that if f is an entire function of exponential type (i.e., order < 1 or order 1 and finite type) then there is a constant C, $0 \le C < \infty$, such that

$$\left|\sum_{|a_k| \le r} \frac{1}{a_k}\right| \le C, \qquad \text{for all } r > 0.$$

Given an example where C can be taken to be zero.

*(ii) Find a similar identity to prove that if f is an entire function of order $n \in \mathbb{N}^*$ and finite type, there is a constant $0 \le C < \infty$ such that

$$\left|\sum_{0 < |a_k| \le r} \frac{1}{a_k^n}\right| \le C, \qquad \text{for all } r > 0,$$

where the sequence $(a_k)_{k \ge 1}$ is the sequence of zeros of f, counted with multiplicities.

10. Let $f(z) = z\prod_{k=1}^{\infty}\left(1 + \frac{z}{k}\right)e^{-z/k}$. What is the order and type of f? (Hint: use the statement of the previous exercise.)

11. Let $(a_k)_{k \ge 1}$ be an infinity sequence of complex numbers $0 < |a_1| \le |a_2| \le \ldots$ with exponent of convergence $\tau = 1$. Show that if the genus $q = 0$, then the canonical product defines a function of order 1 and minimal type.

*12. Let $f(z) = \prod_{k=1}^{\infty}\left(1 + \frac{z}{a_k}\right)$, $a_k > 0$, be a function of order $\rho < 1$. Let ν be the counting function $\nu(t) = \#\{k : |a_k| < t\}$. Show that for $x > 0$ one has

$$\log f(x) = x\int_0^{\infty} \frac{\nu(t)\,dt}{t(x + t)}.$$

Conclude that if $\lim_{t \to \infty} \frac{\nu(t)}{t^{\rho}} = \lambda$, then

$$\lim_{x \to \infty} \frac{\log f(x)}{x^{\rho}} = \frac{\pi\lambda}{\sin \pi\rho}.$$

*13. Let $(a_k)_{k \ge 1}$ be a sequence of complex numbers satisfying $0 < |a_1| \le |a_2| \le \cdots$, of genus 1, and $\nu(r) = \#\{j : |a_j| < r\} \le Ar + B$ for some positive constants A, B. Assume further that the condition of Exercise 4.6.9(i) holds, that is, for some constant $C < \infty$ one has

$$\left|\sum_{|a_k| < r} \frac{1}{a_k}\right| \le C, \qquad \text{for all } r > 0.$$

Show that the canonical product corresponding to this sequence defines a function of order 1 and finite type. (Hint: estimate the partial products $\sum_{k \le \nu(r)} G(z/a_k, 1)$ instead of each term separately.) Generalize this method to the case of order $\rho \in \mathbb{N}^*$.

*14. Let f be a transcendental entire function of order $\rho < 1$, ν the counting function of its zeros $\neq 0$. Show that for $r \to \infty$

(a) $\log M(|f|, r) \leq r \displaystyle\int_0^\infty \frac{\nu(t)}{t(r+t)} dt + O(\log r)$

(b) $\log M(|f|, r) \leq \displaystyle\int_0^r \frac{\nu(t)}{t} dt + r \int_r^\infty \frac{\nu(t)}{t^2} dt + O(\log r)$

(c) There is $\theta = \theta(r)$, $0 \leq \theta \leq 1$, such that

$$\log M(|f|, r) \leq \int_0^r \frac{\nu(t)}{t} dt + \theta r \int_r^\infty \frac{\nu(t)}{t^2} dt + O(\log r).$$

*15. Let f be a transcendental function, $f(0) \neq 0$, such that $\log M(|f|, r) \leq B \log^2 r$ for $r \gg 1$.

(i) Show that for any $\lambda > 1$ one has

$$(\lambda - 1)\nu(r) \log r \leq \int_r^{r^\lambda} \frac{\nu(t)}{t} dt \leq B\lambda^2 \log^2 r$$

and there is $B' \leq 4B$ such that

$$\nu(r) < B' \log r.$$

(ii) Use the previous exercise to show that

$$\log M(|f|, r) \leq \int_0^r \frac{\nu(t)}{t} dt + O(\log r).$$

(iii) Let $\nu(t, \zeta) := \#\{w : 0 < |w| < t, f(w) = \zeta\}$, $N(r, \zeta) := \displaystyle\int_0^r \frac{\nu(t, \zeta)}{t} dt$. Show that

$$\lim_{r \to \infty} \frac{\log M(|f|, r)}{N(r, \zeta)} = 1 \quad \text{(for every } \zeta \in \mathbb{C}\text{)}.$$

*16. Let f be a transcendental function of order 0, $f(0) \neq 0$. Show that

$$\liminf_{r \to \infty} \left(r \int_r^\infty \frac{\nu(t)}{t^2} dt \Big/ \int_0^r \frac{\nu(t)}{t^2} dt \right) = 0.$$

(Hint: let $N(r) := \displaystyle\int_r^\infty \frac{\nu(t)}{t} dt$. If the lim inf is $\alpha > 0$, then $\displaystyle\int_r^\infty N(t) \frac{dt}{t^2} = \kappa r^{1/(\alpha+1)}$, $\kappa > 0$. Conclude that there is a sequence $r_k \to \infty$ and $\beta > 0$ such that $N(r_k) \geq \beta r_k^{\alpha/1+\alpha}$. Show that in this case f cannot have order 0.)

§7. Green Functions and Harmonic Measure

We return to the Dirichlet problem. A nonempty family $\mathscr{F}$ of subharmonic functions defined in an open connected set $\Omega \subseteq \mathbb{C}$ is said to be a ***Perron family*** if it satisfies the following two conditions

(1) $u, v \in \mathscr{F}$ then $\max\{u, v\} \in \mathscr{F}$;
(2) if $\bar{B}(z_0, r) \subseteq \Omega$, $u \in \mathscr{F}$ then the ***Poisson modification*** of u, $\tilde{P}u$, also belongs to $\mathscr{F}$.

Here $\tilde{P}u = \tilde{P}_{z_0,r}u$ is defined as follows:

$$\tilde{P}u(z) = \begin{cases} u(z) & \text{if } z \in \Omega \backslash B(z_0, r) \\ P(u|\partial B(z_0, r))(z) & \text{if } z \in B(z_0, r). \end{cases}$$

The main result about Perron families is the following.

4.7.1. Proposition. *Let $\mathscr{F}$ be a Perron family in Ω. Then, the function $v : \Omega \to]-\infty, +\infty]$ defined by*

$$v(z) = \sup\{u(z) : u \in \mathscr{F}\}$$

is either identically equal to $+\infty$ or it is a harmonic function in Ω.

PROOF. Assume there is a $z_0 \in \Omega$ such that $v(z_0) = +\infty$. We can then find a sequence $\{u_n\}_{n \geq 1} \subseteq \mathscr{F}$, $u_n(z_0) \to \infty$. Replacing u_n by $v_n = \max\{u_1, \ldots, u_n\}$ we get an increasing sequence in $\mathscr{F}$, with $v_n(z_0) \nearrow +\infty$. Fix a disk $\bar{B}(z_0, r) \subseteq \Omega$ and consider $\tilde{P}_{z_0,r}v_n = \tilde{P}v_n \in \mathscr{F}$. We have

$$\tilde{P}v_n(z_0) = \lambda(v_n, z_0, r) \geq v_n(z_0),$$

hence $\tilde{P}v_n(z_0) \to \infty$. Furthermore, the sequence $\tilde{P}v_n$ is also increasing and the functions $\tilde{P}v_n|B(z_0, r)$ are harmonic. By the Harnack inequalities it follows that $\tilde{P}v_n(z) \to \infty$ for every $z \in B(z_0, r)$. It follows that $v \equiv \infty$ in $B(z_0, r)$. In other words, we have just shown that $E = \{z \in \Omega : v(z) = +\infty\}$ is a nonempty open set.

If $z_1 \in \bar{E} \cap \Omega$, let $r_1 > 0$ such that $\bar{B}(z_1, r_1) \subseteq \Omega$. There exists at least one point $z_2 \in B(z_1, r_1/2) \cap E$. Then $z_1 \in B(z_2, r_1/2) \subseteq \bar{B}(z_2, r_1/2) \subseteq \Omega$. But $v(z_2) = +\infty$ and $v \equiv +\infty$ in $B(z_2, r_1/2)$ by the preceding argument. Therefore $v(z_1) = \infty$. That is, $z_1 \in E$. Since E is now open, closed, and nonempty we must have $E = \Omega$.

Let us assume now that $v < \infty$ everywhere in Ω and show that it is harmonic.

Let $z_0 \in \Omega$, $\bar{B}(z_0, r) \subseteq \Omega$. Let $\{u\}_{n \geq 1} \subseteq \mathscr{F}$ be such that $u_n(z_0) \to v(z_0)$. We can also assume here that u_n is an increasing sequence and that $u_n = \tilde{P}u_n$. Therefore, by Harnack's theorem, $\tilde{P}u_n|B(z_0, r) \nearrow u$ which is harmonic in $B(z_0, r)$ and $u(z_0) = v(z_0)$.

Let z_1 be a different point in $B(z_0, r)$. We can similarly find an increasing sequence $\{v_n\}_{n \geq 1} \subseteq \mathscr{F}$ such that $v_n(z_1) \nearrow v(z_1)$. Let $w_n = \tilde{P}(\max\{u_n, v_n\})$ ($\tilde{P} = \tilde{P}_{z_0,r}$). Then w_n is also harmonic in $B(z_0, r)$, $w_n \in \mathscr{F}$ so that $w_n \leq v$ everywhere, and $w_n \geq u_n$, $w_n \geq v_n$ everywhere also. Moreover, since both $\{u_n\}_{n \geq 1}$ and $\{v_n\}_{n \geq 1}$ are increasing sequences, $\{w_n\}_{n \geq 1}$ is an increasing sequence. By Harnack's theorem $w_n|B(z_0, r) \nearrow w$, which is harmonic in $B(z_0, r)$. We also have for $z \in B(z_0, r)$

$$w(z) = \lim_{n\to\infty} w_n(z) \geq \lim_{n\to\infty} u_n(z) = u(z).$$

Therefore $u - w$ is a nonpositive harmonic function in $B(z_0, r)$. But $u(z_0) \leq w(z_0) \leq v(z_0) = u(z_0)$. By the maximum principle, $u \equiv w$. On the other hand

$$v(z_1) \geq w(z_1) \geq \limsup_{n\to\infty} v_n(z_1) = v(z_1).$$

Hence $u(z_1) = v(z_1)$ also, but u is a fixed harmonic function and z_1 was arbitrary, $v \equiv u$ in $B(z_0, r)$. In other words, v is harmonic in $B(z_0, r)$. We conclude that v is harmonic everywhere in Ω. □

4.7.2. Definition. Let Ω be an open subset of Ω and $x_0 \in \partial\Omega$. We say there is a ***barrier*** at x_0 if for every $\delta > 0$, sufficiently small, we can find a function b_δ such that

(1) $-b_\delta$ is subharmonic in Ω.
(2) $b_\delta \geq 0$ in Ω.
(3) $b_\delta \geq 1$ for those $z \in \Omega$ such that $|z - x_0| \geq \delta$, and
(4) $\lim_{\substack{z\in\Omega\\ z\to x_0}} b_\delta(z) = 0$.

Let $h: \partial\Omega \to \mathbb{R}$ be a bounded function. The "Perron family of h" is the family $\mathscr{F}(h)$ of subharmonic functions u in Ω such that of every $\zeta \in \partial\Omega$ we have

$$\limsup_{\substack{z\to\zeta\\ z\in\Omega}} u(z) \leq h(\zeta).$$

We set $v_h(z) := \sup\{u(z) : u \in \mathscr{F}(h)\}$.

4.7.3. Proposition. *The function v_h is harmonic in Ω. If h is continuous at the point $x_0 \in \partial\Omega$ and there is a barrier at x_0, then*

$$\lim_{z\to x_0} v_h(z) = h(x_0).$$

4.7.4. Corollary. *If there is a barrier at each point of $\partial\Omega$, then the Dirichlet problem is solvable in Ω. That is, for every bounded continuous function f on $\partial\Omega$, there is a function u harmonic in Ω and continuous in $\bar{\Omega}$ such that $u|\partial\Omega = f$.*

Proof. To simplify the notation, let $v = v_h$. We can assume $h \geq 0$. Otherwise we add a constant. Each function in $\mathscr{F}(h)$ will be bounded above by $M = \sup\{h(\zeta) : \zeta \in \partial\Omega\} < +\infty$.

It is easy to see that $\mathscr{F}(h)$ is a Perron family and, since $v \leq M$, it follows from §4.7.1 that v is harmonic.

Let $\varepsilon > 0$ and choose $\delta > 0$ such that $|h(x_0) - h(y)| < \varepsilon/2$ if

$$y \in \partial\Omega \cap \bar{B}(x_0, \delta).$$

Let $b = b_\delta$ be a barrier function at x_0 for that δ. Consider the auxiliary function

$$s(z) = h(z_0) - \varepsilon - 2Mb(z) \qquad (z \in \Omega).$$

Then $s(z) \le h(x_0) - \varepsilon$ everywhere in Ω. If $y \in \partial\Omega \cap \bar{B}(x_0, \delta)$ then we have $h(x_0) - \varepsilon < h(y) - \varepsilon/2 < h(y)$, hence

$$\limsup_{z \to y} s(z) < h(y).$$

If $y \in \partial\Omega \cap (\bar{B}(x_0, \delta))^c$, then for $z \in \Omega$ sufficiently close to y we will have $|z - x_0| \ge \delta$, hence $b(z) \ge 1$ and

$$\limsup_{z \to y} s(z) \le h(x_0) - \varepsilon - 2M < h(x_0) - \varepsilon \le h(y).$$

Therefore $s \in \mathscr{F}(h)$. As a consequence we obtain

$$\liminf_{z \to x_0} v(z) \ge \liminf_{z \to x_0} s(z) \ge h(x_0) - \varepsilon,$$

and since $\varepsilon > 0$ is arbitrary,

$$\liminf_{z \to x_0} v(z) \ge h(x_0).$$

It is clear that if $h_1 = h + k$, k constant, then $v_{h_1} = v_h + k$. Therefore, this last inequality does not depend on h being ≥ 0, just on being bounded. We can now apply the same reasoning to $-h$ and define w as $w = -v_{-h}$, that is

$$-w(z) = \sup\{u(z) : u \in \mathscr{F}(-h)\},$$

The function w is harmonic in Ω and

$$\liminf_{z \to x_0} (-w(z)) \ge -h(x_0),$$

or

$$\limsup_{\substack{z \to x_0 \\ z \in \Omega}} w(z) \le h(x_0).$$

To conclude the proof we need the same estimate with w replaced by v. If $u_1 \in \mathscr{F}(h)$, $u_2 \in \mathscr{F}(-h)$ then $u_1 + u_2$ is subharmonic in Ω and for any $y \in \partial\Omega$ we have

$$\begin{aligned}\limsup_{z \to y} (u_1(z) + u_2(z)) &\le \limsup_{z \to y} u_1(z) + \limsup_{z \to y} u_2(z) \\ &\le h(y) + (-h(y)) = 0.\end{aligned}$$

Therefore $u_1 + u_2 \le 0$ in Ω. Hence,

$$(v - w)(z) = \sup\{u_1(z) + u_2(z) : u_1 \in \mathscr{F}(h), u_2 \in \mathscr{F}(-h)\} \le 0,$$

and we can conclude that

$$\limsup_{\substack{z \to x_0 \\ z \in \Omega}} v(z) \le h(x_0).$$

This proves that $\lim_{z \to x_0} v(z) = h(x_0)$. □

One can extend to open subsets of S^2 the notion of subharmonic functions, harmonic functions, Perron families, and barrier functions. For that purpose, one only has to observe that these notions are local and invariant under biholomorphisms. In the neighborhood of $\infty \in S^2$ one considers functions of $w = \frac{1}{z}$, with $\frac{1}{\infty} = 0$.

4.7.5. Proposition. *Let Ω be a connected open subset of S^2 and $x_0 \in \partial\Omega$. If there is a connected compact set K containing more than one point such that $x_0 \in K$ and $K \subseteq \Omega^c$, then there is a barrier at x_0.*

PROOF. Let $x_1 \neq x_0, x_1 \in K$. There is a Moebius transformation that sends x_0 to ∞ and x_1 to 0. We can therefore assume $K \subseteq S^2\backslash\Omega$ and K contains 0 and ∞. Since K is connected, every connected component of $S^2\backslash K$ is simply connected. Let D_0 be the component of $S^2\backslash K$ such that $\Omega \subseteq D_0$. There is a determination $\log_0$ of the logarithm in D_0. Let R_0 be the image of D_0 by the logarithm function. R_0 is biholomorphic to D_0, hence open, connected, and simply connected. We can assume that R_0 intersects the imaginary axis. If not, we replace $\log_0$ by $\log_0 + c$, for some convenient constant c. We have

$$R_0 \cap \{it : t \in \mathbb{R}\} = \bigcup_{j\geq 1}]i\alpha_j, i\beta_j[,$$

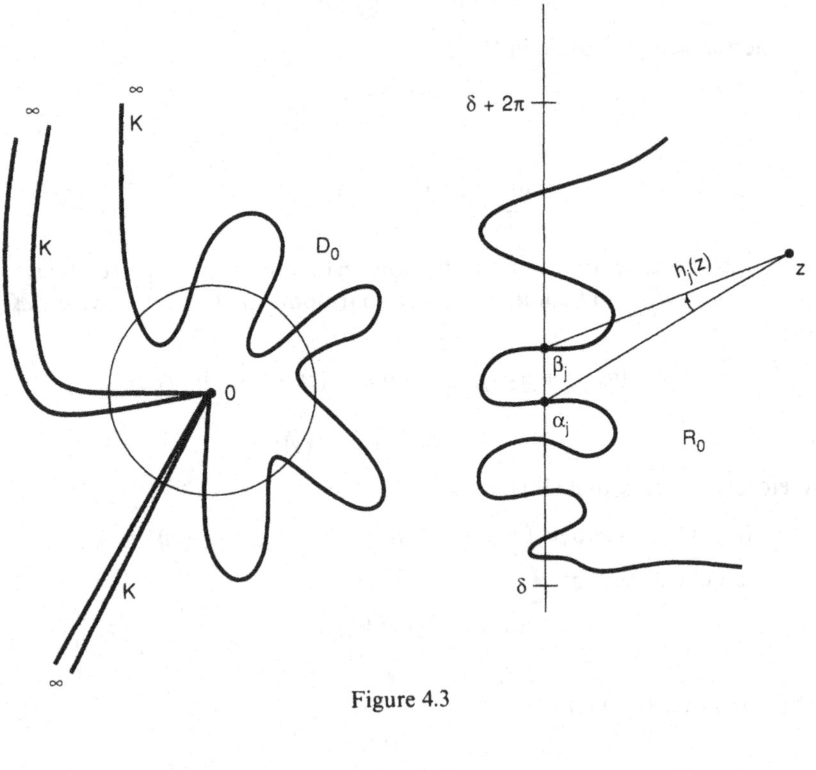

Figure 4.3

where

$$\sum_{j\geq 1} (\beta_j - \alpha_j) \leq 2\pi.$$

Let $h_j(w) = \operatorname{Im}\left(\operatorname{Log}\left(\frac{w - i\alpha_j}{w - i\beta_j}\right)\right) = \operatorname{Arg}\frac{w - i\alpha_j}{w - i\beta_j}$, where $\operatorname{Log}\zeta$ is, as usual, the principal branch of the logarithm for $\operatorname{Re}\zeta > 0$. The functions h_j are harmonic >0 for $\operatorname{Re} w > 0$. If we set

$$\varphi(t) = \operatorname{Im}(\operatorname{Log}(w - i(t\alpha_j + (1 - t)\beta_j))),$$

when $0 \leq t \leq 1$, $\operatorname{Re} w > 0$, then $\varphi(1) - \varphi(0) = h_j(w)$ and

$$\varphi'(t) = \frac{(\operatorname{Re} w)(\beta_j - \alpha_j)}{(\operatorname{Re} w)^2 + (\operatorname{Im} w - (t\alpha_j + (1 - t)\alpha_j))^2} > 0.$$

This implies that $h_j(w) > 0$ and

$$h_j(w) = \int_{\alpha_j}^{\beta_j} \frac{\operatorname{Re} w}{(\operatorname{Re} w)^2 + (\operatorname{Im} w - u)^2}\, du.$$

As a consequence, for any n we have

$$\sum_{1\leq j\leq n} h_j(w) \leq \int_{-\infty}^{\infty} \frac{\operatorname{Re} w}{(\operatorname{Re} w)^2 + (\operatorname{Im} w - u)^2}\, du = \pi.$$

Therefore the series $\sum_{j\geq 1} h_j(w)$ determines a harmonic function in $\operatorname{Re} w > 0$ and $h(w) = -\frac{1}{\pi}\sum_{j\geq 1} h_j(w)$ satisfies $-1 < h(w) < 0$.

It is easy to verify that if $t \in\,]\alpha_j, \beta_j[$ then

$$\lim_{\substack{w\to it\\ \operatorname{Re} w>0}} h_k(w) = \delta_{jk}\pi,$$

where δ_{jk} is the Kronecker delta, $\delta_{jj} = 1$, and $\delta_{jk} = 0$ if $k \neq j$. Hence, if $it \in R_0$,

$$\lim_{\substack{w\to it\\ \operatorname{Re} w>0}} h(w) = -1.$$

Furthermore,

$$\lim_{\substack{|w|\to\infty\\ \operatorname{Re} w>0}} h(w) = 0.$$

Let us now set

$$g(w) = \begin{cases} -1 & \text{if } \operatorname{Re} w \leq 0,\ w \in R_0 \\ h(w) & \text{if } \operatorname{Re} w > 0,\ w \in R_0. \end{cases}$$

The function g is continuous and subharmonic in R_0, $-1 \leq g < 0$, and $\lim_{|w|\to\infty} g(w) = 0$ (when $w \in R_0$, $\operatorname{Re} w > 0$). Let now

$$G(z) = g(\log_0 z) \qquad (z \in D_0.)$$

The function G is subharmonic in D_0, $-1 \leq G < 0$ and $\lim_{\substack{|z|\to\infty \\ z \in D_0}} G(z) = 0$. It is quite possible that $G \to 0$ at certain points in ∂D_0 located in the finite plane. To eliminate this possibility, let t_n be a strictly increasing sequence of positive reals, $0 < t_n \nearrow +\infty$, such that the lines $\operatorname{Re} w = t_n$ intersect R_0 (i.e., the circles of center 0 and radius e^{t_n} intersect D_0). Let g_n be the function constructed by the preceding method so that $g_n = -1$ in $\operatorname{Re} w < t_n$. Set

$$H(z) = \sum_{n \geq 1} \frac{1}{2^n} g_n(\log_0 z) \quad (z \in D_0).$$

This series converges uniformly to a function H continuous in D_0, subharmonic, $-1 \leq H < 0$ and $\lim_{\substack{|z|\to\infty \\ z \in D_0}} H(z) = 0$.

If $y \in \partial\Omega$ then $\log_0 y \in \partial R_0$ and $g_n(\log_0 y) = -1$ for $n \geq n_0(y)$, hence

$$\lim_{\substack{z \to y \\ z \in \Omega}} H(z) < 0.$$

For $M > 0$ sufficiently large, denote by

$$\rho = \sup_{|y| \leq M} \left(\limsup_{\substack{z \to y \\ z \in \Omega}} H(z) \right) < 0.$$

Then the function H/ρ is a barrier at $x_0 = \infty$. □

4.7.6. Corollary. *If no component of $\partial\Omega$ is reduced to a single point then the Dirichlet problem is solvable in Ω.*

Let us recall that in the case of the unit disk $B(0, 1)$ the solution of the Dirichlet problem was done via the Poisson integral representation of the solution:

$$Pf(z) = \frac{1}{2\pi} \int_{|\zeta|=1} P(z, \zeta) f(\zeta) \, ds(\zeta),$$

where $P(z, \zeta)$ is the Poisson kernel for $B(0, 1)$. We will see that such a representation is also valid for every open set for which the Dirichlet problem is solvable; this follows from an abstract functional analysis argument. We will introduce then the ***Green function*** which will allow us, for open sets with C^∞ boundary (or just sufficiently regular boundary), to have a representation of the solution of the Dirichlet problem almost as explicit as in the case of the unit disk.

The starting point of this abstract construction is the observation that $\frac{1}{2\pi} P(z, \zeta) \, ds(\zeta)$ is a measure ≥ 0 in $\partial B(0, 1)$ of total mass 1. Let now Ω be an open set in S^2 for which the Dirichlet problem is solvable and denote $\partial_\infty\Omega$ its boundary in S^2. The space $\mathscr{C}(\partial_\infty\Omega, \mathbb{C})$ of complex valued continuous functions in $\partial_\infty\Omega$ is a Banach space with the uniform norm

$$\|f\| = \sup_{\zeta \in \partial_\infty\Omega} |f(\zeta)|.$$

It is also easy to verify that the space $\mathfrak{h}(\bar{\Omega})$ of continuous functions in $\bar{\Omega}$ which are harmonic in Ω is also a Banach space with the uniform norm. If we denote by $P(f)$ the solution of the Dirichlet problem in Ω with f as boundary data, then

$$P : \mathscr{C}(\partial_\infty\Omega, \mathbb{C}) \to \mathfrak{h}(\bar{\Omega})$$

is an isometry. For a function $f \geq 0$ we also have $Pf \geq 0$. Let $z \in \bar{\Omega}$, the linear functional

$$\varepsilon_z : f \mapsto Pf(z)$$

is continuous and positive in $\mathscr{C}(\partial_\infty\Omega, \mathbb{C})$ and, therefore, by the Riesz' representation theorem there is a unique nonnegative measure in $\partial_\infty\Omega$, denoted ω_z, such that

$$Pf(z) = \int_{\partial_\infty\Omega} f(\zeta)\, d\omega_z(\zeta).$$

Taking $f \equiv 1$ we have $Pf \equiv 1$, which shows that $\omega_z(\partial_\infty\Omega) = 1$. It is also clear that if $z \in \partial_\infty\Omega$ then $\omega_z = \delta_z =$ Dirac measure at z.

4.7.7. Definition. The measure ω_z just introduced is called the ***harmonic measure*** of $\partial_\infty\Omega$ at the point z.

Let us now introduce the Green function of a connected open subset Ω of S^2.

4.7.8. Definition. Let Ω be an open connected subset of S^2 and $z_0 \in \Omega$. The ***Green function*** *of Ω with pole at z_0* is a harmonic function $g(z; z_0, \Omega)$ in $\Omega\backslash\{z_0\}$, which is continuous in $\bar{\Omega}\backslash\{z_0\}$, zero on $\partial_\infty\Omega$, and such that

(i) if $z_0 \neq \infty$, $g(z; z_0, \Omega) + \log|z - z_0|$ is harmonic in Ω.
(ii) if $z_0 = \infty$, $g(z; \infty, \Omega) - \log|z|$ is harmonic in Ω.

4.7.9. Remarks. (1) If $\Omega = \mathbb{C}$ and $z_0 \in \mathbb{C}$ then $g(z; z_0, \mathbb{C}) = \log\dfrac{1}{|z - z_0|}$.

(2) If $\Omega = B(0, 1)$ and $z_0 = 0$ then $g(z; B(0, 1)) = \log\dfrac{1}{|z|}$.

(3) There is at most one Green function $g(z; z_0, \Omega)$ for a given Ω, $z_0 \in \Omega$. This function is >0 by the maximum principle.

(4) If $\varphi : \bar{\Omega}_1 \to \bar{\Omega}_2$ is a homeomorphism which is holomorphic in Ω_1 (and hence a biholomorphic mapping from Ω_1 and Ω_2), then Ω_1 admits a Green function with pole at $z_0 \in \Omega_1$ if and only if Ω_2 admits a Green function with

pole at $\varphi(z_0)$, and

$$g(\varphi(z); \varphi(z_0), \Omega_2) = g(z; z_0, \Omega_1).$$

For instance, a biholomorphic mapping of $B(0, 1)$ onto itself such that z_0 corresponds to 0 is given by the Moebius transformation

$$\varphi(z) = \frac{z - z_0}{1 - \bar{z}_0 z} = w.$$

Therefore,

$$g(z; z_0, B(0, 1)) = g(w; 0, B(0, 1)) = \log\frac{1}{|w|} = \log\left|\frac{1 - \bar{z}_0 z}{z - z_0}\right|.$$

4.7.10. Proposition. *Let Ω be an open connected subset of S^2 for which the Dirichlet problem is solvable. Let z_0 be any point in Ω, then the Green function of Ω with pole at z_0 exists.*

Proof. By means of the transformation $z \mapsto \dfrac{1}{z - z_0}$ one can assume $z_0 = \infty$ and $\partial\Omega$ is a bounded subset of $\mathbb{C}$. Let h be the harmonic function in Ω taking the value $-\log|z|$ on $\partial\Omega$. We let

$$g(z; \infty, \Omega) = h(z) + \log|z| \quad (z \in \Omega).$$

It is easy to see this is the Green function. □

4.7.11. Definition. Let Ω be a connected open set in S^2. A regular exhaustion of Ω is a sequence $\{\Omega_n\}_{n\geq 1}$ of open connected sets Ω_n such that $\Omega_n \subset\subset \Omega_{n+1}$, $\bigcup_{n\geq 1} \Omega_n = \Omega$ and $\partial\Omega_n$ is piecewise regular of class C^∞.

4.7.12. Proposition. Every connected open set in S^2 has a regular exhaustion.

Proof. One can assume $\partial\Omega$ is a compact in $\mathbb{C}$ and $\infty \in \Omega$. We cover $\partial\Omega$ by a finite number of open disks of radius $\frac{1}{n}$, which we can arrange so that none of them is tangent to another one of this finite collection. Take as Ω_n the unbounded component of the complement of the union of the corresponding closed disks. □

4.7.13. Remarks. (1) In the preceding construction, if Ω is simply connected, then the Ω_n are also simply connected.

(2) Since the number of corners of $\partial\Omega_n$ is finite one can modify the construction so that $\partial\Omega_n$ is of class C^∞.

(3) If $\{\Omega_n\}_{n\geq 1}$ is a regular exhaustion and $z_0 \in \Omega_1$, then each Ω_n admits a Green function with pole at z_0, $g_n(z) = g(z; z_0, \Omega_n)$. If $n > m$ then $g_n - g_m$ is

harmonic in Ω_m and >0 in $\partial\Omega_m$, hence $g_n - g_m > 0$ throughout $\bar{\Omega}_m$. Therefore the sequence $\{g_n\}$ converges uniformly throughout every compact of $\Omega\backslash\{z_0\}$ either to the function identically $+\infty$ or to a harmonic function. If Ω has a Green function with pole at z_0, $g(z) = g(z; z_0, \Omega)$, then $g_n < g$ in Ω_n and therefore g_n converges to a harmonic function at most equal to g in $\Omega\backslash\{z_0\}$. But the same inequality forces $\lim_{\substack{z\to\zeta \\ z\in\Omega}}\left(\lim_{n\to\infty} g_n(z)\right) = 0$ if $\zeta \in \partial\Omega$. It follows that $g_n(z; z_0, \Omega) \to g(z; z_0, \Omega)$ uniformly over compact subsets of $\Omega\backslash\{z_0\}$.

4.7.14. Proposition (Symmetry of the Green Function). *Let* $g(z; z_0, \Omega)$ *and* $g(z; z_1, \Omega)$ *be the Green functions of* Ω *with poles at* $z_0 \neq z_1$. *Then*

$$g(z_1; z_0, \Omega) = g(z_0; z_1, \Omega).$$

(*We assume the Dirichlet problem is solvable in* Ω.)

PROOF. Let us assume first that $\partial\Omega$ is piecewise regular of class C^1 and a compact subset of $\mathbb{C}$. The function

$$g(z; z_0, \Omega) + \log|z - z_0|$$

is harmonic in Ω and continuous in $\bar{\Omega}$, with boundary values

$$I(z, z_0) = \log|z - z_0|.$$

Consider the harmonic extension of $I(\cdot, z_0)$

$$P(I(\cdot, z_0))(z) = \int_{\partial\Omega} I(t, z_0)\, d\omega_z(t) \qquad (z \in \Omega).$$

It is clear that for z fixed, this is a harmonic function of $z_0 \in \Omega$. In fact, if $\bar{B}(z_0, r) \subseteq \Omega$ then, applying Fubini's theorem and that $w \mapsto \log|w - t|$ is harmonic in Ω for $t \in \partial\Omega$ fixed, we have

$$\frac{1}{2\pi r}\int_{\partial B(z_0, r)} P(I(\cdot, w))(z)\, ds(w) = \int_{\partial\Omega} \lambda(I(t, \cdot), z_0, r)\, d\omega_z(t)$$

$$= \int_{\partial\Omega} I(t, z_0)\, d\omega_z(t) = P(I(\cdot, z_0))(z),$$

which proves the mean-value property holds for z_0 variable.

On the other hand, we already know one harmonic function in z whose boundary values are precisely $I(\cdot, z_0)$. Therefore we must have the identity

$$P(I(\cdot, z_0))(z) = g(z; z_0, \Omega) + \log|z - z_0|,$$

which implies that for $z \in \Omega$ fixed,

$$z_0 \mapsto g(z; z_0, \Omega) + \log|z - z_0|$$

is harmonic. Therefore $z_0 \mapsto g(z_1; z_0, \Omega)$ is harmonic in $\Omega\backslash\{z_1\}$.

Let D denote the diagonal of $\Omega \times \Omega$. Consider the function

$$\delta(z_1, z_0) = g(z_1; z_0, \Omega) - g(z_0; z_1, \Omega)$$

defined in $\Omega \times \Omega \backslash D$. This function is harmonic in each variable separately. Now, as a function of z_0, $g(z_1; z_0, \Omega)$ has a logarithmic singularity at z_1, and as a function of z_0, $g(z_0; z_1, \Omega)$ also has a logarithmic singularity at z_1, both have opposite signs so that $z_0 \mapsto \delta(z_1, z_0)$ is actually harmonic in a neighborhood of z_1. That is, we have shown δ is actually well defined in $\Omega \times \Omega$ and harmonic in each variable separately.

If $x \in \partial\Omega$ we have

$$\begin{aligned}\liminf_{z_0 \to x} \delta(z_1, z_0) &= \liminf_{z_0 \to x} g(z_1; z_0, \Omega) - \lim_{z_0 \to x} g(z_0; z_1, \Omega) \\ &= \liminf_{z_0 \to x} g(z_1; z_0, \Omega) \geq 0,\end{aligned}$$

hence $\delta \geq 0$ in $\Omega \times \Omega$. Similarly

$$\limsup_{z_1 \to x} \delta(z_1, z_0) = \limsup_{z_1 \to x} (-g(z_1; z_0, \Omega)) \leq 0,$$

hence $\delta \equiv 0$. Therefore $g(z_1; z_0, \Omega) = g(z_0; z_1, \Omega)$ if $z_1 \neq z_0$.

If Ω is an arbitrary open set for which the Dirichlet problem is solvable, let $\{\Omega_n\}_{n \geq 1}$ be a regular exhaustion of Ω. We can suppose z_0, $z_1 \in \Omega_1$. We have $g(z_1; z_0, \Omega_n) = g(z_0; z_1, \Omega_n)$ for every n, therefore this identity holds for their limits, hence by §4.7.13, (3) for the Green function of Ω. □

Our aim is to show how to find the harmonic measure ω_z in terms of the Green function (as in Proposition 4.7.18). We have shown the symmetry of the Green function; we must therefore have some relation between ω_{z_0} and ω_{z_1}.

4.7.15. Proposition. *Let Ω be an open set for which the Dirichlet problem is solvable, z_0, z_1 be two distinct points in Ω. Then ω_{z_0} and ω_{z_1} are mutually absolutely continuous measures and, for every compact set $K \subseteq \Omega$ there is a positive constant M such that*

$$\frac{1}{M}\omega_{z_0}(E) \leq \omega_{z_1}(E) \leq M\omega_{z_0}(E),$$

for every z_0, $z_1 \in K$ and every Borelian set $E \subseteq \partial\Omega$.

PROOF. Let us show first the absolute continuity of ω_{z_1} with respect to ω_{z_0}. Since these are regular measures, it is enough to show that if E is a closed subset of $\partial\Omega$ and $\omega_{z_0}(E) = 0$ then $\omega_{z_1}(E) = 0$ also. Let $u \in \mathscr{C}(\partial\Omega)$ be such that $0 \leq u < 1$ in $\partial\Omega \backslash E$, and $u \equiv 1$ on E. Set $v_n = P(u^n)$. Then $\{v_n\}_{n \geq 1}$ is a decreasing sequence of harmonic functions ≥ 0, and

$$v_n(z_0) = \int_{\partial\Omega} u^n \, d\omega_{z_0} \to \omega_{z_0}(E) = 0.$$

Hence $v_n \to 0$ in Ω and therefore,

$$0 = \lim_{n\to\infty} v_n(z_1) = \lim_{n\to\infty} \int_{\partial\Omega} u^n \, d\omega_{z_1} \to \omega_{z_1}(E),$$

which is what we wanted to prove.

The last part of the proposition is just a little refinement of the preceding argument. From the Harnack Inequality 4.3.9 one can conclude that if v is harmonic positive function in Ω, K a compact subset of Ω, there are constants $c_1 > 0$, $c_2 > 0$ such that for any pair $z_0, z_1 \in K$ we have

$$c_1 v(z_0) \le v(z_1) \le c_2 v(z_0).$$

Let now E be a closed subset of $\partial\Omega$ such that $\omega_{z_0}(E) = \sigma > 0$. Using the preceding argument, we have a sequence $v_n = P(u^n)$, which converges to a harmonic function $v \ge 0$. One has

$$\begin{aligned} v(z_0) &= \lim_{n\to\infty} \int_{\partial\Omega} u^n \, d\omega_{z_0} \\ &= \int_E \omega_{z_0} \\ &= \omega_{z_0}(E) \\ &= \sigma > 0. \end{aligned}$$

Therefore $v > 0$ and we can apply the previous inequality, hence

$$c_1\sigma \le v(z_1) \le c_2\sigma, \qquad \text{for every } z_1 \in K.$$

But

$$v(z_1) = \lim_{n\to\infty} \int_{\partial\Omega} u^n \, d\omega_{z_1} = \omega_{z_1}(E),$$

whence

$$c_1\omega_{z_0}(E) \le \omega_{z_1}(E) \le c_2\omega_{z_0}(E).$$

By the regularity of the measures, these inequalities hold for every Borelian $E \subseteq \partial\Omega$. □

4.7.16. Proposition. *Let Ω_1, Ω_2 be two open connected sets and $\varphi : \bar{\Omega}_1 \to \bar{\Omega}_2$ a homeomorphism which is a biholomorphism from Ω_1 onto Ω_2. If the Dirichlet problem is solvable in Ω_1 (and hence in Ω_2) then for $z_1 \in \Omega_1$ we have*

$$\omega_{\varphi(z_1)} = \varphi_*(\omega_{z_1}).$$

($\varphi_*(\omega_{z_1})$ is the direct image of the measure, i.e., $\int_{\partial\Omega_2} f \, d\varphi_*(\omega_{z_1}) = \int_{\partial\Omega_1} (f \circ \varphi) \, d\omega_{z_1}$.)

PROOF. The proof is left to the reader. □

4.7.17. Proposition. *The harmonic measure ω_{z_0} of a point $z_0 \in \Omega$ has no atoms.*

Proof. We can assume $\infty \in \Omega$, $0 \in \partial\Omega$, and $\partial\Omega \subseteq B(0,1)$. We are going to show $\omega_{z_0}(\{0\}) = 0$ by showing that $\log|z| \in L^1(\partial\Omega, \omega_{z_0})$.

Let $g(z; \infty, \Omega)$ be the Green function of Ω with pole at ∞,

$$u(z) := g(z, \infty, \Omega) - \log|z|.$$

Then u is harmonic Ω, continuous in $\partial\Omega\backslash\{0\}$, and it has the value $-\log|z|$ on $\partial\Omega\backslash\{0\}$. Let $\{f_n\}_{n\geq 1}$ be an increasing sequence of continuous positive functions in $\partial\Omega$ such that $\lim_{n\to\infty} f_n(z) = -\log|z|$, $z \neq 0$, and $\lim_{n\to\infty} f_n(0) = +\infty$. The corresponding harmonic functions $u_n = P(f_n)$ form an increasing sequence in Ω and $u_n(z) \leq u(z)$, since $u_n \leq u$ on $\partial\Omega\backslash\{0\}$. Therefore

$$\begin{aligned} u(z_0) \geq \lim_{n\to\infty} u_n(z_0) &= \lim_{n\to\infty} \int_{\partial\Omega} f_n \, d\omega_{z_0} \\ &= \lim_{n\to\infty} (f_n(0)\omega_{z_0}(\{0\})) + \int_{\partial\Omega} -\log|z| \, d\omega_{z_0}(z), \end{aligned}$$

which proves that $\omega_{z_0}(\{0\}) = 0$ and $\log|z| \in L^1(\partial\Omega, \omega_{z_0})$. □

When the domain Ω has a real analytic boundary then one can give a formula for ω_{z_0} in terms of the Green function.

4.7.18. Proposition. *If Ω is an open connected set, with regular real analytic boundary, $z_0 \in \Omega$ then*

$$d\omega_{z_0}(\cdot) = -\frac{1}{2\pi}\frac{\partial}{\partial n} g(\cdot\,; z_0, \Omega)\, ds,$$

where $g(z; z_0, \Omega)$ is the Green function with pole at z_0, $\dfrac{\partial}{\partial n}$ is the derivative in the direction of the exterior normal and ds is the arc length measure on the boundary.

Proof. We will use Green's formula. Let $h \in C^\infty(\partial\Omega)$ *and, for* $\delta > 0$, *let* $\Omega_\delta = \{z \in \Omega : |z - z_0| > \delta\}$ $(0 < \delta < d(z_0, \Omega^c))$. Let $u = P(h)$ *and*

$$v(z) = g(z; z_0, \Omega).$$

Applying Green's formula to Ω_δ we obtain

$$\int_{\Omega_\delta} (u\Delta v - v\Delta u)\, dx\, dy = \int_{\partial\Omega_\delta} \left(u\frac{\partial v}{\partial n} - v\frac{\partial u}{\partial n} \right) ds,$$

but to justify this formula we need to know that v is a smooth function also in a neighborhood of $\partial\Omega$, This will depend on the following reflection principle.

4.7.19. Lemma (Schwarz's Reflection Principle for Harmonic Functions). *Let v be harmonic in $]a, b[+ i]0, \delta[$, continuous in $]a, b[+ i[0, \delta[$, and zero when*

$\operatorname{Im} z = 0$. *For* $z \in]a,b[+i]-\delta,\delta[$ *we define*

$$V(z) = \begin{cases} v(z) & \text{if } \operatorname{Im} z \geq 0 \\ -v(\bar{z}) & \text{if } \operatorname{Im} z \leq 0, \end{cases}$$

then V *is a harmonic function.*

PROOF. Clearly V is continuous in $G =]a,b[+i]-\delta,\delta[$. If $z_0 \in G$, $\operatorname{Im} z_0 \neq 0$ then it is clear $V(z_0) = \lambda(V, z_0, r)$ for all $r > 0$ sufficiently small. For $z_0 \in]a,b[$ we have by the very definition of V, $\lambda(V, z_0, r) = 0$, but $V(z_0) = 0$ also. Hence V satisfies the mean-value property and it is harmonic. □

Let us return to the proof of §4.7.18. If $\gamma:]-\varepsilon,\varepsilon[\to \mathbb{C}$ is a real analytic function parametrizing locally a component of $\partial\Omega$, $\gamma'(t) \neq 0$ for all t and $\gamma(t) = \sum_{j\geq 0} c_j t^j$ is the series development of γ about $t = 0$, which we can assume converges in the disk of radius ε. Let $\Gamma(z) = \sum_{j\geq 0} c_j z^j$, $z \in B(0,\varepsilon)$, then Γ is a holomorphic injective map in $B(0,r)$ for some r, $0 < r \leq \varepsilon$. Therefore $W = \Gamma(B(0,r))$ is an open neighborhood of $\gamma(0)$ which is biholomorphic to $B(0,r)$. The curve $\gamma(]-r,r[)$ is the image of $B(0,r) \cap \{\operatorname{Im} z = 0\}$. Let us consider therefore the harmonic function $h(z) := g(\Gamma(z); z_0, \Omega)$ in $B(0,r) \cap \{\operatorname{Im} z > 0\}$, which is continuous on $B(0,r) \cap \{\operatorname{Im} z \geq 0\}$ and zero on $B(0,r) \cap \{\operatorname{Im} z = 0\}$. Hence there is a harmonic function H in $B(0,r)$ whose restriction to the upper half of the disk coincides with h. The function Γ^{-1} is well defined in W, $H \circ \Gamma^{-1}$ is now a harmonic function in W which extends $g(\cdot; z_0, \Omega)$ across a segment of $\partial\Omega$. This shows that $g(\cdot; z_0, \Omega)$ is in fact smooth in a neighborhood of $\partial\Omega$, and justifies the use of Green's formula.

Returning to Green's identity, let us recall that since $v = g(\cdot, z_0, \Omega)$ vanishes on $\partial\Omega$, and u, v are both harmonic in Ω_δ, we obtain

$$0 = \int_{\partial\Omega} h \frac{\partial v}{\partial n} ds - \int_{\partial B(0,\delta)} \left(u \frac{\partial v}{\partial n} - v \frac{\partial v}{\partial n} \right) ds.$$

Introducing polar coordinates about z_0 we have $g(z; z_0, \Omega) = -\log r + G(z)$, G harmonic near z_0, $\dfrac{\partial}{\partial n} = \dfrac{\partial}{\partial r}$, hence

$$\frac{\partial v}{\partial n} = -\frac{1}{r} + \frac{\partial G}{\partial n},$$

where $\dfrac{\partial G}{\partial n}$ is continuous near z_0. Therefore

$$\int_{\partial B(0,\delta)} u \frac{\partial v}{\partial n} ds = -\int_{-\pi}^{\pi} u(z_0 + \delta e^{i\theta})\, d\theta + O(\delta) = -2\pi u(z_0) + O(\delta)$$

$$\int_{\partial B(0,\delta)} v \frac{\partial u}{\partial n} ds = -\delta \log \delta \int_{-\pi}^{\pi} \frac{\partial}{\partial r} u(z_0 + \delta e^{i\theta})\, d\theta + O(\delta) = o(1).$$

Letting $\delta \to 0$ we have

$$\int_{\partial\Omega} h\frac{dv}{\partial n}\,ds = -2\pi u(z_0).$$

In other words,

$$u(z_0) = -\frac{1}{2\pi}\int_{\partial\Omega} h(z)\frac{\partial g}{\partial n}(z; z_0, \Omega)\,ds(z),$$

which is the statement of Proposition 4.7.18. □

4.7.20. Remark. Lemma 4.7.19 provides a proof of a stronger Schwarz's reflection principle for holomorphic functions. That is, if $f = u + iv$ is holomorphic in $]a,b[+ i]0,\delta[$, v is continuous in $]a,b[+ i[0,\delta[$ and $v = 0$ on $]a,b[$, then there is a holomorphic function F in G such that

$$F(z) = \begin{cases} f(z) & \text{if } z \in G \cap \{\operatorname{Im} z > 0\} \\ \overline{f(\bar{z})} & \text{if } z \in G \cap \{\operatorname{Im} z < 0\}. \end{cases}$$

($G =]a,b[+ i]-\delta,\delta[$ as in the proof of Lemma 4.7.19.)

In fact, let $g(z) = u(\bar{z}) - iv(\bar{z})$, $z \in G \cap \{\operatorname{Im} z < 0\}$. One verifies without difficulty that $\frac{\partial g}{\partial \bar{z}} = 0$. The difficulty lies in proving the continuity of F across $\operatorname{Im} z = 0$. We will bypass this problem as follows.

The function V from Lemma 4.7.19 is harmonic in G and, since G is simply connected, there is a harmonic function U in G such that $U + iV$ is holomorphic in G. Furthermore, if W is another harmonic function in an open connected subset Ω of G such that $W + iV$ is holomorphic then $W - U$ must be constant in Ω. The choice $W = u$ in $G \cap \{\operatorname{Im} z > 0\}$ leads to $U(z) = u(z) + k_1$ when $\operatorname{Im} z > 0$. The choice of $W(z) = u(\bar{z})$ in $G \cap \{\operatorname{Im} z < 0\}$ leads to $U(z) = u(\bar{z}) + k_2$ when $\operatorname{Im} z < 0$. Since U is continuous we have that for $x_0 \in]a,b[$,

$$U(x_0) - k_1 = \lim_{\varepsilon\to 0^+} U(x_0 + i\varepsilon) - k_1 = \lim_{\varepsilon\to 0^+} u(x_0 + i\varepsilon),$$

which shows the last limit exists. Furthermore

$$U(x_0) - k_2 = \lim_{\varepsilon\to 0^+} U(x_0 - i\varepsilon) - k_2 = \lim_{\varepsilon\to 0^+} u(\overline{x_0 - i\varepsilon}) = \lim_{\varepsilon\to 0^+} u(x_0 + i\varepsilon).$$

hence $k_1 = k_2$, that is, $U + iV - k_1$ is holomorphic in G and coincides with f when $\operatorname{Im} z > 0$ and with g when $\operatorname{Im} z < 0$. Therefore, F is a holomorphic extension of f. □

Assume, as earlier, that Ω is a connected open set with a real analytic boundary. Let $h(z) = h(z; z_0, \Omega)$ be a "harmonic conjugate" of $g(z) := g(z; z_0, \Omega)$, defined in a neighborhood of $\bar{\Omega}\setminus\{z_0\}$ by choosing a path γ_z from a fixed point $z_1 (z_1 \neq z_0)$ to z and letting

$$h(\gamma_z) = 2\operatorname{Re}\int_{\gamma_z} \frac{\partial g}{\partial w}(w)\,dw.$$

The form $2\frac{\partial g}{\partial w}dw$ is closed (cf §4.3.4) but h is not well defined. At every z, there are several determinations depending on the periods, with respect to the holes of $\Omega\backslash\{z_0\}$, of the integral. Nevertheless, locally a determination of h can be chosen, and $Q = g + ih$ is locally holomorphic (though multivalued). Moreover, Q' is well defined, independent of the determination of h. In fact, one has the following.

4.7.21. Proposition. *If Ω is an open connected set with a real analytic boundary and $z_0 \in \Omega$, then*

$$d\omega_{z_0}(z) = \frac{i}{2\pi}Q'(z)\,dz \qquad (z \in \partial\Omega).$$

PROOF. Let n be the exterior normal to $\partial\Omega$ at ζ and τ the unit tangent vector at the same point (with the same orientation as $\partial\Omega$). The Cauchy-Riemann equations and the fact that g vanishes identically on $\partial\Omega$ give

$$0 = \frac{\partial g}{\partial \tau}(\zeta) = -\frac{\partial h}{\partial n}(\zeta).$$

We have $d\zeta = \tau\,ds$, $n = -i\tau$, and we can conclude that

$$\begin{aligned} iQ'(\zeta)\,d\zeta &= i\left(\lim_{t\to 0^+}\frac{Q(\zeta+tn)-Q(\zeta)}{tn}\right)d\zeta = \frac{i\tau}{n}\left(\frac{\partial g}{\partial n}(\zeta) + i\frac{\partial h}{\partial n}(\zeta)\right)ds \\ &= -\frac{\partial g}{\partial n}(\zeta)\,ds = 2\pi\,d\omega_{z_0}, \end{aligned}$$

which proves the proposition. □

The relation

$$d\omega_z = -\frac{1}{2\pi}\frac{\partial}{\partial n}g(\cdot\,; z, \Omega)\,ds$$

shows that the function

$$P(\zeta, z) = -\frac{\ell}{2\pi}\frac{\partial}{\partial n}g(\zeta; z, \Omega),$$

where ℓ = length of $\partial\Omega$, reproduces the harmonic functions in Ω that are continuous up to $\partial\Omega$. That is,

$$u(z) = \int_{\partial\Omega} P(\zeta, z)u(\zeta)\,d\sigma(\zeta),$$

where $d\sigma(\zeta) = \frac{ds}{\ell}$ is the arc-length measure normalized. Therefore this function P has the right to be called the Poisson kernel of Ω. Note that $P \geq 0$ and $\int_{\partial\Omega} P(\zeta, z)\,d\sigma(\zeta) = 1$ for every $z \in \Omega$.

In the next section we will remove the need to have a real analytic boundary for Ω in order to construct a Poisson kernel. The idea is to use §4.7.9 (4), and show that biholomorphic maps have nice boundary regularity properties.

We state now a more or less immediate consequence of Remark 4.7.20 and the proof of Proposition 4.7.18.

4.7.22. Proposition. *If Ω is a Jordan domain with real analytic boundary and f is a conformal map of $B(0, 1)$ onto Ω, then f has an extension to a conformal map defined in a neighborhood of $\bar{B}(0, 1)$.*

PROOF. The proof is left to the reader. □

We can put together the Riesz decomposition theorem and the preceding considerations to obtain a representation formula for subharmonic functions involving the Green function.

4.7.23. Lemma. *Let Ω be a bounded domain with a real analytic boundary. Then for every $\zeta \in \Omega$,*

$$\int_{\partial\Omega} \log|z - z_0|\, d\omega_\zeta(z) = \begin{cases} \log|\zeta - z_0| & \text{if } z_0 \notin \Omega \\ \log|\zeta - z_0| + g(\zeta; z_0, \Omega) & \text{if } z_0 \in \Omega. \end{cases}$$

PROOF. Let us observe first that the formula on the right-hand side is actually a continuous function of z_0. In fact, $g(\zeta; z_0, \Omega) = g(z_0; \zeta, \Omega) \to 0$ when z_0 tends to a point in $\partial\Omega$. For $z_0 \in \Omega$ we already know that $\log|\zeta - z_0| + g(\zeta; z_0, \Omega)$ coincides with a harmonic function of z_0. Secondly, by Proposition 4.7.18 we have that $d\omega_\zeta(z) = -\frac{1}{2\pi}\frac{\partial}{\partial n} g(z; \zeta, \Omega)|dz|$, and for $z_0 \in \partial\Omega$, the function $z \mapsto \log|z - z_0|$ is integrable on the boundary with respect to arc-length measure $|dz|$. Since $\frac{\partial g}{\partial n}$ is bounded, a simple appeal to Lebesgue's dominated convergence theorem shows the integral in the preceding expression is also a continuous function of z_0.

Therefore, it is enough to prove the lemma when $z_0 \notin \partial\Omega$. If $z_0 \notin \bar{\Omega}$, then $\log|z - z_0|$ is a harmonic function of z in a neighborhood of $\bar{\Omega}$, and the formula is correct by definition of the harmonic measure. If $z_0 \in \Omega$, let $h_{z_0}(z)$ be the harmonic function in Ω whose boundary values are $\log|z - z_0|$. We have

$$\int_{\partial\Omega} \log|z - z_0|\, d\omega_\zeta(z) = \int_{\partial\Omega} h_{z_0}(z)\, d\omega_\zeta(z) = h_{z_0}(\zeta) = g(\zeta, z_0, \Omega) + \log|\zeta - z_0|$$

by definition of the Green function. □

4.7.24. Remark. This lemma is valid under very general conditions, certainly we only use that $\frac{\partial g}{\partial n}(z; \zeta, \Omega)$ is bounded and $d\omega_\zeta = -\frac{1}{2\pi}\frac{\partial g}{\partial n}(z; \zeta, \Omega)$. Piecewise smoothness of the boundary suffices for this requirement. (See, e.g., [HK, section 3.7] where this is proved assuming only $m(\partial\Omega) = 0$.) The same remark applies to the following proposition.

4.7.25. Proposition. *Let Ω be as in the previous lemma, u a subharmonic function in an open neighborhood of $\bar{\Omega}$. Then for every $\zeta \in \Omega$ we have*

$$u(\zeta) = \int_{\partial\Omega} u(z)\,d\omega_\zeta(z) - \frac{1}{2\pi}\int_\Omega g(z;\zeta,\Omega)\Delta u(z).$$

PROOF. By Riesz' Decomposition Theorem 4.4.2 (3) and the hypotheses on u we have a relatively compact open set D, $\bar{\Omega} \subseteq D$, and a function u_0 harmonic in D such that for $z \in D$ we have

$$u(z) = u_0(z) + \frac{1}{2\pi}\int_D \log|z - w|\,\Delta u(w).$$

Let us integrate both sides against the harmonic measure $d\omega_\zeta$ of Ω at the point ζ. Then

$$\int_{\partial\Omega} u(z)\,d\omega_\zeta(z) = \frac{1}{2\pi}\int_{\partial\Omega} u_0(z)\,d\omega_\zeta(z) + \frac{1}{2\pi}\int_{\partial\Omega}\left(\int_D \log|z - w|\,\Delta u(w)\right)d\omega_\zeta(z).$$

The total mass of D with respect to the positive measure Δu is finite. The same is true for $\partial\Omega$ with respect to the positive measure $d\omega_\zeta$. The function $\log|z - w|$ is bounded above when $(z, w) \in \partial\Omega \times D$, hence we can apply Fubini's theorem and interchange the order of integration. We obtain

$$\int_{\partial\Omega} u(z)\,d\omega_\zeta(z) = u_0(\zeta) + \frac{1}{2\pi}\int_D \Delta u(w)\left(\int_{\partial\Omega} \log|z - w|\,d\omega_\zeta(z)\right).$$

For $w \in D\backslash\Omega$ the inner integral becomes $\log|\zeta - w|$ by §4.7.23. For $w \in \Omega$ we can use the other part of §4.7.23 and get

$$\int_{\partial\Omega} u(z)\,d\omega_\zeta(z) = u_0(\zeta) + \frac{1}{2\pi}\int_D \log|\zeta - w|\,\Delta u(w) + \frac{1}{2\pi}\int_\Omega g(w;\zeta,\Omega)\Delta u(w)$$

$$= u(\zeta) + \frac{1}{2\pi}\int_\Omega g(w;\zeta,\Omega)\Delta u(w).$$

This is exactly the statement of the proposition. □

We conclude this section by showing that the Riemann mapping theorem can be obtained from the existence of the Green function. We think the argument is instructive.

4.7.26. Proposition. *Let Ω be a proper open subset of $\mathbb{C}$ which is simply connected. Given $z_0 \in \Omega$, there is a conformal map φ of Ω onto $B(0, 1)$ such that $\varphi(z_0) = 0$.*

PROOF. By the same argument as in §2.6.3, we can reduce ourselves to the case where Ω is bounded. Let $g(z; z_0, \Omega)$ be the Green function of Ω with pole at z_0. It exists, since $S^2 \backslash \Omega$ is connected and has more than one point, and therefore the Dirichlet problem is solvable in Ω. Let

$$u(z) = -g(z; z_0, \Omega) - \log|z - z_0| \qquad (z \in \Omega),$$

This function is harmonic in Ω. Let v be the harmonic conjugate function that vanishes at z_0. The holomorphic function

$$\varphi(z) = (z - z_0)\exp(u(z) + iv(z))$$

is the conformal mapping we are looking for.

It is clear that $\varphi(z_0) = 0$ and that $\log|\varphi(z)| = \log|z - z_0| + u(z) = -g(z; z_0, \Omega) \leq 0$. Therefore $\varphi(\Omega) \subseteq B(0, 1)$ and $\lim_{z \to \partial\Omega} |\varphi(z)| = 1$.

Let us now show that φ is surjective. Since φ is an open map, it is enough to prove that φ is also a closed map. If we show that φ is proper, then we will be done with the surjectivity question. For $0 < r < 1$ we claim that the inverse image K_r of $\bar{B}(0, r)$ is a compact subset of Ω. If not, let $(z_n)_{n \geq 1}$ be a sequence in K_r which converges to a point $\zeta \in \partial\Omega$. It would follow that $\lim_{n \to \infty} |\varphi(z_n)| = 1$, which is a contradiction.

To show that φ is injective, we can find a piecewise linear Jordan curve γ in Ω such that $K_r = \varphi^{-1}(\bar{B}(0, r)) \subseteq \operatorname{Int}(\gamma)$ (§1.10.2). The curve $\Gamma = \varphi \circ \gamma$ lies in the annulus $r < |z| < 1$. For a given w_0, $|w_0| < r$, we have that

$$\operatorname{Ind}_\Gamma(w_0) = \frac{1}{2\pi i} \int_\Gamma \frac{dw}{w - w_0} = \frac{1}{2\pi i} \int_\gamma \frac{\varphi'(z)\, dz}{\varphi(z) - w_0}$$

is the number of times the value w_0 is taken. Since $\operatorname{Ind}_\Gamma(w_0) = \operatorname{Ind}_\Gamma(0)$, this is also the number of times the value 0 is taken by φ. From the definition of φ one sees that this happens only once. This concludes the proof of the proposition. □

EXERCISES 4.7

$B = B(0, 1)$.

1. Prove Proposition 4.7.22 in detail.
2. Let Ω be a simply connected open set in $\mathbb{C}$. Show that the level curves $S_\lambda := \{z \in \Omega : g(z; z_0, \Omega) = \lambda > 0\}$ of $g(z; z_0, \Omega)$ are real analytic Jordan curves.
3. Find the Green function of the following domains in $\mathbb{C}$:
   (a) $\Omega = H = \{z \in \mathbb{C} : \operatorname{Im} z > 0\}$
   (b) $\Omega = \{z \in \mathbb{C} : 0 < \operatorname{Arg} z < \alpha < 2\pi\}$
   (c) $\Omega = B(0, R)$
   (d) $\Omega = B(0, R) \cap H$
   (e) $\Omega = \left\{z \in \mathbb{C} : |z| < 1, 0 < \operatorname{Arg} z < \dfrac{\pi}{2}\right\}$.

4. Solve the Dirichlet problem in $H = \{z \in \mathbb{C} : \operatorname{Im} z > 0\}$, where the Dirichlet data $u|_{\mathbb{R}} = f(x)$ is given by
   (a) $f(x) = \chi_{[a,b]}$
   (b) $f(x) = \dfrac{1}{1+x^2}$
   (c) $f(x) = 1$.

5. Solve the Dirichlet problem in the quadrant $\Omega = \left\{z \in \mathbb{C} : 0 < \operatorname{Arg} z < \dfrac{\pi}{2}\right\}$ for the data
   (a) $u|\{0\} \times [0, \infty[= 0,\ u|[0, \infty[\times \{0\} = 1$
   (b) $u|\{0\} \times [0, \infty[= \alpha,\ u|[0, \infty[\times \{0\} = \beta \qquad (\alpha, \beta \in \mathbb{R})$.

6. The Green function $z \mapsto g(z; z_0, \Omega)$ is locally integrable, hence an element of $\mathscr{D}'(\Omega)$. Compute $\Delta g(\cdot\,; z_0, \Omega)$.

7. (i) Show that the domain $\Omega = B\backslash\{0\}$ does not admit a Green function. (Hint: if $g(z; z_0, \Omega)$ is the Green function show that the harmonic function $h(z) = g(z; z_0, \Omega) + \dfrac{1}{2\pi}\log|z - z_0|$ has a removable singularity at $z = 0$. Conclude that the function $G(z) := -g(z; z_0, \Omega)$ for $z \in B\backslash\{0\}$, $G(0) := 0$, is subharmonic. This violates the maximum principle.)
   (ii) Show, using (i), that there could be no conformal map between Ω and any annulus $0 < r_1 < |z| < r_2 < \infty$.

8. Write down explicitly the identity of Proposition 4.7.24, when $u = \log|f|$, f holomorphic in a neighborhood of $\bar{\Omega}$ and has no zeros on $\partial\Omega$. (What happens if there are zeros on $\partial\Omega$?)

*9. (a) Apply Exercise 4.7.7 when $\Omega = \{z \in \mathbb{C} : |z| < R, \operatorname{Im} z > 0\}$. Conclude that if a_k denote the zeros of f in Ω then

$$\log|f(z)| = \sum_{|a_k|<R} \log\left|\frac{z - a_k}{z - \bar{a}_k}\right| + \frac{y}{\pi}\int_{-R}^{R} P_1(t)\log|f(t)|\,dt + \frac{2Ry}{\pi}\int_0^{\pi} P_2(\theta)\log|f(Re^{i\theta})|\,d\theta,$$

where $z = x + iy \in \Omega$, $P_1(t) = \dfrac{1}{t^2 - 2tx + r^2} - \dfrac{R^2}{R^4 - 2tR^2x + |z|^2t^2}$, $P_2(\theta) = \dfrac{(R^2 - |z|^2)\sin\theta}{|R^2e^{2i\theta} - 2Rxe^{i\theta} + |z|^2|^2}$. (Hint: Compute the Green function of Ω using conformal maps.)

   (b) Assume further $f(0) = 1$; then by letting $z \to 0$ conclude that

$$\sum_{r_k<R}\left(\frac{1}{r_k} - \frac{r_k}{R^2}\right)\sin\theta_k = \frac{1}{\pi R}\int_0^{\pi}\log|f(Re^{i\theta})|\sin\theta\,d\theta + \frac{1}{2\pi}\int_0^R\left(\frac{1}{x^2} - \frac{1}{R^2}\right)\log|f(x)f(-x)|\,dx + \frac{1}{2}\operatorname{Im}(f'(0)).$$

Here $a_k = r_ke^{i\theta_k} \in \Omega$.

*10. Use §4.7.20 to prove that if $\Omega =]a, b[+ i]0, \delta]$ and $f \in \mathscr{H}(\Omega)$ is such that as $z_n \to x_0 \in]a, b[$, one has either $\operatorname{Im} f(z_n) \to 0$ or $|f(z_n)| \to \infty$, then f extends to a meromorphic function in $D = \Omega \cup]a, b[\cup \bar{\Omega}$, whose only possible poles lie on the real axis. Here $\bar{\Omega} =]a, b[- i]0, \delta]$. (Hint: consider the function $g(z) = \dfrac{f(z) - i}{f(z) + i}$ in a small half-disk centered about a point $x_0 \in]a, b[$. Show one can assume $g(z) \neq 0$ throughout that half-disk. Apply §4.7.20 to $h(z) = i \log g(z)$.)

11. Let f be a conformal map between $\Omega_1 = \{z : 0 < r_1 < |z| < R_1\}$ and $\Omega_2 = \{w : 0 < r_2 < |w| < R_2\}$. Show that $\dfrac{R_1}{r_1} = \dfrac{R_2}{r_2}$.

12. Let $f \in \mathscr{M}(B)$ be such that $|f(z)| \to 1$ as $|z| \to 1$, show f has an extension to a function F in $\mathscr{M}(S^2)$. For $|z| > 1$ show that $F(z) = 1/\overline{f}(1/\bar{z})$. (Hint: consider $\log|f(z)|$.)

13. Let f be an entire function which is real-valued exactly on the real axis. Show that $f(z) = az + b$ for some $a, b \in \mathbb{R}$, $a \neq 0$. (Hint: f is nonconstant and one can assume that $\operatorname{Im} f(z) > 0$ whenever $\operatorname{Im} z > 0$. Use the argument principle to show that f has exactly one zero. Let $a \in \mathbb{R}$ be that zero, and show the function $\dfrac{f(z)}{z - a}$ must be constant.)

14. Let u be a harmonic function in $H = \{z \in \mathbb{C} : \operatorname{Im} z > 0\}$ and continuous in $\bar{H}$. Assume $u > 0$ in H and $u = 0$ on ∂H. Show that $u(z) = ay$ for some $a > 0$.

15. Let Ω_1, Ω_2 be two domains in $\mathbb{C}$ for which the Dirichlet problem is solvable and $f : \Omega_1 \to \Omega_2$ a holomorphic map. Let

$$u(z) := g(z; z_0, \Omega_1) - g(f(z); f(z_0), \Omega_2).$$

(a) Show that u is harmonic in Ω_1, except possibly at $z = z_0$ and at the points of the discrete set $f^{-1}(\{f(z_0)\})$.

(b) Show that for every singular point $\zeta \in \Omega_1$ of u, there is $k \in \mathbb{N}$ such that $u(z) + k \log|z - \zeta|$ is harmonic in a neighborhood of ζ.

(c) Conclude from (a) and (b) that u is subharmonic in Ω_1 and $u \leq 0$. This is precisely the ***Lindelöf subordination principle***:

$$g(z; z_0, \Omega_1) \leq g(f(z); f(z_0), \Omega_2).$$

(d) Let $m(z_0, f)$ be the multiplicity of $f(z_0)$ as a value of f at $z = z_0$. Show that

$$m(z_0, f) g(z; z_0, \Omega_1) \leq g(f(z); f(z_0), \Omega_2).$$

(e) Let $w_0 \in \Omega_2$ and $f^{-1}(w_0) = \{z_1, \ldots, z_n\}$. Show that

$$\sum_{1 \leq k \leq n} m(z_k, f) g(z; z_k, \Omega_1) \leq g(f(z); w_0, \Omega_2).$$

16. Use Exercise 4.7.15 to show that if $\Omega_1 = B(0, 1)$, $\Omega_2 = \{w \in \mathbb{C} : \operatorname{Re} w > 0\}$, $f : \Omega_1 \to \Omega_2$ holomorphic, and $f(0) = w_0 = u_0 + iv_0$, then

$$|f(z) - iv_0| \leq u_0 \frac{1 + |z|}{1 - |z|}, \qquad |z| < 1.$$

17. Let $f: B \to B$ holomorphic. Show that for every $\zeta \in B$ one has
$$\sum_{z \in f^{-1}(\zeta)} m(z, f)(1 - |z|) < \infty.$$

*18. Let $(z_k)_{k \geq 1} \subseteq B$, $(m_k)_{k \geq 1} \subseteq \mathbb{N}^*$ be such that
$$\sum_{k \geq 1} m_k(1 - |z_k|) < \infty.$$
Construct a holomorphic function $f: B \to B$ that vanishes exactly at the z_k with multiplicity m_k. (Hint: consider $f \in \mathscr{H}(B)$ such that $\log|f(z)| = \sum_{k \geq 1} m_k g(z; z_k, B)$.) This result can be obtained directly using Blaschke products.

19. Let $u \in \mathrm{SH}(B)$, H harmonic in B such that for some ρ, $0 < \rho < 1$, $u(z) \leq H(z)$ for every $z \in \partial B(0, \rho)$. Let $f \in \mathscr{H}(B)$, $f(0) = 0$, $\|f\|_\infty \leq 1$. Show that
$$u(f(z)) \leq H(f(z)) \qquad \text{for every } z \in B(0, \rho)$$
and
$$\lambda(u \circ f, 0, \rho) \leq \lambda(u, 0, \rho).$$

*20. Let $f: B \to \mathbb{C} \backslash [1, \infty[$ be holomorphic, $f(z) = \sum_{n \geq 1} a_n z^n$ for $z \in B$. Using a conformal map from B onto $\mathbb{C} \backslash [1, \infty[$ and the inequality
$$|a_n| = \frac{1}{2\pi} \left| \int_0^{2\pi} f(\rho e^{i\theta}) \rho^{-n} e^{-in\theta} \, d\theta \right| \leq \frac{e}{2\pi} \int_0^{2\pi} \left| f\left(1 - \frac{1}{n}\right) \right| d\theta \left(\rho = 1 - \frac{1}{n}\right),$$
show that there is $c > 0$ such that
$$|a_n| \leq cn.$$

21. Recall that f is a proper map if $f^{-1}(K) \subset\subset B$ whenever $K \subset\subset B$. Show that if $f: B \to B$ is a proper holomorphic map then f is holomorphic in a neighborhood of $\bar{B}$. (Hint: Compare with Exercise 4.7.12.)

§8. Smoothness up to the Boundary of Biholomorphic Mappings

In this section we will consider the possibility of extending up to the boundaries in a C^∞ smooth fashion, a biholomorphic mapping f between two open bounded sets Ω_1, Ω_2 which have C^∞ boundaries. We already showed in Theorem 2.8.8 that when $\bar{\Omega}_1$ and $\bar{\Omega}_2$ are homeomorphic to the closed unit disk $\bar{B}(0, 1)$, i.e., when they are Jordan domains, then without any assumption on the regularity of the boundaries, f has necessarily a continuous extension as a homeomorphism between $\bar{\Omega}_1$ and $\bar{\Omega}_2$. The first proof of this theorem is due to Caratheodory in 1913, but, already in 1885, Paul Painlevé had shown that when $\partial\Omega_1$, $\partial\Omega_2$ are C^∞ regular boundaries then f has a C^∞ extension from $\bar{\Omega}_1$ and $\bar{\Omega}_2$. Until very recently the proofs of this C^∞ extension were extremely intricate. The proof we give here, following [BK], originated in the context of several complex variables. This proof has the additional advantage that (as in Proposition 4.7.22) it provides the C^∞ extension without appealing to Carathéodory's theorem.

Let us start by remarking that if one expects a C^∞ behavior up to the boundary of a biholomorphic map $f: B(0,1) \to \Omega$, we must assume that $\partial\Omega$ be a regular boundary of class C^∞. For instance, for $\Omega = \{z \in \mathbb{C} : |z| < 1, \operatorname{Im} z > 0\}$ the map

$$f(z) = \frac{\sqrt{\dfrac{i(1-z)}{1+z}} - 1}{\sqrt{\dfrac{i(1-z)}{1+z}} + 1}$$

is biholomorphic from $B(0,1)$ onto Ω, and extends to a homeomorphism of the closures, which is not differentiable at $z = \pm 1$.

Recall that if K is an arbitrary set in $\mathbb{C}$, a function $f: K \to \mathbb{C}$ is said to be of class C^∞ on K if there is an open neighborhood U of K and a function $\tilde{f} \in \mathscr{E}(U)$ such that $f = \tilde{f}|K$. In case Ω is an open set with C^∞ regular boundary, $K = \bar{\Omega}$, then f is C^∞ in K if and only if $f|\Omega$ and all its partial derivatives can be continuously extended to K. This fact is not altogether obvious and we will therefore provide a detailed proof.

First, let us observe that one can find two sequences $\{a_k\}_{k\ge 0}$, $\{b_k\}_{k\ge 0}$ of real numbers satisfying

(i) $b_k < 0$

(ii) $\sum_{k\ge 0} |a_n||b_k^n| < \infty$ for every integer $n \ge 0$

(iii) $\sum_{k\ge 0} a_k b_k^n = 1$ for every integer $n \ge 0$

(iv) $\lim_{k\to\infty} b_k = -\infty$.

Namely, let $b_k = -2^k$, and consider for each $N \in \mathbb{Z}^*$ the solutions $a_{k,N}$ of the $N \times N$ system

$$\sum_{0\le k\le N} b_k^n x_k = 1, \qquad 0 \le n \le N.$$

Using Cramer's rule we see that they are given by $a_{k,N} = A_k \cdot B_{k,N}$ where $A_k = \prod_{0\le j\le k-1} \frac{1+2^j}{2^j - 2^k}$, $B_{k,N} = \prod_{k+1\le j\le N} \frac{1+2^j}{2^j-2^k}$. (Recall that an empty product equals 1.) It follows that

$$|A_k| \le \prod_{0\le j\le k-1} 2^{j+2-k} = 2^{-(k^2-3k)/2}$$

$$\log B_{k,N} = \sum_{k+1\le j\le N} \log\left(\frac{1+2^k}{2^j-2^k}\right) \le \sum_{k+1\le j\le N} \frac{1+2^k}{2^j-2^k} \le 4.$$

When N increase toward $+\infty$, $B_{k,N}$ also increases to a limit $B_k \leq e^4$. If we set $a_k = A_k B_k$ we have $|a_k| \leq e^4 2^{-(k^2-3k)/2}$. Therefore, we have $\sum_{k\geq 0} |a_k||b_k^n| < \infty$ for every $n \geq 0$. Let us extend the definition of $a_{k,N}$ by setting $a_{k,N} = 0$ if $k > N$. We have

$$\sum_{k\geq 0} a_{k,N} b_k^n = 1 \qquad \text{if } N \geq n.$$

Since $|a_{k,N}||b_k^n| \leq |a_k||b_k^n|$ and $\sum_{k\geq 0} |a_k||b_k|^n < \infty$ we have, for every $n \geq 0$

$$\lim_{N\to\infty} \sum_{k\geq 0} a_{k,N} b_k^n = \sum_{k\geq 0} a_{k,N} b_k^n = 1.$$

Hence the sequences $\{a_k\}_{k\geq 0}$, $\{b_k\}_{k\geq 0}$ satisfy the four properties (i)–(iv).

Let now $\varphi \in \mathscr{E}(\mathbb{R})$ be any function such that $\varphi(t) = 1$ if $0 \leq t \leq 1$ and $\varphi(t) = 0$ if $t \geq 2$. If f is a C^∞ function in the set $D = \{z \in \mathbb{C} : a < \operatorname{Re} z < b, \operatorname{Im} z > 0\}$ such that it has a continuous extension, together with all its derivatives to the set $D_1 = \{z \in \mathbb{C} : a < \operatorname{Re} z < b, \operatorname{Im} z \geq 0\}$ then we can set for $y < 0$:

$$Ef(x, y) := \sum_{k\geq 0} a_k \varphi(b_k y) f(x, b_k y).$$

For each $y < 0$ the sum is actually finite since $b_k \to -\infty$. Since $\sum_{k\geq 0} |a_k||b_k|^n < \infty$, all the derivatives of Ef converge uniformly when $y \to 0^-$ on compact subsets of $a < x < b$. By (iii) these limits coincide with those obtained for the corresponding derivatives of f when $y \to 0^+$. Therefore the function F defined in $D_2 = \{z \in \mathbb{C} : a < \operatorname{Re} z < b\}$ by

$$F(x, y) = \begin{cases} f(x, y) & \text{for } y > 0 \\ \lim_{t\to 0^+} f(x, t) & \text{for } y = 0 \\ Ef(x, y) & \text{for } y < 0 \end{cases}$$

is a C^∞ function in D_2 that extends f. To simplify the notation, let us call Ef the function F when $a = -1$, $b = 1$.

If Ω is a bounded open set with regular boundary of class C^∞, we can cover $\partial\Omega$ by a finite collection of open sets $U_1, \ldots, U_n$ and find corresponding C^∞ diffeomorphisms $\varphi_j : \Omega_j \to]-1, 1[\times]-1, 1[$ such that $\varphi_j(\partial\Omega \cap U_j)$ lies on the real axis. Let $\Omega = U_0$, $U_{n+1} = \mathbb{C}\backslash\bar{\Omega}$, and $\alpha_0, \ldots, \alpha_{n+1}$ be a C^∞ partition of unity in $\mathbb{C}$ subordinate to $(U_j)_{j=0,\ldots,n+1}$ with $\operatorname{supp} \alpha_j \subset\subset U_j$ if $1 \leq j \leq n$. If f is a C^∞ function in Ω such that all its partial derivatives have a continuous extension to $\bar{\Omega}$, then we let

$$G(f) = \alpha_0 \cdot f + \sum_{1\leq j\leq n} \alpha_j \cdot (E(f \circ \varphi_j^{-1}) \circ \varphi_j),$$

which is C^∞ everywhere and coincides with f in Ω.

If Ω is a bounded simply connected open set with a C^∞ boundary, then $\partial\Omega$ is a Jordan curve having a parametric representation of the form $\gamma : [0, 1] \to \mathbb{C}$,

γ of class C^∞, such that $\left|\frac{d}{dt}\gamma(t)\right| = |\dot{\gamma}(t)| = 1$ for every $t \in [0,1]$ and, $\frac{d^j\gamma}{dt^j}(0) = \frac{d^j\gamma}{dt^j}(1)$ for every $j \geq 0$. We recall that there is also a C^∞ function $\rho: \mathbb{C} \to \mathbb{C}$ such that $\Omega = \{z \in \mathbb{C} : \rho(z) < 0\}$ and the gradient $\nabla\rho(z) \neq 0$ for every $z \in \partial\Omega$ (cf. §1.3.2 (3)). We will also need the following property.

4.8.1. Proposition. *Let Ω be a bounded open set with C^∞ boundary. For every C^∞ function ρ defining Ω as earlier there are two constants $\varepsilon > 0$, $C > 0$, such that if $d(z, \partial\Omega) < \varepsilon$, then*

$$|\rho(z)| \leq C d(z, \partial\Omega).$$

PROOF. We have already pointed out in §1.4.2 that if ρ_1 and ρ_2 are two functions defining Ω there is a function $h \in \mathscr{E}(\mathbb{C})$ such that $\rho_1 = h\rho_2$ and $h > 0$. Hence, the proposition will follow if we find a simple defining function for Ω which satisfies the desired estimate. This is assured by the following lemma.

4.8.2. Lemma. *Let Ω be a bounded open set with regular boundary of class C^∞. There is an $\varepsilon > 0$ such that if $U_\varepsilon = \{z \in \mathbb{C} : d(z, \partial\Omega) < \varepsilon\}$, then the function d_Ω defined in U_ε by*

$$d_\Omega(z) = \begin{cases} d(z, \partial\Omega) & \text{if } z \in U_\varepsilon \cap (\mathbb{C}\setminus\Omega) \\ -d(z, \partial\Omega) & \text{if } z \in U_\varepsilon \cap \Omega \end{cases}$$

is of class C^∞ and such that $\nabla d_\Omega(z) \neq 0$ for every $z \in \partial\Omega$.

PROOF. We can assume without loss of generality that Ω is connected. We will prove the lemma finding $\varepsilon > 0$ corresponding to a neighborhood of Γ, the component of $\partial\Omega$ which is the boundary of the unbounded component of $\mathbb{C}\setminus\Omega$. The reader will convince himself that an analogous argument can be used for the other components of $\partial\Omega$. Since the total number of components of $\partial\Omega$ is finite this will suffice.

Let $\gamma: [0,1] \to \mathbb{C}$ be a C^∞ parameterization of Γ, γ periodic of period 1 and $|\dot{\gamma}(t)| \equiv 1$ as earlier. We can assume $\mathrm{Ind}_\gamma(a) = 1$ for $a \in \Omega$. For $(t,s) \in [0,1] \times]-1,1[$ define H by

$$H(t,s) = \gamma(t) - is\dot{\gamma}(t).$$

Due to the assumption that $\mathrm{Ind}_\gamma(a) = 1$ if $a \in \Omega$, $i\gamma(t)$ is the interior normal to Γ at the point $\gamma(t)$. If $\gamma = \gamma_1 + i\gamma_2$ then

$$H(t,s) = (\gamma_1(t) + s\dot{\gamma}_2(t)) + i(\gamma_2(t) - s\dot{\gamma}_1(t)),$$

and, considered as a map into $\mathbb{R}^2$, its Jacobian matrix is

$$J(t,s) = \begin{pmatrix} \dot{\gamma}_1(t) + s\ddot{\gamma}_2(t) & \dot{\gamma}_2(t) \\ \dot{\gamma}_2(t) - s\ddot{\gamma}_1(t) & -\dot{\gamma}_1(t) \end{pmatrix}.$$

For $s = 0$ we have

$$\det J(t,0) = -(\dot{\gamma}_1(t)^2 + \dot{\gamma}_2(t)^2) = -|\dot{\gamma}(t)|^2 = -1.$$

The inverse function theorem implies that for some $\varepsilon > 0$, the function H is invertible in $[0,1] \times]-\varepsilon, \varepsilon[$. Let $U_\varepsilon^\Gamma = \{z \in \mathbb{C} : d(z,\Gamma) < \varepsilon\}$, $G = H^{-1}$ is defined in U_ε^Γ and it is C^∞. Furthermore, if $G = (G_1, G_2)$ one has

(1) $\gamma(G_1(z))$ is the normal projection of z onto $\partial\Omega$
(2) $G_2(z) = d_\Omega(z)$.

Moreover, the Jacobian matrix of G at $z = \gamma(t) \in \Gamma$ is the inverse to the matrix $J(t,0)$, i.e.,

$$\begin{pmatrix} \dot{\gamma}_1(t) & \dot{\gamma}_2(t) \\ \dot{\gamma}_2(t) & -\dot{\gamma}_1(t) \end{pmatrix}.$$

In particular, the components of $\nabla d_\Omega(z)$ are given by $\dfrac{\partial G_2}{\partial x}(z) = \dot{\gamma}_2(t)$ and $\dfrac{\partial G_2}{\partial y}(z) = -\dot{\gamma}_1(t)$, which shows that $\nabla d_\Omega(z) \neq 0$. This concludes the proof of the lemma and of Proposition 4.8.1. □

4.8.3. Definition. If Ω is a bounded open set in $\mathbb{C}$ with regular boundary of class C^∞ and if $f : \bar{\Omega} \to \mathbb{C}$ is a C^∞ function, then we say that f vanishes to the order k on $\partial\Omega$ if f and all the derivatives of f up to the order $k-1$ vanish on $\partial\Omega$.

4.8.4. Proposition. *If Ω is a bounded open set in $\mathbb{C}$ with regular boundary of class C^∞, ρ is a defining function for Ω, and $f : \bar{\Omega} \to \mathbb{C}$ is a C^∞ function such that for some $C > 0$ and some positive integer k satisfies*

$$|f(z)| \leq C|\rho(z)|^k \qquad (z \in \bar{\Omega})$$

then f vanishes to the order k on $\partial\Omega$.

PROOF. We leave the proof to the reader who, locally can reduce it to the case of a half-plane and use the Taylor formula with integral remainder to conclude the proof. □

We will introduce now two powerful tools in the study of holomorphic maps. They are the Bergman kernel and the Bergman projection. They appear in the context of Hilbert spaces of holomorphic functions. Before proceeding we need to give the reader a brief introduction to the concept of reproducing kernels.

Let H be a Hilbert space whose elements are complex-valued functions in a set S and assume that the point evaluations $f \mapsto f(x)$ $(x \in S)$ are continuous linear functionals on H, i.e., there are constants $M_x > 0$ such that

$$|f(x)| \leq M_x \|f\| \qquad (\forall f \in H, \forall x \in S).$$

The Riesz representation theorem ensures the existence of an element $K_x \in H$ such that

$$f(x) = (f|K_x) \qquad (\forall f \in H).$$

(We have used $\|\cdot\|$ (resp. $(\cdot|\cdot)$) to denote the norm (resp. scalar product) in H.) The function $K: S \times S \to \mathbb{C}$ defined by

$$K(x, y) := (K_y|K_x)$$

is called the ***reproducing kernel*** of H. Note that $K(x, y) = K_y(x)$ and therefore for each y, $x \mapsto K(x, y)$, belongs to the Hilbert space H. Furthermore, $\overline{K(x, y)} = \overline{(K_y|K_x)} = (K_x|K_y) = K(y, x) = K_x(y)$, hence $y \mapsto \overline{K(x, y)}$ belongs to H for every x. Finally, the defining property of the reproducing kernel can be written as

$$f(x) = (f(\cdot)|K(\cdot, x)) \qquad (\forall f \in H).$$

This identity implies that the reproducing kernel is unique.

A typical example of a Hilbert space of functions with a reproducing kernel is the space $H^2(B(0, 1))$ of all holomorphic functions f in the unit disk $B(0, 1)$ whose Taylor developments at the origin, $f(z) = \sum_{n \geq 0} a_n z^n$, satisfy $\sum_{n \geq 0} |a_n|^2 < \infty$. The scalar product is given by

$$(f|g) = \sum_{n \geq 0} a_n \bar{b}_n,$$

if $f(z) = \sum_{n \geq 0} a_n z^n$ and $g(z) = \sum_{n \geq 0} b_n z^n$.

An orthonormal basis of $H^2(B(0, 1))$ is given by $e_n(z) = z^n$ $(n \geq 0)$. This establishes an isomorphism with ℓ^2 and with the subspace of $L^2(\partial B(0, 1), d\theta)$ of the "holomorphic Fourier series" (namely, those whose Fourier coefficients with negative indices are zero).

For $\zeta \in B(0, 1)$ consider the function

$$g_\zeta(z) = \sum_{n \geq 0} \bar{\zeta}^n z^n,$$

$g_\zeta \in H^2(B(0, 1))$, and for any $f \in H^2(B(0, 1))$ we have

$$(f|g_\zeta) = f(\zeta).$$

Therefore, $H^2(B(0, 1))$ has a reproducing kernel and it is given by $K(z, \zeta) = g_\zeta(z)$. That is,

$$K(z, \zeta) = \sum_{n \geq 0} \bar{\zeta}^n z^n = \frac{1}{1 - z\bar{\zeta}}.$$

It is called the ***Szegö kernel.***

The following proposition shows that for a separable Hilbert space with a reproducing kernel, the kernel can always be found as we have done for $H^2(B(0, 1))$ in terms of an orthonormal basis.

4.8.5. Proposition. *Let $\{e_n\}_{n \geq 0}$ be an orthonormal basis for a Hilbert space of functions on S which has a reproducing kernel. This kernel can be computed by* *the formula*

$$K(x, y) = \sum_{n\geq 0} e_n(x)\overline{e_n(y)}.$$

PROOF. For $x \in S$ fixed, the function K_x can be developed in a Fourier series in terms of the basis $\{e_n\}_{n\geq 0}$:

$$K_x = \sum_{n\geq 0} (K_x|e_n)e_n = \sum_{n\geq 0} \overline{(e_n|K_x)}e_n = \sum_{n\geq 0} \overline{e_n(x)}e_n.$$

By Parseval's identity

$$K(x, y) = (K_y|K_x) = \left(\sum_{n\geq 0} \overline{e_n(y)}e_n \middle| \sum_{m\geq 0} \overline{e_m(x)}e_m\right) = \sum_{n\geq 0} e_n(x)\overline{e_n(y)}. \quad \square$$

Another example of a Hilbert space with a reproducing kernel is the following. Let Ω be a bounded open set in $\mathbb{C}$. Recall that dm denotes the Lebesgue measure on Ω. Then $A^2(\Omega)$ is the space of holomorphic functions in Ω such that

$$\|f\|^2 = \int_\Omega |f(z)|^2\, dm(z) < \infty.$$

In other words, $A^2(\Omega) = \mathscr{H}(\Omega) \cap L^2(\Omega, dm)$. In order to prove that $A^2(\Omega)$ is a Hilbert space we need the following lemma.

4.8.6. Lemma. *For every compact set $K \subseteq \Omega$ there is a positive constant C_K such that for every $f \in A^2(\Omega)$*

$$\sup_{z\in K} |f(z)| \leq C_K\|f\|.$$

PROOF. Let $\varphi \in \mathscr{D}(\Omega)$, $\varphi \equiv 1$ in a neighborhood w of K, $K_1 = \operatorname{supp}\left(\frac{\partial\varphi}{\partial\bar{z}}\right)$. For $z \in K$ the representation

$$f(z) = \frac{1}{2\pi i}\int_{K_1} \frac{\partial\varphi}{\partial\bar{\zeta}}(\zeta)\frac{f(\zeta)}{\zeta - z}\, d\zeta \wedge d\bar{\zeta}$$

holds for every $f \in \mathscr{H}(\Omega)$ (apply Pompeiu's formula to φf). Let us denote

$$C_K = \sup_{z\in K} \frac{1}{2\pi}\left(\int_{K_1} \left|\frac{\partial\varphi}{\partial\bar{z}}(\zeta)\right|^2 \frac{1}{|z-\zeta|^2}\, dm(\zeta)\right)^{1/2},$$

which is finite since $K_1 \cap K = \varnothing$. From the Cauchy-Schwarz inequality it follows that for $z \in K$ and $f \in A^2(\Omega)$

$$|f(z)| \leq C_K\left(\int_{K_1} |f(\zeta)|^2\, dm(\zeta)\right)^{1/2} \leq C_K\left(\int_\Omega |f|^2\, dm\right)^{1/2}.$$

Therefore

$$\sup_{z\in K} |f(z)| \leq C_K\|f\|. \quad \square$$

4.8.7. Proposition. *The space $A^2(\Omega)$, with the scalar product*

$$(f|g) = \int_\Omega f\bar{g}\,dm,$$

is a separable Hilbert space of functions with a reproducing kernel.

PROOF. The preceding lemma implies that the point evaluations are continuous. What we have not yet shown is that $A^2(\Omega)$ is complete. It is clear that we can consider $A^2(\Omega)$ as a subspace of $L^2(\Omega, dm)$, hence if we show that it is closed, it will follow that $A^2(\Omega)$ is both complete and separable. If $\{f_j\}_{j\geq 0}$ is a Cauchy sequence in $A^2(\Omega)$ converging to a function $f \in L^2(\Omega, dm)$, then Lemma 4.8.6 shows that $\{f_j\}_{j\geq 0}$ converges uniformly over any compact subset of Ω to a holomorphic function. Therefore f can be taken to be in $\mathscr{H}(\Omega)$. □

The reproducing kernel of $A^2(\Omega)$ will be denoted K_Ω and called the ***Bergman kernel***. Recall that $K_\Omega(z, w) \equiv \overline{K_z(w)}$; hence the reproducing property can be written as

$$h(z) = \int_\Omega K_\Omega(z, w)h(w)\,dm(w),$$

for any $h \in A^2(\Omega)$, $z \in \Omega$.

In the case of the unit disk one can verify that the system

$$e_n(z) = \sqrt{\frac{n+1}{\pi}}\,z^n, \qquad n \geq 0$$

forms an orthonormal basis in $A^2(B(0, 1))$. In fact,

$$\begin{aligned}(e_n, e_m) &= \frac{1}{\pi}\sqrt{n+1}\sqrt{m+1}\int_{|z|<1} z^n\bar{z}^m\,dm(z)\\ &= \frac{1}{\pi}\sqrt{n+1}\sqrt{m+1}\int_{-\pi}^{\pi} e^{i(n-m)\theta}\,d\theta\int_0^1 r^{n+m+1}\,dr\\ &= 2\frac{\sqrt{n+1}\sqrt{m+1}}{n+m+2}\delta_{n,m} = \delta_{n,m}.\end{aligned}$$

Moreover, if $f(z) = \sum_{n\geq 0} a_n z^n$ belongs to $A^2(B(0, 1))$, we have for $0 < r < 1$,

$$\int_{|z|<r} f(z)\bar{z}^n\,dm(z) = 2\pi\sum_{k\geq 0} a_k\delta_{n,k}\frac{r^{n+k+2}}{n+k+2} = \pi a_n\frac{r^{2n+2}}{n+1}.$$

Since $f(z)\bar{z}^n \in L^1(B(0, 1), dm)$ we can let $r \to 1$ and obtain

$$a_n = \sqrt{\frac{n+1}{\pi}}(f|e_n).$$

This shows that $\{e_n\}_{n\geq 0}$ is complete. The Parseval identity becomes

$$\|f\|^2 = \pi \sum_{n\geq 0} \frac{|a_n|^2}{n+1},$$

and the Bergman kernel $K_B(B = B(0,1))$ is:

$$K_B(z,\zeta) = \frac{1}{\pi} \sum_{n\geq 0} (n+1) z^n \bar{\zeta}^n = \frac{1}{\pi(1 - z\bar{\zeta})^2}.$$

Therefore, for $f \in A^2(B(0,1))$ we have

$$f(z) = \frac{1}{\pi} \int_{|\zeta|<1} \frac{f(\zeta)}{(1 - z\bar{\zeta})^2}\, dm(\zeta).$$

This explicit construction of the Bergman kernel was dependent on the density of the polynomials in the space $A^2(B(0,1))$. If Ω is a Jordan domain, the polynomials are dense in $A^2(\Omega)$. We refer to [Mar] and [Ga] for a proof of this fact. For other domains, this question is related to the capacity of the boundary $C(\partial\Omega)$ (to be defined later on), see [Carl] and [Hed].

4.8.8. Proposition. *The map*

$$P_\Omega : f \mapsto \int_\Omega K_\Omega(z,\zeta) f(\zeta)\, dm(\zeta)$$

is the orthogonal projection $L^2(\Omega, dm) \to A^2(\Omega)$. *It is called the* ***Bergman projection***.

PROOF. It is easy to see that $A^2(\Omega)$ is the set of fixed points of P_Ω, and $P_\Omega^2 = P_\Omega$. □

4.8.9. Remark. The preceding theory of the space $A^2(\Omega)$ was made under the assumption that Ω is bounded. This is only imposed to guarantee $A^2(\Omega) \neq \{0\}$. For example, $A^2(\mathbb{C}) = \{0\}$. Whenever we know $A^2(\Omega) \neq \{0\}$ everything else holds. For instance, when $\Omega = \{z \in \mathbb{C} : \operatorname{Im} z > 0\}$.

4.8.10. Proposition (Change of Variables). *Let* $f : \Omega_1 \to \Omega_2$ *be a biholomorphic map between two open subsets of* $\mathbb{C}$. *For any* $\varphi \in L^2(\Omega_2, dm)$, $\psi \in L^2(\Omega_1, dm)$ *we have*

$$\int_{\Omega_1} f'(z)\varphi(f(z))\overline{\psi(z)}\, dm(z) = \int_{\Omega_2} \varphi(w)\overline{(f^{-1})'(w)\psi(f^{-1}(w))}\, dm(w).$$

In other words, the operators

$$\Lambda_1 : L^2(\Omega_2, dm) \to L^2(\Omega_1, dm) \quad \textit{and} \quad \Lambda_2 : L^2(\Omega_1, dm) \to L^2(\Omega_2, dm)$$

which are defined by

$$(\Lambda_1\varphi)(z) := f'(z)\varphi(f(z))$$
$$(\Lambda_2\psi)(w) := (f^{-1})'(w)\psi(f^{-1}(w))$$

are such that

(a) $\|\Lambda_1\varphi\|_{L^2(\Omega_1)} = \|\varphi\|_{L^2(\Omega_2)}$ *(isometry)*
(b) $\|\Lambda_2\psi\|_{L^2(\Omega_2)} = \|\psi\|_{L^2(\Omega_1)}$ *(isometry)*
(c) $(\Lambda_1\varphi|\psi)_{L^2(\Omega_1)} = (\varphi|\Lambda_2\psi)_{L^2(\Omega_2)}$ *(adjointness).*

PROOF. Let us show first that $\Lambda_1\varphi \in L^2(\Omega_1, dm)$. Since $|\varphi|^2 \in L^1(\Omega_2, dm)$ we have that $|\varphi \circ f|^2|J(f)| \in L^1(\Omega_1, dm)$, where $J(f)$ is the determinant Jacobian of the diffemorphism f as a map $\mathbb{R}^2 \to \mathbb{R}^2$. We know that by the Cauchy-Riemann equations $J(f) = |f'|^2$. Therefore, $\Lambda_1\varphi = f'(\varphi \circ f) \in L^2(\Omega_1, dm)$ and $\|\Lambda_1\varphi\| = \|\varphi\|$ by the usual rule of change of variables in the Lebesgue integral. In the same way one sees that $\Lambda_2\psi \in L^2(\Omega_2, dm)$ and $\|\Lambda_2\psi\| = \|\psi\|$.

Finally, since $f'(z) = 1/(f^{-1})'(f(z))$ and $(f^{-1})'(w) = 1/f'(f^{-1}(w))$, we have

$$\begin{aligned}(\Lambda_1\varphi|\psi) &= \int_{\Omega_1} f'(z)\varphi(f(z))\overline{\psi(z)}\,dm(z)\\ &= \int_{\Omega_1} |f'(z)|^2\varphi(f(z))\overline{(f^{-1})'(f(z))\psi(f^{-1}(f(z)))}\,dm(z)\\ &= \int_{\Omega_2} \varphi(w)\overline{(f^{-1})'(w)\psi(f(w))}\,dm(w) = (\varphi|\Lambda_2\psi).\end{aligned}$$ □

4.8.11. Corollary. *Let $f : \Omega_1 \to \Omega_2$ be a biholomorphic map and K_{Ω_1}, K_{Ω_2} the corresponding Bergman kernels. The transformation formula*

$$f'(z)K_{\Omega_2}(f(z), f(\zeta))\overline{f'(\zeta)} = K_{\Omega_1}(z, \zeta)$$

holds.

PROOF. Let $h \in A^2(\Omega_2)$, $a \in \Omega_1$, then

$$\begin{aligned}f'(a)h(f(a)) &= \int_{\Omega_1} K_{\Omega_1}(a, w)f'(w)h(f(w))\,dm(w)\\ &= \int_{\Omega_1} f'(w)h(f(w))\overline{K_{\Omega_1}(w, a)}\,dm(w)\\ &= \int_{\Omega_2} h(\zeta)\overline{(f^{-1})'(\zeta)K_{\Omega_1}(f^{-1}(\zeta), a)}\,dm(\zeta)\\ &= \int_{\Omega_2} h(\zeta)\overline{(f^{-1})'(\zeta)}K_{\Omega_1}(a, f^{-1}(\zeta))\,dm(\zeta).\end{aligned}$$

Replacing a by $f^{-1}(z)$ and multiplying both sides by $(f^{-1})'(z) = 1/f'(a)$, the last identity becomes

$$h(z) = \int_{\Omega_2} h(\zeta)[(f^{-1})'(z)K_{\Omega_1}(f^{-1}(z), f^{-1}(\zeta))\overline{(f^{-1})'(\zeta)}]\,dm(\zeta).$$

The uniqueness of the reproducing kernel allows us to conclude that

$$K_{\Omega_2}(z,\zeta) = (f^{-1})'(z)K_{\Omega_1}(f^{-1}(z), f^{-1}(\zeta))\overline{(f^{-1})'(\zeta)},$$

which is the formula we were looking for with f replaced by f^{-1}. □

4.8.12. Proposition. *Let $f:\Omega_1 \to \Omega_2$ be a biholomorphic map. Then*

$$P_{\Omega_1}(f'(\varphi \circ f)) = f'[(P_{\Omega_2}\varphi)\circ f] \qquad (\varphi \in L^2(\Omega_2, dm)).$$

PROOF.

$$\begin{aligned} P_{\Omega_1}(f'(\varphi \circ f))(z) &= \int_{\Omega_1} K_{\Omega_1}(z,w)f'(w)\varphi(f(w))\,dm(w) \\ &= \int_{\Omega_2} K_{\Omega_1}(z, f^{-1}(\zeta))\overline{(f^{-1})'(\zeta)}\varphi(\zeta)\,dm(\zeta) \\ &= f'(z)[(P_{\Omega_2}\varphi)\circ f](z). \end{aligned}$$ □

After these generalities about the Bergman kernel and projection we return to the main point of this section, the regularity of biholomorphic mappings up to the boundary.

Let Ω be a regular open bounded set of class C^∞ defined by $\rho:\mathbb{C}\to\mathbb{R}$ of class C^∞, let $\Delta = \dfrac{\partial^2}{\partial x^2} + \dfrac{\partial^2}{\partial y^2}$ denote, as always, the Laplace operator.

4.8.13. Lemma. *For every $\theta \in \mathscr{E}(\overline{\Omega})$ we have*

$$P_\Omega(\Delta(\theta\rho^2)) = 0.$$

PROOF. We need to show that for every $h \in A^2(\Omega)$ we have

$$\int_\Omega \bar{h}\Delta(\theta\rho^2)\,dm = 0.$$

In fact, if $h(w) = \overline{K_\Omega(z,w)}$ then we would have

$$0 = \int_\Omega K_\Omega(z,w)\Delta(\theta\rho^2)(w)\,dm(w) = P_\Omega(\Delta(\theta\rho^2))(z).$$

Let $\varepsilon > 0$ sufficiently small, $\rho_\varepsilon(z) = \rho(z) + \varepsilon$ if $\rho(z) \le -\varepsilon$ and $\rho_\varepsilon(z) = 0$ if $\rho(z) > -\varepsilon$. If we let $\Omega_\varepsilon = \{z \in \Omega : \rho_\varepsilon(z) < 0\}$ then Ω_ε is a relatively compact subset of Ω, $\partial\Omega_\varepsilon$ is regular of class C^∞.

We claim that

$$\int_\Omega \bar{h}\Delta(\theta\rho^2)\,dm = \lim_{\varepsilon\to\infty}\int_{\Omega_\varepsilon} \bar{h}\Delta(\theta\rho_\varepsilon^2)\,dm.$$

Namely,

$$\int_\Omega \bar{h}\Delta(\theta\rho^2)\,dm - \int_{\Omega_\varepsilon} \bar{h}\Delta(\theta\rho_\varepsilon^2)\,dm = \int_{\Omega_\varepsilon} \bar{h}\Delta(\theta(\rho^2 - \rho_\varepsilon^2))\,dm + \int_{\Omega\setminus\Omega_\varepsilon} h\Delta(\theta\rho^2)\,dm.$$

For any $\delta > 0$ there is an $\varepsilon_0 > 0$ such that if $0 < \varepsilon \le \varepsilon_0$ one has

$$\left|\int_{\Omega\setminus\Omega_\varepsilon} \bar{h}\Delta(\theta\rho^2)\,dm\right| < \delta/2.$$

This follows from the fact that, since $h \in L^2(\Omega, dm)$, $\bar{h}\Delta(\theta\rho^2) \in L^1(\Omega, dm)$. For any given $\varepsilon > 0$ we have in Ω_ε

$$\Delta(\theta(\rho_\varepsilon^2 - \rho)) = ((\rho + \varepsilon)^2 - \rho^2)\Delta\theta + 2\varepsilon\theta\Delta\rho + 4\varepsilon(\nabla\theta|\nabla\rho),$$

whence the inequality

$$\begin{aligned} &|\Delta(\theta(\rho_\varepsilon^2 - \rho))| \\ &\le 2\varepsilon \max\left(\sup_{z\in\bar{\Omega}} (|\rho| + 1)|\Delta\theta(z)|, \sup_{z\in\bar{\Omega}} |\theta(z)\Delta\rho(z)|, \sup_{z\in\bar{\Omega}} |2(\nabla\theta(z)|\nabla\rho(z))|\right) \\ &= 2\varepsilon M, \end{aligned}$$

holds in Ω_ε. Therefore,

$$\left|\int_{\Omega_\varepsilon} \bar{h}\Delta(\theta(\rho^2 - \rho_\varepsilon^2))\,dm\right| \le 2\varepsilon M \int_\Omega |h|\,dm.$$

Since Ω is bounded, this last quantity can also be made smaller than $\delta/2$ if ε is sufficiently small. The claim is therefore valid.

The integral $\int_{\Omega_\varepsilon} \bar{h}\Delta(\theta\rho_\varepsilon^2)\,dm$ can be computed using Green's formula

$$\int_{\Omega_\varepsilon} \bar{h}\Delta(\theta\rho_\varepsilon^2)\,dm = \int_{\Omega_\varepsilon} \theta\rho_\varepsilon^2\Delta(\bar{h})\,dm + \int_{\partial\Omega_\varepsilon} \left(\frac{\bar{h}\partial(\theta\rho_\varepsilon^2)}{\partial n} - \theta\rho_\varepsilon^2\frac{\partial\bar{h}}{\partial n}\right)ds.$$

The function ρ_ε^2 vanishes to order two on $\partial\Omega_\varepsilon$, therefore we also have $\frac{\partial}{\partial n}(\theta\rho_\varepsilon^2) = 0$. Since h is holomorphic, we have $\Delta\bar{h} = 0$. We conclude the integral on the left-hand side vanishes and, by the claim,

$$\int_\Omega \bar{h}\Delta(\theta\rho^2)\,dm = 0.$$

This concludes the proof of the lemma. □

From now on in this section, for the sake of simplicity we will write

$$d_\Omega(z) = d(z, \partial\Omega)$$

for any $z \in \Omega$. (Note the sign difference with §4.8.2.)

4.8.14. Lemma. *For every positive integer s there is a function* $\varphi_s \in \mathscr{E}(\bar{\Omega})$ *such that* $P_\Omega \varphi_s \equiv 1$ *and a constant* C_s *such that*

$$|\varphi_s(z)| \leq C_s(d_\Omega(z))^s \qquad (z \in \Omega).$$

PROOF. One proceeds by induction on s. For $s = 1$ we take $\varphi_1 = 1 - \Delta(\theta_1 \rho^2)$ for a convenient choice of $\theta_1 \in \mathscr{E}(\bar{\Omega})$. By the previous lemma we have $P_\Omega \varphi_1 = P_\Omega 1 - P_\Omega \Delta(\theta_1 \rho^2) = P_\Omega 1 = 1$.

We need to construct θ_1 so that

$$|\varphi_1(z)| \leq C|\rho(z)|,$$

this is sufficient by §4.8.1. One has

$$\varphi_1 = 1 - \Delta(\theta_1 \rho^2) = 1 - \rho^2 \Delta\theta_1 - 4\rho(\nabla\theta_1|\nabla\rho) - 2\theta_1|\nabla\rho|^2 - 2\rho\theta_1\Delta\rho.$$

One would like to take $\theta_1 = 1/(2|\nabla\rho|^2)$, then we would have

$$|\varphi_1| = \rho|\rho\Delta\theta_1 + 4(\nabla\theta_1|\nabla\rho) + 2\theta_1\Delta\rho| \leq \text{const. } \rho.$$

The problem is that $\nabla\rho$ could vanish, therefore we are compelled to introduce an auxiliary function $\chi \in \mathscr{D}(\mathbb{C})$ with the property that $\chi = 1$ in a neighborhood of $\partial\Omega$ and $\chi = 0$ in a neighborhood of $\{z \in \mathbb{C} : \nabla\rho(z) = 0\}$. Hence we take

$$\theta_1 = \frac{\chi}{|2\nabla\rho|^2},$$

and we have

$$\varphi_1 = 1 - \Delta(\theta_1 \rho^2) = \rho\Phi_1,$$

for some function $\Phi_1 \in \mathscr{E}(\bar{\Omega})$.

The inductive hypothesis is then the following: Assume we have constructed $\varphi_1, \ldots, \varphi_{s-1}$ satisfying

(1) $P_\Omega(\varphi_i) = 1$ and
(2) $\varphi_i = \rho^i \Phi_i$, $\Phi_i \in \mathscr{E}(\bar{\Omega})$.

We will then choose $\theta_s \in \mathscr{E}(\bar{\Omega})$ such that $\varphi_s = \varphi_{s-1} - \Delta(\rho^{s+1}\theta_s)$ verifies (2). Now,

$$\begin{aligned}\varphi_{s-1} - \Delta(\rho^{s+1}\theta_s) = \varphi_{s-1} &- \rho^{s+1}\Delta\theta_s - 2(s+1)\rho^s(\nabla\rho|\nabla\theta_s) \\ &- s(s+1)\rho^{s-1}\theta_s|\nabla\rho|^2 - (s+1)\rho^s\theta_s\Delta\rho.\end{aligned}$$

Let us choose

$$\theta_s = \frac{\chi\Phi_{s-1}}{s(s+1)|\nabla\rho|^2},$$

with the same χ as earlier. Then $\varphi_s = \rho^s\Phi_s$, for some $\Phi_s \in \mathscr{E}(\bar{\Omega})$. This concludes the proof of the lemma. □

We show now that the Bergman kernel K_B for the unit disk satisfies some simple estimates. Recall $K_B(z,\zeta) = \frac{1}{\pi}\frac{1}{(1-z\bar{\zeta})^2}$.

4.8.15. Lemma. *Let s be a positive integer. The Bergman kernel K_B for the unit disk verifies the inequality*

$$\sup_{|z|<1}\left|\frac{\partial^s}{\partial z^s}K_B(z,\zeta)\right| \le \frac{(s+1)!}{\pi}(d_B(\zeta))^{-s-2}.$$

Proof. We have

$$\frac{\partial^s}{\partial z^s}K_B(z,\zeta) = \frac{(s+1)!}{\pi}\frac{\bar{\zeta}^s}{(1-z\bar{\zeta})^{s+2}}.$$

The maximum principle allows us to conclude that

$$\sup_{|z|<1}\left|\frac{\partial^s}{\partial z^s}K_B(z,\zeta)\right| \le \frac{(s+1)!}{\pi}\left(\inf_{|z|=1}|1-z\bar{\zeta}|\right)^{-s-2}.$$

But $\left|\frac{z-\zeta}{1-z\bar{\zeta}}\right| = 1$ if $|z|=1$ and $|\zeta|<1$, hence

$$\inf_{|z|=1}|1-z\bar{\zeta}| = \inf_{|z|=1}|z-\zeta| = d_B(\zeta).$$

Whence the statement of the lemma follows. □

4.8.16. Lemma. *Let $f: B(0,1)\to\Omega$ be a biholomorphic map of the unit disk onto an open simply connected bounded set Ω with a C^∞ regular boundary. There is then a constant $c>0$ such that*

$$d_\Omega(f(z)) \le c\,d_B(z).$$

Proof. Let $\gamma:[0,1]\to\partial\Omega$ be a parameterization of $\partial\Omega$ with $|\dot{\gamma}(t)|=1$ inducing the usual orientation. Let $R_0>0$ be a number smaller than the lower bound of the radius of curvature of every point in $\partial\Omega$. The closed disks of center $\gamma(t)+R_0 i\dot{\gamma}(t)$ and radius R_0, are contained in $\bar{\Omega}$ and touch the boundary at a single point.

Let $p\in\partial\Omega$, $p=\gamma(t)$, $q=\gamma(t)+R_0 i\dot{\gamma}(t)$, and B_p the disk of center q and radius R_0 (See Figure 4.4). The Poisson kernel for B_p is given by

$$P(z,\zeta) = \frac{R_0^2-|z-q|^2}{|\zeta-z|^2} \ge \frac{(R_0-|z-q|)(R_0+|z-q|)}{(2R_0)^2} \ge \frac{1}{4R_0}d_\Omega(z).$$

The function f^{-1} is bounded by 1 on ∂B_p and continuous except possibly at the point p. Therefore, by Poisson's formula,

$$f^{-1}(z) = \int_{\partial B_p} P(z,\zeta)f^{-1}(\zeta)\,d\sigma(\zeta),$$

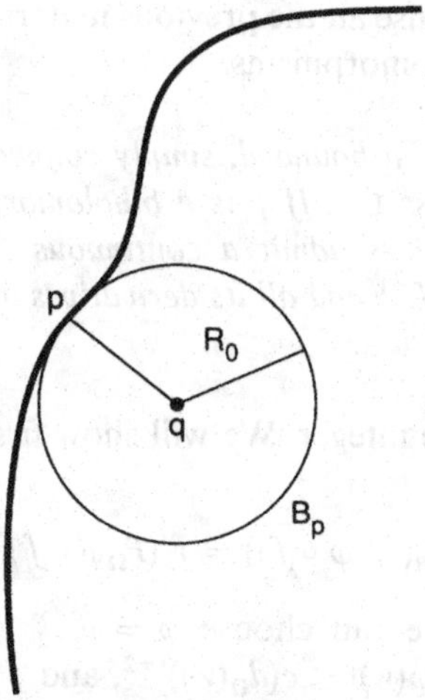

Figure 4.4

for any $z \in B_p$. Since $\int_{\partial B_p} P(z,\zeta)\, d\sigma(\zeta) = 1$, we have

$$d_B(f^{-1}(z)) = 1 - |f^{-1}(z)| \geq \int_{\partial B_p} P(z,\zeta)(1 - |f^{-1}(\zeta)|)\, d\sigma(\zeta)$$

$$\geq \frac{d_\Omega(z)}{8\pi R_0^2} \int_{\partial B_p} (1 - |f^{-1}(\zeta)|)\, |d\zeta|.$$

The function

$$\mu(p) = \frac{1}{8\pi R_0^2} \int_{\partial B_p} (1 - |f^{-1}(\zeta)|)\, |d\zeta|$$

is continuous on the compact set $\partial\Omega$, as one can see by applying Lebesgue's dominated convergence theorem. Since $\mu(p) > 0$ for every $p \in \partial\Omega$, there is a $c > 0$ such that $\mu(p) > \frac{1}{c}$. Therefore

$$d_B(f^{-1}(z)) \geq \frac{1}{c} d_\Omega(z),$$

whenever $z \in \Omega$, $d_\Omega(z) < R_0$. Since the continuous function $d_B(f^{-1}(z))$ is strictly positive on the compact set $\{z \in \Omega : d_\Omega(z) \geq R_0\}$, increasing c, if necessary, we obtain

$$d_B(f^{-1}(z)) \geq \frac{1}{c} d_\Omega(z), \qquad \text{for all } z \in \Omega.$$

Replacing z by $f(w)$ we have

$$d_\Omega(f(w)) \leq c\, d_B(w) \qquad (w \in B(0,1)). \qquad \square$$

We are finally ready to use all the previous material to obtain the theorem on C^∞ extensions of biholomorphisms.

4.8.17. Theorem. *Let Ω be a bounded, simply connected, open set in $\mathbb{C}$, with a regular boundary of class C^∞. If f is a biholomorphism of $B(0,1)$ onto Ω, then f and all its derivatives admit a continuous extension to $\bar{B}(0,1)$ (i.e., $f \in \mathscr{E}(\bar{B}(0,1))$). Moreover, f^{-1} and all its derivatives admit a continuous extension to $\bar{\Omega}$ (i.e., $f^{-1} \in \mathscr{E}(\bar{\Omega})$).*

PROOF. Let s be a positive integer. We will show first that $f^{(s)}$ is a bounded function. We know that

$$P_B(f'(\varphi \circ f)) = f'(P_\Omega \varphi \circ f)$$

for any $\varphi \in L^2(\Omega, dm)$. We can choose $\varphi = \varphi_{s+2}$ given by Lemma 4.8.14. Therefore, $P_\Omega \varphi \equiv 1$ and $|\varphi(w)| \le c(d_\Omega(w))^{s+2}$, and $f' = P_B(f'(\varphi \circ f))$. Let us show now that for some $c_1 > 0$ we have

$$|f'(z)\varphi(f(z))| \le c_1(d_B(z))^{s+1} \qquad (z \in B(0,1)).$$

By the choice of φ we have

$$|\varphi(f(z))| \le c(d_\Omega f(z)))^{s+2} \le c'(d_B(z))^{s+2},$$

where the last inequality is a consequence of §4.8.16. If we can show that

$$|f'(z)| \le c''(d_B(z))^{-1},$$

we will have the desired inequality.

Fix $z \in B(0,1)$ and $0 < r < d_B(z)$ arbitrary, then we can write

$$f'(z) = \frac{1}{\pi r^2} \int_{B(z,r)} f'\, dm.$$

By the Cauchy-Schwarz inequality

$$|f'(z)| \le \frac{1}{\pi r^2} (\pi r^2)^{1/2} \|f'\|_{L^2(B(0,1))} = \sqrt{\frac{m(\Omega)}{\pi}} \frac{1}{r},$$

since $\|f'\|^2_{L^2(B(0,1))} = m(\Omega)$. Given that r was arbitrary we obtain

$$|f'(z)| \le \sqrt{\frac{m(\Omega)}{\pi}} (d_B(z))^{-1},$$

as we wanted.

As we pointed out earlier, $f' = P_B(f'(\varphi \circ f))$, hence

$$f^{(s)}(z) = (P_B(f'(\varphi \circ f)))^{(s-1)}(z) = \frac{d^{s-1}}{dz^{s-1}} \int_B K_B(z,\zeta) f'(\zeta)\varphi(f(\zeta))\, dm(\zeta).$$

The estimates on the derivatives of K_B from Lemma 4.10.15, allow us to take the derivatives under the integral sign

$$f^{(s)}(z) = \int_B \frac{d^{s-1}}{dz^{s-1}} K_B(z,\zeta) f'(\zeta)\varphi(f(\zeta))\, dm(\zeta),$$

and they also show, by the previous estimates of the integrand, that $|f^{(s)}(z)| \leq c_2$ independent of z. For a holomorphic function in $B(0, 1)$ we have that

$$|g(z) - g(w)| \leq \left(\sup_{\zeta \in [z,w]} |g'(\zeta)| \right) |z - w|,$$

hence, if g' is bounded, the function g has a continuous extension to $\bar{B}(0, 1)$. Therefore, f and all its derivatives extend continuously to $\bar{B}(0, 1)$.

Now that we know that $f \in \mathscr{E}(\bar{B}(0, 1))$, let us show that $f^{-1} \in \mathscr{E}(\bar{\Omega})$. This would follow from the inverse function theorem if we knew that $\nabla f(z) \neq 0$ for all $z \in \partial B(0, 1)$. Let $z_0 \in \partial B(0, 1)$ and $w_0 = f(z_0)$. We can assume, to simplify the notation, that $z_0 = 1$. Choose p_0 on the exterior normal to $\partial\Omega$ at w_0 such that the closed disk of center p_0 and radius $|w_0 - p_0|$ touches $\bar{\Omega}$ only at w_0. Let $a \in \mathbb{C}$, $|a| = 1$ be such that

$$\frac{a}{w_0 - p_0} = \frac{1}{|w_0 - p_0|} = M.$$

The function $h(w) := \operatorname{Re}\left(\dfrac{a}{w - p_0}\right)$ is harmonic in Ω, C^∞ in $\bar{\Omega}$ and takes its maximum M at $w = w_0$. Therefore $g(z) = h(f(z)) - M$ is a harmonic function in $B(0, 1)$, C^∞ in $\bar{B}(0, 1)$, $g(z) < 0$ in $\bar{B}(0, 1)\backslash\{1\}$, $g(1) = 0$. Let us show that $\dfrac{\partial g}{\partial r}(1) > 0$. By the chain rule this will imply that $\nabla f(1) \neq 0$. For $0 < r < 1$ we can apply Poisson's formula

$$g(r) = \frac{1}{2\pi} \int_0^{2\pi} \frac{1 - r^2}{|r - e^{i\theta}|^2} g(e^{i\theta})\, d\theta.$$

Hence,

$$\frac{g(1) - g(r)}{1 - r} = \frac{1}{2\pi} \int_0^{2\pi} \frac{1 + r}{|r - e^{i\theta}|^2} (-g(e^{i\theta}))\, d\theta \geq \frac{1}{4\pi} \int_0^{2\pi} (-g(e^{i\theta}))\, d\theta > 0,$$

since $\dfrac{1 + r}{|r - e^{i\theta}|^2} \geq \dfrac{1}{1 + r} \geq \dfrac{1}{2}$. Therefore $\dfrac{\partial g}{\partial r}(1) > 0$. As pointed out earlier, this concludes the proof of the theorem. □

4.8.18. Corollary. *If Ω_1 and Ω_2 are two bounded open sets, simply connected, whose boundaries are regular of class C^∞, and if $f : \Omega_1 \to \Omega_2$ is a biholomorphism, then f and all its derivatives have a continuous extension to $\bar{\Omega}_1$. The same holds for f^{-1} and all its derivatives in $\bar{\Omega}_2$.*

PROOF. We know there are two biholomorphic mappings $\varphi_i : B(0, 1) \to \Omega_i$, which by the preceding theorem have C^∞ extensions up to the boundary of the unit disk, and their inverses φ_i have a C^∞ extension up to the boundary of Ω_i. The same holds for $h = \varphi_2^{-1} \circ f \circ \varphi_1 : B(0,1) \to B(0,1)$ which, by the way, is a Moebius transformation. Therefore, $f = \varphi_2 \circ h \circ \varphi_1^{-1}$ has the announced C^∞ extension properties to the boundary. □

As an application of Theorem 4.8.17 we have the following.

4.8.19. Theorem. *Let Ω be a bounded, simply connected, open set with boundary regular of class C^∞. Let $f : \partial\Omega \to \mathbb{C}$ be a C^∞ function. The solution u of the Dirichlet problem*

$$\begin{cases} \Delta u = 0 & \text{in } \Omega \\ u|_{\partial\Omega} = f \end{cases}$$

is C^∞ in $\bar{\Omega}$.

PROOF. Assume the theorem has been proved for the case $\Omega = B(0,1)$. Let $\Phi : B(0,1) \to \Omega$ be a biholomorphic map. By §4.8.17 it admits a C^∞ extension to $\bar{B}(0,1)$. The function $\tilde{f} = f \circ (\Phi | \partial B(0,1))$ belongs to $\mathscr{E}(\partial B(0,1))$. By the assumption, the solution $\tilde{u}$ of the Dirichlet problem

$$\begin{cases} \Delta \tilde{u} = 0 & \text{in } B(0,1) \\ \tilde{u}|_{\partial B(0,1)} = \tilde{f} \end{cases}$$

is C^∞ on $\bar{B}(0,1)$. Since $\Phi^{-1} \in \mathscr{E}(\bar{\Omega})$ one obtains that $u = \tilde{u} \circ \Phi^{-1} \in \mathscr{E}(\bar{\Omega})$ and solves the original Dirichlet problem. Therefore, to conclude the proof we need to prove the following theorem.

4.8.20. Theorem. *If $f \in \mathscr{E}(\partial B(0,1))$, its Poisson integral Pf is a C^∞ function in $\bar{B}(0,1)$.*

PROOF. If $f(e^{i\theta}) = \sum_{n \in \mathbb{Z}} a_n e^{in\theta}$ is the Fourier series expansion of f, the expression for $Pf(re^{i\theta})$ is

$$Pf(re^{i\theta}) = \sum_{n \in \mathbb{Z}} a_n r^{|n|} e^{in\theta}, \qquad 0 \le r \le 1.$$

Lemma 4.8.21 implies that for every integer $N > 0$ there is a constant $C_N > 0$ such that

$$|a_n| \le C_N (1 + |n|)^{-N} \qquad (n \in \mathbb{Z}).$$

These estimates show that all the partial derivatives of Pf are bounded, which guarantees that Pf has a C^∞ extension to $\bar{B}(0,1)$. In fact, for $0 < r < 1$ we have

$$\frac{\partial^{p+q}}{\partial r^p \partial \theta^q} Pf(re^{i\theta}) = \sum_{\substack{n \in \mathbb{Z} \\ |n| \ge p}} a_n i^q n^q (|n|(|n| - 1) \ldots (|n| - p + 1)) r^{|n| - p} e^{in\theta},$$

which are clearly bounded in $\bar{B}(0,1)$. □

4.8.21. Lemma. *A continuous function $\varphi : \partial B(0, 1) \to \mathbb{C}$ is C^∞ if and only if the sequence $\{b_n\}_{n \in \mathbb{Z}}$ of its Fourier series is rapidly decreasing, i.e., for every integer $N \geq 0$ there is a constant $C_N > 0$ such that*

$$|b_n| \leq C_N(1 + |n|)^{-N} \qquad (n \in \mathbb{Z}).$$

PROOF. If φ is C^∞, then by integration by parts we have

$$b_n = \left(\frac{i}{n}\right)^k \frac{1}{2\pi} \int_{-\pi}^{\pi} u^{(k)}(t) e^{-int}\, dt,$$

if $n \in \mathbb{Z}^*$, $u(t) = \varphi(e^{it})$. The required estimate is then immediate. Conversely, if $|b_n| \leq C_2(1 + |n|)^{-2}$, then the Fourier series $\sum b_n e^{int}$ converges uniformly to $u(t) = \varphi(e^{it})$. The series differentiated term by term are also convergent by the rapid decrease of the sequence. Therefore φ is a C^∞ function. Note that if we had assumed originally only that $\varphi \in L^2$ we would have obtained that φ coincides with a C^∞ function almost everywhere. □

We are going to study now the case of biholomorphic maps between bounded open sets with regular C^∞ boundary but not necessarily simply connected. The first stage consists of proving that such an open set Ω is biholomorphic to an open set Ω', bounded with a regular boundary, the components of the boundary being analytic curves. (For us, this will mean these curves are images of $\partial B(0, 1)$ by maps which are holomorphic on $\partial B(0, 1)$ and injective.) We will also prove that the biholomorphism between Ω and Ω' has a C^∞ extension up to the boundaries.

4.8.22. Proposition. *Every bounded open set Ω in $\mathbb{C}$, with regular boundary of class C^∞, is biholomorphic to a bounded open set Ω', with regular boundary whose components are analytic Jordan curves.*

PROOF. Let $C_0, C_1, \ldots, C_n$ denote the connected components of $\partial\Omega$. (There are only finitely many of them by §1.4.3.) By §1.4.4 they are Jordan curves of class C^∞. We can assume C_0 is the boundary of the unbounded component F_0 of $\mathbb{C} \setminus \Omega$. The open set $U_0 = \mathbb{C} \setminus F_0$ is bounded, simply connected, and with a C^∞ regular boundary. Therefore we can find a biholomorphism h_0 which is C^∞ up to the boundary, from U_0 onto $B(0, 1)$. Denote $C_1^1, \ldots, C_n^1$ the images by h_0 of the curves $C_1, \ldots, C_n$. The open set $U_1 = S^2 \setminus \overline{\text{Int}(C_1^1)}$ is simply connected and with regular C^∞ boundary. Therefore, there exists a biholomorphism h_1, C^∞ up to the boundary, from U_1 onto $S^2 \setminus \bar{B}(0, 1)$. (To see this one needs to apply two inversions to reduce oneself to the case of bounded sets in $\mathbb{C}$.) Under h_1, the image of $\partial B(0, 1)$ is an analytic curve C_0^2. Hence, by $h_1 \circ h_0$ the images of C_0 and C_1 are analytic Jordan curves C_0^2 and $C_1^2 = \partial B(0, 1)$. Let $C_2^2, \ldots, C_n^2$ be the images of $C_1, \ldots, C_n$ under the map $h_1 \circ h_0$. We can now find a biholomorphic map h_2 of $U_2 = S^2 \setminus \overline{\text{Int}(C_2^2)}$, C^∞ up to the boundary, onto $S^2 \setminus \bar{B}(0, 1)$. By $h_2 \circ h_1 \circ h_0$ the images of C_0, C_1, C_2 are analytic Jordan curves $C_0^3, C_1^3, C_2^3 = \partial B(0, 1)$. Continuing in this fashion, we construct the bounded open set Ω', with analytic regular boundary, and the biholomorphism h between Ω and Ω'. □

4.8.23. Proposition. *Let Ω_1, Ω_2 be two open bounded sets with regular boundaries of class C^∞, $h: \Omega_1 \to \Omega_2$ a biholomorphism. Then h and h^{-1} have C^∞ extensions up to the boundaries.*

PROOF. Let Ω'_1, Ω'_2 and φ_1, φ_2 be the open sets with analytic boundaries and the corresponding biholomorphic mappings of $\Omega_i \to \Omega'_i$ given by §4.8.22. The biholomorphism $k = \varphi_2 \circ h \circ \varphi_1^{-1}$ between Ω'_1 and Ω'_2 is in fact the restriction to Ω'_1 of a function holomorphic in a neighborhood U_1 of $\bar{\Omega}'_1$, as follows from §4.7.20 (cf. §4.7.22). The same reasoning holds for k^{-1}. Since $h = \varphi_2^{-1} \circ k \circ \varphi_1$, the proposition is correct. □

4.8.24. Corollary. *If Ω is a bounded open set with C^∞ regular boundary, then its Green function with pole at $z_0 \in \Omega$ is a C^∞ function up to the boundary of Ω and the harmonic measure ω_{z_0} of the point z_0 is given by*

$$d\omega_{z_0}(\cdot) = -\frac{1}{2\pi}\frac{\partial g}{\partial n}(\cdot\,; \zeta_0, \Omega)\, ds.$$

PROOF. If the open set has an analytic boundary these statements hold by §4.7.18. We can now use §4.8.22 and §4.7.16 to finish the proof of the corollary. □

4.8.25. Remark. There is an analogous theorem for domains with regular boundary of class C^k, $1 \leq k < \infty$. The biholomorphisms have only a C^{k-1} extension (in fact, a bit better). The proofs are considerably harder. The interested reader should consult the work of Warschawski in the book [Pom].

One can also use Corollary 4.8.18 to prove an extension of the Schwarz' reflection principle to C^∞ boundaries. The reader will find in [BeL] a proof and further references. The statement is the following:

Let γ_1, γ_2 be two C^∞ curves in $\mathbb{C}$, $z_0 \in \gamma_1$ and D a disk centered at z_0 such that $D \backslash \gamma_1$ has exactly two connected components D_+ and D_-. Suppose there is f holomorphic in D_+, continuous in $\bar{D}_+$ and such that $f(z) \in \gamma_2$ whenever $z \in \gamma_1$. Then f has a C^∞ extension to a neighborhood of z_0. Moreover, if f is not constant, there is $n \in \mathbb{N}^*$ such that $f^{(n)}(z_0) \neq 0$.

Note that a priori f could vanish with infinite order at a boundary point of a domain (give an example!)

EXERCISES 4.8.

1. Compute the Bergman kernel function K_H for the upper half-plane H.
2. Let Ω be a bounded domain in $\mathbb{C}$ with a C^∞ regular boundary. Let $f \in C^\infty(\bar{\Omega})$. Show that the problem

$$\begin{cases} \Delta u = f & \text{in } \Omega \\ u = 0 & \text{on } \partial\Omega,\ u \in C^\infty(\bar{\Omega}) \end{cases} \qquad (*)$$

can be reduced to the Dirichlet problem. (Hint: Extend f to a function $F \in \mathscr{D}(\mathbb{C})$. Solve the equation $\Delta v = F$, $v \in \mathscr{E}(\mathbb{C})$.) Conclude that the problem (*) is solvable in Ω.

3. Method to solve the equation $\dfrac{\partial v}{\partial \bar{z}} = f$ in Ω when $f \in C^\infty(\bar{\Omega})$: Solve the problem (*) of Exercise 4.8.2 and let $v = 4\dfrac{\partial u}{\partial z}$. Show that
   (i) $\dfrac{\partial v}{\partial \bar{z}} = f$ in Ω;
   (ii) $v \in L^2(\Omega, dm)$;
   (iii) v is orthogonal to $A^2(\Omega)$ in $L^2(\Omega, dm)$;
   (iv) $P_\Omega v = 0$;
   (v) v is the unique solution of (i) which is in $L^2(\Omega, dm)$ and $P_\Omega v = 0$.
4. Let Ω be a Jordan domain with a C^∞ boundary. The object of this exercise is to show that the polynomials are dense in $A^2(\Omega)$ with the help of Theorem 4.8.17 and Mergelyan's theorem. We let $f: B(0,1) \to \Omega$ be a conformal map given by the Riemann mapping theorem.
   (a) Show that the map f induces a linear map $A^2(\Omega) \to A^2(B(0,1))$ given by $\varphi \mapsto \varphi \circ f$. This map is continuous and its inverse is continuous.
   (b) Conclude from (a) that the linear combinations of the powers of f^{-1} are dense in $A^2(\Omega)$.
   (c) Show that the polynomials are dense in $A^2(\Omega)$.
5. Let Ω be a bounded domain which is not simply connected and at least one of the components K of Ω^c has more than one point. We want to show that the polynomials cannot be dense in $A^2(\Omega)$.
   (a) Let z_1, z_2 be distinct points of ∂K. Show that the function
   $$f(z) := (z - z_1)^{-1/2}(z - z_2)^{-1/2}$$
   belongs to $A^2(\Omega)$.
   (b) Let Γ be a Jordan curve in Ω such that $K \subseteq \mathrm{Int}(\Gamma)$. Show that if there is a sequence of polynomials $(p_n)_{n\geq 1}$ such that $\|p_n - f\|_{L^2(\Omega)} \to 0$ then $p_n \to f$ uniformly on Γ. Prove that this is impossible.

§9. Introduction to Potential Theory

We have seen that the Riesz decomposition theorem associates to each subharmonic function the logarithmic potential of a positive measure, the Laplacian of this subharmonic function. The study of potentials leads to a notion more delicate than that of zero measure, zero capacity, which allows us to decide which singularities are removable for subharmonic functions. We will see that there are several ways of computing the capacity of a set. In particular, the geometric concept of transfinite diameter of a set will be useful in the second volume to study the arithmetic properties of certain entire functions taking integral values on $\mathbb{Z}$.

For a bounded Borel set E in $\mathbb{C}$, denote by $\mathscr{P} = \mathscr{P}(E)$ the family of all Borel measures $\mu \geq 0$ in $\mathbb{C}$, of total mass equal to 1 and such that $\mu(E) = 1$ (i.e., probability measures on E). For a positive measure μ of finite mass we can define its ***energy*** $I(\mu)$ by

$$I(\mu) := \int_{E\times E} \log\frac{1}{|z-t|}\,d\mu(z)\,d\mu(t).$$

Note that $I(\mu) \in]-\infty, \infty]$ is well defined as the negative of $\int_{E\times E} \log|z-t|\,d\mu(z)\,d\mu(t)$, which has to be interpreted in the sense of §4.2.

4.9.1. Definition. A bounded Borel set E is said to be of ***zero logarithmic capacity*** if $I(\mu) = +\infty$ for every $\mu \in \mathscr{P}(E)$. If not, the ***capacity*** $C(E)$ is defined by

$$C(E) = e^{-V(E)},$$

where

$$V(E) := \inf\{I(\mu) : \mu \in \mathscr{P}(E)\}.$$

(The capacity of the empty set is 0 by definition.)

We remark that $\log\dfrac{1}{|z_1 - z_2|} \geq \log(1/\operatorname{diam}(E)) > -\infty$ when $z_1, z_2 \in E$, where $\operatorname{diam}(E) = \sup_{u,v\in E} |u - v|$ is the diameter of the bounded set E. Hence one has $I(\mu) \geq \log(1/\operatorname{diam}(E))$ for every $\mu \in \mathscr{P}(E)$ and

$$C(E) \leq \operatorname{diam}(E).$$

4.9.2. Proposition. *The function $E \mapsto C(E)$, which to a bounded Borel set assigns its capacity, has the following properties:*

(1) $C(\{a\}) = 0$ $(a \in \mathbb{C})$;
(2) $C(\lambda E + a) = |\lambda| C(E)$ $(\lambda, a \in \mathbb{C})$;
(3) $E_1 \subseteq E_2$ *implies* $C(E_1) \leq C(E_2)$;
(4) *If the Lebesgue measure* $m(E) > 0$, *then* $C(E) > 0$;
(5) $C(E) = \sup\{C(K) : K$ *compact*, $K \subseteq E\}$.

PROOF. (1) $\mu = \delta_a$ is the only measure in $\mathscr{P}(\{a\})$ and $I(\delta_a) = +\infty$.
(2) For $\lambda = 0$ it is (1). Otherwise consider $\varphi(z) = \lambda z + a$, $\varphi : E \to \lambda E + a$. The map $\varphi_* : \mathscr{P}(E) \to \mathscr{P}(\lambda E + a)$ is a bijection and for $\mu \in \mathscr{P}(E)$ one has

$$\begin{aligned} I(\varphi_*\mu) &= \int_{(\lambda E+a)\times(\lambda E+a)} \log\frac{1}{|z_1 - z_2|}\,d(\varphi_*\mu)(\zeta_1)\,d(\varphi_*\mu)(\zeta_2) \\ &= \int_{(E\times E)} \log\frac{1}{|\varphi(z_1) - \varphi(z_2)|}\,d\mu(z_1)\,d\mu(z_2) \\ &= -\log|\lambda| + I(\mu). \end{aligned}$$

Therefore $V(\lambda E + a) = -\log|\lambda| + V(E)$ and $C(\lambda E + a) = |\lambda| C(E)$.

(3) Clearly $\mathscr{P}(E_1) \subseteq \mathscr{P}(E_2)$, hence

$$V(E_1) = \inf\{I(\mu) : \mu \in \mathscr{P}(E_1)\} \geq \inf\{I(\mu) : \mu \in \mathscr{P}(E_2)\} = V(E_2).$$

Therefore $C(E_1) \leq C(E_2)$.

(4) If $m(E) > 0$, let $m|E$ be the Lebesgue measure restricted to E, then $\mu = \dfrac{1}{m(E)} m|E \in \mathscr{P}(E)$ and $I(\mu) < \infty$. This holds because

$$I(\mu) = \frac{2\pi}{(m(E))^2} \int_E U^m \, dm$$

and the logarithmic potential $U^m \in L^1_{\text{loc}}(\mathbb{C})$. Hence $V(E) < \infty$ and $C(E) > 0$.

The last item is a consequence of the regularity of the Borel measures and is left to the reader. □

To clarify the proof of item (4) in Proposition 4.9.2 let us recall that one calls logarithmic potential of a complex measure μ with compact support in $\mathbb{C}$ the function:

$$U^\mu(z) = \frac{1}{2\pi} \int \log|z - t| \, d\mu(t).$$

U^μ can be written, for $|z|$ sufficiently large, as $\dfrac{1}{2\pi} \log|z| + h(z)$, h harmonic near ∞ and zero at ∞. One knows also that $\Delta U^\mu = \mu$ (cf. §4.4.24).

For $\mu \in \mathscr{P}(E)$ the Fubini-Tonelli theorem yields

$$I(\mu) = -2\pi \int \left\{ \frac{1}{2\pi} \int \log|z - t| \, d\mu(t) \right\} d\mu(z) = -2\pi \int U^\mu(z) \, d\mu(z).$$

If $I(\mu) < \infty$ then $\int U^\mu(z) \, d\mu(z) > -\infty$, hence U^μ is finite μ – a.e. One knows, besides, that $U^\mu \in L^1_{\text{loc}}(\mathbb{C})$, and hence, it is finite almost everywhere for the Lebesgue measure.

The same argument used in the proof of §4.9.2, (4), can be used to obtain the following more general result.

4.9.3. Proposition. *Let E_1, E_2 be two compact subsets of $\mathbb{C}$ such that $E_1 \subseteq E_2$, $C(E_1) = 0$ and $C(E_2) > 0$. If $\mu \in \mathscr{P}(E_2)$ satisfies $I(\mu) < \infty$ then $\mu(E_1) = 0$.*

Proof. Applying a homothety (which is allowed by §4.9.2 (2)) we can assume that $\operatorname{diam}(E_2) \leq 1$. Therefore $\log \dfrac{1}{|z_1 - z_2|} \geq 0$ in $E_2 \times E_2$ and

$$\int_{E_1 \times E_1} \log \frac{1}{|z_1 - z_2|} \, d\mu(z_1) \, d\mu(z_2)$$

$$\leq \int_{E_2 \times E_2} \log \frac{1}{|z_1 - z_2|} \, d\mu(z_1) \, d\mu(z_2) = I(\mu) < \infty.$$

If $\mu(E_1) > 0$ then $\mu/\mu(E_1)$ would belong to $\mathscr{P}(E_1)$ and have finite energy which would contradict $C(E_1) = 0$. □

4.9.4. Corollary. *If E_1, E_2 are compact sets with $C(E_1) = C(E_2) = 0$ then $C(E_1 \cup E_2) = 0$.*

PROOF. If μ is a nonnegative measure on $E_1 \cup E_2$ such that $I(\mu) < \infty$ then $\mu(E_1 \cup E_2) \leq \mu(E_1) + \mu(E_2) = 0$. □

4.9.5. Corollary. *If $(E_n)_{n\geq 1}$ is a sequence of compact sets in $\mathbb{C}$ such that $C(E_n) = 0$ and $E = \bigcup_{n\geq 1} E_n$ is compact, then $C(E) = 0$.*

PROOF. If μ is a nonnegative measure on E such that $I(\mu) < \infty$ then $\mu(E) \leq \sum_{n\geq 1} \mu(E_n) = 0$. □

We need to consider other potentials (introduced by M. Riesz), besides the logarithmic potentials.

4.9.6. Definition. For $\alpha \in \mathbb{C}$ and μ a complex measure with compact support in $\mathbb{C}$, we set

$$U_\alpha^\mu(z) = \int \frac{d\mu(t)}{|z - t|^\alpha},$$

which is called the ***Riesz potential of order*** α for the measure μ.

4.9.7. Remark. For z fixed in $\mathbb{C}\backslash K$, $\operatorname{supp}(\mu) = K$, the function $U_\alpha^\mu(z)$ is an entire function of α. For $\alpha \in \mathbb{R}$ and $\mu \geq 0$, U_α^μ is subharmonic in $\mathbb{C}\backslash K$ since $\Delta U_\alpha^\mu = \alpha^2 U_{\alpha+2}^\mu$.

4.9.8. Proposition. *If for some $\alpha \notin -2\mathbb{N}$ there is $R > 0$ such that $U_\alpha^\mu(z) = 0$ for $|z| \geq R$, then $\mu = 0$.*

PROOF. We can assume that $K = \operatorname{supp}(\mu) \subseteq B(0, R)$. One has, for $t \in K$,

$$\frac{1}{|Re^{i\theta} - t|^\alpha} = \frac{1}{R^\alpha\left(1 - \dfrac{te^{-i\theta}}{R}\right)^{\alpha/2}\left(1 - \dfrac{\bar{t}e^{i\theta}}{R}\right)^{\alpha/2}}$$

$$= \frac{1}{R^\alpha} \sum_{m,n\geq 0} (-1)^{m+n} \binom{-\alpha/2}{m}\binom{-\alpha/2}{n} \bar{t}^m t^n e^{-i(n-m)\theta} R^{-(m+n)}.$$

For k an integer we have

$$0 = \int_0^{2\pi} U_\alpha^\mu(Re^{i\theta}) e^{ik\theta}\, d\theta$$

$$= (-1)^k 2\pi R^{-\alpha} \sum_{m\geq 0} \binom{-\alpha/2}{m}\binom{-\alpha/2}{m+k} R^{-(2m+k)} \int_K \bar{t}^m t^{m+k}\, d\mu(t)$$

Since this series is a power series in $1/R$ and $\binom{-\alpha/2}{m}\binom{-\alpha/2}{m+k} \neq 0$ as long as $m + k \geq 0$, we obtain that

$$\int_K \bar{t}^m t^{m+k}\, d\mu(t) = 0$$

for every $m \geq 0$, $m + k \geq 0$. By the Stone-Weierstrass theorem we can conclude that $\int_K f(t)\, d\mu(t) = 0$ for every $f : K \to \mathbb{C}$ continuous. Therefore $\mu = 0$. □

4.9.9. Proposition. *Let μ be a real-valued measure with compact support K. Suppose that $\mu(K) = 0$ and $I(|\mu|) < \infty$. Then U_1^μ exists m—a.e. and*

$$I(\mu) = \frac{1}{2\pi}\int_{\mathbb{C}} (U_1^\mu(z))^2\, dm(z).$$

Hence, $I(\mu) \geq 0$. Moreover, $I(\mu) = 0$ if and only if $\mu = 0$.

Proof. We need first to prove two auxiliary lemmas.

4.9.10. Lemma. *If E is a Lebesgue measurable subset of $\mathbb{C}$, then*

$$\int_E \frac{dm(z)}{|z - t|} \leq 2\sqrt{\pi m(E)} \qquad (t \in \mathbb{C}).$$

Proof of Lemma 4.9.10. This inequality is trivial unless $0 < m(E) < \infty$. Let us consider the disk $D = \bar{B}(t, R)$ such that $\pi R^2 = m(E)$. Then

$$\int_E \frac{dm(z)}{|z - t|} = \int_{E\setminus D} \frac{dm(z)}{|z - t|} + \int_{E\cap D} \frac{dm(z)}{|z - t|} \leq \frac{1}{R}\int_{E\setminus D} dm + \int_{E\cap D} \frac{dm(z)}{|z - t|}$$
$$= \frac{1}{R} m(E\setminus D) + \int_{E\cap D} \frac{dm(z)}{|z - t|}.$$

We have that $m(E\setminus D) = m(E) - m(E \cap D) = m(D) - m(E \cap D) = m(D\setminus E)$. Therefore, since when $z \in D$ we have $|z - t| \leq R$, we obtain

$$\int_E \frac{dm(z)}{|z - t|} = \frac{1}{R} m(D\setminus E) + \int_{E\cap D} \frac{dm(z)}{|z - t|} \leq \int_{E\setminus D} \frac{dm(z)}{|z - t|} + \int_{E\cap D} \frac{dm(z)}{|z - t|}$$
$$= \int_D \frac{dm(z)}{|z - t|} = \int_0^R \int_0^{2\pi} \frac{\rho\, d\rho\, d\theta}{\rho} = 2\pi R = 2\sqrt{\pi m(E)}. \qquad \square$$

4.9.11. Lemma. $\int_E |U_1^\mu(z)|\, dm(z) \leq 2|\mu|(\mathbb{C})\sqrt{\pi m(E)}$, *and* $U_1^\mu \in L^1_{\text{loc}}(\mathbb{C})$.

PROOF OF LEMMA 4.9.11. The theorem of Fubini ensures the inequality. The fact that $U_1^\mu \in L^1_{\text{loc}}(\mathbb{C})$ now follows by letting E be an arbitrary compact set. □

Let us return now the proof of Proposition 4.9.9. We are going to show first that there is a constant C_0 such that if $z_1 \neq z_2$, $\max\{|z_1|,|z_2|\} \le R$, then

$$\frac{1}{2\pi}\int_{|t|\le R}\frac{dm(t)}{|t-z_1||t-z_2|} - \log\frac{1}{|z_1-z_2|} = \log R + C_0 + O(R^{-1}),$$

where $O(R^{-1})$ is a function of R such that in absolute value does not exceed MR^{-1} for some $M > 0$, when $R \to \infty$ but z_1, z_2 remain fixed (or in a compact set).

For that purpose, let us introduce the auxiliary functions

$$I(z_1,z_2,R) = \int_{|t|\le R}\frac{dm(t)}{|t-z_1||t-z_2|}$$

and

$$J(z_1,z_2,R) = \int_{|t|\le R+|z_1|}\frac{dm(t)}{|t-z_1||t-z_2|}.$$

We have (compare with Figure 4.5 in the next page)

$$|I(z_1,z_2,R) - J(z_1,z_2,R)| \le \frac{\pi((R+|z_1|)^2 - R^2)}{(R-|z_1|)(R-|z_2|)}$$

$$= \frac{\pi}{R^2}\frac{2|z_1|R + |z_1|^2}{\left(1-\dfrac{|z_1|}{R}\right)\left(1-\dfrac{|z_2|}{R}\right)}.$$

Using successively the changes of variables $w = t - z_1$ and $s = \dfrac{w}{z_2 - z_1}$ in J we obtain

$$J(z_1,z_2,R) = I\left(0,1,\frac{R+|z_1|}{|z_2-z_1|}\right).$$

In order to obtain the asymptotic behavior of this last integral when $R \to \infty$, we let $A > 2$, and write

$$I(0,1,A) = \int_0^A\int_0^{2\pi}\frac{d\rho\, d\theta}{|1-\rho e^{i\theta}|} = \int_0^2\int_0^{2\pi}\frac{d\rho\, d\theta}{|1-\rho e^{i\theta}|} + \int_2^A\int_0^{2\pi}\frac{d\rho\, d\theta}{|1-\rho e^{i\theta}|}.$$

Since, for $\rho \ge 2$,

$$\frac{1}{|1-\rho e^{i\theta}|} = \frac{1}{\rho}\frac{1}{\left|1-\dfrac{e^{i\theta}}{\rho}\right|} = \frac{1}{\rho}\left|1+\frac{e^{i\theta}}{\rho} + \frac{e^{2i\theta}}{\rho^2\left(1-\dfrac{e^{i\theta}}{\rho}\right)}\right| = \frac{1}{\rho} + O\left(\frac{1}{\rho^2}\right).$$

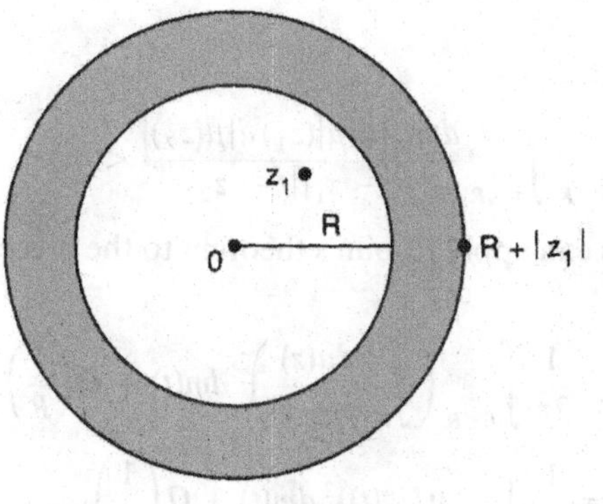

Figure 4.5

One has

$$I(0,1,A)=\int_0^2\int_0^{2\pi}\frac{d\rho\,d\theta}{|1-\rho e^{i\theta}|}+2\pi(\log A-\log 2)+O\left(\frac{1}{A}\right).$$

Let $C_0=\dfrac{1}{2\pi}\displaystyle\int_0^2\int_0^{2\pi}\frac{d\rho\,d\theta}{|1-\rho e^{i\theta}|}-\log 2$, then

$$\begin{aligned}I(z_1,z_2,R)&=J(z_1,z_2,R)+O\left(\frac{1}{R}\right)=I\left(0,1,\frac{R+|z_1|}{|z_2-z_1|}\right)+O\left(\frac{1}{R}\right)\\&=2\pi\left(C_0+\log\frac{R+|z_1|}{|z_2-z_1|}\right)+O\left(\frac{1}{R}\right)\\&=2\pi\left(C_0+\log\frac{1}{|z_2-z_1|}+\log R\right)+O\left(\frac{1}{R}\right),\end{aligned}$$

when $R\to\infty$ and $|z_1|$ remains bounded. This proves the desired asymptotic development. Note that C_0 is independent of z_1, z_2. We use now the hypothesis that $\mu(K)=0$. Then

$$\begin{aligned}I(\mu)&=\int_{K\times K}\log\frac{1}{|z_2-z_1|}\,d\mu(z_1)\,d\mu(z_2)\\&=\frac{1}{2\pi}\int_{K\times K}\left\{\int_{|t|\le R}\frac{dm(t)}{|t-z_1||t-z_2|}\right\}d\mu(z_1)\,d\mu(z_2)+O\left(\frac{1}{R}\right).\end{aligned}$$

The hypothesis that $I(|\mu|)<\infty$ ensures that

$$\int_{K\times K}\log\frac{1}{|z_2-z_1|}\,d|\mu|(z_1)\,d|\mu|(z_2)<\infty.$$

Therefore, one also has

$$\int_{K\times K}\left|\log\frac{1}{|z_2-z_1|}\right|d|\mu|(z_1)\,d|\mu|(z_2)<\infty,$$

and hence, that

$$\int_{K\times K}\int_{|t|\le R}\frac{dm(t)\,d|\mu|(z_1)\,d|\mu(z_2)|}{|t-z_1||t-z_2|}<\infty.$$

As a consequence, we can apply Fubini's theorem to the preceding expression of $I(\mu)$ and obtain

$$\begin{aligned}I(\mu)&=\frac{1}{2\pi}\int_{|t|\le R}\left(\int_K\frac{d\mu(z)}{|t-z|}\right)^2 dm(t)+O\left(\frac{1}{R}\right)\\&=\frac{1}{2\pi}\int_{|t|\le R}(U_1^\mu(t))^2\,dm(t)+O\left(\frac{1}{R}\right).\end{aligned}$$

Let $R\to\infty$, then

$$I(\mu)=\frac{1}{2\pi}\int_{\mathbb{C}}(U_1^\mu(t))^2\,dm(t),$$

which concludes the proof of Proposition 4.9.9. □

4.9.12. Proposition. *If* $V(E)<\infty$ (*i.e.*, $C(E)>0$) *there is a unique measure* $\nu\in\mathscr{P}(E)$ *such that* $I(\nu)=V(E)$.

Proof. By the definition of $V(E)$, one can find a sequence $(\mu_n)_{n\ge1}$ of measures in $\mathscr{P}(E)$ such that $(I(\mu_n))_{n\ge1}$ is a decreasing sequence with limit $V(E)$. Since $\mathscr{P}(E)$ is a metrizable compact space when considered with the topology induced by the weak convergence of measures, we can find a subsequence, also denoted $(\mu_n)_{n\ge1}$, which converges weakly to a measure $\nu\in\mathscr{P}(E)$.

In order to prove that $I(\nu)=V(E)$, let us introduce the continuous functions $g_k(t)=\inf(-\log|t|,k)$, $k\in\mathbb{N}$. We have

$$\begin{aligned}I(\nu)&=\lim_{k\to\infty}\int_{E\times E}g_k(z_1-z_2)\,d\nu(z_1)\,d\nu(z_2)\\&=\lim_{k\to\infty}\lim_{n\to\infty}\int_{E\times E}g_k(z_1-z_2)\,d\mu_n(z_1)\,d\mu_n(z_2).\end{aligned}$$

For k fixed and n arbitrary we have

$$\int_{E\times E}g_k(z_1-z_2)\,d\mu_n(z_1)\,d\mu_n(z_2)\le\int_{E\times E}\log\frac{1}{|z_1-z_2|}\,d\mu_n(z_1)\,d\mu_n(z_2)=I(\mu_n).$$

Therefore

$$I(\nu)\le\liminf_{n\to\infty}I(\mu_n)=V(E).$$

Since $V(E)\le I(\nu)$ by the definition of $V(E)$, we have $I(\nu)=V(E)$.

Let us show there is only one measure in $\mathscr{P}(E)$ with this property. Note that for any pair of measures in $\mathscr{P}(E)$ one has

$$I\left(\frac{\mu_1+\mu_2}{2}\right)+I\left(\frac{\mu_1-\mu_2}{2}\right)=\frac{1}{2}(I(\mu_1)+I(\mu_2)).$$

Assume μ_1, μ_2 are two measures in $\mathscr{P}(E)$ such that $I(\mu_i)=V(E)$. Then $\frac{1}{2}(\mu_1-\mu_2)$ is a real-valued measure supported in E such that

$$\frac{1}{2}(\mu_1-\mu_2)(E)=0.$$

From Proposition 4.9.9 we conclude that $I\left(\frac{1}{2}(\mu_1-\mu_2)\right)\geq 0$. Therefore, the measure $\frac{\mu_1+\mu_2}{2}$ in $\mathscr{P}(E)$ satisfies $I\left(\frac{\mu_1+\mu_2}{2}\right)\leq V(E)$. By the minimality of $V(E)$ we conclude that $I\left(\frac{\mu_1+\mu_2}{2}\right)=V(E)$ and $I\left(\frac{\mu_1-\mu_2}{2}\right)=0$. Appealing again to Proposition 4.9.9, we conclude that $\mu_1=\mu_2$. □

4.9.13. Definition. The measure ν obtained in §4.9.12 satisfying $I(\nu)=V(E)$ is called the ***equilibrium measure*** of E and the corresponding logarithmic potential U^ν is called the ***equilibrium potential*** of E.

We are now going to try to find an effective method to compute the capacity of a compact set. As a corollary of the procedure that follows, we shall be able to conclude that the equilibrium measure of a set is always concentrated in the exterior boundary of that set. (Recall the exterior boundary is the boundary of the unbounded component of $\mathbb{C}\setminus E$.)

Let E be a compact set. The number $\delta_n=\delta_n(E)$ is determined by

$$(\delta_n(E))^{n(n-1)/2}=\sup\left\{\prod_{1\leq j<k\leq n}|z_k-z_j|: z_j\in E, 1\leq j\leq n\right\}.$$

The product $\prod_{1\leq j<k\leq n}(z_k-z_j)$ that appears in the definition of δ_n is precisely the Vandermonde determinant $D(z_1,\ldots,z_n)$

$$D(z_1,\ldots,z_n)=\det\begin{pmatrix}1 & z_1 & z_1^2 & \cdots & z_1^{n-1}\\ 1 & z_2 & z_2^2 & \cdots & z_2^{n-1}\\ \vdots & & & & \vdots\\ 1 & z_n & z_n^2 & \cdots & z_n^{n-1}\end{pmatrix}=\prod_{1\leq j<k\leq n}(z_k-z_j).$$

It is clear that

$$\delta_2(E)=\operatorname{diam}(E)$$

and that, for every n

$$0\leq\delta_n(E)\leq\operatorname{diam}(E).$$

4.9.14. Proposition. *The sequence $(\delta_n(E))_{n\geq 1}$ is decreasing and its limit $\tau(E)$ is called the* ***transfinite diameter*** *of the compact set E.*

PROOF. Since E is compact there exist points $z_1, \ldots, z_{n+1} \in E$ such that

$$(\delta_{n+1}(E))^{n(n+1)/2} = \prod_{1\leq j<k\leq n+1} |z_k - z_j|$$

The right-hand side equals

$$|z_{n+1} - z_1||z_{n+1} - z_2| \cdots |z_{n+1} - z_n| \prod_{1\leq j<k\leq n} |z_k - z_j|$$

which is majored by

$$|z_{n+1} - z_1||z_{n+1} - z_2| \cdots |z_{n+1} - z_n|(\delta_n(E))^{n(n-1)/2}.$$

The same way we obtain, for each index k fixed,

$$(\delta_{n+1}(E))^{n(n+1)/2} \leq \left(\prod_{j\neq k} |z_j - z_k|\right)(\delta_n(E))^{n(n-1)/2}.$$

By multiplying all these inequalities together we obtain

$$\begin{aligned}[(\delta_{n+1}(E))^{n(n+1)/2}]^{n+1} &\leq \left(\prod_{j\neq k} |z_j - z_k|\right)[(\delta_n(E))^{n(n-1)/2}]^{n+1} \\ &\leq (\delta_{n+1}(E))^{n(n+1)}(\delta_n(E))^{(n+1)n(n-1)/2},\end{aligned}$$

which implies

$$\delta_{n+1}(E) \leq \delta_n(E),$$

and proves the proposition. □

4.9.15. Remark. It is clear that if E is a finite set then $\tau(E) = 0$. By Corollary 4.9.4 we also have $C(E) = 0$ in this case. Therefore, for finite sets transfinite diameter and logarithmic capacity coincide. The following proposition shows that this is always the case.

4.9.16. Proposition. *For every compact set E one has $\tau(E) = C(E)$. Moreover, if $C(E) > 0$, the equilibrium measure ν has its support contained in the exterior boundary of E.*

PROOF. By the previous remark we can assume that E is an infinite set. Let $\delta_n = \delta_n(E)$ and $z_1^{(n)}, \ldots, z_n^{(n)} \in E$ be such that

$$\delta_n^{n(n-1)/2} = \prod_{1\leq j<k\leq n} |z_k^{(n)} - z_j^{(n)}|.$$

We remark that we can assume that all these points are in the exterior boundary of E. In fact, let E_∞ be the unbounded component of $\mathbb{C}\backslash E$ and $\hat{E} = \mathbb{C}\backslash E_\infty$. Then $\partial\hat{E} \subseteq E$. Moreover, the function

$z \mapsto \prod_{1 \le j < k < n} |z_k^{(n)} - z_j^{(n)}| \prod_{j=1}^{n-1} |z - z_j^{(n)}|$ is subharmonic and achieves its maximum in $\hat{E}$ at a point of $\partial\hat{E}$. Therefore this maximum must coincide with $\delta_n^{n(n-1)/2}$ and we can assume that $z_n^{(n)} \in \partial\hat{E}$. The same reasoning holds for the other $z_j^{(n)}$.

Let $\mu_j^{(n)}$ be the measure $\dfrac{d\theta}{2\pi n}$ on the circle $\partial B\left(z_j^{(n)}, \dfrac{1}{n}\right)$ and $\mu_n = \sum_{j=1}^{n} \mu_j^{(n)}$, $\mu_n(\mathbb{C}) = 1$. Let us recall that for every $w, \rho \in \mathbb{C}$ one has

$$\frac{1}{2\pi}\int_0^{2\pi} \log|w - \rho e^{i\theta}|\, d\theta \ge \log|w|$$

and, more precisely,

$$\frac{1}{2\pi}\int_0^{2\pi} \log|w - \rho e^{i\theta}|\, d\theta = \begin{cases} \log|w| & \text{if } |w| > |\rho| \\ \log|\rho| & \text{if } |w| \le |\rho|. \end{cases}$$

Hence

$$\begin{aligned} U^{\mu_n}(z) &= \frac{1}{2\pi}\int \log|z - t|\, d\mu_n(t) = \frac{1}{2\pi}\sum_{j=1}^{n}\frac{1}{2\pi n}\int_0^{2\pi} \log\left|z - z_j^{(n)} - \frac{e^{i\theta}}{n}\right| d\theta \\ &\le \frac{1}{2\pi n}\sum_{j=1}^{n} \log|z - z_j^{(n)}| \end{aligned}$$

and

$$\begin{aligned} I(\mu_n) &= -2\pi \int U^{\mu_n}(z)\, d\mu_n(z) \le -\frac{1}{n}\sum_{j=1}^{n}\int \log|z - z_j^{(n)}|\, d\mu_n(z) \\ &= -\frac{1}{n}\sum_{j=1}^{n}\sum_{k=1}^{n}\frac{1}{2\pi n}\int_0^{2\pi} \log\left|z_k^{(n)} + \frac{e^{i\theta}}{n} - z_j^{(n)}\right| d\theta \\ &\le -\frac{1}{n^2}\sum_{\substack{j \ne k \\ 1 \le j,k \le n}} \log|z_k^{(n)} - z_j^{(n)}| - \sum_{j=1}^{n}\frac{1}{2\pi n^2}\int_0^{2\pi} \log\left|\frac{e^{i\theta}}{n}\right| d\theta \\ &\le -\frac{1}{n^2}\sum_{j \ne k} \log|z_k^{(n)} - z_j^{(n)}| + \frac{\log n}{n}. \end{aligned}$$

That is,

$$I(\mu_n) \le \frac{n-1}{n}\log(1/\delta_n) + \frac{\log n}{n}.$$

We can now extract a weakly convergent subsequence $(\mu_{n_k})_{k \ge 1}$ from $(\mu_n)_{n \ge 1}$. Its limit measure $\mu_0 \in \mathscr{P}(E)$ and $\operatorname{supp}\mu_0 \subseteq \partial\hat{E}$, since

$$\operatorname{supp}\mu_n \subseteq \left\{z \in E : d(z, \partial\hat{E}) \le \frac{1}{n}\right\}.$$

By exactly the same reasoning used to show the existence of the equilibrium measure in Proposition 4.9.12, we have

$$I(\mu_0) = \liminf_{k\to\infty} I(\mu_{n_k}) \le \liminf_{k\to\infty} \frac{n_k - 1}{n_k} \log \frac{1}{\delta_{n_k}} = \log \frac{1}{\tau(E)},$$

since $\delta_n \to \tau(E)$.

Conversely, given any set of distinct points $z_1, \ldots, z_n \in E$ we have the inequality

$$\sum_{j\neq k} \log \frac{1}{|z_j - z_k|} \ge n(n-1) \log \frac{1}{\delta_n}.$$

Integrating this inequality against the $n$fold product of a measure $\mu \in \mathscr{P}(E)$ with itself, we obtain

$$n(n-1)\log\frac{1}{\delta_n} \le \int_{E\times\cdots\times E} \sum_{j\neq k} \log \frac{1}{|z_j - z_k|}\, d\mu(z_1)\ldots d\mu(z_n) = n(n-1)I(\mu)$$

If v is the equilibrium measure of E we conclude that

$$V(E) = I(v) \ge \log \frac{1}{\delta_n},$$

and, passing to the limit

$$I(v) \ge \log \frac{1}{\tau(E)} \ge I(\mu_0).$$

Note that if $\tau(E) = 0$ the previous inequality forces $V(E) = \infty$ and $C(E) = 0$. If $\tau(E) > 0$ the last inequality implies $I(\mu_0) < \infty$, and by the minimality $I(v) \le I(\mu_0) < \infty$. Therefore, we can use Proposition 4.9.12 to conclude that $\mu_0 = v$ and $\tau(E) = C(E)$. □

4.9.17. Proposition. *Let Γ be the exterior boundary of the compact set E, then*

$$C(\Gamma) = C(E).$$

PROOF. One has $C(\Gamma) \le C(E)$ by Proposition 4.9.2 (3). The proof of Proposition 4.9.15 shows that $V(E) = I(\mu_0)$ with $\mu_0 \in \mathscr{P}(E)$, $\operatorname{supp}\mu_0 \subseteq \Gamma$. Therefore $\mu_0 \in \mathscr{P}(\Gamma)$ and $V(E) \ge V(\Gamma)$. Hence $C(\Gamma) = C(E)$. □

4.9.18. Corollary. *If E_1 and E_2 are two compact sets having the same exterior boundary then $C(E_1) = C(E_2)$.*

We are going to give now a third way of finding $C(E)$ for a nonempty compact set E in $\mathbb{C}$. Let $\mathscr{P}_n$ denote the space of polynomials of degree $\le n$ and complex coefficients. By $\mathscr{P}_{0,n}$ we denote the closed subset of $\mathscr{P}_n$ formed by the monic polynomials in $\mathscr{P}_n$ (i.e., the coefficient of z^n is exactly equal to 1.) Let

$$M_n(E) = M_n = \inf\left\{\sup_{z\in E} |P(z)| : P \in \mathscr{P}_{0,n}\right\}.$$

If $P \in \mathscr{P}_{0,n}$ has a zero α that does not belong to the convex hull cv(E) of E, then one can find another polynomial $P_1 \in \mathscr{P}_{0,n}$ such that $\sup_{z \in E} |P_1(z)| < \sup_{z \in E} |P(z)|$. In fact, one can separate α from E by a straight line and choose the system of coordinates in $\mathbb{C}$ in such a way that this line is the imaginary axis, E is contained in the left-hand plane and $\alpha > 0$. Therefore, for $z \in E$ we have $|z| < |z - \alpha|$, hence if $P_1(z) = \dfrac{zP(z)}{z - \alpha}$ one has $|P_1(z)| < |P(z)|$ on E. Hence, to find M_n we can restrict ourselves to consider those $P \in \mathscr{P}_{0,n}$ all whose roots lie in cv(E). Since $\sup_{z \in P} |P(z)|$ is a continuous function of the zeros $z_1, \ldots, z_n$ of P, and each of them can be taken independently of the others in the compact set cv(E), we are guaranteed the existence of points $z_1^{(n)}, \ldots, z_n^{(n)} \in \text{cv}(E)$ such that

$$M_n = \sup_{z \in E} \prod_{1 \le j \le n} |z - z_j^{(n)}|.$$

As a first corollary of this observation we can show that $M_{m+n} \le M_n M_m$. Namely, let $z_1^{(n)}, \ldots, z_n^{(n)}$ and $z_1^{(m)}, \ldots, z_m^{(m)}$ be families of points in cv(E) giving M_n and M_m, respectively, then

$$\begin{aligned} M_{m+n} &\le \sup_{z \in E} \left(\prod_{1 \le j \le n} |z - z_j^{(n)}| \prod_{1 \le k \le m} |z - z_k^{(m)}| \right) \\ &\le \left(\sup_{z \in E} \prod_{1 \le j \le n} |z - z_j^{(n)}| \right) \left(\sup_{z \in E} \prod_{1 \le k \le m} |z - z_k^{(m)}| \right) \\ &= M_n M_m. \end{aligned}$$

Let us recall the following simple lemma.

4.9.19. Lemma. *If* $(x_n)_{n \ge 1}$ *is a sequence of real numbers satisfying* $x_{n+m} \le x_n + x_m$ *for every* $n, m \ge 1$, *the* $\lim_{n \to \infty} \dfrac{x_n}{n}$ *exists in* $[-\infty, \infty[$.

PROOF. Let $\alpha > \inf_{n \ge 1} \left\{ \dfrac{x_n}{n} \right\}$. There is an integer s such that $\dfrac{x_s}{s} < \alpha$. For $m > s$ let us write $m = us + v$, $0 \le v < s$, $u, v \in \mathbb{N}$. Then we have

$$x_m = x_{us+v} \le x_{us} + x_v \le u x_s + x_v,$$

hence

$$\frac{x_m}{m} \le \frac{us}{us + v} \frac{x_s}{s} + \frac{x_v}{m}.$$

When $m \to \infty$ we have $u \to \infty$ and, for large m we obtain

$$\frac{x_m}{m} \le \frac{x_s}{s} + \varepsilon < \alpha + \varepsilon.$$

Therefore

$$\inf_{m\geq 1}\frac{x_m}{m}\leq \limsup_{m\to\infty}\frac{x_m}{m}\leq \alpha.$$

It follows that

$$\lim_{m\to\infty}\frac{x_m}{m}=\inf_{m\geq 1}\frac{x_m}{m}. \qquad \square$$

4.9.20. Corollary. *Let $(y_n)_{n\geq 1}$ be a sequence of positive real numbers such that $y_{m+n}\leq y_n y_m$ for every $m, n\geq 1$. Then $\lim_{n\to\infty}\sqrt[n]{y_n}$ exists in $[0,\infty[$ and it is smaller than $y_m^{1/m}$ for any $m\geq 1$.*

Proof. Let $x_n=\log y_n$, then $\frac{x_n}{n}=\log(y_n^{1/n})$ and x_n satisfies the condition of §4.9.19. Therefore $\lim_{n\to\infty} y_n^{1/n}=\exp\left(\lim_{n\to\infty}\frac{x_n}{n}\right)=\exp\left(\inf_{m\geq 1}\frac{x_m}{m}\right)\in[0,\infty[$. $\square$

This last corollary allows us to define the Chebyschev constant $\rho(E)$ of a nonempty compact set E by

$$\rho(E):=\lim_{n\to\infty}(M_n(E))^{1/n}.$$

The function $E\mapsto\rho(E)$ has the following three properties:

(a) If $E_1\subseteq E_2$ then $\rho(E_1)\leq\rho(E_2)$.
(b) If E is a finite set then $\rho(E)=0$.
(c) If $(E_n)_{n\geq 1}$ is a decreasing sequence of compact sets and if $E=\bigcap_{n\geq 1}E_n$, then

$$\rho(E)=\lim_{n\to\infty}\rho(E_n).$$

Property (a) is evidently true. We leave the proof of (b) to the reader. Let us prove (c). Given $\varepsilon>0$ there is an integer $s\geq 1$ such that

$$\rho(E)\leq(M_s(E))^{1/s}\leq\rho(E)+\varepsilon.$$

For this s fixed we have

$$(M_s(E))^{1/s}\leq(M_s(E_n))^{1/n}\leq(M_s(E))^{1/s}+\varepsilon;$$

the last inequality holds for n sufficiently large since then the points of E_n are all very close to E. Therefore, for n sufficiently large we have

$$\rho(E)\leq\rho(E_n)\leq(M_s(E_n))^{1/s}\leq(M_s(E))^{1/s}+\varepsilon\leq\rho(E)+2\varepsilon.$$

The third inequality is a consequence of the last statement in Corollary 4.9.20. It is clear that chain of inequalities proves property (c). $\square$

4.9.21. Proposition. *For any nonempty compact set E in $\mathbb{C}$ we have*

$$\rho(E) = C(E) = \tau(E).$$

PROOF. We can assume E is infinite. Let $z_1, \ldots, z_n$ be the distinct points in E such that

$$\delta_n^{n(n-1)/2} = \prod_{1 \le j < k \le n} |z_j - z_k|,$$

with $\delta_n = \delta_n(E)$. The polynomial $P(z) = \prod_{j=1}^{n} (z - z_j)$ is in $\mathscr{P}_{0,n}$, hence there is a point $z_{n+1} \in E$ such that

$$M_n \le \sup_{z \in E} |P(z)| = |P(z_{n+1})|.$$

Therefore

$$\begin{aligned} \delta_{n+1}^{n(n-1)/2} M_n \le \delta_n^{n(n-1)/2} M_n &\le \left(\prod_{1 \le j < k \le n} |z_j - z_k| \right) |P(z_{n+1})| \\ &\le \prod_{1 \le j < k \le n+1} |z_j - z_k| \le \delta_{n+1}^{n(n+1)/2}. \end{aligned}$$

In other words

$$M_n \le \delta_{n+1}^n,$$

and we can conclude immediately that

$$\rho(E) \le \tau(E) = C(E).$$

Conversely, let Q be a polynomial in $\mathscr{P}_{0,n}$ such that

$$M_n = \sup_{z \in E} |Q(z)|.$$

If $z_1, \ldots, z_{n+1}$ are any family of distinct points in E we have that the Vandermonde determinant $D(z_1, \ldots, z_{n+1})$ can also be expressed as

$$D(z_1, \ldots, z_{n+1}) = \det \begin{pmatrix} 1 & z_1 & z_1^2 & \cdots & z_1^{n-1} & Q(z_1) \\ \vdots & & & & & \vdots \\ 1 & z_{n+1} & z_{n+1}^2 & \cdots & z_{n+1}^{n-1} & Q(z_{n+1}) \end{pmatrix}.$$

Developing this determinant along the last column we obtain

$$D(z_1, \ldots, z_{n+1}) = \sum_{1 \le j \le n+1} (-1)^j Q(z_j) D(z_1, \ldots, \hat{z}_j, \ldots, z_{n+1}),$$

where $D(z_1, \ldots, \hat{z}_j, \ldots, z_{n+1})$ is the Vandermonde $n \times n$ determinant obtained by eliminating the last column and the jth row in the determinant. Of course, $D(z_1, \ldots, \hat{z}_j, \ldots, z_{n+1})$ is exactly $D(z_1, \ldots, z_{j-1}, z_{j+1}, \ldots, z_{n+1})$, that is, the point z_j was removed from the set. We know that by the definition of δ_n

$$|D(z_1, \ldots, \hat{z}_j, \ldots, z_{n+1})| \le \delta_n^{n(n-1)/2},$$

for $1 \le j \le n+1$ and all possible choices of $z_1, \dots, z_{n+1}$. If we choose these points so that

$$|D(z_1,\dots,z_{n+1})| = \delta_{n+1}^{n(n+1)/2}$$

then we have

$$\delta_{n+1}^{n(n+1)/2} \le (n+1)M_n\delta_n^{n(n-1)/2},$$

or, taking the nth root,

$$\delta_{n+1}^{(n+1)/2} \le (n+1)^{1/n}M_n^{1/n}\delta_n^{(n-1)/2}.$$

We take the logarithm of these inequalities and add them for the values $n = 1, 2, \dots, m$. After some simplification we obtain

$$\frac{m+1}{2}\log\delta_{m+1} + \frac{1}{2}\sum_{k=2}^{m}\log\delta_k \le \sum_{k=1}^{m}\frac{\log(k+1)}{k} + \sum_{k=1}^{m}\log(M_k^{1/k}).$$

Recall now that if a sequence $x_n \to a$ then the averages $\dfrac{x_1 + \cdots + x_n}{n} \to a$ also. (This is the Césaro summation procedure.) It is clear that if we divide by m on both sides we have essentially this situation:

$$\frac{1}{2}\left[\frac{m+1}{m}\log\delta_{m+1} + \frac{m-1}{m}\frac{1}{(m-1)}\sum_{k=2}^{m}\log\delta_k\right]$$
$$\le \frac{1}{m}\sum_{k=1}^{m}\frac{\log(k+1)}{k} + \frac{1}{m}\sum_{k=1}^{m}\log(M_k^{1/k}).$$

Taking limits we obtain

$$\log\tau(E) \le \log\rho(E).$$

This gives the other inequality needed to end the proof of the proposition. □

4.9.22. Proposition. *Let E be a nonempty compact set in $\mathbb{C}$ of zero capacity. There exists a measure $\sigma \in \mathscr{P}(E)$ such that*

$$\lim_{z\to\zeta} U^\sigma(z) = -\infty \qquad \textit{for all } \zeta \in E$$

and

$$E = \{z \in \mathbb{C} : U^\sigma(z) = -\infty\}.$$

Proof. The proof of the first part of the previous proposition shows that if we pick points $z_1^{(n)}, \dots, z_n^{(n)}$ in E such that

$$(\delta_n(E))^{n(n-1)/2} = \delta_n^{n(n-1)/2} = \prod_{1\le j<k\le n}|z_j^{(n)} - z_k^{(n)}|$$

then

$$\sup_{z \in E} \left(\prod_{j=1}^{n} |z - z_j^{(n)}| \right)^{1/n} \leq \delta_{n+1}.$$

Let us choose a sequence of integers $(n_k)_{k \geq 1}$ such that

$$\delta_{n_k+1} \leq \exp(-\exp(k)), \qquad k \geq 1.$$

Set

$$\sigma_k = \frac{1}{n_k} \sum_{1 \leq j \leq n_k} \delta_{z_j^{(n_k)}}$$

and

$$\sigma = \sum_{k \geq 1} \frac{\sigma_k}{2^k}.$$

It is clear that $\sigma \in \mathscr{P}(E)$. For $z \in E$ one has

$$U^\sigma(z) = \frac{1}{2\pi} \int \log|z - \zeta| \, d\sigma(\zeta) = \frac{1}{2\pi} \sum_{k \geq 1} \frac{1}{2^k} \left(\frac{1}{n_k} \sum_{1 \leq j \leq n_k} \log|z - z_j^{(n_k)}| \right)$$

$$\leq -\frac{1}{2\pi} \sum_{k \geq 1} \frac{1}{2^k} e^k = -\infty.$$

Since U^σ is harmonic off the support of σ we conclude that

$$E = \{z \in \mathbb{C} : U^\sigma(z) = -\infty\}.$$

The upper semicontinuity of U^σ also proves that

$$\lim_{z \to \zeta} U^\sigma(\zeta) = -\infty$$

for every $\zeta \in E$. □

There is a converse to this proposition. The proof requires the following proposition.

4.9.23. Proposition. *Let E be a compact set with $C(E) > 0$. If ν is the equilibrium measure for E then one has*

$$U^\nu(z) \geq -\frac{V(E)}{2\pi} \qquad \textit{for every } z \in \mathbb{C},$$

and $U^\nu(z) = -\dfrac{V(E)}{2\pi}$ for $z \in E$, with the possible exception of a set $A \subseteq E$ which is the countable union of compact sets of zero capacity, all of them contained in the exterior boundary Γ of E. Moreover, $U^\nu \equiv -\dfrac{V(E)}{2\pi}$ on every bounded component of $\mathbb{C}\backslash\Gamma$.

PROOF. The support F of ν is a compact set contained in the exterior boundary Γ of E. U^ν is harmonic in F^c and subharmonic in $\mathbb{C}$, $\Delta U^\nu = \nu$ and

$$\int U^\nu \, d\nu = -\frac{V(E)}{2\pi}. \qquad (*)$$

Let us consider the set

$$A := \left\{z \in E : U^\nu(z) > -\frac{V(E)}{2\pi}\right\}.$$

We want to show that $C(A) = 0$. By the regularity of the Borel measures we have that $C(A) = \sup\{C(K) : K \text{ compact}, K \subseteq A\}$. It follows from this that, if $C(A) > 0$, then there exists an integer $n_0 \geq 1$ such that $C(A_n) > 0$ for $n \geq n_0$, where

$$A_n := \left\{z \in E : U^\nu(z) \geq -\frac{V(E)}{2\pi} + \frac{1}{n}\right\}.$$

In fact, if K compact, $K \subseteq A$, $C(K) > 0$ then $K = \bigcup_{n \geq 1} (K \cap A_n)$ and by §4.9.5 one has $C(A_{n_0}) \geq C(K \cap A_{n_0}) > 0$ for some $n_0 \geq 1$.

We must also have $\nu(E \setminus A) > 0$. Otherwise the condition $(*)$ could not hold. In fact, $\nu(A) = 1$ also implies that $\nu(A_n) > 0$ for some n, contradicting $(*)$. Therefore there is a compact set B, $B \subseteq E \setminus A$, such that $\nu(B) > 0$. By Proposition 4.9.3 we also have $C(B) > 0$.

To show that $C(A) > 0$ leads to a contradiction we must use the extremality of the equilibrium measure ν, that is, we are going to construct a convenient perturbation of ν.

Let $\mu \in \mathscr{P}(A_{n_0})$ be a measure with finite energy $I(\mu)$. Such a measure exists since $C(A_{n_0}) > 0$.

Every Borel subset of E can be written as a disjoint of union of three Borel sets contained respectively in B, A_{n_0} and $E \setminus (B \cup A_{n_0})$. We can define a real-valued measure σ on E by

$$\sigma(S) = \begin{cases} -\nu(S) & \text{if } S \subseteq B \\ \nu(B)\mu(S) & \text{if } S \subseteq A_{n_0} \\ 0 & \text{if } S \subseteq E \setminus (B \cup A_{n_0}). \end{cases}$$

For $0 < t < 1$ let $\lambda = \nu + t\sigma$. One can verify that $\lambda \geq 0$. (It is obvious that λ is positive in $E \setminus B$ and, in B one uses that $t < 1$.) Moreover, $\lambda(E) = \nu(E) + t(-\nu(B) + \nu(B)\mu(A_{n_0})) = \nu(E) = 1$. It is also easy to show that

$$I(\lambda) < \infty \quad \text{and} \quad I(|\sigma|) < \infty.$$

Therefore, we must have

$$I(\nu) \leq I(\lambda).$$

But

$$I(\lambda) = I(\nu) + t^2 I(\sigma) - 4\pi t \int U^\nu \, d\sigma,$$

and we should then have

$$\int U^\nu \, d\sigma \le 0.$$

On the other hand

$$\begin{aligned}\int U^\nu \, d\sigma &= -\int_B U^\nu \, d\nu + \nu(B) \int_{A_{n_0}} U^\nu \, d\mu \\ &\ge \nu(B)\frac{V(E)}{2\pi} + \nu(B)\left(-\frac{V(E)}{2\pi} + \frac{1}{n_0}\right) \\ &= \frac{\nu(B)}{n_0} > 0,\end{aligned}$$

which is a contradiction. It follows that $C(A) = 0$.

Since $C(A) = 0$ then $C(A_n) = 0$ for all $n \ge 1$, hence $\nu(A_n) = 0$ by Proposition 4.9.3. We still have to show that $A \subseteq \Gamma$, as well as the other properties stated in the proposition. We start by showing that

$$U^\nu(z) \ge -\frac{V(E)}{2\pi} \qquad \text{for all } z \in \mathbb{C}.$$

This will imply that $U^\nu = -\dfrac{V(E)}{2\pi}$ in $E \setminus A$.

We show this inequality in two stages, first on F, the support of ν. Let $z_0 \in F$ such that $U^\nu(z_0) < -\dfrac{V(E)}{2\pi}$. Since U^ν is u.s.c., there is an $\varepsilon > 0$ and a compact neighborhood N of z_0 such that $U^\nu(z) < -\dfrac{V(E)}{2\pi} - \varepsilon$ for $z \in N$. We also must have $\nu(N) > 0$. Otherwise z_0 would not be in F. Hence

$$\begin{aligned}-\frac{V(E)}{2\pi} &= \int_N U^\nu \, d\nu + \int_{(F\setminus N)} U^\nu \, d\nu \\ &\le \left(-\frac{V(E)}{2\pi} - \varepsilon\right)\nu(N) + \nu(F\setminus N)\left(-\frac{V(E)}{2\pi}\right),\end{aligned}$$

since we have already shown that $\nu(A) = 0$ and $U^\nu \le -\dfrac{V(E)}{2\pi}$ in $E \setminus A$. Therefore we have

$$-\frac{V(E)}{2\pi} \le -\varepsilon\nu(N) - \frac{V(E)}{2\pi}$$

which is impossible. That is, we have $U^\nu \ge -\dfrac{V(E)}{2\pi}$ everywhere on F. The second stage is accomplished by the following lemma.

4.9.24. Lemma. *Let E be a compact subset of $\mathbb{C}$, $\mu \in \mathscr{P}(E)$, $c \in \mathbb{R}$. If $U^\mu \geq c$ on E then $U^\mu \geq c$ everywhere.*

Proof of Lemma 4.9.24. Let D be a connected component of $\mathbb{C}\setminus E$. Since $-U^\mu$ is harmonic in D (and tends to $-\infty$ at ∞ if D is the unbounded component of E^c), then it is enough to show that for any $a \in \bar{D} \cap E$ we have

$$\limsup_{\substack{z\to a\\ z\in D}} -U^\mu(z) \leq -c.$$

Let $D_\rho = D \cap B(a,\rho)$, $E_\rho = E \cap B(a,\rho)$. It is not possible that $\mu(\{a\}) > 0$, otherwise by Lemma 4.4.24 we would have $U^\mu(a) = -\infty$. Hence given $\varepsilon > 0$ there exists $\rho > 0$ sufficiently small such that $\mu(E_\rho) < \varepsilon$. Let us choose $z \in D_\rho$ and $z_1 \in E_\rho$ (depending on z) such that $|z_1 - z| \leq |z - t|$ for every $t \in E_\rho$. Then

$$|z_1 - t| \leq |z_1 - z| + |z - t| \leq 2|z - t| \qquad \text{for } t \in E_\rho$$

and

$$\log\frac{1}{|z-t|} \leq \log 2 + \log\frac{1}{|z_1 - t|} \qquad \text{in } E_\rho.$$

This inequality implies

$$\begin{aligned}\int_{E_\rho} \log\frac{1}{|z-t|}\,d\mu(t) &\leq \mu(E_\rho)\log 2 + \int_{E_\rho} \log\frac{1}{|z_1-t|}\,d\mu(t)\\ &\leq \varepsilon\log 2 + \int_E \log\frac{1}{|z_1-t|}\,d\mu(t) - \int_{E\setminus E_\rho} \log\frac{1}{|z_1-t|}\,d\mu(t)\\ &\leq \varepsilon\log 2 - 2\pi c - \int_{E\setminus E_\rho} \log\frac{1}{|z_1-t|}\,d\mu(t),\end{aligned}$$

since $\int_E \log\frac{1}{|z_1-t|}\,d\mu(t) = -2\pi U^\mu(z_1) \leq -2\pi c$.

On the other hand, the integral over $E\setminus E_\rho$ is a continuous function of $z_1 \in B(a,\rho)$. Since $|z_1 - z| \leq |z - a|$ by definition of the point z_1, we have that there is a ρ_0, $0 < \rho_0 < \rho$, such that if $z \in B(a,\rho_0)$ then

$$\int_{E\setminus E_\rho} \log\frac{1}{|z-t|}\,d\mu(t) \leq \int_{E\setminus E_\rho} \log\frac{1}{|z_1-t|}\,d\mu(t) + \varepsilon.$$

Therefore, for $z \in B(a,\rho_0)$ we have

$$\begin{aligned}-2\pi U^\mu(z) &= \int_{E_\rho} \log\frac{1}{|z-t|}\,d\mu(t) + \int_{E\setminus E_\rho} \log\frac{1}{|z-t|}\,d\mu(t)\\ &\leq \int_{E_\rho} \log\frac{1}{|z-t|}\,d\mu(t) + \int_{E\setminus E_\rho} \log\frac{1}{|z_1-t|}\,d\mu(t) + \varepsilon\\ &\leq \varepsilon(1 + \log 2) - 2\pi c.\end{aligned}$$

This proves the lemma. □

The only thing missing to end the proof of Proposition 4.9.23 is to show that $A \subseteq \Gamma$. As in Proposition 4.9.16, let E_∞ be an unbounded component of $\mathbb{C}\backslash E$ and $\hat{E} = \mathbb{C}\backslash E_\infty$. Then $\hat{E}$ is a compact set such that $\partial\hat{E} = \Gamma$, therefore the equilibrium measure of $\hat{E}$ and of E coincide. Let

$$\hat{A} := \left\{ z \in \hat{E} : U^\nu(z) > -\frac{V(E)}{2\pi} \right\}.$$

Then $A \subseteq \hat{A}$ and $C(\hat{A}) = 0$, by the same reasoning as earlier. If $A\backslash\Gamma \neq \varnothing$ then there is a point $z_0 \in \hat{E}$ where $U^\nu(z_0) > -\dfrac{V(E)}{2\pi}$. By the continuity of U^ν in Γ^c there is a neighborhood W of z_0, $W_0 \subseteq \hat{A}$. This is impossible since $m(\hat{A}) = 0$. This concludes the proof of Proposition 4.9.23. Note that the last considerations show that $U^\nu \equiv -\dfrac{V(E)}{2\pi}$ on every component of $\mathbb{C}\backslash\Gamma$ except for the unbounded one. □

The following is the converse to Proposition 4.9.22.

4.9.25. Proposition. *Let μ be a positive measure with compact support E. If E_1 is a compact set contained in the polar set of U^μ then $C(E_1) = 0$.*

Proof. Clearly $E_1 \subseteq E$, since outside U^μ is finite. If $C(E_1) > 0$, let ν be the equilibrium measure of E_1. We know that $U^\nu \geq -\dfrac{V(E)}{2\pi} > -\infty$ everywhere. By Fubini's theorem

$$\begin{aligned}\int_{E_1} U^\mu(z)\,d\nu(z) &= \int_{E_1}\left(\frac{1}{2\pi}\int_E \log|z-\zeta|\,d\mu(\zeta)\right)d\nu(z)\\ &= \int_E\left(\frac{1}{2\pi}\int_{E_1}\log|z-\zeta|\,d\nu(z)\right)d\mu(\zeta)\\ &= \int_{E_1} U^\nu(\zeta)\,d\mu(\zeta) \geq -\frac{V(E_1)\mu(E)}{2\pi} > -\infty,\end{aligned}$$

which contradicts the fact that $U^\mu = -\infty$ on E_1. □

The Riesz decomposition theorem and Propositions 4.9.22 and 4.9.25 allow us to conclude that the compact polar sets are exactly the compacts with zero capacity.

Another remark to be made at this point is that, under very mild restrictions on a compact set E of positive capacity, one can show the exceptional set A of §4.9.23 is empty. This will allow us to relate the capacity of E to the asymptotic behavior at infinite of the Green function of the unbounded component of $S^2\backslash E$. To get there we need first to prove some majoration properties of subharmonic functions. This will be done later.

We study first some elementary conditions on the removal of singularities of harmonic and subharmonic functions. On this account, let us recall that a harmonic function, since it is locally the real part of a holomorphic function, is real analytic. Therefore, if a harmonic function in a connected open set Ω vanishes on a nonempty open set U, $U \subseteq \Omega$, it is identically zero on Ω.

4.9.26. Proposition. *Let Ω be an open set in S^2. A harmonic function h in $\Omega \setminus \{z_0\}$ ($z_0 \in \Omega$) is the restriction of a harmonic function $\tilde{h}$ in Ω if and only if*

$$\lim_{\substack{z \to z_0 \\ z \neq z_0}} \frac{h(z)}{\log|z - z_0|} = 0 \qquad \text{if } z_0 \neq \infty \tag{1}$$

$$\lim_{\substack{z \to \infty \\ z \in \Omega \setminus \{\infty\}}} \frac{h(z)}{\log|z|} = 0 \qquad \text{if } z_0 = \infty. \tag{2}$$

PROOF. That the condition is necessary is evident.

Let $z_0 \neq \infty$ and $\bar{B}(z_0, r) \subseteq \Omega$. Let h_1 be the harmonic function in $B(z_0, r)$ whose boundary values coincide with h. Let us show that $h = h_1$ in $B(z_0, r) \setminus \{z_0\}$. Let $\varepsilon > 0$. Set

$$h_\varepsilon(z) := h(z) - h_1(z) + \varepsilon \log \frac{|z - z_0|}{r},$$

for $z \in \bar{B}(z_0, r) \setminus \{z_0\}$. The function h_ε is zero on $\partial B(z_0, r)$, harmonic in $B(z_0, r) \setminus \{z_0\}$ and, by hypothesis (1), tends to $-\infty$ as $z \to z_0$. Therefore h_ε is subharmonic in $B(z_0, r)$ and, by the maximum principle, $h_\varepsilon \leq 0$ in this disk. Letting ε tend to zero one finds $h \leq h_1$ in $B(z_0, r)$. With the help of $k_\varepsilon = h_1 - h + \varepsilon \log \dfrac{|z - z_0|}{r}$ one proves that h and h_1 coincide. This proves the extension of h to z_0 is possible.

If $z_0 = \infty$ one sets $G_\varepsilon(z) = \pm h(z) \mp h_1(z) + \varepsilon \log|rz|$ if $B(z_0, 1/r)^c \subseteq \Omega$, where h_1 is the solution of the Dirichlet problem in $B(z_0, 1/r)^c$ with h as boundary value. The preceding reasoning can be applied to finish the proof. □

4.9.27. Corollary. (Riemann). *If h is harmonic in $\Omega \setminus \{z_0\}$ and bounded then h admits an extension to a harmonic function in Ω.*

4.9.28. Proposition. *Let Ω be an open set in S^2, $z_0 \in \Omega$, u subharmonic in $\Omega \setminus \{z_0\}$. The function u admits an extension $\tilde{u}$ subharmonic in Ω if and only if*

$$\limsup_{z \to z_0} \frac{u(z)}{\log(1/|z - z_0|)} \leq 0 \qquad \textit{if } z_0 \neq \infty, \tag{1}$$

$$\limsup_{z \to \infty} \frac{u(z)}{\log|z|} \leq 0 \qquad \textit{if } z_0 = \infty. \tag{2}$$

PROOF. We will restrict ourselves to the case $z_0 \neq \infty$. The other case can be reduced to this one (we use the same reasoning).

First we show the condition is necessary. If $\tilde{u}$ exists then

$$\limsup_{\substack{z\to z_0\\ z\neq z_0}} u(z) = \limsup_{\substack{z\to z_0\\ z\neq z_0}} \tilde{u}(z) \leq \tilde{u}(z_0) < \infty.$$

Clearly, $\lim_{\substack{z\to z_0\\ z\neq z_0}} \log \frac{1}{|z - z_0|} = +\infty$. Therefore

$$\limsup_{\substack{z\to z_0\\ z\neq z_0}} \frac{u(z)}{\log(1/|z - z_0|)} \leq 0.$$

To prove the sufficiency let us note that if $\tilde{u}$ exists, then one must have $\tilde{u}(z_0) = \lim_{r\to 0} \lambda(u, z_0, r)$. This proves that the possible extension is unique. We need an auxiliary result to proceed further.

4.9.29. Lemma. *Let u, Ω and z_0 be as in the statement of Proposition 4.9.28. If $\bar{B}(z_0, r) \subseteq \Omega$ and h is a continuous function in $\bar{B}(z_0, r)$, which is harmonic in $B(z_0, r)$ and $h \geq u$ on $\partial B(z_0, r)$, then $h \geq u$ in $B(z_0, r)\setminus\{z_0\}$.*

PROOF. Let $\varepsilon > 0$ and $w_\varepsilon(z) = u(z) - h(z) + \varepsilon \log \frac{|z - z_0|}{r}$ for $z \in \bar{B}(z_0, r)\setminus\{z_0\}$. The function w_ε is subharmonic in $B(z_0, r)\setminus\{z_0\}$. We claim that

$$\lim_{\substack{z\to\zeta\\ 0<|z-z_0|<r}} w_\varepsilon(z) \leq 0 \qquad \text{for } \zeta \in \partial(B(z_0, r)\setminus\{z_0\}).$$

(i) if $\zeta = z_0$ we have

$$w_\varepsilon(z) = \left(\log\frac{1}{|z - z_0|}\right)\left[\frac{u(z)}{\log(1/|z - z_0|)} - \frac{h(z)}{\log(1/|z - z_0|)} - \varepsilon - \frac{\varepsilon \log r}{\log(1/|z - z_0|)}\right].$$

By assumption (1) the expression between brackets is $\leq -\varepsilon/2$ if $0 < |z - z_0|$ is sufficiently small. Therefore

$$\limsup_{\substack{z\to z_0\\ z\neq z_0}} w_\varepsilon(z) = -\infty.$$

(ii) If $\zeta \in \partial B(z_0, r)$ we have

$$\lim_{\substack{z\to\zeta\\ 0<|z-z_0|<r}} w_\varepsilon(z) \leq \left(\limsup_{\substack{z\to\zeta\\ 0<|z-z_0|<r}} u(z)\right) - h(\zeta) \leq u(\zeta) - h(\zeta) \leq 0.$$

Hence the claim is true and we can apply the maximum principle to conclude $w_\varepsilon \leq 0$ in $B(z_0, r)\setminus\{z_0\}$. Letting $\varepsilon \to 0$ we obtain $u(z) \leq h(z)$ in the punctured disk. □

Returning to the proof of Proposition 4.9.28, we let $\tilde{P}_r u$ be the Poisson modification of u with respect to the disk $B(z_0, r)$ (cf. §4.7.1). $\tilde{P}_r u$ is a subharmonic function in Ω. If we let $(r_n)_{n\geq 1}$ be a sequence of positive numbers decreasing to zero, then $(\tilde{P}_{r_n} u)_{n\geq 1}$ is a decreasing sequence of subharmonic functions such that $\tilde{P}_{r_n} u \geq u$ everywhere by Lemma 4.9.29. Therefore the function $\tilde{u} = \inf_n \tilde{P}_{r_n} u$ is subharmonic and, clearly $\tilde{u}|(\Omega\backslash\{z_0\}) = u$. This concludes the proof. □

4.9.30. Corollary. *Let Ω be an open set in S^2, $A = \{z_n\}_{n\geq 1}$ a discrete subset of Ω. Then*

(1) *If u is a subharmonic function in $\Omega\backslash A$ then it admits a subharmonic extension $\tilde{u}$ to Ω if and only if for every n*

(a) $$\limsup_{\substack{z\to a_n \\ z\in\Omega\backslash A}} \frac{u(z)}{\log(1/|z - z_0|)} \leq 0 \qquad \text{if } a_n \neq \infty$$

(b) $$\limsup_{\substack{z\to \infty \\ z\in\Omega\backslash A}} \frac{u(z)}{\log(|z|)} \leq 0 \qquad \text{if } \infty \in A.$$

(2) *If h is harmonic in $\Omega\backslash A$ and satisfies for every n*

(a) $$\limsup_{\substack{z\to a_n \\ z\in\Omega\backslash A}} \frac{h(z)}{\log(1/|z - a_n|)} = 0 \qquad \text{if } a_n \neq \infty$$

(b) $$\lim_{\substack{z\to \infty \\ z\in\Omega\backslash A}} \frac{h(z)}{\log|z|} \leq 0 \qquad \text{if } \infty \in A.$$

then h admits a harmonic extension to Ω.

(3) *The preceding two properties hold if u (resp. h) is bounded above (resp. bounded).*

(4) *If u is subharmonic and bounded above in $S^2\backslash A$, A discrete, then u is a constant.*

4.9.31. Definition. A connected open set Ω in S^2 is said to be ***parabolic*** if every subharmonic function in Ω which is bounded above is constant. If Ω is not parabolic one says it is ***hyperbolic***.

4.9.32. Proposition. *Let Ω be a parabolic open set and $\Omega_1 \subseteq \Omega$ be an open set whose relative boundary $\partial_\Omega(\Omega_1)$ is not empty. Let u be a subharmonic function in Ω_1 bounded above by a constant $M < \infty$. If $\limsup_{\substack{z\to\zeta \\ z\in\Omega_1}} u(z) \leq m$ for every $\zeta \in \partial_\Omega(\Omega_1)$, then $u \leq m$ in Ω_1.*

Conversely, let Ω be a connected open subset of S^2 such that for every open $\Omega_1 \subseteq \Omega$ with $\partial_\Omega(\Omega_1) \neq \varnothing$ and every function u subharmonic in Ω, bounded above in Ω for which $\limsup_{\substack{z\to\zeta \\ z\in\Omega_1}} u(z) \leq m$ for every $\zeta \in \partial_\Omega(\Omega_1)$ one has $u \leq m$ in Ω_1. Then, Ω is parabolic.

Proof. Let us assume first that Ω is parabolic and let

$$v = \begin{cases} \sup(u - M, m - M) & \text{in } \Omega_1 \\ m - M & \text{in } \Omega \backslash \Omega_1. \end{cases}$$

It is clear that v is subharmonic, and hence u.s.c. both in Ω_1 and in $\Omega\backslash\bar{\Omega}_1$. Since in $\partial_\Omega(\Omega_1)$ we have that $v = m - M$, it follows that v is u.s.c. in Ω. If $\zeta \in \partial_\Omega(\Omega_1)$ and $\bar{B}(\zeta, r) \subseteq \Omega$, then from the definition of v we have $v \geq m - M$ on $B(\zeta, r)$, hence $\lambda(v, \zeta, r) \geq m - M = v(\zeta)$. Therefore v is subharmonic in Ω.

Since $v \leq \sup(0, m - M)$ in Ω we conclude that v is constant. This constant must be necessarily $m - M$ since $\partial_\Omega(\Omega_1) \neq \emptyset$. Therefore, $u - M \leq m - M$ in Ω_1, or what is the same, $u \leq m$ in Ω_1.

Conversely, suppose $u \leq M < \infty$ is a subharmonic function in Ω. We want to show u is a constant function. Let $z_0 \in \Omega$, $\Omega_1 = \Omega\backslash\{z_0\}$. Ω_1 is a connected open set with $\partial_\Omega(\Omega_1) = \{z_0\}$. The function $v = u|\Omega_1$ satisfies

$$\limsup_{\substack{z \to z_0 \\ z \in \Omega}} v(z) \leq u(z_0).$$

The hypothesis implies that $u(z) = v(z) \leq u(z_0)$ for every $z \in \Omega\backslash\{z_0\}$. Since z_0 was arbitrary, it follows that u must be a constant. □

4.9.33. Proposition. *Let Ω be a connected open set in S^2, $E \subsetneqq \partial_\infty\Omega$ a countable (or finite) set, and u a subharmonic function in Ω bounded above by $M < \infty$ such that*

$$\limsup_{\substack{z \to \zeta \\ z \in \Omega}} u(z) \leq m < \infty$$

for every $\zeta \in \partial_\infty\Omega\backslash E$. Then $u \leq m$ in Ω.

Proof. We argue by contradiction. Suppose there is a point $z_0 \in \Omega$ such that $u(z_0) > m$. Let $\mu > u(z_0)$. There is a disk $\bar{B}(z_0, r) \subseteq \Omega$ such that $u(z) < \mu$ in it.

For any $w \in S^2\backslash\bar{B}(z_0, r)$ there is a conformal map (in fact, a Moebius transformation) $\varphi : S^2\backslash\bar{B}(z_0, r) \to B(0, 1)$ such that $\varphi(w) = 0$. Choose in this way φ_n conformal mappings such that $\varphi_n(\zeta_n) = 0$ for $E = \{\zeta_n\}_{n \geq 1}$. Hence the functions $\log(1/|\varphi_n(z)|)$ are harmonic and positive in $\Omega\backslash\bar{B}(z_0, r)$.

For a fixed $z_1 \in \Omega\backslash\bar{B}(z_0, r)$ one can find a sequence $\alpha_n > 0$ such that $\sum_{n \geq 1} \alpha_n \log(1/|\varphi_n(z)|) < \infty$. Therefore the function

$$h(z) := \sum_{n \geq 1} \alpha_n \log(1/|\varphi_n(z)|) \not\equiv \infty$$

is harmonic and positive in $\Omega\backslash\bar{B}(z_0, r)$.

For $\varepsilon > 0$ and $z \in \Omega\backslash\bar{B}(z_0, r)$, let

$$v_\varepsilon(z) := u(z) - \varepsilon h(z) - \mu.$$

This function is subharmonic. Let us study its behavior on the boundary of $\Omega\backslash\bar{B}(z_0, r)$.

(i) If $\zeta \in E$ then a term of h tends to $+\infty$ as $z \to \zeta$. Since $u \leq M$ and $\varepsilon > 0$ we have $v_\varepsilon(z) \to -\infty$ as $z \to \zeta$.

(ii) If $\zeta \in \partial_\infty \Omega \setminus E$, $\limsup_{z \to \zeta} u(z) \leq m < u(z_0) < \mu$, and $0 \leq \varepsilon h(z)$, therefore

$$\limsup_{\substack{z \to \zeta \\ z \in \Omega}} v_\varepsilon(z) \leq 0.$$

(iii) If $\zeta \in \partial B(z_0, r)$, one has $|\varphi_n(z)| \to 1$ as $z \to \zeta$, hence $h(z) \to 0$. Nevertheless $h \geq 0$ and $\limsup_{z \to \zeta} u(z) \leq u(\zeta) < \mu$, implies also that $\limsup_{z \to \zeta} v_\varepsilon \leq 0$.

By the maximum principle, $v_\varepsilon \leq 0$ in $\Omega \setminus \bar{B}(z_0, r)$. That is

$$u(z) \leq \varepsilon h(z) + \mu, \qquad z \in \Omega \setminus \bar{B}(z_0, r).$$

Taking the limit $\varepsilon \to 0$, one obtains $u(z) \leq \mu$. Since μ is arbitrary, we have $u(z) \leq u(z_0)$, first in $\Omega \setminus \bar{B}(z_0, r)$, but using that r is arbitrary, one concludes that $u(z) \leq u(z_0)$ for every point $z \in \Omega$. From the maximum principle and the connectivity of Ω, it follows that $u(z) \equiv u(z_0) > m$. This contradicts the fact that $\partial_\infty \Omega \setminus E \neq \varnothing$, and for $\zeta \in \partial_\infty \Omega \setminus E$ one has $\limsup_{z \to \zeta} u \leq m$. □

4.9.34. Corollary. *Let Ω be a connected open set in S^2 with countable boundary. Every subharmonic function in Ω which is bounded above is constant.*

Proof. Let $z_0 \in \Omega$, $\Omega_1 = \Omega \setminus \{z_0\}$. We have $\limsup_{\substack{z \to z_0 \\ z \in \Omega_1}} u(z) \leq u(z_0)$. Therefore, we can apply the previous proposition to the sets

$$\Omega_1, \; E = \partial_\infty \Omega \subsetneqq \partial_\infty \Omega_1 = \partial_\infty \Omega \cup \{z_0\}.$$

We conclude that $u(z) \leq u(z_0)$ in Ω_1 (hence in Ω). Therefore u is a constant function. □

4.9.35. Theorem (Phragmen-Lindelöf Principle—I). *Let Ω be a connected open set in S^2, E a proper and countable subset of $\partial_\infty \Omega$, u a subharmonic function in Ω such that*

(1) $\limsup_{\substack{z \to \zeta \\ \zeta \in \Omega}} u(z) \leq m < \infty$ *for every $\zeta \in \partial_\infty \Omega \setminus E$.*

(2) *There exists a subharmonic function v in Ω such that*

$$\limsup_{\substack{z \to \zeta \\ z \in \Omega}} v(z) \leq 0 \qquad \textit{for every } \zeta \in \partial_\infty \Omega \setminus E.$$

(3) *For every $\varepsilon > 0$,* $\sup_{\zeta \in E} \limsup_{\substack{z \to \zeta \\ z \in \Omega}} (u(z) + \varepsilon v(z)) = A_\varepsilon < \infty$.

Then, $u(z) \leq m$ holds for every $z \in \Omega$.

PROOF. Let $\varepsilon > 0$. The function $w_\varepsilon(z) = u(z) + \varepsilon v(z)$ is subharmonic in Ω and satisfies $\limsup_{z \to \zeta} w_\varepsilon(z) \le m$ for $\zeta \in \partial_\infty \Omega \setminus E$, and $\limsup_{z \to \zeta} w_\varepsilon(z) \le A_\varepsilon$ for $\zeta \in E$. Hence, $\limsup_{z \to \zeta} w_\varepsilon(z) \le \sup(m, A_\varepsilon)$ for every $\zeta \in \partial_\infty \Omega$. It follows from the maximum principle that w_ε is bounded above in Ω. The previous proposition implies that $w_\varepsilon(z) \le m$ for every $z \in \Omega$. Let $B = \{z \in \Omega : v(z) = -\infty\}$, it has zero measure. For $z \in \Omega \setminus B$ we have $u(z) \le m - \varepsilon v(z)$, hence $u(z) \le m$. If $z \in B$, let $r > 0$ such that $\bar{B}(z,r) \subseteq \Omega$. We have $u(z) \le A(u, z, r) \le m$, the last inequality follows from the fact that B has zero measure. □

4.9.36. Corollary. *Let* $\gamma \ge \frac{1}{2}$, $\Omega = \left\{z \in \mathbb{C} : |\operatorname{Arg} z| < \frac{\pi}{2\gamma}\right\}$, *and let* u *be a subharmonic function in* Ω *such that*

(1) $\limsup\limits_{\substack{z \to \zeta \\ z \in \Omega}} u(z) \le m$ *for every* $\zeta \in \partial\Omega$

(2) $\limsup\limits_{\substack{z \to \infty \\ z \in \Omega}} \dfrac{\log u^+(z)}{\log|z|} < \gamma.$

Then $u \le m$ *throughout* Ω.

PROOF. Let δ, η be such that

$$\limsup_{\substack{z \to \zeta \\ z \in \Omega}} \frac{\log u^+(z)}{\log|z|} < \delta < \eta < \gamma,$$

and let $v(z) = \operatorname{Re}(-z^\eta) = -\operatorname{Re}(e^{-\eta \operatorname{Log} z})$ for $z \in \Omega$. (Recall $\operatorname{Log} z = \log|z| + i \operatorname{Arg} z$, $-\frac{\pi}{2\gamma} < \operatorname{Arg} z < \frac{\pi}{2\gamma}$.) One has

$$v(re^{i\theta}) = -r^\eta \cos \eta\theta \qquad \left(|\theta| < \frac{\pi}{2\gamma}\right)$$

and v is harmonic. Since

$$\lim_{\substack{z \to \zeta \\ z \in \Omega}} v(z) = -|\zeta|^\eta \cos \eta \frac{\pi}{2\gamma} \qquad (\zeta \in \partial\Omega)$$

and $0 < \frac{\eta\pi}{2\gamma} < \frac{\pi}{2}$, we have

$$\lim_{\substack{z \to \zeta \\ z \in \Omega}} v(z) \le 0 \qquad (\zeta \in \partial\Omega).$$

Furthermore, there is $R > 0$ such that if $r \ge R$ then

$$u(re^{i\theta}) < r^\delta \qquad \left(|\theta| < \frac{\pi}{2\gamma}\right),$$

hence

$$u(re^{i\theta}) + \varepsilon v(re^{i\theta}) < r^\delta - \varepsilon r^\eta \cos \eta\theta = r^\eta\left(-\varepsilon\cos\frac{\pi\theta}{2\gamma} + \frac{1}{r^{\eta-\delta}}\right)$$

and

$$\limsup_{\substack{z\to\infty\\ z\in\Omega}} (u(z) + \varepsilon v(z)) = -\infty$$

for any $\varepsilon > 0$. We can therefore apply the previous proposition to obtain the inequality $u \le m$ in the angular sector Ω. □

4.9.37. Corollary. *Let* $\gamma \ge \dfrac{1}{2}$, $\Omega = \left\{z \in \mathbb{C} : |\operatorname{Arg} z| < \dfrac{\pi}{2\gamma}\right\}$. *If* f *is a holomorphic function in* Ω *such that*

(1) $\limsup\limits_{\substack{z\to\zeta\\ z\in\Omega}} |f(z)| \le M < \infty$ *for every* $\zeta \in \partial\Omega$,

(2) $\limsup\limits_{\substack{z\to\infty\\ z\in\Omega}} \dfrac{\log(\log^+|f(z)|)}{\log|z|} < \gamma$

then $|f(z)| \le M$ *for every* $z \in \Omega$.

4.9.38. Corollary. *Let* $\rho > \frac{1}{2}$ *and* f *an entire function of order* $\le \rho$. *If* f *is bounded over a family of half-lines* $\arg z = \theta_j$, $1 \le j \le n+1$, $\theta_1 = \theta_{n+1}$, *such that the angle between consecutive half-lines is* $< \pi/\rho$, *then* f *is constant.*

4.9.39. Examples. (1) The function $f(z) = \exp(\alpha z^\gamma)$, $\alpha > 0$, $\gamma > \frac{1}{2}$ is not bounded in $\Omega = \left\{z \in \mathbb{C} : |\operatorname{Arg} z| < \dfrac{\pi}{2\gamma}\right\}$, while $\log M(|f|, r) = \alpha r^\gamma$ and

$$\limsup_{\substack{z\to\infty\\ z\in\Omega}} \frac{\log\log^+|f(z)|}{\log|z|} = \gamma.$$

(2) If $\rho = \frac{1}{2}$ then Corollary 4.9.38 might fail. The entire function $\dfrac{\sin\sqrt{z}}{\sqrt{z}}$ shows there are entire functions of order $\frac{1}{2}$ bounded over a half-line which are not constant. In Corollary 4.9.38, the condition $\rho > \frac{1}{2}$ forces the function f to be bounded on at least two half-lines.

4.9.40. Corollary. *An entire function* f *of order* $\rho < \frac{1}{2}$ *cannot be bounded on any half-line without being constant. If* $\rho = \frac{1}{2}$ *and type* $\tau < \infty$ *then there is at most one half-line where* f *can be bounded unless* f *is constant.*

4.9.41. Proposition. *Let u be a subharmonic function in*

$$\Omega = \{z \in \mathbb{C} : a < \operatorname{Re} z < b\}, \qquad u \le M < \infty \text{ in } \Omega.$$

Let

$$M(x) := \sup_{y \in \mathbb{R}} u(x + iy) \qquad (a < x < b),$$

then $x \mapsto M(x)$ is a convex function in $]a, b[$.

PROOF. Let

$$g(z) = \frac{x-a}{b-a} M(b) + \frac{b-x}{b-a} M(a),$$

where $M(a) = \limsup_{x \to a+0} M(x)$, $M(b) = \limsup_{x \to b-0} M(x)$. Then g is a harmonic function in $\mathbb{C}$, $g(a + iy) = M(a)$, $g(b + iy) = M(b)$. Hence $v(z) = u(z) - g(z)$ is subharmonic in Ω, bounded above and $v \le 0$ on $\partial\Omega$. This ensures that $v \le 0$ in Ω by §4.9.33. Therefore

$$u(z) \le \frac{x-a}{b-a} M(b) + \frac{b-x}{b-a} M(a) \qquad (z = x + iy)$$

and

$$M(x) \le \frac{x-a}{b-a} M(b) + \frac{b-x}{b-a} M(a).$$

Note that this reasoning also allows us to conclude that the function $M(x)$ is a convex function of x. It is enough to replace a and b by any two values x_1, x_2, $a < x_1 < x_2 < b$ to conclude $M((1-t)x_1 + tx_2) \le (1-t)M(x_1) + tM(x_2)$ when $0 \le t \le 1$. □

4.9.42. Corollary. *Let f be a holomorphic function in the strip $\Omega = \{z \in \mathbb{C} : z < \operatorname{Re} z < b\}$, which is bounded in Ω. The function $M(x) := \sup_{y \in \mathbb{R}} |f(x + iy)|$, $a < x < b$, is then a logarithmically convex function in $]a, b[$ and*

$$M(x)^{b-a} \le M(a)^{b-x} M(b)^{x-a}$$

where $M(a) = \limsup_{x \to a+0} M(x)$, $M(b) = \limsup_{x \to b-0} M(x)$.

4.9.43. Proposition. *Let $\Omega = \left\{z \in \mathbb{C} : |\operatorname{Im} z| < \dfrac{\pi}{2}\right\}$ and u a subharmonic function in Ω. If there are two constants $\alpha < 1$ and $A < \infty$ such that*

$$u(z) \le A \exp(\alpha |\operatorname{Re} z|), \qquad (z \in \Omega)$$

$$\limsup_{z \to \zeta} u(z) \le 0 \qquad (\zeta \in \partial\Omega),$$

then $u(z) \le 0$ for every $z \in \Omega$.

Let us remark that $f(z) = \exp(\exp z)$ and $u(z) = \operatorname{Re}(\exp z)$ show that one cannot take $\alpha = 1$ in §4.9.43.

PROOF. Choose β, $0 < \alpha < \beta < 1$, and let $\varepsilon > 0$ be fixed. Let $h_\varepsilon(z) := -\varepsilon \operatorname{Re}(e^{\beta z} + e^{-\beta z}) = -2\operatorname{Re}(\cosh \beta z)$.

For $z \in \bar{\Omega}$ we have

$$\operatorname{Re}(e^{\beta z} + e^{-\beta z}) = (e^{\beta x} + e^{-\beta x})\cos \beta y \geq \delta(e^{\beta x} + e^{-\beta x})$$

with $\delta = \cos\dfrac{\beta\pi}{2} > 0$ (since $0 < \beta < 1$). Hence

$$\limsup_{z\to\zeta}\, (u(z) + h_\varepsilon(z)) \leq 0 \qquad \text{for } \zeta \in \partial\Omega,$$

and, for $z \in \Omega$,

$$u(z) + h_\varepsilon(z) \leq Ae^{\alpha|x|} - \varepsilon\delta(e^{\beta x} + e^{-\beta x}),$$

which shows that

$$\limsup_{\substack{z\to\infty\\ z\in\Omega}}\, (u(z) + h_\varepsilon(z)) \leq 0.$$

Therefore, $u + h_\varepsilon \leq 0$ in Ω, and $u \leq 0$ in Ω by letting $\varepsilon \to 0$. □

Since there exist uncountable compact sets with zero capacity, the following two propositions are more subtle than §4.9.33 and §4.9.35.

4.9.44. Proposition. *Let Ω be a connected open set in $\mathbb{C}$ and $(E_n)_{n\geq 1}$ a locally finite sequence of compact subsets of $\partial\Omega$, all of them of zero capacity, $E = \bigcup_{n\geq 1} E_n$. Let u be a subharmonic function in Ω which is bounded above. If*

$$\limsup_{\substack{z\to\zeta\\ z\in\Omega}}\, u(z) \leq m \qquad \text{for } \zeta \in \partial\Omega\setminus E,$$

then $u \leq m$ in Ω.

PROOF. Given $\mu > m$, without loss of generality we can assume that $0 \in \Omega$, $u(0) < \mu$, and Ω is unbounded. For the construction that follows of a subharmonic function h in $\mathbb{C}$ such that $E \subseteq \{z \in \mathbb{C} : h = -\infty\}$ we can assume $(E_n)_{n\geq 1}$ is an infinite sequence, otherwise we can apply §4.9.22 directly.

For every $n \geq 1$ we can find $\sigma_n \in \mathscr{P}(E_n)$ such that $E_n = \{z \in \mathbb{C} : U^{\sigma_n} = -\infty\}$ (always by §4.9.22)). Since the sequence $(E_n)_{n\geq 1}$ is locally finite we have $d(0, E_n) > 1$ for all n sufficiently large, hence $U^{\sigma_n}(0) > 0$ for those n. Choose a sequence of positive numbers $(\alpha_n)_{n\geq 1}$ such that the two series $\sum_{n\geq 1} \alpha_n$ and $\sum_{n\geq 1} \alpha_n U^{\sigma_n}(0)$ are convergent. The function $h(z) := \sum_{n\geq 1} \alpha_n U^{\sigma_n}(z)$ is locally given as the sum of a finite number of subharmonic functions plus a uniformly convergent series of harmonic functions. Therefore h is subharmonic in $\mathbb{C}$, harmonic in $\mathbb{C}\setminus E$, and $-\infty$ on E. Namely, if $r > 0$ and $n \geq n_0$ we have

$E_n \cap \bar{B}(0, r+1) = \emptyset$; hence for $|z| \le r$ and $n \ge n_0$ we have $U^{\sigma_n}(z) \ge 0$. Now, if $t \ne 0$, $\log|z-t| \le \log|t| + \frac{|z|}{|t|}$, then

$$\sum_{n \ge n_0} \alpha_n U^{\sigma_n}(z) \le \frac{1}{2\pi} \sum_{n \ge n_0} \alpha_n \int_{E_n} \log|t|\, d\sigma_n(t) + \frac{|z|}{2\pi} \sum_{n \ge n_0} \alpha_n \int \frac{d\sigma_n(t)}{|t|}$$

$$\le \sum_{n \ge n_0} \alpha_n U^{\sigma_n}(0) + \frac{r}{2\pi} \sum_{n \ge n_0} \alpha_n < \infty.$$

(We have used that for $n \ge n_0$, $t \in \operatorname{supp} \sigma_n$, implies that $|t| \ge 1$.) Using now that $\log|z-t| \le \log^+|z| + \log^+|t| + \log 2$, the same reasoning shows that there is a positive constant c_0 such that

$$h(z) \le c_0(1 + \log^+|z|).$$

Hence, given $r > 0$ there is a constant $c_1 > 0$ such that

$$h(z) \le c_1\left(1 + \log\left(\frac{|z|}{r}\right)\right) \qquad \text{when } |z| \ge r.$$

Let $r > 0$ be such that $\bar{B}(0,r) \subseteq \Omega$, $u(z) \le \lambda < \mu$ for $z \in \bar{B}(0,r)$. (This is possible because $u(0) < \mu$.) In the domain $\Omega_1 = \Omega \setminus \bar{B}(0,r)$, consider for $\varepsilon > 0$ the auxiliary subharmonic function,

$$v_\varepsilon(z) := u(z) + \varepsilon h(z) - \varepsilon c_1 \log\left(\frac{|z|}{r}\right) - \varepsilon c_1.$$

We claim that

$$\limsup_{\substack{z \to \zeta \\ z \in \Omega_1}} v_\varepsilon(z) \le \mu, \qquad \text{for all } \zeta \in \partial\Omega_1.$$

We note that $v_\varepsilon \le u$ everywhere in Ω_1. Therefore v_ε is bounded in Ω_1 and if $\zeta \in \partial B(0,r)$ or $\zeta \in \partial\Omega \setminus (E \cup \bar{B}(0,r))$ we have

$$\limsup_{\substack{z \to \zeta \\ z \in \Omega_1}} v_\varepsilon(z) \le \limsup_{\substack{z \to \zeta \\ z \in \Omega_1}} u(z) \le \sup(\lambda, m) < \mu.$$

Let $M = \sup_{\Omega} u$. If $\zeta \in \partial\Omega_1 \cap E$, then we have

$$\limsup_{\substack{z \to \zeta \\ z \in \Omega_1}} v_\varepsilon(z) \le M + \limsup_{\substack{z \to \zeta \\ z \in \Omega_1}} h(z) - \varepsilon c_1\left(\log\frac{|\zeta|}{r} + 1\right) = -\infty.$$

It follows that $v_\varepsilon \le \mu$ in Ω_1, a posteriori $u \le \mu$ in Ω_1 since h is finite everywhere in Ω_1. Since $u \le \mu$ in $\bar{B}(0,r)$ we have $u \le \mu$ everywhere in Ω. As μ was an arbitrary quantity $> m$, we can conclude that the inequality $u \le m$ in Ω holds. □

4.9.45. Proposition (Phragmen-Lindelöf Principle II). *Let Ω be an open connected set in $\mathbb{C}$, $E = \bigcup_{n \ge 1} E_n \subseteq \partial\Omega$ be the union of locally finite family of compact* *sets E_n of capacity zero. Let $\hat{E} = E$ if Ω bounded, $\hat{E} = E \cup \{\infty\}$ if not. Let u be a subharmonic function in Ω satisfying:*

(1) $\limsup_{\substack{z \to \zeta \\ z \in \Omega}} u(z) \le m < \infty$, *for every* $\zeta \in \partial\Omega \setminus E$.

(2) *There exists a subharmonic function v in Ω such that*

$$\limsup_{\substack{z \to \zeta \\ z \in \Omega}} v(z) \le 0 \qquad \textit{for every } \zeta \in \partial\Omega \setminus E.$$

(3) *For every* $\varepsilon > 0$, $\sup_{\zeta \in \hat{E}} \left(\limsup_{\substack{z \to \zeta \\ z \in \Omega}} (u(z) + \varepsilon v(z)) \right) = A_\varepsilon < \infty.$

Then $u(z) \le m$ for every $z \in \Omega$.

PROOF. Let $w_\varepsilon(z) = u(z) + \varepsilon v(z)$. By the maximum principle w_ε is bounded above by $\sup(m, A_\varepsilon)$. Proposition 4.9.44 allows us now to conclude that that $w_\varepsilon \le m$ in Ω. Hence $u \le m$ in Ω when we let $\varepsilon \to 0$. □

We return to the question of relating the notion of capacity to that of Green functions.

4.9.46. Proposition. *Let E be a compact set in $\mathbb{C}$, with $C(E) > 0$, and such that every connected component of the exterior boundary Γ has at least two points. Let Ω be the unbounded component of $\mathbb{C} \setminus E$ and $g(z; \infty, \Omega)$ be the Green function of Ω with pole at ∞. Then, if ν denotes the equilibrium measure of E, we have*

$$g(z; \infty, \Omega) = 2\pi U^\nu(z) + V(E).$$

It is standard to call in this context the constant $V(E)$ $(= -\log C(E))$, the ***Robin constant*** of E.

PROOF. The boundary $\partial\Omega$ is Γ. The hypothesis made on it guarantees the existence of the Green function $g(z) := g(z; \infty, \Omega)$. The function on the right-hand side of the equation in the statement of the proposition is subharmonic in $\mathbb{C}$ and harmonic in Ω. At infinity we have the asymptotic behavior

$$2\pi U^\nu(z) = \log|z| + o(1).$$

Moreover, by Proposition 4.9.23, there is an exceptional set A, $A = \bigcup_{n \ge 1} A_n$, each A_n is the compact set of zero capacity given by

$$A_n = \left\{ z \in \Gamma : U^\nu(z) \ge -\frac{V(E)}{2\pi} + \frac{1}{n} \right\},$$

such that $U^\nu(z) \ge -\dfrac{V(E)}{2\pi}$ everywhere and, on $E \setminus A$ we have $U^\nu(z) = -\dfrac{V(E)}{2\pi}$.

Let us consider the function

$$v(z) := 2\pi U^\nu(z) + V(E) - g(z).$$

It is harmonic in Ω and bounded at ∞ since $g(z) = \log|z| + O(1)$ near $z = \infty$. Therefore v is bounded in Ω. For $\zeta \in \Gamma \backslash A_n$ we have

$$\limsup_{\substack{z \to \zeta \\ z \in \Omega}} (2\pi U^\nu(z) + V(E)) \leq 2\pi U^\nu(\zeta) + V(E) \leq \frac{2\pi}{n}.$$

By the continuity of the Green function at the points of Γ we have

$$\limsup_{\substack{z \to \zeta \\ z \in \Omega}} v(z) \leq \frac{2\pi}{n} \qquad (\zeta \in \Gamma \backslash A_n).$$

From Proposition 4.9.44 we obtain

$$v(z) \leq \frac{2\pi}{n} \qquad \text{for every } z \in \Omega.$$

Therefore, $v \leq 0$ in Ω. Since, as pointed out earlier, we also know that $2\pi U^\nu + V(E) \geq 0$, we have

$$0 \leq 2\pi U^\nu + V(E) \leq g \qquad \text{in } \Omega.$$

This ensures that the limit of $2\pi U^\nu(z) + V(E)$ when z approaches any point in Γ from inside Ω is zero. In §4.9.23 we also saw that $2\pi U^\nu + V(E) \equiv 0$ in $\mathbb{C} \backslash \bar{\Omega}$, therefore the function $2\pi U^\nu + V(E)$ is continuous in $\bar{\Omega}$ and has the value zero on the boundary. This proves that $2\pi U^\nu + V(E) \equiv g(z)$ in $\bar{\Omega}$. □

4.9.47. Remarks. (1) It now follows from Proposition 4.9.46 that $U^\nu \equiv -\dfrac{V(E)}{2\pi}$ on Γ when E is a compact set satisfying the hypothesis of §4.9.46. Therefore, the exceptional set A from Proposition 4.9.23 is empty and U^ν is continuous everywhere.

(2) $V(E)$ can be computed as

$$V(E) = \lim_{z \to \infty} g(z; \infty, \Omega) - \log|z|,$$

which is the usual definition of the Robin constant.

(3) These two previous properties hold for any compact set E with positive capacity and such that the Dirichlet problem is solvable in Ω, the unbounded component of $\mathbb{C} \backslash E$.

4.9.48. Examples. (1) As a first example let us consider $E = \bar{B}(0, R)$. We have $g(z; \infty, \Omega) = \log \dfrac{|z|}{R}$. Therefore,

$$V(E) = \lim_{z \to \infty} g(z; \infty, \Omega) - \log|z| = -\log R.$$

Hence $C(E) = R$.

(2) As another example let us take $E = [-1, 1]$. The function $\varphi(z) = \dfrac{1}{2}\left(z + \dfrac{1}{z}\right)$ is a continuous map from $\bar{\Omega}_1$ onto $\bar{\Omega}_2$, where $\Omega_1 = \{z \in \mathbb{C} : |z| > 1\}$

and $\Omega_2 = \mathbb{C}\setminus[-1,1]$. Moreover, φ is a biholomorphism of Ω_1 onto Ω_2 such that $\varphi(\infty) = \infty$. This observation allows us to compute the Green function of Ω_2 with pole at ∞. Namely, if $w = \varphi(z)$,

$$g(w; \infty, \Omega_2) = g(z; \infty, \Omega_1) = \log|z| = \log|w + \sqrt{w^2 - 1}|,$$

where $\sqrt{w^2 - 1}$ is ≥ 0 for $w \in]1, \infty[$. Therefore,

$$V(E) = \lim_{w \to \infty} (g(w; \infty, \Omega_2) - \log|w|) = \lim_{|w| \to \infty} \log\left|1 + \sqrt{1 - \frac{1}{w^2}}\right| = \log 2.$$

Hence $C(E) = \frac{1}{2}$. We conclude that $C([a,b]) = \dfrac{b-a}{4}$ if $a, b \in \mathbb{R}$, $a < b$.

4.9.49. Proposition. *Let Ω be an open connected set in S^2, $\infty \in \Omega$, such that the Dirichlet problem is solvable in Ω. Let ω_∞ be the harmonic measure for ∞ and let γ of the Robin constant of Ω:*

$$\gamma = \lim_{z \to \infty} g(z; \infty, \Omega) - \log|z|.$$

Then the logarithmic potential generated by ω_∞ is a continuous function and verifies:

$$U^{\omega_\infty}(z) = \frac{1}{2\pi} \int \log|z - t| \, d\omega_\infty(t) = \begin{cases} \dfrac{1}{2\pi}((g(z; \infty, \Omega_\infty) - \gamma) & \text{if } z \in \Omega\setminus\{\infty\} \\[2ex] -\dfrac{\gamma}{2\pi} & \text{if } z \notin \bar{\Omega}. \end{cases}$$

In particular, $C(\partial\Omega) > 0$. Moreover, if E verifies the conditions of §4.9.46 and Ω is the unbounded component of E in S^2, then $\gamma = V(E)$ and ω_∞ is the equilibrium measure of E.

Proof. Note that the proposed formula for U^{ω_∞} represents a continuous function in $\mathbb{C}$. For $z_0 \in \Omega$, $z_0 \neq \infty$, and $z \in \Omega\setminus\{z_0\}$, let

$$w(z) := \int_{\partial\Omega} \log|z_0 - \zeta| \, d\omega_z(\zeta) - \log|z_0 - z| + g(z; \infty, \Omega).$$

This function is harmonic in $\Omega\setminus\{z_0\}$, including the point $z = \infty$; it also has a logarithmic pole at z_0 and vanishes on $\partial\Omega$. It follows that $w(z) = g(z; z_0, \Omega)$.
Since $g(z; z_0, \Omega) = g(z_0; z, \Omega)$ we have

$$\int_{\partial\Omega} \log|z_0 - \zeta| \, d\omega_z(\zeta) = g(z_0; z, \Omega) - (g(z; \infty, \Omega) - \log|z - z_0|).$$

Since for u continuous on $\partial\Omega$, $Pu(z) \to Pu(\infty)$ when $z \to \infty$, we can let $z \to \infty$ in this identity and obtain

$$\int_{\partial\Omega} \log|z_0 - \zeta| \, d\omega_\infty(\zeta) = g(z_0; \infty, \Omega) - \gamma.$$

This is the first part of the result we wanted to prove.

Assume now that $z_0 \notin \bar{\Omega}$. Let us define $w(z)$ by the same formula as earlier. This time, w is not only harmonic in Ω but also $\equiv 0$ on $\partial\Omega$. Therefore $w \equiv 0$. Let $z \to \infty$, then

$$0 = w(\infty) = \int_{\partial\Omega} \log|z_0 - \zeta|\, d\omega_\infty(\zeta) + \gamma.$$

On the other hand, the proof of §4.7.17 shows that for $z_0 \in \partial\Omega$ we have

$$U^{\omega_\infty}(z_0) = \lim_{\substack{z \to z_0 \\ z \in \Omega}} U^{\omega_\infty}(z).$$

This implies that the formula holds everywhere.

From the formula just obtained we conclude that

$$I(\omega_\infty) = -2\pi \int U^{\omega_\infty}(z)\, d\omega_\infty(z) = \int_{\partial\Omega} \gamma\, d\omega_\infty(z) = \gamma < \infty.$$

Hence $V(\partial\Omega) < \infty$ and $C(\partial\Omega) > 0$. The rest of the proposition follows from §4.9.46. □

Let Ω be a connected open subset of S^2 such that the Dirichlet problem is solvable in Ω. Let E be a closed subset of $\partial\Omega$. By Urysohn's theorem there is a continuous function $u : \partial\Omega \to [0, 1]$ such that $u^{-1}(\{1\}) = E$. Then, for $z \in \Omega$ we have

$$\omega_z(E) = \int_E d\omega_z(\zeta) = \lim_{n \to \infty} \int_{\partial\Omega} u^n(\zeta)\, d\omega_z(\zeta).$$

Since $u_n := P(u^n)$ is a decreasing sequence of harmonic functions taking values in $[0, 1]$, its limit is a harmonic function. Therefore $z \mapsto \omega_z(E)$ is a harmonic function in Ω. Moreover, if $\zeta \in \partial\Omega$ one has

$$\limsup_{\substack{z \to \zeta \\ z \in \Omega}} \omega_z(E) \le \limsup_{\substack{z \to \zeta \\ z \in \Omega}} u_n(z) = u^n(\zeta) \qquad (n \ge 1).$$

Hence

$$0 \le \limsup_{\substack{z \to \zeta \\ z \in \Omega}} \omega_z(E) \le \chi_E(\zeta) \qquad (\zeta \in \partial\Omega).$$

In particular, $\lim_{z \to \zeta} \omega_z(E) = 0$ if $\zeta \in \partial\Omega \setminus E$.

One can also show that for every $\zeta \in \mathring{E}$ (the relative interior), $\lim_{z \to \zeta} \omega_z(E) = 1$. In fact, let $v \in \mathscr{C}(\partial\Omega)$, $v = 1$ in a neighborhood of ζ and $v = 0$ on $\partial\Omega \setminus E$. Then $v(z) \le u^n(z)$ for every $z \in \partial\Omega$, and therefore $\omega_z(E) \ge Pv(z)$ for every $z \in \Omega$. This shows that $\omega_z(E) \to 1$ as $z \to \zeta$ $(z \in \Omega)$.

One can easily verify, always using the theorem of monotone limits of harmonic functions, that the class $\mathscr{E}$ of Borel sets $E \subseteq \partial\Omega$ for which $z \mapsto \omega_z(E)$ is a harmonic function, is a monotone class. This class contains the closed sets, as we have shown earlier, hence $\mathscr{E} = \mathscr{B}(\partial\Omega)$, the family of all Borel sets in $\partial\Omega$.

Recall that for every compact set $K \subseteq \Omega$ there are constants $c_1, c_2 > 0$ such that

$$c_1 \omega_z(E) \le \omega_{z'}(E) \le c_2 \omega_z(E)$$

for every $z, z' \in K$ (cf. §4.7.15). From this it follows that if f is a Borel measurable function in $\partial\Omega$ which is integrable with respect to the measute ω_{z_0} for some $z_0 \in \Omega$, then f is integrable with respect to every $\omega_z, z \in \Omega$. It is enough to prove it for $f \ge 0$. In this case we know that f is the increasing limit of the step functions $(u_n)_{n \ge 1}$,

$$u_n := \sum_{0 \le k \le n2^n - 1} \frac{k}{2n} \chi_{f^{-1}\left(\left[\frac{k}{2^n}, \frac{k+1}{2^n}\right[\right)} + n\chi_{f^{-1}([n,\infty[)}.$$

The sequence $(Pu_n)_{n \ge 1}$ is an increasing sequence of harmonic functions in Ω that converges at z_0 towards $\int_{\partial\Omega} f(\zeta)\, d\omega_{z_0}(\zeta)$. Hence $Pu_n(z)$ converges everywhere to a finite limit, that is, $f \in L^1(\partial\Omega, d\omega_z)$, $\int_{\partial\Omega} f(\zeta)\, d\omega_z(\zeta) = \lim_{n\to\infty} Pu_n(z)$. Moreover $z \mapsto \int_{\partial\Omega} f\, d\omega_z$ is harmonic in Ω.

Let now u be a subharmonic function in a neighborhood U of $\bar{\Omega}$ and Pu be the Poisson extension of $u|\partial\Omega$,

$$Pu(z) = \int_{\partial\Omega} u(\zeta)\, d\omega_z(\zeta).$$

We want to show that Pu is well defined. Since u is u.s.c., $u|\partial\Omega$ is a decreasing limit of continuous functions $(\varphi_n)_{n \ge 1}$. Hence $P\varphi_n$ is a decreasing sequence of harmonic functions, $Pu = \lim_{n\to\infty} P\varphi_n$ in Ω. Moreover, $(P\varphi_n)|\partial\Omega \ge u$ on $\partial\Omega$, hence we have $P\varphi_n \ge u$ in Ω. Therefore $Pu \ge u$ in Ω, which shows that u is actually integrable with respect to $d\omega_z$ and Pu is well defined.

Suppose now that Ω is bounded and u is a subharmonic function in Ω. Define a u.s.c. function φ on $\bar{\Omega}$ by $\varphi(\zeta) = u(\zeta)$, if $\zeta \in \Omega$, and $\varphi(\zeta) = \limsup_{z\to\zeta} u(z)$, if $\zeta \in \partial\Omega$. Assume that $\sup_{\zeta \in \partial\Omega} \varphi(\zeta) = M < \infty$. The preceding reasoning actually shows that $u \le P(\varphi|\partial\Omega)$ in Ω. Therefore $u \le M$ in Ω (which we would have known anyway using the maximum principle), but if we only know that φ is bounded and $\varphi(\zeta) \le M$ for $\zeta \in \partial\Omega \backslash E$, E a set of zero harmonic measure, then we will still obtain

$$u \le P(\varphi|\partial\Omega) \le M$$

since

$$\int_{\partial\Omega} \varphi\, d\omega_z = \int_{\partial\Omega\backslash E} \varphi\, d\omega_z$$

for every $z \in \Omega$.

A more precise result is the following.

4.9.50 Theorem (of the Two Constants). *Let Ω be a bounded open set of $\mathbb{C}$ such that the Dirichlet problem is solvable in Ω. Let E be a Borel subset of $\partial\Omega$. Let u be a subharmonic function in Ω extended to an upper semicontinuous function φ on $\bar{\Omega}$ by $\varphi(\zeta) = \limsup_{z\to\zeta} u(z)$ for $\zeta \in \partial\Omega$. Assume that $\varphi \leq m$ in E and $\varphi \leq M$ on $\partial\Omega\backslash E$. Then*

$$u(z) \leq m\omega_z(E) + M\omega_z(\partial\Omega\backslash E) \qquad (z \in \Omega).$$

More generally, if φ is bounded and if $\partial\Omega = \left(\bigcup_{1\leq j\leq n} E_j\right)\dot{\bigcup} N$, where $E_1, \ldots, E_n$ are Borel subsets of $\partial\Omega$, N is a Borel subset of zero harmonic measure, and $\varphi \leq m_j < \infty$ on E_j. Then

$$u(z) \leq \sum_{1\leq j\leq n} m_j\omega_z(E_j) \qquad (z \in \Omega).$$

PROOF. $u(z) \leq (P\varphi)(z) \leq \sum_{1\leq j\leq n} m_j\omega_z(E_j)$ since $\varphi|\partial\Omega \leq \sum_{1\leq j\leq n} m_j\chi_{E_j}, \omega_z$ – a.e. for every $z \in \Omega$. □

Let us denote $\omega(z, E, \Omega)$ the value $\omega_z(E)$ for $z \in \Omega$ and E a Borel subset of $\partial\Omega$. It is called the ***harmonic measure*** (at z) of E. This harmonic measure is invariant under conformal mappings.

Let $\varphi: \Omega \to \Omega'$ be a biholomorphic mapping which has an extension to a homeomorphism of $\bar{\Omega}$ onto $\bar{\Omega}'$. Assume the Dirichlet problem is solvable for Ω (hence also for Ω') then

$$\omega(z, E, \Omega) = \omega(\varphi(z), \varphi(E), \Omega').$$

Namely, if h is a continuous function on $\partial\Omega$ one has

$$(P_\Omega h)(z) = \int_{\partial\Omega} h\, d\omega_z$$

and

$$(P_{\Omega'}(h \circ \varphi^{-1}))(\varphi(z)) = \int_{\partial\Omega'} (h \circ \varphi^{-1})\, d\omega_{\varphi(z)}.$$

But

$$P_\Omega h = (P_{\Omega'}(h \circ \varphi^{-1})) \circ \varphi,$$

therefore, if E is a Borel subset of $\partial\Omega$ we will have

$$\omega(z, E, \Omega) = (P_\Omega \chi_E)(z) = (P_{\Omega'} \chi_{\varphi(E)})(\varphi(z)) = \omega(\varphi(z), \varphi(E), \Omega').$$

We exploit this remark in the following examples.

4.9.51. Examples of Harmonic Measures. (1) If $\Omega = \{z \in \mathbb{C} : |z| < R\}$. The usual Poisson formula gives

$$d\omega_z(\mathrm{Re}^{i\theta}) = \frac{1}{2\pi}\frac{R^2 - |z|^2}{|\mathrm{Re}^{i\theta} - z|^2}\,d\theta.$$

For $z = 0$, $E \subseteq \partial\Omega$ one has

$$\omega(0, E, \Omega) = \frac{1}{2\pi}\int_E d\theta.$$

If E is the arc of circle $\zeta_2\zeta_1$ described in the counterclockwise sense then

$$\omega(z, E, \Omega) = \frac{1}{\pi}\left(\operatorname{Arg}\left(\frac{z - \zeta_1}{z - \zeta_2}\right) - \beta\right),$$

β is described by the Figure 4.6 and $\arg\left(\frac{z - \zeta_1}{z - \zeta_2}\right)$ is the determination of the argument whose value at $z = 0$ is 2β.

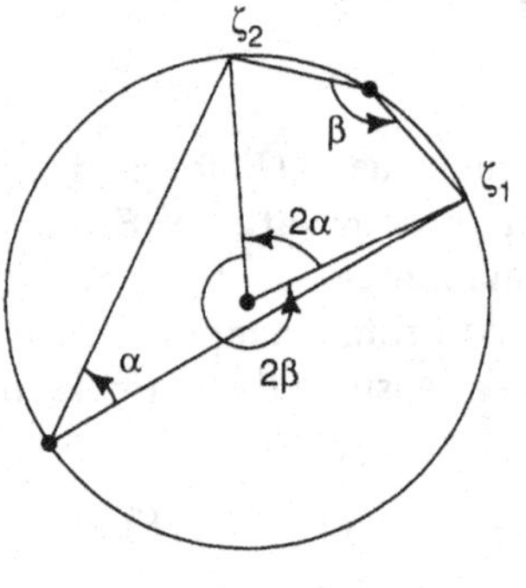

Figure 4.6

(2) If $\Omega = \{z \in \mathbb{C} : |z| < R \text{ and } \operatorname{Im} z > 0\}$, $E = \{z \in \mathbb{C} : |z| = R, \operatorname{Im} z > 0\}$ then $\omega(z, E, \Omega) = \frac{2}{\pi}\operatorname{Arg}\left(\frac{R + z}{R - z}\right) = \frac{2}{\pi}\alpha.$

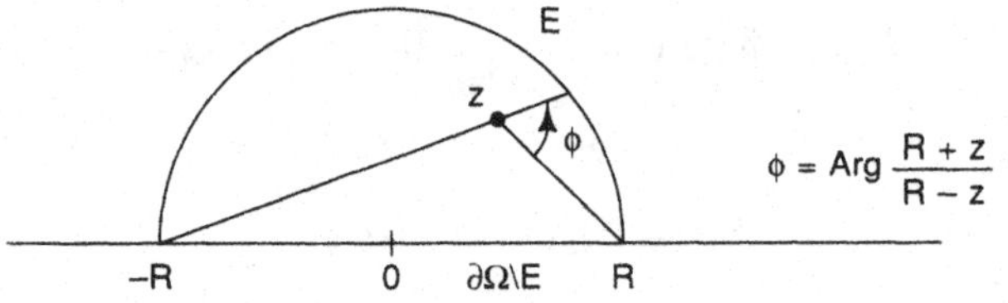

Figure 4.7

(3) $\Omega_1 = \{z \in \mathbb{C} : |z| < 1\}\backslash\{0 \le x \le 1\}$, $E_1 = \{0 \le x \le 1\}$. Then Ω_1 is obtained from the open set Ω of Example 4.9.51(2) (with $R = 1$) by the map $z \mapsto \varphi(z) = z^2$.

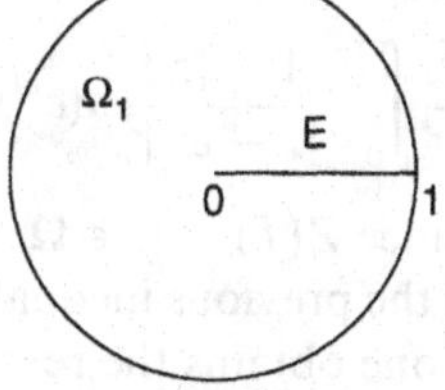

Figure 4.8

Therefore

$$\omega(z, E_1, \Omega_1) = \omega(\sqrt{z}, \partial\Omega \backslash E, \Omega) = 1 - \omega(\sqrt{z}, E, \Omega)$$
$$= 1 - \frac{2}{\pi} \operatorname{Arg}\left(\frac{1 + \sqrt{z}}{1 - \sqrt{z}}\right).$$

(4) $\Omega = \{z \in \mathbb{C} : 0 < r_1 < |z| < r_2 < \infty\}$, $E = \{z \in \mathbb{C} : |z| = r_1\}$. Then

$$\omega(z, E, \Omega) = \frac{\log r_2 - \log|z|}{\log r_2 - \log r_1}$$

$$\omega(z, \partial\Omega \backslash E, \Omega) = \frac{\log|z| - \log r_1}{\log r_2 - \log r_1}.$$

We conclude this chapter with a few results about removable singularities.

4.9.52. Theorem (Radó-Cartan). *Let Ω be an open connected subset of $\mathbb{C}$, $f : \Omega \to \mathbb{C}$ a continuous function such that it is holomorphic in the open subset $\Omega \backslash Z(f)$, $Z(f) = \{z \in \Omega : f(z) = 0\}$. Then f is a holomorphic function in Ω.*

Proof. Let $z_0 \in \Omega$. To simplify the notation we assume $z_0 = 0$ and $B(z_0, R) \subseteq \Omega$ for some $R > 1$. We are going to show that $f | B(0, 1)$ is holomorphic. It is clear that we can assume $f \not\equiv 0$ in $B(0, 1)$.

Let $M = \sup_{|z|=1} |f(z)|$, $p(z) = \log \dfrac{|f(z)|}{M}$. This function is subharmonic in Ω. In fact, p is clearly u.s.c. in Ω. Moreover it is clear that p is harmonic in $\Omega \backslash Z(f)$. If $z \in Z(f)$, one has $-\infty = p(z) \leq A(p, z, r)$ for any $r > 0$ such that $\bar{B}(z, r) \subseteq \Omega$.

Let $f = u + iv$. For $\varepsilon > 0$, the function $u + \varepsilon p$ is subharmonic. It is enough to use the same reasoning that showed that p is subharmonic. Hence, for $|z| < 1$ we have

$$u(z) + \varepsilon p(z) \leq \frac{1}{2\pi} \int_0^{2\pi} \frac{1 - |z|^2}{|z - e^{i\theta}|^2} (u(e^{i\theta}) + \varepsilon p(e^{i\theta}))\, d\theta$$
$$\leq \frac{1}{2\pi} \int_0^{2\pi} \frac{1 - |z|^2}{|z - e^{i\theta}|^2} u(e^{i\theta})\, d\theta,$$

since $p(e^{i\theta}) \leq 0$. If $z \in B(0,1)\backslash Z(f)$ and $\varepsilon \to 0$ we find that

$$u(z) \leq \frac{1}{2\pi} \int_0^{2\pi} \frac{1 - |z|^2}{|z - e^{i\theta}|^2} u(e^{i\theta})\, d\theta.$$

The set $Z(f)$ is a polar set since $Z(f) = \{z \in \Omega : p(z) = -\infty\}$, hence it has measure zero. This implies that the previous inequality is valid throughout by continuity. Using $-f$ and $-u$ one obtains the reverse inequality. Therefore

$$u(z) = \frac{1}{2\pi} \int_0^{2\pi} \frac{1 - |z|^2}{|z - e^{i\theta}|^2} u(e^{i\theta})\, d\theta \qquad (z \in B(0, 1)).$$

This shows that u is harmonic, hence C^∞, in $B(0, 1)$. Clearly the same reasoning applies to v. It follows that f is C^∞. Since $\dfrac{\partial f}{\partial \bar{z}} = 0$ in $B(0,1)\backslash Z(f)$, by continuity we conclude that $\dfrac{\partial f}{\partial \bar{z}} = 0$ in $B(0, 1)$. This proves the Radó-Cartan theorem. □

4.9.53. Corollary. *Let Ω be a connected open subset of $\mathbb{C}$, and ζ_0 a nonisolated point of $\partial\Omega$. Let $f \in \mathscr{H}(\Omega)$ be such that $f(z) \to 0$ whenever $z \to \zeta$ for every $\zeta \in \partial\Omega \cap U$, U some fixed neighborhood of ζ_0. Then $f \equiv 0$.*

Proof. Let $V = B(\zeta_0, r) \subseteq U$, $\Omega_1 = \Omega \cup V$. Hence Ω_1 is connected and the function F defined by

$$F(z) = \begin{cases} f(z) & \text{if } z \in \Omega \\ 0 & \text{if } z \in V\backslash\Omega \end{cases}$$

is continuous in Ω and holomorphic in $\Omega_1\backslash Z(F)$. By the previous theorem, F is holomorphic in Ω_1, hence $F \equiv 0$. □

Let us now prove a result about removable singularities for subharmonic functions.

4.9.54. Lemma. *Let Ω be a connected open set in $\mathbb{C}$, E a closed polar subset of Ω, $E \subseteq \{z \in \Omega : q(z) = -\infty\}$, q a subharmonic function in Ω. For every $a \in \Omega\backslash E$ there is a subharmonic function p in Ω such that $p|E = -\infty$ and $p(z) > -\infty$ for every z sufficiently close to a.*

Proof. Let $\bar{B}(a, R) \subseteq \Omega\backslash E$. We take p to be the Poisson modification of q with respect to $B(a, R)$, that is,

$$p(z) = \begin{cases} q(z) & \text{if } z \in \Omega\backslash B(a, R), \\ P(q|\partial B(a, R))(z) & \text{if } z \in B(a, R). \end{cases}$$

We have already shown that p is subharmonic. The other two properties are clear. □

4.9.55. Theorem. *Let Ω be a connected open set in $\mathbb{C}$, E a closed polar subset of Ω, and u a subharmonic function in $\Omega \setminus E$ such that, for every compact K in Ω, the function $u|(K\setminus E)$ is bounded above. Then, the function $\tilde{u}$ defined on Ω by*

$$\tilde{u}(z) = \begin{cases} u(z) & \text{if } z \in \Omega\setminus E \\ \limsup\limits_{\substack{w\to z \\ w\in\Omega\setminus E}} u(w) & \text{if } z \in E \end{cases}$$

is subharmonic in Ω.

Proof. It is clear that $\tilde{u}$ is a u.s.c. function. Let $w \in \Omega$, $R > 0$ be such that $\bar{B}(w, R) \subseteq \Omega$. We need to show that

$$\tilde{u}(w) \le \lambda(\tilde{u}, w, R).$$

Let $z_0 \in B(w, R)\setminus E$ and p a subharmonic function in Ω such that $p|E = -\infty$ and $p(z) > -\infty$ for z near z_0. There is an $M > 0$ such that $q = p - M \le 0$ in $\bar{B}(w, R)$. For $\varepsilon > 0$, $\tilde{u} + \varepsilon q$ is subharmonic in Ω, hence

$$\begin{aligned}\tilde{u}(z_0) + \varepsilon q(z_0) &\le \frac{1}{2\pi}\int_0^{2\pi} (\tilde{P}_{w,R}(\tilde{u} + \varepsilon q))(w + \mathrm{Re}^{i\theta})\, d\theta \\ &\le (\tilde{P}_{w,R}(\tilde{u}))(z_0),\end{aligned}$$

where $\tilde{P}_{w,R}(g)$ is the Poisson modification of a function g in $\bar{B}(w, R)$. Letting $\varepsilon \to 0$ one obtains

$$\tilde{u}(z_0) \le (\tilde{P}_{w,R}(\tilde{u}))(z_0)$$

for every $z_0 \in B(w, R)\setminus E$. Since $\mathring{E} = \varnothing$, $\tilde{P}_{w,R}(\tilde{u})$ is continuous in $B(w, R)$ and $\tilde{u}$ is u.s.c., we can conclude that

$$\tilde{u}(z) \le (\tilde{P}_{w,R}(\tilde{u}))(z)$$

for every $z \in B(w, R)$. In particular,

$$\tilde{u}(w) \le (\tilde{P}_{w,R}(\tilde{u}))(w) = \lambda(\tilde{u}, w, R).$$

Therefore, the function $\tilde{u}$ is subharmonic in Ω. □

4.9.56. Corollary. *The Dirichlet problem is not solvable for $\Omega = B(0, 1)\setminus\{0\}$.*

Proof. Suppose there is a harmonic function h in Ω which takes the values 1 at $z = 0$ and 1 on $|z| = 1$. By the preceding theorem h would have an extension $\tilde{h}$ harmonic in $B(0, 1)$, $\tilde{h}(0) = 1$, $\tilde{h}(z) = 0$ if $|z| = 1$. This would contradict the maximum principle. □

This example can be generalized to the following.

4.9.57. Proposition. *If E is a compact subset of $\mathbb{C}$, $C(E) = 0$, then the set $D = S^2\setminus E$ cannot have a Green function g with pole at ∞.*

PROOF. If function g exists, then the function $u = -g$ is subharmonic in $\mathbb{C}\setminus E$. Moreover, $\limsup_{\substack{z\to\zeta \\ z\in\mathbb{C}\setminus E}} u(z) = 0$ for every $\zeta \in E$. By Theorem 4.9.55 the function $\tilde{u}$, which extends u to $\mathbb{C}$ being zero at E, is subharmonic in $\mathbb{C}$. There are no subharmonic functions bounded above in $\mathbb{C}$ unless $\tilde{u} =$ constant, hence $u \equiv 0$, impossible. □

EXERCISES 4.9

1. Justify the assertion made after Proposition 4.9.25 that a compact set E has $C(E) = 0$ if and only if it is a polar set.
2. Give a different proof of Proposition 4.9.25 using that $\dfrac{\partial h}{\partial z}$ is holomorphic with an isolated singularity at z_0.
3. Let $f:\Omega \to \Omega_1$ be a biholomorphism between two planar regions. Let E be a compact subset of Ω. Show that $C(E) = 0$ if and only if $C(f(E)) = 0$.
4. Let E_1 be a compact connected set with at least two points. Let E be a compact subset of $\mathbb{C}$, $E_1 \subseteq E$. Show that $C(E) > 0$.
5. Let D_n be an increasing sequence of domains in S^2, $\infty \in D_n$ for all n, $\bigcup_n D_n = D$ and $D = S^2\setminus E$ with $C(E) = 0$. Assume that the Green function g_n of D_n with pole at ∞ exists. Show that $g_n \uparrow \infty$ uniformly on compact subsets of D.
6. Let E, E_1 be two compact subsets of $\mathbb{C}$, Ω, respectively Ω_1, the unbounded component of E^c, resp. E_1^c. Let f be a conformal map of Ω onto Ω_1, $f(z) = z + a_0 + \dfrac{a_1}{z} + \cdots$. Assume further that Ω has a Green function with pole at ∞.
   (a) Show that Ω_1 has also a Green function g_1 with pole at ∞, $g = g_1 \circ f$ and the Robin constants coincide, $\gamma = \gamma_1$. Hence $C(E) = C(E_1)$.
   (b) Let $E_\alpha = \left\{z \in \mathbb{C} : |z| = 1, |\operatorname{Arg} z| \le \dfrac{\alpha}{2}\right\}$, $0 < \alpha < \pi$. Show that the map

$$f(z) = \frac{1}{2}\left[z - 1 + \sqrt{(z - e^{i\alpha/2})(z - e^{-i\alpha/2})}\right],$$

   where the square root is asymptotically equal to z at ∞, maps conformally E_α^c onto $\bar{B}(0,\beta)^c$, for some $\beta > 0$. Show that $\beta = C(E_\alpha)$. Show further that for $|\theta| < \dfrac{\alpha}{2}$, we have

$$\lim_{r\to 1^{\pm}} f(re^{i\theta}) = e^{i\theta/2}\left(i\sin\frac{\theta}{2} \pm \sqrt{\sin^2\frac{\alpha}{2} - \sin^2\frac{\theta}{2}}\right),$$

   and

$$|f(e^{i\theta})| = \sin\frac{\alpha}{4}.$$

   Conclude that $C(E_\alpha) = \sin\dfrac{\alpha}{4}$.

7. Let E be a compact set of positive capacity and assume that the exterior boundary of E is connected and has at least two points. Let $\Omega = E^c$ and ν the equilibrium measure of E. (Recall $g(z, \infty, \Omega) = 2\pi U^\nu + V(E)$.) Show that for every $\lambda > -\dfrac{V(E)}{2\pi}$ the set $S_\lambda = \{z \in \Omega : U^\nu(z) = \lambda\}$ is a real analytic Jordan curve.

8. Let γ_1, γ_2 be two Jordan curves such that γ_1 is interior to γ_2. Show there is a polynomial P and a constant $a > 0$ such that the lemmiscate $\{z \in \mathbb{C} : |P(z)| = a\} \subseteq \operatorname{Int}(\gamma_2)\backslash\overline{\operatorname{Int}(\gamma_1)}$. (Hint; this is a consequence of the Riemann mapping theorem and Runge's theorem.)

   Conclude that given a Jordan curve Γ and $\varepsilon > 0$ there are two lemmiscates L_1 and L_2 such that
$$L_2 \subseteq V_\varepsilon(\overline{\operatorname{Int}(\gamma)})\backslash\overline{\operatorname{Int}(\gamma)},$$
and
$$L_1 \subseteq V_\varepsilon(\overline{\operatorname{Ext}(\gamma)})\backslash\overline{\operatorname{Ext}(\gamma)}.$$
(Recall $V_\varepsilon(K) = \{z \in \mathbb{C} : \operatorname{dist}(z, K) < \varepsilon\}$). Therefore, any simply connected region Ω has an exhaustion $(\Omega_n)_{n\geq 1}$ $\left(\Omega_n \subset\subset \Omega_{n+1}, \bigcup_n \Omega_n = \Omega\right)$, by domains whose boundaries are lemmiscates.

9. Let E be a compact subset of $\mathbb{C}$. We use the notation that follows Definition 4.9.13 and denote $m_1(A)$ the Lebesgue measure of a set $A \subseteq \mathbb{R}$.
   (a) Let $E^* := \{r \geq 0 : \exists z \in E, |z| = r\}$. E^* is compact. (Why?) Show that for any $z_1, \ldots, z_n \in E$ we have
$$|D(z_1, \ldots, z_n)| \geq |D(|z_1|, \ldots, |z_n|)|$$
and conclude that
$$C(E) \geq C(E^*).$$
   (b) Let $\chi_{E^*}(t)$, $t \in \mathbb{R}$, be the characteristic function of E^* and
$$\varphi(s) := \int_0^s \chi_{E^*}(t)\,dt, \qquad s \geq 0.$$
Show that if $0 \leq s_1 \leq s_2$ then
$$0 \leq \varphi(s_2) - \varphi(s_1) \leq s_2 - s_1$$
and, if $0 \leq t_1 < t_2 < \cdots < t_n \leq m_1(E^*)$, there exist $s_1, \ldots, s_n \in E^*$ such that $t_j = \varphi(s_j)$ for $1 \leq j \leq n$. Prove that
$$|D(t_1, \ldots, t_n)| \geq |D(s_1, \ldots, s_n)|.$$
Conclude that if I is an interval of length $m_1(E^*)$, then
$$C(E^*) \geq C(I).$$
   (c) Show that
$$C(E) \geq \frac{1}{4} m_1(E^*).$$

10. Let E be a compact set in $\mathbb{C}$, μ a probability measure on E ($\mu \in \mathscr{P}(E)$), and γ a C^1 Jordan curve such that $E \subseteq \mathrm{Int}(\gamma)$. Compute

$$\int_\gamma \frac{\partial U^\mu}{\partial n}(z)|dz|.$$

(Hint: replace γ by $\partial B(0, R)$ with $R \gg 1$ and use Gauss' Theorem 4.4.28.)

*11. Let $E_1, \ldots, E_n$ be closed subsets of $\bar{B}\left(0, \frac{1}{2}\right)$ and $E = \bigcup_{j=1}^{n} E_j$. Show that

$$\frac{1}{\log C(E)} \geq \sum_{j=1}^{n} \frac{1}{\log C(E_j)}.$$

(Hint: let ν_j be the equilibrium measures of the E_j. For any convex combination $\mu = \sum_{j=1}^{n} \lambda_j \nu_j$, $\lambda_j \geq 0$, $\sum_{j=1}^{n} \lambda_j = 1$, show that

$$\sup U^\mu \leq -\frac{1}{2\pi} \min_{1 \leq j \leq n} \lambda_j V(E_j),$$

hence $V(E) \geq \min_{1 \leq j \leq n} \lambda_j V(E_j)$. Choose λ_j so that $\lambda_j V(E_j)$ is independent of j to finish the proof.)

12. Let f be a meromorphic function in the unit disk which has a simple pole at $z = 0$. Let $E = S^2 \backslash f(B(0, 1))$. Why is it compact? Why does the Green function g of $\mathbb{C} \backslash E$ with pole at ∞ exist? Show that the function

$$u(z) := -g(f(z)) - \log|z|$$

is subharmonic in $B(0, 1)$ and $u \leq 0$. Moreover,

$$u(0) = -\log|\mathrm{Res}(f, 0)| - V(E).$$

Conclude that

$$C(E) \leq |\mathrm{Res}(f, 0)|.$$

Use this inequality to find the relation between capacity and analytic capacity (cf. §2.7.9).

13. Prove Corollary 4.9.40.

14. Let u be a subharmonic function in the half-plane $\mathrm{Im}\, z > 0$. Define $M(r) := \sup_{0<\theta<\pi} u(re^{i\theta})$ for $r > 0$ and $u(x) := \limsup_{z \to x} u(z)$, for $x \in \mathbb{R}$. Assume that $u(x) \leq 0$ for all $x \in \mathbb{R}$ and

$$\limsup_{r \to \infty} \frac{M(r)}{r} \leq 0.$$

Show that $u(z) \leq 0$ for every z with $\mathrm{Im}\, z \geq 0$. (Proceed as follows: use the Two-Constants Theorem 4.9.50 and Example 4.9.51.2, to show that

(i) Given $\varepsilon > 0$ and $B > 0$, let $R > B$ be such that $M(R) < \varepsilon R$. In the half-disk $\Omega := \{z \in \mathbb{C} : |z| < R, \mathrm{Im}\, z > 0\}$, show that

$$u(z) \leq \frac{2\varepsilon R}{\pi} \mathrm{Arg} \frac{R + z}{R - z}.$$

(ii) Conclude from (i) that for any z with $\operatorname{Im} z \geq 0$, you have

$$u(z) \leq \frac{4\varepsilon}{\pi} \operatorname{Im} z.$$

(iii) Finish the proof.)

15. Let f be a holomorphic function in the angular region $\alpha < \arg z < \beta$, $2\pi > \beta - \alpha > 0$, f continuous up to the boundary of this region, and $\rho = \dfrac{\pi}{\beta - \alpha}$. Assume $|f(z)| \leq M$ on the sides of this angular region and that

$$\lim_{r \to \infty} r^{-\rho} \sup_{\alpha < \theta < \beta} \log|f(re^{i\theta})| = 0.$$

Show that $|f(z)| \leq M$ everywhere. (Hint: assume $\alpha = -\beta$, $0 < \beta < \pi$, and consider the auxiliary function $g(z) := f(z)e^{-\varepsilon z^\rho}$.)

16. Let f be holomorphic in $\operatorname{Re} z \geq 0$ and vanish at every $n \in \mathbb{N}$. Assume further that for some $\alpha < \pi$, $M, A > 0$ one has

$$|f(z)| \leq M \exp(A \operatorname{Re} z + \alpha|\operatorname{Im} z|).$$

Show that $f \equiv 0$. (Hint: consider the auxiliary function $g(z) := \dfrac{f(z)}{\sin \pi z} e^{-Az}$ and apply Exercise 4.4.25. Alternatively, apply a convenient variation of Exercise 4.7.8(b).)

*17. Let f be holomorphic and bounded in the angle $\Omega := \{z \in \mathbb{C} : |\operatorname{Arg} z| < \alpha < \pi\}$, and continuous in $\bar{\Omega}$. Assume $f(re^{\pm i\alpha}) \to L$ as $r \to \infty$. Show that $f(z) \to L$ when $z \in \Omega$, $|z| \to \infty$. (Hint: consider the auxiliary function $F(z) := \dfrac{z}{z + \lambda} f(z)$ for a convenient choice of $\lambda > 0$.) Show that the condition $\alpha < \pi$ can be removed by the change of variable $z = \zeta^2$.

18. Let f be an entire function satisfying the inequalities

$$|f(z)| \leq Ae^{B|z|}, \qquad \text{for every } z \in \mathbb{C}$$

and

$$|f(x)| \leq M, \qquad \text{for every } x \in \mathbb{R}.$$

Show that f satisfies the inequality

$$|f(z)| \leq Me^{B|\operatorname{Im} z|}, \qquad \text{for every } z \in \mathbb{C}.$$

*19. Let f be a holomorphic function of order at most ρ in the angle $\alpha < \arg z < \beta$, i.e., for any $\varepsilon > 0$

$$\limsup_{r \to \infty} \frac{\log|f(re^{i\theta})|}{r^{\rho+\varepsilon}} = 0$$

uniformly for $\alpha < \theta < \beta$. The Phragmen-Lindelöf indicator function h is defined by

$$h(\theta) = \limsup_{r \to \infty} \frac{\log|f(re^{i\theta})|}{r^\rho}.$$

(a) Compute h in case $f(z) = \exp(P(z))$, P a polynomial of degree exactly ρ (which then must be an integer).

(b) Let $\alpha < \theta_1 < \theta_2 < \beta, \theta_2 - \theta_1 < \frac{\pi}{\rho}$. Assume $h(\theta_1) \leq h_1, h(\theta_2) \leq h_2, h_j \in \mathbb{R}$, and let $H(\theta) := (h_1 \sin\rho(\theta_2 - \theta) + h_2 \sin(\theta - \theta_1))/\sin\rho(\theta_2 - \theta_1)$. Show that

$$h(\theta) \leq H(\theta) \qquad \text{for } \theta_1 \leq \theta \leq \theta_2.$$

What happens if h_1 or $h_2 = -\infty$?

(c) Let $\alpha < \theta_1 < \theta_2 < \theta_3 < \beta$, $\theta_2 - \theta_1 < \frac{\pi}{\rho}$, $\theta_3 - \theta_2 < \frac{\pi}{\rho}$, and $H(\theta) = a\cos\rho\theta + b\sin\rho\theta$ be such that $h(\theta_1) \leq H(\theta_1)$ and $h(\theta_2) \geq H(\theta_2)$, then use (b) to show that

$$H(\theta_3) \leq h(\theta_3).$$

Conclude that

$$h(\theta_1)\sin\rho(\theta_3 - \theta_2) + h(\theta_2)\sin\rho(\theta_1 - \theta_3) + h(\theta_3)\sin\rho(\theta_2 - \theta_1) \geq 0.$$

(d) Let h be finite in the interval $\theta_1 \leq \theta \leq \theta_3$, $\theta_3 - \theta_1 < \frac{\pi}{\rho}$. Let $\theta_2 \in]\theta_1, \theta_3[$ and $H_1(\theta) = a_1\cos\rho\theta + b_1\sin\rho\theta$, $H_2(\theta) = a_2\cos\rho\theta + b_2\sin\rho\theta$ be chosen so that

$$H_1(\theta_1) = h(\theta_1), \qquad H_1(\theta_2) = h(\theta_2), \qquad H_2(\theta_2) = h(\theta_2), \qquad H_2(\theta_3) = h(\theta_3).$$

Show that

$$H_2(\theta) \leq h(\theta) \leq H_1(\theta) \qquad \text{if } \theta_1 \leq \theta \leq \theta_2$$

$$H_1(\theta) \leq h(\theta) \leq H_2(\theta) \qquad \text{if } \theta_2 \leq \theta \leq \theta_3.$$

Conclude that if $\theta \in [\theta_1, \theta_3]\backslash\{\theta_2\}$, then

$$\frac{H_1(\theta) - H_1(\theta_2)}{\theta - \theta_2} \leq \frac{h(\theta) - h(\theta_2)}{\theta - \theta_2} \leq \frac{H_2(\theta) - H_2(\theta_2)}{\theta - \theta_2}.$$

Use these inequalities to prove that h is continuous at θ_2. Hence h is continuous in the interval $]\theta_1, \theta_3[$.

(c) Show that if h is finite in $[\alpha, \beta]$, $\alpha < \theta_1 < \theta_2 < \beta$, $\varepsilon > 0$, then there is $r_0 = r_0(\varepsilon) > 0$ such that

$$|f(re^{i\theta})| \leq \exp[r^\rho(h(\theta) + \varepsilon)]$$

for $r \geq r_0$ and $\theta_1 \leq \theta \leq \theta_2$.

20. Let f be a holomorphic function of order ρ and type τ in the angle $|\text{Arg}\, z| \leq \frac{2\pi}{\rho}$. Assume f is bounded on the sides of the angle. Show its indicator function h (defined in the previous exercise) satisfies the inequality

$$h(\theta) \leq \tau r^\rho \cos\rho\theta.$$

21. Let f be holomorphic in the half-strip $S := \{z = x + iy : x \geq 0, |y| \leq \alpha\}$ satisfying $|f(x \pm i\alpha)| \leq A$ on ∂S and $|f(x + iy)| \leq B\exp(e^{\beta x})$, with $0 \leq \beta < \frac{\pi}{2\alpha}$. Show that

f is bounded by A everywhere in S. $\left(\text{Hint: consider the auxiliary function } f(z)\exp(-e^{\sigma z}), \text{ with } \varepsilon > 0, \beta < \sigma < \frac{\pi}{2\alpha}.\right)$

Notes to Chapter 4

1. The characterization of harmonic functions in terms of the mean-value property for all small disks (Proposition 4.3.3) can be substantially improved. Essentially, it is enough to consider averages for two fixed radii. We will discuss problems of this kind in the second volume. We refer to [Za5] and [BG] for different kinds of mean-value theorems.

2. In the second volume we will consider in detail ideals, interpolation, and related problems for spaces of entire functions of finite order. These spaces arise naturally in harmonic analysis when one applies the Fourier transform to study integro-differential equations of the convolution type. We shall take up again there the construction of entire functions with given zeros following the ideas of Remark 4.6.16. The use of the canonical potentials as done in §4.6 owes much to [HK] and [GL].

3. The interpolation problems of the type that appear in Exercise 4.5.11 and Exercise 4.9.16 are related to questions of analytic continuation and functional equations that will also be considered in the next volume.

4. Besides its clear interest in the solvability of the Dirichlet problem and removability of singularities of harmonic functions, the notions of Green function and capacity play a role in the study of functional equations and arithmetical functions, as will be seen in the next volume. There is an extensive literature in potential theory and on the especially interesting questions of the relation between capacity and the Hausdorff dimension of a set and the solvability of the Dirichlet problem and Neumann problem for sets with very rough boundaries. The reader will find the small monographs [Fu2] and [Carl] excellent introductions to this subject. In particular, in [Carl] one finds a criterion to decide exactly when a Cantor set has capacity zero. An essentially definitive contribution to the relation between capacity of a subset E of a rectifiable curve and the length of the set E can be found in [BJ]. There is a very large literature on potential theory in $\mathbb{R}^n$. We mention [Lan] or [Hel] as a possible starting point for further study.

5. The Riemann mapping theorem is not valid when one replaces $\mathbb{C}$ by $\mathbb{C}^n$, $n \geq 2$. It was of considerable interest and difficulty to prove that biholomorphic mappings between domains with smooth boundaries, extend smoothly to the boundary. This theorem was first proved by C. Fefferman. The argument followed in §4.8 was carried out in the case of $\mathbb{C}^n$, $n \geq 2$, by S. Bell and others. We refer to [BK] for the particulars. Similarly, the C^∞ version of the Schwartz reflection principle mentioned at the end of §4.8 seems to have been discovered first in the context of several variables.

CHAPTER 5

Analytic Continuation and Singularities

§1. Introduction

When we say "given a holomorphic function in an open set Ω," we are already making a choice of the domain of the function. Sometimes it is evident that the function is in fact the restriction to Ω of a holomorphic function defined on a larger open set. The obvious example of a removable isolated singularity comes to mind. Another example occurs when we define the function by a power series expansion, for instance, for $f(z) = \sum_{n\geq 0} z^n$ in $B(0, 1)$, we can sum the series and find that the function $z \mapsto (1 - z)^{-1}$, holomorphic in $\mathbb{C}\setminus\{1\}$, extends the function f to this larger open set.

Regretfully, the intuitive concept of extension of a function across a point in the boundary of Ω quickly leads to bothersome difficulties, multivalued functions may appear. For instance, the function $z \mapsto \operatorname{Log} z$ in $\mathbb{C}\setminus]-\infty,0]$ admits one-sided extensions at the nonzero boundary points, but they do not coincide with the original function. Another typical example of this behavior arises when the functions are defined as Cauchy transforms. In fact, let $\gamma: [0, 1] \to \mathbb{C}$ be a Jordan arc, piecewise C^1, and f a holomorphic function in a neighborhood of $\gamma([0, 1])$. The function $F: \mathbb{C}\setminus\gamma([0, 1]) \to \mathbb{C}$ defined by

$$F(z) = \frac{1}{2\pi i}\int_\gamma \frac{f(t)}{t - z}\,dt$$

is holomorphic and has, in a neighborhood of each point of γ different from the endpoints, the same type of behavior as the logarithm. Namely, let $\zeta \in \gamma(]0, 1[)$, $\zeta = \gamma(t_0)$. One can choose $r > 0$ and two values t_1, t_2 such that $0 < t_1 < t_0 < t_2 < 1$, $\gamma(]t_1, t_2[) \Subset B(\zeta, r)$, $\gamma(t_j) \in \partial B(\zeta, r)$ $(j = 1, 2)$, and f is holomorphic in a neighborhood of $\bar{B}(\zeta, r)$. If C_1, C_2 are the arcs of $\partial B(\zeta, r)$ shown in Figure 5.1, then the curve $(\gamma|[t_1, t_2])\cdot C_2$ is a Jordan curve of index 1

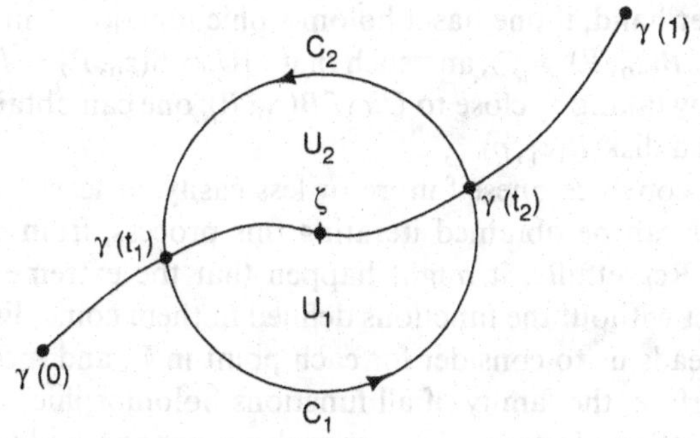

Figure 5.1

with respect to the points of U_2, and $C_1 \cdot \overline{(\gamma | [t_1, t_2])}$ is also a Jordan curve with index 1 with respect to U_1. Let $\gamma_1 = \gamma | [0, t_1]$, $\gamma_2 = \gamma | [t_2, 1]$. Then, the function

$$F_1(z) = \frac{1}{2\pi i} \int_{\gamma_1 \cdot \bar{C}_2 \cdot \gamma_2} \frac{f(t)}{t - z}\, dt$$

is holomorphic in $\mathbb{C} \backslash (\gamma_1 \cdot \bar{C}_2 \cdot \gamma_2)$ and coincides with F in U_1. One defines F_2 similarly using $\gamma_1 \cdot C_1 \cdot \gamma_2$. It follows that the limits

$$\ell_j(\zeta) = \lim_{\substack{z \to \zeta \\ z \in U_j}} F(z) \qquad (j = 1, 2)$$

exist, and, using once more Cauchy's theorem,

$$\ell_2(\zeta) - \ell_1(\zeta) = f(\zeta).$$

(This is usually called Plemelj's formula, cf. Exercise 3.6.9.)

In view of these examples, one needs to formalize the intuitive concept of analytic continuation, eliminating the pathology of multivalued functions. To do this, we must "leave" the complex plane and use the concept of a Riemann surface; one obtains this way a maximal surface (in a sense that will be defined precisely) which will be the natural domain of definition of the given function as a holomorphic function. It is precisely in this context that the notion of Riemann surfaces arose.

The origin of the methods that we will introduce is the following: Given a holomorphic function f in a disk $B(z_0, R)$, consider a point z_1 in that disk which lies close to the boundary. We can develop the function f in a Taylor series about z_1. This series,

$$\sum_{n \geq 0} \frac{1}{n!} f^{(n)}(z_1)(z - z_1)^n$$

converges to f in $B(z_1, R - |z_1 - z_0|)$, but it is possible that the radius of convergence r of this series is strictly bigger than $R - |z_1 - z_0|$. We have, therefore, a function defined in a neighborhood of a certain arc of $\partial B(z_0, R)$, through which the function f admits an analytic continuation.

On the other hand, if one has a holomorphic function F in an open set U such that $U \cap \partial B(z_0, R) \neq \varnothing$, and such that $f|U \cap B(z_0, R) = F|U \cap B(z_0, R)$, one sees that, by taking z_1 close to $U \cap \partial B(z_0, R)$, one can obtain F as the sum of the series in a disk $B(z_1, \rho)$.

As one can convince oneself more or less easily, at least locally, the continuations of f can be obtained iterating this process, from disk to disk to form a chain. Regretfully, it might happen that the extreme disks of these chains intersect without the functions defined in them coinciding in the intersection. This leads us to consider for each point in $\mathbb{C}$, and more generally in a Riemann surface, the family of all functions holomorphic near that point, that is, to the notion of germs of holomorphic functions, and to its associated sheaf. This will be the tool that will help us to study the question of analytic continuation. We start by considering first some elementary aspects of the problem of analytic continuation.

Several of the subjects of this chapter, like overconvergence and Dirichlet series, will reappear in the second volume.

§2. Elementary Study of Singularities and Dirichlet Series

5.2.1. Definition. A point z_1 in the boundary of the disk $B(0, R)$ of convergence of a power series $s(z) = \sum_{n \geq 0} a_n z^n$ is said to be a ***regular point*** (for s) if there is $r > 0$ and a function f holomorphic in $B(z_1, r)$ such that f and s coincide in $B(0, r) \cap B(z_1, r)$. A point of $\partial B(0, R)$ which is not regular for s is said to be ***singular***.

Singular points always exist.

5.2.2. Proposition. *Let $s(z) = \sum_{n \geq 0} a_n z^n$ be a power series with radius of convergence R, $0 < R < \infty$. The set of singular points is a nonempty closed subset of $\partial B(0, R)$.*

Proof. It is clear from their definition that the set of regular points is open in $\partial B(0, R)$. Assume there are no singular points. Then, for every $a \in \partial B(0, R)$, there is a disk $B(a, \rho_a)$ ($\rho_a > 0$) and a holomorphic function F_a in that disk such that $s|(B(0, R) \cap B(a, \rho_a)) = F_a|(B(0, R) \cap B(a, \rho_a))$. Whenever $B(a, \rho_a) \cap B(b, \rho_b) \neq \varnothing$ we have that $F_a \equiv F_b$ in $B(a, \rho_a) \cap B(b, \rho_b)$, since this intersection is connected and both F_a and F_b coincide with s in the nonempty open set $B(0, R) \cap B(a, \rho_a) \cap B(b, \rho_b)$. Therefore, there is a function F holomorphic in the open set

$$\Omega = B(0, R) \cup \left(\bigcup_{a \in \partial B(0, R)} B(a, \rho_a) \right)$$

extending s. Ω contains a disk $B(0, R + \varepsilon)$ for some $\varepsilon > 0$. Hence the radius of convergence of the Taylor series of F about 0 is at least $R + \varepsilon$; since this is exactly the original series s, we have a contradiction. □

Let us see now several examples showing that every point of the boundary of the disk could be a singular point. Furthermore, this phenomenon could occur even if the power series converges to a C^∞ function in the closed disk.

5.2.3. Examples. (1) If a power series $f(z) = \sum_{n\geq 0} a_n z^n$ with radius of convergence $R, 0 < R < \infty$, has all its coefficients $a_n \geq 0$, the point $z = R$ is a singular point. If not, the series

$$\sum_{n\geq 0} \frac{1}{n!} f^{(n)}\left(\frac{R}{2}\right)\left(z - \frac{R}{2}\right)^n$$

would converge in $B(R/2, R_1)$ for some $R_1 > \dfrac{R}{2}$. By the positivity of the coefficients we have $\left|f^{(n)}\left(\dfrac{R}{2}e^{i\theta}\right)\right| \leq f^{(n)}\left(\dfrac{R}{2}\right)$, for every $\theta \in [0, 2\pi]$. Therefore the series

$$\sum_{n\geq 0} \frac{1}{n!} f^{(n)}\left(\frac{R}{2}e^{i\theta}\right)\left(z - \frac{R}{2}e^{i\theta}\right)^n$$

will also converge in the disks $B\left(\dfrac{R}{2}e^{i\theta}, R_1\right)$, and f will have no singular points in $\partial B(0, R)$.

(2) The Weierstrass series

$$f(z) = \sum_{n\geq 0} a_n z^{2^n},$$

$a_n \geq 0$, $\limsup_{n\to\infty} \sqrt[2^n]{a_n} = 1$, has 1 as a radius of convergence and every point in $\partial B(0, 1)$ is a singular point. Namely, for any integer N we have

$$f(z) = f_N(z) + \sum_{n\geq N} a_n z^{2^n} = f_N(z) + \sum_{k\geq 0} a_{N+k}(z^{2^N})^{2^k}$$

with $f_N(z) = \sum_{0\leq n\leq N-1} a_n z^{2^n}$, and the second series admits $z = 1$ as a singular point by the previous example. On the other hand, the function defined by this second series is invariant under the rotations $z \mapsto e^{2\pi i p/2^N} z$, $p \in \mathbb{N}$, hence the points $e^{2\pi i p/2^N} z$ are singular points for f. This guarantees that the set of singular points is dense in $\partial B(0, 1)$. Since it is closed, it coincides with $\partial B(0, 1)$.

(3) The series $f(z) = \sum_{n\geq 0} e^{-2^{n/2}} z^{2^n}$ has radius of convergence 1 since

$$\limsup_{n\to\infty} (e^{-2^{n/2}})^{1/2^n} = 1.$$

The series for the derivatives

$$f^{(k)}(z) = \sum_{k\leq 2^n} 2^n(2^n - 1)\dots(2^n - k + 1)e^{-2^{n/2}} z^{2^n - k},$$

converge uniformly in the closed disk. Therefore f is C^∞ in $\bar{B}(0,1)$. On the other hand, by the previous example, every point of the boundary is a singular point.

A generalization of this example is the following theorem due to Hadamard.

5.2.4. Theorem. *Let λ, p_k, q_k $(k \geq 1)$ be positive integers such that:*

$$p_1 < p_2 < p_3 < \cdots, \quad q_1 < q_2 < q_3 < \cdots, \qquad \text{and } \lambda q_k > (\lambda + 1)p_k.$$

*If the series $f(z) = \sum_{n\geq 0} a_n z^n$ has a radius of convergence equal to 1 and, if $a_n = 0$ for $p_k < n < q_k$ (**lacunarity condition**), then for every regular point $\beta \in \partial B(0,1)$, the subsequence*

$$s_{p_k}(z) = \sum_{0\leq n\leq p_k} a_n z^n$$

*of the sequence of partial sums of f converges uniformly in a neighborhood of β (**overconvergence**).*

PROOF. If $g(z) = f(\beta z)$, the function g also satisfies the lacunarity condition. We can therefore assume that $\beta = 1$. This point being regular means that f is in fact holomorphic in an open set Ω containing $B(0,1) \cup \{1\}$. Consider the auxiliary polynomial φ of degree $\lambda + 1$ given by

$$\varphi(w) := \tfrac{1}{2}(w^\lambda + w^{\lambda+1})$$

and the auxiliary function $F(w) = f(\varphi(w))$ defined for $w \in \varphi^{-1}(\Omega)$. If $|w| \leq 1$, $w \neq 1$, we have $|\varphi(w)| < 1$ since $|1 + w| < 2$. We also have $\varphi(1) = 1$. Hence there exists $\varepsilon > 0$ such that $\varphi(B(0, 1 + \varepsilon)) \subseteq \Omega$. It follows that the series

$$F(w) = \sum_{n\geq 0} b_n w^n$$

converges in $|w| < 1 + \varepsilon$. Note that the largest power appearing in $(\varphi(w))^n$ is $(\lambda + 1)n$, the smallest is λn. Therefore, the polynomial $(\varphi(w))^{p_k}$ has no terms in common with $(\varphi(w))^{q_k}$. Since we have $F(w) = \sum_{n\geq 0} a_n(\varphi(w))^n (|w| < 1)$ then the partial sums

$$\sum_{0\leq n\leq p_k} a_n(\varphi(w))^n = \sum_{0\leq m\leq(\lambda+1)p_k} b_m w^m, \qquad k = 1, 2, 3, \ldots$$

converge in $|w| < 1 + \varepsilon$ when $k \to \infty$. Therefore $(s_{p_k}(z))_{k\geq 1}$ converges for every $z \in \varphi(B(0, 1 + \varepsilon))$. This is the neighborhood of 1 mentioned in the theorem. □

5.2.5. Corollary. *Let λ be an integer > 0, $(p_k)_{k\geq 1}$ a sequence of integers >0 satisfying*

$$p_{k+1} > \left(1 + \frac{1}{\lambda}\right)p_k, \qquad k = 1, 2, 3, \ldots.$$

Assume that the power series

$$f(z) = \sum_{k \geq 1} a_k z^{p_k}$$

has radius of convergence 1. *Then every point of* $\partial B(0, 1)$ *is singular.*

PROOF. We note that in this case the sequence $(s_{p_k})_{k \geq 1}$ of Theorem 5.2.4 is, up to repetitions, exactly the sequence of partial sums of the series of f. If $\beta \in \partial B(0, 1)$ were a regular point, then the series would converge in a neighborhood of β. It follows from the properties of power series that the radius of convergence is strictly bigger than 1. This contradiction proves the corollary. □

We recommend [Di] and [Hil, vol. 2] to the reader who wants to know more about singularities of holomorphic functions.

Power series are not the only kind of series expansions usually considered for holomorphic functions. In analytic number theory one makes frequent use of Dirichlet series. We shall study now these series a little bit, and see that their properties about singularities and analytic continuation are similar to those we have just seen for power series. The reader will find a very nice introduction to the subject in [HR]. For deeper properties of Dirichlet series, see [Man] and [Ber].

5.2.6. Definition. We call ***Dirichlet series*** a series of the form $\sum_{n \geq 1} a_n e^{-\lambda_n s}$, $s \in \mathbb{C}$, $a_1, a_2, \ldots$ complex numbers, and $\lambda_1 < \lambda_2 < \cdots$ a sequence of real numbers converging to $+\infty$.

The a_n are called the **coefficients** and the λ_n the ***frequencies*** of the Dirichlet series.

For $\lambda_n = n$, the Dirichlet series becomes a power series in the variable e^{-s}. Another particular case corresponds to $\lambda_n = \log n$, the Dirichlet series can be written then as $\sum_{n \geq 1} \frac{a_n}{n^s}$. The series corresponding to $a_n = 1$ for every $n \geq 1$, defines the Riemann ζ-function.

We remark that in the rest of this section we shall use consistently the notation $\sigma = \operatorname{Re} s$, $\tau = \operatorname{Im} s$, $s = \sigma + i\tau$.

To study properties of Dirichlet series we can always assume $\lambda_1 \geq 0$. Consider first the question of absolute convergence.

5.2.7. Proposition. *If a Dirichlet series converges absolutely at the point* $s_0 = \sigma_0 + i\tau_0$, *then it converges absolutely and uniformly in the closed half-plane* $\operatorname{Re} s \geq \sigma_0$.

PROOF. $|a_n e^{-\lambda_n s}| \leq |a_n| e^{-\lambda_n \sigma_0}$ in the half-plane. □

5.2.8. Proposition. *For every Dirichlet series there is a value* $\alpha \in [-\infty, \infty]$ *such that the series is absolutely convergent for* $\operatorname{Re} s > \alpha$ *and is not absolutely convergent for* $\operatorname{Re} s < \alpha$.

PROOF. Simply set $\alpha = \inf\left\{\alpha \in \mathbb{R} : \sum_{n=1}^{\infty} |a_n|e^{-\lambda_n \sigma} \text{ is convergent}\right\}$. □

5.2.9. Definition. The value α is called the ***abscissa of absolute convergence*** and denoted σ_a. The vertical line $\operatorname{Re} s = \sigma_a$ is called the ***line of absolute convergence*** and the half-plane $\operatorname{Re} s > \sigma_a$, the ***half-plane of absolute convergence.***

On the line $\operatorname{Re} s = \sigma_a$, the Dirichlet series could be absolutely convergent or not. For instance, both series $\sum_{n\geq 2} \frac{1}{(\log n)^2} \frac{1}{n^s}$ and $\sum_{n\geq 1} \frac{1}{n^s}$ have $\sigma_a = 1$. The first converges absolutely on that line and the second does not. The cases $\sigma_a = \pm\infty$ can occur: consider $\sum_{n\geq 1} \frac{n!}{n^s}$ and $\sum_{n\geq 1} \frac{1}{n!n^s}$.

A very useful identity in the study of Dirichlet series is the Abel summation formula: Let $(a_n)_{1\leq n\leq N}$, $(b_n)_{1\leq n\leq N}$ be two finite sequences of complex numbers and $B_k = b_1 + \cdots + b_k$, $1 \leq k \leq N$. Then

$$\sum_{1\leq k\leq N} a_k b_k = \sum_{1\leq k\leq N-1} (a_k - a_{k+1})B_k + a_N B_N.$$

Its verification has already been left as an exercise to the reader (Exercise 2.2.8).

Let us turn now to the convergence (not necessarily absolute) of Dirichlet series.

5.2.10. Proposition. *For every Dirichlet series $\sum_{n\geq 1} a_n e^{-\lambda_n s}$ there is a value $\beta \in [-\infty, \infty]$ such that the series converges for $\operatorname{Re} s > \beta$ and diverges for $\operatorname{Re} s < \beta$. Furthermore, the sum $f(s)$ of the Dirichlet series is holomorphic in the half-plane $\operatorname{Re} s > \beta$.*

PROOF. The proof follows from the following lemma.

5.2.11. Lemma. *If the series $\sum_{n\geq 1} a_n e^{-\lambda_n s}$ converges for $s = s_0$, then it converges in $\operatorname{Re} s > \operatorname{Re} s_0 = \sigma_0$. Furthermore, it converges uniformly in every sector $|\operatorname{Arg}(s - s_0)| \leq \frac{\pi}{2} - \delta$, $\delta > 0$.*

PROOF. We can assume that $s_0 = 0$ and hence $\sum_{n\geq 1} a_n$ is convergent.

Let us show first that the sum of the series

$$|e^{-\lambda_1 s}| + \sum_{n\geq 1} |e^{-\lambda_{k+1} s} - e^{-\lambda_k s}|$$

is uniformly bounded in every sector of the form $|\operatorname{Arg} s| \leq \frac{\pi}{2} - \delta$ $(\delta > 0)$.

Since

$$e^{-\lambda_{k+1}s} - e^{-\lambda_k s} = -s \int_{\lambda_k}^{\lambda_{k+1}} e^{-\lambda s}\, d\lambda$$

we have

$$|e^{-\lambda_{k+1}s} - e^{-\lambda_k s}| \le |s| \int_{\lambda_k}^{\lambda_{k+1}} e^{-\lambda\sigma}\, d\lambda = \frac{|s|}{\sigma}(e^{-\lambda_{k+1}\sigma} - e^{-\lambda_k\sigma}).$$

Recalling that $0 \le \lambda_1 < \lambda_2 < \cdots$, the series we are considering is bounded by

$$1 + \frac{|s|}{\sigma} \sum_{k\ge 0} (e^{-\lambda_{k+1}\sigma} - e^{-\lambda_k\sigma}) \le 1 + \frac{|s|}{\sigma} \le 1 + \frac{1}{\sin\delta}.$$

In order to prove the uniform convergence of the original Dirichlet series in the sector we apply the Abel summation procedure. Let $S_N(s) = \sum_{0\le n\le N} a_n e^{-\lambda_n s}$ be a partial sum of the Dirichlet series, $b_{p,m} = \sum_{p\le j\le m} a_j$, and $M_p = \max_{m\ge p} |b_{p,m}|$. The sequence M_p is decreasing and $M_p \to 0$ when $p \to \infty$, due to the fact that $\sum_{n\ge 1} a_n$ is convergent. Therefore

$$\begin{aligned} |S_q(s) - S_{p-1}(s)| &= |b_{p,p}(e^{-\lambda_p s} - e^{-\lambda_{p+1}s}) + \cdots \\ &\quad + b_{p,q-1}(e^{-\lambda_{q-1}s} - e^{-\lambda_q s}) + b_{p,q}e^{-\lambda_q s}| \\ &\le M_p\left(1 + \frac{1}{\sin\delta} + e^{-\lambda_q\sigma}\right) \\ &\le M_p\left(2 + \frac{1}{\sin\delta}\right) \to 0 \qquad \text{as } p \to \infty. \end{aligned}$$

This proves the lemma and Proposition 5.2.10. □

5.2.12. Definition. The value β from Proposition 5.2.10 is called ***abscissa of convergence*** of the Dirichlet series and is denoted by σ_c. The line $\operatorname{Re} s = \sigma_c$ is called **line of convergence**, and the half-plane $\operatorname{Re} s > \sigma_c$ is **half-plane of convergence**. One has clearly $\sigma_c \le \sigma_a$ and one can have $\sigma_c < \sigma_a$. For instance, for the series

$$1 - \frac{1}{2^s} + \frac{1}{3^s} - \frac{1}{4^s} + \cdots$$

we have $\sigma_a = 1$ and $\sigma_c = 0$ (by the alternating series criteria). The strip $\sigma_c < \operatorname{Re} s < \sigma_a$ is called strip of conditional convergence.

5.2.13. Remark. Even though the half-plane of convergence of a Dirichlet series is a concept analogous to the disk of convergence of a power series, we see that there is an essential difference between the two. The disk of convergence and the disk of absolute convergence coincide, while the corresponding half-planes might not.

The question of when a function holomorphic in a half-plane of the form $\operatorname{Re} s > a$ is represented by a Dirichlet series is a difficult one (cf. [Leo]). On the other hand, the uniqueness of such an eventual representation is easy to prove. It follows from the fact that, if $\sum_{n\geq 1} a_n e^{-\lambda_n s}$ converges in $\operatorname{Re} s > \sigma_c$ and it has infinitely many zeros $s_j \to \infty$ in a sector $|\operatorname{Arg}(s - \sigma_c)| \leq \frac{\pi}{2} - \delta$ $(\delta > 0)$, then $a_1 = a_2 = \cdots = 0$. In fact, we note that

$$\lim_{\substack{s\to\infty \\ |\arg(s-\sigma_c)|\leq \pi/2-\delta}} e^{\lambda_p s}\left(\sum_{n\geq p} a_n e^{-\lambda_n s}\right) = a_p.$$

Therefore, we can prove that the coefficients $a_1, a_2, \ldots$ vanish by a simple induction.

5.2.14. Proposition. *The following relations hold*:

(1) $\sigma_a - \sigma_c \leq \limsup_{n\to\infty} \dfrac{\log n}{\lambda_n}$

(2) *If* $\sigma_c > 0$ *then* $\sigma_c = \limsup_{n\to\infty} \dfrac{\log|a_1 + \cdots + a_n|}{\lambda_n}$

(3) *If* $\sigma_c < 0$ *then* $\sigma_c = \limsup_{n\to\infty} \dfrac{\log\left|\sum_{j\geq n+1} a_j\right|}{\lambda_n}$.

Proof. Let $k = \limsup_{n\to\infty} \dfrac{\log n}{\lambda_n}$. We can assume that $k < \infty$. Let $k_1 > k$, $\varepsilon > 0$. We shall show that if the Dirichlet series converges for $\sigma_0 \in \mathbb{R}$, then it converges absolutely for $s = \sigma_1 = \sigma_0 + (1 + \varepsilon)k_1$. This assertion clearly implies (1). Now, since $\dfrac{\log n}{\lambda_n} < k_1$ for $n \geq n_0$, we have

$$|a_n|e^{-\lambda_n\sigma_1} = |a_n e^{-\lambda_n\sigma_0}|e^{-\lambda_n(1+\varepsilon)k_1} \leq |a_n e^{-\lambda_n\sigma_0}|\frac{1}{n^{1+\varepsilon}}.$$

The convergence of the Dirichlet series at σ_0 implies that the sequence $(|a_n e^{-\lambda_n\sigma_0}|)_{n\geq 1}$ is bounded, hence the absolute convergence at σ_1 is immediate.

Let us now prove (2). Let

$$\ell = \limsup_{n\to\infty} \frac{\log|a_1 + \cdots + a_n|}{\lambda_n} \geq 0.$$

(If $\ell < 0$ then we will have convergence of the series at $s = 0$, to the value 0, contradicting $\sigma_c > 0$.) We claim that $\sigma_c \le \ell$. Clearly we can assume $\ell < \infty$, and given $\varepsilon > 0$ it is enough to show convergence at $s = \sigma = \ell + 2\varepsilon$. We have $\log|a_1 + \cdots + a_n| \le (\ell + \varepsilon)\lambda_n$ when $n \ge n_0$. We use once again the Abel summation procedure and the notation from §5.2.11. For $q > p > n_0$ we have

$$\sum_{p\le j\le q} a_j e^{-\lambda_j s} = \sum_{p\le j\le q-1} b_{0,j}(e^{-\lambda_j s} - e^{-\lambda_{j+1}s}) - b_{0,p-1}e^{-\lambda_p s} + b_{0,q}e^{-\lambda_q s}.$$

Hence

$$\left|\sum_{p\le j\le q} a_j e^{-\lambda_j s}\right| \le s \sum_{p\le j\le q-1} e^{(\ell+\varepsilon)\lambda_j}\int_{\lambda_j}^{\lambda_{j+1}} e^{-st}\,dt + e^{(\ell+\varepsilon)\lambda_{p-1}-\lambda_p s} + e^{(\ell+\varepsilon)\lambda_q - \lambda_q s}$$

$$\le s \sum_{p\le j\le q-1}\int_{\lambda_j}^{\lambda_{j+1}} e^{(\ell+\varepsilon-s)t}\,dt + e^{-\varepsilon\lambda_p} + e^{-\varepsilon\lambda_q}$$

$$\le s\int_{\lambda_p}^{\infty} e^{-\varepsilon t}\,dt + 2e^{-\varepsilon\lambda_p},$$

where we have used $s = \ell + 2\varepsilon$ in the last two inequalities. It is clear that the last term tends to 0 as $p \to \infty$, since $\varepsilon > 0$. Therefore the Dirichlet series converges for $s = \ell + 2\varepsilon$, which shows that $\sigma_c \le \ell$. Let us show now that $\sigma_c \ge \ell$. Let $s_0 \in \mathbb{R}$ be a point at which the Dirichlet series converges. We have $s_0 > 0$ by the hypothesis. Let $b_j = a_j e^{-\lambda_j s_0}$ and $B_p = b_1 + \cdots + b_p$. The convergence also ensures that there is an $M \ge 0$ such that $|B_n| \le M$ for all n. The Abel summation formula shows that

$$\left|\sum_{1\le j\le n} a_j\right| = \left|\sum_{1\le j\le n} b_j e^{\lambda_j s_0}\right| = \left|\sum_{1\le j\le n-1} B_j(e^{\lambda_j s_0} - e^{\lambda_{j+1}s_0}) + B_n e^{\lambda_n s_0}\right|$$

$$\le M\left(\sum_{1\le j\le n-1}(e^{\lambda_{j+1}s_0} - e^{\lambda_j s_0}) + e^{\lambda_n s_0}\right) \le 2Me^{\lambda_n s_0}.$$

This clearly implies that $\ell \le s_0$, hence $\ell \le \sigma_c$.

The case $\sigma_c < 0$ is left to the reader. □

5.2.15. Corollary. *If the abscissa of absolute convergence σ_a is strictly positive then*

$$\sigma_a = \limsup_{n\to\infty}\frac{\log(|a_1| + \cdots + |a_n|)}{\lambda_n}.$$

PROOF. One applies the preceding result to the series $\sum_{n\ge 0}|a_n|e^{-\lambda_n s}$. □

5.2.16. Corollary. *If the Dirichlet series is of the form $\sum_{n\ge 0}\frac{a_n}{n^s}$, then one has*

$$\sigma_a - \sigma_c \le 1.$$

PROOF. We have $\lambda_n = \log n$ in this case. □

5.2.17. Corollary. *If the Dirichlet series* $\sum_{n\geq 0} a_n e^{-\lambda_n s}$ *is such that* $\lim_{n\to\infty} \frac{\log n}{\lambda_n} = 0$, *then*

$$\sigma_a = \sigma_c = \limsup_{n\to\infty} \frac{\log|a_n|}{\lambda_n}.$$

PROOF. Let $\gamma = \limsup_{n\to\infty} \frac{\log|a_n|}{\lambda_n}$. Let $s_0 = \sigma_0 + it_0$ be a point in the half-plane $\operatorname{Re} s > \gamma$. Pick $\varepsilon \in]0, (\sigma_0 - \gamma)/3[$. There is $n_0 \in \mathbb{N}^*$ such that for $n \geq n_0$ one has

$$|a_n e^{-\lambda_n s_0}| < e^{-\lambda_n(\sigma_0 - \gamma - \varepsilon)} < \frac{1}{n^2}.$$

Namely, it is enough to assume that $\frac{\log|a_n|}{\lambda_n} < \gamma + \varepsilon$ and $\frac{\log n}{\lambda_n} < \frac{\sigma_0 - \gamma - \varepsilon}{2}$. We conclude that the series converges in $\operatorname{Re} s > \gamma$. That is, $\sigma_a = \sigma_c \leq \gamma$.

On the other hand, if $\sigma_0 < \gamma$, for any $\varepsilon \in]0, \gamma - \sigma_0[$, we can find a sequence n_k of indices such that

$$|a_{n_k} e^{-\lambda_{n_k}\sigma_0}| > e^{-\lambda_{n_k}(\sigma_0 - \gamma - \varepsilon)} > 1.$$

This shows that the series diverges for $\operatorname{Re} s < \gamma$. □

We have shown, by an elementary argument, that for a function represented by a Dirichlet series, one has uniqueness of the coefficients a_n, for the given set of frequencies. We could ask whether we could expand the function using two distinct sets of frequencies. We will see later that the answer is also negative. This will follow from a better knowledge of the behavior of a Dirichlet series in the half-planes $\operatorname{Re} s \geq \delta > \sigma_c$ (assuming, of course, that $\sigma_c < \infty$).

5.2.18. Proposition. *For every* $\delta > \sigma_c$ *there is a constant* $M \geq 0$ *such that if* $\operatorname{Re} s > \delta$, $n \in \mathbb{N}^*$, *then*

$$\left|\sum_{k=1}^{n} a_k e^{-\lambda_k s}\right| \leq M\left(\frac{|s-\delta|}{\sigma - \delta}(e^{-\lambda_1(\sigma-\delta)} - e^{-\lambda_n(\sigma-\delta)}) + e^{-\lambda_n(\sigma-\delta)}\right).$$

In particular, for $f(s) = \sum_{k=1}^{\infty} a_k e^{-\lambda_k s}$ *we have*

$$|f(s)| \leq M\frac{|s-\delta|}{\sigma-\delta} e^{-\lambda_1(\sigma-\delta)} \qquad (\operatorname{Re} s > \delta).$$

PROOF. Since the series converges for $s = \delta$, there is a constant $M \geq 0$ such that $\left|\sum_{k=1}^{n} a_k e^{-\lambda_k \delta}\right| \leq M$ for every $n \geq 1$. Let $\operatorname{Re} s = \sigma > \delta$. Set $U_p = \sum_{k=1}^{p} a_k e^{-\lambda_k s}$, and apply once more the Abel summation formula; we obtain

$$\sum_{k=1}^{n} a_k e^{-\lambda_k s} = \sum_{p=1}^{n-1} U_p(e^{-\lambda_p(s-\delta)} - e^{\lambda_{p+1}(s-\delta)}) + U_n e^{-\lambda_n(s-\delta)}.$$

Recalling that

$$e^{-\lambda_p(s-\delta)} - e^{\lambda_{p+1}(s-\delta)} = (s-\delta)\int_{\lambda_p}^{\lambda_{p+1}} e^{-t(s-\delta)}\,dt$$

we have

$$\left|\sum_{k=1}^{n} a_k e^{-\lambda_k s}\right| \le M|s-\delta| \int_{\lambda_1}^{\lambda_n} e^{-t(\sigma-\delta)}\,dt + Me^{-\lambda_n(\sigma-\delta)}$$
$$\le M\left(\frac{|s-\delta|}{\sigma-\delta} e^{-\lambda_1(\sigma-\delta)} + e^{-\lambda_n(\sigma-\delta)}\right).$$

Letting $n \to \infty$, we find

$$|f(s)| \le M\frac{|s-\delta|}{\sigma-\delta} e^{-\lambda_1(\sigma-\delta)}. \qquad \square$$

5.2.19. Proposition. *If* $f(s) = \sum_{k=1}^{\infty} a_k e^{-\lambda_k s}$, *then for every* $\sigma_1 > \sigma_c$ *the following two properties hold*:

(1) *There is a constant* $M \ge 0$ *such that*

$$|f(s)| \le M|\operatorname{Im} s|$$

whenever $\operatorname{Re} s \ge \sigma_1$, $|\operatorname{Im} s| \ge 1$.

(2) $$\lim_{|\tau|\to\infty} \frac{|f(\sigma + i\tau)|}{|\tau|} = 0$$

uniformly in σ, $\operatorname{Re} s = \sigma \ge \sigma_1$.

PROOF. To prove the first statement, we apply the preceding proposition with $\delta \in]\sigma_c, \sigma_1[$ and obtain $M_0 \ge 0$ such that for $\operatorname{Re} s \ge \sigma \ge \sigma_1$ we have that

$$|f(s)| \le M_0 \frac{|\sigma - \delta + i\tau|}{\sigma - \delta} e^{-\lambda_1(\sigma-\delta)} \le M_0\left(1 + \frac{|\tau|}{\sigma-\delta}\right) \le M|\tau|$$

for a convenient choice of M, once we assume that $|\tau| \ge 1$. (Recall that $\lambda_1 \ge 0$.) The proof of the second property follows from examination of the proof of Proposition 5.2.18. In fact, the constant M such that

$$\left|\sum_{k\ge n} a_k e^{-\lambda_k s}\right| \le M\left(1 + \frac{|\tau|}{\sigma-\delta}\right) e^{-\lambda_n(\sigma-\delta)}$$

can be chosen independent of n. For $\sigma \ge \sigma_1$ we have therefore

$$\frac{|f(\sigma+i\tau)|}{|\tau|} \le \frac{1}{|\tau|}\sum_{k=1}^{n-1} |a_k| e^{-\sigma_1\lambda_k} + M\left(\frac{1}{|\tau|} + \frac{1}{\sigma_1-\delta}\right) e^{-\lambda_n(\sigma_1-\delta)}.$$

Hence

$$\limsup_{|\tau|\to\infty} \frac{|f(\sigma+i\tau)|}{|\tau|} \le \inf_{n\ge 1}\left(\frac{M}{\sigma_1-\delta}e^{-\lambda_n(\sigma_1-\delta)}\right) = 0.$$

This ends the proof of the proposition. □

We are now closer to the proof of the uniqueness of the frequencies in a Dirichlet series. We still need a few technical lemmas.

5.2.20. Lemma. *If $\gamma > 0$, one has*

$$\frac{1}{2\pi i}\int_{\gamma-i\infty}^{\gamma+i\infty} \frac{e^{us}}{s}\,ds = \begin{cases} 1 & \text{if } u > 0 \\ 0 & \text{if } u < 0, \end{cases}$$

where the notation $\int_{\gamma-i\infty}^{\gamma+i\infty}$ *indicates* $\lim\limits_{\substack{k\to\infty\\ h\to\infty}}\int_{\gamma-ik}^{\gamma+ih}$, *both here and in what follows, unless explicitly mentioned.*

PROOF. We integrate $\frac{e^{us}}{s}$ along the boundary of the rectangle

$$-\delta \le \operatorname{Re} s \le \gamma, \quad -k \le \operatorname{Im} s \le h \qquad \text{if } u > 0$$

and

$$\gamma \le \operatorname{Re} s \le \delta, \quad -k \le \operatorname{Im} s \le h \qquad \text{if } u < 0,$$

and we let $h, k, \delta \to +\infty$.

(a) If $u > 0$ we find, by the residue theorem,

$$\frac{1}{2\pi i}\int_{\gamma-ik}^{\gamma+ih} e^{us}\frac{ds}{s} - 1 = \frac{1}{2\pi i}\int_{-\infty}^{\gamma} \frac{e^{u(\sigma+ih)}}{\sigma+ih}\,d\sigma - \frac{1}{2\pi i}\int_{-\infty}^{\gamma}\frac{e^{u(\sigma-ik)}}{\sigma-ik}\,d\sigma,$$

since $\lim\limits_{\delta\to\infty}\int_{-\delta-ik}^{-\delta+ih} e^{us}\frac{ds}{s} = 0$, as it follows from the inequality

$$\left|\int_{-\delta-ik}^{-\delta+ih} e^{us}\frac{ds}{s}\right| \le \frac{e^{-u\delta}}{\delta}(h+k).$$

We have as a consequence the inequality

$$\left|\frac{1}{2\pi i}\int_{\gamma-ik}^{\gamma+ih} e^{us}\frac{ds}{s} - 1\right| \le \frac{1}{2\pi h}\int_{-\infty}^{\gamma} e^{u\sigma}\,d\sigma + \frac{1}{2\pi k}\int_{-\infty}^{\gamma} e^{u\sigma}\,d\sigma$$
$$\le \frac{1}{2\pi}\frac{e^{u\gamma}}{u}\left(\frac{1}{h}+\frac{1}{k}\right).$$

This inequality allows us to draw the desired conclusion when $u > 0$ by letting $h \to \infty$, $k \to \infty$.

(b) The case $u < 0$ is completely similar and we leave it as an exercise to the reader. □

5.2.21. Lemma. *If $\gamma > 0$ one has*

$$\lim_{h\to\infty} \frac{1}{2\pi i} \int_{\gamma-ih}^{\gamma+ih} \frac{ds}{s} = \frac{1}{2}.$$

PROOF. The proof is a routine computation. □

We are now ready to prove a theorem that yields immediately the uniqueness of the frequencies and the coefficients of a convergent Dirichlet series.

5.2.22. Theorem. *Let $f(s) = \sum_{k=1}^{\infty} a_k e^{-\lambda_k s}$ be a Dirichlet series with $\sigma_c < \infty$. For $u \in]\lambda_n, \lambda_{n+1}[$ and $\gamma > \max(\sigma_c, 0)$ we have*

$$\frac{1}{2\pi i} \int_{\gamma-i\infty}^{\gamma+i\infty} f(s) \frac{e^{us}}{s} ds = \sum_{k=1}^{n} a_k.$$

PROOF. Let

$$g(s) = e^{us} f(s) - \sum_{k=1}^{n} a_k e^{(u-\lambda_k)s} = \sum_{k=n+1}^{\infty} a_k e^{-(\lambda_k - u)s}.$$

By Lemma 5.2.20 we have

$$\frac{1}{2\pi i} \int_{\gamma-i\infty}^{\gamma+i\infty} \left(\sum_{k=1}^{n} a_k e^{(u-\lambda_k)s} \right) \frac{ds}{s} = \sum_{1\le k\le n} a_k.$$

To finish the proof of the theorem we need to show that

$$\int_{\gamma-i\infty}^{\gamma+i\infty} \frac{g(s)}{s} ds = 0.$$

Note that $g(s)$ is a Dirichlet series whose smallest frequency is strictly positive; we can rewrite it as $g(s) = e^{-(\lambda_{n+1}-u)s} F(s)$, where $F(s) = \sum_{k\ge n+1} a_k e^{-(\lambda_k - \lambda_{n+1})s}$ has abscissa of convergence σ_c. Integrating $\frac{g(s)}{s} = e^{-(\lambda_{n+1}-u)s} \frac{F(s)}{s}$ along the boundary of the rectangle: $\gamma \le \operatorname{Re} s \le \delta$, $-k \le \operatorname{Im} s \le h$, one obtains 0 since $\frac{g(s)}{s}$ is holomorphic in a neighborhood of this rectangle. Hence

$$\left| \int_{\gamma-ik}^{\gamma+ih} \frac{F(s)e^{-(\lambda_{n+1}-u)s}}{s} ds \right| \le \int_{\gamma}^{\infty} \left(\frac{|F(\sigma - ik)|}{k} + \frac{|F(\sigma + ih)|}{h} \right) e^{-(\lambda_{n+1}-u)\sigma} d\sigma,$$

since

$$\lim_{\delta\to\infty} \int_{\delta-ik}^{\delta+ih} F(s)e^{-(\lambda_{n+1}-u)s} \frac{ds}{s} = 0,$$

by the inequality

$$\left|\int_{\delta-ik}^{\delta+ih} F(s)e^{-(\lambda_{n+1}-u)s}\frac{ds}{s}\right| \leq \int_{-k}^{h} \frac{|F(\delta+i\tau)|}{\delta} e^{-(\lambda_{n+1}-u)\delta}\,d\tau.$$

The last quantity tends to zero by §5.2.19 (1). By §5.2.19 (2), for a given $\varepsilon > 0$ there is $R > 0$, independent of $\sigma \geq \gamma$ such that

$$\frac{|F(\sigma+ih)|}{h} \leq \varepsilon \quad \text{and} \quad \frac{|F(\sigma+ik)|}{k} \leq \varepsilon,$$

whenever $h \geq R$ and $k \geq R$. We conclude that

$$\left|\int_{\gamma-ik}^{\gamma+ih} F(s)e^{-(\lambda_{n+1}-u)s}\frac{ds}{s}\right| \leq \frac{2\varepsilon}{\lambda_{n+1}-u}.$$

This inequality concludes the proof of the theorem. □

5.2.23. Corollary. *Let* $f(s) = \sum_{k=1}^{\infty} a_k e^{-\lambda_k s} = \sum_{j=1}^{\infty} b_j e^{-\mu_j s}$, *for* $\operatorname{Re} s > \max(\sigma_c, \sigma_c')$, *where* σ_c, σ_c' *are the respective abscissas of convergence and* $\max(\sigma_c, \sigma_c') < \infty$. *Then, for any* $\lambda_k \notin \{\mu_j\}_{j\geq 1}$ *we have* $a_k = 0$, *for any* $\mu_j \notin \{\lambda_k\}_{k\geq 1}$, *we have* $b_j = 0$, *and, if* $\lambda_k = \mu_j$, *then* $a_k = b_j$.

5.2.24. Corollary. *Under the same hypothesis as in Theorem* 5.2.2 *we have*

$$\lim_{R\to\infty} \frac{1}{2\pi i}\int_{\gamma-iR}^{\gamma+iR} \frac{f(s)e^{\lambda_n s}}{s}\,ds = \sum_{k=1}^{n-1} a_k + \frac{a_n}{2}.$$

PROOF. By Theorem 5.2.2, we have

$$\frac{1}{2\pi i}\int_{\gamma-i\infty}^{\gamma+i\infty} \frac{f(s) - a_n e^{-\lambda_n s}}{s} e^{\lambda_n s}\,ds = \sum_{k=1}^{n-1} a_k.$$

Let us write

$$\frac{1}{2\pi i}\int_{\gamma-iR}^{\gamma+iR} \frac{f(s) - a_n e^{-\lambda_n s}}{s} e^{\lambda_n s}\,ds = \frac{1}{2\pi i}\int_{\gamma-iR}^{\gamma+iR} \frac{f(s)e^{\lambda_n s}}{s}\,ds - \frac{a_n}{2\pi i}\int_{\gamma-iR}^{\gamma+iR} \frac{ds}{s},$$

passing to the limit when R tends to infinity and using Lemma 5.2.21 we obtain the conclusion of the corollary. □

5.2.25. Theorem. *Let* $f(s) = \sum_{k=1}^{\infty} a_k e^{-\lambda_k s}$ *have abscissa of convergence* $\sigma_c < \infty$. *For* $u \in]\lambda_n, \lambda_{n+1}[$, $\gamma > \max(0, \sigma_c)$, *and* p *integer* ≥ 1 *one has*

$$\frac{p!}{2\pi i}\int_{\gamma-i\infty}^{\gamma+i\infty} \frac{f(s)e^{us}}{s^{p+1}}\,ds = \sum_{k=1}^{n} a_k(u-\lambda_k)^p.$$

PROOF. Note that the case $p = 0$ is just Theorem 5.2.22. To prove this theorem we observe first that integrating by parts several times yields

$$\frac{1}{2\pi i}\int_{\gamma-i\infty}^{\gamma+i\infty}\frac{f(s)e^{us}}{s^{p+1}}\,ds = \frac{1}{2\pi i}\frac{1}{p!}\int_{\gamma-i\infty}^{\gamma+i\infty}\frac{(f(s)e^{us})^{(p)}}{s}\,ds,$$

where the contributions of the endpoints vanish, thanks to §5.2.18. For every integer p we have

$$(f(s)e^{us})^{(p)} = \sum_{k=1}^{\infty} a_k(u-\lambda_k)^p e^{(u-\lambda_k)s}$$

in the half-plane of convergence. Applying now Theorem 5.2.22 to this last Dirichlet series we obtain the identity we were looking for. □

5.2.26. Theorem. *Let $f(s) = \sum_{n\geq 1} a_n e^{-\lambda_n s}$ be a Dirichlet series with $-\infty < \sigma_c < \infty$, and $a_n \geq 0$ for every n. Then the point $s = \sigma_c$ is a singular point for the function f.*

PROOF. If not, there is $\sigma_0 > \sigma_c$ and a disk centered at σ_0 and radius $R > \sigma_0 - \sigma_c$ in which the Taylor series

$$f(s) = \sum_{k\geq 0}\frac{f^{(k)}(\sigma_0)}{k!}(s-\sigma_0)^k$$

converges. Since the convergence of the Dirichlet series $\sum_{n\geq 1} a_n e^{-\lambda_n s}$ is uniform in a neighborhood of σ_0, we have for $k \in \mathbb{N}$

$$f^{(k)}(\sigma_0) = (-1)^k \sum_{n\geq 1} a_k \lambda_n^k e^{-\lambda_n \sigma_0},$$

hence, for $|s-\sigma_0| < R$ we have

$$f(s) = \sum_{k\geq 0}\left(\sum_{n\geq 1} a_n\lambda_n^k e^{-\lambda_n\sigma_0}\right)\frac{(\sigma_0-s)^k}{k!}.$$

If we now consider s real, $\sigma_0 - R < s < \sigma_c < \sigma_0$, then $\sigma_0 - s > 0$ and not only the series converges, but the order of summation can be interchanged using that every term is positive. This way we obtain

$$\begin{aligned} f(s) &= \sum_{n\geq 1} a_n e^{-\lambda_n\sigma_0}\left(\sum_{k\geq 0}\frac{\lambda_n^k(\sigma_0-s)^k}{k!}\right) \\ &= \sum_{n\geq 1} a_n e^{-\lambda_n\sigma_0} e^{\lambda_n(\sigma_0-s)} \\ &= \sum_{n\geq 1} a_n e^{-\lambda_n s}. \end{aligned}$$

Therefore the Dirichlet series converges at $s < \sigma_c$, which is impossible. □

As we mentioned earlier, we shall return to the study of general Dirichlet series after we introduce the Mellin transform in the second volume. More general series (with complex frequencies) will appear there also, in the context of mean periodicity.

EXERCISES 5.2

1. Let $f(z) = \sum_{n\geq 0} a_n z^n$ have radius of convergence 1 and let $F(\zeta) = \frac{1}{1-\zeta} f\left(\frac{\zeta}{1-\zeta}\right)$.
   (a) Show that F is holomorphic in the half-plane $\operatorname{Re}\zeta < \frac{1}{2}$, and $z = 1$ is a singular point for f if and only if $\zeta = \frac{1}{2}$ is a singular point for F.
   (b) Show that $F(\zeta) = \sum_{n\geq 0} b_n \zeta^n$, with $b_n = \sum_{k=0}^{n} \binom{n}{k} a_k$.
   (c) Prove that $z = 1$ is a singular point for f if and only if $\liminf_{n\to 0} |b_n|^{-1/n} = \frac{1}{2}$.

2. Prove by induction that for any $k \in \mathbb{N}^*$ there is a polynomial P_k of degree k, with coefficients in $\mathbb{Z}$ such that

$$\sum_{n\geq 1} n^k z^n = \frac{P_k(z)}{(1-z)^{k+1}} \qquad \text{for } |z| < 1.$$

(In fact, one can also show the coefficients of P_k are ≥ 0.) Conclude that if Q is a polynomial, then the series

$$f(z) = \sum_{n\geq 0} Q(n) z^n$$

has radius of convergence exactly one but it is the restriction to the unit disk of a rational function whose only pole occurs at $z = 1$. What is the relation between the order of the pole and the degree of Q?

*3. Let f be a meromorphic function in a neighborhood of the closed disk $\bar{B}(0, R)$ whose only pole lies at $z = z_0$, $|z_0| = R$. Let $f(z) = \sum_{n\geq 0} a_n z^n$ be the Taylor series expansion of f about $z = 0$. Show that there is an $A \in \mathbb{C}^*$ and $\rho > R$ such that

$$a_n = \frac{An^{\nu-1}}{z_0^n} + O(\rho^{-n}), \qquad n \to \infty,$$

where ν is the order of the pole z_0. (Hint: let $P\left(\frac{1}{z-z_0}\right)$ be the principal part of f about $z = z_0$ and consider the holomorphic function $F(z) = f(z) - P\left(\frac{1}{z-z_0}\right)$ to compute the a_n.)

Conclude that

$$\lim_{n\to\infty} \frac{(n+1)f^{(n)}(0)}{f^{(n+1)}(0)} = z_0.$$

4. Prove statement (3) of Proposition 5.2.14.
5. Prove Corollary 5.2.23.
6. Find σ_a, σ_c (when appropriate) for the following Dirichlet series:

   (i) $\sum_{n\geq 1} \frac{n!}{n^s}$

(ii) $\sum_{n\geq 1} \frac{1}{n^s}$

(iii) $\sum_{n\geq 1} \frac{(-1)^{n+1}}{n^s}$

(iv) $\sum_{n\geq 1} \frac{1}{n!n^s}$

(v) $\sum_{n\geq 2} (-1)^n e^{-(\log\log n)s}$

(vi) $\sum_{n\geq 2} \frac{(-1)^n}{n} e^{-(\log\log n)s}$

7. Show that if $\lambda_n = n\log n$ then $\sigma_a = \overline{\lim_{n\to\infty}} \frac{\log|a_n|}{\lambda_n}$ for the Dirichlet series $\sum_{n\geq 1} a_n e^{-\lambda_n s}$.

8. Let $f(s) = \sum_{n\geq 1} a_n e^{-\lambda_n s}$, $\sigma = \operatorname{Re} s > \sigma_a$, and $M(\sigma) = \sup_{t\in\mathbb{R}} |f(\sigma + it)|$. Show that
$$a_n e^{-\lambda_n \sigma} = \lim_{T\to\infty} \frac{1}{T} \int_0^T f(\sigma + it) e^{i\lambda_n t}\, dt.$$
Conclude that
$$|a_n| e^{-\lambda_n \sigma} \leq M(\sigma).$$

*9. Let $f(s) = \sum_{n\geq 1} a_n e^{-\lambda_n s}$, $g(s) = \sum_{n\geq 1} b_n e^{-\lambda_n s}$ converge absolutely for $s = \alpha$, $s = \beta$, respectively. Show that
$$\lim_{T\to\infty} \frac{1}{T} \int_0^T f(\alpha + it) g(\beta - it)\, dt = \sum_{n\geq 1} a_n b_n e^{-\lambda_n(\alpha+\beta)}.$$
What is the limiting value of the average $\frac{1}{2T} \int_{-T}^{T} |f(\alpha + it)|^2\, dt$ as $T \to \infty$?

10. Let $f(s) = \sum_{n\geq 1} \frac{a_n}{n^s}$ have $\sigma_c < \infty$. If $x > 0$ is not an integer, $\gamma > 0$, and $\operatorname{Re} s > \sigma_c + \gamma$, then
$$\sum_{n<x} \frac{a_n}{n^s} = \frac{1}{2\pi i} \int_{\gamma - i\infty}^{\gamma + i\infty} f(s + z) x^z \frac{dz}{z}$$
What is the corresponding statement for a general Dirichlet series $\sum_{n\geq 1} a_n e^{-\lambda_n s}$? (Compare with Theorem 5.2.22.)

11. Let φ be an entire function of exponential type A and $f(s) = \sum_{n\geq 1} a_n e^{-\lambda_n s}$ a Dirichlet series with $\sigma_a < \infty$. Show that the Dirichlet series $\sum_{n\geq 1} a_n \varphi(\lambda_n) e^{-\lambda_n s}$ has abscissa of absolute convergence $\leq \sigma_a + A$.

12. Let f be a function holomorphic in a neighborhood of the origin that satisfies the functional equation $f(2z) = \frac{d}{dz}(f(z))^2$. Show that f must be an entire function.

Find a solution of the form $f(z) = e^{\lambda z}$. (Hint: use the equation to define f in larger and larger domains.)

*13. (a) Choose $N \in \mathbb{N}^*$ such that $\sum_{n \geq N} n e^{-2^{n/2}} < 1$.

(b) Let $f(z) = z + \sum_{n \geq N} e^{-2^{n/2}} z^n$. Show that $f \in \mathscr{H}(B(0,1)) \cap C^\infty(\bar{B}(0,1))$.

(c) Show that f is injective in $\bar{B}(0,1)$.

(d) Let $\Gamma = f(\partial B(0,1))$. It is a C^∞ Jordan curve (why?). Show there is no point $w_0 \in \Gamma$ for which one can find a function $\varphi : [-1,1] \to \mathbb{C}$, real analytic, injective, $\varphi(0) = w_0$, $\dot{\varphi}(0) \neq 0$, $\varphi([-1,1]) \subseteq \Gamma$. In other words, Γ is nowhere real analytic.

14. For $n \in \mathbb{N}$ denote by $\Delta_n = \{z \in \mathbb{C} : \operatorname{Re} z \geq n, |\operatorname{Im} z| \leq \pi\}$ and $\gamma_n = \partial \Delta_n$ with its usual orientation. For $z \notin \gamma_n$, let

$$h_n(z) := -\frac{1}{2\pi i} \int_{\gamma_n} \frac{\exp(\exp \zeta)}{\zeta - z} \, d\zeta.$$

(i) Show that h_n is holomorphic in $\mathbb{C} \setminus \gamma_n$. Moreover, when $n \geq m$,

$$h_n | \Delta_m^c = h_m | \Delta_m^c.$$

Conclude that there is an entire function h such that $h | \Delta_n^c = h_n$ and $h(\bar{z}) = \overline{h(z)}$.

(ii) Show that

$$-\frac{1}{2\pi i} \int_{\gamma_n} \exp(\exp \zeta) \, d\zeta = 1$$

for every $n \in \mathbb{N}$.

(iii) Let $g(z) = zh(z) + 1$. Prove that $|g(z)| \leq \dfrac{10}{d(z, \gamma_0)}$ whenever $z \notin \gamma_0$.

(iv) For $0 < x < n$, $n \in \mathbb{N}^*$, show that $h(x) = e^{e^x} + h_0(x)$, and conclude that $\lim_{x \to \infty} h(x) = \infty$.

(v) Prove that for every $\theta \in]0, 2\pi[$, $\lim_{r \to \infty} h(re^{i\theta}) = 0$. Show that $H(z) = h(z) e^{-h(z)}$ has the property

$$\lim_{r \to \infty} H(re^{i\theta}) = 0$$

for every $\theta \in \mathbb{R}$. (The same is true for $e^{-h(z)} - e^{-h(2z)}$.)

(vi) Let $B > 0$, find a nonzero entire function f such that $|f(z)| \leq \dfrac{C}{|z|}$ when $|\operatorname{Im} z| \geq B$, for some constant $C > 0$.

15. Show that if f is an entire function that satisfies the inequality

$$|f(z)| \leq \frac{1}{|\operatorname{Im} z|}$$

everywhere, then $f \equiv 0$. (Hint. Consider the auxiliary function

$$g(z) := (\operatorname{Im} z) f(z),$$

and develop it in series de Fourier to show $g \equiv 0$.) See Exercise 3.5.10 for a different proof.

§3. A Brief Study of the Functions Γ and ζ

We have already seen the Euler Γ-function appear several times in this book, e.g., in Exercises 3.2.4 and 3.3.11. It has a number of remarkable properties and we recommend the book [Cam], which is entirely dedicated to this function. We shall see later that the Γ-function appears intertwined with another famous function, the Riemann ζ-function, which, in fact, was also originally defined by Euler.

We have up to now given two different definitions of the Γ-function. The first one is

$$\Gamma(z) = \int_0^\infty e^{-t} t^{z-1}\, dt, \tag{$*$}$$

and we indicated in Exercise 3.2.4 how this function, originally defined and holomorphic in the half-plane $\operatorname{Re} z > 0$, has an analytic continuation to a meromorphic function in $\mathbb{C}$ with simple poles at $z = 0, -1, -2, \ldots$. Namely, for $\operatorname{Re} z > 0$ we rewrite ($*$) in the form

$$\begin{aligned}\Gamma(z) &= \int_0^1 e^{-t} t^{z-1}\, dt + \int_1^\infty e^{-t} t^{z-1}\, dt\\ &= \sum_{n\geq 0} \frac{(-1)^n}{n!} \int_0^1 t^{n+z-1}\, dt + \int_1^\infty e^{-t} t^{z-1}\, dt\\ &= \sum_{n\geq 0} \frac{(-1)^n}{n!} \frac{1}{z+n} + \int_1^\infty e^{-t} t^{z-1}\, dt.\end{aligned}$$

The first term is clearly a meromorphic funtion with simple poles at $z = 0, -1, -2, \ldots$, and the second term is an entire function of z.

The second definition we gave was in Example 4.5.7 (2), where we *defined* Γ *indirectly* by saying that $\frac{1}{\Gamma}$ is the entire function given by the infinite product expansion

$$\frac{1}{\Gamma(z)} = z e^{\gamma z} \prod_{n=1}^{\infty} \left(1 + \frac{z}{n}\right) e^{-z/n}. \tag{†}$$

A priori, it is not clear that the definitions ($*$) and (†) coincide. If they do, one can draw unexpected consequences, e.g., Γ has no zeros in $\mathbb{C}$, since $\frac{1}{\Gamma}$ is entire. There are several ways of conciliating the two definitions. We start with ($*$), which clearly implies that $\Gamma(x) > 0$ for $x \in]0, \infty[$ and show that $\frac{1}{\Gamma(x)}$ has the form (†) for $x \in]0, \infty[$. Since we already know that the function Γ is meromorphic in $\mathbb{C}$ and hence $\frac{1}{\Gamma}$ is also meromorphic in $\mathbb{C}$, this will prove that (†) holds.

5.3.1. Lemma. *The functions* $\varphi_n(z) := \int_0^n \left(1 - \frac{t}{n}\right)^n t^{z-1}\, dt$ $(n \in \mathbb{N}^*)$ *converge to* Γ *locally uniformly in the half-plane* $\operatorname{Re} z > 0$.

PROOF. (Compare with Exercise 3.3.11). We start by proving that the auxiliary functions $\psi_n(t) := e^{-t} - \left(1 - \frac{t}{n}\right)^n$, defined for $0 \le t \le n$, satisfy the inequalities

$$0 \le \psi_n(t) \le \frac{t^2 e^{-t}}{n}. \qquad (**)$$

The proof of the left-hand side inequality is very easy. The function $\chi_n(t) := \left(1 - \frac{t}{n}\right)^n e^t$ satisfies $\chi_n(0) = 1$ and $\chi_n'(t) = e^t \left(1 - \frac{t}{n}\right)^{n-1} \left(-\frac{t}{n}\right) \le 0$ in $0 \le t \le n$. Hence $\chi_n(t) \le 1$, or equivalently $\psi_n(t) \ge 0$.

The other inequality is considerably harder. It is enough to show it for large n. Consider first the function

$$g_n(t) := e^t \left(1 - \frac{t}{n}\right)^{n-1} - 2.$$

Then $g_n'(t) = e^t \left(1 - \frac{t}{n}\right)^{n-2} \frac{1-t}{n} = 0$ only if $t = 1$, which is a local maximum of g_n. The value $g_n(1) = e\left(1 - \frac{1}{n}\right)^{n-1} - 2 \to -1$, hence $g_n(t) \le 0$ in $0 \le t \le n$, for all large n. To continue the proof, let $\theta_n(t) := 1 - e^t \left(1 - \frac{t}{n}\right)^n - \frac{t^2}{n}$. We want to show that $\theta_n \le 0$. We have $\theta_n(0) = 0$ and

$$\theta_n'(t) = \frac{t}{n}\left[e^t \left(1 - \frac{1}{n}\right)^{n-1} - 2\right] \le 0$$

for all large n. Hence θ_n is decreasing and both inequalities in (**) hold, at least for large n.

Note that if $x = \operatorname{Re} z$, $0 < x \le 1$

$$\left|\Gamma(z) - \int_0^n e^{-t} t^{z-1}\, dt\right| \le \int_n^\infty e^{-t} t^{x-1}\, dt \le n^{x-1} \int_n^\infty e^{-t}\, dt \le \frac{e^{-n}}{n^{1-x}}.$$

If $x > 1$ we can integrate by parts and obtain

$$\int_n^\infty e^{-t} t^{x-1}\, dt = e^{-n} n^{x-1} + (x-1) \int_n^\infty t^{x-2} e^{-t}\, dt.$$

It is clear that for any $0 < A < B < 0$ there is a constant C such that whenever $A \le \operatorname{Re} z \le B$ we have

$$\left|\Gamma(z) - \int_0^n e^{-t} t^{z-1}\, dt\right| \le C n^{B-1} e^{-n} \to 0 \qquad \text{as } n \to \infty.$$

On the other hand, the inequalities (**) show that

$$\left|\varphi_n(z) - \int_0^n e^{-t}t^{z-1}\,dt\right| \leq \int_0^n t^{x-1}\left(e^{-t} - \left(1 - \frac{t}{n}\right)^n\right) dt$$
$$\leq \frac{1}{n}\int_0^n t^{x+1}e^{-t}\,dt \leq \frac{\Gamma(x+2)}{n}.$$

This concludes the proof of the lemma. □

We can describe the functions φ_n more explicitly by integration by parts

$$\varphi_n(z) = \int_0^n \left(1 - \frac{t}{n}\right)^n t^{z-1}\,dt = \frac{1}{z}\int_0^n \left(1 - \frac{t}{n}\right)^{n-1} t^z\,dt.$$

Iterating this procedure we obtain

$$\varphi_n(z) = \frac{n!n^z}{z(z+1)\dots(z+n)} = \frac{n^z}{z(1+z)(1+z/2)\dots(1+z/n)}.$$

Since these functions have no zeros in $\operatorname{Re} z > 0$, we can conclude by Hurwitz's theorem that the same holds for Γ. Moreover,

$$\prod_{k=1}^n \left(1 + \frac{z}{k}\right) = \left[\prod_{k=1}^n \left(1 + \frac{z}{k}\right)e^{-z/k}\right]\exp\left(\left[1 + \frac{1}{2} + \cdots + \frac{1}{n}\right]z\right)$$
$$= n^z \exp\left(\left[1 + \frac{1}{2} + \cdots + \frac{1}{n} - \log n\right]z\right)\prod_{k=1}^n \left(1 + \frac{z}{k}\right)e^{-z/k}.$$

Since $\lim_{n\to\infty}\left(\sum_{k=1}^n \frac{1}{k} - \log n\right) = \gamma$, it is clear that

$$\Gamma(z) = \lim_{n\to\infty} \frac{n!n^z}{z(z+1)\dots(z+n)} = \frac{1}{e^{\gamma z}z\prod_{k=1}^\infty \left(1 + \frac{z}{k}\right)e^{-z/k}},$$

as we wanted to show. From the first identity, sometimes called Gauss' formula, it is easy to derive the functional equation for the Γ-function

$$\Gamma(z+1) = \lim_{n\to\infty} \frac{n!n^{z+1}}{(z+1)\dots(z+n+1)}$$
$$= z\lim_{n\to\infty}\left[\frac{n}{z+n+1}\right]\lim_{n\to\infty} \frac{n!n^z}{z(z+1)\dots(z+n)} = z\Gamma(z).$$

We remind the reader that we have proved this under the assumption that $\operatorname{Re} z > 0$, but since both sides are meromorphic functions of z, it holds for every value $z \neq 0, -1, -2, \dots$. Furthermore, it is easy to see from the proof that Gauss' formula for the Γ-function is valid everywhere in $\mathbb{C}\setminus\{0, -1, -2, \dots\}$.

An important property of the Γ-function whose proof we shall omit is ***Stirling's formula***: For any $\varepsilon > 0$, in the region $|\operatorname{Arg} z| \leq \pi - \varepsilon$ we have

$$\lim_{|z|\to\infty} \frac{\Gamma(z)}{\sqrt{2\pi e^{-z} z^{z-1/2}}} = 1,$$

where $z^{z-1/2} = \exp((z-\frac{1}{2})\log z)$. (See [Cam], [Mar], [WW], or Exercise 5.5.35 for a proof.) In particular, for $n \in \mathbb{N}$, we have

$$n! \sim \sqrt{2\pi n}\left(\frac{n}{e}\right)^n,$$

where the symbol $\sim$ indicates the quotient of the two sides has limit 1 when $n \to \infty$.

There are several relations between the Γ-function and the ζ-function. One type of relation is obtained by considering the logarithmic derivative of Γ. From (†) we have

$$\psi(z) := \frac{\Gamma'(z)}{\Gamma(z)} = -\gamma - \frac{1}{z} + \sum_{n=1}^{\infty}\left(\frac{1}{n} - \frac{1}{z+n}\right), \tag{††}$$

hence

$$\psi'(z) = \frac{1}{z^2} + \sum_{n=1}^{\infty} \frac{1}{(z+n)^2}.$$

Evaluating this series at $z = 1$, we get $\psi'(1) = \zeta(2)$. Proceeding this way, one can evaluate $\psi^{(n)}(1)$ in terms of $\zeta(n+1)$. A more interesting relationship arises as follows. Let $a \in \mathbb{C}^*$ be fixed and define the generalized ζ-function

$$\zeta(s, a) := \sum_{n=0}^{\infty} \frac{1}{(a+n)^s},$$

which is holomorphic for $\operatorname{Re} s > 1$. Note that $\zeta(s, 1) = \zeta(s)$. We observe that for $0 < a$ and $\operatorname{Re} s > 0$

$$\int_0^{\infty} x^{s-1} e^{-(a+n)x}\, dx = \frac{\Gamma(s)}{(a+n)^s}$$

and, therefore,

$$\Gamma(s)\zeta(s, a) = \sum_{n=0}^{\infty} \int_0^{\infty} x^{s-1} e^{-(a+n)x}\, dx = \int_0^{\infty} \frac{x^{s-1} e^{-ax}}{1 - e^x}\, dx,$$

which is a representation very similar to (*) for the Γ-function. Let us consider now a contour C as in the adjacent Figure 5.2, which does not contain any of the points $2\pi i n$, $n \in \mathbb{Z}^*$, and the corresponding integral

$$\int_{\infty}^{(0+)} \frac{(-z)^{s-1} e^{-az}}{1 - e^{-z}}\, dz,$$

where it is assumed that $|\arg(-z)| \le \pi$, $\arg(-z) = 0$ when $z < 0$, $0 < a \le 1$. (It is standard to symbolize the integral over C by $\int_{\infty}^{(0+)}$. This path C is called

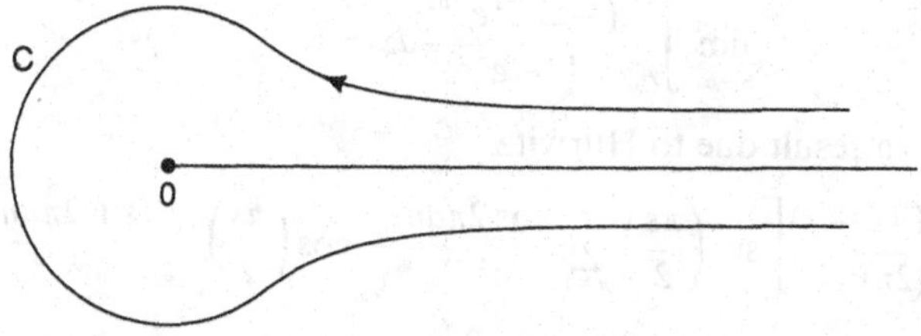

Figure 5.2

a Hankel contour.) As a function of s, this is an entire function. Shrinking the contour C to the positive real axis we obtain

$$\int_{\infty}^{(0+)} \frac{(-z)^{s-1}e^{-az}}{1-e^{-z}}\,dz = -2i\sin\pi s \int_0^{\infty} \frac{x^{s-1}e^{-az}}{1-e^{-x}}\,dx.$$

In Exercise 3.3.12 it was proved that

$$\frac{1}{\Gamma(s)\Gamma(1-s)} = \frac{\sin\pi s}{\pi}.$$

Using this identity, we obtain for $\operatorname{Re} s > 1$, $s \notin \mathbb{N}$, that

$$\zeta(s,a) = -\frac{\Gamma(1-s)}{2\pi i}\int_{\infty}^{(0+)} \frac{(-z)^{s-1}e^{-az}}{1-e^{-z}}\,dz.$$

This shows that $\zeta(s,a)$ has an analytic continuation to the whole plane, which is a meromorphic function and whose only possible singularities are those of $\Gamma(1-s)$, i.e., at most simple poles at $s = 1, 2, \ldots$. On the other hand, $\zeta(s,a)$ is a convergent Dirichlet series for $\operatorname{Re} s > 1$. Hence $\zeta(s,a)$ and the function $\zeta(s)$ have only one singularity, a simple pole at $s = 1$. In fact

$$\lim_{\zeta\to 1} \frac{\zeta(s,a)}{\Gamma(1-s)} = -\frac{1}{2\pi i}\int_{\infty}^{(0+)} \frac{e^{-az}}{1-e^{-z}}\,dz = -1.$$

Therefore, the residue is $\operatorname{Res}(\zeta(s,a), s = 1) = 1$.

Let us denote by C_N the circle $\partial B(0,(2N+1)\pi)$ in the positive direction, and by $((2N+1)\pi, 0+)$ the portion of the contour C lying inside C_N. Here $N \in \mathbb{N}^*$. A computation of residues allows us to calculate

$$\frac{1}{2\pi i}\int_{C_N} \frac{(-z)^{s-1}e^{-az}}{1-e^{-z}}\,dz = \frac{1}{2\pi i}\int_{(2N+1)\pi}^{(0+)} \frac{(-z)^{s-1}e^{-az}}{1-e^{-z}}\,dz$$
$$= 2\sum_{n=1}^{N} (2\pi n)^{s-1}\sin\left(\frac{1}{2}s\pi + 2\pi an\right).$$

The function $e^{-az}(1-e^{-z})^{-1}$ is uniformly bounded over all the circles C_N, hence if $0 < a \le 1$ and $\operatorname{Re} s < 0$, we have

$$\lim_{n\to\infty} \int_{C_N} \frac{(-z)^{s-1}e^{-az}}{1-e^{-z}}\,dz = 0$$

and one obtains a result due to Hurwitz:

$$\zeta(s, a) = \frac{2\Gamma(1-s)}{(2\pi)^{1-s}}\left[\sin\left(\frac{\pi s}{2}\right)\sum_{n=1}^{\infty}\frac{\cos 2\pi a n}{n^{1-s}} + \cos\left(\frac{\pi s}{2}\right)\sum_{n=1}^{\infty}\frac{\sin 2\pi a n}{n^{1-s}}\right].$$

We set $a = 1$ to find the functional equation of the ζ-function

$$\zeta(s) = \frac{2\Gamma(1-s)}{(2\pi)^{1-s}}\sin\left(\frac{\pi s}{2}\right)\zeta(1-s),$$

which is usually rewritten as

$$2^{1-s}\Gamma(s)\zeta(s)\cos\left(\frac{\pi s}{2}\right) = \pi^s\zeta(1-s).$$

This formula holds for every value of s, with due consideration for the poles of the factors, and forces ζ to have simple zeros at $s = -2, -4, -6, \ldots$, and nowhere else, except possibly in the strip $0 \le \operatorname{Re} s \le 1$. The still unsolved Riemann hypothesis states that in this strip all the zeros lie in the vertical line $\operatorname{Re} s = \frac{1}{2}$.

The relation of the ζ-function with number theory arises out of a relation discovered by Euler

$$\zeta(s)\prod_p\left(1 - \frac{1}{p^s}\right) = 1, \qquad (\operatorname{Re} s > 1),$$

where the product takes place over all the primes $p = 2, 3, 5, \ldots$. The existence of a pole for ζ at $s = 1$ forces the relation

$$\sum_p \frac{1}{p} = \infty.$$

This is Euler's proof of the existence of infinitely many primes. This already shows that the primes could not be very sparse, otherwise the series would converge. The prime number theorem asserts that if p_n represents the nth prime then $p_n \sim n \log n$. It turns out that this is a consequence of a theorem of Hadamard–de la Vallée–Poussin that $\zeta(s)$ does not vanish when $\operatorname{Re} s = 1$. The Riemann hypothesis furnishes further information on the distribution of the prime numbers. We refer to [Ti2] and [Pa] for an up-to-date account of this subject.

Exercises 5.3

1. Let $\psi(z) = \Gamma'(z)/\Gamma(z)$.

   (a) Show that for $k \ge 2$, $\psi^{(k-1)}(z) = (-1)^{k-1}(k-1)!\zeta(k, z)$.

   (b) Use (a) to show that $\Gamma'(1) = -\gamma$, $\Gamma''(1) = \gamma^2 + \zeta(2)$, and that, in general, $\Gamma^{(n)}(1)$ is a polynomial of degree n in γ with coefficients computed in terms of the values $\zeta(k)$, $k \in \mathbb{N}$.

(c) Show that $\psi'(z) + \psi'(z + \frac{1}{2}) = 2\psi'(2z)$. Use this to prove

$$\Gamma(2z) = 2^{2z-1}\Gamma(z)\Gamma(z + \tfrac{1}{2}).$$

(d) Prove $f(x) = \log \Gamma(x)$ is a convex function of x for $x > 0$. (It is a theorem of Bohr-Mollerup that the four conditions $\Gamma(x) > 0$, $\Gamma(x + 1) = x\Gamma(x)$, $\Gamma(1) = 1$, and convexity of $\log \Gamma$, characterize the Γ-function.)

2. Use Stirling's formula to check that

$$\int_x^{x+1} \log \Gamma(t)\,dt = x(\log x - 1) + \frac{1}{2}\log 2\pi \qquad \text{for } x > 0.$$

3. Show for $x > 0$ not an integer that

$$\Gamma(x) = -\frac{1}{2i \sin \pi x}\int_\infty^{(0+)} (-t)^{x-1}e^{-t}\,dt,$$

where $(-t)^{x-1}$ is defined by taking $\arg(-t) = 0$ on the negative real axis and $|\arg(-t)| \le \pi$. The meaning of the path is that it does not pass through $t = 0$, otherwise starts at ∞, encloses 0 counterclockwise, and returns to ∞, cf. Figure 5.2. Show this identity still holds for any $x \in \mathbb{C}\backslash\mathbb{Z}$. Conclude that

$$\frac{1}{\Gamma(z)} = \frac{i}{2\pi}\int_\infty^{(0+)} (-t)^{-z}e^{-t}\,dt \qquad (z \in \mathbb{C}).$$

4. (Alternative derivation of the inequalities in Lemma 5.3.1).

(a) Show that for $0 \le y < 1$, $1 + y \le e^y \le \dfrac{1}{1 - y}$.

(b) Replacing y by $\dfrac{t}{n}$, obtain $0 \le e^{-t} - \left(1 - \dfrac{t}{n}\right)^n \le e^{-t}\left\{1 - \left(1 - \dfrac{t^2}{t^2}\right)\right\}$.

(c) Use that for $0 \le \eta \le 1$, $(1 - \eta)^n \ge 1 - n\eta$ to obtain the inequality

$$0 \le e^{-t} - \left(1 - \frac{t}{n}\right)^n \le e^{-t}\frac{t^2}{n} \qquad \text{if } 0 \le t \le n, \quad n \in \mathbb{N}^*$$

*5. (Proof of Stirling's asymptotic formula).

(a) Show that when $|\operatorname{Arg} z| \le \pi - \varepsilon$ ($\varepsilon > 0$), we have

$$\operatorname{Log}\Gamma(z) = -\gamma z - \operatorname{Log} z - \sum_{n=1}^{\infty}\left[\frac{z}{n} - \operatorname{Log}\left(1 + \frac{z}{n}\right)\right].$$

(b) Let $[t]$ denote the integral part of the real number t. Show that for $n \in \mathbb{N}$, $n \ge 2$, and $\operatorname{Re} z > 0$,

$$\begin{aligned}\int_0^n \frac{[t] - t + 1/2}{t + z}\,dt &= \sum_{k=1}^{n-1}\int_k^{k+1}\left(\frac{k + 1/2 + z}{t + z} - 1\right)dt \\ &= \sum_{k=1}^{n-1}\left[\frac{z}{k} - \operatorname{Log}\left(1 + \frac{z}{k}\right)\right] - \log[(n-1)!] \\ &\quad - z\left(1 + \frac{1}{2} + \cdots + \frac{1}{n-1}\right) - \left(z + \frac{1}{2}\right)\operatorname{Log} z \\ &\quad + \left(n - \frac{1}{2} + z\right)\log(n + 2) - n.\end{aligned}$$

(c) Use that for $n \to \infty$ and z fixed with $\operatorname{Re} z > 0$, one has

$$1 + \frac{1}{2} + \cdots + \frac{1}{n-1} = \log n + \gamma + O(1),$$

$$\operatorname{Log}(n + z) = \log n + \frac{z}{n} + O\left(\frac{1}{n^2}\right),$$

to show that for $\operatorname{Re} z > 0$

$$\operatorname{Log} \Gamma(z) = \left(z - \frac{1}{2}\right)\operatorname{Log} z - z + \frac{1}{2}\log 2\pi + \int_0^\infty \frac{[t] - t + 1/2}{t + z}\,dt.$$

Why is this formula valid for all $z \notin]-\infty, 0]$?

(d) Let $\varphi(u) = \displaystyle\int_0^u \left([t] - t + \frac{1}{2}\right) dt$. Show that φ is periodic of period 1 and, hence, bounded. Integrate by parts to obtain

$$\int_0^\infty \frac{[t] - t + 1/2}{t + z}\,dt = \int_0^\infty \frac{\varphi(u)\,du}{(u + z)^2}.$$

Show that when $|\operatorname{Arg} z| \le \pi - \varepsilon$, the last integral is $O\left(\dfrac{1}{|z|}\right)$ *as* $|z| \to \infty$. This concludes the proof of Stirling's asymptotic formula for the Γ-function.

6. Prove Euler's relation

$$\zeta(s) = \prod_p \left(1 - \frac{1}{p^s}\right)^{-1} \qquad \text{for } \operatorname{Re} s > 1.$$

Why does it imply that the infinite product $\prod_p \left(1 - \dfrac{1}{p}\right)$ is divergent? Why does this imply that $\sum_p \dfrac{1}{p} = \infty$?

7. Let $\xi(s) := s(s - 1)\pi^{-1/2s}\zeta(s)\Gamma\left(\dfrac{s}{2}\right)$. Show that ξ is an entire function and satisfies the equation $\xi(s) = \xi(1 - s)$.

*8. Use Exercise 5.3.6 to prove that for $\operatorname{Re} s > 1$ we have

$$\frac{\zeta'(s)}{\zeta(s)} = -\sum_{n=2}^\infty \frac{\Lambda(n)}{n^s},$$

where $\Lambda(n) = \log p$ if n is a power of a prime p, $\Lambda(n) = 0$ otherwise.

9. Show that for $\operatorname{Re} s > 1$,

$$\sum_{n=1}^\infty \frac{(-1)^{n+1}}{n^s} = (1 - 2^{1-s})\zeta(s).$$

Conclude that the Dirichlet series $\displaystyle\sum_{n=1}^\infty \frac{(-1)^{n+1}}{n^s}$ which has $\sigma_a = 1$, $\sigma_c = 0$ (from Exercise 5.2.6) is the restriction of an entire function to the half-plane $\operatorname{Re} s > 0$.

10. (a) Show that if χ is a C^1 function in the interval $[\alpha, \beta]$, $0 \le \alpha < \beta < \infty$, then

$$\sum_{\alpha < n \le \beta} \chi(n) = \int_\alpha^\beta \chi(x)\,dx + \int_\alpha^\beta \left(x - [x] - \frac{1}{2}\right)\chi'(x)\,dx$$
$$+ \left(\alpha - [\alpha] - \frac{1}{2}\right)\chi(\alpha) - \left(\beta - [\beta] - \frac{1}{2}\right)\chi(\beta).$$

(Compare with Exercise 5.3.5.)

(b) Let α, β be integers, $\operatorname{Re} s > 1$ and $\chi(x) = x^{-s}$. Use part (a) to show that

$$\zeta(s) = \frac{1}{2} + \frac{1}{s-1} + s\int_1^\infty \frac{[x] - x + 1/2}{x^{s+1}}\,dx.$$

(c) Show that

$$\int_1^n \frac{[x] - x}{x^2}\,dx = \sum_{k=2}^n \frac{1}{k} - \log n.$$

(d) Use parts (b) and (c) to conclude that

$$\lim_{s\to 1}\left(\zeta(s) - \frac{1}{s-1}\right) = \gamma.$$

(e) Use the functional equation to obtain that $\zeta(0) = -\frac{1}{2}$.

(f) Show that the functional equation implies that

$$\frac{\zeta'(1-s)}{\zeta(1-s)} = \log 2\pi + \frac{\pi}{2}\tan\frac{s\pi}{2} - \frac{\Gamma'(s)}{\Gamma(s)} - \frac{\zeta'(s)}{\zeta(s)}.$$

(g) Use parts (d) and (e) to conclude that

$$\frac{\zeta'(0)}{\zeta(0)} = \log 2\pi.$$

11. Show that the function $f(z) = \int_0^1 t^{z-1}e^t\,dt$ is holomorphic for $\operatorname{Re} z > 0$ and admits an analytic continuation to a function meromorphic in the whole complex plane.

12. Same as Exercise 5.3.11, for $f(z) = \int_1^\infty \frac{t^{z-1}}{1+t^2}\,dt$.

13. Let $\sigma > 1$, use Exercise 5.2.9 to compute

$$\lim_{T\to\infty} \frac{1}{2T}\int_{-T}^T |\zeta(\sigma + it)|^2\,dt.$$

14. Show that for $x > 0$

$$\frac{1}{2\pi i}\int_W \exp[(e^t - 1 - t)x]\,dt = \frac{e^{x\log x - x}}{\Gamma(x+1)},$$

where W is a path in the half-strip $\{\operatorname{Re} t > 0, |\operatorname{Im} t| < \pi\}$, asymptotic to the rays $\operatorname{Im} t = \pm\pi$, starts at $\infty - i\pi$ and ends at $\infty + i\pi$. (Hint: Consider the transformation $t = \operatorname{Log}(-z)$, then W is the image of a Hankel contour $(\infty, 0+)$ used in the text. See next page.)

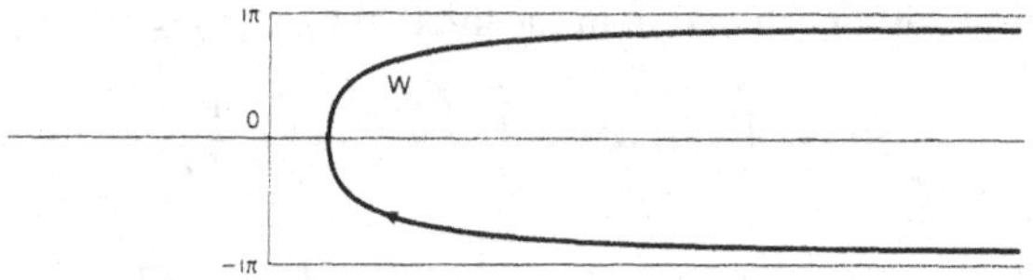

§4. Covering Spaces

5.4.1. Definitions

(1) If X and Y are two topological spaces and $p : X \to Y$ is a continuous map we say $p^{-1}(y)$ is *the* ***fiber*** *of p over y.*

(2) If $p : X_1 \to Y$, $q : X_2 \to Y$ are continuous maps, we say that a continuous map $f : X_1 \to X_2$ ***preserves the fibers*** if $p = q \circ f$.

(3) If X, Y, Z are topological spaces, and $p : Y \to X$ and $f : Z \to X$ are continuous maps, a ***lifting*** of f is any continuous map $g : Z \to Y$ such that $p \circ g = f$.

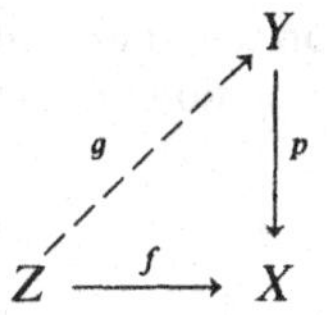

5.4.2. Proposition. *Let X and Y be Hausdorff topological spaces and $p : Y \to X$ a local homeomorphism. Let Z be a connected topological space and $f : Z \to X$ a continuous map. If g_1, g_2 are two liftings of f such that $g_1(z_0) = g_2(z_0)$ for some $z_0 \in Z$, then $g_1 = g_2$.*

PROOF. Let $A = \{z \in Z : g_1(z) = g_2(z)\}$. This set is a nonempty closed set by the hypotheses. We will show it is open. Let $z \in A$ and $y = g_1(z) = g_2(z)$. Since p is a local homeomorphism, there is an open neighborhood U of y which is homeomorphic by p to an open neighborhood V of $p(y) = f(z)$. By the continuity of g_1 and g_2, there is a neighborhood W of z such that $g_1(W) \cup g_2(W) \subseteq V$. Let $\varphi = (p|V)^{-1} : V \to U$. From $p \circ g_j = f$ we deduce $g_j|W = \varphi \circ (f|W)$, $j = 1, 2$. Hence $W \subseteq A$ and the proposition holds. □

The concepts of loop, path, homotopy, and fundamental group that we have introduced in Chapter 1 for open subsets of $\mathbb{C}$ have an immediate extension to the case of topological spaces. We have also proved properties of the liftings of homotopies with respect to the map $\mathbb{C} \to \mathbb{C}^*$, $z \mapsto e^z$. We are now going to extend the validity of these properties.

5.4.3. Proposition (Lifting of Homotopic Curves. Abstract Form of the Monodromy Theorem). *Let X and Y be two Hausdorff spaces $p : Y \to X$ a local* *homeomorphism. Let $a, b \in X$ and $a_1 \in p^{-1}(a)$. Let $H: [0,1] \times [0,1] \to X$ be a continuous map such that $H(0,s) = a$ and $H(1,s) = b$ for every $s \in [0,1]$. Let $\gamma_s(t) = H(t,s)$ $(0 \le t \le 1)$. If every path γ_s can be lifted to a path C_s starting at a_1, then C_0 and C_1 end at the same point and they are homotopic with fixed endpoints.*

Proof. We redo here (for the convenience of the reader) the proof already given in the case of $\mathbb{C}$ (cf. Proposition 1.6.28).

Define $K: [0,1] \times [0,1] \to Y$ by $K(s,t) = C_s(t)$.

5.4.4. Lemma. *There is an $\varepsilon_0 > 0$ such that K is continuous in $[0, \varepsilon_0] \times [0,1]$.*

Proof. There is a neighborhood V of a_1 and a neighborhood U of a such that $p|V: V \to U$ is a homeomorphism. Since $H(\{0\} \times [0,1]) = \{a\}$, there is $\varepsilon_0 > 0$ such that $H([0,\varepsilon_0] \times [0,1]) \subseteq U$. The uniqueness of the lifting of curves that follows from Proposition 5.4.2 implies that $C_s|[0,\varepsilon_0] = \varphi \circ (\gamma_s|[0,\varepsilon_0])$ for $0 \le s \le 1$, where $\varphi = (p|V)^{-1}$. Therefore

$$K|[0,\varepsilon_0] \times [0,1] = \varphi \circ (H|[0,\varepsilon_0] \times [0,1]),$$

which shows that K is continuous in $[0,\varepsilon_0] \times [0,1]$. □

5.4.5. Lemma. *The map K is continuous in $[0,1] \times [0,1]$.*

Proof. Let us assume that $(t_0, \sigma) \in [0,1] \times [0,1]$ is a point of discontinuity for K. Let $\tau = \inf\{t \in [0,1] : K \text{ is not continuous at } (t,\sigma)\}$. Clearly $\varepsilon_0 \le \tau \le t_0$. Let $x = H(\tau,\sigma)$, $y = K(\tau,\sigma) = C_\sigma(\tau)$. Let V, U be neighborhoods of y and x, respectively, such that $p|V: V \to U$ is a homeomorphism, and set $\varphi = (p|V)^{-1}$. Since H is continuous there is an $\varepsilon > 0$ such that $H(I_\varepsilon(\tau) \times I_\varepsilon(\sigma)) \subseteq U$, where $I_\varepsilon(\xi) = \{t \in [0,1] : |t - \xi| < \varepsilon\}$. In particular, $\gamma_\sigma(I_\varepsilon(\tau)) \subseteq U$. Hence

$$C_\sigma|I_\varepsilon(\tau) = \varphi \circ (\gamma_\sigma|I_\varepsilon(\tau)).$$

Let us choose $t_1 \in I_\varepsilon(\tau)$ such that $t_1 < \tau$. Then

$$K(t_1,\sigma) = C_\sigma(t_1) \in V.$$

Since K is continuous at the point (t_1,σ) there is a δ, $0 < \delta \le \varepsilon$ such that

$$K(t_1,s) = C_s(t_1) \in V \qquad \text{for } s \in I_\delta(\sigma).$$

The uniqueness of the lifting shows that, for $s \in I_\delta(\sigma)$, one has

$$C_s|I_\varepsilon(\tau) = \varphi \circ (\gamma_s|I_\varepsilon(\tau)).$$

Hence $K = \varphi \circ H$ in $I_\varepsilon(\tau) \times I_\varepsilon(\sigma)$, which contradicts the definition of τ. Therefore K is continuous in $[0,1] \times [0,1]$. □

To finish the proof of Proposition 5.4.3 we observe that $H(\{1\} \times [0,1]) = \{b\}$ and $H = p \circ K$ implies that the connected compact set $K(\{1\} \times [0,1])$ is contained in the discrete set $p^{-1}(b)$. (Why is $p^{-1}(b)$ discrete?) This clearly implies that $K(\{1\} \times [0,1])$ is a single point. That is, the endpoint $C_s(1)$ of all the paths coincide. This proves both statements in Proposition 5.4.3. □

5.4.6. Definitions

(1) Let X, Y be two topological spaces. A map $p: Y \to X$ is said to be a ***covering map*** if for every $x \in X$ there is an open neighborhood U of x such that $p^{-1}(U) = \dot{\bigcup}_{i \in I} V_i$, where the V_i are open sets in Y, pairwise disjoint, $p|V_i: V_i \to U$ are homeomorphisms, and I is an index set depending on x.
Such an open set U is said to be ***trivializing*** (*for* p). If we consider I with the discrete topology, one sees that $p^{-1}(U)$ is homeomorphic to $U \times I$ via the trivializing map φ, $\varphi: p^{-1}(U) \to U \times I$, defined by $\varphi(y) = (p(y), i(y))$, where $i(y)$ is the unique $i \in I$ such that $y \in V_i$.

(2) Let $p_1: \tilde{X}_1 \to X$ and $p_2: \tilde{X}_2 \to X$ be two covering maps of the same topological space X. A ***morphism*** between these covering maps is any continuous maps $h: \tilde{X}_1 \to \tilde{X}_2$ such that $p_2 \circ h = p_1$. (More precisely, one should say a morphism from $(\tilde{X}_1, p_1)$ to $(\tilde{X}_2, p_2)$.)

(3) A morphism is said to be an ***isomorphism*** if the map h is also a homeomorphism (it follows that h^{-1} is also a morphism).

(4) When $\tilde{X}_1 = \tilde{X}_2 = \tilde{X}$ and $p_1 = p_2 = p$, the isomorphisms are called ***automorphisms*** of the covering map $p: \tilde{X} \to X$. They form a group $G(p)$ under composition.
For instance, if $\tilde{X} = X \times F$, F discrete and X connected, $p := pr_1$, then every element of $G(p)$ is of the form $\xi_\sigma: (x, y) \mapsto (x, \sigma(y))$, with $\sigma \in \mathfrak{S}_F$, the group of permutations of F. Hence $G(p)$ can be identified to $\mathfrak{S}_F$.

(5) If $p: Y \to X$ is a covering map, X is locally connected, and Y' is an open connected component of Y, then $p': Y' \to X$, $p' := p|Y'$, is also a covering map.

Examples of covering maps

(1) $p: \mathbb{C}^* \to \mathbb{C}^*$, $p(z) = z^k$, $k \in \mathbb{N}^*$;
(2) $p: \mathbb{C} \to \mathbb{C}^*$, $p(z) = \exp z$.

One can prove for covering maps a uniqueness property similar to that of Proposition 5.4.2 without the Hausdorff hypothesis.

5.4.7. Proposition. *Let $p: Y \to X$ be a covering map, Z a connected topological space, $f: Z \to X$ continuous. If g_1 and g_2 are two liftings of f that coincide at a point $z_0 \in Z$, then $g_1 = g_2$.*

PROOF. Let $E = \{z \in Z : g_1(z) = g_2(z)\}$. It is not empty by hypothesis. We need to show that E is both open and closed. (Why is E not automatically closed?)

Let us show first that E is open. Let $z \in E$ and U a trivializing neighborhood of $f(z)$. Denote $p^{-1}(U) = \dot{\bigcup}_{i \in I} V_i$, $\sigma_i = (p|V_i)^{-1}$. We have $g_1(z) = g_2(z) \in V_{i_0}$ for some $i_0 \in I$. By continuity, there is an open neighborhood W of z such that $g_i(W) \cup g_2(W) \subseteq V_{i_0}$ and $f(W) \subseteq U$. It follows that $g_1|W = \sigma_{i_0} \circ (f|W) = g_2|W$. Therefore E is open.

We shall now show that E is closed. Let $z \in \bar{E}$, U a trivializing open neighborhood of $f(z)$, $p^{-1}(U) = \dot{\bigcup} V_i$, $\sigma_i = (p|V_i)^{-1}$ as before. If $g_1(z) \in V_{i_1}$, $g_2(z) \in V_{i_2}$, then we can find a neighborhood W of z such that $g_1(W) \subseteq V_{i_1}$, $g_2(W) \subseteq V_{i_2}$, and $f(W) \subseteq U$. Hence $g_1|W = \sigma_{i_1} \circ (f|W)$ and $g_2|W = \sigma_{i_2} \circ (f|W)$. But $E \cap W$ is not empty and for $\zeta \in E \cap W$ we have $g_1(\zeta) = g_2(\zeta)$, which is only possible if $i_1 = i_2$ (otherwise $V_{i_1} \cap V_{i_2} \neq \varnothing$). Therefore we conclude that $g_1|W = g_2|W$. In particular, $z \in E$. □

5.4.8. Definition. A continuous map $p: Y \to X$ is said to have the ***lifting property for paths*** if for every $x_0 \in X$, every $y_0 \in p^{-1}(x_0)$, and every continuous path $\gamma: [0, 1] \to X$ starting at x_0, there is a lifting C of γ, $C: [0, 1] \to \gamma$ continuous, and $C(0) = y_0$.

5.4.9. Proposition. *Every covering map $p: Y \to X$ has the lifting property for paths.*

PROOF. Let $\gamma: [0, 1]$ be a continuous path in X starting at x_0, $y_0 \in p^{-1}(x_0)$. By the compactness of $[0, 1]$, there is a partition $0 = t_0 < t_1 < \cdots < t_n = 1$ of $[0, 1]$ and trivializing open sets $U_k \subseteq X$ $(1 \le k \le n)$ such that $\gamma([t_k, t_{k+1}]) \subseteq U_k$ and $p^{-1}(U_k) = \dot{\bigcup}_{i \in I_k} V_{k,i}$.

We show by induction on k, $k = 0, 1, \ldots, n$, that there is lifting $C: [0, t_k] \to Y$, continuous, $C(0) = y_0$, $p \circ C = \gamma|[0, t_k]$. For $k = 0$, it is evident. Let us assume $k \ge 1$ and that $C|[0, t_{k-1}]$ has already been found. Denote $y_{k-1} = C(t_{k-1})$. Since $p(y_{k-1}) = p(C(t_{k-1})) = \gamma(t_{k-1}) \in U_k$, there exists a unique $i \in I_k$ such that $y_{k-1} \in V_{k,i}$. Let $\varphi = (p|V_{k,i})^{-1}: U_k \to V_{k,i}$. If we let

$$C|[t_{k-1}, t_k] = \varphi \circ (\gamma|[t_{k-1}, t_k])$$

we get a lifting C of γ, continuous in $[0, t_k]$. □

Remarks. Let $p: Y \to X$ be a covering map.

(1) The lifting of a closed curve is not necessarily a closed curve. For instance if $p: \mathbb{C} \to \mathbb{C}^*$, $p(z) = e^z$ and γ is the closed curve given by $\gamma(t) = e^{2\pi i t}$, then a lifting starting at $k \in \mathbb{Z}$ ends at $k + 1$.

(2) The liftings of two continuous paths, which are homotopic with fixed endpoints in X, are homotopic with fixed endpoints in Y, if they have the same starting point. In particular, if γ is a loop with base point a, which is homotopic in X to the constant loop ε_a, any lifting of γ to Y starting at $a_1 \in p^{-1}(a)$ will be a loop homotopic to the constant loop ε_{a_1}.

5.4.10. Proposition. *Let X, Y be two topological spaces, X connected, and $p: Y \to X$ a covering map. For every $x_0, x_1 \in X$, the sets $p^{-1}(x_0)$ and $p^{-1}(x_1)$ have the same cardinality.*

PROOF. Let $E = \{x \in X : p^{-1}(x)$ have the same cardinal as $p^{-1}(x_0)\}$. We show that E is open and closed. If $x \in E$ and U is a trivializing neighborhood of x, $U = \dot{\bigcup}_{i \in I} V_i$, then it is clear that the cardinal $\# p^{-1}(x') = \# I$ for every $x' \in U$. Hence $U \subseteq E$, and E is open. The same reasoning shows that if $x \in \bar{E}$ one must have $x \in E$. Simply use that $U \cap E \neq \varnothing$. □

Remark. The cardinal $\# p^{-1}(x)$ is called indistinctly ***multiplicity** of p* or the ***number of sheets of p***. In general, there is no canonical way of enumerating the sheets.

5.4.11. Definition. A topological space X is said to be ***simply connected*** if it is arcwise connected and if, for every $a \in X$, every loop of base a is homotopic to ε_a.

For instance, every open set in $\mathbb{C}$ biholomorphic to $B(0, 1)$ is simply connected.

5.4.12. Proposition. *Let X, Y be topological spaces, $p: Y \to X$ a covering map. Let Z be a simply connected locally arcwise connected space and $f: Z \to X$ a continuous map. For any choice $z_0 \in Z$, $y_0 \in Y$ such that $f(z_0) = p(y_0)$ there is a unique lifting g of f, $g: Z \to Y$ such that $g(z_0) = y_0$.*

PROOF. Let $z \in Z$ and $\gamma: [0, 1] \to Z$ be a continuous path joining z_0 to z. Hence $\alpha = f \circ \gamma$ is a curve in X of origin $f(z_0)$ and endpoint $f(z)$. Let $C: [0, 1] \to Y$ be the unique lifting of α starting at y_0. We define $g(z) := C(1)$. This definition does not depend on the choice of γ since, if γ_1 is another curve joining z_0 to z, then γ_1 is homotopic to γ (because $\gamma_1 \cdot \bar{\gamma}$ is homotopic to ε_{z_0}). Hence, $\alpha_1 = f \circ \gamma_1$ is homotopic to α by the homotopy $f \circ H$, where H is the homotopy between γ and γ_1. Therefore the liftings C_1 of α_1 such that $C_1(0) = C(0) = y_0$ will verify $C_1(1) = C(1)$.

It is clear that $p \circ g = f$.

We need to show that g is a continuous map. Let $z \in Z$, $y = g(z)$, and V a neighborhood of y such that $p|V: V \to U := p(V)$ is a homeomorphism onto the open neighborhood U of $p(y) = f(z)$. Since f is continuous and Z is locally arcwise connected, there is an arcwise connected neighborhood W of z such that $f(W) \subseteq U$. It is clear that if we show tht $g(W) \subseteq V$ then the continuity of g at z follows. Let γ, α be defined as earlier; for $z' \in W$ let γ' be a path joining z to z' in W. The curve $\alpha' = f \circ \gamma'$ is contained in U and $c' = (p|V)^{-1} \circ \alpha'$ is a lifting of α' with origin y. Therefore, the composite path $C \cdot C'$ lifts $\alpha \cdot \alpha' = f \circ (\gamma \cdot \gamma^{-1})$ with origin z_0, and $g(z') = C \cdot C'(1) = C'(1) \in V$. □

A familiar example of application of this proposition is the case $p(z) = e^z$ between $Y = \mathbb{C}$ and $X = \mathbb{C}^*$, Z simply connected open subset of $\mathbb{C}$, and $f: Z \to \mathbb{C}^*$ a continuous function. Its lifting is $\log f$, the branch of the logarithm being determined by the choice of $z_0 \in Z$ and $y_0 \in \mathbb{C}$ such that $f(z_0) = e^{y_0}$.

Remark. The previous proof remains valid if we assume only that p is a local homeomorphism which has the lifting property for paths.

We introduce now a class of topological spaces X for which the local homeomorphisms $p: X \to Y$ that have the lifting property for paths are necessarily covering maps.

5.4.13. Definition (Topological Manifold of Dimension 2). A ***topological manifold*** of dimension 2 is a Hausdorff space X such that for every $a \in X$ there is an open neighborhood U of a and a homeomorphism φ of U onto an open subset of $\mathbb{R}^2$.

Such a pair (U, φ) is called a ***coordinate patch*** (or ***local chart***) of X.

Remark. One can replace 2 by $n \geq 1$ without any difficulty in Definition 5.4.13 and talk about a topological manifold of dimension n. The following statements will still hold.

5.4.14. Proposition. *Let X be a topological manifold, Y a Hausdorff space, and $p: Y \to X$ a local homeomorphism having the lifting property for paths. Then p is a covering map.*

PROOF. Let $x_0 \in X$ and $\{y_i\}_{i \in I}$ be the distinct elements of $p^{-1}(x_0)$. Let (U, φ) be a chart in X with $x_0 \in U$ and U homeomorphic to a ball. Therefore U is simply connected and locally arcwise connected. Let $j: U \to X$ be the canonical injection. For every $i \in I$ there is a lifting $f_i: U \to Y$ of j such that $f_i(x_0) = y_i$. If $V_i = f_i(U)$ then we leave to the reader the verification that V_i is open, we have $p^{-1}(U) = \dot{\bigcup_{i \in I}} V_i$, and $p|V_i: V_i \to U$ is a homeomorphism. This concludes the proof. □

The following is a simple condition implying that a local homeomorphism is indeed a covering map.

5.4.15. Lemma. *Let X, Y be two locally compact spaces and $p: Y \to X$ a continuous proper map with discrete fibers. Then*
(1) *For every $x \in X$, $p^{-1}(x)$ is a finite set.*
(2) *If $x \in X$ and V is an open neighborhood of $p^{-1}(x)$, there exists an open neighborhood U of x such that $p^{-1}(U) \subseteq V$.*

PROOF. Let us recall that a map p is proper if p^{-1}(compact set) is a compact set. It is immediate now that (1) holds. To prove (2), observe that $Y \setminus V$ is closed and hence $p(Y \setminus V)$ is closed because p is a proper map. The set $U = X \setminus p(Y \setminus V)$ is an open neighborhood of x such that $p^{-1}(U) \subseteq V$. □

5.4.16. Proposition. *Let X, Y be two locally compact spaces and $p: Y \to X$ a continuous proper map which is also a local homeomorphism. Then p is a covering map.*

PROOF. Let $x \in X$, $p^{-1}(x) = \{y_1, \ldots, y_n\}$. For every j, $1 \le j \le n$, there is an open neighborhood W_j of y_j and an open neighborhood U_j of x such that $p|W_j : W_j \to U_j$ is a local homeomorphism. We can also assume that the W_j are pairwise disjoint. By Lemma 5.4.15 there is an open neighborhood U of x such that $U \subseteq U_1 \cap \cdots \cap U_n$ and $p^{-1}(U) \subseteq W_1 \cup \cdots \cup W_n$. Let $V_j = W_j \cap p^{-1}(U)$. Then $p^{-1}(U) = \dot{\bigcup}_{1 \le j \le n} V_j$ and $p|V_j : V_j \to U$ is a homeomorphism. Therefore p is a covering map. □

5.4.17. Example. The map $z \mapsto \tan z$ from $\mathbb{C}$ into $S^2\backslash\{\pm i\}$ is a meromorphic map. Let us show it is a covering map. We have

$$\tan z = \frac{1}{i}\frac{1 - w^2}{1 + w^2}, \qquad w = e^{-iz}.$$

Since $z \mapsto e^{iz}$ is a covering map of $\mathbb{C}$ onto $\mathbb{C}^*$, it is enough to show that $w \mapsto \frac{1}{i}\frac{1 - w^2}{1 + w^2}$ from $\mathbb{C}^*$ into $S^2\backslash\{\pm i\}$ is a covering map. For any $\zeta \in S^2\backslash\{\pm i\}$ the equation

$$\zeta = \frac{1}{i}\frac{1 - w^2}{1 + w^2}$$

has two roots: if $\zeta = \infty$ they are $\pm i$, if $\zeta \ne \infty$, they are the two solutions of

$$w^2 = \frac{-\zeta + i}{\zeta - i} \in \mathbb{C}^*.$$

It is evident that $p : w \mapsto \frac{1}{i}\frac{1 - w^2}{1 + w^2}$ is a local homeomorphism as one sees by computing the derivative $p'\left(\text{or}\left(\frac{1}{p}\right)' \text{ near } w = \pm i\right)$. Finally, if K is a compact subset of $S^2\backslash\{\pm i\}$, there is an $\varepsilon > 0$ such that $|\zeta \pm i| \ge \varepsilon$ for any $\zeta \in K$. This immediately implies that there are $0 < a < b$ such that $a \le |w| \le b$ if $w^2 = \frac{-\zeta + i}{\zeta - i}$. Hence p is a proper map and, by Proposition 5.4.16, a covering map.

EXERCISES 5.4

1. Is the map $\mathbb{C}\backslash\{0\} \to \mathbb{C}\backslash\{0, 1\}$, given by $z \mapsto e^z$, a covering map?
2. Is the map $z \mapsto e^z$, from $\mathbb{C}\backslash 2\pi i\mathbb{Z}$ into $\mathbb{C}\backslash\{0, 1\}$, a covering map?
3. Let $pr_1 : \mathbb{C}^2 \to \mathbb{C}$, $\mathrm{pr}_1(z, w) = z$, and $p = pr_1|X$, $X := \{(z, w) \in \mathbb{C}^2 : w^2 - z = 0\}$. Is $p : X \to \mathbb{C}$ a covering map?
4. Consider $X_1 = \{(z, w) \in \mathbb{C}^* \times \mathbb{C} : w^2 - z = 0\}$, $pr_1 : \mathbb{C}^* \times \mathbb{C} \to \mathbb{C}^*$ the projection onto the first coordinate, and $p_1 = pr_1|X_1 : X_1 \to \mathbb{C}^*$. Is p_1 a covering map?
5. Let $X = \{(z, w) \in \mathbb{C}^2 : zw - \sin(zw) = 0\}$, $p = pr_1|X$ (same notation as earlier). Is $X \to \mathbb{C}$ a covering map?

6. Let $P \in \mathbb{C}[z]$ be a polynomial with complex coefficients of degree n. Its critical points are those $z \in \mathbb{C}$ such that $P'(z) = 0$, its critical values are the values that P takes at the critical points. Let V be the set of critical values of P, i.e., $V = P(\{z \in \mathbb{C} : P'(z) = 0\})$. Let
$$X := \mathbb{C} \setminus P^{-1}(V), \qquad Y := \mathbb{C} \setminus V.$$
(a) Show that $P : X \to Y$ is a covering map of multiplicity n.
(b) Let $\sigma : X \to X$ be a continuous map such that $P \circ \sigma = P$. Show that σ is holomorphic in X and σ is a proper map from X into X. Show now that σ is the restriction to X of an affine map of the form $z \mapsto \lambda z + \mu$, $\lambda, \mu \in \mathbb{C}$, and $\lambda^q = 1$ for some integer q.

7. Let $p : \mathbb{C} \setminus \{1\} \to \mathbb{C} \setminus \{2\}$ given by $p(z) = z^2 - 2z + 3$. Show that p is a covering map. Which is its multiplicity? Show there is only one nontrivial homeomorphism $\sigma : \mathbb{C} \setminus \{1\} \to \mathbb{C} \setminus \{1\}$ such that $p \circ \sigma = p$. Show it has the form $z \mapsto az + b$. Find a and b.

8. Let $p : X \to Z$ be a covering map of multiplicity 2. For every $x \in X$, denote by $f(x)$ the other element of the fiber over $p(x)$. Show that $f : X \to X$ is a homeomorphism.

9. In the vector space $\mathbb{R}^4$, let us denote the canonical basis by $\{1, i, j, k\}$. The space of quaternions $\mathbb{H}$ is the vector space $\mathbb{R}^4$ as an $\mathbb{R}$-algebra, whose multiplication table satisfies
$$i^2 = j^2 = k^2 = -1, \quad i \cdot j = -j \cdot i = k, \quad j \cdot k = -k \cdot j = i, \quad k \cdot i = -i \cdot k = j$$
and 1 is the identity.
(a) Show that $\mathbb{H}$ is a noncommutative field.
(b) Let $q = x + yi + zj + tk \in \mathbb{H}$, $N(q) = (x^2 + y^2 + z^2 + t^2)^{1/2}$ be the Euclidean norm. Show $N(q_1 \cdot q_2) = N(q_1)N(q_2)$.
(c) Show that $S^3 = \{q \in \mathbb{H} : N(q) = 1\}$ is a compact group with the induced multiplication.
(d) Show that for $q \in S^3$, the map $x \in \mathbb{H} \mapsto qxq^{-1} \in \mathbb{H}$ is a linear map of $\mathbb{R}^4$ given by a matrix in $SO(3)$.
(e) Show that the map $m : S^3 \to SO(3)$ defined by $m(q)(x) = qxq^{-1}$, is a two-sheeted covering map.

§5. Riemann Surfaces

5.5.1. Definitions. Let X be a topological manifold of dimension 2.

(1) Two charts (U_1, φ_1) and (U_2, φ_2) are said to be ***holomorphically compatible*** if either $U_1 \cap U_2 = \varnothing$ or if not, the map
$$\varphi_2 \circ \varphi_1^{-1} : \varphi_1(U_1 \cap U_2) \to \varphi_2(U_1 \cap U_2)$$
is a biholomorphic map between the two open sets in $\mathbb{C}$.

(2) An ***atlas*** is a family $\mathfrak{U} = \{(U_j, \varphi_j)\}_{j \in J}$ of charts, pairwise holomorphically compatible, such that $\bigcup_{j \in J} U_j = X$.

(3) Two atlases are equivalent if every chart of one is compatible with every chart of the other, i.e., if their reunion is also an atlas.

Remarks

(1) If $\varphi : U \to V$ is a chart and U_1 is an open subset of U, then $\varphi | U_1 : U_1 \to V_1 = \varphi(U_1)$ is a new chart compatible with (U, φ).
(2) The notion of equivalence of atlases is in fact an equivalence relation.

5.5.2. Definition. A structure of ***Riemann surface*** on a topological variety of dimension 2 is a choice of an equivalence class of atlases.

5.5.3. Definition. Let X be a Riemann surface and $Y \subseteq X$ an open subset. A function $f: Y \to \mathbb{C}$ is said to be of class C^k ($k \in \mathbb{N} \cup \{\infty\}$) (respectively holomorphic) if for every chart (U, φ), the function defined in the open subset $\varphi(U \cap Y)$ of $\mathbb{C}$ given by

$$f \circ \varphi^{-1} : \varphi(U \cap Y) \to \mathbb{C}$$

is C^k (respectively holomorphic) in the usual sense.

The family of functions of class C^k in Y is denoted $\mathscr{E}_k(Y)$ (if $k = \infty$ we suppress the index) and the family of holomorphic functions is denoted $\mathscr{H}(Y)$.

5.5.4. Remarks

(1) The families $\mathscr{E}_k(Y)$ and $\mathscr{H}(Y)$ are $\mathbb{C}$-algebras with the usual laws of addition and product.
(2) Every chart (U, φ) determines a function $\varphi \in \mathscr{H}(U)$.
(3) If $a \in Y$ and $f \in \mathscr{H}(Y \setminus \{a\})$ is bounded in a neighborhood of a, then f has a unique extension to a function $\tilde{f} \in \mathscr{H}(Y)$.
(4) Y has a natural structure of Riemann surface by taking charts $(U \cap Y, \varphi | (U \cap Y))$ whenever (U, φ) is a chart in X.

5.5.5. Definitions

(1) Let X_1, X_2 be two Riemann surfaces. A continuous function $f: X_1 \to X_2$ is said to be ***holomorphic*** if, for every pair of charts (U_1, φ_1) of X_1 and (U_2, φ_2) of X_2, respectively, such that $f(U_1) \subseteq U_2$, the function

$$\varphi_2 \circ f \circ \varphi_1^{-1} : \varphi_1(U_1) \to \varphi_2(U_2)$$

is holomorphic.
(2) A map $f: X_1 \to X_2$ is called ***biholomorphic***, or a ***conformal map***, if it is a homeomorphism such that both f and f^{-1} are holomorphic. We also say that X_1 and X_2 are ***biholomorphic***, or ***conformally equivalent***.

5.5.6. Remarks

(1) If $f: X \to Y$ and $g: Y \to Z$ are holomorphic maps, then $g \circ f$ is also a holomorphic map.
(2) If $f: X \to Y$ is a holomorphic map and U an open subset of Y, then the map $f^*: \mathscr{H}(U) \to \mathscr{H}(f^{-1}(U))$, $g \mapsto g \circ f$, is an algebra homomorphism.

5.5.7. Examples. (1) $\mathbb{C}$, or any nonempty open subset $X \subseteq \mathbb{C}$, are Riemann surfaces. The atlas is (X, id).

(2) The Riemann sphere S^2: $S^2 = \mathbb{C} \cup \{\infty\}$ is a Riemann surface for the following atlas: $U_1 = \mathbb{C}$, $U_2 = \mathbb{C}^* \cup \{\infty\}$, $\varphi_1 = id$, $\varphi_2(z) = z^{-1}$ for $z \in \mathbb{C}^*$ and $\varphi_2(\infty) = 0$. The map $\varphi_2 \circ \varphi_1^{-1} : \mathbb{C}^* \to \mathbb{C}^*$ is $z \mapsto 1/z$.

(3) Complex torus: Let w_1, w_2 be two complex numbers, linearly independent over $\mathbb{R}$. Let Γ be the lattice $\Gamma = \mathbb{Z}w_1 \oplus \mathbb{Z}w_2$. Consider $\mathbb{C}/\Gamma$ with the quotient topology. It is a compact Hausdorff space. The canonical map $\mathbb{C} \to \mathbb{C}/\Gamma$ is open since the equivalence relation "modulo Γ" is open. Namely, if V is open in $\mathbb{C}$, then

$$\pi^{-1}(\pi(V)) = \bigcup_{w \in \Gamma} (w + V),$$

which is an open set.

We can define in $\mathbb{C}/\Gamma$ a structure of Riemann surface as follows: Let V be an open subset of $\mathbb{C}$ which intersects each Γ-coset in at most one point. Then $U = \pi(V)$ is an open set and $\pi|V : V \to U$ is a homeomorphism. The pair $(U, (\pi|V)^{-1})$ is a chart. Any two such charts are holomorphically compatible since $\psi = \varphi_2 \circ \varphi_1^{-1}$ has the property that $\psi(z) - z \in \Gamma$, hence $z \mapsto \psi(z) - z$ is constant in each component of $\varphi_1(U_1 \cap U_2)$ and, therefore, ψ is holomorphic. Let us also point out that $\pi : \mathbb{C} \to \mathbb{C}/\Gamma$ is a covering map.

(4) The Riemann surface of the square root: Let $U_1 = \{z \in \mathbb{C} : 0 < \arg z < 2\pi\}$, $U_2 = \{z \in \mathbb{C} : -\pi < \arg z < \pi\}$. In U_1 there is a determination of the "function square root," namely,

$$f_1(z) = \exp(\tfrac{1}{2}(\log|z| + i \arg z)) \qquad (0 < \arg z < 2\pi)$$

and in U_2 there is another determination f_2

$$f_2(z) = \exp(\tfrac{1}{2}(\log|z| + i \arg z)) \qquad (-\pi < \arg z < \pi).$$

Evidently the sign chosen for f_1 and f_2 was a bit arbitrary. We are going to construct a two-sheeted covering space $p : \tilde{X} \to \mathbb{C}^*$ which will be a Riemann surface and, on this surface, we will have a holomorphic function $\tilde{f}$ which takes at the two points of $p^{-1}(z)$, $z \in \mathbb{C}^*$, the two possible values of $\sqrt{z}$.

Let Y be the disjoint union

$$Y = (\{1\} \times U_1 \times \{-1, 1\}) \dot{\cup} (\{2\} \times U_2 \times \{-1, 1\})$$

considered as a subset of $\{1, 2\} \times \mathbb{C}^* \times \{-1, 1\}$. Let $\mathfrak{G} = \{id, \sigma\}$ be the group of permutations of $\{-1, 1\}$ ($\sigma(t) = -t$ for $t \in \{-1, 1\}$). Consider now the maps $g_{11} : U_1 \to \mathfrak{G}$, $g_{12}, g_{21} : U_1 \cap U_2 \to \mathfrak{G}$, $g_{22} : U_2 \to \mathfrak{G}$, defined by $g_{11}(z) = g_{22}(z) = id$, $g_{12}(z) = g_{21}(z) = id$ if $\operatorname{Im} z > 0$ and $g_{12}(z) = g_{21}(z) = \sigma$ if $\operatorname{Im} z < 0$.

Introduce the equivalence relation ρ in Y given by $(i, z, \varepsilon)\rho(j, \zeta, \varepsilon')$ if and only if $z = \zeta$, $\varepsilon' = g_{i,j}(z)(\varepsilon)$. The reader will be able to verify easily:

(i) The relation ρ is open.

(ii) If $\tilde{X} = Y/\rho$ denotes the quotient topological space and $q : Y \to \tilde{X}$ is the canonical projection, then q is an open map.

(iii) The map $(i, z, \varepsilon) \mapsto z$ passes to the quotient and induces a map $p: \tilde{X} \to \mathbb{C}^*$ surjective, open, which is a homeomorphism of each open set $V_{i,\alpha,\varepsilon} = q(\{i\} \times U_\alpha \times \{\varepsilon\})$ onto U_α. From this it follows that p is a covering map with multiplicity equal to two.

Figure 5.3 in the following page illustrates the identification of Y by the number of crosses representing the corresponding points. The solid line represents the "first sheet" and the dotted one the "second" one.

(iv) If $p_{i,\alpha,\varepsilon} = p|V_{i,\alpha,\varepsilon}$, then any two charts are holomorphically compatible since $p_{i,\alpha,\varepsilon} \circ (p_{j,\beta,\varepsilon'})^{-1}(z) = z$.

(v) Therefore $\tilde{X}$ is a Riemann surface and $p: \tilde{X} \to \mathbb{C}^*$ is holomorphic in the sense that for each $p_{i,\alpha,\varepsilon}$, the function $p \circ p_{i,\alpha,\varepsilon}^{-1}$ is holomorphic in an open subset of $\mathbb{C}$.

On Y we have a function F defined by

$$F(i, z, \varepsilon) = \varepsilon f_i(z)$$

This function is compatible with ρ since one can easily verify

$$F(i, z, \varepsilon) = F(j, z, g_{ij}(z)\varepsilon).$$

Hence we can define a holomorphic function $\tilde{f}: \tilde{X} \to \mathbb{C}$ by letting

$$\tilde{f}(q(i, z, \varepsilon)) = F(i, z, \varepsilon).$$

It can be seen that if $\sigma_1 : U_1 \to \tilde{X}$, $\sigma_2 : U_2 \to \tilde{X}$ are the two sections

$$\sigma_1(z) = q(1, z, +1)$$

$$\sigma_2(z) = q(2, z, -1)$$

the $\tilde{f} \circ \sigma_1(z)$ and $\tilde{f} \circ \sigma_2(z)$ are the two values of $\sqrt{z}$.

One can "pluck the holes" 0 and ∞ in $\tilde{X}$ and obtain a compact Riemann surface, cf. §5.12.

(5) The Riemann surface of the logarithm: As before, let $U_1 = \{z \in \mathbb{C} : 0 < \arg z < 2\pi\}$ and $U_2 = \{z \in \mathbb{C} : -\pi < \arg z < \pi\}$. In U_1 and U_2 we have countably many determinations of the "function log"

$$f_{1,k}(z) = \log|z| + i \arg z + 2\pi i k \qquad (0 < \arg z < 2\pi), \quad k \in \mathbb{Z},$$

$$f_{2,k}(z) = \log|z| + i \arg z + 2\pi i k \qquad (-\pi < \arg z < \pi), \quad k \in \mathbb{Z}.$$

We are going to construct a covering space $\tilde{X}$ with countably many sheets, $p: \tilde{X} \to \mathbb{C}^*$, which will be a Riemann surface, and we will also construct a holomorphic function $\tilde{f}$ on $\tilde{X}$ such that at the points of $p^{-1}(z)$, $z \in \mathbb{C}^*$, it takes all the possible values of $\log z$.

As earlier, we leave the justification of all the assertions to the reader.

Let Y be the disjoint union:

$$Y = (\{1\} \times U_1 \times \mathbb{Z}) \dot{\cup} (\{2\} \times U_2 \times \mathbb{Z}),$$

as a subspace of $\{1, 2\} \times \mathbb{C}^* \times \mathbb{Z}$. Let $\mathfrak{G} = \mathfrak{G}_\mathbb{Z} =$ the group of permutations (bijections) of the integers. Consider the maps $g_{11} : U_1 \to \mathfrak{G}$, $g_{22} : U_2 \to \mathfrak{G}$,

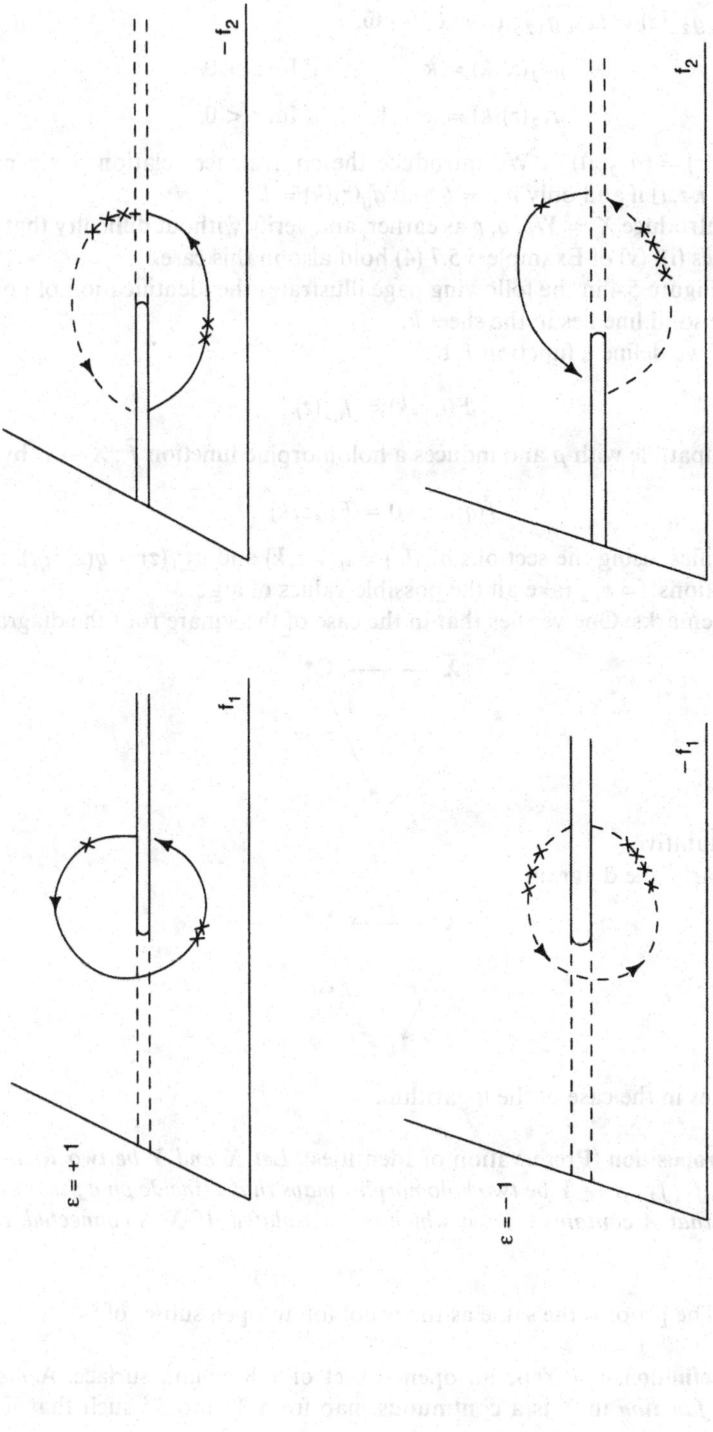

Figure 5.3

$$g_{11}(z) = g_{22}(z) = id_Z, g_{12} : U_1 \cap U_2 \to \mathfrak{G},$$

$$g_{12}(z)(k) = k \qquad \text{if } \operatorname{Im} z > 0,$$

$$g_{12}(z)(k) = k + 1, \qquad \text{if } \operatorname{Im} z < 0,$$

and $g_{21}(z) = (g_{12}(z))^{-1}$. We introduce the equivalence relation ρ given by $(i, z, k)\rho(j, z, l)$ if and only if $z = \zeta$ and $g_{ij}(z)(k) = l$.

We introduce $\tilde{X} = Y/\rho$, q, p as earlier, and verify without difficulty that the properties (i)–(v) of Example 5.5.7 (4) hold also in this case.

The Figure 5.4 in the following page illustrates the identification of points in Y, the solid line lies in the sheet k.

On Y we define a function F by

$$F(i, z, k) = f_{i,k}(z).$$

It is compatible with ρ and induces a holomorphic function $\tilde{f} : \tilde{X} \to \mathbb{C}$ by

$$\tilde{f}(q(i, z, k)) = F(i, z, k).$$

One verifies, using the sections $\sigma_{1,k}(z) = q(1, z, k)$ and $\sigma_{2,k}(z) = q(2, z, k)$, that the functions $f \circ \sigma_{i,k}$ take all the possible values of $\log z$.

(6) Remarks: One verifies that in the case of the square root the diagram

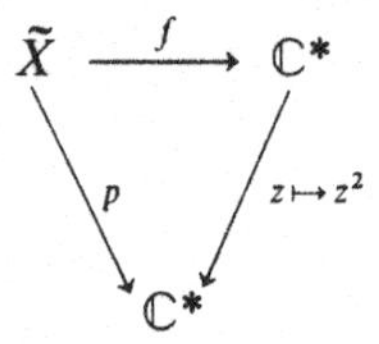

is commutative.

Similarly, the diagram

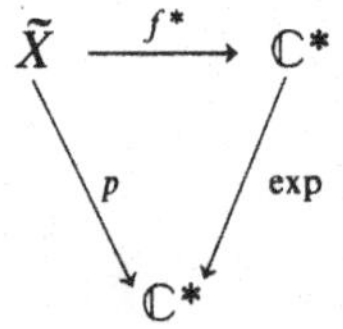

commutes in the case of the logarithm.

5.5.8. Proposition (Preservation of Identities). *Let X and Y be two Riemann surfaces, $f_1, f_2 : X \to Y$ be two holomorphic maps that coincide on a set $A \subseteq X$. Assume that A contains a point which is not isolated. If X is connected, then $f_1 \equiv f_2$.*

PROOF. The proof is the same as the proof for an open subset of $\mathbb{C}$. □

5.5.9. Definition. Let Y be an open subset of a Riemann surface. A ***meromorphic function*** in Y is a continuous map from Y into S^2 such that if we

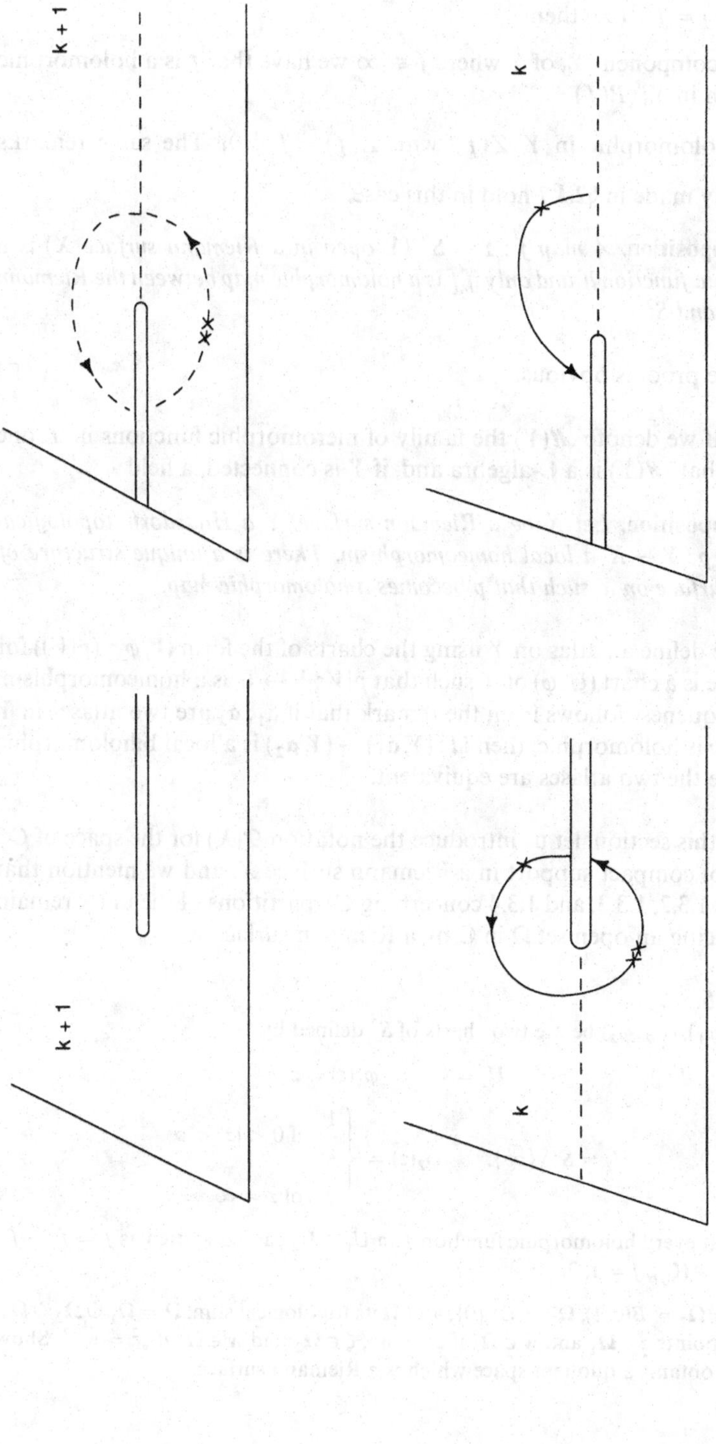

Figure 5.4

denote $P(f) = f^{-1}(\infty)$ then

(1) In any component Y_0 of Y where $f \not\equiv \infty$ we have that f is a holomorphic function in $Y_0 \backslash P(f)$.

(2) $\dfrac{1}{f}$ is holomorphic in $Y \backslash Z(f)$, with $Z(f) = f^{-1}(0)$. The same remarks already made in §2.4.7 hold in this case.

5.5.10. Proposition. *A map $f: Y \to S^2$ (Y open in a Riemann surface X) is a meromorphic function if and only if f is a holomorphic map between the Riemann surfaces Y and S^2.*

PROOF. The proof is obvious. □

Remarks. If we denote $\mathcal{M}(Y)$ the family of meromorphic functions in Y, one can show that $\mathcal{M}(Y)$ is a $\mathbb{C}$-algebra and, if Y is connected, a field.

5.5.11. Proposition. *Let X be a Riemann surface, Y a Hausdorff topological space, and $p: Y \to X$ a local homeomorphism. There is a unique structure of Riemann surface on Y such that p becomes a holomorphic map.*

PROOF. We define an atlas on Y using the charts of the form $(V, \varphi \circ (p|V))$ for which there is a chart (U, φ) of X such that $p|V: V \to U$ is a homeomorphism.

The uniqueness follows from the remark that if $\mathfrak{a}_1$, $\mathfrak{a}_2$ are two atlases in Y for which p is holomorphic, then $id: (Y, \mathfrak{a}_1) \to (Y, \mathfrak{a}_2)$ is a local biholomorphic map, hence the two atlases are equivalent. □

To end this section, let us introduce the notation $\mathcal{D}(X)$ for the space of C^∞ functions of compact support in a Riemann surface X, and we mention that statements 1.3.2, 1.3.3, and 1.3.4 concerning C^∞ partitions of the unity remain valid replacing an open set Ω in $\mathbb{C}$ by a Riemann surface X.

EXERCISES 5.5

1. Let (U_1, φ_1), (U_2, φ_2) be the two charts of S^2 defined by

$$U_1 = \mathbb{C}, \qquad \varphi_1(z) = z$$

$$U_2 = S^2 \backslash \{\infty\}, \qquad \varphi_2(z) = \begin{cases} \dfrac{1}{z} & \text{if } 0 < |z| < \infty \\ 0 & \text{of } z = \infty. \end{cases}$$

Show that every holomorphic function f in $U_1 \cap U_2$ can be written as $f = f_1 - f_2$, with $f_j \in \mathcal{H}(U_j)$, $j = 1, 2$.

2. Let $\Omega_1 = \Omega_2 = B(0, 1)$, $\Omega_3 = \mathbb{C} \backslash \{0\}$, and Ω its topological sum, $\Omega = \Omega_1 \dot\cup \Omega_2 \dot\cup \Omega_3$. Identify points $z \in \Omega_1$ and $w \in \Omega_3$ if $z^2 = w^2$, $\zeta \in \Omega_2$ and $w \in \Omega_3$ if $\zeta^2 = w^{-2}$. Show that one obtains a quotient space which is a Riemann surface.

§6. The Sheaf of Germs of Holomorphic Functions

Let $a \in \mathbb{C}$ and consider the family of all pairs (U, f), where U is an open subset of $\mathbb{C}$ containing the point a, and $f \in \mathscr{H}(U)$. We define an equivalence relation in this family by setting $(U, f) \sim (V, g)$ if and only if there is an open neighborhood W of a such that $W \subseteq U \cap V$ and $f|W = g|W$. The class of equivalence of (U, f) is denoted by f_a and called the ***germ of the holomorphic function*** f *at the point* a. We denote by $\mathcal{O}_a$ the set of all germs of holomorphic functions at a.

Clearly the same construction could have been done with C^∞ functions instead of holomorphic functions. We denote $\mathscr{E}_a$ the set of all germs of C^∞ functions at the point a.

Denote by $f_a(a)$ the ***value*** *at the point* a of the germ f_a. It is defined by $f_a(a) = f(a)$, which is independent of the choice of representative f of the germ f_a. In the same way, we can define the value at a of the derivatives of f_a, $f_a^{(n)}(a)$, $n \geq 0$.

5.6.1. Proposition. *The set $\mathcal{O}_a$ has a natural structure of principal local ring, where the unique maximal ideal is $M_a = \{f_a \in \mathcal{O}_a : f_a(a) = 0\}$. As a ring, it is isomorphic to the ring $\mathbb{C}\{\zeta\}$ of convergent power series in a single variable ζ. The residue field $\mathcal{O}_a/M_a$ is isomorphic to $\mathbb{C}$.*

PROOF. If f_a and g_a have as representatives (U, f) and (V, g), respectively, one defines $f_a + g_a$ and $f_a \cdot g_a$ as the classes of $(U \cap V, f + g)$ and $(U \cap V, f \cdot g)$, respectively. If $\lambda \in \mathbb{C}$, λf_a is the class of $(U, \lambda f)$. We leave to the reader the verification that these operations are well defined in $\mathcal{O}_a$. It is also easy to see that $\mathcal{O}_a$ is an integral domain.

Let M_a be the set of noninvertible elements in $\mathcal{O}_a$. If $f_a(a) \neq 0$ then f_a is invertible in $\mathcal{O}_a$ with inverse $(1/f)_a$. Conversely, if f_a has an inverse g_a, then $g_a(a) f_a(a) = 1$ and it follows that $f_a(a) \neq 0$. Therefore $M_a = \{f_a \in \mathcal{O}_a : f_a(a) = 0\}$.

Let $f_a \in \mathcal{O}_a$, then the power series $\sum_{n \geq 0} \frac{1}{n!} f_a^{(n)}(a)(z - a)^n$ has a positive radius of convergence, since it is the Taylor expansion of a representative f of f_a in a disk of center a. The map assigning to f_a this power series is clearly a ring isomorphism between $\mathcal{O}_a$ and $\mathbb{C}\{(z - a)\}$, the ring of convergent power series in the variable $(z - a)$. It now follows that the only proper ideals in $\mathcal{O}_a$ are the ideals M_a^k, $k \in \mathbb{N}^*$. In fact, if I is a proper ideal in $\mathbb{C}\{\zeta\}$, let

$$k = \inf\left\{\nu \in \mathbb{N} : \exists s(\zeta) = \sum_{n \geq 0} a_n \zeta^n \in I \text{ with } a_\nu \neq 0\right\}.$$

Clearly $k \geq 1$, if not s will be invertible and $I = \mathbb{C}\{\zeta\}$. Similarly, $k < \infty$, since $I \neq \{0\}$. Let $s \in I$ be such that the infimum is attained at s, then

$$s(\zeta) = \zeta^k(a_k + a_{k+1}\zeta + \cdots) = \zeta^k \sigma(\zeta).$$

Since $a_k \neq 0$, the series σ is invertible in $\mathbb{C}\{\zeta\}$. Hence I is the ideal generated by ζ^k. In other words, the only proper ideals in $\mathcal{O}_a$ are M_a^k and $\mathcal{O}_a$ is a local ring.

The last assertion of the proposition is obvious. □

Consider now the set $\mathcal{O} = \dot{\bigcup}_{a \in \mathbb{C}} \mathcal{O}_a$, and introduce in it a topology as follows. Let (U, f) be a representative of $f_a \in \mathcal{O}_a$. Denote

$$N(U, f) = \{f_z \in \mathcal{O}_z : z \in U\},$$

and let a set $\Omega \subseteq \mathcal{O}$ be open if and only if it contains a subset of the form $N(U, f)$ whenever $f_a \in \Omega$. One verifies without difficulty that $\mathcal{O}$ becomes a topological space, and the sets $N(U, f)$ form a fundamental system of open neighborhoods of f_a.

In $\mathcal{O}$ we have a natural continuous map $p : \mathcal{O} \to \mathbb{C}$, $p(f_a) = a$.

5.6.2. Definition. The space $\mathcal{O}$ with the projection map $p : \mathcal{O} \to \mathbb{C}$ is called the ***sheaf of germs of holomorphic functions*** on $\mathbb{C}$.

Remark. One can define in the same way the sheaf $\mathscr{E}$ of germs of C^∞ functions on $\mathbb{C}$. In both cases ($\mathcal{O}$ and $\mathscr{E}$) one can also replace $\mathbb{C}$ by an arbitrary Riemann surface X; we denote by $\mathcal{O}_X$ and $\mathscr{E}_X$ the corresponding sheaves (we suppress the index when no confusion could arise).

5.6.3. Proposition. *The topological space $\mathcal{O}$ is Hausdorff.*

Proof. Let $f_a \neq g_b$. If $a \neq b$ then one can choose representatives (U, f) and (V, g) such that $U \cap V = \varnothing$. Hence $N(U, f) \cap N(V, f) = \varnothing$. If $a = b$, then since $f_a \neq g_a$ we can find a disk $B(a, \rho)$ and holomorphic functions f, g representing f_a and g_a, respectively, such that $N(B(a, \rho), f) \cap N(B(a, \rho), g) = \varnothing$. If this last claim were not true, then there is a germ $h_z \in N(B(a, \rho), f) \cap N(B(a, \rho), g)$. That is, $h_z = f_z = g_z$, which means that f and g coincide in an open subset of $B(a, \rho)$ and therefore, $f = g$ in $B(a, \rho)$. Hence $f_a = g_a$, which is a contradiction. □

Remark. It is easy to verify that $\mathscr{E}$ is not Hausdorff. It is also clear that the same proof shows that $\mathcal{O}_X$ is Hausdorff for any Riemann surface X.

5.6.4. Proposition. *The map $p : \mathcal{O}_X \to X$ is a local homeomorphism.*

Proof. If $N(U, f)$ is a neighborhood of f_a then $p : N(U, f) \to U$ is a homeomorphism. It is clearly bijective. It is continuous and open because if V is an open neighborhood of $z_0 \in U$, then $p(N(V, f)) = V$ and $N(V, f)$ is an open neighborhood of f_{z_0}. □

5.6.5. Corollary. *$\mathcal{O}_X$ is a Riemann surface and $p : \mathcal{O}_X \to X$ is a holomorphic map.*

5.6.6. Definition. Let $f_a \in \mathcal{O}_a$ and $\gamma : [0, 1] \to X$ a continuous curve originating at $\gamma(0) = a$. The ***analytic continuation*** *of f_a along γ* is the lifting $\tilde{\gamma}$ of γ to $\mathcal{O}_X$ such that $\tilde{\gamma}(0) = f_a$ (if it exists). The germ $\tilde{\gamma}(t)$ is called the germ at the point $\gamma(t)$ obtained by the analytic continuation of f_a along γ.

Due to the uniqueness of the lifting, it follows from this definition that if the analytic continuation of the germs of a holomorphic function exists along a curve, it is uniquely determined. But it could well occur, even for $X = \mathbb{C}$, that the lifting does not exist: for instance if $\gamma : [0, 1] \to \mathbb{C}$ is the straight line segment $\gamma(t) = t$, then γ cannot be lifted to $\mathcal{O}$ having as origin the germ at $z = 0$ of the function $\dfrac{1}{1 - z}$.

In particular, $p : \mathcal{O}_X \to X$ is never a covering map. Nevertheless, we have the following.

5.6.7. Theorem (Monodromy Theorem). *Let X be a Riemann surface, γ_0, γ_1 be two paths homotopic with fixed endpoints in X. Let $a = \gamma_0(0)$, $b = \gamma_0(1)$, H be a homotopy between γ_0, γ_1, and γ_s be the curves $\gamma_s(t) = H(t, s)$. If $f_a \in \mathcal{O}_a$ has an analytic continuation $\tilde{\gamma}_s$ along every curve γ_s then $\tilde{\gamma}_0(1) = \tilde{\gamma}_1(1)$ in $\mathcal{O}_b$.*

PROOF. It is an immediate translation of §5.4.3 to this setting since $\mathcal{O}_X$ is a Hausdorff space. □

5.6.8. Corollary. *Let X be a simply connected Riemann surface, $a \in X$, $f_a \in \mathcal{O}_a$ a germ of a holomorphic function that admits analytic continuation along any path in X starting at a. There is then a unique holomorphic function $f \in \mathscr{H}(X)$ such that its germ at the point a coincides with f_a.*

PROOF. For $z \in X$, let f_z be the germ at z arising from f_a by analytic continuation along a curve joining a to z. It does not depend on the curve due to the simple connectedness of X. Let $f(z) = f_z(z)$; one sees that this defines a holomorphic function in X such that its germ at the point a coincides with f_a. □

Remarks

(1) In general, even if the analytic continuation is possible along every curve that starts at a and ends at b, the germ at b of the analytic continuations do not coincide (except, of course, if the curves are homotopic).

(2) If $\gamma : [0, 1] \to X$ is a curve in a Riemann surface, $f_a \in \mathcal{O}_a$ ($a = \gamma(0)$), and we have a family of germs $f_{\gamma(t)} \in \mathcal{O}_{\gamma(t)}$ ($0 \leq t \leq 1$) such that $f_{\gamma(0)} = f_a$ and such that, also, for every $\tau \in [0, 1]$ there is a neighborhood I_τ of τ in $[0, 1]$, an open set $U_\tau \subseteq X$, $\gamma(I_\tau) \subseteq U_\tau$ and $\tilde{f} \in \mathscr{H}(U_\tau)$ such that the germ at $\gamma(t)$ of $\tilde{f}$ coincides with $f_{\gamma(t)}$ for every $t \in I_\tau$, then $f_{\gamma(1)}$ is the analytic continuation of f_a along γ. Usually, the open sets U_τ are coordinate patches centered at $\gamma(\tau)$.

Let us assume that X and Y are two Riemann surfaces, $\mathcal{O}_X$, $\mathcal{O}_Y$ the corresponding sheaves of germs of holomorphic functions. Assume further that there is a local biholomorphism $p: Y \to X$. Then, for every $y \in Y$ the pull back is an isomorphism $p^*: \mathcal{O}_{X,p(y)} \to \mathcal{O}_{Y,y}$, $p^*f = f \circ p$, and we denote by $p_*: \mathcal{O}_{Y,y} \to \mathcal{O}_{X,p(y)}$ its inverse.

5.6.9. Definition. Assume X is a Riemann surface, $a \in X$, and $f_a \in \mathcal{O}_a$. A quadruple $(Y, p, \tilde{f}, b)$ is ***an analytic continuation*** *of* f_a if:

(1) Y is a connected Riemann surface and $p: Y \to X$ is a local biholomorphism,
(2) $\tilde{f}$ is a holomorphic function in Y, and
(3) $b \in Y$ is such that $p(b) = a$ and $p_*(\tilde{f}_b) = f_a$.

An analytic continuation $(Y, p, \tilde{f}, b)$ is said to be ***maximal*** if it satisfies the following universal property: If (Z, q, g, c) is another analytic continuation of f_a, then there is a holomorphic map $F: Z \to Y$ such that $F(c) = b$ and $F^*(\tilde{f}) = \tilde{f} \circ F = g$.

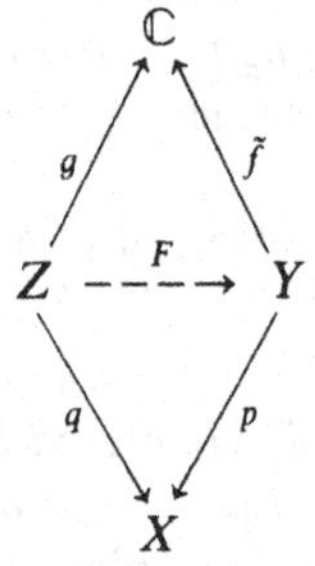

Such maximal analytic continuation, if it exists, is unique up to biholomorphic mappings; if $(Y, p, \tilde{f}, b)$ and (Z, q, g, c) are maximal, then there are $F: Z \to Y$, $G: Y \to Z$ such that $f(c) = b$, $G(b) = c$, $\tilde{f} \circ F = g$, and $g \circ G = \tilde{f}$. Hence $F \circ G: Y \to Y$ is such that $p \circ F \circ G = p$ and $(F \circ G)(b) = b = id_Y(b)$. By the uniqueness of the liftings we have $F \circ G = id_Y$. Similarly $G \circ F = id_Z$ and hence, F and G are biholomorphic mappings inverse to each other.

5.6.10 Lemma. *Let X be a Riemann surface, $a \in X$, $f_a \in \mathcal{O}_a$, and $(Y, p, \tilde{f}, b)$ be an analytic continuation of f_a. If $\gamma: [0,1] \to Y$ is a curve such that $\gamma(0) = b$ and $\gamma(1) = y$, then the germ $p_*(\tilde{f}_y) \in \mathcal{O}_{p(y)}$ is the analytic continuation of f_a along $\alpha = p \circ \gamma$.*

PROOF. For $t \in [0,1]$, let $f_{\alpha(t)} = p_*(\tilde{f}_{\gamma(t)}) \in \mathcal{O}_{\alpha(t)}$. One has $f_{\alpha(0)} = f_a$ and $f_{\alpha(1)} = p_*(\tilde{f}_y)$. Let $t_0 \in [0,1]$. Since p is a local biholomorphic map, there are open neighborhoods $V \subseteq Y$ and $U \subseteq X$ of $\gamma(t_0)$ and $\alpha(t_0)$, respectively, such that $p|V: V \to U$ is a biholomorphic map. Let $q = (p|V)^{-1}$ and $g = q^*(\tilde{f}|V) = (\tilde{f}|V) \circ q \in \mathscr{H}(U)$. Hence $p_*(\tilde{f}_\zeta) = g_{p(\zeta)}$ for every $\zeta \in V$. On the other hand, there is a neighborhood I_{t_0} of t_0 in $[0,1]$ such that $\gamma(I_{t_0}) \subseteq V$ and $\alpha(I_{t_0}) \subseteq U$. For $t \in I_{t_0}$ we have

$$g_{\alpha(t)} = p_*(\tilde{f}_{\gamma(t)})) = f_{\alpha(t)}.$$

It follows from the preceding remarks that $p_*(\tilde{f}_y)$ is the analytic continuation of f_a along α. □

5.6.11. Theorem. *Let X be a Riemann surface, $a \in X$, and $f_a \in \mathcal{O}_a$. There is a maximal analytic continuation $(Y, P, \tilde{f}, b)$ of f_a, where Y is the connected component of $\mathcal{O}_X$ containing f_a and $P: Y \to X$ is the restriction to Y of the canonical map $p: \mathcal{O}_X \to X$.*

PROOF. Let Y be the connected component of $\mathcal{O}_X$ containing f_a. First we remark that Y is an open subset of $\mathcal{O}_X$. In fact, when U is a disk, the open sets $N(U, g)$ in $\mathcal{O}_X$ are connected by arcs, hence $\mathcal{O}_X$ is locally connected (by arcs) and it follows that its connected components are open. Therefore, Y is a Riemann surface such that $P: Y \to X$ is a local biholomorphic map.

Define now $\tilde{f}$ as follows. Every $\zeta \in Y$ is in fact a germ of a holomorphic function at the point $p(\zeta) \in X$. Let $\tilde{f}(\zeta) = \zeta(p(\zeta))$. It is easy to verify that $\tilde{f}$ is holomorphic on Y and $p_*(\tilde{f}_\zeta) = \zeta$ for every $\zeta \in Y$. If we set $b = f_a$ we see that $(Y, P, \tilde{f}, b)$ is an analytic continuation of f_a.

Assume there is another analytic continuation of f_a, (Z, q, g, c). Define $F: Z \to Y$ as follows: if $\zeta \in Z$ and $g(\zeta) = z$, the germ $q_*(g_\zeta) \in \mathcal{O}_{X,z}$ is an analytic continuation of f_a along a certain path joining a to z, and we know that Y is the set of all germs that can be obtained by analytic continuation of f_a along curves. Therefore, there is a unique $w \in Y$ such that $q_*(g_\zeta) = w$. Let $F(\zeta) = w$. One can verify that F is a holomorphic map from Z to Y, $P \circ F = q$, $F(c) = b$, and $F^*(\tilde{f}) = g$. This concludes the proof of the theorem. □

5.6.12. Proposition. *Let X be a connected Riemann surface, $a \in X$. Suppose that $f_a \in \mathcal{O}_a$ can be analytically continued along any path in X starting at a. Let $(Y, p, \tilde{f}, b)$ be the maximal analytic continuation of f_a. Then $p: Y \to X$ is a covering map.*

PROOF. In fact, Y is a topological manifold of dimension 2, $p: Y \to X$ a local homeomorphism having the lifting property for curves. The proposition is then a corollary of 5.4.13.

Y is sometimes called the ***Riemann surface of the germ*** f_a.

The concept of sheaf plays an important role in complex analysis, especially in the case of several variables. We recommend [Gu], [God], and [Bred] for further study.

EXERCISES 5.6

1. Let X be a connected Riemann surface, $z_0 \in X$, and $f, g \in \mathcal{O}_{z_0}$, which admit an analytic continuation along a path γ starting at z_0 and ending at z_1. Let $f_\gamma, g_\gamma \in \mathcal{O}_{z_1}$ be those analytic continuations. What can you say about the germs of $\lambda f + \mu g$ ($\lambda, \mu \in \mathbb{C}$) and fg?

2. Let X be an open subset of $\mathbb{C}$, $z_0 \in X$, $f \in \mathcal{O}_{z_0}$. Assume f admits an analytic

continuation f_γ along a path γ in X which starts at z_0 and ends at z_1, $f_\gamma \in \mathcal{O}_{z_1}$. Show that the germ $g = \dfrac{df}{dz}$ also admits an analytic continuation g_γ along γ and that $g_\gamma = \dfrac{df_\gamma}{dz}$ in $\mathcal{O}_{z_1}$.

*3. Let $\theta \in \mathbb{R}$, L_θ be the ray of direction θ, i.e., $L_\theta = \{re^{i\theta} : r > 0\}$, and H_θ be the half-plane "perpendicular" to $\bar{L}_\theta$, i.e., $H_\theta = \{z \in \mathbb{C} : \operatorname{Re}(ze^{i\theta}) < 0\}$. Let

$$f_\theta(z) = \int_{L_\theta} \exp\left(zt + \frac{(\log t)^2}{4\pi i}\right) dt,$$

where $\log t = \log|t| + i\theta$.

(a) Show that f_θ is holomorphic in H_θ.

(b) Show that if $\theta < \theta' < \theta + \dfrac{\pi}{2}$ then $f_\theta = f_{\theta'}$ in $H_\theta \cap H_{\theta'}$. In particular, if we let f denote the restriction of $f_{-\pi}$ to the disk $B(1, 1)$, $f(z) = \sum_{n \geq 0} a_n(z-1)^n$, then f has an analytic continuation along any path in $\mathbb{C}^*$ starting at $z = 1$.

Let γ be the unit circle $\partial B(0, 1)$ in the counterclockwise direction.

(c) Show that f_γ concides with f_π.

(d) Show that for $z \in B(1, 1)$ (or even for $\operatorname{Re} z > 0$)

$$f_\gamma(z) = \int_{L_{-\pi}} t \exp\left(zt + \frac{(\log t)^2}{4\pi i}\right) dt.$$

Conclude that

$$f_\gamma(z) = \frac{d}{dz} f(z) \qquad \text{for } z \in B(1, 1)$$

or

$$f_\gamma(z) = \sum_{n \geq 1} n a_n (z-1)^{n-1} \qquad \text{for } |z - 1| < 1.$$

4. Let $X = \mathbb{C} \setminus \{-1, 1\}$, $a = 0 \in X$, f_0 the germ at $z = 0$ of the determination of $\log(1 - z^2)$ such that $f_0(0) = 0$. Show that f_0 admits an analytic continuation along any path in X that starts at the origin. Let $(Y, p, \tilde{f}, b)$ be a maximal analytic continuation of f_0 such that $p(b) = a = 0$. Show that if $U = \mathbb{C} \setminus (]-\infty, -1] \cup [1, \infty[)$ then

$$p | p^{-1}(U) : p^{-1}(U) \to U$$

is a trivial covering map. Find all the determinations of $\tilde{f}$ over U.

5. Same as Exercise 5.6.4 but for the following cases:

(i) $X = \mathbb{C} \setminus \{-1, 0, 1\}$, $a = \dfrac{1}{2}$, $f_a(z) = (1 - z^2)^{1/2} \log z$, $f_a\left(\dfrac{1}{2}\right) = -\dfrac{\sqrt{3} \log 2}{2}$, $U = \mathbb{C} \setminus (]-\infty, -1] \cup [1, \infty[)$;

(ii) $X = \mathbb{C} \setminus \{0\}$, $a = 1$, $f_a(z) = (1 + z^2) z^{1/3}$, $f_a(1) = 1$, $U = \mathbb{C} \setminus]-\infty, 0[$;

(iii) $X = \mathbb{C} \setminus \{2, 3\}$, $a = 0$, $f_a(z) = ((z^2 - 2)(z^2 - 3))^{1/2}$, $f_a(0) = \sqrt{6}$,

$$U = \mathbb{C} \setminus (\{2 - it : t \geq 0\} \cup \{3 + it : t \geq 0\});$$

(iv) $X = \mathbb{C} \setminus \{0, 1\}$, $a = -1$, $f_a(z) = \log(z^{1/2} + (z-1)^{1/2})$, $f_a(-1) = (1 + \sqrt{2})i$,

$$U = \mathbb{C} \setminus (]-\infty, 0] \cup [1, \infty[);$$

(v) $X = \mathbb{C} \setminus \{-1, 1\}$, $a = 2$, $f_a(z) = (1 - z^2)^{1/2}$, $f_a(2) = \sqrt{3} i$, $U = \mathbb{C} \setminus [-1, 1]$.

(vi) $X = \mathbb{C}\backslash\{1,2,3,\ldots,2n\}$, $a = 0$, $f_a(z) = ((z-1)(z-2)\ldots(z-2n))^{1/2}$, $f_a(0) = (2n!)^{1/2}$, $U = \mathbb{C}\backslash\left(\bigcup_{j=1}^{n}[2j-1,2j]\right)$.

6. Let $X = \mathbb{C}\backslash\{-ti : t \geq 2\}$, $a = 1 - i$, $f_a(z) = \log(z + 2i)$. Find all the possible values at $b = 1 + i$ of the analytic continuations of f_a along paths in X joining $a = 1 - i$ to $b = 1 + i$, assuming that $f_a(1 - i) = \sqrt{2} + i\dfrac{17\pi}{4}$.

7. Same as in Exercise 5.6.6 in the following cases:

(i) $X = \mathbb{C}\backslash(]-\infty,-1] \cup [1,\infty[)$, $a = 0$, $f_a(z) = \log(1 - z^2)$, $f_a(0) = 0$, $b = \dfrac{1+i}{\sqrt{2}}$;

(ii) $X = \mathbb{C}\backslash([1,\infty[\cup \{-1 + it : t \geq 0\})$, $a = 0$, $f_a(z) = (1 - z^2)^{1/2}$, $f_a(0) = -1$, $b = -5$;

(iii) $X = \mathbb{C}\backslash([-5,-3] \cup [3,5])$, $a = 0$, $f_a(z) = ((z^2 - 9)(z^2 - 25))^{1/2}$, $f_a(0) = 15$, $b = i$.

8. Find the analytic continuation along the given paths in the following cases:

(i) $f(z) = z^{1/5}$, $f(1) = 1$ (see Figure 5.5)

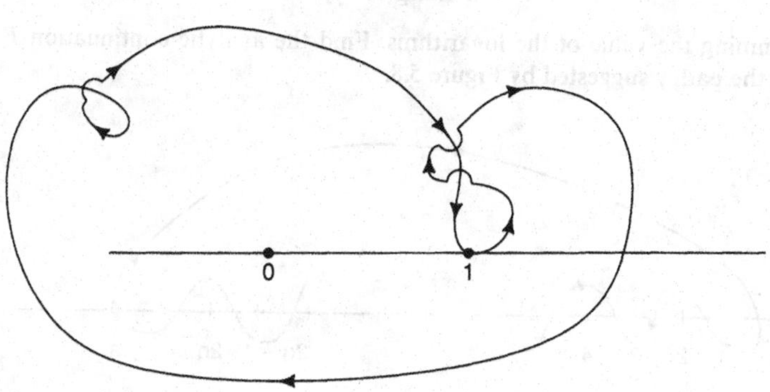

Figure 5.5

(ii) $f(z) = (1 + z^{1/2})\log z$, $f(1) = 4\pi i$ (see Figure 5.6)

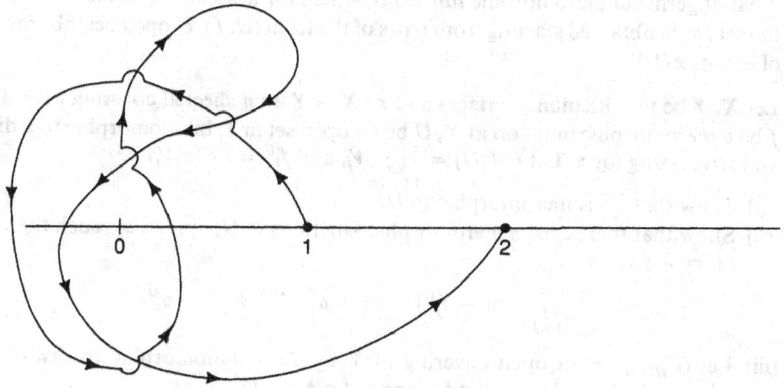

Figure 5.6

(iii) $f(z) = ((z^2 - 9)(z^2 - 25))^{1/2}$, $f(0) = 15$ (see Figure 5.7)

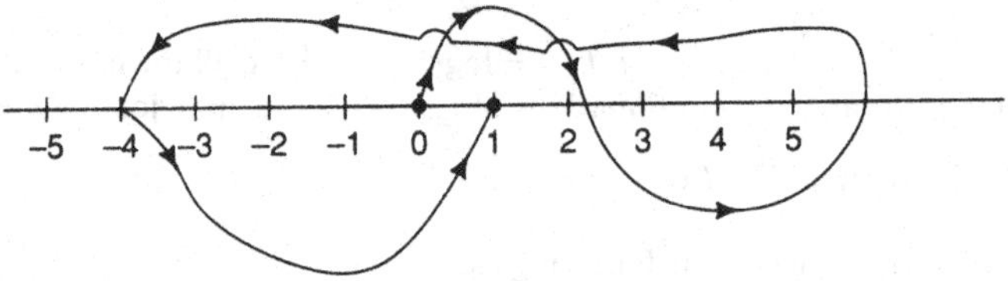

Figure 5.7

9. Let $f_1, \ldots, f_{2n}$ be entire functions and

$$f(z) := \sum_{1 \le j \le 2n} f_j(z) \log(z - j),$$

considered as a germ in the neighborhood of a point $a > 2n$, with

$$-\pi < \arg(z - j) < \pi$$

determining the value of the logarithms. Find the analytic continuation F of f along the path γ suggested by Figure 5.8.

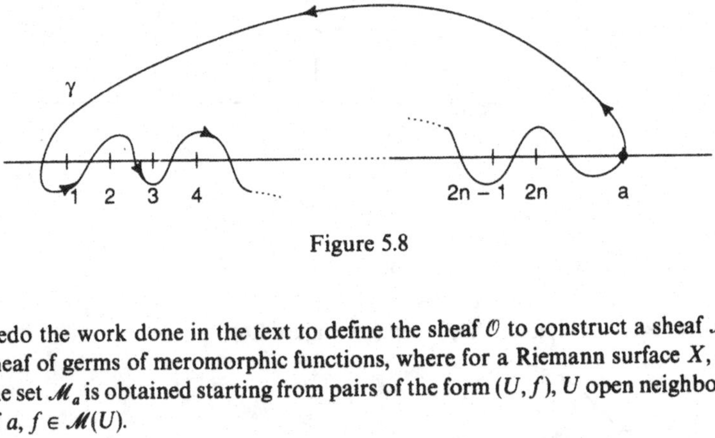

Figure 5.8

10. Redo the work done in the text to define the sheaf $\mathcal{O}$ to construct a sheaf $\mathcal{M}$, the sheaf of germs of meromorphic functions, where for a Riemann surface X, $a \in X$, the set $\mathcal{M}_a$ is obtained starting from pairs of the form (U, f), U open neighborhood of a, $f \in \mathcal{M}(U)$.

11. Let X, Y be two Riemann surfaces and $\pi: X \to Y$ an n-sheeted covering map. Let f be a meromorphic function in X, U be an open set in Y, biholomorphic to a disk and trivializing for π. Let $\pi^{-1}(U) = \bigcup_{1 \le j \le n} V_j$, and $f_j^U = f \circ (\pi | V_j)^{-1}$.
   (i) Show that f_j^U is meromorphic in U.
   (ii) Show that there exist n meromorphic functions in U, $a_1^U, \ldots, a_n^U$, such that for every $w \in \mathbb{C}$

$$\prod_{1 \le j \le n} (w - f_j^U) = w^n + a_1^U w^{n-1} + \cdots + a_n^U.$$

   (iii) Let $(U_l)_{l \in \Lambda}$ be an open covering of Y by sets biholomorphic to disks and trivializing for π. Show that for any $k, l \in \Lambda, j \in \{1, \ldots, n\}$

$$a_j^{U_k}|U_k \cap U_l = a_j^{U_l}|U_k \cap U_l.$$

Conclude that there exist functions $a_1, \dots, a_n \in \mathscr{M}(Y)$ such that

$$a_j|U_k = a_j^{U_k}$$

for any j, k.

12. Let $f(z) = \displaystyle\int_1^2 e^{-1/t} \frac{dt}{t - z}$. Show it is holomorphic in $\mathbb{C}\backslash[1, 2]$. Find all the analytic continuations of the germ of f at 0 along paths in $\mathbb{C}\backslash\{1, 2\}$, joining 0 to z, $z \neq 1$, $z \neq 2$.

13. Find the analytic continuation of the germ $f(z) = \log\log\sqrt{z}$, $f(e^{2\pi}) = \log \pi$ along the paths in Figure 5.9.

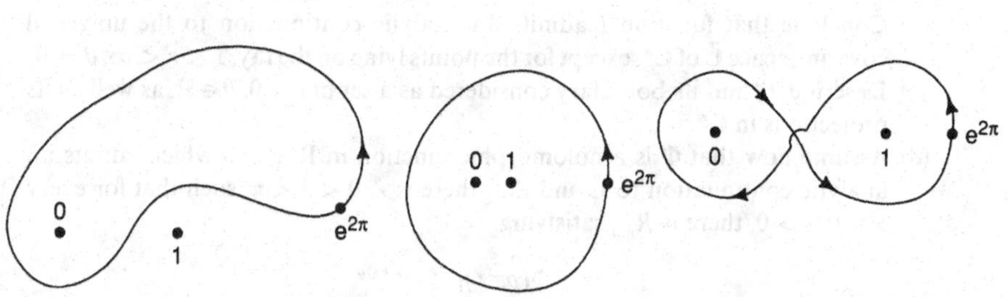

Figure 5.9

14. Let φ be an entire function, $f(z) = \dfrac{1}{2\pi i}\displaystyle\int_0^1 \frac{\varphi(t)}{t - z}$, $z \in \mathbb{C}\backslash[0, 1]$. Let γ be a closed path in $\mathbb{C}\backslash\{0, 1\}$, starting at $z \in \mathbb{C}\backslash[0, 1]$. Show the analytic continuation f_γ of f along γ exists and it is given by

$$f_\gamma(z) = f(z) + (\operatorname{Ind}_\gamma(1) - \operatorname{Ind}_\gamma(0))\varphi(z).$$

15. Let D be a simply connected domain in $\mathbb{C}$ and $f \in \mathscr{H}(D)$ be such that $f(D)$ is a simply connected domain and $f'(z) \neq 0$ for every $z \in D$. It is often "proved" using the monodromy theorem that f is a biholomorphic map. The following is a simple counterexample.

Let $f(z) := \displaystyle\int_0^z e^{\zeta^2}\, ds$, $z \in \mathbb{C}$.

(a) Show that $f'(z) \neq 0$ everywhere.
(b) Show f is not injective.
(c) f is an odd function.
(d) $f: \mathbb{C} \to \mathbb{C}$ is surjective. (Hint: if f omits the value a, then it must also omit $-a$.)

16. Let φ be a holomorphic function in a neighborhood of the closed half-plane $\operatorname{Re} z \geq 0$ such that for every $\varepsilon > 0$, $0 < \alpha \leq \dfrac{\pi}{2}$ there is $R_{\varepsilon,\alpha}$ such that

$$|\varphi(z)| \leq e^{\varepsilon|z|},$$

whenever $|z| \geq R_{\varepsilon,\alpha}$ and $|\operatorname{Arg} z| < \alpha$.

(a) Let $\tilde{B}$ denote the universal covering space of $B(0,1)\setminus\{0\}$. Show that the function

$$I(\zeta) := \int_0^\infty \varphi(x)\zeta^x\,dx,$$

$\zeta^x = \exp(x \operatorname{Log} \zeta)$, is holomorphic in $B(0,1)\setminus]-1,0[$ and admits an analytic continuation to $\tilde{B}$.

(b) Let $\psi \in]-\alpha,\alpha[$ and L_ψ be the ray $\{z = \rho e^{i\psi}, \rho > 0\}$. Set

$$I_\psi(\zeta) := \int_{L_\psi} \varphi(z)\zeta^z\,dz.$$

Show that I_ψ is holomorphic in the set

$$T_\psi := \{\zeta = re^{i\theta} : (\log r)\cos\psi - \theta\sin\psi < 0, r > 0\}.$$

Conclude that function I admits an analytic continuation to the universal covering space $\tilde{\mathbb{C}}$ of $\mathbb{C}^*$, except for the points lying on the ray, $1 \le r < \infty, \theta = 0$.

(c) Describe T_ψ and its boundary considered as a set in $r > 0$, $\theta \in \mathbb{R}$, as well as its projections in $\mathbb{C}^*$.

(d) Assume now that Φ is a holomorphic function in $\operatorname{Re} z \ge 0$ which admits an analytic continuation to $\tilde{\mathbb{C}}$ and that there is λ, $0 < \lambda < \pi$, such that for every $\alpha > 0$, $\varepsilon > 0$, there is $R_{\varepsilon,\alpha}$ satisfying

$$|\Phi(\rho e^{i\psi})| \le e^{(\lambda+\varepsilon)\rho}$$

if $|\psi| \le \alpha$, $\rho > R_{\varepsilon,\alpha}$. Define

$$h(\psi) := \limsup_{\rho\to\infty} \frac{\log|\Phi(\rho e^{i\psi})|}{\rho}.$$

Show that the integral I, defined as in part (a) but with respect to Φ, admits an analytic continuation to $\tilde{\mathbb{C}}\setminus\sigma(h)$, where

$$\sigma(h) = \bigcap_{\psi\in\mathbb{R}} \{\zeta = re^{i\theta} : (\log r)\cos\psi - \theta\sin\psi + h(\psi) \ge 0, r > 0\}.$$

Therefore, all the possible singular points of (the analytic continuation of) I project into the "interior" of the curve $(\log r)^2 = \lambda^2 - \theta^2$.

17. Let $u_0 \in \mathbb{C}\setminus\{0,1\}$, γ_0, γ_1, γ_∞ be straight line segments that start at u_0 and end, respectively, at 0, 1, ∞ and do not intersect, except at u_0. By the angle between γ_0 and γ_1 we mean the angular sector that does not contain γ_∞. Let us also fix three values $a > -1$, $b > -1$, $\lambda > -1$, such that $a + b + \lambda < -1$. If $z \in \mathbb{C}\setminus(\gamma_0 \cup \gamma_1 \cup \gamma_\infty)$, let γ_z be a path starting at u_0, otherwise disjoint from $\gamma_0 \cup \gamma_1 \cup \gamma_\infty$ and ending at z. We can choose a determination of the function

$$\varphi(u) := u^a(u-1)^b(u-z)^\lambda$$

for u in a neighborhood of u_0, continue it along γ_0, γ_1, γ_∞, γ_z (except for the endpoints) and integrate this analytic continuation. This defines four functions of z

$$v_0 := \int_{\gamma_0} \varphi(u)\,du, \quad v_1 := \int_{\gamma_1} \varphi(u)\,du, \quad v_\infty := \int_{\gamma_\infty} \varphi(u)\,du, \quad v_z := \int_{\gamma_z} \varphi(u)\,du.$$

(a) Show that the integrals defining $v_0, \ldots, v_z$ are absolutely convergent. Why is v_z independent of the path γ_z? Show the four functions are holomorphic functions of z in $\mathbb{C}\setminus\{\gamma_0 \cup \gamma_1 \cup \gamma_\infty\}$.

(b) Using the contour from the following figure show that

$$(1 - e^{2\pi i a})v_0 + (e^{2\pi i a} - e^{2\pi i(a+\lambda)})v_z$$
$$+ (e^{2\pi i(a+\lambda)} - e^{2\pi i(a+b+\lambda)})v_1 + (e^{2\pi i(a+b+\lambda)} - 1)v_\infty = 0.$$

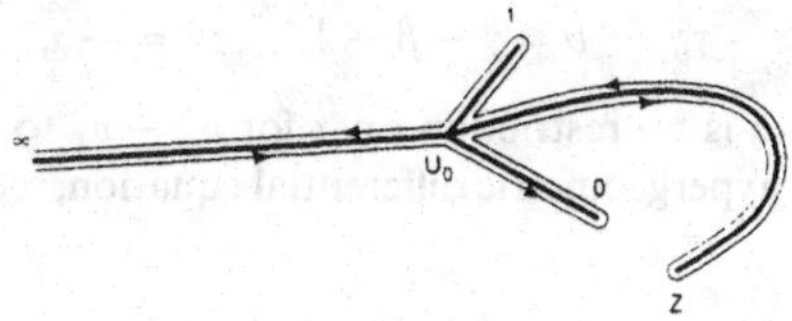

(c) Show that if z approaches a point ζ in $\gamma_0\setminus\{0\}$ from inside the angle between γ_0 and γ_1, i.e., from the left, the values $v_0, \ldots, v_\infty$ admit a limit value which we denote by the symbols $v_0, v_1, v_\infty, v_\zeta$. If z approaches ζ from the right, i.e., from outside the angle, they also admit a limit which we denote $v_0', \ldots, v_\zeta'$. Show these limits satisfy the relations

$$v_1' - v_1 = 0, \qquad v_\infty' = v_\infty$$
$$(1 - e^{2\pi i a})(v_0' - v_0) + (e^{2\pi i a} - e^{2\pi i(a+\lambda)})(v_\zeta' - v_\zeta) = 0.$$

Note also that this observation allows us to define the analytic continuation of $v_0, \ldots, v_z$ when γ_z is a path in $\mathbb{C}\setminus\{0, 1\}$.

(d) Consider the functions $w_1 := v_z - v_0$, $w_2 = v_z - v_1$, and the corresponding values w_1', w_2' from (c). Show that

$$\begin{pmatrix} w_1' \\ w_2' \end{pmatrix} = \begin{pmatrix} e^{-2\pi i(a+\lambda)} & 0 \\ [e^{-2\pi i(a+\lambda)} - e^{-2\pi i\lambda}] & 1 \end{pmatrix} \begin{pmatrix} w_1 \\ w_2 \end{pmatrix}.$$

(e) Let $v_0'', \ldots, v_z''$ the values when z lies on the left side of γ_1, i.e., when z crosses from the inside to the outside of the angle between γ_0 and γ_1. Show that

$$v_0'' - v_0 = 0, \qquad v_\infty'' - v_\infty = 0$$
$$(e^{2\pi i a} - e^{2\pi i(a+\lambda)})(v_z'' - v_z) + (e^{2\pi i(a+\lambda)} - e^{2\pi i(a+b+\lambda)})(v_1'' - v_1) = 0.$$

And giving w_1'', w_2'' the corresponding values, we have

$$\begin{pmatrix} w_1'' \\ w_2'' \end{pmatrix} = \begin{pmatrix} 1 & [e^{2\pi i(b+\lambda)} - e^{2\pi i\lambda}] \\ 0 & e^{2\pi i(b+\lambda)} \end{pmatrix} \begin{pmatrix} w_1 \\ w_2 \end{pmatrix}.$$

(f) Find the corresponding transformation law when z follows the following path

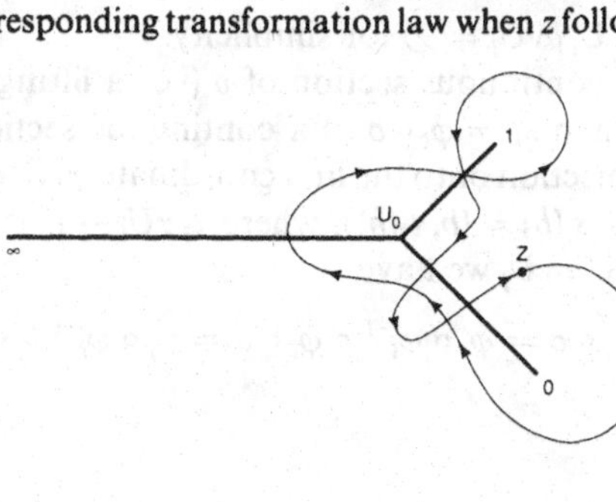

The above integrals are called hypergeometric (see [Pi], vol. 2, p. 257 ff. and vol. 3, p. 321 ff., [Hen]). The function $v_\infty - v_1$ admits a power series expansion of the form

$$\text{const.}\left(1+\frac{\alpha\beta}{1\gamma}z+\cdots+\frac{\alpha(\alpha+1)\ldots(\alpha+n-1)}{1.2\ldots n}\cdot\frac{\beta(\beta+1)\ldots(\beta+n-1)}{\gamma(\gamma+1)\ldots(\gamma+n-1)}z^n+\cdots\right)$$

when $|z|<1$, α, β, γ are parameters related to a, b, λ by the expressions

$$a=\alpha-\gamma, \qquad b=\gamma-\beta-1, \qquad \lambda=-\alpha, \qquad \gamma>\beta>0.$$

Moreover, there is no restriction on α for $v_\infty - v_1$ to be well defined. It is a solution of the hypergeometric differential equation, see Exercise 5.15.7 below.

§7. Cocycles

Let X, Y be two topological spaces, X connected, and $p: Y \to X$ a covering map such that every fiber has the same cardinality as a certain set F (which will be considered with the discrete topology). Let $(U_i)_{i\in I}$ be a covering of X by trivializing open sets and, for every $i \in I$, let φ_i be a trivialization $\varphi_i : p^{-1}(U_i) \to U_i \times F$.

If $U_i \cap U_j$ is not empty we have, for $(b, x) \in (U_i \cap U_j) \times F$,

$$(\varphi_i \circ \varphi_j^{-1})(b, x) = (b, g_{ij}(b)x),$$

where g_{ij} is a locally constant map from $U_i \cap U_j$ into the group $\mathfrak{S}_F$ of permutations of F (also considered with its discrete topology).

If $U_i \cap U_j \cap U_k \neq \varnothing$, we have

$$g_{ij}(b) \circ g_{jk}(b) \circ g_{ki}(b) = id_F, \qquad b \in U_i \cap U_j \cap U_k,$$

since

$$(\varphi_i \circ \varphi_j^{-1}) \circ (\varphi_j \circ \varphi_k^{-1}) \circ (\varphi_k \circ \varphi_i^{-1}) = id_{(U_i \cap U_j \cap U_k)\times F}.$$

In particular,

$$g_{ii}(b) = id_F \qquad \text{for } b \in U_i$$

$$g_{ij}(b) = g_{ji}(b))^{-1}, \qquad \text{for } b \in U_i \cap U_j.$$

We will set $g_{ij} = id_F$ if $U_i \cap U_j = \varnothing$ for simplicity.

Let $\sigma : X \to Y$ be a continuous section of p (i.e., a lifting of p, if it exists). Over the set U_i, the map $s_i := \varphi_i \circ \sigma$ is a continuous section of the "trivial covering map," the projection onto the first coordinate, $pr_1 : U_i \times F \to U_i$. This map has the form $b \mapsto s_i(b) = (b, \tau_i(b))$, where $\tau_i : U_i \to F$ is locally constant. Over the intersection $U_i \cap U_j$ we have

$$s_j = \varphi_j \circ \sigma = \varphi_j \circ \varphi_i^{-1} \circ \varphi_i \circ \sigma = \varphi_j \circ \varphi_i^{-1} \circ s_i,$$

whence

$$s_j(b) = (b, \tau_j(b)) = (b, g_{ji}(b)(\tau_i(b))),$$

that is,

$$\tau_j(b) = g_{ji}(b)(\tau_i(b)), \qquad (b \in U_i \cap U_j).$$

Conversely, given maps $\tau_i : U_i \to F$ which are locally constant and verify these last relations, we can construct a section σ of p by $\sigma(b) = \varphi_i^{-1}(b, \tau_i(b)) = \varphi_i^{-1} \circ s_i(b)$, whenever $b \in U_i$.

Let us remark that since F and $\mathfrak{S}_F$ are given the discrete topology, the maps τ_i and g_{ij} are continuous. Hence the section σ thus defined is a continuous section.

5.7.1. Definitions

(1) Let X be a topological space, G a group (endowed with the discrete topology), e the identity element of G, and $(U_i)_{i \in I}$ a covering of X. A ***cocycle for the covering*** $(U_i)_{i \in I}$ ***with values in G*** is a family $(g_{ij})_{(i,j) \in I \times I}$ such that, if $U_i \cap U_j \neq \varnothing$, then g_{ij} is a locally constant map of $U_i \cap U_j$ into G and, if $U_i \cap U_j = \varnothing$, $g_{ij} \equiv e$. We assume they verify the cocycle condition:

$$g_{ij}(b) g_{jk}(b) g_{ki}(b) = e, \qquad \text{for every } b \in U_i \cap U_j \cap U_k.$$

(2) We say that a cocycle (g_{ij}) for the covering $(U_i)_{i \in I}$ of X with values in G is a ***trivial cocycle*** if there is a family of locally constant maps $f_i : U_i \to G$ such that, if $U_i \cap U_j \neq \varnothing$,

$$g_{ij}(b) = f_i(b) f_j(b)^{-1}, \qquad \text{for every } b \in U_i \cap U_j.$$

We have just seen that given a covering map with connected base space, we can associate to it a cocycle with values in the group of permutations of the fiber. Conversely, we have the following theorem.

5.7.2. Theorem. *Let X be a topological space, $(U_i)_{i \in I}$ an open covering of X, F a set, and (g_{ij}) a cocycle for this covering with values in the group $\mathfrak{S}_F$ of permutations of F. There is a covering map $p : \tilde{X} \to X$ for which the g_{ij} are the elements of the cocycle found by the previous procedure. Two covering spaces having the same property are isomorphic.*

Proof. Let Y be the topological sum of the $U_i \times F$:

$$Y = \dot{\bigcup_{i \in I}} (\{i\} \times U_i \times F) = \{(i, b, \xi) \in I \times X \times F : b \in U_i\},$$

where I and F are considered with the discrete topology.

Let ρ be the equivalence relation in Y which identifies (i, b, ξ) to (j, c, η) if $b = c$ and $\xi = g_{ij}(b)(\eta)$. If $U_i \cap U_j \neq \varnothing$ we can define a map

$$h_{ij} : \{j\} \times (U_i \cap U_j) \times F \to \{i\} \times (U_i \cap U_j) \times F$$
$$(j, b, \eta) \mapsto (i, b, g_{ij}(\eta)).$$

The maps h_{ij} are homeomorphisms, with inverse h_{ji}. Furthermore, the equivalence class of (j, b, η) is given by $\{(i, b, g_{ij}(\eta)) : \text{for every } i \in I \text{ such that } b \in U_i\}$. It follows that the equivalence relation ρ is open; if $\omega \subseteq \{j\} \times U_j \times F$ we have

$$\text{saturated of } \omega = \text{sat}(\omega) = \bigcup_{i : U_i \cap U_j \neq \varnothing} h_{ij}(\omega \cap (\{j\} \times (U_i \cap U_j) \times F)).$$

Let $\tilde{X} := Y/\rho$ with the quotient topology and $q : Y \to \tilde{X}$ the canonical projection. Then q is an open map. The projection $pr_2 : I \times X \times F \to X, (i, b, \xi) \mapsto b$, restricted to Y passes to the quotient and induces a continuous surjective map $p : \tilde{X} \to X$, such that $p \circ q = pr_2$.

This map p is also open. Namely, pr_2 is open on Y because

$$pr_2(\{i\} \times U \times \{\xi\}) = U,$$

for every open subset of U_i, and $p(\Omega) = pr_2(q^{-1}(\Omega))$, for every open $\Omega \subseteq X$.

The map p is locally injective; since q is open, the sets $q(\{i\} \times U_i \times \{\xi\})$ from an open covering of $\tilde{X}$. On such a set, p is necessarily injective. In fact, if $p(q(i, b, \xi)) = q(p(i, b', \xi))$ with $b, b' \in U_i$, we have $b = b'$ since $p \circ q = pr_2$.

Therefore we have shown that p is a surjective local homeomorphism.The maps $\sigma_{i,\xi} : U_i \to \tilde{X}$ defined, for $i \in I$, $\xi \in F$ fixed, by $\sigma_{i,\xi}(b) := q(i, b, \xi)$, are evidently continuous sections of p and $\sigma_{i,\xi}(U_i) = q(\{i\} \times U_i \times \{\xi\})$ is an open subset of $\tilde{X}$. If $\xi \neq \eta$ we have that the open sets $\sigma_{i,\xi}(U_i)$ and $\sigma_{i,\eta}(U_i)$ are disjoint. If not, let $x \in \sigma_{i,\xi}(U_i) \cap \sigma_{i,\eta}(U_i)$ and $b = p(x)$. Then $q(i, b, \xi) = q(i, b, \eta)$ and $\eta = g_{ii}(b)\xi = \xi$. Finally, $p^{-1}(U_i) = \dot{\bigcup}_{\xi \in F} \sigma_{i,\xi}(U_i) = \dot{\bigcup}_{\xi \in F} q(\{i\} \times U_i \times \{\xi\})$. Hence, p is a covering map.

We see now that $q_i := q|\{i\} \times U_i \times F$ is a homeomorphism of $\{i\} \times U_i \times F$ onto $p^{-1}(U_i)$ such that $(p \circ q_i)(i, b, \xi) = b$. The map

$$\varphi_i := q_i^{-1} : p^{-1}(U_i) \to \{i\} \times U_i \times F \simeq U_i \times F$$

is a trivialization of p over U_i. One can verify without difficulty that

$$(\varphi_i \circ \varphi_j^{-1})(b, \xi) = (b, g_{ij}(b)(\xi)).$$

Finally, let $p_1 : \tilde{X}_1 \to X$ be a covering map with trivializations $\varphi_i' : p_1^{-1}(U_i) \to U_i \times F$ such that $(\varphi_i' \circ (\varphi_j')^{-1})(b, \xi) = (b, g_{ij}(b)(\xi))$. Then the continuous and open map $\theta : Y \to \tilde{X}_1$ equal to $(\varphi_i')^{-1}$ on $\{i\} \times U_i \times F$, passes to the quotient and determines an isomorphism of covering spaces from $\tilde{X}$ to $\tilde{X}_1$. This concludes the proof of the theorem. □

5.7.3. Corollary. *Let G be a group and (g_{ij}) a cocycle with values in G for a covering $(U_i)_{i \in I}$ of a topological space X. Then there is a covering map $p : \tilde{X} \to X$, which is unique up to isomorphism, such that the fibers are exactly G and the canonical cocycle associated can be identified to (g_{ij}). Moreover, there is a group* *monomorphism* $\varphi: G \to G(p)$ *such that, via this monomorphism,* G *acts freely and transitively in every fiber.*

PROOF. We take $F = G$ (hence $\mathfrak{S}_F \equiv \mathfrak{S}_G$) in the preceding theorem. By the canonical injection $\mathrm{j}: G \to \mathfrak{S}_G$, $\mathrm{j}(\xi)(\eta) = \xi\eta$, the given cocycle can be considered taking values in $\mathfrak{S}_G$. Therefore Theorem 5.7.2 is applicable to this case and we retain the notation from its proof.

The group G acts on $I \times X \times G$ by

$$s \cdot (i, b, \xi) = (i, b, \xi s^{-1}), \qquad s \in G.$$

Under this action the set Y, defined in the proof of the previous theorem, is stable ($s \cdot Y \subseteq Y$ for every $s \in G$) and this action is compatible with the equivalence relation ρ. If $(i, b, \xi)\rho(j, b, \eta)$ we have $\xi = g_{ij}(b)\eta$ and hence,

$$[s \cdot (i, b, \xi)]\rho[s \cdot (j, b, \eta)],$$

since $\xi s^{-1} = g_{ij}(b)\eta s^{-1}$. It follows that we can let G act on $\tilde{X}$ via

$$s \cdot q(i, b, \xi) = q(i, b, \xi s^{-1}).$$

The morphisms $\varphi(s)$ of $\tilde{X}$ constructed this way are automorphisms of the covering map $p: \tilde{X} \to X$. In fact, since $p \circ q = pr_2$, we have

$$p(s \cdot q(i, b, \xi)) = p(q(i, b, \xi s^{-1})) = b.$$

It is easy to see that the map $\varphi: G \to G(p)$ defined by the preceding construction is a group monomorphism.

Finally, let $b_0 \in X$. We have $p^{-1}(b_0) = q(\{i : b_0 \in U_i\} \times \{b_0\} \times G) \simeq G$. We want to check that the action of G is free and transitive on the fiber above b_0. Let $I_0 = \{i : b_0 \in U_i\}$ and choose $i_0 \in I_0$. Let $\alpha: I_0 \times \{b_0\} \times G \to G$ be given by $\alpha(i, b_0, \xi) = g_{i_0 i}(b_0)\xi$. We have $\alpha(i, b_0, \xi) = \alpha(j, b_0, \eta)$ if and only if $q(i, b_0, \xi) = q(j, b_0, \eta)$. Let $z_1 = q(i, b_0, \xi)$ and $z_2 = q(j, b_0, \eta)$, with $i, j \in I_0$. If there is an $s \in G$ such that $s \cdot z_1 = z_2$, that is, $s \cdot z_1 = q(i, b_0, \xi s^{-1}) = q(j, b_0, \eta)$, we have $\xi s^{-1} = g_{ij}(b_0)\eta$, and it follows that $s = \eta^{-1} \cdot g_{ji}(b_0) \cdot \xi$. Conversely, this s will definitely send z_1 into z_2. This shows that G acts freely and transitively on the fiber above b_0. □

It is clear that the definition of cocycles depends on the covering. On the other hand, if we have another covering $(V_m)_{m \in M}$, such that each $V_m \subseteq U_l$ for some l, we can define a cocycle element g_{mn} on $V_m \cap V_n$ by restriction. In this way for instance, when X is a Riemann surface, we can assume that every U_l is homeomorphic to a disk. An example of application of this type of condition on the covering appears in the following proposition.

5.7.4. Proposition. *Let* X *be a connected topological space that is locally connected by arcs. Let* $(U_i)_{i \in I}$ *be an open covering of* X *by connected open sets,* F

a set, and (g_{ij}) *a cocycle for the covering with values in* $\mathfrak{S}_F$. *Let* $\tilde{X}$ *be the covering space constructed in Theorem 5.7.2. Keeping the notation of the proof of §5.7.2, a sufficient condition for the connectivity of* $\tilde{X}$ *is the following:*

For every (i, b, ξ) *and* (j, c, η) *in* Y *there is a finite sequence of indices* $i_1, i_2, \ldots, i_n$, *a sequence* $b_1, \ldots, b_n \in X$, *and a sequence* $\xi_1, \ldots, \xi_n \in F$ *such that*

(1) $i_1 = i, \xi_1 = \xi, b_1 = b$;
(2) $i_n = j, \xi_n = \eta, b_n = c$; *and*
(3) *for* $1 \leq k \leq n+1$, $b_{k+1} \in U_{i_k} \cap U_{i_{k+1}}, g_{i_{k+1}, i_k}(b_{k+1})(\xi_k) = \xi_{k+1}$.

PROOF. Given $z_1, z_2 \in \tilde{X}$ it is enough to find a continuous path γ in $\tilde{X}$ starting at z_1 and ending at z_2. Let $b = p(z_1)$, $c = p(z_2)$, and $i, j \in I$ such that $b \in U_i$, $c \in U_j$. Choose any $\xi, \eta \in F$ such that $q(i, b, \xi) = z_1$, $q(j, c, \eta) = z_2$, and corresponding finite sequences (i_k), (ξ_k), (b_k). For each k there is an arc γ_k joining b_k to b_{k+1} in U_{i_k} since the U_i are arcwise connected. One can then define a path $\gamma: [1, n] \to \tilde{X}$ by

$$\gamma(t) := q(i_k, U_{i_k}, \gamma_k(t-k)) \qquad \text{for } k \leq t \leq k+1. \qquad \square$$

As an application of the concept of cocycles we give the construction of the Riemann surface of a germ $f \in \mathcal{O}_{z_0}$, $z_0 \in X$, X connected Riemann surface, when f admits an analytic continuation along any path in X starting at z_0 (cf. Proposition 5.6.12).

Let X be a connected Riemann surface $(U_i)_{i \in I}$ a covering of X by simply connected open sets, $z_0 \in U_{i_0}$, and $f \in \mathcal{O}_{z_0}$ a germ of a holomorphic function which has an analytic continuation along any continuous path γ starting at z_0.

5.7.5. Lemma. *Let* $i \in I$, $z_i \in U_i$ *and* γ *be a path starting at* z_0 *and ending at* z_i. *There is a function* $f_\gamma \in \mathscr{H}(U_i)$ *which can be obtained as the analytic continuation of* f. *This function* f_γ *can be obtained by analytic continuation in* U_i *of the analytic continuation of* f *along* γ. *If we replace* z_i *and* γ *by* $\gamma' = \gamma \cdot \alpha$, $z_i' =$ *endpoint of* γ', α *a path in* U_i *starting at* z_i, *then the function* $f_{\gamma'} \in \mathscr{H}(U_i)$ *thus obtained coincides with* f_γ. *Moreover,* f_γ *depends only on the homotopy class of* γ *in* X.

PROOF. By hypothesis, f has an analytic continuation along γ. Let g be the germ of a holomorphic function at $\gamma(1)$ obtained this way. By hypothesis g can also be analytically continued along any path of origin $\gamma(1)$, which is entirely contained in U_i. Hence by §5.6.8, there is a function $f_\gamma \in \mathscr{H}(U_i)$ whose germ at $\gamma(1)$ is exactly g. The other statements in the lemma are clear from the definition of g_γ. $\square$

Recall that $\mathscr{C}(X, z_0, z_i)$ is the set of all paths in X joining z_0 to $z_i \in U_i$, and $\pi_1(X, z_0, z_1)$ the set of their homotopy classes in X (under homotopies with fixed endpoints z_0, z_i). For every path $\alpha \in \mathscr{C}(U_i, z_i, z_i')$ the map $[\gamma] \to [\gamma \cdot \alpha]$ from $\pi_1(X, z_0, z_i)$ into $\pi_1(X, z_0, z_i')$ is a bijection. For this reason we denote by $\pi_1(X, z_0, U_i)$ one arbitrary choice among these sets. Introduce the equivalence relation ρ_i in $\pi_1(X, z_0, U_i)$ given by $[\gamma]\rho_i[\gamma']$ if and only if $f_\gamma = f_{\gamma'}$ (with the notations of §5.7.5). Let E_i be the quotient set $\pi_1(X, z_0, U_i)/\rho_i$, denote its elements by $[[\gamma]]$ and f_γ the function in $\mathscr{H}(U_i)$ corresponding to $[[\gamma]]$. We want to show that all the sets E_i are equipotent.

Let β be a path in X joining $z_i \in U_i$ to $z_j \in U_j$. Every other path in X joining z_0 to z_j is homotopic to a path of the form $\gamma \cdot \beta$, where γ joins z_0 to z_i; if c joins z_0 to z_j, then c is homotopic to $(c\overline{\beta})\beta$.

5.7.6. Lemma. *The map $\theta_{ij}: [\alpha] \in \pi_1(X, z_0, z_i) \mapsto [\alpha\beta] \in \pi_1(X, z_0, z_j)$ is compatible with the relations ρ_i and ρ_j and determines a bijection $\varphi_{ij}: E_i \to E_j$ which depends only on the homotopy class of β. The inverse bijection is defined in an analogous way using $\overline{\beta}$ instead of β.*

PROOF. It is enough to show that θ_{ij} is compatible with ρ_i and ρ_j. Now if $[\alpha_1]\rho[\alpha_2]$, it means that $f_{\alpha_1} = f_{\alpha_2} \in \mathscr{H}(U_i)$. Hence $f_{\alpha_1\beta} = f_{\alpha_2\beta}$ and we have $[\alpha_1\beta]\rho_j[\alpha_2\beta]$. □

In what follows let F be a set equipotent to all the E_i and identify any φ_{ij} with an element from $\mathfrak{S}_F$.

With the help of the φ_{ij} we would like now to construct a cocycle (g_{ij}) with values in $\mathfrak{S}_F$. For $i \in I$, $\zeta \in U_i$, let $\alpha_{i,\zeta}$ be a path in U_i starting at z_i and ending at ζ.

Assume now that $\zeta \in U_i \cap U_j$. We have a bijection $g_{ij}(\zeta): F \to F$ which is induced by the map $[[\gamma]] \in E_j \mapsto [[\gamma \cdot \alpha_{j,\zeta} \cdot \overline{\alpha_{i,\zeta}}]] \in E_i$.

5.7.7. Lemma. *The map $g_{ij}: U_i \cap U_j \to \mathfrak{S}_F$, given by $\zeta \mapsto g_{ij}(\zeta)$ is well defined, independent of the choices of $\alpha_{i,\zeta}$ and $\alpha_{j,\zeta}$, and locally constant. The family (g_{ij}) is a cocycle with values in $\mathfrak{S}_F$ for the covering $(U_i)_{i \in I}$ of X. (As usual, we set $g_{ij} = id_F$ if $U_i \cap U_j = \varnothing$.)*

PROOF. The main observation is that if V is a simply connected open subset of a component of $U_i \cap U_j$, then the paths $\alpha_{j,\zeta} \cdot \overline{\alpha_{i,\zeta}}$ for different $\zeta \in V$ are homotopic to each other. This implies that g_{ij} is constant in V. Similarly one can see that the cocycle condition $g_{ij}(\zeta) \cdot g_{jk}(\zeta) \cdot g_{ki}(\zeta) = id_F$ for $\zeta \in U_i \cap U_j \cap U_k$ is satisfied because $\gamma\alpha_{i,\zeta} \cdot \overline{\alpha_{k,\zeta}} \cdot \alpha_{k,\zeta} \cdot \overline{\alpha_{j,\zeta}} \cdot \alpha_{j,\zeta} \cdot \overline{\alpha_{i,\zeta}}$ and γ are homotopic to each other in X. □

Before proceeding with the construction we state two almost obvious observations.

5.7.8. Lemma

1. *It is possible to choose the covering* $(U_i)_{i\in I}$ *so that the cocycle* (g_{ij}) *of* §5.7.7 *satisfies the connectivity condition of Proposition* 5.7.4.
2. *If two coverings* $(U_i)_{i\in I}$, $(U_j')_{k\in K}$, *by open simply connected sets, are such that the corresponding cocycles* (g_{ij}), $(g_{k,l}')$ *satisfy the condition of* §5.7.4, *then the associated covering spaces are homeomorphic.*

PROOF. The covering of X by all the coordinate patches satisfies this condition. The second statement is an easy exercise for the reader. We note that the covering spaces are actually biholomorphic. □

By §5.7.2 and Lemma 5.7.7 we conclude that there is a covering map $p:\tilde{X}\to X$ defined by the cocycle (g_{ij}), $\tilde{X}$ is a Riemann surface, and p is a local biholomorphism. We want to show now the existence of a holomorphic function $\tilde{f}:\tilde{X}\to\mathbb{C}$ which will give us the maximal analytic continuation of the germ f. For $(i,\zeta,[[\gamma]])\in\{i\}\times U_i\times E_i$ we set

$$\varphi(i,\zeta,[[\gamma]])=f_\gamma(\zeta).$$

If $(i,\zeta,[[\alpha]])$ and $(j,\zeta,[[\beta]])$ are ρ-equivalent in $Y=\dot{\bigcup\limits_{i\in J}}\{i\}\times U_i\times E_i$ (with the notation of §5.7.2) that means that $[[\alpha]]=g_{ij}(\zeta)[[\beta]]=[[\beta\cdot\alpha_{j,\zeta}\cdot\overline{\alpha_{i,\zeta}}]]$. It follows that the functions f_α and f_β coincide in a neighborhood of ζ and hence, the function $\tilde{f}:\tilde{X}\to\mathbb{C}$ defined by $\tilde{f}(q(i,\zeta,[[\gamma]]))=\varphi(i,\zeta,[[\gamma]])$ is actually holomorphic in $\tilde{X}$.

Let $V=q(\{i_0\}\times U_{i_0}\times[[\varepsilon_{z_0}]])$ ($\varepsilon_{z_0}=$ constant loop at z_0), then $p|V:V\to V_{i_0}$ is a biholomorphism, $p(q(i_0,z_0,[[\varepsilon_{z_0}]]))=z_0$ and $p_*(\tilde{f}_{z_0})=f$. All of this assures that $(\tilde{X},p,\tilde{f},q(i_0,z_0,[[\varepsilon_{z_0}]]))$ is an analytic continuation of f.

Finally, let us show this analytic continuation is maximal. We know that the manifold $\tilde{X}$ is arcwise connected. Assume now that (Z,r,g,c) is another analytic continuation of f. Let us define $f:Z\to X$ as follows: if $\zeta\in Z$ and $r(\zeta)=z$, the germ $r_*(g_\zeta)\in\mathcal{O}_{X,\zeta}$ is an analytic continuation of f along a curve joining z_0 to z (see §5.6.10). Hence, there is a unique $w\in\tilde{X}$, $w=q(i,z,[[\gamma]])$ such that $r_*(g_\zeta)=(f_\gamma)_\zeta$. We set $F(\zeta)=w$. It is not hard to show that F is a holomorphic map such that $p\circ F=r$, $F(c)=q(i_0,z_0,[[\varepsilon_{z_0}]])$ and $F^*(\tilde{f})=g$.

We have in this way shown that §5.6.12 can be obtained using cocycles.

Remark. The reader will note that the constructions of the Riemann surfaces of the square root and of the logarithm given in §5.5.7 are particular cases of the preceding recipe, which allows us to construct more concretely the Riemann surfaces of multivalued functions.

We recommend [JS] for more details on this subject and constructions that are akin to ours.

EXERCISES 5.7

1. Let $F = \{0, 1, 2\}$ and U_1, U_2 be the open subsets of $\mathbb{C}$ given by

$$U_1 = \{z \in \mathbb{C} : 0 < \arg z < 2\pi\}, \qquad U_2 = \{z \in \mathbb{C} : -\pi < \arg z < \pi\}.$$

Let $Y = (\{1\} \times U_1 \times \{0,1,2\}) \dot{\cup} (\{2\} \times U_2 \times \{0,1,2\})$, $g_{11} : U_1 \to \mathfrak{G}_F$, $g_{11} = id_F$, $g_{22} : U_2 \to \mathfrak{G}_F$, $g_{22} = id_F$,

$$g_{12}(z) = \begin{cases} id_F & \text{if } \operatorname{Im} z > 0 \\ g_{12}(z)(0) = 1 & \\ g_{12}(z)(1) = 2 & \text{if } \operatorname{Im} z < 0 \\ g_{12}(z)(2) = 0 & \end{cases}$$

and $g_{21} = g_{12}^{-1}$.

Show that the construction from §5.7.7 of the Riemann surface of $z^{1/3}$ and the abstract construction of §5.4.7 lead to this same cocycle. In this way one obtains a Riemann surface X, $p : X \to \mathbb{C}\backslash\{0\}$ a three-sheeted covering map, and a holomorphic function $F : X \to \mathbb{C}$ taking at the points $p^{-1}(z)$ the three values of $z^{1/3}$.

2. Repeat the construction of Exercise 5.7.1 for the Riemann surface of $z^{1/n}$, $n \geq 2$.

3. Let X_k, X_l denote the Riemann surfaces of $z^{1/k}$ and $z^{1/l}$, $p_k : X_k \to \mathbb{C}^*$ and $p_l : X_l \to \mathbb{C}^*$ the corresponding covering maps ($k, l \in \mathbb{N}^*$). Show there is a biholomorphic map $\sigma_{kl} : X_k \to X_l$ such that $p_l \circ \sigma_{kl} = \sigma_k$. (Hint: show first that the function $z^{1/k}$ on X_k establishes a biholomorphism between X_k and $\mathbb{C}^*$.)

4. Describe the Riemann surfaces of the functions $\sqrt{z(z-1)}$, $\sqrt{z(z-1)(z-2)}$, etc. Construct a two-sheeted covering map $p : X \to \mathbb{C}\{0, 1\}$ and a holomorphic function $F : X \to \mathbb{C}$ whose values at the two points of $p^{-1}(z)$ $(z \neq 0, 1)$ are the two possible values of $\sqrt{z(z-1)}$. Similarly, for $\sqrt{z(z-1)(z-2)}$, etc.

5. Let Ω be an open subset of $\mathbb{C}$, ω a closed 1-form of class C^1 in Ω. For every U open, simply connected subset of Ω we let f^U be a C^2-function in U such that $df = \omega | U$. Let $(U_i)_{i \in I}$ be an open covering of Ω by simply connected sets. Construct a cocycle associated to this covering by $g_{ij}(z) = f^{U_j} - f^{U_i}$, with values in $\mathbb{C}$ (as an additive group). Use Corollary 5.7.3 to obtain a covering map $p : X \to \Omega$.

What properties does this map have? What happens when Ω is simply connected? What if all the periods of ω are zero? (Compare also with §1.7.)

§8. Group Actions and Covering Spaces

We will first develop a few elementary aspects of the theory of continuous group actions on topological spaces, among other things to permit a better understanding of the following property ($\mathscr{D}$), a property which will have important consequences for covering spaces.

Let X be a topological space and G a topological group that acts continuously (on the left) on X. One says that this action satisfies the property

($\mathscr{D}$) if every $x \in X$ has a neighborhood U such that

$$\{g \in G : gU \cap U \neq \varnothing\} = \{e\}$$

(that is, the sets gU are pairwise disjoints).

One says the action of G on X is ***discrete*** when ($\mathscr{D}$) is satisfied.

5.8.1. Proposition. *Let X be a topological space and G a topological group acting continuously on the left on X. Let $p : X \to X/G$ be the canonical projection. If the condition ($\mathscr{D}$) is satisfied then p is a covering map. An analogous result holds for actions on the right.*

PROOF. The equivalence relation ρ defining X/G is, of course, $x\rho x'$ if and only if there is $g \in G$ such that $x' = g \cdot x$. It is an open relation since $\operatorname{sat}(\Omega) = \bigcup_{g \in G} g \cdot \Omega$, and $g \cdot \Omega$ is open in X if Ω is open in X. Therefore, the canonical projection π is both continuous and open.

We claim that p is also locally injective. Let $x \in X$, let U_x be the open neighborhood of x whose existence is guaranteed by ($\mathscr{D}$). It follows that p must be injective on U_x. If not, there are two distinct points $x_1, x_2 \in U_x$ and $g \in G$ such that $x_2 = gx_1$. This clearly contradicts ($\mathscr{D}$).

Let us show now that for every $x \in X$, the open set $V_x = p(U_x)$ trivializes p. It is clear that $p|U_x : U_x \to V_x$ is a homeomorphism. Clearly the same is true for $p|(gU_x) : gU_x \to V_x$. Furthermore, it is absolutely obvious that $p^{-1}(V_x) = \dot{\bigcup}_{g \in G} (gU_x)$. This concludes the proof of the proposition. □

5.8.2. Corollary. *Let G be a topological group and H a discrete subgroup of G. Then, the canonical projection $\pi : G \to G/H$ is a covering map. (Here G/H denotes the space of left cosets of H, $\{Hg : g \in G\}$.)*

PROOF. The group H acts on G in the obvious manner: $(h, g) \mapsto hg$ ($h \in H, g \in G$). The equivalence relation ρ from §5.8.1 says $g_1 \rho g_2$ if and only if $g_1 g_2^{-1} \in H$, i.e., $g_1 \in Hg_2$. Hence, once we verify the property ($\mathscr{D}$), we will be done. Since H is discrete, there is a neighborhood V of e in G such that $V \cap H = \{e\}$. Since G is a topological group, there is another neighborhood U of e such that $U \cdot U^{-1} \subseteq V$. If $h \in H$, $h \neq e$ we see that $hU \cap U = \varnothing$. One can see that this implies ($\mathscr{D}$). □

5.8.3. Proposition. *Let G be a topological group acting continuously on a connected topological space X in such a way that ($\mathscr{D}$) holds. Then, the group $G(p)$ of automorphisms of the covering map $p : X \to X/G$ is isomorphic to G.*

PROOF. Every $g \in G$ determines an automorphism $\tilde{g}$ by $x \mapsto g \cdot x$. Conversely, if $s \in G(p)$ and $x \in X$ there is $g \in G$ such that $s(x) = g \cdot x$. Hence $s = \tilde{g}$, by the uniqueness of the liftings of p taking the value $s(x) = g \cdot x$ at x. □

Let now $p: \tilde{X} \to X$ be a covering map. The discrete group $G(p)$ acts on the left on $\tilde{X}$. It acts on $\tilde{X}$ and any fiber $p^{-1}(x)$ in a continuous fashion. Moreover, if $\tilde{X}$ is connected, then $G(p)$ acts freely. Namely, if $s \in G$ is such that s and $id_{\tilde{X}}$ coincide at a point x, then s and $id_{\tilde{X}}$ are two liftings of p which coincide at a point, hence $s = id_{\tilde{X}}$. More generally we have the following theorem.

5.8.4. Theorem. *If $p: \tilde{X} \to X$ is a covering map, $\tilde{X}$ is connected, and X is locally connected, then the group $G(p)$ acts freely and $(\mathscr{D})$ holds.*

Proof. Let $y \in \tilde{X}$ and $x = p(y)$. There is an open connected neighborhood of x trivializing p and a unique continuous section σ of p over U such that $\sigma(x) = y$. Let $V = \sigma(U)$. If $s \in G(p)$ is such that $sV \cap V \neq \varnothing$, then the two sections $s \circ \sigma$ and s verify $(s \circ \sigma)(U) \cap \sigma(U) \neq \varnothing$. Since U was trivializing, this implies $s \circ \sigma = \sigma$. The free action proved before implies $s = id_{\tilde{X}}$. □

5.8.5. Corollary. *If $p: \tilde{X} \to X$ is a covering map, $\tilde{X}$ connected, X locally connected, then the canonical projection $q: \tilde{X} \to \tilde{X}/G(p)$ is a covering map such that $G(q) = G(p)$.*

5.8.6. Proposition. *Let $p: \tilde{X} \to X$ be a covering map with $\tilde{X}$ connected and X locally connected. Let $\pi: \tilde{X}/G(p) \to X$ induced by p. Then π is a covering map such that, for every $x \in X$, $\pi^{-1}(x)$ is isomorphic to the homogeneous space $p^{-1}(x)/G(p)$.*

Proof. The projection p passes to the quotient and induces $\pi: \tilde{X}/G(p) \to X$ such that $\pi \circ q = p$, where $q: \tilde{X} \to \tilde{X}/G(p)$ is a covering map by §5.8.5.

$$\begin{array}{ccc} \tilde{X} & \xrightarrow{p} & X \\ & {\scriptstyle q}\searrow \quad \nearrow{\scriptstyle \pi} & \\ & \tilde{X}/G(p) & \end{array}$$

Let $x \in X$, V open connected neighborhood of x which trivializes p. Let $(s_z)_{z \in p^{-1}(x)}$ be the family of all sections of p over V. They verify:

(a) $s_z(x) = z$ and $s_z(V)$ open in $\tilde{X}$;
(b) $s_{z_1}(V) \cap s_{z_2}(V) \neq \varnothing$ if and only if $z_1 = z_2$;
(c) $p^{-1}(V) = \bigcup_{z \in p^{-1}(x)} s_z(V)$.

For every $z \in p^{-1}(x)$, $q \circ s_z$ is a homeomorphism of V onto $s_z(V)$. This set is open in $\pi^{-1}(V)$, since $q \circ p_z$ is injective and open on V.

Let us choose now for every $y \in \pi^{-1}(x)$ a point $z(y) \in q^{-1}(y)$. This way we obtain a family of sections $\sigma_y = q \circ s_{z(y)}$ $(y \in \pi^{-1}(x))$ of π above V, and

$\pi^{-1}(V) = \bigcup_{y \in \pi^{-1}(x)} \sigma_y(V)$. Hence $\pi : \tilde{X}/G(p) \to X$ is a covering map. It is also clear that $\pi^{-1}(x)$ can be identified to $p^{-1}(x)/G(p)$. This shows the proposition is correct. □

To finish this section let us mention the following.

5.8.7. Corollary. *Let $p : \tilde{X} \to X$ be a covering map with $\tilde{X}$ connected and X locally connected. The group $G(p)$ acts transitively on every fiber $p^{-1}(x)$ if and only if it acts transitively on one fiber.*

§9. Galois Coverings

5.9.1. Definition. A covering map $p : \tilde{X} \to X$, where $\tilde{X}$ is connected and X is locally connected, is called a ***Galois covering*** if its group of automorphisms $G(p)$ acts transitively on each fiber $p^{-1}(x)$ $(x \in X)$.

Under these conditions $\pi : \tilde{X}/G(p) \to X$ is a homeomorphism and the covering $p : \tilde{X} \to X$ is isomorphic to the covering $q : \tilde{X} \to \tilde{X}/G(p)$.

Remarks

(1) Every endomorphism of a Galois covering is an automorphism: let h be a morphism from the Galois covering, $p : \tilde{X} \to X$ into itself, and let $z \in \tilde{X}$. There is a $g \in G(p)$ such that $h(z) = g \cdot z = \tilde{g}(z)$. It follows that $\tilde{g} = h$.
(2) The following coverings are Galois:
   (a) $p : \mathbb{C}^* \to \mathbb{C}$, $p(z) = z^k$, $k \in \mathbb{N}^*$, $G(p) \cong \mathbb{Z}/k\mathbb{Z}$,
   (b) $p : \mathbb{C} \to \mathbb{C}^*$, $p(z) = \exp(z)$, $G(p) \cong \mathbb{Z}$.

The theorem that follows is one of the most important of the theory. The reader will recognize that our previous construction of the integral of closed differential forms along continuous paths is just a particular case of this theorem. It is just the fact that the interval $[0, 1]$ is both locally connected and simply connected, hence it satisfies the requirements of Theorem 5.9.2.

5.9.2. Theorem. *Let X be a connected, locally connected space. The following conditions are equivalent:*

(a) *Every covering map $p : \tilde{X} \to X$ is trivial.*
(b) *Every connected covering of X is a homeomorphism.*
(c) *Every Galois covering of X is a homeomorphism.*
(d) *For every group G and every open covering $(U_i)_{i \in I}$ of X, every cocycle with values in G for the covering $(U_i)_{i \in I}$ is trivial.*

Proof. (a) ⇒ (b). The connectedness of $\tilde{X}$, X and the existence of a global section of $p : \tilde{X} \to X$ implies the uniqueness of this section. It follows that p is a homeomorphism.

(b) ⇒ (c). By definition of Galois coverings, $\tilde{X}$ is connected.

(c) ⇒ (d). Let $p : \tilde{X} \to X$ be the covering map associated to the given cocycle. Let C be a connected component of $\tilde{X}$. We have the following lemma.

5.9.3. Lemma. *Let $p : \tilde{X} \to X$ be a covering map whose base X is connected and locally connected. Let C be a connected component of $\tilde{X}$. The map $p|C : C \to X$ is a covering map.*

PROOF. Let $x \in X$ and V be a connected open neighborhood of x trivializing p. Let $(s_z)_{z \in p^{-1}(x)}$ be the family of sections of p over V such that $p^{-1}(V) = \dot{\bigcup}_{z \in p^{-1}(x)} s_z(V)$. For every $z \in p^{-1}(x)$ we have either $s_z(V) \cap C = \varnothing$ or $s_z(V) \subseteq C$. Let $J := \{z \in p^{-1}(x) : s_z(V) \subseteq C\}$. Then $(p|C)^{-1}(V) = \dot{\bigcup}_{z \in J} s_z(V)$, hence $p|C : C \to X$ is a covering map. □

We come back to the proof of (c) ⇒ (d) by proving that the covering map $p|C : C \to X$ is Galois. In fact, $G(p|C)$ acts transitively on each fiber since $G(p)$ already had that property (by §5.7.4) and since, for $s \in G(p)$, $s(C) \cap C \neq \varnothing$ implies $s(C) = C$.

Therefore $p|C : C \to X$ is a homeomorphism. Let $\tau = (p|C)^{-1}$, then τ is a global section of p. With the notations of §5.7.2, for every $i \in I$ let $\sigma_{i,e}$ be the section $x \mapsto q(i, x, e)$ defined on U_i. On the intersection $U_i \cap U_j$ we have (cf. §5.7.4)

$$\sigma_{i,e}(x) = q(i, x, e) = q(j, x, g_{ji}(x)e) = (g_{ji}(x))^{-1} \cdot q(j, x, e) = g_{ij}(x) \cdot \sigma_{j,e}(x),$$

by definition of ρ and the action of G. In other words

$$\sigma_{i,e}(x) = g_{ij}(x)\sigma_{j,e}(x) \qquad (x \in U_i \cap U_j).$$

For every $i \in I$, $\sigma_{i,e}(x)$ and $\tau(x)$ are two points in $p^{-1}(x)$ ($x \in U_i$). Since G acts transitively, there is a function $f_i : U_i \to G$ such that

$$\sigma_{i,e}(x) = f_i(x)\tau(x) \qquad (x \in U_i).$$

The function f_i is locally constant since the sections $\sigma_{i,e}$ and $f_i(x)\tau(x)$ coincide necessarily on a neighborhood of x.

On $U_i \cap U_j$ we now have

$$\sigma_{i,e}(x) = f_i(x)\tau(x) = g_{ij}(x)\sigma_{j,e}(x) = g_{ij}(x)f_j(x)\tau(x).$$

Since G acts freely, we obtain

$$g_{ij}(x) = f_i(x)(f_j(x))^{-1} \qquad (x \in U_i \cap U_j),$$

hence the cocycle is trivial.

(d) implies (a): The base X being connected allows us to construct a cocycle (g_{ij}) with values in $\mathfrak{G}_F$, where F is the fiber of $p : \tilde{X} \to X$ such that the covering map considered is, up to isomorphism, the covering map associated to this cocycle. By (d) the cocycle is trivial. We could have assumed to start with that the open sets of $(U_i)_{i\in I}$ were connected. Let now $f_i: U_i \to \mathfrak{G}_F$ be the constant functions such that $g_{ji} = f_j f_i^{-1}$ on $U_i \cap U_j$. For every $i \in I$ and $f \in F$ let $\sigma_{i,f}$ be the section of p over U_i such that if $\varphi_i: p^{-1}(U_i) \to U_i \times F$ is the trivialization over U_i then we have

$$(\varphi_i \circ \sigma_{i,f})(x) = (x, f) = s_{i,f}(x).$$

On $U_i \cap U_j$ we have (use that the g_{ij} are constant in this case)

$$\begin{aligned}\sigma_{i,f}(x) &= (\varphi_i^{-1} \circ s_{i,f})(x) = (\varphi_j^{-1} \circ \varphi_j \circ \varphi_i^{-1} \circ s_{i,f})(x)\\ &= \varphi_j^{-1}((x, g_{ji} \cdot f)) = \sigma_{j, g_{ji}\cdot f}(x).\end{aligned}$$

If when $x \in U_i$ we replace f by $f_i \cdot f$, then

$$\sigma_{i, f_i \cdot f}(x) = \sigma_{j, g_{ji} f_i \cdot f}(x) = \sigma_{j, f_j \cdot f}(x).$$

We can set therefore

$$\sigma_f(x) = \sigma_{i, f_i \cdot f}(x) \qquad \text{when } x \in U_i,$$

and so determine a family of global sections of p, $(\sigma_f)_{f\in F}$, such that

(1) $\sigma_f(X) \cap \sigma_g(X) = \varnothing$ if $f \neq g$. If not, there are $i \in I$, $x \in U_i$, such that $\sigma_{i, f_i\cdot f}(x) = \sigma_{i, f_i \cdot g}(x)$, which by definition means $(x, f_i \cdot f) = (x, f_i \cdot g)$. Therefore $f = g$.

(2) $p^{-1}(X) = \dot{\bigcup}_{f\in F} \sigma_f(X)$. Namely, if $x_0 \in U_i$ and $z \in p^{-1}(x_0)$ there is $f \in F$ such that $z = \sigma_{i, f_i \cdot f}(x_0)$. If $\varphi_i(z) = (x_0, f_0)$ it is enough to take $f = f_i^{-1} \cdot f_0$.

These two conditions show that the covering map p was trivial.

This concludes the proof of the theorem. Its applications will become clear in the next two sections. □

§10. The Exact Sequence of a Galois Covering

Let us start with some elementary remarks about the lifting of homotopies. If $p: \tilde{X} \to X$ is a covering map, $x_0 \in X$, $z_0 \in p^{-1}(x_0)$. The map

$$\begin{gathered}\pi_1(p): \pi_1(\tilde{X}, z_0) \to \pi_1(X, x_0),\\ [\alpha] \mapsto [p \circ \alpha]\end{gathered}$$

is an injective group homomorphism. The verification that $\pi_1(p)$ is a group homomorphism is elementary. On the other hand, if $\pi_1(p)([\alpha]) = [\varepsilon_{x_0}]$, when we lift the two homotopic loops with base point x_0, $p \circ \alpha$, and ε_{x_0}, to paths with common origin z_0, we find by the uniqueness of the lifting that their liftings are α and ε_{z_0}. We know these two paths are homotopic, hence $[\alpha] = [\varepsilon_{z_0}]$, thus showing the injectivity of $\pi_1(p)$.

It is also true that $\pi_1(p)$ is surjective if and only if $\mathscr{C}(\tilde{X}, z_0, z) = \varnothing$ for every

$z \in p^{-1}(x_0)\backslash\{z_0\}$. In fact, if $\pi_1(p)$ is surjective and $\gamma \in \mathscr{C}(\tilde{X}, z_0, z)$, then $p \circ \gamma$ is a loop at x_0, which admits a lifting to a loop α with base point z_0 (by the surjectivity of $\pi_1(p)$). The uniqueness of the lifting implies that $\gamma = \alpha$, hence $z = \gamma(1) = \gamma(0) = z_0$. Therefore, if $z_0 \neq z$ we will have $\mathscr{C}(\tilde{X}, z_0, z) = \varnothing$. Conversely, if $\mathscr{C}(\tilde{X}, z_0, z) = \varnothing$ for every $z \in p^{-1}(x_0)\backslash\{z_0\}$, whenever we lift a loop α with base point x_0 to a path γ starting at z_0 such that $p \circ \gamma = \alpha$, we find that $\gamma(1) = z_0$. Hence $\pi_1(p)$ is surjective.

In particular, if $\tilde{X}$ is arcwise connected, $p^{-1}(x_0)$ reduces to a point once $\pi_1(p)$ is known to be surjective. If, for instance, X is simply connected then the surjectivity is true; hence p is a homeomorphism. More generally, the following proposition holds.

5.10.1. Proposition. *Let $p: \tilde{X} \to X$ be a covering map, $x \in X$, $z_1, z_2 \in p^{-1}(x)$. If z_1, z_2 can be connected by a continuous path in $\tilde{X}$, the groups $\pi_1(p)(\pi_1(\tilde{X}, z_1))$ and $\pi_1(p)(\pi_1(\tilde{X}, z_2))$ are conjugate in $\pi_1(X, x)$.*

PROOF. Let γ be a path in $\tilde{X}$ joining z_1 to z_2. Let $c = p \circ \gamma$. Then $[c] \in \pi_1(X, x)$. The map

$$[p \circ \alpha] \mapsto [c]^{-1}[p \circ \alpha][c]$$

is the conjugation we use to prove the proposition. □

This proposition admits a converse. If H is a subgroup of $\pi_1(X, x)$ which is conjugate to $\pi_1(p)(\pi_1(\tilde{X}, z_1))$ for some $z_1 \in p^{-1}(x)$, then there is $z_2 \in p^{-1}(x)$ such that z_1 and z_2 can be joined by a path in $\tilde{X}$ and $H = \pi_1(p)(\pi_1(\tilde{X}, z_2))$. Namely, if $H = [c]^{-1}\pi_1(p)(\pi_1(\tilde{X}, z_1))[c]$, let γ be the lifting of c starting at z_1. If we set $z_2 = \gamma(1)$ we see that $H = \pi_1(p)(\pi_1(\tilde{X}, z_2))$.

5.10.2. Proposition. *Let $p: \tilde{X} \to X$ be a covering map, with $\tilde{X}$ arcwise connected. Let $x \in X$, $z \in p^{-1}(x)$, γ_1, γ_2 two loops with base x, and α_1, α_2 their liftings with origin z. In order that α_1 and α_2 have the same endpoint, it is necessary and sufficient that $[\gamma_1][\gamma_2]^{-1} \in \pi_1(p)(\pi_1(\tilde{X}, z))$.*

PROOF. If $\alpha_1(1) = \alpha_2(1)$ then $\alpha_1\bar{\alpha}_2$ a loop with base z whose projection by p is $\gamma_1\bar{\gamma}_2$. This shows that $[\gamma_1][\gamma_2]^{-1} \in \pi_1(p)(\pi_1(\tilde{X}, z))$.

Conversely, if $[\gamma_1][\gamma_2]^{-1} \in \pi_1(p)(\pi_1(\tilde{X}, z))$, there is a loop α with base point z which is a lifting of $\gamma_1\bar{\gamma}_2$. From the uniqueness of the lifting, it follows that $\alpha(1/2) = \alpha_1(1) = \alpha_2(1)$. □

5.10.3. Proposition. *Let $p: \tilde{X} \to X$ be a covering map, $\tilde{X}$ arcwise connected. For every $x \in X$ and $z_1, z_2 \in p^{-1}(x)$, there is a loop in x with base x whose lifting starts at z_1 and ends at z_2.*

In particular, the fiber $p^{-1}(x)$ has the same cardinal as the homogeneous space $\pi_1(X, x)/\pi_1(p)(\pi_1(\tilde{X}, z))$.

PROOF. It is enough to consider the loop $\gamma = p \circ \alpha$ in X, where α is a path in $\tilde{X}$ joining z_1 to z_2, and apply the preceding result. □

It follows that under the preceding conditions, $p: \tilde{X} \to X$ is an n-sheeted covering if and only if $\pi_1(p)(\pi_1(\tilde{X}, z))$ has index n.

In this order of ideas we have the following important theorem.

5.10.4. Theorem. *Let X be a connected topological space, locally arcwise connected and simply connected. Every covering of X is trivial.*

PROOF. Let $p: \tilde{X} \to X$ be a covering map. Then $\tilde{X}$ is locally arcwise connected and its connected components are then arcwise connected. These connected components are then covering spaces of multiplicity one of X by the preceding remark. □

5.10.5. Corollary. *Let X be a connected, locally arcwise connected and simply connected topological space. For every group G and every open covering of X, every cocycle with values in G for that covering must be trivial.*

Remarks

(1) This last corollary gives a new justification of the theory of integration of closed one-forms along continuous paths.
(2) Corollary 5.6.10 is also included in this corollary.

We are now going to let the group $\pi_1(X, x_0)$ act on the fiber $p^{-1}(x_0)$ of a covering $p: \tilde{X} \to X$, $x_0 \in X$, in order to find an important relation between $\pi_1(X, x_0)$ and $G(p)$.

We have already seen that, given $[\alpha] \in \pi_1(X, x_0)$ and $z \in p^{-1}(x_0)$, one can determine a unique point $z' \in p^{-1}(x_0)$ as follows: choose a representative loop α of $[\alpha]$, lift it to a path γ of origin z, and let $z' = \gamma(1)$. We write $z' = z \cdot [\alpha]$, since it depends only on z and the homotopy class $[\alpha]$. The following identities are evident:

(1) $z \cdot [\varepsilon_{x_0}] = z$;
(2) $(z \cdot [\alpha]) \cdot [\beta] = z \cdot ([\alpha][\beta])$.

They express the fact that the group $\pi_1(X, x_0)$ acts on the right on the fiber $p^{-1}(x_0)$. The trajectory of $z \in p^{-1}(x_0)$ under this action of $\pi_1(X, x_0)$ is precisely the trace in $p^{-1}(x_0)$ of the arcwise connected component of $\tilde{X}$ containing z.

The stabilizer of z is $\pi_1(p)(\pi_1(\tilde{X}, z))$. In fact, if $[\alpha]$ stabilizes z and γ is the lifting of α such that $\gamma(0) = z$, then $\gamma(1) = z \cdot [\alpha] = z$ implies that γ is a loop. Hence $[\alpha] \in \pi_1(p)(\pi_1(\tilde{X}, z))$.

Let us assume now the $\tilde{X}$ and X are connected, locally arcwise connected, and that $p: \tilde{X} \to X$ is a Galois covering. For fixed $x_0 \in X$, $z_0 \in p^{-1}(x_0)$, one can construct a group homomorphism

$$\rho : \pi_1(X, x_0) \to G(p)$$

as follows: $\rho([\alpha])(z_0) = z_0 \cdot [\alpha]$. That is, $\rho([\alpha])$ is the unique element of $G(p)$ that transforms z_0 into $z_0 \cdot [\alpha]$.

5.10.6. Theorem. *Let $p : \tilde{X} \to X$ be a Galois covering where $\tilde{X}, X$ are connected and locally arcwise connected. The map ρ just introduced is a surjective group homomorphism whose kernel is $\pi_1(p)(\pi_1(\tilde{X}, z_0))$. In other words, the sequence*

$$\{[\varepsilon_{z_0}]\} \longrightarrow \pi_1(\tilde{X}, z_0) \xrightarrow{\pi_1(p)} \pi_1(X, x_0) \xrightarrow{\rho} G(p) \longrightarrow \{id_{\tilde{X}}\}$$

is exact. (It will be called the exact sequence of the Galois covering $p : \tilde{X} \to X$.) It also follows that the group $G(p)$ is isomorphic to $\pi_1(X, x_0)/\pi_1(p)(\pi_1(\tilde{X}, z_0))$.

PROOF. To show that $\rho([\alpha][\beta]) = \rho([\alpha]) \circ \rho([\beta])$, it is enough to show that these two automorphisms take the same value at z_0. Now

$$\rho([\alpha][\beta])(z_0) = z_0 \cdot ([\alpha][\beta]) = (z_0 \cdot [\alpha])[\beta] = \rho([\alpha])(z_0) \cdot [\beta]$$

and

$$(\rho([\alpha]) \circ \rho([\beta]))(z_0) = \rho([\alpha])(\rho([\beta])(z_0)) = \rho([\alpha])(z_0 \cdot [\beta]).$$

Therefore, we need to prove the following.

5.10.7. Lemma. *For every $s \in G(p)$, every $\beta \in \pi_1(X, x_0)$ and $z_0 \in p^{-1}(x_0)$ we have*

$$s(z_0 \cdot [\beta]) = s(z_0) \cdot [\beta].$$

PROOF. Let β be a representative loop of $[\beta]$ and γ be its lifting starting at z_0 and ending at $z_0 \cdot [\beta]$. Then $s \circ \gamma$ is the lifting of β which starts at $s(z_0)$ and ends at $s(z_0 \cdot [\beta])$. By definition of the action of $\pi_1(X, x_0)$, this endpoint is $s(z_0) \cdot [\beta]$. □

Therefore ρ is a group homomorphism whose kernel is the stabilizer of z_0 under the action of $\pi_1(X, x_0)$, i.e., $\pi_1(p)(\pi_1(\tilde{X}, z_0))$. Finally the surjectivity of ρ follows from the fact that $\tilde{X}$ is arcwise connected. Namely, if $z \in p^{-1}(x_0)$ let γ be a path in $\tilde{X}$ starting at z_0 and ending at z, then $z_0 \cdot [\alpha] = z$, with $[\alpha] = [p \circ \gamma]$. To finish the proof of the theorem we need only to recall that we know already that $\pi_1(p)$ is injective. □

5.10.8. Corollary. *Let $p : \tilde{X} \to X$ be a covering map with X, $\tilde{X}$ connected and locally arcwise connected. Assume further that $\tilde{X}$ is simply connected. Then the covering is Galois and ρ is a group isomorphism identifying $\pi_1(X, x_0)$ to $G(p)$.*

PROOF. Since $\tilde{X}$ is simply connected, ρ is an isomorphism between $\pi_1(X, x_0)$ and $G(p)$. This implies also that the covering is Galois, since if $z_0, z_1 \in p^{-1}(x_0)$

there is definitely an $[\alpha] \in \pi_1(X, x_0)$ such that $z_1 = z_0 \cdot [\alpha]$. If $f = \rho([\alpha]) \in G(p)$, then $f(z_0) = z_1$. □

§11. Universal Covering Space

5.11.1. Definition. Let X and Y be two connected topological spaces. A covering $p : Y \to X$ is said to be a ***universal covering space*** *of* X if it satisfies the following property:

For every covering map $q : Z \to X$ with Z connected, and for every $y_0 \in Y$, $z_0 \in X$ such that $p(y_0) = q(z_0)$, there is a unique lifting f of p with respect to q, $f : Y \to Z$, $q \circ f = p$ (i.e., a morphism of covering maps).

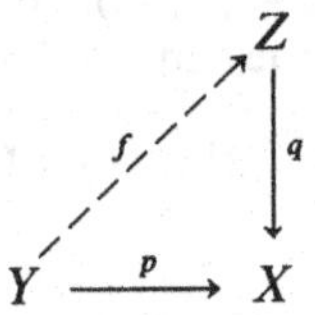

A connected space has, up to isomorphism, at most one universal covering space. Note that we should really say universal covering map, but the accepted terminology is justified by the following proposition.

5.11.2. Proposition. *Let X, Y be two connected spaces, Y simply connected. Every covering map $p : Y \to X$ is a universal covering of X.*

PROOF. The lifting of p with respect to q is assured by the hypothesis. □

One can characterize the connected, locally arcwise connected spaces X which have a universal covering space. For us it suffices the following case.

5.11.3. Theorem. *Let X be a connected topological manifold. There is a topological manifold $\tilde{X}$ connected and simply connected, and a covering map $p : \tilde{X} \to X$. (Hence $\tilde{X}$ is "the" universal covering space of X.)*

PROOF. It is enough to construct a covering map $p : \tilde{X} \to X$, with $\tilde{X}$ connected and simply connected. Let $(U_i)_{i \in I}$ be the open covering of X by all open sets which are connected and simply connected. For every $i \in I$ choose $z_i \in U_i$. As observed in §5.7, for every continuous path α in U_i which joins z_i to z_i', the map $[\gamma] \mapsto [\gamma\alpha]$ is a bijection of $\pi_1(X, z_0, z_i)$ onto $\pi_1(X, z_0, z_i')$. For this reason we will simply denote E_i one of these sets. We are going to show that all the sets E_i are equipotent. Let γ be a path joining z_i to z_j. Every path c joining z_0 to z_j is homotopic in X to a path of the form $\alpha\gamma$, where α is a continuous path joining z_0 to z_i, namely, $c\bar{\gamma}\gamma$. The following lemma is clear.

5.11.4. Lemma. *The map φ_{ji} which associates to $[\alpha] \in \pi_1(X, z_0, z_i)$ the class $[\alpha\gamma] \in \pi_1(X, z_0, z_j)$ is a bijection $\varphi_{ji}: E_i \to E_j$, which depends only on the homotopy class of γ. Its inverse is defined analogously using $\bar{\gamma}$ instead of γ.*

In what follows we denote F a set of the same cardinal as all the E_i and identify φ_{ji} to an element of $\mathfrak{G}_F$. Using the method of §5.7.7 we will construct a cocycle with values in $\mathfrak{G}_F$ for the covering $(U_i)_{i \in I}$. For every $i \in I$ and $\zeta \in U_i$ choose a path $\alpha_{i,\zeta}$ joining z_i to ζ in U_i, and for $\zeta \in U_i \cap U_j$ consider the bijection $g_{ij}(\zeta): F \to F$ which arises, modulo the preceding identifications, from the map $[\gamma] \in E_j \mapsto [\gamma\alpha_{j,\zeta}\overline{\alpha_{i,\zeta}}] \in E_i$. We have the following statement analogous to 5.7.7.

5.11.5. Lemma. *The map $g_{ij}: U_i \cap U_j \to \mathfrak{G}_F$ is well defined, independent of the choice of $\alpha_{i,\zeta}$ and $\alpha_{j,\zeta}$, and locally constant. The collection (g_{ij}) is a cocycle with values in $\mathfrak{G}_F$ for the covering $(U_i)_{i \in I}$ of X. (As always, $g_{ij} = id_F$ if $U_i \cap U_j \neq \varnothing$.)*

As a consequence of this lemma we can apply §5.7.2 and obtain a covering map $p: \tilde{X} \to X$ defined by the cocycle (g_{ij}). Keeping the notation of §5.7.2, we will show now that $\tilde{X}$ is simply connected. Let us verify first that $\tilde{X}$ is arcwise connected; it is done by the same construction as in §5.7.4. If $q(i, \zeta, [\gamma])$ is given with $\gamma(1) = \zeta$, then there is a partition $0 = t_0 < t_1 < \cdots < t_n = 1$ of $[0, 1]$ and open sets $U_{i_0}, \ldots, U_{i_n}$ of the covering such that $\gamma([t_{k-1}, t_k]) \subseteq U_{i_k}$. The curve $\tilde{\gamma}: [0, 1] \to \tilde{X}$ defined by $\tilde{\gamma}(t) = q(i_k, \gamma(t), [\gamma_t])$ if $t \in [t_{k-1}, t_k[$ for $k < n$ and $t \in [t_{n-1}, t_n]$ if $k = n$, where $\gamma_t(s) = \gamma(ts)$ $(0 \leq s \leq 1)$, is a lifting of γ with origin $q(i_0, z_0, [\varepsilon_{z_0}])$ and endpoint $q(i, \zeta, [\gamma])$, as desired. Let now α be a loop in $\tilde{X}$ with basepoint $q(i_0, z_0, [\varepsilon_{z_0}])$. Then $\gamma = p \circ \alpha$ is a loop in X with base point z_0. The previous lifting $\tilde{\gamma}$ of γ with origin $q(i_0, z_0, [\varepsilon_{z_0}])$ and endpoint $q(i_0, z_0, [\gamma])$ is, by the uniqueness of the liftings, equal to α. Therefore $\tilde{\gamma}(1) = q(i_0, z_0, [\gamma]) = q(i_0, z_0, [\varepsilon_{z_0}])$. It follows that $[\gamma] = [\varepsilon_{z_0}]$, and, by the liftings of homotopies, $[\alpha] = [\varepsilon_{q(i_0, z_0, [\varepsilon_{z_0}])}]$. This shows that $\tilde{X}$ is simply connected. □

5.11.6. Examples

(1) $\exp: \mathbb{C} \to \mathbb{C}^*$ is the universal covering space of $\mathbb{C}^*$. For $n \in \mathbb{Z}$ let $\tau_n: \mathbb{C} \to \mathbb{C}$ be the translation by $2\pi i n$. Then $\exp(\tau_n(z)) = \exp(z)$ and, hence, $\tau_n \in G(\exp)$. If $\sigma \in G(\exp)$, then $\exp(\sigma(0)) = \exp(0) = 1$ and therefore there is $n \in \mathbb{Z}$ such that $\sigma(0) = \tau_n(0)$. It follows that $\sigma = \tau_n$ and $G(\exp) \cong \mathbb{Z} = \pi_1(\mathbb{C}^*)$.

(2) If $H = \{z \in \mathbb{C} : \operatorname{Re} z < 0\}$ and $D^* = \{z \in \mathbb{C} : 0 < |z| < 1\}$, then $\exp: H \to D^*$ is the universal covering space as earlier, and one has $\pi_1(D^*) \cong \mathbb{Z}$.

5.11.7. Theorem. *Let X, Y be two connected topological manifolds, $q: Y \to X$ a covering map, and $p: \tilde{X} \to X$ the universal covering of X. Let $f: \tilde{X} \to Y$ be a lifting of p. Then f is a covering map and there is a subgroup G of $G(p)$ such that (up to homeomorphism) $Y = \tilde{X}/G$. Moreover, G is isomorphic to $\pi_1(Y)$.*

PROOF. We know that $G(p)$ acts freely and condition ($\mathscr{D}$) is satisfied (by §5.8.4). Therefore, if G is a subgroup of $G(p)$, $\tilde{X}$ is a covering space of $\tilde{X}/G$ since G also acts freely and ($\mathscr{D}$) holds for G (§5.8.1). Hence, to prove the theorem is sufficient to show that $f: \tilde{X} \to Y$ can be identified to the canonical map $\tilde{X} \to \tilde{X}/G$ for a convenient subgroup G of $G(p)$. By §5.8.3 it will also follow that $G \cong \pi_1(Y)$.

We first observe that if $\sigma \in G(p)$ then $f \circ \sigma$ is also a lifting of p. Consider $G = \{\sigma \in G(p): f \circ \sigma = f\}$, the group of isotropy of f. If $z \in \tilde{X}$, $\sigma \in G$, and $z' = \sigma(z)$ then it is obvious that $f(z) = f(z')$. Conversely, let z, $z' \in \tilde{X}$ in the same fiber $p^{-1}(x)$, $x \in X$, be such that $f(z) = f(z')$. Since $G(p)$ is transitive, there is $\sigma \in G(p)$ such that $\sigma(z) = z'$. It follows $f \circ \sigma = f$, since the two liftings of p coincide in z. To conclude the proof we need to show that f is open and its image is closed. This will imply that f is surjective and Y is homeomorphic to $\tilde{X}/G$. First, f is open. Namely, if W is a connected open set on which p is injective and $p(W)$ is trivializing for q, then on W we have that f is injective, since $p = q \circ f$. Furthermore, $f(W)$ is a connected subset of $q^{-1}(p(W))$, but $f(W)$ intersects every fiber of q above $p(W)$. Hence $f(W)$ must be a connected component of the open set $q^{-1}(p(W))$. It follows that $f(W)$ is open in Y. Second, let us show that $f(\tilde{X})$ is closed in Y. If $y \in \overline{f(\tilde{X})}$ and $x = q(y)$, choose a connected open neighborhood of x trivializing q. There is a unique component U_1 of $q^{-1}(U)$ that contains y. Let $V \subseteq U$ be a connected open neighborhood of x which trivializes p and such that $f(V) \subseteq q^{-1}(U)$. Every component of $p^{-1}(V)$ has its image under f contained in a component of $q^{-1}(U)$. Therefore U_1 contains the image of a component V_1 of $p^{-1}(V)$. Since there is $z_1 \in V_1$ such that $p(z_1) = x$ we have $f(z_1) \in f(\tilde{X}) \cap U_1$ and $q(f(z_1)) = x$. This implies that $f(z_1) = y$ since already $q(y) = x$ and $q|U_1$ is injective. This proves entirely the theorem. □

If one analyzes the proof of the existence of a universal covering space for a topological variety and the one of the existence of a maximal analytic continuation for a genus admitting analytic continuation along every curve, both obtained as applications of the notion of cocycle, one realizes that the manifold Y for which the maximal analytic continuation exists is a quotient of the universal covering space $\tilde{X}$ of the manifold X on which the germ f was defined at $x_0 \in X$. In fact, one should consider the subgroup H of $\pi_1(X, x_0)$ of those loops $[\alpha]$ such that the analytic continuation of f along α coincides with the original germ f. Then $Y \cong \tilde{X}/H$. More precisely we have the following proposition.

5.11.8. Proposition. *Let X be a connected Riemann surface, $p: \tilde{X} \to X$ a universal covering space of X, $z_0 \in X$, and $f \in \mathcal{O}_{X,z_0}$ a germ of a holomorphic function admitting analytic continuation along any curve in X which starts at z_0. Let as in §5.6, $p_*: \mathcal{O}_{\tilde{X},z_0} \to \mathcal{O}_{X,z_0}$ be the inverse of the map $p^*: \mathcal{O}_{X,z_0} \to \mathcal{O}_{\tilde{X},z_0}$ given by $p^*(\phi_{z_0}) = (\phi \circ p)_{\zeta_0}$. Then, there is a holomorphic function $\tilde{f}$ on $\tilde{X}$ and a point $\zeta_0 \in \tilde{X}$ such that $p(\zeta_0) = z_0$, $p_*(\tilde{f}_{\zeta_0}) = f_{z_0}$ and, moreover,* *there is a maximal analytic continuation* (X^*, s, f^*, ξ_0) *of* f_{z_0} *with* $S(\xi_0) = z_0$ *and*

$$\tilde{f} = f^* \circ r,$$

where $r: \tilde{X} \to X^*$ *is the lifting of* p *such that* $r(\xi_0) = \zeta_0$.

PROOF. We use the notation from the construction of $\tilde{X}$ in §5.11.3. We see that for every path γ joining z_0 to $z_i \in U_i$ there is $f_\gamma \in \mathscr{H}(U_i)$, obtained as in §5.7.5 by analytic continuation along γ. The function ψ defined on $\tilde{Y} := \dot{\bigcup}_{i \in I} \{i\} \times U_i \times \tilde{E}_i$ (where $\tilde{E}_i$ is the set $\pi_1(X, z_0, z_i)$, $z_i \in U_i$) by

$$\psi(i, \zeta, [\gamma]) := f_\gamma(\zeta)$$

induces a holomorphic function $\tilde{f}: \tilde{X} \to \mathbb{C}$ given by

$$\tilde{f}(\tilde{q}(i, \zeta, [\gamma])) := \psi(i, \zeta, [\gamma]) = f_\gamma(\zeta),$$

with $\tilde{q}: \tilde{Y} \to \tilde{X}$ as in §5.7.2.

In the same way, if we recapitulate the construction in §§5.7.5, 5.7.6, 5.7.7 and 5.7.8 of X^*, there is a function ϕ defined on $Y^* := \dot{\bigcup}_{i \in I} \{i\} \times U_i \times E_i$ where here E_i is $\tilde{E}_i$ modulo the equivalence relation ρ_i, $[\gamma] \rho_i [\gamma']$ if and only if $f_\gamma = f_{\gamma'}$), ϕ defined by

$$\phi(i, \zeta, [[\gamma]]) := f_\gamma(\zeta)$$

($[[\gamma]]$ the class of $[\gamma]$ modulo ρ_i). It induces a holomorphic function f^* on X^* by

$$f^*(q^*(i, \zeta, [[\gamma]])) := \phi(i, \zeta, [[\gamma]]) = f_\gamma(\zeta),$$

with $q^*: Y^* \to X^*$ as in §5.7.2.

It is now immediate that if r is a lifting of p and we choose $\xi_0 \in r^{-1}(\zeta_0)$, then $f^* \circ r = \tilde{f}$. □

The Riemann mapping theorem characterized $B(0, 1)$ as the "only" simply connected plane domain. A generalization of it is the characterization of the universal covering spaces of every Riemann surface. This is the content of the following theorem, usually called the uniformization theorem. The reader will find a proof using many of the concepts developed in this book in [Fo]. A rather different proof appears in [Car].

5.11.9. Uniformization Theorem. *The universal covering space* $\tilde{X}$ *of a connected Riemann surface* X *is biholomorphic to one of the three surfaces* S^2, $\mathbb{C}$ *or* $B(0, 1)$.

5.11.10. Definition. The connected Riemann surface X is said to be ***elliptic***, ***parabolic***, or ***hyperbolic***, if $\tilde{X}$ is biholomorphic to S^2, $\mathbb{C}$, or $B(0, 1)$, respectively.

5.11.11. Proposition. *An elliptic Riemann surface is biholomorphic to* S^2. *A parabolic Riemann surface is biholomorphic to one and only one of the three surfaces* $\mathbb{C}$, $\mathbb{C} \setminus \{0\}$ *or a complex torus* $\mathbb{C}/\Gamma$ *(see §5.5.7).*

PROOF. If X is elliptic then $p: S^2 \to X$ is a covering map and X is biholomorphic to $S^2/G(p)$, $G(p)$ a subgroup of the group $\operatorname{Aut}(S^2)$. All the automorphisms of S^2 are of the form

$$z \mapsto \frac{az + b}{cz + d}, \qquad \begin{pmatrix} a & b \\ c & c \end{pmatrix} \in GL(2, \mathbb{C}).$$

Such a map has necessarily one fixed point. On the other hand, two elements of $G(p)$ that coincide at one point coincide everywhere, hence $G(p) = \{id_{S^2}\}$ and $X \cong S^2$.

If X is parabolic then it is conformally equivalent to $\mathbb{C}/G(p)$, $G(p)$ a subgroup of $\operatorname{Aut}(\mathbb{C})$. Hence, we need to find out all the discrete subgroups of $\operatorname{Aut}(\mathbb{C})$ that do not have fixed points. Since any automorphism of $\mathbb{C}$ has the form

$$z \mapsto az + b, \qquad a \in \mathbb{C}^*, \qquad b \in \mathbb{C},$$

the only way it would not have a fixed point is that $a = 1$. The discreteness condition implies that $G(p)$ can only be one of two types:

(i) there is $\omega \in \mathbb{C}^*$ such that $G(p) = \{z \mapsto z + n\omega : n \in \mathbb{Z}\}$;
(ii) there are $\omega_1, \omega_2 \in \mathbb{C}^*$, $\omega_1/\omega_2 \notin \mathbb{R}$ such that

$$G(p) = \{z \mapsto z + m\omega_1 + n\omega_2 : m, n \in \mathbb{Z}.\}$$

In the second case $X \cong \mathbb{C}/\Gamma$, $\Gamma = \omega_1\mathbb{Z} \oplus \omega_2\mathbb{Z}$. In the first case, $p(z) = \exp\left(\frac{\omega}{2\pi i} z\right)$ is a concrete realization of the covering map $p: \mathbb{C} \to \mathbb{C}^*$ with group $G(p)$ isomorphic to $\omega\mathbb{Z}$. So X is conformal to $\mathbb{C}^*$ in the first case. □

Note that the four cases that appear in the last proposition, S^2, $\mathbb{C}$, $\mathbb{C}^*$, and $\mathbb{C}/\Gamma$ are topologically different. For instance, S^2 and $\mathbb{C}/\Gamma$ are both compact, but one is simply connected ($\pi_1(S^2) = 0$) and the other is not. The first corollary of this observation is the following.

5.11.12. Corollary (Little Picard Theorem). *Let $f \in \mathscr{H}(\mathbb{C})$. Then, either f is constant or $\mathbb{C} \setminus f(\mathbb{C})$ contains at most one point.*

PROOF. Let $a, b \in \mathbb{C} \setminus f(\mathbb{C})$, $a \neq b$. The Riemann surface $X = \mathbb{C} \setminus \{a, b\}$ is not homeomorphic to either $\mathbb{C}$ or $\mathbb{C}^*$, since $\pi_1(X) \neq \{0\}$, $\mathbb{Z}$, and it is clearly not compact. Therefore $\tilde{X} = B(0, 1)$. Let $p: B(0, 1) \to X$ be the covering map. Since $f: \mathbb{C} \to X$ and $\mathbb{C}$ is simply connected, there is a holomorphic lifting $g: \mathbb{C} \to B(0, 1)$, $f = p \circ g$. By Liouville's theorem, g is constant, a fortiori, f is also constant. □

Compact connected Riemann surfaces are topologically characterized by their genus g, the number of "handles" one has to add to S^2 to obtain (up to homeomorphism) X. Then $g = 0$ for S^2, $g = 1$ for a complex torus. Another immediate consequence of §5.11.11 is that a compact Riemann surface X with genus $g \geq 2$ is hyperbolic. In fact, one can also prove that if genus $g \leq 1$, then X is not hyperbolic (see [Fo]). For instance, if X is compact connected and $g = 0$ then X is conformal to S^2.

For noncompact Riemann surfaces the classification is harder to accomplish. Even for a seemingly simple example like the Riemann surface defined by the equation

$$e^z + e^w = 1,$$

takes considerable effort to show it is parabolic [De]. Good references on this subject are [AS] and [Nev].

Exercises 5.11

1. Show that two covering maps of $X = B(0,R)\backslash\{0\}$ $(0 < R \leq \infty)$ having the same multiplicity are necessarily isomorphic. Is $X = B(0,R)\backslash\{0\}$ essential for the correctness of this result?

2. In $\mathbb{R}^3$ consider the segments A_1B_1, A_2B_2, A_3B_3 such that

$$A_1 = (0,-1,1) \qquad B_1 = (0,2,1)$$
$$A_2 = (0,2,1) \qquad B_2 = (1,2,0)$$
$$A_3 = (1,2,0) \qquad B_3 = (1,0,0)$$

Let Z be the union of those segments and the set $\left\{\left(x, \sin\frac{\pi}{x}, 0\right) : 0 < x < \pi\right\}$ and $Z_n = Z + (0,0,n)$, $n \in \mathbb{Z}$. Let $p: \mathbb{R}^3 \to \mathbb{R}^2$ be the projection $(x,y,z) \mapsto (x,y)$, $X = p(Z)$, and $Y = \bigcup_{n \in \mathbb{Z}} Z_n$. Show that X is arcwise connected, Y is a connected covering space of X whose connected components by arcs are the sets Z_n, but no Z_n is a covering space for X (with the map $p|Z_n$). Show further that Y is not trivializable, even though X is simply connected (compare with §5.9.2 and §5.10.4).

3. Is the cocycle associated to the covering map $\mathbb{C}^* \to \mathbb{C}^*$, $z \mapsto z^2$ and the covering $U_1, \ldots, U_4$ of $\mathbb{C}^*$ given by

$$U_1 = \{z \in \mathbb{C} : \operatorname{Re} z > 0\}, \qquad U_2 = \{z \in \mathbb{C} : \operatorname{Re} z < 0\},$$
$$U_3 = \{z \in \mathbb{C} : \operatorname{Im} z > 0\}, \qquad U_4 = \{z \in \mathbb{C} : \operatorname{Im} z < 0\}$$

trivial?

Same question for the covering map $z \mapsto z^n$ and an open covering of $\mathbb{C}^*$ defined similarly.

4. Show that the two-sheeted covering map $m: S^3 \to SO(3)$ constructed in Exercise 5.4.9 is a Galois covering. Determine the group $G(m)$.

5. Are the covering maps associated to the functions $z^{1/n}$ in $\mathbb{C}^*$ Galois coverings? (See Exercises 5.7.1 and 5.7.2.) Same question for $\sqrt{z(z-1)}$. Determine the group of automorphisms of those coverings.

6. Let X be a Riemann surface and $\mathscr{C}(X)$ the space of complex-valued continuous functions in X.
   (i) Let $f \in \mathscr{C}(X)$ be such that $f(x) \neq 0$ for every $x \in X$.
      (a) Show there is an open covering $(U_i)_{i\in I}$ of X and continuous functions $g_i : U_i \to \mathbb{C}$ such that $f_i|U_i = e^{g_i}$.
      (b) Show that the functions $h_{k,l}$ defined by $h_{k,l} = \frac{1}{2\pi i}(g_l - g_k)$ in $U_l \cap U_k \neq \varnothing$ and $h_{k,l} = 0$ if $U_l \cap U_k = \varnothing$, determine a cocycle of the open covering $(U_i)_{i\in I}$ with values in $\mathbb{C}$.
      (c) For every $i \in I$, let g_i' be another continuous function in U_i such that $f|U_i = e^{g_i'}$, $(h_{k,l}')_{k,l}$ the corresponding cocycle. What can you say about the cocycle $(h_{k,l} - h_{k,l}')_{k,l}$?
      (d) What does it mean that the cocycle $(h_{k,l})_{k,l}$ is trivial? What can you say about $(h_{k,l})_{k,l}$ if X is simply connected?
   (ii) Assume that X is compact and let $\mathscr{U} = (U_1, \ldots, U_N)$ be a finite open covering of X, $(\alpha_i)_{1\le i\le N}$ a C^∞ partition of unity subordinate to the covering $\mathscr{U}$ $\left(\text{i.e., } \operatorname{supp}(\alpha_i) \subseteq U_i, \sum_i \alpha_i = 1\right)$. Let $(h_{k,l})_{1\le k,l\le N}$ be any cocycle for the covering $\mathscr{U}$ with values in $\mathbb{Z}$. For each k, $1 \le k \le N$, let

$$g_k = 2\pi i \sum_{l=1}^{N} \alpha_l h_{l,k} \qquad \text{in } U_k.$$

      (a) Show that g_k is continuous in U_k.
      (b) Show that $\frac{1}{2\pi i}(g_l - g_k) = h_{k,l}$ in $U_k \cap U_l$.
      (c) Show that there is $f \in \mathscr{C}(X)$, never zero, whose associated cocycle (i.e., the one constructed in (i)) is the given cocycle $(h_{k,l})_{k,l}$.

7. Let $\Gamma_1 = \omega_1\mathbb{Z} \oplus \omega_2\mathbb{Z}$, $\Gamma_2 = \alpha_1\mathbb{Z} \oplus \alpha_2\mathbb{Z}$ be two lattices of rank two in $\mathbb{C}$ (i.e., $\{\omega_1, \omega_2\}$ and $\{\alpha_1, \alpha_2\}$ are two $\mathbb{R}$-basis of $\mathbb{C}$). Assume there is a complex number $\tau \neq 0$ such that $\tau\omega_1, \tau\omega_2 \in \Gamma_2$.
   (i) Show that the function $f: \mathbb{C} \to \mathbb{C}$, $f(z) = \tau z$ induces a covering map $p: \mathbb{C}/\Gamma_1 \to \mathbb{C}/\Gamma_2$ such that $G(p)$ isomorphic to the group $\Gamma_2/\tau\Gamma_1$.
   (ii) Find all the holomorphic maps $g: \mathbb{C}/\Gamma_1 \to \mathbb{C}/\Gamma_2$ which map 0 into 0.
   (iii) Find all the biholomorphic maps of $\mathbb{C}/\Gamma_1$ into itself.

8. Let $p: Y \to X$ be a Galois covering, with both X and Y connected and locally arcwise connected. Assume $\pi_1(X)$ is generated by N loops $\gamma_1, \ldots, \gamma_N$. For $x_0 \in X$, $y \in p^{-1}(x_0)$ let $y \cdot [\gamma_j]$ represent the endpoint of the lifting of γ_j which starts at y. Show that knowing the N transformations $p^{-1}(x_0) \to p^{-1}(x_0)$ given by $y \mapsto y \cdot [\gamma_j]$ completely determines the group homomorphism $\rho: \pi_1(X, x_0) \to G(p)$ (see §5.10.6 for the notation).

   As an application, find $G(p)$ for the coverings of $\mathbb{C}\setminus\{0\}$ associated to $z^{1/n}$ and of $\mathbb{C}\setminus\{0,1\}$ associated to $\sqrt{z(z-1)}$.

9. Let X, B be two connected locally connected spaces, $p: X \to B$ a covering map, A a topological space, $f: A \to B$ a continuous map. Let

$$Y := \{(a, x) \in A \times X : f(a) = p(x)\},$$

assumed to be $\neq \varnothing$ and with the topology induced by $A \times X$. Let $p_A: Y \to A$ be the map $p_A(a, x) = a$.

(i) If U is a connected open neighborhood of $b = f(z)$ which trivializes p, show that $V = f^{-1}(U)$ is an open set trivializing p_A. (Hint: if $\mathscr{S}(U)$ is the set of sections s of p over U consider the set $\mathscr{S}_1(V)$ of maps

$$\sigma(s) : \alpha \in V \to (\alpha, s(f(\alpha)))$$

and show that

(a) $\sigma(s)(V)$ is an open subset of $p_A^{-1}(V)$;

(b) If $s_1 \neq s_2$, then $s_1(U) \cap s_2(U) = \varnothing$ and $\sigma(s_1)(V) \cap \sigma(s_2)(V) = \varnothing$;

(c) $p_A^{-1}(V) = \bigcup_{s \in \mathscr{S}(U)} \sigma(s)(V)$.

Conclude that (i) is correct.

(ii) Show that $p_A : Y \to A$ is a covering map (called ***inverse image*** *of* $p : X \to B$ *by* f).

(iii) Show that there is a unique morphism H of covering maps, $H : Y \to X$, such that if $q : D \to A$ is any covering map and $r : D \to X$ is a continuous map that satisfies $p \circ r = f \circ q$, then there is a unique continuous map $\sigma : D \to Y$ verifying $p_A \circ \sigma = q$ and $H \circ \sigma = r$.

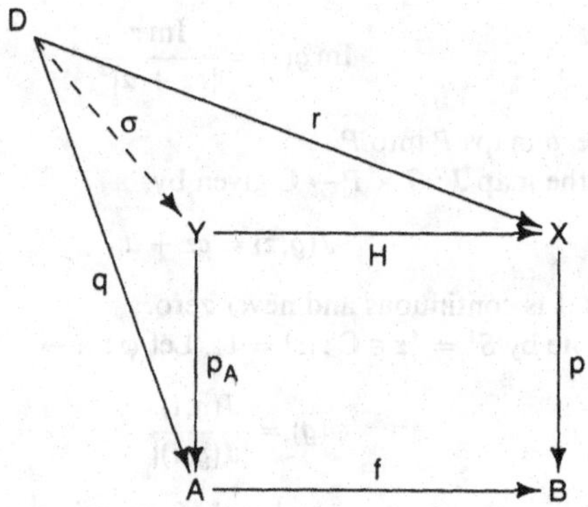

Figure 5.10

This problem will be used in the following one.)

10. (Universal covering space of $SL(2, \mathbb{R})$): Consider the topological group $G = SL(2, \mathbb{R})$ defined by

$$G = \left\{ \begin{pmatrix} a & b \\ c & d \end{pmatrix} : a, b, c, d \in \mathbb{R}, ad - bc = 1 \right\}$$

with matrix multiplication as the operation of G. It has two subgroups

$$B = \left\{ \begin{pmatrix} 1/t & u \\ 0 & t \end{pmatrix} : t > 0, u \in \mathbb{R} \right\}$$

$$K = \left\{ \begin{pmatrix} \cos\theta & -\sin\theta \\ \sin\theta & \cos\theta \end{pmatrix} : \theta \in \mathbb{R} \right\},$$

all of them considered with the topology induced by $\mathbb{R}^4$.

(i) (a) Show that the map $\mu : B \times K \to B$, $\mu(b, k) = bk$, is bijective and continuous.

(b) Let L be a compact subset of G. Show there exist two constants $\alpha > 0$,

$\beta > 0$ such that

$$\max(|a|, |b|, |c|, |d|) \le \beta$$
$$c^2 + d^2 \ge \alpha$$

for every $\begin{pmatrix} a & b \\ c & d \end{pmatrix} \in L$. Conclude that μ is a homeomorphism from $B \times K$ onto G.

(c) Show that G is connected and, using that the fundamental group of a product is the direct product of the fundamental groups, determine $\pi_1(G)$.

(ii) (a) Let $P = \{z \in \mathbb{C} : \operatorname{Im} z > 0\}$ and to $g \in G$ associate a Moebius transformation as follows: if $g = \begin{pmatrix} a & b \\ c & d \end{pmatrix}$ then

$$g(z) := \frac{az + b}{cz + d}.$$

Recall that

$$\operatorname{Im} g(z) = \frac{\operatorname{Im} z}{|cz + d|^2}$$

and, hence, g maps P into P.

(b) Consider the map $J : G \times P \to \mathbb{C}$ given by

$$J(g, z) = cz + d.$$

Show that J is continuous and never zero.

(iii) As usual, denote by $S^1 = \{z \in \mathbb{C} : |z| = 1\}$. Let $\varphi : G \to S^1$ defined by

$$\varphi(g) = \frac{J(g, i)}{|J(g, i)|}$$

and $p : \tilde{G} \to G$ the covering map which is the inverse image by φ of the covering map $q : \mathbb{R} \to S^1$, $q(x) = e^{ix}$. (See preceding exercise.) Then

$$\tilde{G} = \{(g, \theta) \in G \times \mathbb{R} : \varphi(g) = q(\theta)\}, \qquad p(g, \theta) = g.$$

(a) Let A be a subset of $B \times S^1 \times \mathbb{R}$ of the triplets (b, z, θ) such that $z = q(\theta)$. Show that the map $\psi : B \times S^1 \times \mathbb{R} \to G \times \mathbb{R}$ induced by μ determines a homeomorphism from A onto $\tilde{G}$. Conclude that $\tilde{G}$ is homeomorphic to $B \times \mathbb{R}$.

(b) Find the fundamental group of $\pi_1(\tilde{G})$.

(iv) Consider the map $\chi : G \times P \to S^1$ defined by

$$\chi(g, z) = \frac{J(g, z)}{|J(g, z)|}.$$

(a) Show there is a unique continuous map $\omega : \tilde{G} \times P \to \mathbb{R}$ such that

$$\chi(p(\gamma), z) = e^{i\omega(\gamma, z)}$$

satisfying $\omega(\tilde{e}, i) = 0$, where $\tilde{e} = \begin{pmatrix} 1 & 0 \\ 0 & 1 \end{pmatrix}$.

(b) Find $\omega(\gamma, i)$ in terms of $\gamma = (g, \theta)$.

(v) Denoted by $\mathfrak{m}: G \times G \to G$ the group multiplication $(g_1, g_2) \mapsto g_1 g_2$.
   (a) Show that there is a unique continuous map $\tilde{\mathfrak{m}}: \tilde{G} \times \tilde{G} \to \tilde{G}$ such that

$$p(\tilde{\mathfrak{m}}(\gamma_1, \gamma_2)) = \mathfrak{m}(p(\gamma_1), p(\gamma_2)),$$

   and

$$\tilde{\mathfrak{m}}(\tilde{e}, \tilde{e}) = \tilde{e}.$$

   (b) Compare the two maps of $\tilde{G} \times \tilde{G} \times \tilde{G}$ into $\tilde{G}$

$$(\gamma_1, \gamma_2, \gamma_3) \mapsto \tilde{\mathfrak{m}}(\tilde{\mathfrak{m}}(\gamma_1, \gamma_2), \gamma_3)$$

$$(\gamma_1, \gamma_2, \gamma_3) \mapsto \tilde{\mathfrak{m}}(\gamma_1, \tilde{\mathfrak{m}}(\gamma_2, \gamma_3)).$$

   Compare also the maps from $\tilde{G}$ into itself

$$\gamma \mapsto \tilde{\mathfrak{m}}(\gamma, \tilde{e})$$

$$\gamma \mapsto \tilde{\mathfrak{m}}(\tilde{e}, \gamma).$$

   (c) Do the same as in part (a) to lift to $\tilde{G}$ the map $\mathfrak{i}: G \to G$, $g \mapsto g^{-1}$.
   (d) Prove that there is on $\tilde{G}$ a unique structure of (topological) group such that $p: \tilde{G} \to G$ becomes a group homomorphism and $\tilde{e}$ is the group identity.

(vi) Let V be an open connected neighborhood of $\tilde{e}$ in $\tilde{G}$ and $h: V \to \mathbb{R}$ a continuous local homomorphism (i.e., if $\gamma_1, \gamma_2 \in V$ are such that $\gamma_1, \gamma_2 \in V$ then $h(\gamma_1 \gamma_2) = h(\gamma_1) + h(\gamma_2)$).
   (a) Show there is an open neighborhood U of $\tilde{e}$ in $\tilde{G}$ such that $U = U^{-1}$, $U^2 \subseteq V$, and $\tilde{G} = \bigcup_{n \geq 1} U^n$.
   (b) For $\gamma \in \tilde{G}$ consider the map

$$h_\gamma: U \cdot \gamma \to \mathbb{R}$$

$$h_\gamma(\delta) = h(\delta\gamma^{-1}).$$

   Construct a cocycle $(g_{\gamma,\delta})$ for the covering $(U \cdot \gamma)_{\gamma \in \tilde{G}}$ with values in $\mathbb{R}$, $\mathbb{R}$ considered with the discrete topology. Use this cocycle to show there is a unique group homomorphism $\tilde{h}: \tilde{G} \to \mathbb{R}$ which extends h.

§12. Algebraic Functions. I

We discuss in this section the construction of the ***complete Riemann surface*** of $\sqrt{f(z)}$, f a polynomial in $\mathbb{C}[z]$. In the next section we will consider the general case of algebraic functions, i.e., functions $w = w(z)$ defined by equations of the form $P(z, w) = 0$, with $P \in \mathbb{C}[z, w]$. The previous case corresponds to the equation $w^2 - P(z) = 0$.

In the previous sections, by way of exercises at least, we have seen how to obtain a Riemann surface for $\sqrt{z}$, $\sqrt{z(z-a)}$, etc., i.e., a covering surface of $S^2 \backslash \{\infty, 0\}$, $S^2 \backslash \{\infty, 0, a\}$, etc., where the function $\sqrt{z}$, resp. $\sqrt{z(z-a)}$, is well defined. In order to "pluck the holes" we need the concept of branch point of a holomorphic map.

5.12.1. Definition. Let X, Y be two Riemann surfaces, $p: X \to Y$ a nonconstant holomorphic map. A point $z \in X$ is called a ***branch point*** (or ***ramification point***) if, for every neighborhood V of z, $p|V$ is not injective. It is easy to see that this is equivalent to the statement that there are charts (U, φ) of X about z and (W, ψ) of Y about $p(z)$, and an integer $k \geq 2$ such that

(i) $\varphi(z) = \psi(p(z)) = 0$
(ii) $p(U) \subseteq W$
(iii) $(\psi \circ p \circ \varphi^{-1})(u) = u^k$ for $u \in \varphi(U)$.

One says that *z is a* ***branch point of order*** $k - 1$.

When X and Y are connected Riemann surfaces and $p: X \to Y$ is a nonconstant holomorphic and proper map, then the set A of branch points of p is a closed discrete subset of X. Moreover, $B = p(A)$ is also closed in Y. The latter assertion follows from the fact that p is proper, hence a closed transformation. It is clear that if A had an accumulation point z, the function $\psi \circ p \circ \varphi^{-1}: \varphi(U) \to \mathbb{C}$ would have $u = 0$ as an accumulation point of zeros of $\frac{d}{du}(\psi \circ p \circ \varphi^{-1})$. It follows that p would be constant on U, hence everywhere.

The set B is called the ***set of critical values*** *of* p.

5.12.2. Definition. Let X, Y be two Riemann surfaces. One says that a nonconstant holomorphic map $p: X \to Y$ is a ***branched covering map*** if the three sets, A of branch points, $B = p(A)$ of critical values, and $p^{-1}(B)$, are all closed discrete sets, and, furthermore, $p: X \backslash p^{-1}(B) \to Y \backslash B$ is a covering map.

Remark. In case one also assumes p is proper and Y connected, then the number of sheets n of $p: X \backslash p^{-1}(B) \to Y \backslash B$ is a well-determined integer. One says then that p is an ***n-sheeted branched covering***.

We are going to study first the Riemann surface of $\sqrt{f(z)}$,

$$f(z) = (z - a_1)\dots(z - a_n),$$

where the a_j are distinct complex numbers.

5.12.3. Proposition. *There is a compact connected Riemann surface X and a two-sheeted branched covering map $\pi: X \to S^2$ such that the set of critical values $B \subseteq \{a_1, \dots, a_n, \infty\}$, the set of branch points A coincides with $\pi^{-1}(B)$, and there is a function $F \in \mathcal{M}(X)$ such that $F^2 = f \circ \pi$ over $X \backslash A$ and for any $z \in S^2 \backslash B$, if $\pi^{-1}(z) = \{\zeta_1, \zeta_2\}$, then $F(\zeta_1) = -F(\zeta_2)$.*

Furthermore, if $(\tilde{X}, \tilde{\pi}, \tilde{F})$ is another triple with the same properties, then there is a biholomorphic map $\sigma: \tilde{X} \to X$ such that $\pi \circ \sigma = \tilde{\pi}$ and $F \circ \sigma = \tilde{F}$.

PROOF. For every $z \in \Omega = S^2 \backslash \{a_1, \dots, a_n, \infty\}$, the polynomial $w \mapsto w^2 - f(z)$ has two distinct roots which determine two germs $\varphi_z, \psi_z \in \mathcal{O}_z$ such that $\varphi_z^2 = \psi_z^2 = f_z$.

Let X_1 be the subset of the sheaf $\mathcal{O}_\Omega$ of all the germs $\varphi_z \in \mathcal{O}_z$, $z \in \Omega$, such that $\varphi_z^2 = f_z$. Denote by π the restriction to X_1 of the canonical projection $\mathcal{O}_\Omega \to \Omega$ (i.e., $\pi(\varphi_z) = z$).

5.12.4. Lemma. *For every $z \in \Omega$, let $B(z, \delta) \subseteq \Omega$. Then there exist two holomorphic functions $f_1, f_2 \in \mathscr{H}(B(z, \delta))$ such that*

(i) $w^2 - f(\zeta) = (w - f_1(\zeta))(w - f_2(\zeta))$ for every $\zeta \in B(z, \delta)$;
(ii) $\pi^{-1}(B(z, \delta)) = N(B(z, \delta), f_1) \dot\cup N(B(z, \delta), f_2)$ (notation of §5.6);
(iii) $\pi | N(B(z, \delta), f_j) : N(B, z, \delta), f_j) \to B(z, \delta)$ is a biholomorphic map, $j = 1, 2$;
(iv) $\pi : \pi^{-1}(B(z, \delta)) \to B(z, \delta)$ is a two-sheeted covering map.

PROOF. There are exactly two determinations, f_1 and $f_2 = -f_1$, of $\sqrt{f}$ on the simply connected set $B(z, \delta)$. The rest is obvious. □

This lemma ensures that X_1 is a Riemann surface and $\pi : X_1 \to \Omega$ is a two-sheeted covering map.

Let now $F : X_1 \to \mathbb{C}$ be defined by $F(\varphi_z) = \varphi_z(z)$. The funtion F is holomorphic (as already shown in §5.6) since locally, $f = \varphi \circ \pi$, with φ holomorphic. Moreover, $F(z)^2 = (\varphi_z(z))^2 = f(\pi(\varphi_z)) = f(z)$. To end the proof of the proposition we only need to pluck the holes lying over $a_1, \ldots, a_n, \infty$.

(a) For every $1 \le j \le n$, choose $r_j > 0$ sufficiently small so that the closed disks $\bar{B}(a_j, r_j)$ are disjoint.

The function $g_j(z) = \prod_{k \ne j} (z - a_k)$ does not vanish in $B(a_j, r_j)$, hence there is $h_j \in \mathscr{H}(B(a_j, r_j))$ such that $h_j^2 = g_j$ in that disk. Hence

$$f(z) = (z - a_j)(h_j(z))^2, \qquad z \in B(a_j, r_j).$$

If $0 < \rho < r_j$, $\theta \in \mathbb{R}$, let $\zeta = a_j + \rho e^{i\theta}$. There is a unique $\varphi_\zeta \in \mathcal{O}_\zeta$ such that $\varphi_\zeta^2 = f_\zeta$ and $\varphi_\zeta(\zeta) = \sqrt{\rho} e^{i\theta/2} h_j(\zeta)$. Performing the analytic continuation along the circle $\partial B(a_j \rho)$ of the initial germ at $\theta = 0$, one obtains the opposite germ when $\theta = 2\pi$. Let $U_j = B(a_j, r_j) \backslash \{a_j\}$, $V_j = \pi^{-1}(U_j)$, then $\pi : V_j \to U_j$ is a covering map with two sheets.

(b) Let $r > \max\{|a_j| : 1 \le j \le n\}$. Then $U = \{z \in \mathbb{C} : |z| > r\} \cup \{\infty\}$ is an open neighborhood of ∞ in S^2, biholomorphic to a disk, and does not contain any a_j. On $U \backslash \{\infty\}$ one can write

$$f(z) = z^n h(z),$$

with h holomorphic in U and h does not vanish.

If n is odd there is then a meromorphic function k in U such that $f(z) = z(k(z))^2$ for $z \in U \backslash \{\infty\}$, and the only possible pole of k is at $z = \infty$.

If n is even there is a meromorphic function k on U such that $f(z) = (k(z))^2$ in $U \backslash \{\infty\}$, and the only pole of k occurs at $k = \infty$.

As before, we let $V = \pi^{-1}(U \backslash \{\infty\})$ and obtain that $\pi : V \to U \backslash \{\infty\}$ is a two-sheeted covering map. If n is odd, V is connected, but, if n is even, it is not. The reason is that a germ φ_ζ with $\varphi_\zeta^2 = f_\zeta$, $\zeta \in U \backslash \{\infty\}$, can be taken as the germ k_ζ (or $-k_\zeta$), and analytic continuation along the circle $\partial B(0,|\zeta|)$ returns to the same germ and not to its opposite.

To construct X we are going to "glue" a disk $\tilde{V}_j$ over V_j and one or two disks $\tilde{V}$ over V, depending on whether V is connected or not. As a set X shall be the union of X_1 and $\{a_1,\ldots,a_n,\infty\}$ or $\{a_1,\ldots,a_n,\infty,\infty'\}$.

Assume first that n is odd. Consider the topological space X_2 which is the disjoint union of X_1, $\tilde{V}_1, \tilde{V}_2, \ldots, \tilde{V}_n, \tilde{V}$, where $\tilde{V}_j$ is a copy of $B(a_j, r_j)$, $1 \le j \le n$, and $\tilde{V}$ is a copy of U. We introduce in X_2 an equivalence relation $\mathscr{R}$ defined as follows:

(i) For $1 \le j \le n$, we consider the two coverings of $U_j = B(a_j, r_j)\backslash\{a_j\}$

$$\pi : V_j \to U_j$$

and

$$p_j : U_j \to U_j, \qquad p_j(z) = a_j + \frac{(z-a_j)^2}{r_j}$$

and show these two covering maps are isomorphic. Namely, let us define $\ell_j : V_j \to U_j$ as follows: fix $\xi_0 \in V_j$ and $\zeta_0 \in U_j$ such that $\pi(\xi_0) = p_j(\zeta_0)$. For $\xi \in V_j$ there is a path γ_1 in V_j joining ξ_0 to ξ. Let γ_2 be the lifting, starting at ζ_0, of the path $\pi \circ \gamma_1$. Set $\ell_j(\xi) = \gamma_2(1) =$ endpoint of γ_2. It is easy to verify that $\ell_j : V_j \to U_j$ is a biholomorphic map such that $p_j \circ \ell_j = \pi$.

(ii) The same way one constructs an isomorphism of coverings $\ell : V \to U\backslash\{\infty\}$ between the connected two-sheeted covering maps $\pi : V \to U\backslash\{\infty\}$ and $p : U\backslash\{\infty\} \to U\backslash\{\infty\}$, $p(z) = \dfrac{z^2}{r}$.

Now, the relation $\mathscr{R}$ consists in identifying the points in X_2 that correspond to each other by isomorphisms $\ell_1, \ldots, \ell_n, \ell$. We leave to the reader the easy task of verifying that $X = X_2/\mathscr{R}$ is a connected compact Riemann surface.

X_1 is a dense open subset of X, $X\backslash X_1 = \{\tilde{a}_1, \ldots, \tilde{a}_n, \tilde{\infty}\}$, where $\tilde{a}_j$, resp. $\tilde{\infty}$, is the class of a_j, resp. ∞, modulo $\mathscr{R}$. The original map $\pi : X_1 \to \mathbb{C}$ can be extended by continuity to X setting $\pi(\tilde{a}_1) = a_j$, $\pi(\tilde{\infty}) = \infty$, obtaining a map $\pi : X \to S^2$ which is a holomorphic, proper, branched covering map. The branch point set $A = \{\tilde{a}_1, \ldots, \tilde{a}_n, \tilde{\infty}\}$, the order of every branch point is exactly 1, and $\pi(A) = B = (a_1, \ldots, a_n, \infty\}$.

We already know that the function F defined on X_1 by $F(\varphi_z) = \varphi_z(z)$ is a holomorphic function satisfying $F^2 = f \circ \pi$. We define it at $\tilde{a}_j$ by $F(\tilde{a}_j) = 0$ and $F(\tilde{\infty}) = \infty$. It is easy to verify that it is a continuous map $X \to S^2$, hence $F \in \mathscr{M}(X)$ with a pole at ∞.

In case n is even, the construction is the same, except for the fact we have two copies of U, say $\tilde{V}$ and $\tilde{V}'$, in X_2. The corresponding maps p, p' are just the identity. The only difference is that $X\backslash X_1 = \{\tilde{a}_1, \ldots, \tilde{a}_n, \tilde{\infty}, \tilde{\infty}'\}$, $A = \{\tilde{a}_1, \ldots, \tilde{a}_n)$, and $B = \{a_1, \ldots, a_n\}$, i.e., the points at infinity are not branch points any longer. We let $F(\tilde{a}_j) = 0$, $F(\tilde{\infty}) = F(\tilde{\infty}') = \infty$ to obtain the corresponding meromorphic map.

Finally, let us verify the essential uniqueness of the construction. There is an isomorphism of two-sheeted covering maps $\sigma : \tilde{X}_1 \to X_1$,

$$\tilde{X}_1 := \tilde{X} \backslash \tilde{\pi}^{-1}(\{a_1, \ldots, a_n, \infty\}),$$

defined as follows: For $z \in \tilde{X}_1$, let U_z be an open neighborhood of z in $\tilde{X}_1$, biholomorphic to a disk, such that $\tilde{\pi} | U_z$ is injective. The function $\tilde{F} \circ (\tilde{\pi} | U_z)^{-1}$ is holomorphic in the neighborhood $\tilde{\pi}(U_z)$ of the point $\tilde{\pi}(z)$, let φ_z denote the corresponding germ in $\mathcal{O}_{\tilde{\pi}(z)}$. We set $\sigma(z) = \varphi_z$. It is now easy to verify that $\sigma : \tilde{X}_1 \to X_1$ is a biholomorphism which extends to a biholomorphism in $\tilde{X} \to X$ with the required properties. □

5.12.5. Definition. The Riemann surface X just constructed is called the ***complete Riemann surface*** of the algebraic function $\sqrt{f(z)}$, $f(z) = (z - a_1) \ldots (z - a_n)$, $a_1, \ldots, a_n$ distinct.

EXERCISES 5.12

1. Let $a_1, a_2 \in \mathbb{C}$, $a_1 \neq a_2$. Using the method of cocycles, construct a two-sheeted covering map $\pi : X_1 \to \mathbb{C} \backslash \{a_1, a_2\}$ and a holomorphic function $F : X_1 \to \mathbb{C}$ such that for every $z \in \mathbb{C} \backslash \{a_1, a_2\}$, $\pi^{-1}(\{z\}) = \{\zeta_1, \zeta_2\}$ one has $F(\zeta_1) = -F(\zeta_2)$ and $F^2 = f \circ \pi$, with $f(z) = (z - a_1)(z - a_2)$. We suggest using the following two simply connected open sets U_1, U_2 that verify $\mathbb{C} \backslash \{a_1, a_2\} = U_1 \cup U_2$, ($D_1$, D_2, Δ_1, Δ_2 are parallel rays).

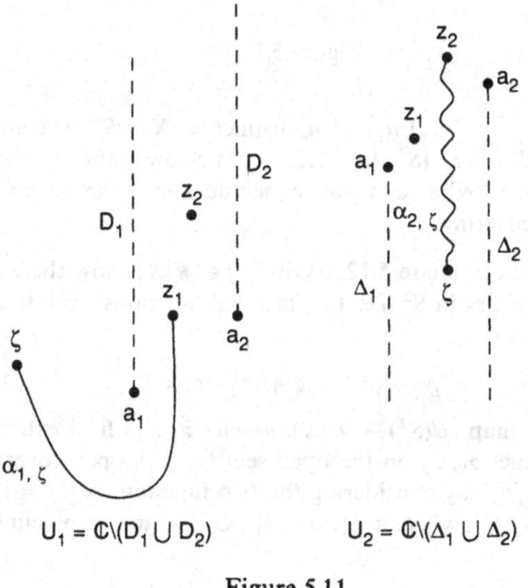

Figure 5.11

2. Same as in Exercise 5.12.1 with n distinct points.
3. Let $a_1, \ldots, a_n$ be n distinct complex numbers. Let $f(z) = (z - a_1) \ldots (z - a_n)$,

$$X_1 = \{(z, w) \in (\mathbb{C}\backslash\{a_1,\ldots,z_n\}) \times \mathbb{C} : w^2 - f(z) = 0\},$$

and $\pi(z, w) = z$. Show that $\pi : X_1 \to \mathbb{C}\backslash\{a_1,\ldots,z_n\}$ is a two-sheeted covering map and there is a holomorphic function $F : X_1 \to \mathbb{C}$ such that for every $z \in \mathbb{C}\backslash\{a_1,\ldots,a_n\}$, $\pi^{-1}(\{z\}) = \{\zeta_1, \zeta_2\}$, one has $F(\zeta_1) = -F(\zeta_2)$ and $F^2 = f \circ \pi$.

4. Let X be the complete Riemann surface of $\sqrt{z - a}$ and $F : X \to S^2$ the meromorphic function such that $F(a) = 0$, $F(\infty) = \infty$, and $F(\zeta)^2 = \pi(\zeta) - a$ given in Proposition 5.12.3. Show that F is a biholomorphic map. In other words, X is conformally equivalent to S^2.

5. Let $f(z) = (z - a_1) \ldots (z - a_n)$, $a_1, \ldots, a_n$ distinct. Let $a_0 \in \mathbb{C}\backslash\{a_1,\ldots,a_n\}$ and the γ_j that appear in the Figure 5.12, the generators of $\pi_1(\mathbb{C}\backslash\{a_1,\ldots,a_n\}, a_0)$ (see Proposition 1.6.27). Let $\varphi_{a_0} \in \mathcal{O}_{a_0}$ be a solution of $\varphi_{a_0}^2 = f_{a_0}$. For a loop $\gamma \in \pi_1(\mathbb{C}\backslash\{a_1,\ldots,z_n\}, a_0)$ describe the action of γ on φ_{a_0} in terms of the expression of γ in the basis $\gamma_1, \ldots, \gamma_n$.

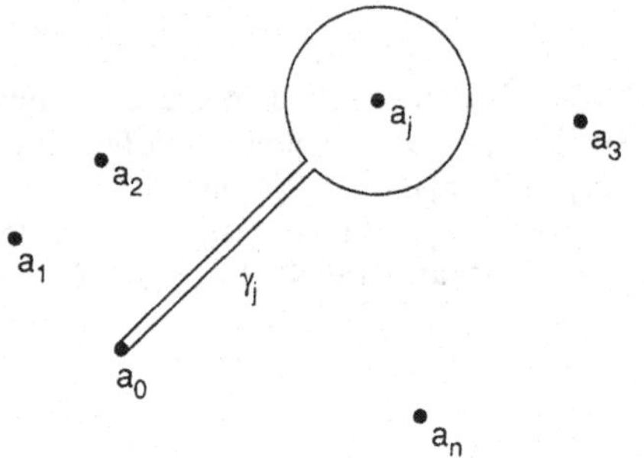

Figure 5.12

6. Let $f(z) = (z - a_1) \ldots (z - a_n)$, $a_1, \ldots, a_n$ distinct, $\pi : X \to S^2$ the complete Riemann surface of $\sqrt{f}$ and $X_1 = \pi^{-1}(S^2\backslash\{a_1,\ldots,a_n,\infty\})$. Show that $\pi : X_1 \to \mathbb{C}\backslash\{a_1,\ldots,a_n\}$ is a Galois covering. What can you conclude about the group $G(\pi)$ of automorphisms of the covering?

7. Let f and X be as in Exercise 5.12.6. Given $g \in \mathcal{M}(X)$ show there are two meromorphic functions c_1, c_2 in S^2 (i.e., two rational functions) such that g verifies the equation

$$g^2 + (c_1 \circ \pi)g + (c_2 \circ \pi) = 0.$$

Conclude that the map $\mathcal{M}(S^2) \to \mathcal{M}(X)$, $h \mapsto h \circ \pi$, is a field extension of degree 2. (One can construct c_1, c_2 on the open sets U_i, $(U_i)_i$ open covering trivializing $\pi : X_1 \to \mathbb{C}\backslash\{a_1,\ldots,a_n\}$ by considering the two functions $g_{i,j}$, $j = 1, 2$, defined by $g_{i,j} = (g|V_{i,j}) \circ (\pi|V_{i,j})^{-1}$, where $\pi^{-1}(U_i) = V_{i,1} \dot{\cup} V_{i,2}$, and then defining $c_{i,1}, c_{i,2}$ by the formula

$$w^2 + c_{i,1}w + c_{i,2} = (w - g_{i,1})(w - g_{i,2}).)$$

8. Let X be the complete Riemann surface of $\sqrt{(z - a_1)(z - a_2)(z - a_3)}$, a_1, a_2, a_3 distinct. Prove that X cannot be biholomorphic to S^2. (One can consider the loops

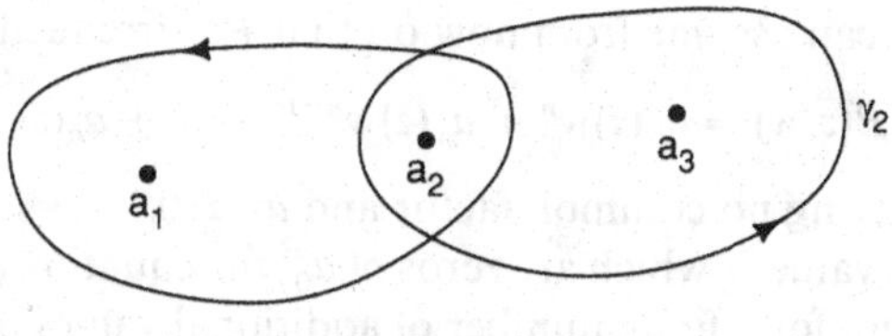

Figure 5.13

γ_1, γ_2 in $\mathbb{C}\setminus\{a_1, a_2, a_3\}$ suggested in Figure 5.13, and lift them to two nonhomotopic loops in X.)

§13. Algebraic Functions. II

We proceed to construct the complete Riemann surface of an algebraic function along the same lines as was done for $\sqrt{f}$ in the last section.

5.13.1. Definition. Let $P(z, w) \in \mathbb{C}[z, w]$ be a nontrivial polynomial in two variables. The (unbranched) Riemann surface X_P of the equation $P = 0$ is the largest open subset of the sheaf $\mathcal{O}_{\mathbb{C}}$ formed by the germs φ_z of holomorphic functions satisfying the equation $P(\zeta, \varphi_z(\zeta)) = 0$ in a neighborhood of z.

Remark. One could define an analogous notion for an equation $A = 0$, with A a holomorphic function of two variables. For instance, $A(z, w) = e^w - z = 0$. In this case, the holomorphicity in ζ of $A(\zeta, \varphi_z(\zeta))$ for any germ φ_z is clear. The surface X_A can be cononically identified to the unbranched Riemann surface of the logarithm constructed in §5.5.7.

Given a polynomial $P \in \mathbb{C}[z, w]$ one can write it as

$$P = P_1^{n_1} \dots P_N^{n_N}$$

with P_j irreducible, i.e., prime in the factorial ring $\mathbb{C}[z, w]$. We claim that the surface X_P is the disjoint union of the surfaces X_{P_j}. It is clear that if $\varphi_z \in X_{P_j}$ then $P(\zeta, \varphi_z(\zeta)) = 0$ in a neighborhood of z. On the other hand, the set of points in $\mathbb{C}^2$ defined by $P_i(z, w) = P_j(z, w) = 0$, $i \neq j$, is finite. Therefore, a germ φ_z cannot belong to $X_{P_i} \cap X_{P_j}$. This shows that $X_P \supseteq \dot{\bigcup}_{1 \le j \le N} X_{P_j}$. Finally, let $\varphi_z \in X_P$ be defined in $B(z, r)$ and let $z_0 \in B(z, r)$ be such that there is no w_0 and a pair $i \neq j$ with $P_i(z_0, w_0) = P_j(z_0, w_0) = 0$. Since $P(z_0, \varphi_z(z_0)) = 0$, there is only one index, say i_0, for which $P_{i_0}(z_0, \varphi_z(z_0)) = 0$. In a neighborhood of z_0 we have therefore $P_j(\zeta, \varphi_z(\zeta)) \neq 0$ if $j \neq i_0$. Hence $P_{i_0}(\zeta, \varphi_z(\zeta)) \equiv 0$ for all points in a neighborhood of z_0 and we can conclude $P_{i_0}(\zeta, \varphi_z(\zeta)) \equiv 0$ in $B(z, r)$. In other words, $\varphi \in X_{P_{i_0}}$ and, hence, $X_P = \dot{\bigcup}_j X_{P_j}$.

Therefore, we can assume from now on that P is irreducible and write

$$P(z, w) = a_0(z)w^n + a_1(z)w^{n-1} + \cdots + a_n(z),$$

with $a_i \in \mathbb{C}[z]$ having no common factor and $a_0 \not\equiv 0$.

Except for the values z which are zeros of a_0, the equation (in w) $P(z, w) = 0$ has n roots. Except for a finite number of additional values of z, these n roots are distinct. We denote by $\Delta(P)$ the set of those exceptional values of z; this is called the set of critical points of P. For simplicity we add ∞ to $\Delta(P)$. In this way, $\Delta(P)$ is the union of ∞, the roots of a_0, and the $z \in \mathbb{C}$ such that the two polynomials in w, $P(z, \cdot)$ and $\frac{\partial P}{\partial w}(z, \cdot)$, have a common root. ($P$, $\frac{\partial P}{\partial w}$ have only finitely many common zeros, thanks to the irreducibility of P.)

5.13.2. Lemma. *Let $z_0 \in S^2 \setminus \Delta(P)$ and $w_1, \ldots, w_n$ be the distinct roots of $w \mapsto P(z_0, w)$. There are germs $\varphi_1, \ldots, \varphi_n \in \mathcal{O}_{z_0}$ such that $\varphi_j(z_0) = w_j$ and*

$$a_0(z) \prod_{j=1}^{n} (w - \varphi_j(z)) = P(z, w)$$

for every z in a neighborhood of z_0 and $w \in \mathbb{C}$.

Proof. Let $\varepsilon > 0$ be such that the closed disks $\bar{B}(w_j, \varepsilon)$ are disjoint. Let $r > 0$ be such that $P(z, w) \neq 0$ when $|z - z_0| \leq r$ and $|w - w_j| = \varepsilon$ for some j. By Rouche's theorem, for each fixed z, $|z - z_0| < r$, there is only one zero of $P(z, w) = 0$ in $B(w_j, \varepsilon)$, and it is given by

$$\varphi_j(z) = \frac{1}{2\pi i} \int_{\partial B(w_j, \varepsilon)} w \frac{\partial P}{\partial w}(z, w) \frac{dw}{P(z, w)}.$$

The functions $\varphi_j \in \mathcal{H}(B(w_j, \varepsilon))$ and clearly satisfy the other conditions of the lemma. □

Let us consider now the situation in a neighborhood of a point $z_0 \in \Delta(P) \setminus \{\infty\}$. Let $\varepsilon > 0$ be such that $B(z_0, \varepsilon) \cap \Delta(P) = \{z_0\}$, and define $V_{z_0} = \pi^{-1}(B(z_0, \varepsilon) \setminus \{z_0\}) \cap X_P$, where $\pi : \mathcal{O}_{\mathbb{C}} \to \mathbb{C}$ is the canonical projection, $\pi(\varphi_z) = z$. For every open simply connected subset U of $B(z_0, \varepsilon) \setminus \{z_0\}$ we can now construct, thanks to the lemma, n holomorphic function $f_1^U, \ldots, f_n^U$ in U, such that for each $z \in U$ the distinct roots $w_1, \ldots, w_n$ of $P(z, w) = 0$ are given by the values $f_1^U(z), \ldots, f_n^U(z)$. Hence

$$\pi^{-1}(U) \cap X_P = \dot{\bigcup_{1 \leq j \leq n}} N(U, f_j^U)$$

and $\pi | N(U, f_j^U) : N(U, f_j^U) \to U$ is a biholomorphism for every j. This shows that

$$\pi : V_{z_0} \to B(z_0, \varepsilon) \setminus \{z_0\}$$

is an n-sheeted covering map.

Let γ be a generator of $\pi_1(B(z_0,\varepsilon)\setminus\{z_0\},a)$ of index 1 with respect to z_0. The n elements $f_{1,a},\ldots,f_{n,z}\in\mathcal{O}_a$, which are germs of distinct roots of $P(z,w)=0$ near $z=a$, can be continued along γ into a new system of roots which are a permutation σ of the original ones. We denote them by $f_{\sigma(1),a},\ldots,f_{\sigma(n),a}$. One can decompose σ into disjoint cycles. For instance, up to renumbering, we could have $\sigma(1)=2$, $\sigma(2)=3,\ldots,\sigma(q-1)=q$ and $\sigma(q)=1$ for some $q\le n$. This means that if we start with $f_{1,a}$ then continuing along γ we obtain $f_{2,a}$, continuing once more we obtain $f_{3,a}$, and after q times we come back to $f_{1,a}$. In this situation, the following property holds.

5.13.3. Lemma. *Let $f_a\in\mathcal{O}_a$ be a germ admitting analytic continuation along any path in $U=B(z_0,\varepsilon)\setminus\{z_0\}$ that starts at $z=a$. Let γ be the generator of $\pi_1(U,a)$ that has index 1 with respect to z_0. Assume that the analytic continuation of f_a along γ^q is equal to f_a. Then there is a Laurent series convergent in $B(0,\varepsilon^{1/q})\setminus\{0\}$ such that*

$$f_a(z_0+(\zeta-z_0)^q)=\sum_{n\in\mathbb{Z}}c_n(\zeta-z_0)^n.$$

PROOF. We can assume without loss of generality that $z_0=0$, $\varepsilon=1$, $U=B(0,1)\setminus\{0\}$. Let $b\in U$ be such that $b^q=a$. The map $\varphi:\zeta\to\zeta^q$ of U into U has the property that the germ g_b of $f_a\circ\varphi$ at $\zeta=b$, when continued along γ, returns to itself. By the monodromy theorem, g_b determines a holomorphic function in U. This function can be expanded in a Laurent series $\sum_{n\in\mathbb{Z}}a_n\zeta^n$ in U, hence

$$f_a(\zeta^q)=\sum_{n\in\mathbb{Z}}c_n\zeta^n,$$

as we wanted to show. □

Applying the lemma to $f_{1,a}$ one finds

$$f_{1,a}(z_0+(\zeta-z_0)^q)=\sum_{n\in\mathbb{Z}}c_n(\zeta-z_0)^n$$

for a series $f(t)=\sum_{n\in\mathbb{Z}}c_nt^n$ convergent in $B(0,\varepsilon^{1/q})\setminus\{0\}$. It is usual to write $\xi=z_0+(\zeta-z_0)^q$, hence

$$f_{1,a}(\xi)=\sum_{n\in\mathbb{Z}}c_n(\xi-z_0)^{n/q},$$

which is called a ***Puiseux series*** representation for $f_{1,a}$. In terms of f we have $f_{1,a}(\xi)=f((\xi-z_0)^{1/q})$. With this notation the different branches $f_{1,a},\ldots,f_{q,a}$ can also be written as

$$f_{j,a}(\xi)=f((\xi-z_0)^{1/q}),\qquad 1\le j\le q,$$

where one just takes into account the different choices of the root. If $(\xi-z_0)^{1/q}$ is the choice of $f_{1,a}$, the others are $e^{2\pi i(j-1)/q}(\xi-z_0)^{1/q}$, i.e.,

$$f_{j,a}(\xi) = f(e^{2\pi i(j-1)/q}(\xi - z_0)^{1/q}) = \sum_{n\in\mathbb{Z}} c_n[e^{2\pi i(j-1)/q}(\xi - z_0)^{1/q}]^n$$

or

$$f_{j,a}(z_0 + (\zeta - z_0)^q) = \sum_{n\in\mathbb{Z}} c_n[e^{2\pi i(j-1)/q}(\zeta - z_0)]^n.$$

If $q < n$, then one can consider the different cycles of $f_{q+1,a}, \ldots, f_{n,a}$, and find corresponding Puiseux series expansions for the different roots.

If we are in the situation where $a_0(z_0) \neq 0$, then the solutions w of $P(z, w) = 0$ remain in a bounded set when z is near z_0. In fact, the functions $a_j(z)/a_0(z)$ are bounded in a neighborhood of z_0, and from the expression

$$w^n = -\left(\frac{a_1(z)}{a_0(z)} w^{n-1} + \cdots + \frac{a_n(z)}{a_0(z)}\right)$$

we see that either $|w| \leq 1$ or

$$|w| = \left|\frac{a_1(z)}{a_0(z)} + \frac{a_2(z)}{a_0(z)}\frac{1}{w} + \cdots + \frac{a_n(z)}{a_0(z)}\frac{1}{w^{n-1}}\right| \leq \left|\frac{a_1(z)}{a_0(z)}\right| + \left|\frac{a_2(z)}{a_0(z)}\right| + \cdots + \left|\frac{a_n(z)}{a_0(z)}\right|.$$

In other words, if $\alpha > 0$ is sufficiently small so that $a_0(z) \neq 0$ in $|z - z_0| \leq \alpha$, we have that any root w of $P(z, w) = 0$ satisfies

$$|w| \leq \max\left\{1, \max_{|z-z_0|\leq\alpha} \frac{|a_1(z) + \cdots + |a_n(z)||}{|a_0(z)|}\right\} \leq M < \infty.$$

Therefore, the different branches $f_{j,a}$ of the first cycle, or if one prefers, $f_{1,a}(z_0 + (\zeta - z_0)^q)$, remain bounded when $\zeta \to z_0$. This implies that f is holomorphic even at $t = 0$, so that

$$f_{j,a}(\xi) = \sum_{n\geq 0} c_n(\xi - z_0)^{n/q}, \qquad 1 \leq j \leq q.$$

The same happens for every cycle of the permutation σ. What does this mean for the covering map $\pi: V_{z_0} \to B(z_0, \varepsilon)\setminus\{z_0\}$?

If $q = 1$, there is only one branch in the cycle, which is then the restriction to $B(z_0, \varepsilon)\setminus\{z_0\}$ of a holomorphic function in $B(z_0, \varepsilon)$. This corresponds to a connected component of V_{z_0}, which has only one sheet.

If $q \geq 2$, the cycle corresponds to a connected component of V_{z_0} of exactly q sheets.

If $a_0(z_0) = 0$ with multiplicity $k \geq 1$, then, by the same reasoning as earlier, if w is a root of $P(z, w) = 0$, $z \in B(z_0, z)\setminus\{z_0\}$, we have

$$|z - z_0|^k|w| \leq \max\left\{1, \frac{|a_1(z)| + \cdots + |a_n(z)|}{|a_0(z)|/|z - z_0|^k}\right\} \leq M < \infty$$

as long as $0 < |z - z_0| \leq \alpha$, α sufficiently small. In terms of the function f associated to the first cycle this means that

$$t^k f(t) = \sum_{n\in\mathbb{Z}} c_n t^{n+k}, \qquad 0 < |t| < \varepsilon^{1/q}$$

remains bounded as $t \to 0$. So f has at most a pole of order k and

$$f_{j,a}(\xi) = \sum_{n \geq -k} c_n(\xi - z_0)^{n/q}.$$

The description of the connected components of the covering $\pi : V_{z_0} \to B(z_0, \varepsilon)\backslash\{z_0\}$ and their number of sheets remains the same.

It remains to consider the point $z_0 = \infty \in \Delta(P)$. Let $d = \max\{\deg a_j, 0 \leq j \leq n\}$. Then we can consider the polynomial equation

$$z^d P\left(\frac{1}{z}, w\right) = 0$$

about $z = 0$ and the local situation is the same as earlier.

Now we are ready to construct the complete Riemann surface of the irreducible equation $P = 0$, proceeding in the same fashion as we did for $\sqrt{f(z)}$ in §5.12, i.e., for $P(z, w) = w^2 - \prod_{1 \leq j \leq n} (z - a_j)$.

For that purpose, let us denote $z_1, \ldots, z_N$ the points in $\Delta(P)$. For each j, denote by $1 \leq k \leq m_j$ the different connected components of the coverings $\pi : V_j \to U_j\backslash\{z_j\}$, where the U_j are open disks with disjoint closures in S^2 (for $z_N = \infty$ we mean $U_N = \{z \in \mathbb{C} : |z| > r\} \cup \{\infty\}$), and $V_j = \pi^{-1}(U_j\backslash\{z_j\}) \cap X_P$. V_j has connected components $V_{j,k}$ of multiplicity q_{jk}. We have $\sum_{1 \leq k \leq m_j} q_{jk} = n$.

Consider now the topological space $\tilde{X}$ which is the disjoint union of X_P and $\sum_{1 \leq j \leq N} m_j$ disks $\tilde{V}_{j,k}$. One glues V_{jk} to $\tilde{V}_{jk}$ by introducing isomorphisms σ_{jk} between the covering maps

$$\pi | V_{jk} : V_{jk} \to U_j\{z_j\}$$

and

$$p_{jk} : \tilde{V}_{jk}\backslash\{c_{jk}\} \to U_j\backslash\{z_j\},$$

where c_{jk} is the center of $\tilde{V}_{jk}$ and p_{jk} is a covering map with q_{jk} sheets given by the q_{jk}th power.

To construct σ_{jk}, choose $\xi_0 \in X_{jk}$, $\zeta_0 \in \tilde{V}_{jk}\backslash\{c_{jk}\}$ such that $\pi(\xi_0) = p_{jk}(\zeta_0)$, and if ξ is the endpoint of a path γ in V_{jk} starting at ζ_0, let γ_1 be the lifting to $\tilde{V}_{jk}\backslash\{c_{jk}\}$ of $\pi \circ \gamma$, starting at ζ_0. We define $\alpha_{jk}(\xi) = \gamma_1(1)$. One verifies without difficulty that σ_{jk} is a biholomorphic map from V_{jk} onto $\tilde{V}_{jk}\backslash\{c_{jk}\}$ such that $p_{jk} \circ \sigma_{jk} = \pi | V_{jk}$.

Introduce in $\tilde{X}$ the equivalence relation $\mathscr{R}$ which identifies the points that correspond to each other via some σ_{jk}. The quotient space $\tilde{X}_P = \tilde{X}/\mathscr{R}$ is now a Riemann surface in which X_P appears as a dense open set. (In fact, $\tilde{X}_P\backslash X_P$ is a finite collection of points.)

The covering map $\pi : X_P \to \mathbb{C}$ can be extended to a holomorphic map $\pi : \tilde{X}_P \to S^2$ setting $\pi(\tilde{c}_{jk}) = z_j$, where $\tilde{c}_{jk}$ is the class of c_{jk} modulo $\mathscr{R}$. It is a branched covering map.

The global holomorphic function $F: X_P \to \mathbb{C}$ defined by $F(\varphi_z) = \varphi_z(z)$ can be extended by continuity to the points $\tilde{c}_{jk}$ to a finite or ∞ value. According to whether the corresponding Laurent series $\sum_{n \in \mathbb{Z}} a_n t^n$ is holomorphic or has a pole at $t = 0$. The meromorphic function $F: \tilde{X}_P \to S^2$ thus obtained verifies $P(\pi(\varphi), f(\varphi)) = 0$ for every $\varphi \in X_P$.

5.13.4. Definition. The triple $(\tilde{X}_P, \pi, F)$ associated by the preceding construction to the irreducible polynomial $P \in \mathbb{C}[z, w]$ is called the ***(complete) Riemann surface of the equation*** $P = 0$. The function F is the ***algebraic function associated to the equation*** $P = 0$.

If P is not irreducible we denote $\tilde{X}_P$ the (disjoint) union of the $\tilde{X}_{P_j}$ for the different irreducible factors of P.

5.13.5. Proposition. *If P is irreducible then $\tilde{X}_P$ is connected.*

Proof. It is enough to prove that X_P is connected. For that, it is enough to show that for a fixed $z_0 \in \Omega = S^2 \setminus \Delta(P)$, the n germs $\varphi_1, \ldots, \varphi_n$ which are solutions of $P(z, w) = 0$ in a neighborhood of z_0, are in the connected component of the germ φ_1 in $\mathcal{O}_\Omega$. We argue by contradiction.

Let us assume that $\varphi_1, \ldots, \varphi_k$ are in the same component and $\varphi_{k+1}, \ldots, \varphi_n$ in other components. Consider the symmetric functions σ_i of the φ_j, $1 \leq j \leq k$, defined by

$$\prod_{1 \leq j \leq k} (w - \varphi_j) = w^k - \sigma_1 w^{k-1} + \cdots + (-1)^k \sigma_k.$$

If γ is a loop in $\pi_1(\Omega, z_0)$, the analytic continuation along γ preserves the set $\{\varphi_j : 1 \leq j \leq k\}$. Hence the σ_j are well-defined holomorphic functions in Ω. The only possible singularities of the σ_j are poles at the points of $\Delta(P)$. The reason is that, from the Puiseux series expansions of the solutions of $P = 0$ near a point $c \in \Delta(P)$, one sees that all of them are bounded by $|z - c|^{-M}$ for some fixed power M if $c \neq \infty$, and by $|z|^M$ if $c = \infty$. Therefore the $\sigma_j \in \mathcal{M}(S^2)$. Let $Q(z)$ be the least-common multiple of the denominators of the rational function σ_j. Then

$$R(z, w) := Q(z) \prod_{1 \leq j \leq k} (w - \varphi_j)$$

is a polynomial in (z, w) of degree k in w. It follows that R and P are relatively prime, since P is irreducible and cannot divide R. Therefore, there are only finitely many common roots $(z, w) \in \mathbb{C}^2$ of the equations

$$R(z, w) = P(z, w) = 0.$$

This contradicts the fact that $R(z, \varphi_1(z)) \equiv 0$ and $P(z, \varphi_1(z)) \equiv 0$ in a neighborhood of z_0.

We conclude that $k < n$ is impossible, hence $\tilde{X}_P$ is connected. □

5.13.6. Proposition. *$\tilde{X}_P$ is compact.*

Proof. We can assume P is irreducible. $\tilde{X}_P$ is the union of finitely many relatively compact regions. Namely, the images under the equivalence $\mathscr{R}$ of the $\tilde{V}_{jk}$ and of $X_P\left(\bigcup_{j,k} V_{jk}\right)$. □

Remark. It follows that $\pi : \tilde{X}_P \to S^2$ is surjective, since π is an open map and $\tilde{X}_P$ is compact.

One can prove that the converse of Proposition 5.13.6 is also true. Any compact Riemann surface can be defined as $\tilde{X}_P$ for a convenient polynomial $P \in \mathbb{C}[z, w]$ (see [Fo]).

We have only scratched the surface of the subject of algebraic functions, or equivalently, compact Riemann surfaces. The reader should consult [Fo], [JS], [Gri2], [ACG], [Si], [FK], and [Cl] for this beautiful and rich theory.

Another interesting question is to decide in an effective way which are possible Puiseux expansions of algebraic function around the critical points. This can be done in terms of the Newton diagram (cf. [Hil]).

Exercises 5.13

1. Describe the complete Riemann surfaces associated to the following algebraic equations.
   (a) $P(z, w) = w^n - z = 0$ (Riemann surface of $\sqrt[n]{z}$)
   (b) $P(z, w) = w^3 - 3w + 2z = 0$
   (c) $P(z, w) = w^3 - 3w + z^6 = 0$
   (d) $P(z, w) = w^4 - 2w^2 + 1 - z = 0$.

   Find exactly the ramification points and the behavior at those points of the algebraic function F defined by the equation. Study the action of $\pi_1(S^2\backslash\Delta(P))$ on the fibers $\pi^{-1}(\{z\})$ for points $z \in S^2\backslash\Delta(P)$. Find out, if possible, the subgroup of the group of permutations of $\pi^{-1}(\{z\})$ generated by the action of $\pi_1(S^2\backslash\Delta(P))$.

2. Prove that if $P \in \mathbb{C}[z, w]$ is an irreducible polynomial and, if (X_1, π_1, F_1) and (X_2, π_2, F_2) are two triples satisfying the same conditions that satisfies the complete Riemann surfaces $(\tilde{X}_P, \pi, F)$ of $P = 0$, then there is a biholomorphic map $\sigma : X_1 \to X_2$ such that $\pi_2 \circ \sigma = \pi_1$ and $F_2 \circ \sigma = F_1$.

3. Study the Riemann surface $\sqrt{f}$, for $f(z) = (z - a_1)^{\alpha_1} \dots (z - a_n)^{\alpha_n}$, where $a_1, \dots, a_n \in \mathbb{C}$ are distinct and the α_j are integers ≥ 1.

4. Construct the noncomplete Riemann surface X_p associated to an irreducible $P \in \mathbb{C}[z, w]$, using $\{(z, w) \in \mathbb{C}^2 : P(z, w) = 0\}$, as in Exercise 5.12.3.

5. Let X, $\tilde{X}$ be the noncomplete, resp. complete, Riemann surface of $w^4 - 2w^2 + 1 - z = 0$.
   (a) X is a four-sheeted covering space of $\mathbb{C}\backslash\{0, 1\}$ (cf. Exercise 5.13.1).
   (b) Let $z_0 = \frac{1}{2}$, $\pi^{-1}(\{z_0\}) = \{\xi_1, \xi_2, \xi_3, \xi_4\}$, with $\xi_1 = \sqrt{1 + \sqrt{\frac{1}{2}}}$, $\xi_2 = \sqrt{1 - \sqrt{\frac{1}{2}}}$, $\xi_3 = -\sqrt{1 + \sqrt{\frac{1}{2}}}$, $\xi_4 = -\sqrt{1 - \sqrt{\frac{1}{2}}}$. Show that the action on $\{\xi_1, \xi_2, \xi_3, \xi_4\}$ of a loop $\gamma_1 \in \pi_1(\mathbb{C}\backslash\{0, 1\}, \frac{1}{2})$ with $\operatorname{Ind}_{\gamma_1}(0) = 1$, $\operatorname{Ind}_{\gamma_1}(1) = 0$ is given by (notations from §5.10.6)

$$\begin{cases} \xi_1[\gamma_1] = \xi_2 \\ \xi_2[\gamma_1] = \xi_1 \end{cases} \quad \text{and} \quad \begin{cases} \xi_3[\gamma_1] = \xi_4 \\ \xi_4[\gamma_1] = \xi_3 \end{cases}$$

(c) The same as (b) when $\gamma_2 \in \pi_1(\mathbb{C}\setminus\{0,1\}, \frac{1}{2})$ has $\mathrm{Ind}_{\gamma_2}(0) = 0$, $\mathrm{Ind}_{\gamma_2}(1) = 1$. (Answer: $\xi_1[\gamma_2] = \xi_1$, $\xi_2[\gamma_2] = \xi_4$, $\xi_3[\gamma_2] = \xi_3$, $\xi_4[\gamma_2] = \xi_2$).

(d) Let $\gamma_3 \in \pi_1(\mathbb{C}\setminus\{0,1\}, \frac{1}{2})$ with $\mathrm{Ind}_{\gamma_3}(0) = \mathrm{Ind}_{\gamma_3}(1) = 1$. Show that $[\gamma_3] = [\gamma_1][\gamma_2]^{-1}$ and characterize the action on $\pi^{-1}(\{z_0\})$.

(e) Show that the covering map $\pi : X \to \mathbb{C}\setminus\{0,1\}$ is Galois and determine $G(\pi)$.

6. Redo Exercise 5.13.5 for the other Riemann surfaces of Exercise 5.13.1.

7. Show that if F is an algebraic function which is single-valued on $S^2\setminus B$, $\#(B) < \infty$, then F is a rational function.

8. Either using Exercise 5.13.7 or directly, show that if F is an algebraic function then F'/F is a rational function. Compare with Exercise 2.8.19 for the case $\sqrt{(z - a_1)\dots(z - a_n)}$.

§14. The Periods of a Differential Form

The purpose of this section is to conclude the proof of Proposition 1.7.9. The missing part was that given a set of periods one could find a closed 1-form with those periods. We need to develop a bit the calculus of differential forms in a Riemann surface. This is exactly analogous to what was done in §1.2. We recommend [We] for a thorough development of this subject and its relation to the geometry of complex manifolds (Riemann surfaces being manifolds of complex dimension 1).

Let (U, φ), $\varphi = u + iv$, be a coordinate chart on a Riemann surface X (u, v real-valued). One can define the differential operators $\frac{\partial}{\partial u}, \frac{\partial}{\partial v}, \frac{\partial}{\partial \varphi}, \frac{\partial}{\partial \bar{\varphi}}$ from $\mathscr{E}(U)$ into itself by the formulas

$$\frac{\partial f}{\partial u}(\zeta) = \left(\frac{\partial}{\partial x}(f \circ \varphi^{-1})\right)(\varphi(\zeta)),$$

$$\frac{\partial f}{\partial v}(\zeta) = \left(\frac{\partial}{\partial y}(f \circ \varphi^{-1})\right)(\varphi(\zeta)),$$

$$\frac{\partial f}{\partial \varphi}(\zeta) = \left(\frac{\partial}{\partial z}(f \circ \varphi^{-1})\right)(\varphi(\zeta)),$$

$$\frac{\partial f}{\partial \bar{\varphi}}(\zeta) = \left(\frac{\partial}{\partial \bar{z}}(f \circ \varphi^{-1})\right)(\varphi(\zeta)),$$

where $z = x + iy$ represents the coordinates in $\mathbb{C}$, $z = \varphi(\zeta) = u(\zeta) + iv(\zeta)$, $\zeta \in U$.

Let $a \in X$. The ring $\mathscr{E}_a$ of germs at the point a of C^∞ functions is a local ring whose only maximal ideal $\mathscr{N}_a$ is the set of germs f such that $f(a) = 0$. We denote $\mathscr{N}_a^k$ its successive powers. One sees easily, using Taylor's expansion, that for any chart (U, φ) about a the ideal $\mathscr{N}_a^2$ can be characterized as the set of $f \in \mathscr{E}_a$ such that

$$f(a) = \frac{\partial f}{\partial u}(a) = \frac{\partial f}{\partial v}(a) = 0.$$

5.14.1. Definition. The complex vector space $T_a^*(X)$ defined as the quotient

$$T_a^*(X) := \mathscr{N}_a/\mathscr{N}_a^2$$

is called the ***cotangent space*** to X at the point a. If V is an open set containing a and $f \in \mathscr{E}(V)$, its differential at a is the element $df(a) \in T_a^*(X)$ defined by

$$df(a) = \dot{\overline{f - f(a)}},$$

where $f(a)$ is the germ of the constant function and the notation $\dot{g}$ denotes the class of $g \in \mathscr{N}_a$ modulo $\mathscr{N}_a^2$.

5.14.2. Proposition. *Let X be a Riemann surface, $a \in X$, (U, φ) a chart about a, $\varphi = u + iv$, $z = x + iy = \varphi(\zeta) = u(\zeta) + iv(\zeta)$, for $\zeta \in U$. The differentials $du(a)$ and $dv(a)$ form a basis of $T_a^*(X)$. The same is true for $d\varphi(a)$ and $d\bar{\varphi}(a)$. We have the relations $d\varphi(a) = du(a) + idv(a)$ and $d\bar{\varphi}(a) = du(a) - idv(a)$. Moreover, if f is C^∞ in a neighborhood of a, we can write*

$$df(a) = \frac{\partial f}{\partial u}(a)\,du(a) + \frac{\partial f}{\partial v}(a)\,dv(a)$$

$$= \frac{\partial f}{\partial \varphi}(a)\,d\varphi(a) + \frac{\partial f}{\partial \bar{\varphi}}(a)\,d\bar{\varphi}(a).$$

PROOF. Let us show first that $du(a)$ and $dv(a)$ span $T_a^*(X)$. Let $\theta \in T_a^*(X)$ be represented by $\alpha \in \mathscr{N}_a$. Taylor's formula gives

$$\alpha(\zeta) - \alpha(a) = \alpha(\zeta) = (u(\zeta) - u(a))\left(\frac{\partial}{\partial x}(\alpha \circ \varphi^{-1})\right)(\varphi(a))$$

$$+ (v(\zeta) - v(a))\left(\frac{\partial}{\partial x}(\alpha \circ \varphi^{-1})\right)(\varphi(a)) + \psi(\zeta)$$

where $\psi \in \mathscr{N}_a^2$. Passing to the quotient we have

$$\theta = \dot{\alpha} = \frac{\partial \alpha}{\partial u}(a)\,du(a) + \frac{\partial \alpha}{\partial v}(a)\,dv(a).$$

Moreover, if $\theta = 0$, $\alpha \in \mathscr{N}_a^2$ and $\dfrac{\partial \alpha}{\partial u}(a) = \dfrac{\partial \alpha}{\partial v}(a) = 0$. Therefore $\{du(a), dv(a)\}$ is a basis for $T_a^*(X)$. One verifies that

$$\frac{\partial}{\partial \varphi} = \frac{1}{2}\left(\frac{\partial}{\partial u} + \frac{1}{i}\frac{\partial}{\partial v}\right), \qquad \frac{\partial}{\partial \bar{\varphi}} = \frac{1}{2}\left(\frac{\partial}{\partial u} - \frac{1}{i}\frac{\partial}{\partial v}\right),$$

which shows that $\{d\varphi(a), d\bar{\varphi}(a)\}$ is another basis for $T_a^*(X)$.

If f is smooth near a, we have

$$f - f(a) = \frac{\partial f}{\partial u}(a)(u - u(a)) + \frac{\partial f}{\partial v}(a)(v - v(a)) + \psi,$$

$\psi \in \mathcal{N}_a^2$. Hence

$$df(a) = \frac{\partial f}{\partial u}(a)\,du(a) + \frac{\partial f}{\partial v}(a)\,dv(a) = \frac{\partial f}{\partial \varphi}(a)\,d\varphi(a) + \frac{\partial f}{\partial \bar{\varphi}}(a)\,d\bar{\varphi}(a). \qquad \square$$

Clearly we do not need $f \in C^\infty$ near a to be able to define $df(a)$. Differentiability is enough.

We are now going to define the differential forms of type $(1, 0)$ and $(0, 1)$. For that we need to find out the relations that appear when we change coordinate charts. Let (U, φ) and (V, ψ) be two charts about a. Let $h = \varphi \circ \psi^{-1}$. If $\varphi(\zeta) = z$ and $\psi(\zeta) = z'$, then

$$\frac{\partial \bar{h}}{\partial \bar{z}'}(\psi(a)) = \overline{\frac{\partial h}{\partial z'}(\psi(a))} \qquad \text{and} \qquad \frac{\partial h}{\partial \bar{z}'} = \frac{\partial \bar{h}}{\partial z'} = 0.$$

Hence,

$$\frac{\partial \varphi}{\partial \psi}(a) = \left(\frac{\partial}{\partial z'}(\varphi \circ \psi^{-1})\right)(\psi(a)) = \frac{\partial h}{\partial z'}(\psi(a)) \neq 0, \qquad \frac{\partial \varphi}{\partial \bar{\psi}}(a) = 0,$$

$$d\varphi(a) = \frac{\partial \varphi}{\partial \psi}(a)\,d\psi(a) + \frac{\partial \varphi}{\partial \bar{\psi}}\,d\bar{\psi}(a) = \frac{\partial \varphi}{\partial \psi}(a)\,d\psi(a),$$

and

$$d\bar{\varphi}(a) = \overline{\frac{\partial \varphi}{\partial \psi}(a)}\,d\bar{\psi}(a).$$

These formulas allow us to decompose the space $T_a^*(X)$ as a direct sum $T_a^*(X) = T_a^{1,0}(X) \oplus T_a^{0,1}(X)$, of the two subspaces

$$T_a^{1,0}(X) = \mathbb{C}\,d\varphi(a),$$
$$T_a^{0,1}(X) = \mathbb{C}\,d\bar{\varphi}(a),$$

which do not depend on the chart considered. The elements of $T_a^{1,0}(X)$ are called differential forms at a of type $(1, 0)$, those of $T_a^{0,1}(X)$ are said to be of type $(0, 1)$. If f is continuously differentiable in a neighborhood of a one can write, in a unique fashion,

$$df(a) = \partial f(a) + \bar{\partial} f(a),$$

with $\partial f(a) \in T_a^{1,0}(X)$ and $\bar{\partial} f(a) \in T_a^{0,1}(X)$. These formulas reveal that in terms of a coordinate chart (U, φ) we have

$$\partial f(a) = \frac{\partial f}{\partial \varphi}(a)\,d\varphi(a),$$

and

$$\bar{\partial} f(a) = \frac{\partial f}{\partial \bar{\varphi}}(a)\, d\bar{\varphi}(a).$$

5.14.3. Definition. Let U be an open subset of the Riemann surface X. We denote by $T^*(U) = \dot{\bigcup\limits_{a \in U}} T_a^*(X)$. A differential form ω of degree one is a map

$$\omega : U \to T^*(U)$$

such that $\omega(a) \in T_a^*(X)$ for any $a \in U$. If $\omega(a) \in T_a^{1,0}(X)$ for every $a \in U$ we say ω is of type $(1, 0)$, while if $\omega(a) \in T_a^{0,1}(X)$ for every $a \in U$ we say it is of type $(0, 1)$. We sometimes abbreviate and say that ω is 1-form, a (1,0)-form, or a $(0, 1)$-form.

Example. If $f \in \mathscr{E}(U)$, then df, ∂f, $\bar{\partial} f$ are differential forms of degree 1, the last two being, respectively, of type $(1, 0)$ and $(0, 1)$. As it is easy to verify, $f \in \mathscr{E}(U)$ is holomorphic in U if and only if $\bar{\partial} f = 0$. The usual Leibniz rule for derivatives of a product imply $d(fg) = f\,dg + g\,df$ and analogous relations for $\partial(fg)$ and $\bar{\partial}(fg)$ hold, where $f, g \in \mathscr{E}(U)$.

Remarks. (1) Let (U, φ) be a chart in X, $\varphi = u + iv$. A differential form ω of degree 1 in U can be written as

$$\omega = f\,du + g\,dv = k\,d\varphi + l\,d\bar{\varphi},$$

where f, g, k, l are well-defined complex-valued functions in U.

(2) If (U, φ) and (V, ψ) are two charts such that $U \cap V \neq \varnothing$, and if ω is a differential form of degree 1 in $U \cap V$, then

$$\omega = f\,d\varphi + g\,d\bar{\varphi} = k\,d\psi + l\,d\bar{\psi}.$$

The formulas $d\varphi = \dfrac{\partial \varphi}{\partial \psi} d\psi$ and $d\bar{\varphi} = \dfrac{\overline{\partial \varphi}}{\partial \psi} d\bar{\psi}$ imply the relations

$$k = f \frac{\partial \varphi}{\partial \psi} \qquad \text{and} \qquad l = g \frac{\overline{\partial \varphi}}{\partial \psi}.$$

(3) If (U, φ) is a chart in X, $\varphi(\zeta) = z$, and $\omega = f\,d\varphi + g\,d\bar{\varphi}$ is a differentiable form of degree 1 in U, we introduce the "local version" ω_1 of ω in the open subset $U_1 = \varphi(U)$ of the complex plane by

$$\omega_1 = (f \circ \varphi^{-1})\,dz + (g \circ \varphi^{-1})\,d\bar{z}.$$

(4) When we are in the situation that we have two charts (U, φ), (V, ψ), $U \cap V \neq \varnothing$, $z = \varphi(\zeta)$, $z' = \psi(\zeta)$, and ω a differential form of degree 1 in $U \cap V$, $\omega = f\,d\varphi + g\,d\bar{\varphi} = k\,d\psi + l\,d\bar{\psi}$, we have two local versions

$$\omega_1 = (f \circ \varphi^{-1})\,dz + (g \circ \varphi^{-1})\,d\bar{z}, \qquad \text{in} \qquad U_1 = \varphi(U \cap V),$$

$$\omega_2 = (k \circ \psi^{-1})\,dz' + (l \circ \psi^{-1})\,d\bar{z}', \qquad \text{in} \qquad U_2 = \psi(U \cap V).$$

Since we know that $k \circ \psi^{-1} = (f \circ \psi^{-1})\left(\frac{\partial \varphi}{\partial \psi} \circ \psi^{-1}\right)$ and $l \circ \psi^{-1} = (g \circ \psi^{-1}) \times \overline{\left(\frac{\partial \varphi}{\partial \psi} \circ \psi^{-1}\right)}$, if we let $h = \varphi \circ \psi^{-1} : U_2 \to U_1$, and consider the pullback of ω_1 by h, $h^*(\omega_1)$, we find

$$\begin{aligned} h^*(\omega_1) &= (f \circ \varphi^{-1} \circ h)\,dh + (g \circ \varphi^{-1} \circ h)\,d\bar{h} \\ &= (f \circ \varphi^{-1})\left(\frac{\partial \varphi}{\partial \psi} \circ \psi^{-1}\right) dz' + (g \circ \psi^{-1})\overline{\left(\frac{\partial \varphi}{\partial \psi} \circ \psi^{-1}\right)}\,\overline{dz'} \\ &= \omega_2. \end{aligned}$$

So that the two local versions of ω are given by $h^*(\omega_1) = \omega_2$, this allows us to talk about the continuity, smoothness, etc., of a differential form of degree 1 in a way that is independent of the coordinate chart.

5.14.4. Definition. Let W be an open subset of the Riemann surface X, ω a differential form of degree 1 in W. We say that ω is of class C^k in W ($k \in \mathbb{N} \cup \{\infty\}$) if for any chart (U, φ), $U \subseteq W$, $\omega = f\,d\varphi + g\,d\bar{\varphi}$ with f, $g \in \mathscr{E}_k(U)$. We say ω is a holomorphic differential form of degree 1 if, for any chart (U, φ), $U \subseteq W$, $\omega = h\,d\varphi$, with $h \in \mathscr{H}(U)$.

We will denote $\mathscr{E}_k^1(W)$ the family of all differential forms of degree 1 of class C^k in W. Similarly, $\mathscr{E}_k^{1,0}(W)$, $\mathscr{E}_k^{0,1}(W)$, $\Omega^1(W) = \Omega(W)$, are those of type $(1,0)$ and class C^k, type $(0,1)$ and class C^k, and holomorphic forms in W. One can define in a natural way the associated sheaves of germs $\mathscr{E}_k^1$, $\mathscr{E}_k^{1,0}$, $\mathscr{E}_k^{0,1}$, Ω. If $k = \infty$, we will suppress the index k.

In the same way as was done in §1.2, we can define the wedge product of vectors. If $\zeta, \eta \in T_a^*(X)$ we obtain an element $\zeta \wedge \eta$ of a vector space denoted indistinctly $\Lambda^2 T_a^*(X)$ or $T_a^{*2}(X)$. If (U, φ) is a coordinate chart at the point a, $\varphi = u + iv$, we have that both $d\varphi(a) \wedge d\bar{\varphi}(a)$ and $du(a) \wedge dv(a)$ form a basis for the one-dimensional complex vector space $T_a^{*2}(X)$. Moreover,

$$d\varphi(a) \wedge d\bar{\varphi}(a) = -2i\,du(a) \wedge dv(a).$$

5.14.5. Definition. Let W be an open subset of X, a differential form of degree 2 on W is a map $\omega : W \to \dot{\bigcup}_{a \in W} T_a^{*2}(X) = T^{*2}(X)$ such that $\omega(a) \in T_a^{*2}(X)$ for every $a \in W$. We also say ω is a 2-form in W.

Let (U, φ) be a chart in X, $\varphi = u + iv$, a differential form ω of degree 2 in U can be written as

$$\omega = f\,d\varphi \wedge d\bar{\varphi} = g\,du \wedge dv,$$

for some uniquely determined functions $f, g : U \to \mathbb{C}$. One can also introduce the local version ω_1 of ω in $\varphi(U)$ by

$$\omega_1 = (f \circ \varphi^{-1})\,dz \wedge d\bar{z},$$

where $z = \varphi(\zeta)$, $\zeta \in U$. Different local versions of ω are related by the pullback map.

5.14.6. Definition. Let W be an open subset of the Riemann surface W, ω a differential form of degree 2 in W. We say that

(1) ω is of class C^k in W ($k \in \mathbb{N} \cup \{\infty\}$), if, for any chart (U, φ) such that $U \subseteq W$, $\omega = f\, d\varphi \wedge d\bar{\varphi}$ with $f \in \mathscr{E}_k(U)$;
(2) ω is holomorphic in W if, for any such chart (U, φ), $\omega = h\, d\varphi \wedge d\bar{\varphi}$ with $h \in \mathscr{H}(U)$.

We will denote by $\mathscr{E}_k^2(W)$ the set of differential forms of degree 2 and class C^k in W (with k suppressed if $k = \infty$). By $\Omega^2(W)$ we denote the set of all holomorphic differential forms of degree 2 in W. The corresponding sheaves shall be denoted $\mathscr{E}_k^2$ and Ω^2.

The wedge product induces a map

$$\mathscr{E}_k^1(W) \times \mathscr{E}_l^1(W) \to \mathscr{E}_m^2(W), \qquad m = \inf\{k, l\}$$
$$(\alpha, \beta) \mapsto \alpha \wedge \beta,$$

where $(\alpha \wedge \beta)(a) = \alpha(a) \wedge \beta(a)$.

We shall now define the operator $d : \mathscr{E}^1(W) \to \mathscr{E}^2(W)$. Locally (i.e., in a chart) a C^∞ differential form ω of degree 1 can be written as

$$\omega = \sum_k f_k\, dg_k,$$

$f_k, g_k \in C^\infty$ (see, for instance, the alternative ways we write in terms of $d\varphi$, $d\bar{\varphi}$ or du, dv when the chart is (U, φ)). We define

$$d\omega = \sum_k df_k \wedge dg_k.$$

One can easily show this 2-form is independent of the particular representation chosen for ω. It is also immediate that if $f \in \mathscr{E}_2(X)$, then

$$d(df) = 0.$$

5.14.7. Definition. Let W be an open set in a Riemann surface. A differential form $\omega \in \mathscr{E}_k^1(W)$, $k \geq 1$, is said to be closed if $d\omega = 0$. The form is said to be exact in W if there is a continuously differentiable function f such that $\omega = df$ in W.

Remark. The previous remark implies that every exact form is closed, since $d(df) = 0$.

5.14.8. Proposition. *Let W be an open set in a Riemann surface.*

(a) *Every* $\omega \in \Omega(W)$ is closed.
(b) *Every* $\omega \in \mathscr{E}^{1,0}(W)$ *that is closed is holomorphic.*

PROOF. Let (U, φ) be a coordinate chart with $U \subseteq W$ and $\omega = f\,d\varphi$ in W. Then $d\omega = df \wedge d\varphi = -\dfrac{\partial f}{\partial \bar{\varphi}} d\varphi \wedge d\bar{\varphi}$. If f is holomorphic then $\dfrac{\partial f}{\partial \bar{\varphi}} = 0$ and, hence, $d\omega = 0$. The converse is equally clear. □

5.14.9. Definition. Let $f: X \to Y$ be a C^∞ map between two Riemann surfaces. For every open set W of Y, f induces $\mathbb{C}$-linear maps, ***pullbacks***, $f^*: \mathscr{E}^j(W) \to \mathscr{E}^j(f^{-1}(W))$ $(j = 0, 1, 2, \mathscr{E}^0(W) = \mathscr{E}(W))$, where the pullback is defined as follows:

(1) If $g \in \mathscr{E}(W)$, then $f^*g = g \circ f$.
(2) If $\alpha \in \mathscr{E}^1(W)$ is written locally as $\alpha = \sum_j f_j\,dg_j$, then $f^*\alpha = \sum_j (f^*f_j)\,d(f^*g_j)$.
(3) If $\beta \in \mathscr{E}^2(W)$ is written locally as $\beta = \sum_j f_j\,dg_j \wedge dh_j$, then $f^*\beta = \sum_j (f^*f_j) d(f^*g_j) \wedge d(f^*h_j)$.

It is necessary to verify the independence of the definition of pullback with respect to the local representations. This is done easily; it depends only on the chain rule. For any $g \in \mathscr{E}(W)$ one has

$$f^*\,dg = d(f^*g).$$

Similarly, for any $\omega \in \mathscr{E}^1(W)$ one has

$$f^*\,d\omega = d(f^*\omega).$$

It is also immediate that if $f: X \to Y$ and $g: Y \to Z$ are C^∞ maps, then $(g \circ f)^* = f^* \circ g^*$.

We now note that the procedures explained in Chapter 1 to integrate 1-forms of class C^1 along a piecewise-C^1 path and closed 1-forms along continuous paths are valid in the case of forms defined in an open set W of a Riemann surface. We have therefore at our disposal the analogue of statements 1.7.6, 1.7.7, and 1.7.8. One can define $B^1(W)$, $Z^1(W)$, and $H^1(W)$ in a natural fashion and one arrives to §1.7.9 which remains valid, except that the surjectivity of the map $j: Z^1(W) \to \operatorname{Hom}(\pi_1(W, a), \mathbb{C})$ remains still to be proven. It is to this proof that we are consecrating the remainder of this section. Once this is done the consequences 1.7.10, 1.7.11, and 1.7.12 remain valid for Riemann surfaces.

Let X be a connected Riemann surface and $p: \tilde{X} \to X$ its universal covering space. Recall that the group $G(p)$ isomorphic to $\pi_1(X)$. If $\sigma \in G(p)$ and $f: \tilde{X} \to \mathbb{C}$ is any function we define $\sigma \cdot f: \tilde{X} \to \mathbb{C}$ by $\sigma \cdot f = f \circ \sigma^{-1}$. It is clear that if g is another function $g: \tilde{X} \to \mathbb{C}$ we have

$$\sigma \cdot (f + g) = \sigma \cdot f + \sigma \cdot g$$

$$\sigma \cdot (fg) = (\sigma \cdot f)(\sigma \cdot g).$$

Moreover, if $\sigma, \tau \in G(p)$, $(\sigma \circ \tau) \cdot f = \sigma \cdot (\tau \cdot f)$.

5.14.10. Definition. A function $f : \tilde{X} \to \mathbb{C}$ is said to be ***additively automorphic*** if for every $\sigma \in G(p)$ there is a constant $a_\sigma \in \mathbb{C}$ such that

$$f - \sigma \cdot f = a_\sigma.$$

These constants a_σ are called the ***automorphic constants*** of f.

The obvious example of such a function is one which is invariant under the action of $G(p)$, that is, $a_\sigma = 0$ for every $\sigma \in G(p)$.

5.14.11. Proposition. *Let $f : \tilde{X} \to \mathbb{C}$ be an additively automorphic function. Then the map $\sigma \mapsto a_\sigma$ from $G(p) = \pi_1(X)$ into $\mathbb{C}$ is an element of* $\operatorname{Hom}(\pi_1(X), \mathbb{C})$.

PROOF.

$$a_{\sigma \circ \tau} = f - \sigma \cdot (\tau \cdot f) = f - \sigma \cdot f + \sigma \cdot (f - \tau \cdot f) = a_\sigma + \sigma \cdot a_\tau = a_\sigma + a_\tau. \quad \square$$

5.14.12. Proposition. *Let X be a connected Riemann surface and $p : \tilde{X} \to X$ its universal covering space. Then*

(a) *If $\omega \in Z^1(X)$ and f is a primitive of $p^*\omega$, then f is additively automorphic and its constants a_σ, $\sigma \in \pi_1(X)$, are precisely the periods of ω.*

(b) *Conversely, if $f \in \mathscr{E}(\tilde{X})$ is additively automorphic there is $\omega \in Z^1(X)$ such that $df = p^*\omega$.*

PROOF. (a) If $\sigma \in G(p) = \pi_1(X)$, since $p \circ \sigma^{-1} = p$, then function $\sigma \cdot f$ is also a primitive of $p^*\omega$. Therefore $f - \sigma \cdot f = \text{constant} \equiv a_\sigma$. Let $x_0 \in X$, $z_0 \in p^{-1}(x_0)$. For $\sigma \in G(p)$, the corresponding element $\tilde{\sigma} \in \pi_1(X, x_0)$ can be represented by taking a curve $\gamma : [0, 1] \to \tilde{X}$ such that $\gamma(0) = y_0 = \sigma^{-1}(z_0)$ and $\gamma(1) = z_0$. Then $\alpha = p \circ \gamma$ is a loop in X with base point x_0 such that $\tilde{\sigma} = [\alpha]$. The periods of ω are then computed by

$$\int_\alpha \omega = \int_\gamma p^*\omega = f(\gamma(1)) - f(\gamma(0)) = f(z_0) - f(\sigma^{-1}(z_0)) = a_\sigma.$$

(b) If f is additively automorphic with constants a_σ, $\sigma \in G(p)$, we have $\sigma^*(df) = d(\sigma^* f) = d(f + a_\sigma) = df$. Since p is a local biholomorphism we always have locally $\omega \in \mathscr{E}^1(X)$ such that $p^*\omega = df$, but since df is invariant under $G(p)$ we obtain that ω is globally well defined. Furthermore, $p^*(d\omega) = dp^*\omega = ddf = 0$, but p^* is injective, hence $d\omega = 0$. $\square$

Let now X, Y be two Riemann surfaces, $p : Y \to X$ a covering map. The group $G(p)$ of automorphisms of p acts on $\mathscr{E}(Y)$ by $\sigma \cdot f = f \circ \sigma^{-1}$. The differences $a_\sigma = f - \sigma \cdot f \in \mathscr{E}(Y)$, $\sigma \in G(p)$, are then called the automorphic factors associated to f. The map $\sigma \in G(p) \mapsto a_\sigma \in \mathscr{E}(Y)$ satisfies the following:

$$a_{\sigma \circ \tau} = a_\sigma + \sigma \cdot a_\tau,$$

by the same argument of Proposition 5.14.2. Namely, $a_{\sigma\circ\tau} = f - (\sigma\circ\tau)\cdot f = f - \sigma\cdot f + \sigma\cdot(f - \tau\cdot f) = a_\sigma + \sigma\cdot a_\tau$. The only difference is that a_τ is not constant in general.

If the covering map p is Galois and the factors a_σ are all identically zero then $f \in p^*\mathscr{E}(X) \subseteq \mathscr{E}(Y)$, and one can identify it to a function on X.

Let $x \in X$, U an open connected neighborhood of x which trivializes p. Let $p^{-1}(U) = \dot{\bigcup}_{i\in I} V_i$. Considering $G(p)$ as usual with the discrete topology, one can construct a trivializing map

$$\varphi : p^{-1}(Y) \to U \times G(p)$$

by choosing $i_0 \in I$ since, in that case, for every $i \in I$ there is a unique $\sigma_i \in G(p)$ such that $\sigma_i(V_{i_0}) = V_i$. We set $\varphi(y) = (p(y), \sigma_i)$ if $y \in V_i$. The diagram

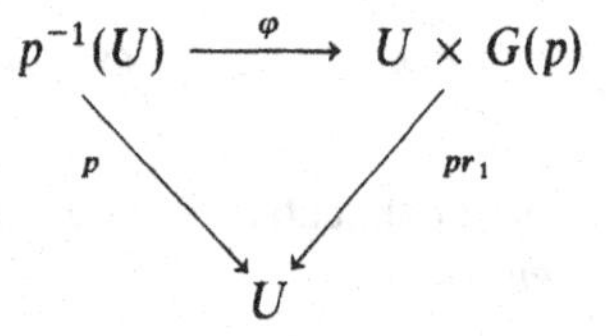

is then clearly commutative. Furthermore, φ is compatible with the action of $G(p)$: if $\varphi(y) = (x, \sigma)$, we have $\varphi(\tau(y)) = (x, \tau\circ\sigma)$ for every $\tau \in G(p)$. This trivialization of φ can therefore be written $\varphi = (p, \eta)$ with $\eta : p^{-1}(U) \to G(p)$ satisfying $\eta(\tau(y)) = \tau\circ\eta(y)$. Moreover, η is continuous, or what is the same, locally constant.

This construction will be used in the following proposition.

5.14.13. Proposition. *Let X, Y be two Riemann surfaces and $p : Y \to X$ a Galois covering. For every map $a : G(p) \to \mathscr{E}(Y)$, $\sigma \mapsto a_\sigma$, satisfying $a_{\sigma\circ\tau} = a_\sigma + \sigma\cdot a_\tau$ for every $\sigma, \tau \in G(p)$, there exist $f \in \mathscr{E}(Y)$ having the functions a_σ as automorphic factors.*

Proof. It consists of three steps:

(i) Choose an open covering $(U_i)_{i\in I}$ of X by connected open sets trivializing p. On $Y_i = p^{-1}(U_i)$ let $\varphi_i : p^{-1}(U_i) \to U_i \times G(p)$ be a trivialization like the one just constructed, $\varphi_i = (p, \eta_i)$. Define a function $f_i : Y_i \to \mathbb{C}$ by

$$f_i(y) = a_{\eta_i(y)}(y),$$

which is C^∞ in Y_i.

(ii) We verify that $f_i - \sigma\cdot f_i = a_\sigma$ on Y_i for every $\sigma \in G(p)$. In fact, if $y \in Y_i$ we have, by definition

$$\begin{aligned}(\sigma\cdot f_i)(y) &= f_i(\sigma^{-1}(y)) = a_{\eta_i(\sigma^{-1}(y))}(\sigma^{-1}(y)) \\ &= a_{\sigma^{-1}\circ\eta_i(y)}(\sigma^{-1}(y)).\end{aligned}$$

The relation $a_{\sigma\circ\tau} = a_\sigma + \sigma\cdot a_\tau$, with $\tau = \sigma^{-1}\circ\eta_i(y)$, becomes

$$a_\sigma(y) = a_{\sigma \circ \tau}(y) - a_{\sigma^{-1} \circ \eta_i(y)}(\sigma^{-1}(y)) = a_{\eta_i(y)}(y) - (\sigma \cdot f_i)(y)$$
$$= f_i(y) - (\sigma \cdot f_i)(y),$$

which is the relation we wanted to verify.

(iii) The differences $g_{i,j} = f_i - f_j \in \mathscr{E}(Y_i \cap Y_j)$ can be considered as being elements of $\mathscr{E}(U_i \cap U_j)$, since for any $\sigma \in G(p)$

$$f_i - \sigma \cdot f_i = a_\sigma = f_j - \sigma \cdot f_j$$

leads to

$$f_i - f_j = \sigma \cdot (f_i - f_j).$$

Furthermore, it is clear that $g_{i,j} + g_{j,k} + g_{k,i} = 0$ on $U_i \cap U_j \cap U_k$.

Let now $(\alpha_v)_v$ be a partition of unity by elements of $\mathscr{D}(X)$ such that for every v there is $j(v) \in I$ with $\alpha_v \in \mathscr{D}(U_{j(v)})$. Set $g_i = \sum_v \alpha_v g_{i,j(v)}$. It is easy to see that $g_i \in \mathscr{E}(U_i)$ and $g_{i,j} = g_i - g_j$. Hence, the functions

$$\tilde{f}_i = f_i - p^* g_i \in \mathscr{E}(Y_i),$$

satisfy $\tilde{f}_i - \sigma \cdot \tilde{f}_i = a_\sigma$ and, whenever $Y_i \cap Y_j \neq \varnothing$, we have on $Y_i \cap Y_j$

$$\tilde{f}_i - \tilde{f}_j = f_i - f_j - (p^* g_i - p^* g_j) = g_{i,j} - g_{i,j} = 0.$$

Therefore, the $\tilde{f}_i$ define a global function $f \in \mathscr{E}(Y)$ satisfying $f - \sigma \cdot f = a_\sigma$ for every $\sigma \in G(p)$. □

We are finally ready to prove the surjectivity missing in §1.7.9.

5.14.14. Proposition. *Let X be a connected Riemann surface. The map $j : H^1(X) \to \operatorname{Hom}(\pi_1(X), \mathbb{C})$ which associates to every cohomology class of a closed 1-form its periods, is surjective.*

Proof. Let $p : \tilde{X} \to X$ be the universal covering space of X. We have $G(p) = \pi_1(X)$. Let $\sigma \mapsto a_\sigma$ be an element of $\operatorname{Hom}(\pi_1(X), \mathbb{C})$ considered as a map from $G(p)$ into $\mathscr{E}(\tilde{X})$ (all its values being constant functions). It satisfies $a_{\sigma \circ \tau} = a_\sigma + \sigma \cdot a_\tau$, since $\sigma \cdot a_\tau = a_\tau \circ \sigma^{-1} = a_\tau$ and $a_{\sigma \circ \tau} = a_\sigma + a_\tau$. The preceding proposition ensures the existence of $f \in \mathscr{E}(\tilde{X})$ with the a_σ as automorphic constants. Hence $\omega = df$ induces on X a closed differential 1-form having periods a_σ by §5.14.12. □

The reader will find in [Fo] the proof of the theorem of Behnke-Stein proving that there is a 1-form $\omega \in \Omega(X)$ with given periods a_σ *if X is not* compact.

We abandon now the theory of Riemann surfaces. For the reader who wants to delve more deeply into this wonderful subject, we recommend the beautiful book [Fo], which has inspired much of our exposition on the subject. For a more geometric point of view, we recommend [JS] and [Gri2].

EXERCISES 5.14

1. Show that $\Omega(S^2) = \{0\}$.

2. Let X be the complete Riemann surface of $\sqrt{z-a}$. Show that $\Omega(X) = \{0\}$.

3. Let X be a connected Riemann surface. Denote by $\mathcal{M}^{1,0}(X)$ the 1-forms ω in X whose expression in any local chart (U, φ) is $\omega = h\,d\varphi$, $h \in \mathcal{M}(U)$. Show that $\mathcal{M}^{1,0}(X)$ is an $\mathcal{M}(X)$-vector space.

   Find the dimension of $\mathcal{M}^{1,0}(S^2)$ over $\mathcal{M}(S^2)$. Answer the same question when X is the complete Riemann surface of $\sqrt{z-a}$.

4. Let X be a Riemann surface. Show that a differential form $\omega \in \mathscr{E}^1(X)$, which is closed, has a primitive if and only if all its periods are zero.

5. Let $\Gamma = \omega_1\mathbb{Z} \oplus \omega_2\mathbb{Z}$ be a lattice of rank two in $\mathbb{C}$, $X = \mathbb{C}/\Gamma$ and $p : \mathbb{C} \to X$ the canonical projection map. Show there is a differential form $\alpha \in \Omega(X)$ such that $p^*(\alpha) = dz$. Find the periods of α.

   Let $\alpha_j : [0, 1] \to \mathbb{C}$, $\alpha_j(t) = t\omega_j$ and $\gamma_j := p \circ \alpha_j$, $j = 1, 2$. Then $[\gamma_1]$, $[\gamma_2]$ are the generators for $\pi_1(X)$ $(= \pi_1(X, 0))$.

   Let $\beta \in \Omega(X)$. Prove that there is a function $g \in \mathscr{H}(\mathbb{C})$ such that $g(0) = 0$ and $dg = p^*\beta$.

   Let R be the parallelogram suggested by Figure 5.14. Prove the identity

$$0 = \int_{\alpha_1} g'(z)\,dz + \int_{\alpha_3} g'(z)\,dz - \int_{\alpha_4} (g'(z) + \omega_2)\,dz$$

$$- \int_{\alpha_2} (g'(z) - \omega_1)\,dz = \omega_1 g(\omega_2) - \omega_2 g(\omega_1).$$

   Conclude that there is a $\lambda \in \mathbb{C}$ such that β and $\lambda\alpha$ have the same periods. Prove that $\dim_{\mathbb{C}}\Omega(X) = 1$.

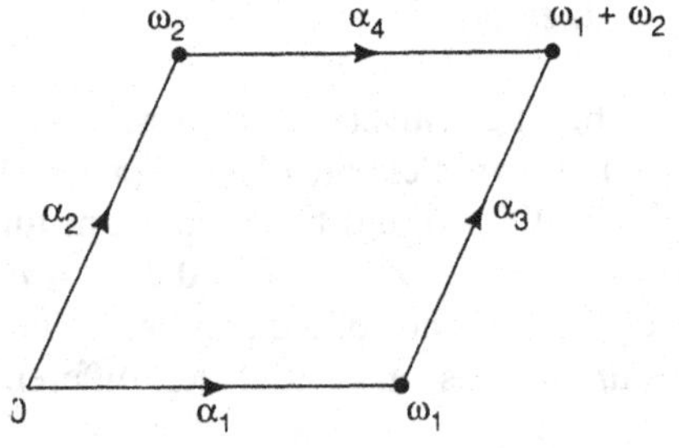

Figure 5.14

6. Compute $\dim_{\mathbb{C}}\Omega(X)$, when X is the complete Riemann surface of $\sqrt{(z-a)(z-b)}$, $a \neq b$.

7. Let $\pi : X \to S^2$ be the branched covering map corresponding to complete Riemann surface of $\sqrt{f}$, $f(z) = (z - a_1)\ldots(z - a_{2n+1})$, where $a_1, a_2, \ldots, a_{2n+1}$ are all distinct

complex numbers. Show that the formula

$$\omega_j = \frac{\pi^{j-1}\,d\pi}{\sqrt{f\circ\pi}}, \qquad 1 \le j \le n$$

defines elements of $\Omega(X)$. Moreover, they are $\mathbb{C}$-linearly independent. (See [Fo] for a proof that $\{\omega_1,\ldots,\omega_n\}$ forms a basis of $\Omega(X)$.)

§15. Linear Differential Equations

We are going to consider differential equations of the form

$$y' = Ay + B,$$

where $y:\Omega\to\mathbb{C}^n$ is a holomorphic map to be determined, $A:\Omega\to M(n,\mathbb{C})$ and $B:\Omega\to\mathbb{C}^n$ are holomorphic maps (i.e., each component of y, A, and B is holomorphic). The meaning of the derivative y' is that each entry of the vector y is differentiated. Ω is assumed to be a simply connected open set in $\mathbb{C}$. In general, one imposes an initial condition $y(z_0) = x_0 \in \mathbb{C}^n$ for some $z_0\in\Omega$. We shall show there is a unique solution to this problem.

We start by solving the following problem: Given Ω simply connected open subset of $\mathbb{C}$, $z_0\in\Omega$, and $A:\Omega\to M(n,\mathbb{C})$ holomorphic map, we look for a holomorphic map $f:\Omega\to M(n,\mathbb{C})$ such that

$$\begin{cases} f' = Af \\ f(z_0) = id_{\mathbb{C}^n} = id. \end{cases} \tag{$*$}$$

We identify $M(n,\mathbb{C})$ to $\mathscr{L}(\mathbb{C}^n)$, the algebra of linear endomorphisms of $\mathbb{C}^n$, which then carries a natural norm $\|\cdot\|$. (For $x\in\mathbb{C}^n$, $\|x\| = (|x_1|^2+\cdots+|x_n|^2)^{1/2}$. For $A\in M(n,\mathbb{C})$, $\|A\| = \max\{\|Ax\| : \|x\| = 1\}$.)

If $n = 1$, $\Omega = \mathbb{C}$, $A = a\in\mathbb{C}$, and $z_0 = 0$, a solution to $(*)$ is $f(z) = e^{az}$, which can be represented by the power series

$$e^{az} = \sum_{k=0}^{\infty}\frac{a^k z^k}{k!}.$$

One remarks that every term of this series can be obtained from the preceding one by multiplying it by a and then taking the primitive of the term thus obtained, normalized by the condition that it vanishes at the origin. We copy this procedure in the general case $(*)$. Introduce a sequence of functions in Ω with values in $M(n,\mathbb{C})$ by the recurrence formulas:

$$\varphi_0(z) = id$$

$$\varphi_1(z) = \int_{\gamma_z} A(u)\varphi_0(u)\,du$$

$$\varphi_2(z) = \int_{\gamma_z} A(u)\varphi_1(u)\,du$$

$$\vdots$$

$$\varphi_{n+1}(z) = \int_{\gamma_z} A(u)\varphi_n(u)\,du,$$

where γ_z is an arbitrary path joining z_0 to z in Ω. The sequence $(\varphi_n)_{n\geq 0}$ is well defined since one can verify by induction that all the φ_n are holomorphic in Ω. (Note that the integration takes place in each entry of the matrix $A\varphi_n$.)

5.15.1. Lemma. *The series $\sum_{n\geq 0} \varphi_n$ is absolutely and uniformly convergent on every compact subset of Ω. Its sum $f(z) = \sum_{n\geq 0} \varphi_n(z)$ is a holomorphic function in Ω solving the problem (*). The solution to (*) is unique. Moreover, f takes values in $GL(n, \mathbb{C})$.*

PROOF. (i) Let us prove the normal convergence of the series over any compact subset of Ω. It is sufficient to do it for closed disks $\bar{B}(z_1, R) \subseteq \Omega$. Fix a piecewise C^1 path γ joining z_0 to z_1 in Ω, we will take $\gamma_z = \gamma\alpha_z$, where α_z is the line segment joining z_1 to z in $\bar{B}(z_1, R)$. Let M be a majorant of $\|A(u)\|$ on $\bar{B}(z_1, R) \cup \gamma$. Let M_1 be a majorant of $M|\gamma_z'(u)|$ when $0 \leq u \leq 1$, $z \in \bar{B}(z_1, R)$.

For a fixed $z \in \bar{B}(z_1, R)$ we introduce now the auxiliary sequence of functions $\tilde{\varphi}_n : [0, 1] \to M(n, \mathbb{C})$ defined by

$$\tilde{\varphi}_1(t) = id$$

$$\tilde{\varphi}_1(t) = t \int_0^1 A(\gamma_z(ts))\tilde{\varphi}_0(ts)\gamma_z'(ts)\,ds$$

$$\vdots$$

$$\tilde{\varphi}_{n+1}(t) = t \int_0^1 A(\gamma_z(ts))\tilde{\varphi}_n(ts)\gamma_z'(ts)\,ds.$$

Using that $s \mapsto \gamma_z(ts)$ is a path joining z_0 to $\gamma_z(t)$, one proves by induction that

$$\tilde{\varphi}_n(t) = \varphi_n(\gamma_z(t)), \qquad 0 \leq t \leq 1.$$

We can estimate the norm of the $\tilde{\varphi}_n(t)$ for $0 \leq t \leq 1$ as follows:

$$\|\tilde{\varphi}_0(t)\| = 1$$

$$\|\tilde{\varphi}_1(t)\| \leq tM_1$$

$$\|\tilde{\varphi}_2(t)\| \leq t \int_0^t (ts)M_1^2\,ds = \frac{t^2M_1^2}{2}$$

$$\|\tilde{\varphi}_3(t)\| \leq \frac{t^3M_1^3}{2} \int_0^1 s^2\,ds = \frac{(tM_1)^3}{3!}$$

and, by induction,

$$\|\tilde{\varphi}_n(t)\| \leq \frac{(tM_1)^n}{n!}.$$

Therefore,

$$\|\varphi_n(z)\| \leq \frac{M_1^n}{n!},$$

since $\varphi_n(z) = \tilde{\varphi}_n(1)$. This implies that the series $\sum_{n\geq 0} \varphi_n$ converges normally in $\bar{B}(z_1, R)$. Hence, $f(z) = \sum_{n\geq 0} \varphi_n(z)$ is a holomorphic function in Ω with values in $M(n, \mathbb{C})$.

(ii) We also have that the series $A(z)f(z) = \sum_{n\geq 0} A(z)\varphi_n(z)$ converges normally in every compact subset of Ω. Therefore, it can be integrated term by term over any path γ_z:

$$\int_{\gamma_z} A(u)f(u)\,du = \sum_{n\geq 0} \int_{\gamma_z} A(u)\varphi_n(u)\,du = \sum_{n\geq 0} \varphi_{n+1}(z) = f(z) - \varphi_0(z).$$

Hence

$$f(z) = id + \int_{\gamma_z} A(u)f(u)\,du,$$

which ensures that f solves the problem $(*)$.

(iii) The same reasoning shows that there are solutions in Ω to the (adjoint) problem

$$\begin{cases} g'(z) = -g(z)A(z) \\ g(z_0) = id. \end{cases} \qquad (**)$$

(Just note that introducing $h = g^t$ = the transpose of g, and $B = -A^t$, one turns $(**)$ into the problem $(*)$ for h and B instead of f and A.)

Let now f be any solution of $(*)$ and g any solution of $(**)$. Then

$$(gf)' = g'f + gf' = -gAf + gAf = 0.$$

Therefore, the matrix-valued holomorphic function gf must be constant. Evaluating at z_0 we obtain

$$gf = id.$$

This implies that f takes values in $GL(n, \mathbb{C})$. It also implies that the solution is unique. Namely, fix a solution g of $(**)$, then any solution of $(*)$ must be pointwise the inverse of $g(z)$. □

5.15.2. Definition. The ***resolvent*** of the differential equation $(*)$ is the function $R : \Omega \times \Omega \to GL(n, \mathbb{C})$ defined as follows: for $z_0 \in \Omega$ fixed, the function

$z \mapsto R(z, z_0)$ is the unique holomorphic function in Ω with values in $M(n, \mathbb{C})$ such that $R(z_0, z_0) = id$ and $\dfrac{d}{dz} R(z, z_0) = A(z) R(z, z_0)$.

5.15.3. Proposition. *The resolvent satisfies the following properties:*

(1) $R(z_0, z_0) = id, \qquad (z_0 \in \Omega)$;
(2) $R(z_2, z_1) R(z_1, z_0) = R(z_2, z_0) \qquad (z_0, z_1, z_2 \in \Omega)$;
(3) $(R(z_1, z_0))^{-1} = R(z_0, z_1) \qquad (z_0, z_1 \in \Omega)$;
(4) $\dfrac{d}{dz}(R(z, z_0)) = A(z) R(z, z_0) \qquad (z, z_0 \in \Omega)$.

The function $z_0 \mapsto R(z_1, z_0)$ is holomorphic in Ω for every z_1 fixed in Ω. Furthermore R is a continuous function of both variables.

Proof. (1) and (4) are just the definition of R. (3) is an immediate consequence of (2) and (1) since R is invertible. To show (2), one only needs to verify that $f(z) = R(z, z_1)\,(R(z_0, z_1))^{-1}$ is a solution of $(*)$, which is immediate. Hence $f(z) = R(z, z_0)$ and (2) holds.

It follows from (3) that for z_1 fixed the function $z_0 \mapsto R(z_1, z_0)$ is also holomorphic in Ω. Namely, $R(z_1, z_0) = (R(z_0, z_1))^{-1}$, $R(z_0, z_1)$ is holomorphic in z_0 by definition and its inverse is obtained by taking rational operations in the entries of $R(z_0, z_1)$.

Finally, the continuity in both variables is also a consequence of (2) and (3). Namely, fixed $z_0 \in \Omega$. Then

$$R(z_1, z_2) = R(z_1, z_0)(R(z_2, z_0))^{-1},$$

and both functions are continuous in their first variables. In fact, this argument shows that R is a holomorphic function in both variables simultaneously, i.e., its entries can be given locally as power series in two variables. □

We are now ready to solve the original inhomogeneous differential equation.

5.15.4. Theorem. *Let Ω be a simply connected open subset of $\mathbb{C}$, $A : \Omega \to M(n, \mathbb{C})$ a holomorphic matrix-valued function, $B : \Omega \to \mathbb{C}^n$ a holomorphic vector-valued function, $z_0 \in \Omega$, and $x_0 \in \mathbb{C}^n$. Then, there exists a unique vector-valued holomorphic function $f : \Omega \to \mathbb{C}^n$ solving the linear differential equation*

$$\begin{cases} f'(z) = A(z) f(z) + B(z) & (z \in \Omega) \\ f(z_0) = x_0. \end{cases} \tag{†}$$

This function is given by the ***Green-Lagrange formula***

$$f(z) = R(z, z_0) x_0 + \int_{\gamma_z} R(z, u) B(u)\, du,$$

where γ_z is a path joining z_0 to z in Ω.

PROOF. Let us assume (†) has a holomorphic solution f. Define an auxiliary vector-valued function g by

$$g(z) = R(z_0, z)f(z),$$

which is holomorphic in Ω by the previous proposition. We also have $g(z_0) = f(z_0) = x_0$, and, using item 3 in Proposition 5.15.3,

$$f(z) = R(z, z_0)g(z).$$

Hence

$$f'(z) = \left(\frac{d}{dz}R(z, z_0)\right)g(z) + R(z, z_0)g'(z) = A(z)R(z, z_0)g(z) + R(z, z_0)g'(z)$$

$$= A(z)f(z) + R(z, z_0)g'(z).$$

Therefore,

$$g'(z) = (R(z, z_0))^{-1}B(z) = R(z_0, z)B(z).$$

Integrating this equation, we obtain

$$g(z) = x_0 + \int_{\gamma_z} R(z_0, u)B(u)\,du.$$

Conversely, if we define g by the last integral, then the reasoning can be reversed and $f(z) = R(z, z_0)g(z)$ solves (†). Moreover, it is now immediate that f is represented by the Green-Lagrange formula. □

Remarks. (1) If $B = 0$, we say that the equation (†) is homogeneous. The solution f such that $f(z_0) = x_0$ is then simply given by $f(z) = R(x, z_0)x_0$.

Finding all solutions of the homogeneous equation now becomes the question of finding the resolvent $R(z, z_0)$ as a function of z for a fixed-value $z_0 \in \Omega$. For $z_1 \in \Omega$, we have

$$R(z, z_1) = R(z, z_0)(R(z_1, z_0))^{-1}.$$

(2) The function $\int_{\gamma_z} R(z, u)B(u)\,du$ is just the solution of (†) for the case $x_0 = 0$. In this form, the solution f of (†) appears as a sum of a particular solution with initial condition zero and a solution of the homogeneous equation. If we have another particular solution, we reduce the problem (†) to a different homogeneous problem.

(3) A system of solutions $f_1, \ldots, f_n$ of the equation $f' = Af$ whose initial conditions $f_1(z_0) = x_{0,1}, \ldots, f_n(z_0) = x_{0,n}$ form a vector space basis of $\mathbb{C}^n$, is said to be a ***fundamental system of solutions***. The matrix $W(z) = (f_1(z), \ldots, f_n(z))$ whose columns are the f_j is called a ***fundamental matrix***.

Since for any $z \in \Omega$ we have

$$W(z) = (f_1(z), \ldots, f_n(z)) = R(z, z_0)(f_1(z_0), \ldots, f_n(z_0))$$

we conclude that the matrix W is invertible everywhere in Ω. It follows that the vector space of solutions of $f' = Af$ has dimension exactly n. Furthermore, for any fundamental matrix we have

$$R(z, z_0) = W(z)W(z_0)^{-1}.$$

Hence, the Green-Lagrange formula for the solution f of (†) can be written

$$f(z) = W(z)W(z_0)^{-1}x_0 + W(z)\int_{\gamma_z} (W(u))^{-1}B(u)\,du.$$

(4) If we let $\Delta(z, z_0) = \det(R(z, z_0))$ we have the relation

$$\Delta(z, z_0) = \exp\left(\int_{\gamma_z} (T_r(A(u))\,du\right).$$

In fact, let $\varphi(z) = R(z, z_0)$ and $\delta(z) = \Delta(z, z_0)$. For $z \in \Omega$ and $|h|$ small we have

$$\delta(z + h) = \delta(z)\det(\varphi(z + h)\varphi(z)^{-1}),$$

and

$$\varphi(z + h) = \varphi(z) + h\varphi'(z) + o(h) = \varphi(z) + hA(z)\varphi(z) + o(h).$$

Hence

$$\det(\varphi(z + h)\varphi(z)^{-1}) = \det(id + hA(z) + o(h)) = 1 + h\operatorname{Tr}(A(z)) + o(h),$$

the last formula just follows from the properties of the determinant. Therefore,

$$\delta(z + h) = \delta(z) + h\operatorname{Tr}(A(z))\delta(z) + o(h).$$

Letting $h \to 0$ we conclude that δ satisfies the scalar differential equation

$$\begin{cases} \delta'(z) = \operatorname{Tr}(A(z))\delta(z) \\ \delta(z_0) = 1. \end{cases}$$

It follows that

$$\Delta(z, z_0) = \delta(z) = \exp\left(\int_{\gamma_z} \operatorname{Tr}(A(u))\,du\right).$$

In terms of a fundamental matrix W this can be rewritten as

$$\det(W(z)) = \det(W(z_0))\exp\left(\int_{\gamma_z} \operatorname{Tr}(A(u))\,du\right).$$

We can now show some elementary consequences of these remarks for scalar linear differential equations of order n in a simply connected open set $\Omega \subseteq \mathbb{C}$. This means an equation of the form

$$f^{(n)} + a_1 f^{(n-1)} + \cdots + a_{n-1}f' + a_n f = b, \qquad (\rhd)$$

where $a_n, \ldots, a_n$, b are holomorphic functions in Ω, and the holomorphic function f is the unknown we are looking for.

To such a function f one associates the vector-valued function $F := (f, f', \ldots, f^{(n-1)})$, where we consider F as a column vector, $F : \Omega \to \mathbb{C}^n$. Clearly $F' = (f', f'', \ldots, f^{(n)})$, hence the equation ($\rhd$) is equivalent to

$$F' = AF + B, \qquad (\rhd\rhd)$$

with $A \in M(n, \mathbb{C})$,

$$A := \begin{pmatrix} 0 & 1 & 0 & \ldots & 0 \\ 0 & 0 & 1 & \ldots & 0 \\ 0 & 0 & & \ldots & 1 \\ \cdot & \cdot & & & \cdot \\ -a_n & -a_{n-1} & & & -a_1 \end{pmatrix}$$

and $B \in \mathbb{C}^n$,

$$B := \begin{pmatrix} 0 \\ \vdots \\ 0 \\ b \end{pmatrix}.$$

It is easy to see that if $\Phi = (u_1, \ldots, u_n)$ is a solution of ($\rhd\rhd$), then u_1 is a solution of ($\rhd$) and $u_2 = u_1', \ldots, u_n = u^{(n-1)}$.

Hence, every solution f of ($\rhd$) is completely determined by its initial conditions

$$(f(z_0), f'(z_0), \ldots, f^{(n-1)}(z_0)) \in \mathbb{C}^n,$$

which can be chosen arbitrarily. Therefore, as in the real case, we have that the space of solutions of the corresponding homogeneous equation (i.e., ($\rhd$) with $b = 0$) has dimension n. As earlier, the problem of solving ($\rhd$) reduces to the study of the resolvent $R(z, z_0)$, or what is the same, to a fundamental matrix W for the solutions of the homogeneous equation. In this case we have that

$$W = \begin{pmatrix} f_1 & \ldots & f_n \\ f_1' & & f_n' \\ \vdots & & \vdots \\ f_1^{(n-1)} & \ldots & f_n^{(n-1)} \end{pmatrix},$$

for a system of solutions $f_1, \ldots, f_n$ of ($\rhd$) whose initial conditions at a point z_0 form a basis of $\mathbb{C}^n$. Its determinant, $w = \det W$, is usually called the ***Wronskian*** of the functions $f_1, \ldots, f_n$. From part (4) of the previous remark, we have the relation

$$w(z) = w(z_0) \exp\left(-\int_{\gamma_z} a_1(u)\, du \right),$$

since $\mathrm{Tr}(A) = -a_1$ in this case.

The formula of Green-Lagrange specializes to the following: a solution φ

of ($\rhd$) with given initial conditions can be written as

$$\varphi(z) = \varphi_0(z) + \frac{1}{w(z_0)} \sum_{1 \le j \le n} f_j(z) \int_{\gamma_z} b(s) w_j(s) \exp\left(\int_{\gamma_s} a_1(u)\, du\right) ds,$$

where:

(i) φ_0 is the solution of the homogeneous equation satisfying the given initial conditions;
(ii) $f_1, \ldots, f_n$ is a fundamental system of solutions and w is its Wronskian;
(iii) $w_j(s)$ is the determinant of the matrix obtained replacing the jth column of W by the vector $(0, \ldots, 0, 1)$.

This formula can be derived very simply by specializing the discussion in (1), (2), and (3) of the previous remark. In fact, from (2) we know that the solution F of ($\rhd\rhd$) with zero initial conditions is given by

$$\int_{\gamma_z} R(z, s) B(s)\, ds = W(z) \int_{\gamma_z} (W(s))^{-1} B(s)\, ds.$$

To compute $(W(s))^{-1} B(s)$ we observe that $B(s) = b(s)(0, \ldots, 0, 1) = b(s)E$. We can apply Cramer's rule to compute $(W(s))^{-1}E$ which is precisely the solution C of the equation $WC = E$. This observation yields $C_j(s) = w_j(s)/w(s) = w_j(s) \exp\left(\int_{\gamma_s} a_1(u)\, du\right) \Big/ w(z_0)$. Finally, we recall that we only need to compute the first component of F to obtain the term in φ involving the f_j. In practice one just writes $\varphi = y + \varphi_1$ with

$$y(z) = \sum_{1 \le k \le n} C_k(z) f_k(z),$$

and φ_1 a convenient solution of the homogeneous equation. The unknown functions C_k are such that

$$W(s) \begin{pmatrix} C_1'(s) \\ \vdots \\ C_n'(s) \end{pmatrix} = \begin{pmatrix} 0 \\ \vdots \\ 0 \\ b(s) \end{pmatrix}.$$

One solves this linear system of equations in the unknowns $C_1', \ldots, C_n'$ and then integrates. This is usually called ***Euler's method of variation of parameters.***

We will say now a few words about linear differential equations of first order on a Riemann surface.

5.15.5. Definition. On a Riemann surface X we call a linear differential homogeneous equation of first order a relation of the form

$$df = A \cdot f,$$

where $A \in M(n, \Omega(X))$ is a matrix of holomorphic 1-forms, $A = (a_{ij})_{ij}$, $a_{ij} \in \Omega(X)$ and $f : X \to \mathbb{C}^n$ is the holomorphic solution of this equation.

Remark. In a local chart (U, φ) this equation becomes

$$\frac{df}{d\varphi} = Bf,$$

where $A = B\,d\varphi$, B is a holomorphic matrix.

5.15.6. Proposition. *Let X be a simply connected Riemann surface, $A \in M(n, \Omega(X))$, $z_0 \in X$. For every $x_0 \in \mathbb{C}^n$ there is a unique holomorphic map $f: X \to \mathbb{C}^n$ such that*

$$\begin{cases} df = Af \\ f(z_0) = x_0. \end{cases}$$

The family S_A of all holomorphic solutions of the equation $df = Af$ is a complex vector space of dimension n.

PROOF. By the previous remark and Theorem 5.15.4 there is a simply connected U_0 open neighborhood of z_0 and a unique holomorphic solution $f: U_0 \to \mathbb{C}^n$ of $df = Af$ with $f(z_0) = x_0$. This already shows that if there is a solution in X to our equation, it must be unique. We need only to show that the solution f defined in U_0 can be analytically continued along any path γ in X starting at z_0. Once this is done, the corresponding global function, still denoted f, will satisfy $df = Af$ everywhere by the principle of preservation identities under analytic continuation.

We can find simply connected coordinate patches U_j and a partition $0 = t_0 < t_1 < \cdots < t_n = 1$ such that $\gamma([t_{j-1}, t_j]) \subseteq U_{j-1}$ for $j = 1, \ldots, n$. In such U_j, for every $x_j \in \mathbb{C}^n$ there is a unique solution f_j of $df = Af$ in U_j such that $f_j(\gamma(t_j)) = x_j$. Therefore, starting with $f_0 = f$ we can construct the f_j by recurrence in such a way that $f_j(\gamma(t_j)) = f_{j-1}(\gamma(t_j))$. Therefore f_j and f_{j-1} coincide in the connected component of $U_j \cap U_{j-1}$ containing $\gamma(t_j)$. This shows the existence of the analytic continuation of f along γ.

The map $f \in S_A \mapsto f(z_0)$ is linear and bijective by the preceding part, hence $\dim S_A = n$. □

5.15.7. Corollary. *Let $p: \tilde{X} \to X$ be the universal covering space of a Riemann surface X. Let $A \in M(n, \Omega(X))$, $z_0 \in X$, $y_0 \in p^{-1}(z_0)$, $x_0 \in \mathbb{C}^n$. There exists a unique holomorphic solution $f: \tilde{X} \to \mathbb{C}^n$ of the equation $df = (p^*A)f$ such that $f(y_0) = x_0$.*

Let X be a Riemann surface. $A \in M(n, \Omega(X))$ and $p: \tilde{X} \to X$ the universal covering space of X. Let S_{p^*A} be the space of solutions of $df = (p^*A)f$ in $\tilde{X}$. Let $f_1, \ldots, f_n$ be a basis of S_{p^*A}. The fundamental matrix $\Phi = (f_1, \ldots, f_n)$ satisfies the differential equation $d\Phi = (p^*A)\Phi$. For $\sigma \in G(p) \cong \pi_1(X)$ we set $\sigma \cdot \Phi := \Phi \circ \sigma^{-1}$. Then $d(\sigma \cdot \Phi) = (p^*A)(\sigma \cdot \Phi)$ and it is also a fundamental matrix. Hence, there is a matrix $T(\sigma) \in GL(n, \mathbb{C})$ such that $\sigma \cdot \Phi = \Phi T(\sigma)$

(product of matrices on the right-hand side of this equation). Now, let σ, $\tau \in G(p)$, then

$$\Phi T(\sigma \circ \tau) = (\sigma \circ \tau)\cdot\Phi = \sigma\cdot(\tau\cdot\Phi) = \sigma\cdot(\Phi T(\tau)) = (\sigma\cdot\Phi)T(\tau) = \Phi T(\sigma)T(\tau),$$

hence $\sigma \mapsto T(\sigma)$ is a group homomorphism from $G(p)$ into $GL(n, \mathbb{C})$.

Conversely, if $T: G(p) \to GL(n, \mathbb{C})$ is a group homomorphism and $\Phi: \tilde{X} \to GL(n, \mathbb{C})$ is a holomorphic map such that

$$\sigma\cdot\Phi = \Phi T(\sigma) \tag{††}$$

for every $\sigma \in G(p)$, then the matrix $(d\Phi)\Phi^{-1}$ is invariant under $G(p)$, since

$$\sigma\cdot(d\Phi)\Phi^{-1} = (d\Phi T(\sigma))(\Phi T(\sigma))^{-1} = (d\Phi)\Phi^{-1},$$

hence it is of the form p^*A, with $A \in M(n, \Omega(X))$. Therefore, Φ is a fundamental solution for the differential equation $df = (p^*A)f$.

A matrix satisfying the equation (††) for the representation T of $G(p)$ into $GL(n, \mathbb{C})$ is called an ***automorphic matrix.***

Let us consider now the special case

$$X = \{z \in \mathbb{C} : 0 < |z| < R\},$$

where $0 < R \leq \infty$. We take $\tilde{X} = \{w \in \mathbb{C} : \operatorname{Re} w < \log R\}$ and $p(w) = \exp(w)$ as a concrete realization of the universal covering space of X. The group $G(p) \cong \pi_1(X)$ is isomorphic to $\mathbb{Z}$. On $\tilde{X}$ the function $L(w) = w$ is such that $\exp \circ L = p$. We choose the isomorphism between $G(p)$ and $\mathbb{Z}$ in such a way that $\sigma_0 = 1 \in \mathbb{Z}$ acts by

$$\sigma_0\cdot L = L \circ \sigma_0^{-1} = L + 2\pi i,$$

so that σ_0^{-1} arises out of a loop going once counterclockwise in X around the origin.

Let now $A \in M(n, \Omega(X))$ and Φ a fundamental matrix of solutions of $df = (p^*A)f$ in $\tilde{X}$. The behavior of Φ as an automorphic matrix is determined by the matrix $T \in GL(n, \mathbb{C})$ such that

$$\sigma_0\cdot\Phi = \Phi T$$

In this case, T is called the ***automorphic factor***. If Ψ is another fundamental matrix there is $S \in GL(n, \mathbb{C})$ such that $\Psi = \Phi S$, Hence

$$\sigma_0\cdot\Psi = \sigma_0\cdot(\Phi S) = (\sigma_0\cdot\Phi)S = \Phi TS = \Psi(S^{-1}TS).$$

Therefore, by a convenient change of fundamental matrix we can assume that the automorphic factor T has the Jordan canonical form.

5.15.8. Proposition. *Let $T \in GL(n, \mathbb{C})$ be given and $B \in M(n, \mathbb{C})$* such that

$$\exp(2\pi i B) = T.$$

Then the matrix-valued function $\Phi_0 = \exp(BL)$ is a fundamental matrix in $\tilde{X} = \{w \in \mathbb{C} : \operatorname{Re} w < \log R\}$ for the equation

$$df = p^*\left(\frac{B}{z}dz\right)f,$$

associated to the differential equation

$$f' = \frac{B}{z}f$$

on $X = \{z \in \mathbb{C} : 0 < |z| < R \leq \infty\}$.

The matrix Φ_0 *has* T *as automorphic factor*

$$\sigma_0 \cdot \Phi_0 = \Phi_0 T,$$

with $\sigma_0 = 1 \in \mathbb{Z} \cong G(p)$, $\sigma_0 . L = L + 2\pi i$, $L(w) = w$.

PROOF. It is very simple. In the coordinates w, the differential equation in $\tilde{X}$ is just $f'(w) = Bf(w)$. We have $\Phi_0(w) = \exp(Bw)$, hence $\Phi_0' = B\Phi_0$, which shows that Φ_0 is a fundamental matrix. Furthermore,

$$(\sigma \cdot \Phi_0)(w) = \exp((w + 2\pi i)B) = \exp(wB)\exp(2\pi iB) = \Phi_0(w)T. \qquad \square$$

Remark. If $T \neq id_{\mathbb{C}^n}$, there is no holomorphic function $f : X \to \mathbb{C}$, $f \not\equiv 0$ such that $f' = \frac{B}{z}f$. If not, we would have

$$p^* df = d(f \circ p) = p^*\left(\frac{B}{z}dz\right)(p^*f) = (Bdw)(p^*f),$$

hence the function $g = p^*f$ is $2\pi i$-periodic, invariant under σ_0, and would solve $g' = Bg$. This is impossible since it implies $\exp(2\pi iB) = id_{\mathbb{C}^n}$.

5.15.9. Proposition. *Let* $X = \{z \in \mathbb{C} : 0 < |z| < R\}$ $(0 < R \leq \infty)$, $p : \tilde{X} \to X$ *its universal covering space,* $\tilde{X} = \{w \in \mathbb{C} : \operatorname{Re} w < \log R\}$, $p(w) = \exp(w)$, *and* $A : X \to M(n, \mathbb{C})$ *a holomorphic map. The differential equation in* X

$$f' = Af$$

has a fundamental matrix Φ *in* $\tilde{X}$ *of the form* $\Phi = \Psi\Phi_0$, *with* $\Phi_0(w) = \exp(wB)$ *for some* $B \in M(n, \mathbb{C})$ *and* $\Psi : X \to GL(n, \mathbb{C})$ *holomorphic.*

PROOF. Let Φ be a fundamental matrix for the solutions in $\tilde{X}$ of $df = (p^*(A\,dz))f$. (That is, $df(w) = A(e^w)e^w f(w)\,dw$.) We have $\sigma_0 \cdot \Phi = \Phi T$ for some $T \in GL(n, \mathbb{C})$ and $\sigma_0 = 1 \in \mathbb{Z}$ the generator of $G(p)$. We can find $B \in M(n, \mathbb{C})$ such that $T = \exp(2\pi iB)$ and hence, the function $\Phi_0(w) = \exp(wB)$ satisfies $\sigma_0 \cdot \Phi_0 = \Phi_0 T$. Let $\Psi = \Phi\Phi_0^{-1}$, then $\sigma \cdot \Psi = \Psi$. This proves the proposition.

$\square$

Remark. In other words, any fundamental matrix Φ of the preceding equation can be written as a product of a very simple matrix, $\Phi_0(w) = \exp(wB)$, and a matrix Ψ that can be developed in a Laurent series in X about $z = 0$.

5.15.10. Definition. The origin is said to be a ***regular singular point*** or a ***Fuchsian type singularity*** for the differential equation $f' = Af$ in X, $A \in M(n, \mathscr{H}(X))$, if the matrix Ψ from Proposition 5.15.9 has at most a pole at $z = 0$.

5.15.11. Proposition. *Let* $X = \{z \in \mathbb{C} : 0 < |z| < R\}$ $(0 < R \leq \infty)$. *If the matrix* $A \in M(n, \mathscr{H}(X))$ *has at most a pole of order* 1 *at* $z = 0$, *then the origin is a regular singular point for the differential equation*

$$f' = Af.$$

PROOF. The proof needs two lemmas.

5.15.12. Lemma. *Let* $k \geq 0$, $F :]0, r_0] \to]0, \infty[$ *be a function of class* C^1 *such that*

$$|F'(r)| \leq \frac{kF(r)}{r}, \qquad 0 < r \leq r_0.$$

Then

$$F(r) \leq F(r_0)/(r/r_0)^k, \qquad 0 < r \leq r_0.$$

PROOF. By hypothesis

$$\frac{d}{dr}(\log F(r)) = \frac{F'(r)}{F(r)} \geq -\frac{k}{r}.$$

Integrating in $[r, r_0]$ we obtain

$$\log \frac{F(r_0)}{F(r)} \geq -k \log\left(\frac{r_0}{r}\right),$$

which is equivalent to the conclusion of the lemma. □

5.15.13. Lemma. *Let* $f \in \mathscr{H}(X)$. *Then*

$$\left|\frac{\partial}{\partial r}|f(re^{i\theta})|^2\right| \leq 2|f(re^{i\theta})||f'(re^{i\theta})|.$$

PROOF. We have $f' = \dfrac{df}{dz} = \dfrac{\partial f}{\partial r}e^{-i\theta}$ and $\left|\dfrac{\partial f}{\partial r}\right| = |f'|$. Furthermore, $\dfrac{\partial \bar{f}}{\partial r} = \overline{\dfrac{\partial f}{\partial r}}$, hence $\left|\dfrac{\partial \bar{f}}{\partial r}\right| = |f'|$ also. This gives

$$\left|\frac{\partial}{\partial r}|f|^2\right| = \left|\bar{f}\frac{\partial f}{\partial r} + f\frac{\partial \bar{f}}{\partial r}\right| \leq 2|f||f'|.$$ □

We can now return to the proof of Proposition 5.15.11.

We already know there is a fundamental matrix Φ in $\tilde{X}$ for the equation $f' = Af$ in X, and that this fundamental matrix has the form $\Phi = \Psi\Phi_0$, $\Phi_0(w) = \exp(wB)$ for some $B \in M(n, \mathbb{C})$ and $\Psi : X \to GL(n, \mathbb{C})$ holomorphic. We have that the matrix Φ satisfies the equation

$$\frac{d}{dw}\Phi(w) = zA(z)\Phi(w) \qquad (z = e^w).$$

That is,

$$zA(z)\Psi(z)\Phi_0(w) = \frac{d}{dw}(\Psi(z)\Phi_0(w)) = z\frac{d\Psi(z)}{dz}\Phi_0(w) + \Psi(z)B\Phi_0(w).$$

Dividing by Φ_0 we obtain a differential equation in X for the matrix-valued holomorphic function Ψ

$$\Psi' = A\Psi - \frac{\Psi}{z}B.$$

Since the matrix A has at most a pole of order 1 at $z = 0$, there is a matrix-valued holomorphic function $A_1 : B(0, R) \to M(n, \mathbb{C})$ such that $A = \frac{1}{z}A_1$. That is, we have

$$\Psi'(z) = \frac{1}{z}(A_1(z)\Psi(z) - \Psi(z)B).$$

Let, for the moment, the norm of a matrix (C_{ij}) be $\|C\| = \left(\sum_{i,j}|C_{ij}|^2\right)^{1/2}$. With this definition of norm, there is an $r_0 > 0$ and $M \geq 0$ such that

$$\|\Psi'(z)\| \leq \frac{M}{|z|}\|\Psi(z)\| \qquad \text{for } 0 < |z| \leq r_0.$$

If we now let $\theta \in \mathbb{R}$ be fixed, $r \in]0, r_0]$, $\Psi = (\Psi_{ij})$ and

$$F(r) = \|\Psi(re^{i\theta})\|^2 = \sum_{i,j}|\Psi_{kj}(re^{i\theta})|^2,$$

we conclude from Lemma 5.15.13 that

$$\begin{aligned}|F'(r)| &\leq 2\sum_{k,j}|\Psi_{kj}(re^{i\theta})||\Psi'_{kj}(re^{i\theta})| \\ &\leq 2\|\Psi(re^{i\theta})\|\,\|\Psi'(re^{i\theta})\| \\ &\leq \frac{2}{r}M\|\Psi(re^{i\theta})\|^2 = \frac{2M}{r}F(r).\end{aligned}$$

Applying Lemma 5.15.12 we obtain

$$F(r) \leq F(r_0)\left(\frac{r}{r_0}\right)^{-2M}.$$

That is,

$$\|\Psi(re^{i\theta})\| \leq \|\Psi(r_0 e^{i\theta})\| \left(\frac{r}{r_0}\right)^{-M},$$

which implies that Ψ has a pole of order at most M at the origin. In other words, the origin is a regular singular point for $f' = Af$ in X. $\square$

Propositions 5.15.9 and 5.15.11 have the following consequence.

5.15.14. Proposition. *Let $A_1 \in M(n, \mathbb{C})$, $A_2 : B(0, R) \to M(n, \mathbb{C})$ holomorphic, $X = B(0, R)\setminus\{0\}$. Then the differential equation*

$$f'(z) = \left(\frac{A_1}{z} + A_2(z)\right) f(z) \qquad (z \in X)$$

has n vector-valued linearly independent solutions of the form $f_\lambda = (f_{\lambda,1}, \ldots, f_{\lambda,n})$

$$f(z) = \sum_{k \geq 0} P_k(\log z) z^{\lambda+k},$$

where λ is an eigenvalue of the matrix A_1 and P_k is a vector-valued polynomial of degree $\leq n-1$, which is not identically zero. The series is absolutely and uniformly convergent in every closed disk in X, for every determination of the logarithm (and corresponding determination of $z^\lambda = e^{\lambda \log z}$).

PROOF. The singular point $z = 0$ is a regular point by §5.15.11. Hence the equation in $\tilde{X}$

$$dg = p^*\left(\left(\frac{A_1}{z} + A_2(z)\right) dz\right) g = (A_1 + A_2(e^w)e^w) g\, dw$$

has a fundamental matrix Φ of the form

$$\Phi(w) = \Psi(e^w) e^{wB}, \qquad B \in M(n, \mathbb{C})$$

where Ψ is holomorphic in X. From §5.15.3 we know that Ψ has a Laurent expansion of the form

$$\Psi(z) = \sum_{k \geq m} z^k \Psi_k = z^{-m} \tilde{\Psi}(z)$$

where $\tilde{\Psi}(z)$ is holomorphic at $z = 0$ and $m \in \mathbb{N}$. In other words we have

$$\Phi(w) = \tilde{\Psi}(e^w) e^{w(B - mI)}.$$

We can change the fundamental matrix in such a way that B is in Jordan canonical form. Then, for each Jordan block of size $d \times d$ of the Jordan canonical form, there are d columns of the fundamental matrix which will be of the form $\varphi_l = (\varphi_{l,1}, \ldots, \varphi_{l,n})$, $0 \leq l \leq d-1$,

$$\varphi_{l,j}(w) = e^{(v-m)w} \sum_{k=0}^{\infty} P_{l,j,k}(w)e^{kw},$$

where v is an eigenvalue of B. In the z-plane (using the same symbol $\varphi_{l,j}$ by abuse of language)

$$\varphi_{l,j}(z) = z^{v-m} \sum_{k=0}^{\infty} P_{l,j,k}(\log z)z^k.$$

The coefficients $P_{l,j,k}$ are polynomials of degree $\leq l \leq d-1 \leq n-1$. Let us consider first the case $l = 0$. We have that the vector $\varphi_0(z)$ has the form $\varphi_0(z) = z^{v-m}h(z)$, h is a holomorphic vector at $z = 0$, $h \not\equiv 0$. Assume $h(0) \neq 0$ and set $\lambda = v - m$. Replacing into the differential equation we obtain

$$z^{\lambda-1}(\lambda I - A_1)h + z^{\lambda}(h' - A_2 h) = 0.$$

Dividing by z^λ and letting $z \to 0$ we obtain

$$(\lambda I - A_1)h(0) = 0,$$

so that λ is an eigenvalue of A_1. If $h(0) = 0$, $h'(0) \neq 0$, then we can write $\varphi_0(z) = z^{\lambda+1}\tilde{h}(z)$ and the same reasoning indicates that $\lambda + 1$ is definitively an eigenvalue of A_1. In any case, φ_0 can be written in such a way that the leading power is z^{v+s}, $s \in \mathbb{Z}$, v eigenvalue of B, $v + s$ eigenvalue of A_1. Note also that, if above $\lambda + 1$ was not an eigenvalue of A_1, then $h(0) \neq 0$. The general case φ_l, $l \neq 0$, follows the same lines. □

5.15.15. Remarks. (1) A slightly more careful analysis of the proof of the last proposition leads to the following: for every φ_l, $0 \leq l \leq d-1$, corresponding to the same Jordan block of B one has

$$\varphi_l(z) = z^\lambda \sum_{k=0}^{\infty} P_{l,k}(\log z)z^k,$$

with the same eigenvalue λ of A_1 and degree $P_{l,0} = l$.

In fact, let us say that we are considering the Jordan block J_1 of B corresponding to the eigenvalue v_1, i.e.

$$B = \begin{pmatrix} J_1 & 0 & \\ & J_2 & \\ & 0 & \ddots \end{pmatrix},$$

$$J_1 = \begin{pmatrix} v_1 & 1 & 0 & \cdots & 0 \\ 0 & \ddots & \ddots & & \vdots \\ \vdots & & & & 0 \\ & & & & 1 \\ 0 & & & & v_1 \end{pmatrix} = \begin{pmatrix} v_1 & 0 & \cdots & 0 \\ \vdots & \ddots & & \vdots \\ 0 & & & v_1 \end{pmatrix} + \begin{pmatrix} 0 & 1 & 0 & \cdots & 0 \\ & & \ddots & \ddots & \vdots \\ \vdots & & & & 1 \\ 0 & & \cdots & & 0 \end{pmatrix},$$

J_1 is a $d \times d$ matrix. Then

$$e^{wJ_1} = e^{\nu_1 w}\begin{pmatrix} 1 & w & \frac{w^2}{2} & \cdots & \frac{w^{d-1}}{(d-1)!} \\ 0 & 1 & w & & \\ & 0 & 1 & & \\ \vdots & & 0 & \ddots & \vdots \\ & \vdots & \vdots & & w \\ 0 & & & & 1 \end{pmatrix}.$$

Therefore, in the computation of φ_0 we see that it is the order of the origin as a zero of the entry $\tilde{\psi}_{1,1}(z)$ of the matrix $\tilde{\psi}(z)$ that counts. If this were k, then $\tilde{\psi}_{1,1}(z) = a_0 z^k(1 + \cdots)$, $a_0 \neq 0$, and

$$\varphi_0 = z^{\nu_1 - m + k}\begin{pmatrix} a_0(1 + \cdots) \\ 0 \\ \vdots \end{pmatrix}.$$

The first entry of φ_1, $\varphi_{1,1}$ will now have the form

$$\begin{aligned}\varphi_{1,1}(z) &= z^{\nu_1 - m + k}[(a_0 \log z(1 + \cdots)) + \text{holomorphic function of } z] \\ &= z^{\nu_1 - m + k}[(a_0 \log z + b_0) + (a_1 \log z + b_1)z + \cdots].\end{aligned}$$

The entry $\varphi_{1,2}$ is given as

$$\varphi_{1,2}(z) = z^{\nu_1 - m + k}(\text{holomorphic function of } z)$$

and the following ones are zero.

The same reasoning holds for $\varphi_2, \ldots, \varphi_{d-1}$ and for the other Jordan blocks.

It also follows that once we know the form of the solution we can determine the shape of the Jordan blocks of B. The eigenvalues ν of B are then given as $\lambda + m - k$, with λ eigenvalue of A_1 and some $k \in \mathbb{N}$. Since B is only determined up to a matrix C such that $\exp(2\pi i C) = id$, we see that we can take B with eigenvalues $\lambda + m$, λ eigenvalue of A_1, and the Jordan structure of B obtained from the solutions. In the example of the second-order equations which is discussed at the end of this section, there is a case where this is not the most convenient choice of B. What is much harder is to determine m and the Jordan structure of B *a priori* (see [Ju] and [CoL]).

(2) One can also prove that if a formal series

$$f(z) = z^{\lambda} \sum_{k \geq 0} P_k(\log z) z^k$$

is a solution of the equation from Proposition 5.15.14, then it is convergent in every closed disk in $0 < |z| < r$, for r sufficiently small. (See, e.g., [Hi2], [Hen], and [In].)

5.15.16. Proposition. *Let* $A = \dfrac{A_1}{z} + A_2$, *where* $A_1 \in M(n, \mathbb{C})$ *and* $A_2 : B(0, \rho) \to M(n, \mathbb{C})$ *is a holomorphic map. Assume that the endomorphism* *of $M(n, \mathbb{C})$ given by $Z \mapsto A_1 Z - ZA_1$ does not have any eigenvalue which is a positive integer. Then, there is a unique holomorphic map $H : B(0, \rho) \to GL(n, \mathbb{C})$ such that*

$$\begin{cases} H'(z) = \dfrac{1}{z}(A_1 H(z) - H(z)A_1) + A_2(z)H(z), \quad z \in B(0, \rho)\setminus\{0\}, \\ H(0) = id_{\mathbb{C}^n} = I. \end{cases}$$

Moreover, if a formal proven series $\sum_{k\geq 0} H_k z^k$ satisfies the differential equation and the initial condition, then it is automatically convergent and it is the Taylor series of H.

Proof. Let us assume that a formal power series $\sum_{k\geq 0} H_k z^k$ with $H_k \in M(n, \mathbb{C})$, verifies the preceding equation and the initial condition. Then $H_0 = I$ and

$$H' = \sum_{k\geq 1} kH_k z^{k-1},$$

while

$$\frac{1}{z}(A_1 H - HA_1) = \sum_{k\geq 1} (A_1 H_k - H_k A_1)z^{k-1}.$$

Let us write $A_2(z) = \sum_{k\geq 0} B_k z^k$, with $B_k \in M(n, \mathbb{C})$, convergent in $B(0, \rho)$. Then we must have for $k \geq 1$

$$A_1 H_k - H_k A_1 - kH_k = -(B_{k-1} + B_{k-2}H_1 + \cdots + B_0 H_{k-1}).$$

The hypothesis made earlier on A_1 implies that this recurrence relation determines uniquely the matrices $H_1, H_2, \ldots$.

We want to show that the formal power series that are obtained from the recurrence relations converges in $B(0, \rho)$ to a holomorphic map with values in $GL(n, \mathbb{C})$. From the recurrence we obtain the inequality (with the operator norm, so that $\|H_0\| = 1$)

$$(k - 2\|A_1\|)\|H_k\| \leq \|B_{k-1}\| + \|B_{k-2}\|\,\|H_1\| + \cdots + \|B_0\|\,\|H_{k-1}\|.$$

From the Cauchy inequalities for A_2, we know that there is a positive function $M(r)$, $0 < r < \rho$, such that

$$\|B_k\| \leq \frac{1}{3}\frac{M(r)}{r^k}.$$

Therefore, for any $k \geq 3\|A_1\|$ we have

$$\|H_k\| \leq \frac{M(r)}{k}\left(\frac{1}{r^{k-1}} + \frac{\|H_1\|}{r^{k-2}} + \cdots + \frac{\|H_{k-2}\|}{r} + \|H_{k-1}\|\right).$$

Let us fix now r, let $0 \leq t < r$. From this inequality we can easily derive

$$\|H_1\| + 2t\|H_2\| + \cdots + kt^{k-1}\|H_k\|$$
$$\leq \frac{M(r)}{(1 - t/r)}(\|H_0\| + t\|H_1\| + \cdots + t^k\|H_k\|).$$

Introducing the auxiliary function

$$I_k(t) = \|H_0\| + t\|H_1\| + \cdots + t^k\|H_k\|,$$

then we have

$$I_k'(t) \leq \frac{M(r)}{(1 - t/r)} I_k(t), \qquad 0 \leq t < r.$$

Integrating in t, this leads to

$$I_k(t) \leq M(r)\log(1 - t/r).$$

This inequality clearly implies that $\sum_{k\geq 0} H_k z^k$ converges uniformly for $|z| \leq t < r$. Hence it is a holomorphic function in $B(0, \rho)$. This proves the existence of a holomorphic solution to the differential equation with the initial value $H(0) = I$. It also proves its uniqueness. What we still need to show is that H takes values in $GL(n, \mathbb{C})$.

Consider the system

$$\begin{cases} K' = \dfrac{1}{z}(A_1 K - KA_1) - KA_2 \\ K(0) = I. \end{cases}$$

The same argument shows this system has a holomorphic solution $K: B(0, \rho) \to M(n, \mathbb{C})$. Also note that the uniqueness part implies that if $A_2 = 0$, then $X \equiv I$ is the unique solution to

$$\begin{cases} X' = \dfrac{1}{z}(A_1 X - XA_1) \\ X(0) = I. \end{cases}$$

Consider now the matrix-valued holomorphic function KH. It satisfies $KH(0) = I$, and

$$\begin{aligned} (KH)' &= K'H + KH' \\ &= \frac{1}{z}(A_1 K - KA_1)H - KA_2 H + K\frac{1}{z}(A_1 H - HA_1) + KA_2 H \\ &= \frac{1}{z}(A_1(KH) - (KH)A_1). \end{aligned}$$

From the previous observation it follows that $KH \equiv I$. Hence $H(z) \in GL(n, \mathbb{C})$ for every $z \in B(0, \rho)$. □

5.15.17. Remarks. (1) Consider now the equation

$$R' = AR = \left(\frac{A_1}{z} + A_2\right)R,$$

with $A_1 \in M(n, \mathbb{C})$, $A_2 : B(0, \rho) \to M(n, \mathbb{C})$ holomorphic. We assume A_1 is such that $Z \mapsto A_1 Z - Z A_1$ has no eigenvalues in $\mathbb{N}^*$.

Let us write $R = H\Gamma$, where H is the unique holomorphic function in $B(0, \rho)$ with values in $GL(n, \mathbb{C})$ that solves the equation from Proposition 5.15.16. Therefore we have

$$R' = H\Gamma' + \frac{1}{z}(A_1 H - H A_1)\Gamma + A_2 H\Gamma = AR + H\Gamma' - \frac{1}{z} H A_1 \Gamma,$$

which imposes on Γ the differential equation

$$H\Gamma' = \frac{1}{z} H A_1 \Gamma,$$

and, since H is invertible, this is equivalent to

$$\Gamma' = \frac{A_1}{z}\Gamma.$$

From §5.15.9, we know that Γ is of the following form. Fix some $z_0 \in X = B(0, \rho)\backslash\{0\}$, then

$$\Gamma(z) = \exp(\log(z/z_0)A_1)\Gamma(z_0)$$

and has $T = \exp(2\pi i A_1)$ as automorphic factor.

If we cut X along a ray which does not contain z_0, we obtain a simply connected domain where the resolvent $R(z, z_0)$ must be

$$R(z, z_0) = H(z)\Gamma(z),$$

with

$$R(z_0, z_0) = H(z_0)\Gamma(z_0) = I,$$

that is,

$$\Gamma(z_0) = H(z_0)^{-1}.$$

Hence

$$R(z, z_0) = H(z)\exp(\log(z/z_0)A_1)H(z_0)^{-1}.$$

Let $\tilde{X} = \{w \in \mathbb{C} : \operatorname{Re} w < \log \rho\}$, $p : \tilde{X} \to X$, $p(w) = e^w$. The differential equation in $\tilde{X}$

$$df = p^*\left(\left(\frac{A_1}{z} + A_2\right)dz\right)f$$

admits as a fundamental matrix

$$\Phi = p^*(R(z, z_0)H(z_0)e^{(\log z_0)A_1}),$$

which is of the form

$$\Phi(w) = H(e^w)e^{A_1 w} = \Psi(w)e^{A_1 w}.$$

Moreover, for $\sigma_0 = 1 \in \mathbb{Z} \cong G(p)$ we have

$$\begin{aligned}(\sigma_0 \cdot \Phi)(w) &= (\Phi \circ \sigma_0^{-1})(w) = \Psi(w)e^{(2\pi i + w)A_1} \\ &= \Phi(w)e^{2\pi i A_1} = \Phi(w)T,\end{aligned}$$

where T is the automorphic factor.

A fundamental matrix in X is

$$W(z) = H(z)e^{A_1 \log z}.$$

(2) We refer the reader to the classical literature [Hi2], [CoL], and [In] for the study of systems with regular singular points of the form $f' = Af$, $A = \dfrac{A_1}{z} + A_2$, where A_1 is such that $Z \mapsto A_1 Z - Z A_1$ has eigenvalues in $\mathbb{N}^*$.

We take the opportunity to mention here how the eigenvalues of A_1 as a linear transformation in $\mathbb{C}^n$ and those of the map $ad(A_1): Z \mapsto A_1 Z - Z A_1$ as a linear transformation in $M(n, \mathbb{C}) \cong \mathbb{C}^{n^2}$ are related. Let $\lambda_1, \ldots, \lambda_n$ be the eigenvalues of A_1 (counted with multiplicities), then for the map $ad(A_1)$ the eigenvalues are all the possible differences $\lambda_j - \lambda_k$, $1 \le k, j \le n$. The proof is very simple and relegated to the exercises.

We end this section with a glance into the theory of scalar equations of order n with a singularity of Fuchsian type at the origin.

Let $a_1, a_2, \ldots, a_n$ be holomorphic functions in the disk $B(0, \rho)$. A differential equation of order n is said to be of ***Fuchsian type*** if it has the form

$$f^{(n)}(z) + \frac{a_1(z)}{z} f^{(n-1)}(z) + \frac{a_2(z)}{z^2} f^{(n-2)}(z) + \cdots + \frac{a_{n-1}(z)}{z^{n-1}} f'(z) + \frac{a_n(z)}{z^n} f = 0.$$

Note that it could happen that $z = 0$ is a regular point for this equation if the a_j vanish to sufficiently high order. Its associated first-order system can be written as

$$F' = \frac{1}{z^n} AF,$$

with $F = (f, f', \ldots, f^{(n-1)})$ and

$$A = \begin{pmatrix} 0 & z^n & 0 & & 0 \\ 0 & 0 & z^n & \cdots & 0 \\ \vdots & \vdots & \vdots & & \vdots \\ 0 & 0 & 0 & & z^n \\ -a_n & -za_{n-1} & -z^2 a_{n-2} & \cdots & -z^{n-1} a_1 \end{pmatrix}.$$

The equation for the resolvent is

$$R' = \frac{1}{z^n} AR.$$

A priori this system does not seem to fall into the case previously considered, but one can reduce it to that case introducing matrices Δ and Γ such that $R = \Delta\Gamma$ and

$$\Delta(z) = \begin{pmatrix} z^{n-1} & 0 & & 0 \\ 0 & z^{n-2} & \dots & 0 \\ 1 & 0 & & \\ & & z & 0 \\ 0 & 0 & 0 & 1 \end{pmatrix}.$$

One has $\det \Delta(z) \neq 0$ if $z \neq 0$. We obtain

$$R' = \Delta'\Gamma + \Delta\Gamma' = \frac{1}{z^n} A\Delta\Gamma,$$

and hence Γ satisfies the differential equation

$$\Gamma' = \frac{1}{z^n}\Delta^{-1}(A\Delta - z^n\Delta')\Gamma$$

in $B(0, \rho)\backslash\{0\}$. Now

$$\Delta^{-1} = \begin{pmatrix} z^{1-n} & 0 & \dots & 0 \\ 0 & z^{2-n} & & \\ 1 & & \ddots & \\ & & z^{-1} & \\ 0 & & \dots & 1 \end{pmatrix}$$

and

$$\Delta' = \begin{pmatrix} (n-1)z^{n-2} & 0 & \dots & 0 \\ 0 & (n-2)z^{n-3} & & \\ \vdots & & \ddots & \\ & & 1 & \\ 0 & \dots & & 0 \end{pmatrix}.$$

Hence

$$\Delta^{-1}\Delta' = \frac{1}{z}\begin{pmatrix} (n-1) & 0 & \dots & 0 \\ 0 & (n-2) & & 0 \\ \vdots & & \ddots & \vdots \\ & & 1 & \\ 0 & \dots & & 0 \end{pmatrix}$$

and

$$\Delta^{-1}A\Delta = \begin{pmatrix} 0 & z^{n-1} & 0 & \dots & 0 \\ 0 & 0 & z^{n-1} & \dots & 0 \\ \vdots & & & & \\ 0 & & \dots & & z^{n-1} \\ -a_n z^{n-1} & & \dots & & -a_1 z^{n-1} \end{pmatrix}.$$

Hence, Γ verifies the differential equation

$$\Gamma' = \frac{1}{z}\begin{pmatrix} (n-1) & 1 & 0 & \dots & 0 \\ 0 & -(n-2) & & \dots & 0 \\ 0 & & \dots & -1 & 1 \\ -a_n & -a_{n-1} & \dots & -a_2 & -a_1 \end{pmatrix}\Gamma,$$

which is of the preceding type

$$\Gamma' = \left(\frac{1}{z}A_1 + A_2\right)\Gamma,$$

with A_2 holomorphic in $B(0, \rho)$ and

$$A_1 = \begin{pmatrix} -(n-1) & 1 & 0 & \dots & 0 \\ 0 & -(n-2) & & \dots & 0 \\ \vdots & & & & \\ 0 & \dots & & -1 & 1 \\ -a_n(0) & \dots & & -a_2(0) & -a_1(0) \end{pmatrix}.$$

The characteristic polynomial of A_1 is

$$\begin{aligned} \det(\lambda I - A_1) = (\lambda + n - 1)(\lambda + n - 2)\dots\lambda \\ + (\lambda + n - 1)(\lambda + n - 2)\dots(\lambda + 1)a_1(0) + \cdots \\ + (\lambda + n - 1)a_{n-1}(0) + a_n(0). \end{aligned}$$

Therefore, if none of the differences $\lambda_i - \lambda_j$ between the eigenvalues $(\lambda_k)_{k=1}^n$ of A_1 is in $\mathbb{N}^*$, we can apply the preceding study. Letting $\lambda = r - n + 1$ allows us to rewrite the characteristic equation of A_1 in the more classical form of the ***indicial equation***

$$r(r-1)\dots(r-n+1) + a_1(0)r(r-1)\dots(r-n+2) + \cdots + a_{n-1}(0)r + a_n(0) = 0.$$

What we need then is that none of the differences $r_j - r_k$ of the roots $(r_k)_{k=1}^n$ of the indicial equation belongs to $\mathbb{N}^*$.

The reader will easily find that the indicial equation is the equation that has to be satisfied by $r \in \mathbb{C}$ in order for the original Fuchsian differential equation to have a solution of the form $f(z) = z^r\varphi(z)$, φ holomorphic in $B(0, \rho)$, $\varphi(0) \neq 0$, $z^r = \exp(r \log z)$.

If certain differences between roots of the indicial equation are integers (zero or not) one looks for solutions of the form $\sum_{k \geq 0} P_k(\log z) z^{\lambda+k}$, $\lambda = r - n + 1$, where P_k is a polynomial of degree $\leq n - 1$ (cf. §5.15.14 and §5.15.15).

As an example, let us study in more detail the extremely important case of Fuchsian type differential equations of second order.

Let a_1, a_2 be two holomorphic functions in $B(0, \rho)$, $X = B(0, \rho)\backslash\{0\}$. The differential equation in X

$$y'' + \frac{a_1(z)}{z} y' + \frac{a_2(z)}{z^2} y = 0$$

has a fundamental system of solutions (φ_1, φ_2) of one of the two following forms (where Ψ_1, Ψ_2 are holomorphic in $B(0, \rho)$):

In the first case

$$\begin{cases} \varphi_1(z) = z^{r_1}\Psi_1(z) \\ \varphi_2(z) = z^{r_2}\Psi_2(z). \end{cases} \tag{1}$$

This case occurs, for instance, if the matrix

$$A_1 = \begin{pmatrix} -1 & 0 \\ -a_2(0) & -a_1(0) \end{pmatrix}$$

is diagonalizable with eigenvalues λ_1, λ_2, which do not differ by a non-zero integer. The values r_1, r_2 are the roots of the indicial equation $r^2 + r(a_1(0) - 1) + a_2(0) = 0$.

In the other case

$$\begin{cases} \varphi_1(z) = z^r\Psi_1(z) \\ \varphi_2(z) = z^r(\Psi_1(z)\log z + \Psi_2(z)), \end{cases} \tag{2}$$

Ψ_2 with at most a pole at 0. This case occurs, for instance, if the matrix A_1 is not diagonalizable and r is the (double) root of the indicial equation.

In order to see this, one writes the associated first-order system

$$\begin{pmatrix} w_1' \\ w_2' \end{pmatrix} = \begin{pmatrix} 0 & 1 \\ -\dfrac{a_2}{z^2} & -\dfrac{a_1}{z} \end{pmatrix} \begin{pmatrix} w_1 \\ w_2 \end{pmatrix}$$

$$= \frac{1}{z^2}\begin{pmatrix} 0 & z^2 \\ -a_2 & -a_1 z \end{pmatrix}\begin{pmatrix} w_1 \\ w_2 \end{pmatrix} = \frac{A(z)}{z^2}\begin{pmatrix} w_1 \\ w_2 \end{pmatrix}.$$

According to the previous remark we look for a solution (w_1, w_2) of the form

$$\begin{pmatrix} w_1 \\ w_2 \end{pmatrix} = \begin{pmatrix} z & 0 \\ 0 & 1 \end{pmatrix}\begin{pmatrix} u_1 \\ u_2 \end{pmatrix} = \Delta(z)\begin{pmatrix} u_1 \\ u_2 \end{pmatrix}.$$

This leads to the new system

$$\begin{pmatrix} u_1' \\ u_2' \end{pmatrix} = \frac{1}{z^2}\begin{pmatrix} z & 0 \\ 0 & 1 \end{pmatrix}^{-1}\left(\begin{pmatrix} 0 & z^2 \\ -a_2 & -a_1 z \end{pmatrix}\begin{pmatrix} z & 0 \\ 0 & 1 \end{pmatrix} - z^2\begin{pmatrix} 1 & 0 \\ 0 & 0 \end{pmatrix}\right)\begin{pmatrix} u_1 \\ u_2 \end{pmatrix}$$
$$= \frac{1}{z^2}\Delta^{-1}(A\Delta - z^2\Delta')\begin{pmatrix} u_1 \\ u_2 \end{pmatrix}.$$

Now

$$\Delta^{-1}\Delta' = \begin{pmatrix} 1/z & 0 \\ 0 & 1 \end{pmatrix}\begin{pmatrix} 1 & 0 \\ 0 & 0 \end{pmatrix} = \begin{pmatrix} 1/z & 0 \\ 0 & 1 \end{pmatrix},$$
$$\Delta^{-1}A\Delta = \begin{pmatrix} 0 & z^2 \\ -a_2 & -a_1 z \end{pmatrix},$$

and hence,

$$\begin{pmatrix} u_1' \\ u_2' \end{pmatrix} = \frac{1}{z}\begin{pmatrix} -1 & 1 \\ -a_2 & -a_1 \end{pmatrix}\begin{pmatrix} u_1 \\ u_2 \end{pmatrix} = \frac{\tilde{A}}{z}\begin{pmatrix} u_1 \\ u_2 \end{pmatrix},$$

which shows that this system has $z = 0$ as a regular singular point. Let $A_1 = \tilde{A}(0)$,

$$A_1 = \begin{pmatrix} -1 & 1 \\ -a_2(0) & -a_1(0) \end{pmatrix},$$

whose characteristic polynomial is

$$\det(\lambda I - A_1) = \lambda^2 + (a_1(0) + 1)\lambda + (a_2(0) + a_1(0)),$$

and the corresponding indicial equation is obtained by letting $r = \lambda + 1$,

$$r^2 + r(a_1(0) - 1) + a_2(0) = 0.$$

We consider now the two possible cases.

First case: The roots r_1, r_2 of the indicial equation do not differ by a nonzero integer.

Then a fundamental matrix of solutions of

$$U' = \frac{1}{z}\tilde{A}U$$

is of the form

$$U(z) = H(z)e^{A_1 \log z},$$

which leads to a fundamental matrix W of the original system

$$W' = \frac{1}{z^2}AW$$

by setting

$$W(z) = \begin{pmatrix} z & 0 \\ 0 & 1 \end{pmatrix} H(z) e^{A_1 \log z}.$$

Subcase 1(a): If A_1 is diagonalizable, with $\lambda_1 = r_1 - 1$, $\lambda_2 = r_2 - 1$ as eigenvalues, up to a similarity, we have

$$e^{A_1 \log z} = \begin{pmatrix} z^{\lambda_1} & 0 \\ 0 & z^{\lambda_2} \end{pmatrix}.$$

Since a similarity corresponds to a different choice of basis of solutions, we find that we have a fundamental system (φ_1, φ_2) of the type (1) for the original second-order scalar equation.

Subcase 1(b): If A_1 is not diagonalizable we must have $r_1 = r_2 = r$, $\lambda_1 = \lambda_2 = r - 1$. Always up to a change of basis we have

$$\exp(A_1 \log z) = \exp\left(\begin{pmatrix} \lambda & 1 \\ 0 & \lambda \end{pmatrix} \log z\right) = \exp\left(\begin{pmatrix} \lambda & 0 \\ 0 & \lambda \end{pmatrix} \log z\right) \exp\left(\begin{pmatrix} 0 & 1 \\ 0 & 0 \end{pmatrix} \log z\right)$$

$$= \begin{pmatrix} z^{\lambda} & 0 \\ 0 & z^{\lambda} \end{pmatrix} \begin{pmatrix} 1 & \log z \\ 0 & 1 \end{pmatrix} = \begin{pmatrix} z^{\lambda} & z^{\lambda} \log z \\ 0 & z^{\lambda} \end{pmatrix}.$$

This leads to a fundamental system of the form (2).

Second case: The two roots of the indicial equation differ by a nonzero integer.

In this case A_1 is diagonalizable but all we can assert from §5.15.4(i) is that there is a matrix B with a double eigenvalue being one of the eigenvalues of A_1 and such that

$$U(z) = H(z) e^{B \log z}.$$

Then we have a fundamental system as in (1) or (2) according to B being diagonalizable or not.

As an example of application, let us consider in $X = \mathbb{C} \setminus \{0\}$, the *Bessel equations of order* $p \in \mathbb{C}$,

$$y'' + \frac{y'}{z} + \left(1 - \frac{p^2}{z^2}\right) y = 0.$$

The indicial equation in this case is

$$r^2 - p^2 = 0$$

with roots $r = \pm p$. The matrix A_1 is

$$A_1 = \begin{pmatrix} -1 & -1 \\ p^2 & 1 \end{pmatrix}.$$

If $2p \notin \mathbb{Z}$, then we are in subcase 1(a) since the matrix A_1 is diagonalizable and we have two independent solutions of the form $z^p \Psi_1(z)$, $z^{-p} \Psi_2(z)$, with

Ψ_1, Ψ_2 entire functions. In the other case, p integer or p half-integer differ in the existence of a logarithmic term in the first case and not in the second. To show this in detail we use the fact that even if $2p \in \mathbb{Z}$, there is definitely a solution of the form (cf. Remark 5.15.15)

$$\varphi(z) = z^p \sum_0^\infty c_n z^n, \qquad c_0 \neq 0,$$

and we can find the coefficients by substitution into the equation. The equations for the coefficients are

$$(r^2 - p^2)c_0 = 0$$

$$((r+1)^2 - p^2)c_1 = 0$$

$$((r+n)^2 - p^2)c_n + c_{n-2} = 0, \; n \geq 2.$$

The first equation is simply the indicial equation since $c_0 \neq 0$. The second leads to $c_1 = 0$, and hence to $c_{2n+1} = 0$ for every $n \in \mathbb{N}$ by the third equation. With $n = 2k$, the third equation becomes

$$4k + (r+k)c_{2k} + c_{2k-2} = 0, \qquad k \geq 1,$$

or

$$c_{2k} = (-1)^k \left(\frac{1}{2}\right)^{2k} \frac{c_0}{k!(r+1)\ldots(r+k)}.$$

This procedure could only break down if $r + k = 0$ for some $k \in \mathbb{N}^*$. Hence, if $p \notin \mathbb{Z}$ we can take either choice $r = \pm p$; if $p \in \mathbb{Z}$ we can only take the choice $r = p \geq 0$. With this proviso, we normalize the coefficients by choosing $c_0 = 1/\Gamma(r+1)$. We obtain the functions, called ***Bessel functions of order*** p,

$$J_p(z) = \left(\frac{z}{2}\right)^p \sum_{k=0}^\infty \frac{(-1)^k}{2^k k!\,\Gamma(p+k+1)} z^{2k}.$$

It is immediate that the series represents an entire function, and when $p \notin \mathbb{N}$ we have two linearly independent solutions of the Bessel equation. (The linear independence is an immediate corollary of the obvious linear independence of z^p, z^{-p} for $p \notin \mathbb{Z}$. What happens when $p \in \mathbb{N}$? The relation

$$4k(-p+k)c_{2k} + c_{2k-2} = 0,$$

leads to $c_{2k} \neq 0$ for $k < p$, but for $k = p$, it yields $0.\, c_{2p} + c_{2p-2} = 0$, which is a contradiction. Therefore, there can only be one independent solution of the form $z^p h(z)$, h entire. It must be a multiple of $J_p(z)$ defined earlier. The other one will be of the form $\varphi(z) = J_p(z)\log z + g(z)$, g with at most a pole at 0. Replacing into the Bessel equation we have

$$g''(z) + \frac{1}{z}g'(z) + \left(1 - \frac{p^2}{z^2}\right)g(z) = -\frac{2}{z}J_p'(z),$$

which admits a solution

$$g(z) = z^{-p} \sum_{k=0}^{\infty} a_k z^k.$$

For the case of the Bessel equation of order 1/2 we have that the independent solutions of

$$z^2 y'' + zy + (z^2 - \tfrac{1}{4})y = 0$$

are

$$J_{1/2}(z) = \frac{\cos z}{z^{1/2}}$$

and

$$J_{-1/2}(z) = \frac{\sin z}{z^{1/2}}.$$

We have not touched at all the relations between solutions of differential equations in the complex domain and analytic continuation, asymptotic developments, special functions, and group representations, Nevanlinna theory. Many of these questions are currently being studied in a completely different way than what the reader will find in classical textbooks [Hi2], [CoL], [In], and [Hen], we return to this subject in the next and last section of this book, as well as in the next volume. For the moment, we offer a few other references of interest: [Ju], [Del], [Ol], [Wa], [Vil], [Fi], and [Rub2].

Exercises 5.15

1. Given $A \in M(n, \mathbb{C})$ we denote $ad(A): M(n, \mathbb{C}) \to M(n, \mathbb{C})$ the map defined by

$$ad(A)M = AM - MA.$$

We want to show that if $\lambda_1, \ldots, \lambda_n$ are the eigenvalues of A, counted with multiplicities, then $\lambda_j - \lambda_k$, $1 \leq k, j \leq n$, are all the eigenvalues of $ad(A)$.

Denote by x a column vector in $\mathbb{C}^n$. Then x^t, its transpose, is a row vector. We let e_j be the column vector representing the jth element of the canonical basis of $\mathbb{C}^n$. The matrices $M_{jk} = e_j \cdot e_k^t$ form a basis of $M(n, \mathbb{C})$.

(a) Show that if A is a diagonal matrix, then the matrices M_{jk} are eigenvectors of $ad(A)$ with eigenvalues $\lambda_j - \lambda_k$. Hence $ad(A)$ is a diagonal matrix with respect to the basis $(M_{jk})_{j,k}$.

(b) Recall that A is diagonalizable if there is a diagonal matrix D and an invertible matrix P such that $A = PDP^{-1}$ $(D, P \in M(n, \mathbb{C}))$. Show that in case A is diagonalizable, $ad(A)$ is also diagonalizable.

(c) Let x, y be eigenvectors of A with eigenvalues λ, μ, respectively. Show that

(i) xy^t is a nonzero matrix. (Hint: apply this matrix to the column vector $\bar{y}$, conjugate to y.)

(ii) $ad(A)xy^t = (\lambda - \mu)xy^t$.

(d) Let P be an invertible matrix such that $A = PTP^{-1}$, $T = (t_{jk})$ an upper triangular matrix.

(i) If $M \in M(n, \mathbb{C})$ is an eigenvector of $ad(A)$ with eigenvalue α, show that $N = P^{-1}MP$ is an eigenvector of $ad(T)$ with eigenvalue α.
(ii) Let k be the smallest index such that $Ne_k \neq 0$. Show that Ne_k is an eigenvector of T with eigenvalue $(t_{kk} + \alpha)$.
(iii) Use (ii) to conclude that $\alpha = \lambda_i - \lambda_j$, for some pair of eigenvalues λ_i, λ_j of A.

(e) What are the eigenvalues of the map $S(A): M(n, \mathbb{C}) \to M(n, \mathbb{C})$,

$$S(A)M = AM + MA?$$

2. Let Ω be an open set in $\mathbb{C}$, $a \in \Omega$, $f_1, \ldots, f_n \in \mathcal{O}_a$, which are linearly independent. Assume further that
(i) for each $\gamma \in \pi_1(\Omega, a)$, and each j, the germ f_j admits an analytic continuation along the loop γ to a germ $\gamma f_j \in \mathcal{O}_a$;
(ii) for every γ, j as in (i), we have that

$$\gamma f_j = \mu_1 f_1 + \cdots + \mu_n f_n \quad \text{in} \quad \mathcal{O}_a$$

for some constants $\mu_1, \ldots, \mu_n$.
Let W be the $n \times (n + 1)$ matrix

$$W = \begin{bmatrix} f_1 & f_1' & \cdots & f_1^{(n)} \\ \vdots & \vdots & & \vdots \\ f_n & f_n' & \cdots & f_n^{(n)} \end{bmatrix}$$

and denote by w_j the $n \times n$ minors obtained by deleting the jth column of W. Clearly $w_j \in \mathcal{O}_a$ for $1 \leq j \leq n + 1$.
(a) Show that w_{n+1} is not identically zero;
(b) Show that property (i) still holds for every γ and every w_j;
(c) Let $F = \begin{bmatrix} f_1 \\ \vdots \\ f_n \end{bmatrix}$. Show that for $\gamma \in \pi_1(\Omega, a)$ there is a nonsingular matrix T_γ such that

$$\gamma F = T_\gamma F.$$

(d) Show that if we denote by $\tilde{w}_{n+1}$ the multivalued holomorphic function in Ω defined by analytic continuation of w_{n+1}, then the zeros of $\tilde{w}_{n+1}$ and its multiplicities are independent of the branch of $\tilde{w}_{n+1}$ lying above a point $z_0 \in \Omega$. Let V be multiplicity variety of the zeros of $\tilde{w}_{n+1}$.
(e) Show that $a_j = w_j/w_{n+1}$, $1 \leq j \leq n$, are germs in $\mathcal{M}_a$ and admit analytic continuation to a (single-valued) meromorphic function in Ω.
(f) Let $g \in \mathcal{H}(\Omega)$ define the multiplicity variety V and denote

$$b_j := ga_j \qquad (1 \leq j \leq n).$$

Show that if D denotes the following differential operator with holomorphic coefficients in Ω:

$$D := \sum_{j=0}^{n} b_{n-j} \frac{d^n}{dz^n}$$

with $b_0 = g$, then every f_j is a solution of the homogeneous equation $Du = 0$.

3. Find the automorphic factor for the fundamental matrix of the system

$$f'(z) = \left[\frac{1}{z}\begin{pmatrix}0 & 0\\ 0 & -1\end{pmatrix} + \begin{pmatrix}0 & 1\\ 0 & 0\end{pmatrix}\right] f(z), \qquad 0 < |z| < \infty.$$

Show it is not similar to $\exp\left(2\pi i\begin{pmatrix}0 & 0\\ 0 & -1\end{pmatrix}\right)$ as predicted by Remark 5.15.7(1) Why?

4. Let r_1, r_2 be the roots of the indicial equation of the Fuchsian type differential equation (with singularity at $z = 0$)

$$f''(z) + \frac{a_1(z)}{z} f'(z) + \frac{a_2(z)}{z^2} f(z) = 0.$$

(a) Introduce a change of variable $z = \zeta^k, k \in \mathbb{N}, g(\zeta) = f(\zeta^k)$. Show that g satisfies a Fuchsian type equation with corresponding roots kr_1, kr_2 for the indicial equation.
(b) What happens when the change of variable is of the form $z = \varphi(\zeta)$, $\varphi(0) = 0$, $\varphi'(0) \neq 0$?
(c) What happens in the case of an equation of order n?

5. For a system $f' = Af$, A an $n \times n$ matrix holomorphic in the region $0 < R < |z| < \infty$, we say that the system is regular or it has a Fuchsian type singularity at $z = \infty$ if the system $g' = Bg$ obtained by the change of variables $z \mapsto \dfrac{1}{z}$ has $z = 0$ as a regular point or, respectively, it has a Fuchsian type singularity.

Show for either of these two possibilities to occur it is necessary and sufficient that $z = \infty$ is a removable singularity for A and $A(\infty) = 0$.

6. Consider the differential equation

$$f^{(n)} + p_1 f^{(n-1)} + \cdots + p_n f = 0,$$

where $z = \infty$ is an isolated singularity for all the coefficients p_j. We say that $z = \infty$ is a singular point of the Fuchsian type if the system obtained by the substitution $z = \dfrac{1}{w}$ has a Fuchsian type singularity at $w = 0$. Show that this can happen if and only if each coefficient p_j is holomorphic at $z = \infty$ and vanishes at least to the order j at that point. (Hint: it is easier to first write $p_j(z) = q_j(z)/z^j$.)

7. Use the definition and criteria given in the text and in Exercise 5.15.6 to classify all the singularities of the following equations:
(a) $(1 - z^2)f''(z) - 2zf'(z) + \nu(\nu + 1)f(z) = 0$ (Legendre equation)
(b) $(1 - z^2)f''(z) - 2zf'(z) + \left[\nu(\nu + 1) - \dfrac{\mu^2}{1 - z^2}\right] f(z) = 0$ (associated Legendre equation)
(c) $y'' + \dfrac{y'}{z} + \left(1 - \dfrac{p^2}{z^2}\right) y = 0$ (Bessel equation)
(d) $u'' + \left(\dfrac{a}{z} + \dfrac{b}{z - 1}\right) u' + \dfrac{c}{z(z - 1)} u = 0$ (hypergeometric equation)

(ν, μ, p, a, b, c are complex parameters).

8. Let A be an $n \times n$ matrix holomorphic except for a finite number of singularities at $z_1, \dots, z_m \in \mathbb{C}$, and, possibly $z = \infty$. Show that the system $f' = Af$ has only Fuchsian type singularities if and only if there are constant nonzero matrices $A_1, \dots, A_m$ such that

$$A(z) = \frac{A_1}{z - z_1} + \cdots + \frac{A_m}{z - z_m}.$$

Moreover, the point $z = \infty$ is not a singular point if and only if $A_1 + \cdots + A_m = 0$. Could there be a system with precisely one singular point?

9. For a differential equation

$$f^{(n)} + p_1 f^{(n-1)} + \cdots + p_n f = 0,$$

where the coefficients have singularities only at $z_1, \dots, z_m$, and possibly, $z = \infty$, show that all the singularities are of Fuchsian type if and only if the p_j are rational functions of the form

$$p_j(z) = \frac{q_j(z)}{\left(\prod_{k=1}^{m} (z - z_k)\right)^j},$$

q_j a polynomial, $\deg q_j \leq (m-1)j$.

10. (a) Can there be a second-order equation $f'' + p_1 f' + p_2 f = 0$ without singular points?
(b) Show that if there is precisely one Fuchsian type singular point at $z = z_1 \neq \infty$, then the equation is

$$f'' + \frac{z}{z - z_1} f' = 0.$$

Find a fundamental system.
(c) Which is the only equation satisfying (b) when the only singular point is $z = \infty$?

*11. Write down all the differential equations of second order which have exactly two singular points of the Fuchsian type at $z = z_1$ and $z = z_2$, and $z = \infty$ is not a singularity. (Ans: $f'' + \frac{2z + a}{(z - z_1)(z - z_2)} f + \frac{b}{(z - z_1)^2 (z - z_2)^2} f = 0$, with a, b arbitrary constants.)

*12. Show that every second-order differential equation all whose singularities are of Fuchsian type and lie at $z = 0, 1, \infty$ are of the form

$$u'' + \left(\frac{a}{z} + \frac{b}{z - 1}\right) u' + \frac{c}{z(z - 1)} u = 0$$

(see Exercise 5.15.7(d)).

13. What are the conditions on the parameters a, b, c of the hypergeometric equation in order that there is a solution which is a polynomial?

(Remark: There are only seven choices of triplets (a, b, c) such that all the solutions of the hypergeometric equation are algebraic (see [Hi2]).)

14. Recall that the Schwarzian derivative $\{w, z\}$ of a holomorphic function w with respect to the variable z is given by $\{w, z\} = \left(\frac{w''}{w'}\right)' - \frac{1}{2}\left(\frac{w''}{w'}\right)^2$. Show that if u_1, u_2 are two linearly independent solutions of the differential equation
$$u'' + qu = 0$$
in a simply connected region Ω, and $w = u_1/u_2$, then
$$\{w, z\} = zq(z)$$
(whenever $u_2(z) \neq 0$).

15. A method, due to Laplace, to solve the differential equation with linear coefficients
$$(a_0 z + b_0)u''(z) + (a_1 z + b_1)u'(z) + (a_2 z + b_2)u = 0,$$
is to assume u is of the form
$$u(z) = \int_C e^{zt} v(t)\, dt$$
for a convenient holomorphic function v and contour C. Show that formally v must satisfy the first-order equation
$$P(t)\frac{dv}{dt} - Q(t)v(t) = 0.$$
Find P, Q in terms of $a_0, b_0, \ldots, b_2$. Rewrite this equation as
$$\frac{1}{v}\frac{dv}{dt} = \mu + \frac{\lambda_1}{t - \alpha_1} + \frac{\lambda_2}{t - \alpha_2}.$$
In case λ_1, λ_2 satisfy $\operatorname{Re} \lambda_j > -1$, show that if we take C to be any finite segment, then one can define a solution u_1 of the original equation by this method. Use Euler's trick of writing $u_2 = wu_1$ to find a second linearly independent solution (cf. [In] for generalizations).

§16. The Index of Differential Operators

We conclude this volume proving some index theorems for linear differential equations with holomorphic coefficients in several spaces of holomorphic functions and of germs of holomorphic functions. Surprisingly these results seem to have escaped the classical textbooks [Hi2], [CoL], [Ha], and [In], and only some of the corollaries have found their way there ([In] has the most complete account). We follow the article of [Mal]. A slightly different account appears in [Schm]. One should consider this section only as an introduction to the present renewal of interest on the theory of singular points by authors with very different perspectives on the subject, e.g., [Del], [BV], and [Be].

In order not to make this section too long we will assume known the main facts about the index of operators [Ka] and sheaf cohomology [Bre] and [God]. We will recall here some definitions and basic results.

5.16.1. Definition. Let E, F be two $\mathbb{C}$-vector spaces and $u : E \to F$ a linear map. We say that u is an ***operator with index*** if $\ker u$ and $\operatorname{coker} u$ have finite dimension. The ***index*** $\chi(u)$ is the integer defined by

$$\chi(u) = \dim \ker u - \dim \operatorname{coker} u.$$

If $u : E \to F$ and $v : F \to G$ are operators with index, then $v \circ u$ also is an operator with index and

$$\chi(v \circ u) = \chi(v) + \chi(u).$$

If E and F are Banach spaces, $u : E \to F$ is an operator with index and $v : E \to F$ is a compact operator, then $u + v$ is an operator with index and

$$\chi(u + v) = \chi(u).$$

Finally, if we have a commutative diagram of $\mathbb{C}$-vector spaces and linear applications

$$\begin{array}{ccc}
0 & & 0 \\
\downarrow & & \downarrow \\
F_1 & \xrightarrow{u} & E_2 \\
{\scriptstyle i_1}\downarrow & & \downarrow{\scriptstyle i_2} \\
F_1 & \xrightarrow{v} & E_2 \\
{\scriptstyle \pi_1}\downarrow & & \downarrow{\scriptstyle \pi_2} \\
F_1/E_1 & \overset{w}{\dashrightarrow} & F_2/E_2 \\
\downarrow & & \downarrow \\
0 & & 0
\end{array}$$

where the i_k are injective, π_k the canonical projections, and u, v are operators with index, then the unique linear map w that makes the diagram commutative is also an operator with index

$$\chi(w) = \chi(v) - \chi(u).$$

Since we are going to use several times this type of diagram, let us indicate here how to prove the last identity.

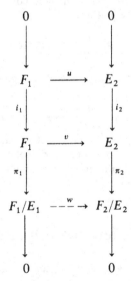

One completes the diagram to the following one

$$\begin{array}{ccccccccc}
 & & 0 & & 0 & & 0 & & \\
 & & \downarrow & & \downarrow & & \downarrow & & \\
0 & \longrightarrow & \ker u & \longrightarrow & E_1 & \xrightarrow{u} & E_2 & \longrightarrow & \operatorname{coker} u & \longrightarrow & 0 \\
 & & \downarrow & & \downarrow & & \downarrow & & \\
0 & \longrightarrow & \ker v & \longrightarrow & F_1 & \xrightarrow{v} & F_2 & \longrightarrow & \operatorname{coker} v & \longrightarrow & 0 \\
 & & \downarrow & & \downarrow & & \downarrow & & \\
0 & \longrightarrow & \ker w & \longrightarrow & F_1/E_1 & \xrightarrow{w} & F_2/E_2 & \longrightarrow & \operatorname{coker} w & \longrightarrow & 0 \\
 & & \downarrow & & \downarrow & & \downarrow & & \\
 & & 0 & & 0 & & 0 & &
\end{array}$$

In this diagram all the rows and all the columns are exact. A classical lemma of homological algebra (the five lemma) allows us to construct in this situation a linear map $\delta : \ker w \to \operatorname{coker} u$ in such a way that the sequence

$$0 \to \ker u \to \ker v \to \ker w \xrightarrow{\delta} \operatorname{coker} u \to \operatorname{coker} v \to \operatorname{coker} w \to 0$$

is exact. All these vector spaces are finite dimensional, hence one knows that

$$\dim \ker u - \dim \ker v + \dim \ker w$$
$$- \dim \operatorname{coker} u + \dim \operatorname{coker} v - \dim \operatorname{coker} w = 0,$$

which yields immediately the equality $\chi(w) = \chi(v) - \chi(u)$.

In this section we denote by $\mathcal{O}$ the ring of germs of holomorphic functions at the origin of $\mathbb{C}$, and $\hat{\mathcal{O}}$ the ring of $\mathbb{C}[[z]]$ of formal power series. That is,

$$\hat{\mathcal{O}} = \mathbb{C}[[z]] = \left\{ \sum_{n \geq 0} a_n z^n : a_n \in \mathbb{C} \right\}.$$

This is a local ring with the usual sum and multiplication of power series. Let $\hat{\mathcal{M}}$ denote the maximal ideal of $\hat{\mathcal{O}}$ consisting of all the formal power series such that $a_0 = 0$. Its kth power $\hat{\mathcal{M}}^k$ is therefore the principal ideal

$$\hat{\mathcal{M}}^k = z^k \hat{\mathcal{O}}.$$

The corresponding maximal ideal in $\mathcal{O}$ is denoted $\mathcal{M} = \hat{\mathcal{M}} \cap \mathcal{O}$.

If $f = \sum_{n \geq 0} a_n z^n \in \hat{\mathcal{O}}$, we denote by $v(f)$ the smallest integer n such that $a_n \neq 0$. That is,

$$f = \sum_{n \geq v(f)} a_n z^n, \qquad a_{v(f)} \neq 0.$$

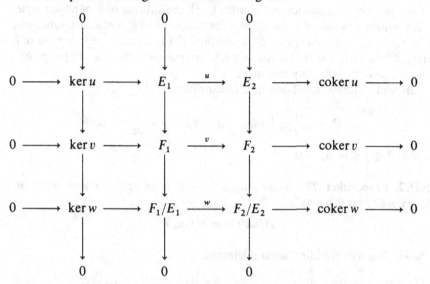

It certainly exists if $f \neq 0$. The integer $v(f)$ is called the ***valuation*** of f. We set $v(0) = \infty$. Since we can identify $\mathcal{O}$ with $\mathbb{C}\{z\}$, the subring of $\hat{\mathcal{O}}$ of power series with nonzero radius of convergent, every element of $\mathcal{O}$ has also a valuation. Clearly, $v(f) = 0$, $f \in \mathcal{O}$ (resp. $\hat{\mathcal{O}}$) if and only if it is an invertible element of $\mathcal{O}$ (resp. $\hat{\mathcal{O}}$). In fact, $p = v(f)$ is the smallest integer such that $f \in \mathcal{M}^p$ (resp $\hat{\mathcal{M}}^p$), the pth power of the maximal ideal.

In what follows D will denote a linear differential operator

$$D = a_m \frac{d^m}{dz^m} + a_{m-1} \frac{d^{m-1}}{dz^{m-1}} + \cdots + a_1 \frac{d}{dz} + a_0,$$

$a_j \in \mathcal{O}$, $0 \leq j \leq m$, $a_m \neq 0$.

5.16.2. Proposition. *The linear map $D : \mathcal{O} \to \mathcal{O}$ is an operator with index, its index $\chi(D, \mathcal{O})$ is given by*

$$\chi(D, \mathcal{O}) = m - v(a_m).$$

PROOF. We will need first several lemmas.

5.16.3. Lemma. *For $0 < r < \infty$ and $p \in \mathbb{N}$, the complex vector space*

$$E_p(r) := \{f \in \mathcal{H}(B(0,r)) \cap \mathcal{E}_p(\bar{B}(w_0, r))\}$$

can be given the norm

$$\|f\| = \sum_{0 \leq j \leq p} \max_{|z| \leq r} |f^{(j)}(z)|,$$

which turns it into a Banach space.

(Here $\mathcal{E}_p(\bar{B}(w_0, r))$ is the space of functions whose partial derivatives with respect to z and $\bar{z}$ exist up to order p and are continuous in the closed disk. It coincides with those functions which are restrictions to the closed disk of functions of class C^p in a larger open disk. The subspace $E_p(r)$ consists of those holomorphic functions f in $B(0,r)$ whose derivatives $f^{(j)}$, $0 \leq j \leq p$ have a continuous extension to the closed disk.)

PROOF. If $(f_n)_{n \geq 0}$ is a Cauchy sequence in $E_p(r)$, then for $0 \leq k \leq p$, the sequence $(f_n^{(k)})_{n \geq 0}$ clearly converges uniformly in $\bar{B}(0,r)$ to a function $g_k \in E_0(r)$, i.e., a function continuous in the closed disk and holomorphic inside. Furthermore, it is clear that for $z \in B(0,r)$ we have $g_k' = g_{k+1}$ if $0 \leq k \leq p - 1$. This implies that $g_0 \in E_p(r)$ and that $(f_n)_{n \geq 0}$ converges towards g_0 in $E_p(r)$. □

5.16.4. Lemma. *Let $m \geq 1$, $D' = \sum_{0 \leq j \leq m-1} a_j \frac{d^j}{dz^j}$, $a_j \in E_1(r)$. Then*

$$D' : E_m(r) \to E_0(r)$$

is a compact operator.

PROOF. Let $(f_n)_{n\geq 0}$ be a sequence of elements in the unit ball of $E_m(r)$. Let $g_n = D'f_n$, $z, \zeta \in \bar{B}(0,r)$, then

$$g_n(z) - g_n(\zeta) = (z - \zeta)\int_0^1 g_n'(\zeta + t(z-\zeta))\,dt,$$

hence

$$|g_n(z) - g_n(\zeta)| \leq M|z - \zeta|,$$

where M is a positive constant independent of n, due to the hypothesis on D' and $(f_n)_{n\geq 0}$. Hence the sequence $(g_n)_{n\geq 0}$ is bounded in $\bar{B}(0,r)$ and equicontinuous. Therefore, by Ascoli's theorem, it has a convergent subsequence in $E_0(r)$. □

From the remarks at the beginning of this section we can now conclude that, when $a_j \in E_1(r)$, to show that $D: E_m(r) \to E_0(r)$ is an operator with index, it is enough to show this holds for $a_m \dfrac{d^m}{dz^m}$. Furthermore, $\chi(D: E_m(r) \to E_0(r)) = \chi\left(a_m \dfrac{d^m}{dz^m}: E_m(r) \to E_0(r)\right)$ (here we assume $a_m \in E_0(r)$). Using the factorization of $a_m \dfrac{d^m}{dz^m}$:

$$E_m(r) \xrightarrow{\frac{d}{dz}} E_{m-1}(r) \xrightarrow{\frac{d}{dz}} \cdots E_1(r) \xrightarrow{\frac{d}{dz}} E_0(r) \xrightarrow{a_m} E_0(r),$$

where the last operator is just multiplication by a_m, we have reduced the question to prove the following lemma.

5.16.5. Lemma.

(1) *The operator* $\dfrac{d}{dz}: E_{p+1}(r) \to E_p(r)$ *has an index equal to* 1.

(2) *The operator* $a_m: E_0(r) \to E_0(r)$ *has an index if* a_m *does not vanish in* $\bar{B}(0,r)\setminus\{0\}$, *its index is* $-v(a_m)$.

(3) *The operator* $D: E_m(r) \to E_0(r)$ *has index* $m - v(a_m)$.

PROOF. (1) It is clear that $\dim\ker\left(\dfrac{d}{dz}\right) = 1$ since $\bar{B}(0,r)$ is connected. On the other hand, $\dfrac{d}{dz}$ is surjective. Namely, if $f \in E_r(r)$, let

$$g(z) = z\int_0^1 f(tz)\,dt = \int_0^z f(\zeta)\,d\zeta.$$

Then $g \in E_{p+1}(r)$ and $\dfrac{d}{dz}g = f$.

(2) If $a_m(z) = z^k b(z)$, $k = v(a_m)$, and $b(z) \neq 0$ in $\bar{B}(0,r)$, then it is clear that

the operator of multiplication by a_m is injective, and that its cokernel is generated by $1, z, \ldots, z^{k-1}$.

(3) This is just a consequence of the preceding two parts. □

To end the proof of Proposition 5.16.2 we will exploit the fact that for any fixed $p \geq 0$ and $0 < r_0 < \infty$ one has

$$\mathcal{O} = \varinjlim_{0<r<r_0} E_p(r).$$

The symbol $\varinjlim_{0<r<r_0} E_p(r)$ represents the vector space "inductive limit" of the $E_p(r)$. This space is the union $\bigcup_{0<r<r_0} E_p(r)$, modulo the equivalence relation: $f \in E_p(r_2)$, $g \in E_p(r_2)$ are equivalent if $f = g$ in $\bar{B}(0, \inf(r_1, r_2))$. With this definition, the fact that $\mathcal{O}$ coincides with the inductive limit is obvious.

Given the original operator D, its coefficients were only assumed to be holomorphic in a neighborhood of the origin. Let us choose $r_0 > 0$ such that all the a_j are holomorphic in a neighborhood of $\bar{B}(0, r_0)$ and a_m does not vanish in $\bar{B}(0, r_0) \backslash \{0\}$. This is possible since a_m is not identically zero.

For any r, $0 < r < r_0$ we denote

$$D_r : E_m(r) \to E_0(r)$$

the operator D acting on $E_m(r)$. We must have $\dim \ker D_r \leq m$; if not in every simply connected open subset of $B(0, r) \backslash \{0\}$ we will have more than m independent solutions of the homogeneous equation

$$f^{(m)} + \frac{a_{m-1}}{a_m} f^{(m-1)} + \cdots + \frac{a_0}{a_m} f = 0,$$

which is impossible by the results of the previous section. It follows that, after taking r_0 smaller if necesssary, we can assume that $\dim \ker D_r = N = \text{constant}$ for $0 < r < r_0$. In fact, we can even assume we have a basis of N functions in $E_m(r_0)$. Under these conditions the $\dim \operatorname{coker} D_r$ is also stationary, namely, $\dim \operatorname{coker} D_r = N - \chi(D_r) = N - m + v(a_m)$.

Consider now the diagram

$$\begin{array}{ccccccccc}
 & & 0 & & 0 & & 0 & & \\
 & & \downarrow & & \downarrow & & \downarrow & & \\
0 & \longrightarrow & \ker D_r & \longrightarrow & E_m(r) & \xrightarrow{D_r} & E_0(r) & \xrightarrow{\pi_r} & \operatorname{coker} D_r & \longrightarrow & 0 \\
 & & \downarrow \rho & & \downarrow \rho & & \downarrow \rho & & \downarrow \bar{\rho} & & \\
0 & \longrightarrow & \ker D & \longrightarrow & \mathcal{O} & \xrightarrow{D} & \mathcal{O} & \xrightarrow{\pi} & \operatorname{coker} D & \longrightarrow & 0
\end{array}$$

where ρ denotes the restriction map and $\bar{\rho}$ the map obtained from ρ after passing to the quotient. The restriction maps ρ are clearly injective.

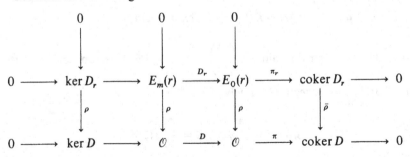

We must have $\dim \ker D = \dim \ker D_r = N$ for all sufficiently small r. If not, let $u_1, \ldots, n_{N+1} \in \ker D$ which are linearly independent. Then for $0 < r \ll 1$ we have that all $u_j \in E_m(r)$ and $D_r u_j = Du_j = 0$, hence $\dim \ker D_r \geq N + 1$. This is impossible, hence $\dim \ker D \leq N$. On the other hand the injectivity of ρ implies the reverse inequality.

We also know that $\dim \operatorname{coker} D_r = M$ is constant for r small. We would like to show that $\dim \operatorname{coker} D = M$. Let us show first of all that $\dim \operatorname{coker} D \leq M$. Suppose there are $u_1, \ldots, u_{M+1} \in \mathcal{O}$ such that $\pi(u_1), \ldots, \pi(u_{M+1})$ are linearly independent. This means that for any choice $\lambda_1, \ldots, \lambda_{M+1} \in \mathbb{C}$, not all zero, we have $\sum_{j=1}^{M+1} \lambda_j u_j \notin \operatorname{Im} D$. On the other hand for $0 < r$ sufficiently small, all the $u_j \in E_0(r)$, hence the set $\{\pi_r(u_1), \ldots, \pi_r(u_{M+1})\} \subseteq \operatorname{coker} D_r$ and it is necessarily linearly dependent. Hence there are scalars $\lambda_1, \ldots, \lambda_{M+1}$, not all zero, and $f \in E_m(r)$ such that $D_r f = \sum_{j=1}^{M+1} \lambda_j u_j$ in $E_0(r)$. This clearly implies that $\sum_{j=1}^{M+1} \lambda_j u_j \in \operatorname{Im} D$, a contradiction. To obtain $\dim \operatorname{coker} D \geq M$ we will show that $\bar{\rho}$ is injective. Let us assume there is $u \in E_0(r)$ such that $\bar{\rho}(\pi_r(u)) = 0$. This means that there is $f \in \mathcal{O}$ such that $u = Df$ in some disk $B(0, r')$, $0 < r' \leq r$. Let us cover $B(0, r) \setminus \{0\}$ with three open angular sectors such that the intersection of any two of them is nonempty and connected. If S is one of those sectors, then it is simply connected and, taking initial conditions at a point $z_0 \in S \cap B(0, r')$ which coincide with $f(z_0), \ldots, f^{(m-1)}(z_0)$, we have a unique holomorphic function f_S in S solving $Df_S = u$ and satisfying those initial conditions. Therefore, $f_S = f$ in $S \cap B(0, r')$, hence f_S is an analytic continuation of f to the sector. If S' is another sector, we have a corresponding function $f_{S'}$, $f_{S'} = f = f_S$ in $S \cap S' \cap B(0, r')$. This shows that f has an analytic continuation (still called f) to $B(0, r)$, solving $Df = u$. If we can prove that $f \in C^m$ in $\bar{B}(0, r)$ then we will have $\pi_r(u) = 0$, as we wanted to show. Let $z_1 \in \partial B(0, r)$, take $z_0 \in B(0, r)$ very close to z_1 so that for some $\rho > 0$ we have $z_1 \in B(z_0, \rho)$ and a_m has no zeros in $B(z_0, \rho)$ (it is enough to take z_0 and ρ so that $|z_0| + \rho < r_0$ and $|z_0| - \rho > 0$). Then the linear system associated to the equation

$$\frac{1}{a_m} Df = f^{(m)} + \frac{a_{m-1}}{a_m} f^{(m-1)} + \cdots + \frac{a_0}{a_m} f = \frac{u}{a_m}$$

has a resolvent $R(z, z_0)$, and if we call F (respectively, B) the vector corresponding to f (resp. u/a_m), the Green-Lagrange formula can be written as

$$F(z) = R(z, z_0)F(z_0) + (z - z_0) \int_0^1 R(z, z_0 + t(z - z_0))B(z_0 + t(z - z_0))\, dt.$$

Since u is continuous in $B(z_0, \rho) \cap \bar{B}(0, r)$, the same holds for the vector-valued function B, hence for F. It we start with $F(z_0) = (f(z_0), \ldots, f^{(m-1)}(z_0))$, this shows that $f, \ldots, f^{(m-1)}$ are continuous up to $\partial B(0, r)$. The equation itself shows the same holds for $f^{(m)}$. Hence $f \in E_m(r)$.

Now that we know that dim coker D = dim coker D_r, we can conclude that

$$\chi(D, \mathcal{O}) = \chi(D_r : E_m(r) \to E_0(r)) = m - v(a_m). \qquad \square$$

The differential operator D also operates in the space of formal power series $\hat{\mathcal{O}}$. It will be denoted by the same symbol D. This operator D still has an index but it is not necessarily the same as in $\mathcal{O}$.

5.16.6. Proposition. *The map $D : \hat{\mathcal{O}} \to \hat{\mathcal{O}}$ is an operator with index and*

$$\chi(D, \hat{\mathcal{O}}) = \max_{0 \le k \le m} (k - v(a_k)).$$

Proof. Let $n = \max\{k - v(a_k) : 0 \le k \le m\}$, $P = \{p : p - v(a_p) = n\}$. For every $p \in P$ we have $a_p(z) = z^{p-n} b_p(z)$, $b_p \in \mathcal{O}$, and $b_p(0) \neq 0$. If $p \notin P$, we can still write $a_p(z) = z^{p-n} b_p(z)$, but then $b_p(0) = 0$. Let k be an integer $\ge q = \max\{p : p \in P\}$, then we have

$$a_p(z) \frac{d^p}{dz^p}(z^k) = k(k-1)\dots(k-p+1) b_p(0) z^{k-n} + O(z^{k-n+1}).$$

Hence

$$D(z^k) = \left(\sum_{p \in P} k(k-1)\dots(k-p+1) b_p(0) \right) z^{k-n} + O(z^{k-n+1}).$$

The coefficient of z^{k-n} is a polynomial $Q(k)$ of degree q in k with leading term $b_p(0) \neq 0$. It follows that there is $k_0 \in \mathbb{N}^*$ such that this coefficient is $\neq 0$ whenever $k \ge k_0$.

To continue the proof we need the following lemma.

5.16.7. Lemma. *Let $k \ge k_0$ and let $g \in \hat{\mathcal{O}}$ be such that $v(g) \ge k - n$. There is a unique $f \in \hat{\mathcal{O}}$ such that $v(f) \ge k$ and $Df = g$. In other words, D induces a bijection $\tilde{D} : \hat{\mathcal{M}}^k \to \hat{\mathcal{M}}^{k-n}$ for every $k \ge k_0$.*

Proof. Let us write $g = c_{k-n} z^{k-n} + c_{k-n+1} z^{k-n+1} + \cdots \in \hat{\mathcal{M}}^{k-n}$ and assume there is $f = d_k z^k + d_{k+1} z^{k+1} + \cdots \in \hat{\mathcal{M}}^k$ such that $Df = g$. From the previous discussion we conclude that

$$Df = Q(k) d_k z^{k-n} + O(z^{k-n+1}).$$

Hence, d_k is uniquely determined by the equation $d_k = c_{k-n} Q(k)$, since $Q(k) \neq 0$ whenever $k \ge k_0$. Writing now

$$f = d_k z^k + f_{k+1}, \qquad \text{with} \qquad f_{k+1} = d_{k+1} z^{k+1} + \cdots \in \hat{\mathcal{M}}^{k+1},$$

we have

$$\begin{aligned} D(f_{k+1}) &= D(f - d_k z^k) = g - (c_{k-n} z^{k-n} + O(z^{k-n+1})) \\ &= c'_{k-n+1} z^{k-n+1} + \cdots \in \hat{\mathcal{M}}^{k-n+1}. \end{aligned}$$

The previous reasoning implies that $d_{k+1} = c'_{k-n+1}/Q(k+1)$. In this way, we can determine uniquely all the coefficients of f. Moreover, it is clear that with this choice of coefficients we can define an element $f \in \hat{\mathcal{O}}$ such that $\hat{D}(f) = g$. Hence, the lemma is correct. □

To conclude the proof of Proposition 5.16.6 let us consider the diagram

$$\begin{array}{ccccccccccc}
 & & 0 & & 0 & & 0 & & 0 & & \\
 & & \downarrow & & \downarrow & & \downarrow & & \downarrow & & \\
0 & \longrightarrow & \ker \tilde{D} = 0 & \longrightarrow & \hat{\mathcal{M}}^k & \xrightarrow{\tilde{D}} & \hat{\mathcal{M}}^{k-n} & \longrightarrow & \operatorname{coker} \tilde{D} = 0 & \longrightarrow & 0 \\
 & & \downarrow & & \downarrow & & \downarrow & & \downarrow & & \\
0 & \longrightarrow & \ker D & \longrightarrow & \hat{\mathcal{O}} & \xrightarrow{D} & \hat{\mathcal{O}} & \longrightarrow & \operatorname{coker} D & \longrightarrow & 0 \\
 & & \downarrow & & \downarrow & & \downarrow & & \downarrow & & \\
0 & \longrightarrow & \ker \bar{D} & \longrightarrow & \hat{\mathcal{O}}/\hat{\mathcal{M}}^k & \xrightarrow{\bar{D}} & \hat{\mathcal{O}}/\hat{\mathcal{M}}^{k-n} & \longrightarrow & \operatorname{coker} \bar{D} & \longrightarrow & 0 \\
 & & \downarrow & & \downarrow & & \downarrow & & \downarrow & & \\
 & & 0 & & 0 & & 0 & & 0 & &
\end{array}$$

where $\bar{D}$ is obtained by passing to the quotient. The columns and rows being exact, we conclude, with the help of the five lemma, that the sequence

$$0 \to \ker \tilde{D} \to \ker D \to \ker \bar{D} \to \operatorname{coker} \tilde{D} \to \operatorname{coker} D \to \operatorname{coker} \bar{D} \to 0$$

is also exact. This one breaks into two short exact sequences (isomorphisms)

$$0 \to \ker D \to \ker \bar{D} \to 0$$

$$0 \to \operatorname{coker} D \to \operatorname{coker} \bar{D} \to 0.$$

The spaces $\hat{\mathcal{O}}/\hat{\mathcal{M}}^k$ and $\hat{\mathcal{O}}/\hat{\mathcal{M}}^{k-n}$ are finite dimensional, of respective dimensions k and $k-n$. Hence $\bar{D}$ has an index and, by the last two isomorphisms, D also has an index and $\chi(\bar{D}) = \chi(D)$. To compute $\chi(\bar{D})$ we use that

$$\dim \ker \bar{D} + \dim \operatorname{Im} \bar{D} = k,$$

and

$$\dim \operatorname{coker} \bar{D} = k - n - \dim \operatorname{Im} \bar{D}.$$

Therefore,

$$\chi(D) = \chi(\bar{D}) = \dim \ker \bar{D} - \dim \operatorname{coker} \bar{D} = n = \max_{0 \le p \le m} (p - v(a_p)).$$

This concludes the proof of Proposition 5.16.6. □

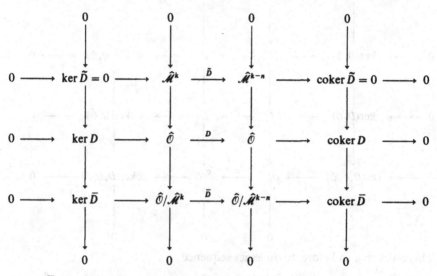

Let us consider now the exact sequence

$$0 \to \mathcal{O} \to \hat{\mathcal{O}} \to \hat{\mathcal{O}}/\mathcal{O} \to 0,$$

from which we obtain the following diagram. The last row was obtained by passing to the quotient, and the notation is, we hope, clear.

$$\begin{array}{ccccccccccc}
 & & & & 0 & & 0 & & 0 & & \\
 & & & & \downarrow & & \downarrow & & \downarrow & & \\
0 & \longrightarrow & \ker(D, \mathcal{O}) & \longrightarrow & \mathcal{O} & \xrightarrow{D} & \mathcal{O} & \longrightarrow & \operatorname{coker}(D, \mathcal{O}) & \longrightarrow & 0 \\
 & & \downarrow & & \downarrow & & \downarrow & & \downarrow & & \\
0 & \longrightarrow & \ker(D, \hat{\mathcal{O}}) & \longrightarrow & \hat{\mathcal{O}} & \xrightarrow{D} & \hat{\mathcal{O}} & \longrightarrow & \operatorname{coker}(D, \hat{\mathcal{O}}) & \longrightarrow & 0 \\
 & & \downarrow & & \downarrow & & \downarrow & & \downarrow & & \\
0 & \longrightarrow & \ker(D, \hat{\mathcal{O}}/\mathcal{O}) & \longrightarrow & \hat{\mathcal{O}}/\mathcal{O} & \xrightarrow{D} & \hat{\mathcal{O}}/\mathcal{O} & \longrightarrow & \operatorname{coker}(D, \hat{\mathcal{O}}/\mathcal{O}) & \longrightarrow & 0 \\
 & & & & \downarrow & & \downarrow & & \downarrow & & \\
 & & & & 0 & & 0 & & 0 & &
\end{array}$$

This gives rise, as before, to an exact sequence

$$0 \to \ker(D, \mathcal{O}) \to \ker(D, \hat{\mathcal{O}}) \to \ker(D, \hat{\mathcal{O}}/\mathcal{O}) \to \operatorname{coker}(D, \mathcal{O}) \to \operatorname{coker}(D, \hat{\mathcal{O}})$$
$$\to \operatorname{coker}(D, \hat{\mathcal{O}}/\mathcal{O}) \to 0.$$

The third and sixth terms, $\ker(D, \hat{\mathcal{O}}/\mathcal{O})$ and $\operatorname{coker}(D, \hat{\mathcal{O}}/\mathcal{O})$, are the obstructions to the maps

$$\ker(D, \mathcal{O}) \to \ker(D, \hat{\mathcal{O}})$$
$$\operatorname{coker}(D, \mathcal{O}) \to \operatorname{coker}(D, \hat{\mathcal{O}})$$

being isomorphisms. For instance, if the first map were an isomorphism, then every formal power series solution f of the homogeneous equation $Df = 0$ would be automatically convergent in a neighborhood of the origin.

We can prove the following comparison theorem.

5.16.8. Theorem. *One always has*

(1) $\operatorname{coker}(D, \hat{\mathcal{O}}/\mathcal{O}) = 0$, *and*

(2) $\dim \ker(D, \hat{\mathcal{O}}/\mathcal{O}) = \left\{ \max\limits_{0 \le p \le m} (p - v(a_p)) \right\} - (m - v(a_m))$.

Proof. The meaning of assertion (1) is that every $f \in \hat{\mathcal{O}}$ has a decomposition of the form $f = Dg + h$, with $g \in \hat{\mathcal{O}}$ and $h \in \mathcal{O}$. To prove this, choose any $h \in \mathcal{O}$ (for instance a polynomial) such that $v(f - h) \ge k_0 - n$. By Lemma 5.16.7, there is $g \in \hat{\mathcal{O}}$ such that $Dg = f - h$.

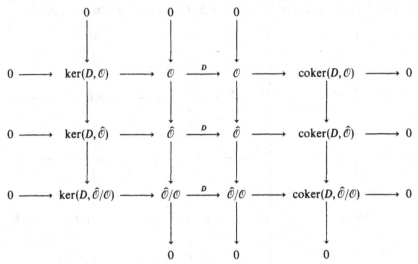

To prove assertion (2), recall that we already know that

$$\chi(D, \mathcal{O}) = m - v(a_m),$$
$$\chi(D, \hat{\mathcal{O}}) = \max_{0 \le p \le m} (p - v(a_p)).$$

From the exact sequence

$$0 \to \ker(D, \mathcal{O}) \to \ker(D, \hat{\mathcal{O}}) \to \ker(D, \hat{\mathcal{O}}/\mathcal{O}) \to \operatorname{coker}(D, \mathcal{O}) \to \operatorname{coker}(D, \hat{\mathcal{O}}) \to 0$$

we obtain

$$0 = \dim \ker(D, \mathcal{O}) - \dim \ker(D, \hat{\mathcal{O}}) + \dim \ker(D, \hat{\mathcal{O}}/\mathcal{O}) - \dim \operatorname{coker}(D, \mathcal{O}) + \dim \operatorname{coker}(D, \hat{\mathcal{O}}) = \dim \ker(D, \hat{\mathcal{O}}/\mathcal{O}) + \chi(D, \mathcal{O}) - \chi(D, \hat{\mathcal{O}}).$$

Therefore, the assertion (2) holds. □

Hence, in order that $\ker(D, \hat{\mathcal{O}}/\mathcal{O}) = 0$ and that, as a consequence, $\ker(D, \mathcal{O}) \cong \ker(D, \hat{\mathcal{O}})$ and $\operatorname{coker}(D, \mathcal{O}) \cong \operatorname{coker}(D, \hat{\mathcal{O}})$, it is necessary and sufficient that $m - v(a_m) = \max\{p - v(a_p) : 0 \le p \le m\}$. In other words, one should have, for every p ($0 \le p \le m$)

$$v(a_p) \ge v(a_m) + p - m.$$

We express this fact by means of the following definitions.

5.16.9. Definitions

(1) We say the origin is a ***regular singular point*** for D if, for every p, $0 \le p \le m$,

$$v(a_p) \ge v(a_m) + p - m$$

(2) The ***index of irregularity*** of D at the origin is the nonnegative integer

$$i(D) = \max_{0 \le p \le m} (p - v(a_p)) - (m - v(a_m)).$$

Remarks. (1) A classical example, due to Euler, of an irregular singularity is the following:

$$Df = z^2 \frac{df}{dz} - f.$$

One has $i(D) = 1$ and one observes that $D\left(\sum_{n \ge 0} n! z^{n+1}\right) = z$. Therefore $\sum_{n \ge 0} n! z^{n+1}$ is a basis of $\ker(D, \hat{\mathcal{O}}/\mathcal{O})$.

(2) One always has $\dim \ker(D, \mathcal{O}) \ge m - v(a_m)$, which is a statement of Perron giving a lower bound for the number of independent convergent power series which solve the homogeneous equation $Df = 0$.

(3) If we consider the equivalent matrix form of the operator $\frac{d^m}{dz^m} + \frac{a_{m-1}}{a_m} \frac{d^{m-1}}{dz^{m-1}} + \ldots + \frac{a_1}{a_m} \frac{d}{dz} + \frac{a_0}{a_m}$, we have the matrix

$$A = \begin{pmatrix} 0 & 1 & 0 & \dots & 0 \\ 0 & 0 & 1 & \dots & 0 \\ & & & & \vdots \\ 0 & 0 & & \dots & 1 \\ -\dfrac{a_0}{a_m} & -\dfrac{a_0}{a_m} & & & -\dfrac{a_{m-1}}{a_m} \end{pmatrix}.$$

The condition $v(a_p) \geq v(a_m) + p - m$ means that a_p/a_m has a pole at origin of order $v(a_m) - v(a_p) \leq m - p$. In particular, the matrix A has a pole of order at most m. If we compare with the discussion in the previous section we see that a regular singularity in the sense of Definition 5.16.9 is a regular Fuchsian singularity in the sense of §5.15.

We turn now to a global index theorem.

5.16.10. Theorem. *Let Ω be a connected open subset of $\mathbb{C}$ and $D = \sum_{0 \leq k \leq m} a_k \dfrac{d^k}{dz^k}$ a differential operator of order m with coefficients in $\mathscr{H}(\Omega)$. We assume that the Betti number $b_1 = \dim_{\mathbb{C}} H^1(\Omega) < \infty$ and that the number $v(a_m, \Omega)$ of zeros of a_m in Ω (counted with multiplicity) is also finite. Then $D : \mathscr{H}(\Omega) \to \mathscr{H}(\Omega)$ is an operator with index and its index is*

$$\chi(D, H(\Omega)) = m(1 - b_1) - v(a_m, \Omega).$$

Proof. Let Z be the set of zeros of a_m in Ω and let $\Omega' = \Omega \backslash Z$. Let $\operatorname{Ker} D_{\Omega'}$ denote the sheaf of solutions of D in Ω', that is, its fiber $(\operatorname{Ker} D)_a := \ker\{D : \mathcal{O}_a \to \mathcal{O}_a\}$ and $\operatorname{Ker} D_{\Omega'} = \dot{\bigcup}_{a \in \Omega'} (\operatorname{Ker} D)_a$. It is a sheaf of complex vector spaces. In Ω' we have the exact sequence of sheaves

$$0 \to \operatorname{Ker} D_{\Omega'} \to \mathcal{O} \xrightarrow{D} \mathcal{O} \to 0.$$

The existence and uniqueness theorem ensures that $\operatorname{Ker} D_{\Omega'}$ is locally isomorphic to $\mathbb{C}^m$.

5.16.11. Lemma. *Let $D = a_m \dfrac{d^m}{dz^m} + \cdots + a_0$ be a differential operator of order m with holomorphic coefficients in a connected open subset U of $\mathbb{C}$ such that $\dim_{\mathbb{C}} H^1(U) = \beta_1 < \infty$ and a_m does not vanish anywhere in U. Then the operator $D : \mathscr{H}(U) \to \mathscr{H}(U)$ has an index equal to*

$$\chi(D, \mathscr{H}(U)) = m(1 - \beta_1).$$

Proof. We will use here freely the main results and notation from the theory of cohomology with supports ([God], [Bre]). We shall denote by D_W the operator D acting in the space $\mathscr{H}(W)$.

We can assume $0 \in U$. Let U^* be the open subset of S^2 obtained from U by the inversion $z \mapsto w = \dfrac{1}{z}$. Using the same reasoning as in Theorem 3.3.1 we can find β_1 closed straight line segments in $\mathbb{C}$, $L_1, \ldots, L_{\beta_1}$, pairwise disjoint, joining a point of the boundary of the holes of U^* to a point of the exterior boundary of U^* in such a way that $U^*\backslash(L_1 \cup \cdots \cup L_{\beta_1})$ is simply connected. Furthermore, every L_j is entirely contained in U^* except for its endpoints. We can also find simply connected open sets $U_1^*, \ldots, U_{\beta_1}^*$, pairwise disjoint, such that $L_j \cap U^* \subseteq U_j^* \subseteq U^*$ and each $U_j^*\backslash L_j$ has exactly two connected components, which are also simply connected. It follows that we can find smooth Jordan arcs (in fact, arcs of circle or straight line segments) $F_1, \ldots, F_{\beta_1}$, pairwise disjoint, contained in U except for their endpoints, $U\backslash(F_1 \cup \cdots \cup F_{\beta_1})$ is a simply connected open set, and such that the corresponding simply connected open subsets U_j of U are such that $F_j \cap U \subseteq U_j$ and $U_j\backslash F_j$ has exactly two connected components which are also simply connected. From the exact sequence of sheaves of vector spaces over U

$$0 \to \operatorname{Ker} D_U \to \mathcal{O} \xrightarrow{D_U} \mathcal{O} \to 0,$$

with $\operatorname{Ker} D_U$ locally isomorphic to $\mathbb{C}^m$, we can obtain the following commutative diagram whose columns represent exact sequences of cohomology with supports in $F = \bigcup_{1 \le j \le \beta_1} F_j$.

$$\begin{array}{ccccccc}
 & & 0 & & 0 & & 0 \\
 & & \downarrow & & \downarrow & & \downarrow \\
0 & \longrightarrow & \Gamma_F(U, \operatorname{Ker} D_U) & \longrightarrow & \Gamma_F(U, \mathcal{O}) & \xrightarrow{D_U} & \Gamma_F(U, \mathcal{O}) \\
 & & \downarrow & & \downarrow & & \downarrow \\
0 & \longrightarrow & \Gamma(U, \operatorname{Ker} D_U) & \longrightarrow & \Gamma(U, \mathcal{O}) & \xrightarrow{D_U} & \Gamma(U, \mathcal{O}) \\
 & & \downarrow & & \downarrow & & \downarrow \\
0 & \longrightarrow & \Gamma(U\backslash F, \operatorname{Ker} D_U) & \longrightarrow & \Gamma(U\backslash F, \mathcal{O}) & \xrightarrow{D_U} & \Gamma(U\backslash F, \mathcal{O}) \\
 & & \downarrow & & \downarrow & & \downarrow \\
 & & H_F^1(U, \operatorname{Ker} \bar{D}_U) & \longrightarrow & H_F^1(U, \mathcal{O}) & \xrightarrow{\bar{D}_U} & H_F^1(U, \mathcal{O}) \\
 & & \downarrow & & \downarrow & & \downarrow \\
 & & H^1(U, \operatorname{Ker} \bar{D}_U) & \longrightarrow & H^1(U, \mathcal{O}) & \xrightarrow{\bar{D}_U} & H^1(U, \mathcal{O})
\end{array}$$

Theorem 3.2.2 can be stated as saying that $H^1(U, \mathcal{O}) = 0$. As usual, $\Gamma(U, \mathcal{O})$ (resp. $\Gamma_F(U, \mathcal{O})$) represents the sections in U of the sheaf $\mathcal{O}$ (resp. sections with support in F). It is clear that $\Gamma_F(U, \operatorname{Ker} D_U) = \Gamma_F(U, \mathcal{O}) = 0$, since no nonzero holomorphic function can have support in F. Introducing these facts in the previous diagram, we obtain

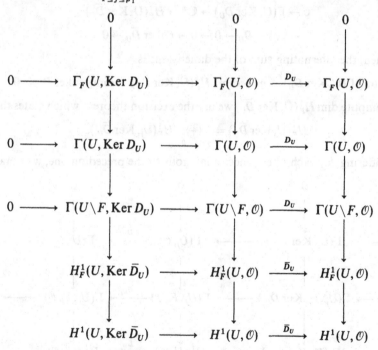

$$\begin{array}{ccccccc}
 & & 0 & & 0 & & 0 \\
 & & \downarrow & & \downarrow & & \downarrow \\
0 & \longrightarrow & \Gamma(U, \operatorname{Ker} D_U) & \longrightarrow & \Gamma(U, \mathcal{O}) & \xrightarrow{D_U} & \Gamma(U, \mathcal{O}) \\
 & & \downarrow & & \downarrow & & \downarrow \\
0 & \longrightarrow & \Gamma(U \backslash F, \operatorname{Ker} D_U) & \longrightarrow & \Gamma(U \backslash F, \mathcal{O}) & \xrightarrow{D_U} & \Gamma(U \backslash F, \mathcal{O}) \\
 & & \downarrow & & \downarrow & & \downarrow \\
 & & H_F^1(U, \operatorname{Ker} \bar{D}_U) & \longrightarrow & H_F^1(U, \mathcal{O}) & \xrightarrow{\bar{D}_U} & H_F^1(U, \mathcal{O}) \\
 & & & & \downarrow & & \downarrow \\
 & & & & 0 & & 0
\end{array}$$

Since $U \backslash F$ is simply connected, we have that $\Gamma(U \backslash F, \operatorname{Ker} D_U) \cong \mathbb{C}^m$. We can apply now the five lemma, as done several times before, and obtain the exact sequence

$$\begin{aligned} 0 \to \Gamma(U, \operatorname{Ker} D_U) \to \mathbb{C}^m \to H_F^1(U, \operatorname{Ker} \bar{D}_U) \\ \to \operatorname{coker} D_U \to 0 \to 0 \to \operatorname{coker} \bar{D}_U \to 0. \end{aligned}$$

Then, the alternating sum of the dimensions is

$$\dim_{\mathbb{C}} \Gamma(U, \operatorname{Ker} D_U) - m + \dim_{\mathbb{C}} H_F^1(U, \operatorname{Ker} \bar{D}_U) - \dim_{\mathbb{C}} \operatorname{coker} D_U = 0.$$

To compute $\dim H_F^1(U, \operatorname{Ker} \bar{D}_U)$ we use the excision theorem which states that

$$H_F^1(U, \operatorname{Ker} \bar{D}_U) = \bigoplus_{1 \le j \le \beta_1} H_F^1(U_j, \operatorname{Ker} \bar{D}_{U_j}).$$

Considering, for each j, sequences analogous to the preceding one, we obtain

$$\begin{array}{ccccccccc}
 & & 0 & & 0 & & 0 & & \\
 & & \downarrow & & \downarrow & & \downarrow & & \\
0 & \longrightarrow & \Gamma(U_j, \operatorname{Ker} D_{U_j}) & \longrightarrow & \Gamma(U_j, \mathcal{O}) & \xrightarrow{D_{U_j}} & \Gamma(U_j, \mathcal{O}) & \longrightarrow & 0 \\
 & & \downarrow & & \downarrow & & \downarrow & & \\
0 & \longrightarrow & \Gamma(U_j \backslash F_j, \operatorname{Ker} D_{U_j}) & \longrightarrow & \Gamma(U_j \backslash F_j, \mathcal{O}) & \xrightarrow{D_{U_j}} & \Gamma(U_j \backslash F_j, \mathcal{O}) & \longrightarrow & 0 \\
 & & \downarrow & & \downarrow & & \downarrow & & \\
0 & \longrightarrow & H_{F_j}^1(U_j, \operatorname{Ker} \bar{D}_{U_j}) & \longrightarrow & H_{F_j}^1(U_j, \mathcal{O}) & \xrightarrow{\bar{D}_{U_j}} & H_{F_j}^1(U_j, \mathcal{O}) & & \\
 & & & & \downarrow & & \downarrow & & \\
 & & & & 0 & & 0 & &
\end{array}$$

and the exact sequence

$$0 \to \Gamma(U_j, \operatorname{Ker} D_{U_j}) \cong \mathbb{C}^m \to \Gamma(U_j \setminus F_j, \operatorname{Ker} D_{U_j}) \cong \mathbb{C}^{2m} \to H^1_{F_j}(U_j, \operatorname{Ker} \bar{D}_{U_j}) \to 0.$$

(Recall that $U_j \setminus F_j$ has exactly two components.) Therefore,

$$\dim_{\mathbb{C}} H^1_{F_j}(U_j, \operatorname{Ker} \bar{D}_{U_j}) = m,$$
$$\dim_{\mathbb{C}} H^1_F(U, \operatorname{Ker} \bar{D}_U) = m\beta_1,$$

and

$$\chi(D, H(U) = \dim \Gamma(U, \operatorname{Ker} D_U) - \dim \operatorname{coker} D_U = m - m\beta_1 = m(1 - \beta_1).$$

Hence the statement of the lemma is true. □

Let us denote by $\zeta_1, \ldots, \zeta_M$ the points in $Z = \{z \in \Omega : a_m(z) = 0\}$. Let Δ_j be disks centered at ζ_j, pairwise disjoint, $\Delta_j \subseteq \Omega$. Let $\Delta'_j = \Delta_j \cap \Omega'$ be the corresponding punctured disks. The map

$$R : \mathscr{H}(\Omega') \to \bigoplus_{1 \le j \le M} (\mathscr{H}(\Delta'_j)/\mathscr{H}(\Delta_j))$$

which to every $u \in \mathscr{H}(\Omega')$ associates the collection of classes $(\tilde{u}_1, \ldots, \tilde{u}_M)$ of $u|\Delta'_j$ modulo $\mathscr{H}(\Delta_j)$ is a surjective map; if $u_j \in \mathscr{H}(\Delta'_j)$, $1 \le j \le M$ are given, the Mittag-Leffler theorem implies that there is a function $u \in \mathscr{H}(\Omega')$ and functions $v_j \in \mathscr{H}(\Delta_j)$ such that $u_j = u - v_j$ in Δ'_j. Hence $R(u) = (\tilde{u}_1, \ldots, \tilde{u}_M)$.

It is clear that $\ker R = \mathscr{H}(\Omega)$, hence

$$\mathscr{H}(\Omega')/\mathscr{H}(\Omega) \cong \bigoplus_{1 \le j \le M} (\mathscr{H}(\Delta'_j)/\mathscr{H}(\Delta_j))$$

via the map R.

The previous lemma implies that $\chi(D, \mathscr{H}(\Delta'_j)) = m(1 - 1) = 0$. We also have the following lemma.

5.16.12. Lemma. $\chi(D, \mathscr{H}(\Delta_j)) = m - v(a_m, \zeta_j)$.

Proof. Clearly $v(a_m, \zeta_j)$ indicates the multiplicity of ζ_j as zero of a_m. Let φ_j be the radius of Δ_j. To simplify the notation, we set $\zeta_j = 0$ and eliminate the index j whenever it is clear. By definition of the disks Δ_j, a_m does not vanish in $\Delta' = \Delta'_j = B(0, r_j) \setminus \{0\}$. For $0 < r < r_j$, consider the diagram (cf. §5.16.2)

$$\begin{array}{ccccccccc}
0 & \longrightarrow & \ker D_r & \longrightarrow & E_m(r) & \xrightarrow{D_r} & E_0(r) & \longrightarrow & \operatorname{coker} D_r & \longrightarrow & 0 \\
 & & \uparrow \rho & & \uparrow \rho & & \uparrow \rho & & \uparrow \bar{\rho} & & \\
0 & \longrightarrow & \ker D & \longrightarrow & \mathscr{H}(\Delta) & \xrightarrow{D} & \mathscr{H}(\Delta) & \xrightarrow{\pi} & \operatorname{coker} D & \longrightarrow & 0 \\
 & & \uparrow & & \uparrow & & \uparrow & & & & \\
 & & 0 & & 0 & & 0 & & & &
\end{array}$$

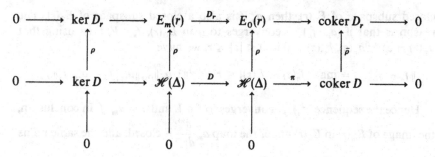

where ρ is the restriction map and $\bar{\rho}$ is the induced map. Since the first three columns are exact, we have $\dim \ker D \leq \dim \ker D_r$. Therefore, it is enough to prove the reverse inequality $\dim \ker D_r \leq \dim \ker D$ to conclude that $\rho : \ker D \to \ker D_r$ is bijective. If the last inequality were not true, let $N = \dim \ker D$ and $u_1, \ldots, u_{N+1}$ be linearly independent functions in $E_m(r)$ satisfying $D(u_j) = 0$. These solutions have an analytic continuation to $N + 1$ linearly independent solutions of $\mathscr{H}(\Delta)$ by the reasoning of §5.16.5. Hence we have the contradiction that $\dim \ker D \geq N + 1$.

We are now going to show that

$$\dim \operatorname{coker} D = \dim \operatorname{coker} D_r.$$

Let us start by showing that $\bar{\rho} : \operatorname{coker} D \to \operatorname{coker} D_r$ is injective: If $\bar{\rho}(\pi(u)) = 0$ for $u \in \mathscr{H}(\Delta)$, it means that $\rho(u) = D_r(f)$ for some $f \in E_m(r)$. As earlier, f has an analytic continuation to the whole disk Δ and $Df = u$. Hence $\pi(u) = 0$. Therefore $\dim \operatorname{coker} D \leq \dim \operatorname{coker} D_r$.

To prove the equality of the dimensions we need to know that $\operatorname{Im} D_r$, the image of D_r, is closed in $E_0(r)$ (and of finite codimension). Since D_r is a compact perturbation of $a_m \dfrac{d^m}{dz^m}$, it is enough to show that $a_m \dfrac{d^m}{dz^m} : E_m(r) \to E_0(r)$ has a closed image, with kernel and cokernel of finite dimension (cf. §5.16.5). Classical functional analysis then shows the same holds for D_r.

Let us now show that $a_m \dfrac{d^m}{dz^m}$ has a closed image. Recall that if g is a holomorphic function in $E_0(r)$, then the integral over the straight-line segment $[0, z]$

$$F(z) := \int_{[0,z]} \frac{(z - t)^{m-1}}{(m - 1)!} g(t)\, dt$$

represents the mth primitive of g, vanishing, together with its $(m - 1)$ first derivatives at $z = 0$. Suppose now that a sequence $g_n = \dfrac{d^m f_n}{dz^m}$ converges in $E_0(r)$ to a function g. The corresponding functions F_n converge in $E_m(r)$ to F. Hence, the image of $E_m(r)$ in $E_0(r)$ under the map $\dfrac{d^m}{dz^m}$ is closed. Now, if V is a closed subspace of $E_0(r)$, then $a_m V$ is also a closed subspace of $E_0(r)$. The reason is that if $(a_m \cdot f_n)_{n \geq 1}$ converges to g in $E_0(r)$, $f_n \in V$, then using that $a_m(z) = z^{v(a_m)} b_m(z)$, $b_m(z) \neq 0$ in $0 \leq |z| \leq r$, we have

$$\|f_n - f_k\|_{0,r} = \max_{|z|=r} |f_n(z) - f_k(z)| \leq r^{-v(a_m)} \|1/b_m\|_{0,r} \|a_m \cdot f_n - a_m \cdot f_k\|_{0,r}.$$

Hence the sequence $(f_n)_{n \geq 1}$ converges to $f \in V$ and $g = a_m \cdot f$. In conclusion, the image of $E_m(r)$ in $E_0(r)$ under the map $a_m \dfrac{d^m}{dz^m}$ is closed, and the same holds for $\operatorname{Im} D_r$. Moreover, D_r is an operator with index

$$\chi(D_r) = \chi\left(a_m \frac{d^m}{dz^m}\right) = m - v(a_m).$$

Note that $\rho(\mathscr{H}(\Delta))$ is dense in $E_m(r)$. This can be shown the following way. Let $f \in E_m(r)$. For $0 < s < 1$ let $f_s(z) = f(sz)$. The function f_s is holomorphic in $B(0, r/s)$, hence $f_s \in E_m(r)$. One has $f_s^{(j)}(z) = s^j f^{(j)}(sz)$. By the uniform continuity of $f^{(j)}$, $0 \le j \le m$, for every $\varepsilon > 0$ there is an $s_\varepsilon \in]0, 1[$ such that if $s_\varepsilon < s < 1$ we have

$$|f_s^{(j)}(z) - f^{(j)}(z)| \le \varepsilon \qquad (\text{for every } |z| \le r, 0 \le j \le m).$$

This implies that f_s converges to f in $E_m(r)$ when $s \to 1$. Fix now $s \in]s_\varepsilon, 1[$. The Taylor series of f_s converges uniformly in $|z| \le r$, hence there is an n_ε such that

$$\left\| f_s(z) - \sum_{n=0}^{n_\varepsilon} \alpha_n(s) z^n \right\|_{m,r} \le \varepsilon,$$

where $\|\cdot\|_{m,r}$ denotes the norm in $E_m(r)$. Altogether we have

$$\left\| f - \sum_{n=0}^{n_\varepsilon} \alpha_n(s) z^n \right\|_{m,r} \le 2\varepsilon,$$

which shows that even the polynomials are dense in $E_m(r)$.

We are now ready to return to the proof of the inequality $\dim \operatorname{coker} D_r \le \dim \operatorname{coker} D$. Let us assume the opposite. Then, with $v = \dim \operatorname{coker} D$, we have $\pi_r(u_1), \ldots, \pi_r(u_{v+1})$ in $\operatorname{coker} D_r$ which are linearly independent. Since $\operatorname{Im} D_r$ is closed and $u_j \notin \operatorname{Im} D_r$, we can find $\delta > 0$ such that the balls $B(u_j, \delta) = \{u \in E_0(r) : \|u - u_j\|_{m,r} < \delta\}$ are pairwise disjoint and do not intersect $\operatorname{Im} D_r$. The topology of $\operatorname{coker} D_r$ is Hausdorff since $\operatorname{Im} D_r$ is closed. Moreover, $\operatorname{coker} D_r$ is finite dimensional, hence its topology is precisely the usual Euclidean topology. Therefore we can also choose δ sufficiently small so that if $v_j \in B(u_j, \delta)$ are chosen arbitrarily, the system $\pi(v_1), \ldots, \pi(v_{v+1})$ is still a basis for $\operatorname{coker} D_r$. By the density of $\rho(\mathscr{H}(\Delta))$ in $E_m(r)$ we can assume $v_j \in \mathscr{H}(\Delta)$. None of these functions could be in $\operatorname{Im} D$, if not $v_j = Df_j$ and $D_r \rho(f_j) = \rho(v_j) \in \operatorname{Im} D_r \cap B(u_j, \delta)$ which is impossible. Moreover, there cannot exist constants $\lambda_1, \ldots, \lambda_{v+1}$ and $f \in \mathscr{H}(\Delta)$ such that $Df = \sum_{j=1}^{v+1} \lambda_j v_j$. If not, we would have $\sum_1^{v+1} \lambda_j \pi_r(\rho(v_j)) = 0$, which is also impossible. Therefore, the only possibility left is that $\dim \operatorname{coker} D_r = \dim \operatorname{coker} D$.

It follows that $D : \mathscr{H}(\Delta) \to \mathscr{H}(\Delta)$ has an index and

$$\chi(D, \mathscr{H}(\Delta)) = \chi(D_r) = m - v(a_m). \qquad \square$$

Let us conclude the proof of Theorem 5.16.10. Since

$$\mathscr{H}(\Omega')/\mathscr{H}(\Omega) \cong \bigoplus_{1 \le j \le m} (\mathscr{H}(\Delta_j')/\mathscr{H}(\Delta_j)),$$

we have

$$\chi(D, \mathscr{H}(\Omega')/\mathscr{H}(\Omega)) = \sum_{1 \le j \le M} \chi(D, \mathscr{H}(\Delta_j')/\mathscr{H}(\Delta_j)).$$

Consider the following diagram, whose rows and columns are exact:

$$\begin{array}{ccccccccc}
 & & 0 & & 0 & & 0 & & \\
 & & \downarrow & & \downarrow & & \downarrow & & \\
0 \to & \ker(D, \mathscr{H}(\Omega)) & \to & \mathscr{H}(\Omega) & \overset{D}{\to} & \mathscr{H}(\Omega) & \to & \operatorname{coker}(D, \mathscr{H}(\Omega)) & \to 0 \\
 & \downarrow & & & & \downarrow & & \downarrow & \\
0 \to & \ker(D, \mathscr{H}(\Omega')) & \to & \mathscr{H}(\Omega') & \overset{D}{\to} & \mathscr{H}(\Omega') & \to & \operatorname{coker}(D, \mathscr{H}(\Omega')) & \to 0 \\
 & \downarrow & & \downarrow & & \downarrow & & \downarrow & \\
0 \to & \ker(D, \mathscr{H}(\Omega')/\mathscr{H}(\Omega)) & \to & \mathscr{H}(\Omega')/\mathscr{H}(\Omega) & \overset{D}{\to} & \mathscr{H}(\Omega')/\mathscr{H}(\Omega) & \to & \operatorname{coker}(D, \mathscr{H}(\Omega')/\mathscr{H}(\Omega')) & \to 0 \\
 & & & \downarrow & & \downarrow & & \downarrow & \\
 & & & 0 & & 0 & & 0 &
\end{array}$$

It leads to the exact sequence

$$\begin{aligned}
0 &\to \ker(D, \mathscr{H}(\Omega)) \to \ker(D, \mathscr{H}(\Omega')) \to \ker(D, \mathscr{H}(\Omega')/\mathscr{H}(\Omega)) \\
&\to \operatorname{coker}(D, \mathscr{H}(\Omega)) \to \operatorname{coker}(D, \mathscr{H}(\Omega')) \to \operatorname{coker}(D, \mathscr{H}(\Omega')/\mathscr{H}(\Omega)) \to 0.
\end{aligned}$$

From the corresponding relation among dimensions

$$\begin{aligned}
0 = {} & \dim \ker(D, \mathscr{H}(\Omega)) - \dim \ker(D, \mathscr{H}(\Omega')) + \dim \ker(D, \mathscr{H}(\Omega')/\mathscr{H}(\Omega)) \\
& - \dim \operatorname{coker}(D, \mathscr{H}(\Omega)) - \dim \operatorname{coker}(D, \mathscr{H}(\Omega')) \\
& + \dim \operatorname{coker}(D, \mathscr{H}(\Omega')/\mathscr{H}(\Omega)),
\end{aligned}$$

we obtain,

$$\begin{aligned}
\chi(D, \mathscr{H}(\Omega)) &= \chi(D, \mathscr{H}(\Omega')) - \chi(D, \mathscr{H}(\Omega')/\mathscr{H}(\Omega)) \\
&= \chi(D, \mathscr{H}(\Omega')) - \sum_{1 \le j \le M} \chi(D, \mathscr{H}(\Delta_j')/\mathscr{H}(\Delta_j)) \\
&= m(1 - \beta_1) - \sum_{1 \le j \le M} (\chi(D, \mathscr{H}(\Delta_j')) - \chi(D, \mathscr{H}(\Delta_j)) \\
&= m(1 - \beta_1) + \sum_{1 \le j \le M} (m - v(a_m, \zeta_j)) \\
&= m(1 - \beta_1 + M) - v(a_m, \Omega).
\end{aligned}$$

Now $b_1 = \dim H^1(\Omega)$ is the number of holes of Ω, and $\beta_1 = \dim H^1(\Omega')$ is the number of holes of Ω'. It is clear that $b_1 = \beta_1 + M$. Hence

$$\chi(D, \mathscr{H}(\Omega)) = m(1 - b_1) - v(a_m, \Omega)$$

as asserted by the theorem. □

5.16.13. Example. Let $D = z^2\frac{d^2}{dz^2} + z\frac{d}{dz} + \left(z^2 - \frac{1}{4}\right)$ the operator in $\Omega = \mathbb{C}^*$ defining the Bessel functions of order 1/2. Since we have two independent solutions $J_{1/2}(z) = \frac{\cos z}{z^{1/2}}$ and $J_{-1/2}(z) = \frac{\sin z}{z^{1/2}}$, which are not single-valued, and no linear combination of them is single-valued either, the space of holomorphic solutions of $Df = 0$ in $\mathbb{C}^*$ must be $\Gamma(\mathbb{C}^*, \ker D) = \{0\}$. The index of the operator is, in this case, zero since $b_1 = 1$ and $v(a_2, \mathbb{C}^*) = v(z^2, \mathbb{C}^*) = 0$. One can conclude that $D: \mathscr{H}(\mathbb{C}^*) \to \mathscr{H}(\mathbb{C}^*)$ is surjective. One can verify this fact directly by the method of variation of parameters as follows.

The Wronskian $w(z)$ of $J_{1/2}$ and $J_{-1/2}$ is z^{-1}. The Green-Lagrange formula for a particular solution of the equation $Df = g$ gives

$$f(z) = -\frac{\cos z}{\sqrt{z}}\int^z \frac{\sin t}{t\sqrt{t}}g(t)\,dt + \frac{\sin z}{\sqrt{z}}\int^z \frac{\cos t}{t\sqrt{t}}g(t)\,dt,$$

where $\int^z$ designates a primitive of the integrand. We want to show that f is a single-valued function in $\mathbb{C}^*$. Change variables $t = s^2$, then

$$G_1(z) = \int^z \frac{\sin t}{t\sqrt{t}}g(t)\,dt = 2\int^{\sqrt{z}} \frac{\sin(s^2)}{s^2}g(s^2)\,ds,$$

$$G_2(z) = \int^z \frac{\cos t}{t\sqrt{t}}g(t)\,dt = 2\int^{\sqrt{z}} \frac{\cos(s^2)}{s^2}g(s^2)\,ds,$$

and both integrands are Laurent series containing only even powers of s. Therefore $G_1(z) = \sqrt{z}H_1(z)$, $G_2(z) = \sqrt{z}H_2(z)$, where H_1, H_2 are single-valued holomorphic functions in $\mathbb{C}^*$. It follows that

$$f(z) = -\cos z H_1(z) + \sin z H_2(z),$$

which is single-valued, holomorphic in $\mathbb{C}^*$, and solves $Df = g$. In other words, we have shown directly that $D: \mathscr{H}(\mathbb{C}^*) \to \mathscr{H}(\mathbb{C}^*)$ is surjective.

Exercises 5.16

1. Let $\Omega = \mathbb{C}^*$, $D = \frac{d^2}{dz^2} + \frac{1}{z}\frac{d}{dz} + \left(1 - \frac{p^2}{z^2}\right)$ the Bessel operator. For which values of the parameter p is $\dim\operatorname{coker} D = 0, 1, 2$?

2. Let $\Omega = \mathbb{C}$, $D = z^2\frac{d^2}{dz^2} - (4z + \lambda z^2)\frac{d}{dz} + (u - \kappa z)$. For which values of κ do we have $\dim\ker D = 2$? (Hint: Use Frobenius method from the previous chapter. Ans: $\kappa = -\lambda$, $\kappa = -2\lambda$, and $\kappa = -3\lambda$.)

3. Let $\Omega = \mathbb{C}$, find $\dim\ker D$ and $\dim\operatorname{coker} D$ when $D = z^2\frac{d^2}{dz^2} + z^2\frac{d}{dz} - 2$.

Notes to Chapter 5

The four subjects treated in this chapter, analytic continuation, Dirichlet series, Riemann surfaces, and differential equations in the complex domain, have long and separate histories, but their freshness and interest as sources of research problems and applications remain unabated. Moreover, as we have tried to convey, they are quite related to each other. We take the opportunity here to mention a few of these relationships and suggest further subjects of study.

1. The classical theory of Dirichlet series, their analytic continuation, and location of singularities, is richer in the case where the exponents $\lambda_n \approx n$, i.e., then the Dirichlet series is approximately a power series (see e.g., [Di], [Man], [Ber], and [Po]). This is an unnecessary restriction, as we shall show in the following volume, where we explore the relations between singularities, analytic continuation, functional equations, and interpolation.

2. Trying to give the reader a definite direction to proceed from the introduction to the theory of Riemann surfaces we gave in this chapter is almost impossible; there are too many choices. Nevertheless, we mention a few topics we feel the reader should be able to understand with the background provided by this book: (i) a deeper study of Riemann surfaces [Fo], [FK], and [Gu]; (ii) Teichmüller spaces [Leh]; (iii) algebraic curves as an introduction to algebraic geometry [Gri2], [Cl], and [GH1]; (iv) differential geometry of complex manifolds [We] and [Na2]; (v) discrete groups [Bea] and [Mas]; and (vi) the study of meromorphic functions in complex tori, i.e., elliptic functions and their relation to number theory [Lang]. We will come back to this last item in the next volume.

3. The Schwarz-Christoffel transformations that we studied in Chapter 2 can be understood as part of the general fact that a branched covering, e.g., $\pi : B(0, 1) \to S^2$ leads to a differential equation $w'' + \varphi w = 0$, written in terms of the Schwarzian derivative $\{\pi, z\}$, $\varphi = \frac{1}{2}\{\pi, z\}$. The local sections of π are quotients of a pair of independent solutions of the equation (see [Hi2], chapter 10).

4. The study of differential equations with singular points leads naturally to quite a few subjects that we have barely mentioned in the text. For asymptotic developments of solutions there is the excellent textbook [Ol]. There is a myriad of reasons to study special functions, especially hypergeometric functions and their properties. Two very different approaches are [Vil] and [Fi]. In the next volume we shall consider hyperfunctions and infinite-order differential operators, as well as the relations between asymptotic developments and the analytic continuation of solutions of differential equations. The concept of resurgence, which has been developed in the last few years by [Ec] and others, arises also naturally when considering the analytic continuation of solutions of differential equations and functional equations. Finally, let us mention that differential equations in the complex domain also have very interesting relations with Nevanlinna theory [Hi2], [Ban], and [Ro] and with dynamical systems [Ar].

5. The proofs of the index theorems should convince the reader of the value of sheaf theory and homological algebra in complex analysis. Further illustrations appear in the theory of several complex variables, especially in the study of analytic varieties (i.e., complex manifolds with singularities, see [Ho1]).

6. We are also confident that at the end of this volume the reader can turn to complex analysis in several variables and find out by himself the close relation between this subject and what we presented here.

References

The books [He1], [Hen], and especially [Bu] contain extensive bibliographies. Here we only list those books and manuscripts we have consulted while preparing the present volume or suggested for further reading in the text.

[Ab] M. Abate: *Iteration Theory of Holomorphic Maps on Taut Manifolds.* Mediterranean Press, 1989.

[Ah1] L. V. Ahlfors: *Complex Analysis.* McGraw-Hill, 1953.

[Ah2] L. V. Ahlfors: *Conformal Invariants.* McGraw-Hill, 1973.

[Ah3] L. V. Ahlfors: *Lectures on Quasiconformal Mappings.* Van Nostrand, 1966.

[AS] L. V. Ahlfors and L. Sario: *Riemann Surfaces.* Princeton University Press, 1960.

[ACG] E. Arbarello, M. Cornalba, and P. A. Griffiths: *Topics in the Theory of Algebraic Curves.* Springer-Verlag, 1985.

[Arm] M. A. Armstrong: *Basic Topology.* Springer-Verlag, 1983.

[Av] V. Avanissian: Fonctions p.s.h., différence de deux fonctions p.s.h. de type exponentiel. Comptes Rendus Acad. Sci. Paris 225 (1961), 499–500.

[Ba] A. Baernstein II, et al., ed: *The Bieberbach Conjecture.* Amer. Math. Soc., 1986.

[BGHT] D. Barrett, R. Gail, S. Hantler, and B. A. Taylor: Varieties in a two-dimensional polydisk with univalent projection at the boundary. IBM Research Report, RC 15848 (1990).

[Bea] A. Beardon: *The Geometry of Discrete Groups.* Springer-Verlag, 1983.

[BK] S. Bell and S. Krantz: Smoothness to the boundary of conformal maps. Rocky Mountain J. Math. 17 (1987), 23–40.

[BL] S. Bell and L. Lémpert: A C^∞ Schwarz reflection principle in one and several variables. J. Diff. Geom. (in press).

[BG] C. A. Berenstein and R. Gay: A local version of the 2-circles theorem. Israel J. Math. 55 (1986), 267–288.

[BT] C. A. Berenstein and B. A. Taylor: A new look at interpolation theory for entire functions of one variable. Advances in Math. 3 (1979), 109–143.

[BH] C. A. Berenstein and D. Hamilton: Et la conjecture de Bieberbach devint le théorème de Louis de Branges. La Recherche 16 (1985), 691–693.

[Ber] V. Bernstein: *Séries de Dirichlet.* Gauthier-Villars, 1933.

[Be] D. Bertrand: Travaux récents sur les points singuliers des équations différentielles linéaires. *Sem. Bourbaki* 31 (1978/79), 538.01–538.16, Springer-Verlag, 1980.

[Bie] L. Bieberbach: *Conformal Mappings.* Chelsea Publishing, 1953.

[BJ] C. J. Bishop and P. W. Jones: Harmonic measures and arc length. Ann. Math. (in press).
[BR1] C. Blair and L. A. Rubel: A universal entire function. Amer. Math. Monthly 90 (1983), 331–332.
[BR2] C. Blair and L. A. Rubel: A triply universal entire function. Enseign. Math. 30 (1984), 269–274.
[Bl] P. Blanchard: Complex dynamics on the Riemann sphere. Bull. Amer. Math. Soc. 11 (1984), 85–111.
[Bo] H. P. Boas and R. P. Boas: Short proofs of three theorems of harmonic functions. Proc. Amer. Math. Soc. 102 (1988), 906–908.
[De B] L. de Branges: A proof of the Bieberbach conjecture. Acta Math. 154 (1985), 137–152.
[Bre] G. E. Bredon: *Sheaf Theory*. McGraw-Hill, 1967.
[Brem] H. Bremerman: *Distributions, Complex Variables and Fourier Transform*. Addison-Wesley, 1965.
[Bu] R. B. Burckel: *An Introduction to Classical Complex Analysis*. Academic Press, 1979.
[Cam] R. Campbell: *Les Intégrales Euleriennes et Leurs Applications*. Dunod, 1966.
[Car] C. Caratheodory: *Conformal Representation*. Cambridge University Press, 1958.
[Carle] T. Carleman: Zur Theorie der Minimalflächen. Math. Z. 9 (1921), 154–160.
[Carl] L. Carleson: *Selected Problems on Exceptional Sets*. Van Nostrand, 1967.
[Ca] H. Cartan: *Differential Forms*. Hermann, 1970.
[Ch] P. Charpentier: Formules explicites pour les solutions minimales de l'équation $\bar{\partial} u = f$ dans la boule et dans le polydisque de $\mathbb{C}^n$. Ann. Inst. Fourier 30 (1980), 125–154.
[Cl] C. H. Clemens: *A Scrapbook of Complex Curve Theory*. Plenum Press, 1980.
[CoL] E. A. Coddington and N. Levinson: *Theory of Ordinary Differential Equations*. McGraw-Hill, 1955.
[CL] E. F. Collingwood and A. J. Lohwater: *The Theory of Cluster Sets*. Cambridge University Press, 1966.
[Co] J. B. Conway: *Functions of One Complex Variable*. Springer-Verlag, 1986.
[De] P. Deligne: *Equations Différentielles à Points Singuliers Reguliers*. Springer-Verlag, 1970.
[Dem] J.-P. Demailly: Fonctions holomorphes à croissance polynomiale sur la surface de equation $e^x + e^y = 1$. Bull. Soc. Math. France 103 (1979), 179–191.
[Dev] R. L. Devaney and L. Keen, ed: *Chaos and Fractals*. Amer. Math. Soc., 1989.
[Di] P. Dienes: *The Taylor Series*. Dover, 1957.
[Die] J. Dieudonné: *Calcul Infinitésimal*. Hermann, 1980.
[Du1] P. L. Duren: *Univalent Functions*. Springer-Verlag, 1983.
[Du2] P. L. Duren: *Theory of H^p Spaces*. Academic Press, 1970.
[E] J. Ecalle: *Les Fonctions Resurgents*. Publ. University Orsay, 1971.
[Ev] M. Evgrafov: *Recueil de Problemes sur la Théorie des Fonctions Analytiques*. Mir, 1974.
[FK] H. Farkas and I. Kra: *Riemann Surfaces*. Springer-Verlag, 1980.
[Fi] N. J. Fine: *Basic Hypergeometric Series and Applications*. Amer. Math. Soc., 1988.
[Fo] O. Forster: *Lectures on Riemann Surfaces*. Springer-Verlag, 1981.
[Fu1] W. H. J. Fuchs: *Théorie de l'Approximation des Fonctions d'Une Variable Complexe*. Presses de l'Univ. de Montreal, 1968.
[Fu2] W. H. J. Fuchs: *The Theory of Functions of One Complex Variable*. Van Nostrand, 1967.
[Gab1] R. M. Gabriel: An inequality concerning integrals of positive subharmonic functions along certain circles. J. London Math. Soc. 5 (1930), 129–131.
[Gab2] R. M. Gabriel: Some results concerning integrals of moduli of regular functions along curves of certain types. Proc. London Math. Soc. 28 (1926), 121–127.

[Ga1] D. Gaier: *Lectures on Complex Approximation.* Birkhauser, 1987.
[Ga2] D. Gaier: *Konstruktive Methoden der konformen Abbildung.* Springer-Verlag.
[Gak] F. D. Gakhov: *Boundary Value Problems.* Pergamon, 1966.
[Gam] T. W. Gamelin: *Uniform Algebras.* Prentice Hall, 1969.
[GK] T. W. Gamelin and D. Khavinson: The isoperimetric inequality and rational approximation. Amer. Math. Monthly 96 (1989), 18–30.
[Gay] R. Gay: Thése 3e. cycle. Strasbourg, 1970.
[GS] I. M. Gelfand and G. E. Shilov: *Generalized Functions.* Academic Press, 1968.
[Ge] A. O. Gelfond: *The Calculus of Finite Differences* (in Russian). GITTL, 1952.
[Go] C. Godbillon: *Elements de Topologie Algébrique.* Hermann, 1971.
[God] R. Godement: *Topologie Algébrique et Théorie des Faisceaux.* Hermann, 1958.
[Gre] M. J. Greenberg: *Algebraic Topology, a First Course.* Harper, 1981.
[Gri1] P. A. Griffiths: *Entire Holomorphic Mappings in One and Several Complex Variables.* Princeton University Press, 1976.
[Gri2] P. A. Griffiths: *Introduction to Algebraic Curves.* Amer. Math. Soc., 1989.
[GH] P. A. Griffiths and J. Harris: *Principles of Algebraic Geometry.* John Wiley & Sons, 1978.
[Gu] R. C. Gunning: *Lectures on Riemann Surfaces.* Princeton University Press, 1966.
[HR] G. H. Hardy and M. Riesz: *The General Theory of Dirichlet Series.* Cambridge University Press, 1915.
[Har] P. Hartman: *Ordinary Differential Equations.* John Wiley & Sons, 1964.
[Ha] W. K. Hayman: *Meromorphic Functions.* Oxford University Press, 1964.
[HK] W. K. Hayman and P. B. Kennedy: *Subharmonic Functions.* Academic Press, 1976.
[Hed] L. I. Hedberg: Approximation in the mean by analytic functions. Trans. Amer. Math. Soc. 163 (1972), 157–171.
[He1] M. Heins: *Complex Function Theory.* Academic Press, 1968.
[He2] M. Heins: *Selected Topics in the Classical Theory of Functions of a Complex Variable.* Holt, Rinehart and Winston, 1962.
[Hel] L. Helms: *Introduction to Potential Theory.* John Wiley & Sons, 1969.
[Hem] J. A. Hempel: Precise bounds in the theorems of Schottky and Picard. S. London Math. Soc. 21 (1980), 279–286.
[Hen] P. Henrici: *Applied and Computational Complex Analysis.* John Wiley & Sons, 1974.
[HS] E. Hewitt and K. Stromberg: *Analytic Function Theory.* Springer-Verlag, 1965.
[Hi1] E. Hille: *Analytic Function Theory.* Chelsea, 1959.
[Hi2] E. Hille: *Ordinary Differential Equations in the Complex Domain.* John Wiley & Sons, 1964.
[Ho1] L. Hörmander: *An Introduction to Complex Analysis in Several Variables.* North-Holland, 1973.
[Ho2] L. Hörmander: *The Analysis of Linear Partial Differential Operators.* Springer-Verlag, 1983.
[Hor] J. Horvath: *Topological Vector Spaces and Distributions.* Addison-Wesley, 1966.
[HC] A. Hurwitz and R. Courant: *Funktionentheorie.* Springer-Verlag, 1964.
[In] E. L. Ince: *Ordinary Differential Equations.* Dover, 1956.
[JS] G. A. Jones and D. Singerman: *Complex Functions.* Cambridge University Press, 1987.
[Jo] C. Jordan: *Cours d'Analyse.* Gauthier-Villars, 1893–6.
[Ju] W. B. Jurkat: *Meromorphe Differentialgleichungen.* Springer-Verlag, 1978.
[Ka] T. Kato: *Perturbation Theory for Linear Operators.* Springer-Verlag, 1966.
[Kn] K. Knopp: *Theory and Applications of Infinite Series.* Blackie & Son, 1961.
[Kob] H. Kober: *Dictionary of Conformal Representations.* Dover, 1957.
[Ko] P. Koosis: *Introduction to H^p Spaces.* Cambridge University Press, 1980.
[Kr] J. Krzyz: *Problems in Complex Function Theory.* American Elsevier, 1971.

[La] N. S. Landkof: *Foundation of Modern Potential Theory*. Springer-Verlag, 1972.
[Leb] N. N. Lebedev: *Special Functions and Their Applications*. Dover, 1972.
[Le] O. Lehto: *Univalent Functions and Teichmüller Spaces*. Springer-Verlag, 1987.
[LG] P. Lelong and L. Gruman: *Entire Functions of Several Complex Variables*. Springer-Verlag, 1986.
[Leo] A. F. Leont'ev: *Exponential Series* (in Russian). Nauka, 1976.
[Lev] B. J. Levin: *Distribution of Zeros of Entire Functions*. Amer. Math. Soc., 1964.
[LR] D. H. Luecking and L. A. Rubel: *Complex Analysis. A Functional Analysis Approach*. Springer-Verlag, 1984.
[MOS] W. Magnus, F. Oberhettinger and R. P. Soni: *Formulas and Theorems for the Special Functions of Mathematical Physics*. Springer-Verlag, 1966.
[Mal] B. Malgrange: Sur les points singuliers des équations différentielles. Enseign. Math. 20 (1974), 147–176.
[Man] S. Mandelbrojt: *Dirichlet Series*. Reidel, 1972.
[Mar] A. I. Markushevich: *Theory of Functions of a Complex Variable*. Chelsea, 1977.
[Mas] B. Maskit: *Kleinian Groups*. Springer-Verlag, 1987.
[Me] S. N. Mergelyan: Uniform approximation to functions of a complex variable. Amer. Math. Soc. Transl. 3 (1954), 294–391.
[MK] D. S. Mitrinović and J. D. Kečkić: *The Cauchy Method of Residues*. Reidel, 1984.
[Mo] S. Momm: Lower bounds for the modulus of analytic functions. Bull. London Math. Soc. (in press).
[Na1] R. Narasimhan: *Analysis on Real and Complex Manifolds*. Masson, 1973.
[Na2] R. Narasimhan: *Complex Analysis in One Variable*. Birkhäuser, 1985.
[Neh] Z. Nehari: *Conformal Mapping*. Dover, 1975.
[Nev] R. Nevanlinna: *Analytic Functions*. Springer-Verlag, 1970.
[Ol] F. W. Olver: *Asymptotics and Special Functions*. Academic Press, 1972.
[Pa] S. J. Patterson: *An Introduction to the Theory of the Riemann Zeta Function*. Cambridge University Press, 1978.
[Po] G. Polya: *Collected Papers, Volume 1*. MIT Press, 1974.
[PS] G. Polya and G. Szegö: *Problems and Theorems in Analysis*, 2 volumes. Springer-Verlag, 1983.
[Pom] C. Pommerenke: *Univalent Functions*. Vanderhoeck & Ruprecht, 1975.
[Ra] T. Radó: *Subharmonic Functions*. Chelsea, 1949.
[RR] J.-P. Rosay and W. Rudin: Arakelian's approximation theorem. Amer. Math. Monthly 96 (1989), 432–434.
[Rub1] L. A. Rubel: Four counterexamples to Bloch's principle. Proc. Amer. Math. Soc. 98 (1986), 257–260.
[Rub2] L. A. Rubel: Some research problems in algebraic differential equations. Trans. Amer. Math. Soc. 280 (1983), 43–52.
[Ru] W. Rudin: *Real and Complex Analysis*. McGraw-Hill, 1966.
[SZ] S. Saks and A. Zygmund: *Analytic Functions*. Elsevier, 1971.
[Sa] G. Sansone and J. Gerretsen: *Lectures on the Theory of Functions of a Complex Variable*. Noordhoff, 1960.
[Sch] H. H. Schaeffer: *Topological Vector Spaces*. Springer-Verlag, 1970.
[Schm] J. Schmets: *Equations Différentieles et Espaces d'Hyperfonctions*. Université de Liege, 1981.
[S] L. Schwartz: *Théorie des Distributions*. Hermann, 1966.
[Se] R. E. Seeley: Extension of C^∞ functions defined in a half-space. Proc. Amer. Math. Soc. 15 (1964), 625–626.
[Shi] A. Shields: On fixed points of commuting analytic functions. Proc. Amer. Math. Soc. 15 (1964), 703–706.
[Sh] B. Shiffman: Introduction to Carlson-Griffiths equidistribution theory, in *Value Distribution Theory*, I. Laine and S. Richman, eds. Springer-Verlag, 1983.
[Si] C. L. Siegel: *Topics in Complex Function Theory*. John Wiley & Sons, 1969.

[Sp] M. Spivak: *Calculus on Manifolds*. Benjamin, 1965.

[Ti1] E. C. Titchmarsh: *The Theory of Functions*. Oxford University Press, 1939.

[Ti2] E. C. Titchmarsh: *The Theory of the Riemann Zeta Function*. Oxford University Press, 1951.

[Tv] H. Tverberg: A proof of the Jordan curve theorem. Bull. London Math. Soc. 12 (1980), 34–38.

[Va1] G. Valiron: *Fonctions Entières et Fonctions Méromorphes d'Une Variable*. Gauthier-Villars, 1925.

[Va2] G. Valiron: *Fonctions Analytiques*. Presses Université de France, 1954.

[Vil] N. J. Vilenkin: *Fonctions Espéciales et Théorie de la Representation des Groupes*. Dunod, 1969.

[Vi] A. G. Vitushkin: Analytic capacity of sets in problems of approximation theory. Russian Math. Surveys 22 (1967), 135–200.

[Wa] J. L. Walsh: *Interpolation and Approximation by Rational Functions in the Complex Domain*. Amer. Math. Soc., 1969.

[Was] W. Wasow: *Linear Turning Point Theory*. Springer-Verlag, 1985.

[We] R. O. Wells, Jr: *Differential Analysis on Complex Manifolds*. Prentice Hall, 1973.

[Wer] J. Werner: *Banach Algebras and Several Complex Variables*. Springer-Verlag, 1976.

[WW] E. T. Whittaker and G. N. Watson: *A Course on Modern Analysis*. Cambridge University Press, 1963.

[Za1] L. Zalcman: *Analytic Capacity and Rational Approximation*. Springer-Verlag, 1968.

[Za2] L. Zalcman: A heuristic principle in complex function theory. Amer. Math. Monthly 82 (1975), 813–817.

[Za3] L. Zalcman: Picard's theorem without tears. Amer. Math. Monthly 85 (1978), 265–268.

[Za4] L. Zalcman: Real proof of complex theorems (and vice versa). Amer. Math. Monthly 81 (1974), 115–137.

[Za5] L. Zalcman: Offbeat integral geometry. Amer. Math. Monthly 87 (1980), 161–175.

[Za6] L. Zalcman: Uniqueness and non-uniqueness for the Radon transform. Bull. London Math. Soc. 14 (1982), 241–245.

Notation and Selected Terminology

$:=$	This symbol is used to indicate that the left hand side is defined by the right hand side.
$\equiv$	We use this occasionally to indicate equality throughout a set.
$\#(A)$	Denotes the number of points of the set A. Sometimes we write simply $\# A$.
$A \times B := \{(a, b) : a \in A, b \in B\}$	Cartesian product of two sets A and B.
$\prod_{i \in I} A_i$; pr_i	Cartesian product of a family A_i of sets indexed by $i \in I$; projection onto the ith coordinate.
$\mathbb{N} := \{0, 1, 2, \ldots\}$	Set of non-negative integers.
$\mathbb{N}^* := \mathbb{N} \backslash \{0\}$	Set of natural numbers.
$\mathbb{Z}$	Set of all integers.
$\mathbb{Q}$	Set of all rational numbers.
$\mathbb{R}$	Set of all real numbers.
$\mathbb{R}^* := \mathbb{R} \backslash \{0\}$	Nonzero real numbers.
$\mathbb{C}$	Set of all complex numbers.
$\mathbb{C}^* := \mathbb{C} \backslash \{0\}$	Nonzero complex numbers.
$(z_n)_{n \geq 1}$	Sequence indexed by $\mathbb{N}^*$.
$\log^+ x := \max(\log x, 0)$	For $x > 0$.
$\dot{\bigcup}_{i \in I} A_i$	Disjoint union of a collection of sets. If all $A_i \subseteq X$, it means that $A_i \cap A_j = \varnothing$ whenever $i \neq j$ and $\dot{\bigcup}_{i \in I} A_i = \bigcup_{i \in I} A_i$.
$A + B := \{a + b : a \in A, b \in B\}$	We use this when A, B are subset of $\mathbb{C}$. It makes sense in any additive group.

$]a,b[:= \{x \in \mathbb{R} : a < x < b\}$	Open interval in the real line with end-points a, b. We allow $a = -\infty$ or $b = \infty$.
$[a,b] := \{x \in \mathbb{R} : a \leq x \leq b\}$	Closed interval in $\mathbb{R}$. Here $-\infty < a \leq b < \infty$.
$]a,b] := \{x \in \mathbb{R} : a < x < b\}$ $[a,b[:= \{x \in \mathbb{R} : a \leq x < b\}$	Semi-closed intervals in $\mathbb{R}$.
$\operatorname{Re} z$, $\operatorname{Im} z$	Denotes the real and imaginary parts of the complex number z.
$\bar{z}$, $\lvert z\rvert$	Denotes the conjugate (resp. absolute value) of $z \in \mathbb{C}$.
$B(z_0,r) := \{z \in \mathbb{C} : \lvert z - z_0\rvert < r\}$ $\bar{B}(z_0,r) := \{z \in \mathbb{C} : \lvert z - z_0\rvert \leq r\}$	The open (resp. closed) disk of center z_0 and radius r, $0 < r < \infty$. (Generally in the complex plane. In chapter 1 we use the same notation in $\mathbb{R}^n$).
	For $A \subseteq \mathbb{C}$ these symbols denote:
$\mathring{A}$	Interior of A
$\bar{A}$	Closure of A
$\partial A := \bar{A} \backslash A$	Boundary of A
$A^c := \mathbb{C} \backslash A$	Complement of A
	The symbol $\bar{A}$ is also used to denote closure when A is a subset of some topological space. In every instance this more general meaning is used, it will be clear from the context.
$S^2 := \{(x,y,z) \in \mathbb{R}^3 : x^2 + y^2 + z^2 = 1\}$	It is the unit sphere in $\mathbb{R}^3$. We identify it to $\mathbb{C} \cup \{\infty\}$ via stereographic projection.
$\partial_\infty A$	When $A \subseteq \mathbb{C}$ is considered as a subset of S^2, this indicates the boundary of A relative to S^2 (Chapter 4).
$\partial_\Omega A$	Also in Chapter 4, when $A \subseteq \Omega \subseteq \mathbb{C}$, this indicates the relative boundary of A in Ω, i.e., $(\bar{A} \cap \Omega) \backslash \Omega$.
Exterior boundary of Ω	Boundary of unbounded component of Ω^c.
A discrete in Ω	This means that A has no acumulation points in Ω.
Domain	An open connected subset of $\mathbb{C}$.
$d(z,A) := \inf\{\lvert z - a\rvert : a \in A\}$	Distance from the point z to the set A.
$d(A,B) := \inf\{\lvert a - b\rvert : a \in A, b \in B\}$	Distance between two sets A and B.
$V(K,\varepsilon) := \{z \in \mathbb{C} : d(z,K) < \varepsilon\}$	ε-neighborhood of the set K.
$cv(A)$	Convex hull of a subset A of $\mathbb{C}$, i.e., the smallest convex set containing A.

$A \subset\subset B$	A is relatively compact in B, i.e., $\bar{A}$ is compact, $\bar{A} \subseteq \mathring{B}$.
Exhaustion $(K_n)_{n\geq 1}$ of Ω	A collection of sets such that $K_n \subset\subset K_{n+1}$, $\bigcup_{n\geq 1} K_n = \Omega$.
	When Γ is a Jordan curve:
$\mathrm{Int}(\Gamma)$	Interior of Γ, i.e., bounded component of Γ^c.
$\mathrm{Ext}(\Gamma)$	Exterior of Γ, i.e., unbounded component of Γ^c.
$f: A \to B$	Denotes a function with domain A and values in B.
$\mathrm{Im}(f) := \{f(a) : a \in A\}$	When f is a function as above, the image of f.
$f \mid C$	The restriction of f to a subset C of A.
$f: a \mapsto b$	Denotes the specific assignment given by the function f. Sometimes this symbol is used to define the function f in terms of a concrete formula.
$f(z) = O(g(z))$ when $z \to a$	For numerically valued functions, it means that there is a constant C such that $\|f(z)\| \leq C\|g(z)\|$ for all values of z near a.
$f(z) = o(g(z))$ when $z \to a$	It means that $\lim_{z\to a} \frac{f(z)}{g(z)} = 0$.
$r \gg 1$	It means r sufficiently large, i.e., $r > r_0$ for some value r_0.
f is u.s.c. in Ω	It means $f: \Omega \to [-\infty, \infty[$ is upper semicontinuous.
f is l.s.c. in Ω	It means $f: \Omega \to [-\infty, \infty]$ is lower semicontinuous.
f^*, f_*	Upper (resp. lower) regularized of a real valued function, defined in 4.4.38, only used in Chapter 4.
$\mathscr{C}(A, B)$	Set of all continuous maps from A into B.
$\mathscr{C}(\Omega) := \mathscr{C}(\Omega, \mathbb{C})$	Set of all complex valued continuous maps in Ω.
$\mathscr{C}_{\mathbb{R}}(\Omega)$	Set of all real valued continuous maps in Ω.
f is of class $C^k(\Omega)$	For $1 \leq k \leq \infty$, it means that f has k continuous partial derivatives in an open set Ω. For $k = 0$, it means f is continuous.

$f \in C^k(\bar{\Omega})$	It means that there is an open set $V \supseteq \bar{\Omega}$, $f \in C^k(V)$, V depends on f.
$\mathscr{E}_k^i(\Omega)$, $\mathscr{E}_k^{1,0}(\Omega)$, $\mathscr{E}_k^{0,1}(\Omega)$	Spaces of functions and differential forms with coefficients in $C^k(\Omega)$, defined in §1.2. When $k = \infty$, the index k is omitted.
$f^*\omega$	Pullback (or inverse image) of a form ω by a map f, see §1.2.
$f^*g = g \circ f$	Composition of two functions.
$\operatorname{supp}(\mu)$	Denotes the support of a function, measure, or distribution.
$\mathscr{D}(\Omega)$	Denotes the space of functions f of class C^∞ in the open set Ω such that $\operatorname{supp}(f) \subset\subset \Omega$.
$\mathscr{D}_k^i(\Omega)$	Spaces of functions and differential forms in Ω with coefficients in $\mathscr{D}(\Omega)$, defined in §1.5.
d	Exterior differential, see §1.2.
$\wedge$	Wedge product, see §1.2
$\dfrac{\partial}{\partial z} := \dfrac{1}{2}\left(\dfrac{\partial}{\partial x} - i\dfrac{\partial}{\partial y}\right)$	Cauchy-Riemann operator, see §1.2.
$\dfrac{\partial}{\partial \bar{z}} := \dfrac{1}{2}\left(\dfrac{\partial}{\partial x} + i\dfrac{\partial}{\partial y}\right)$	
$\partial, \bar{\partial}$	Complex differentials, see §1.2.
$\lVert x \rVert := \sqrt{x_1^2 + \cdots + x_n^2}$	Euclidean length of a vector $x = (x_1, \ldots, x_n) \in \mathbb{R}^n$.
$(x \mid y) := x_1 y_1 + \cdots + x_n y_n$	Scalar product of two vectors in $\mathbb{R}^n$.
$\operatorname{grad} \varphi := \left(\dfrac{\partial \varphi}{\partial x_1}, \ldots, \dfrac{\partial \varphi}{\partial x_n}\right)$	Gradient of a function of n variables. Also denoted $\nabla\varphi$.
$\dfrac{\partial \varphi}{\partial n} := (\nabla\varphi \mid n)$	Normal derivative. Here n denotes the exterior unit normal on a boundary $\partial\Omega$.
$\operatorname{div} F := \dfrac{\partial F_1}{\partial x_1} + \ldots + \dfrac{\partial F_n}{\partial x_n}$	Divergence of a vector field $F = (F_1, \ldots, F_n)$.
$\Delta := \dfrac{\partial^2}{\partial x^2} + \dfrac{\partial^2}{\partial y^2}$	Laplace operator in $\mathbb{R}^2$.
dm	Lebesgue measure in $\mathbb{R}^2$. In Chapter 4 we use m_1 to denote Lebesgue measure in $\mathbb{R}$.
$m(A)$	Lebesgue measure of a measurable subset A of $\mathbb{R}^2$.

ds, $\lvert dz\rvert$	Arc length measure on a rectifiable curve. The second notation is used exclusively in $\mathbb{C}$.
$\ell(\gamma)$	Length of a curve γ.
$d\sigma := ds/\ell(\gamma)$	Normalized arc length measure (4.7.9).
$\mathscr{C}(\Omega; a, b)$	Family of all paths in Ω starting at a and ending at b, §1.6.
$\mathscr{C}(\Omega; a)$	Family of all loops in Ω, with base point a, §1.6
$\pi_1(\Omega; a)$, $\pi_1(\Omega)$	Fundamental group of Ω (1.6.8).
$\alpha\beta$	Composition of paths: α followed by β.
$\bar{\alpha}$	Inverse of a path.
$[\alpha]$	Homotopy class of a path (1.6.2).
ε_a	Constant path at a.
$d(\gamma)$	Degree of the path γ (1.6.23).
$\mathrm{Ind}_\gamma(a)$	Index of γ with respect to a (1.8.1).
$Z^j(\Omega)$, $B^j(\Omega)$, $H^j(\Omega)$	de Rham spaces of cocycles, coboundaries, cohomology (1.5.9).
$Z_1(\Omega; \mathbb{Z})$, $B_1(\Omega; \mathbb{Z})$, $H_1(\Omega; \mathbb{Z})$	Spaces of 1-cycles, boundaries and homology in Ω with values in $\mathbb{Z}$, §1.9.
$\mathrm{Ind}_\delta(T)$	Index of $\delta \in Z_1(\Omega; \mathbb{Z})$ with respect to the hole T of Ω (1.10.6).
$\mathrm{Res}_\omega(T)$	Residue of $\omega \in Z^1(\Omega)$ with respect to the hole T of Ω (1.10.9).
$I_\omega(\gamma)$	Integral of $\omega \in Z^1(\Omega)$ along a continuous path γ (1.7.5).
$j_\omega(\gamma)$	Period map. Restriction of I_ω to $\mathscr{C}(\Omega; a)$ (1.7.9).
$L(A)$	Free group generated by A (1.6.17).
$[G, G]$	Commutator subgroup of a group G (1.6.17).
Abelianized version of G	$G/[G, G]$ (1.6.17).
$\mathbb{Z}^{(A)}$	Family of all $f: A \to \mathbb{Z}$ such that $f(a) \neq 0$ only for finitely many values of A (1.6.19).
$\mathscr{L}_{\mathbb{R}}(\mathbb{C}, \mathbb{R})$	Space of all $\mathbb{R}$-linear maps from $\mathbb{C}$ into $\mathbb{R}$ (§1.1).
$\mathscr{L}_{\mathbb{R}}(\mathbb{C})$	Space of all $\mathbb{R}$-linear maps from $\mathbb{C}$ into itself (§1.1).
$\mathscr{L}_{\mathbb{C}}(\mathbb{C})$	Space of all $\mathbb{C}$-linear maps from $\mathbb{C}$ into itself (§1.1).
$\mathscr{B}(\mathbb{R}^2 \times \mathbb{R}^2, \mathbb{C})$	Space of all alternating $\mathbb{R}$-bilinear maps (§1.1) of $\mathbb{R}^2 \times \mathbb{R}^2$ into $\mathbb{C}$ (§1.1).

$M(n, \mathbb{C})$	Space of all n × n matrices with complex entries. Coincides with the space $\mathscr{L}_{\mathbb{C}}(\mathbb{C}^n)$ of $\mathbb{C}$-linear maps of $\mathbb{C}^n$ into itself.
$GL(n, \mathbb{C}) := \{A \in M(n, \mathbb{C}), \det A \neq 0\}$	Invertible n × n matrices.
$SL(2, \mathbb{R})$	Space of all 2 × 2 matrices with real entries and determinant 1.
$SU(1,1)$	See Exercise 2.3.15.
$\mathscr{H}(\Omega)$ (Ω open)	Algebra of holomorphic functions in the open set $\Omega \subseteq \mathbb{C}$.
$\mathscr{H}(K)$ (K closed)	Algebra of functions holomorphic in a neighborhood of the closed set K, §1.11.
$\mathscr{M}(\Omega)$	Space of meromorphic functions on Ω (1.4.7)
$Z(f)$, $P(f)$	Set of zeros (resp. poles) of $f \in \mathscr{M}(\Omega)$ (2.4.7).
$\mathscr{H}(\Omega)\|K := \{f\|K : f \in \mathscr{H}(\Omega)\}$	Space of restrictions of functions holomorphic in Ω to the (closed) set K, §3.1.
$\frac{df}{dz}, f'$	Derivative of a holomorphic function (2.1.13).
$f^{(n)}$	nth derivative of a holomorphic function (2.1.14).
$f^{[n]}$	nth iterate of a function (2.2.9).
$f^{\#}$	Spherical derivative of a meromorphic function (Exercise 2.3.1).
$\sigma(z_1, z_2)$	Chordal distance between two points in S^2 (Exercise 2.3.1).
$\mathrm{Aut}(\Omega)$	Group of (conformal) automorphisms of Ω (2.3.9).
$\mathscr{M}$	Moebius group (2.3.12).
$M(u, r) := \sup_{\|z\|=r} u(z)$,	Notation used for subharmonic functions, (mainly chapter 4).
$M(u, z_0, r) := \sup_{\|z-z_0\|=r} u(z)$	
$\\|f\\|_\infty = \\|f\\|_{L^\infty(K)}$	L^∞ norm on a set $K \subseteq \mathbb{C}$.
$\\|f\\|_{L^p(K)} := \left(\int_K \|f(z)\|^p \, dm\right)^{1/p}$	L^p norm on a set $K \subseteq \mathbb{C}$.
$\\|\mu\\|_K$	Total variation of a Radon measure on a (closed) set K, §3.1.
$\|\mu\|$	Total variation measure of a Radon measure, see [Ru (6.1)].
$\hat{\mu}$	Cauchy transform of a measure (resp. distribution) (2.1.5, 3.6.17).

$A(\bar{\Omega}) := \mathscr{H}(\Omega) \cap \mathscr{C}(\bar{\Omega})$	Space of holomorphic functions, continuous to the boundary of Ω (3.7.1).
$H^{\perp}, \mu \perp H$	Orthogonal space to a subspace H; linear functional orthogonal to H, §3.1.
$\hat{K}_{\Omega} = \hat{K}$	Holomorphically convex hull (in Ω) of compact set $K \subset\subset \Omega$, §3.1.
$A^2(\Omega) = \mathscr{H}(\Omega) \cap L^2(\Omega, dm)$	Bergman space (4.8.7).
$SH(\Omega)$	Space of subharmonic functions in Ω (Chapter 4).
$\mathscr{D}'(\Omega)$	Space of distributions in Ω, §3.8.
T_f	Distribution associated to a function $f \in L^1_{loc}(\Omega)$, §3.8.
$\langle T, \varphi \rangle$ $(T \in \mathscr{D}'(\Omega), \varphi \in \mathscr{D}(\Omega))$	See §3.8.
$pv(f)$	Cauchy principal value of $f \in \mathscr{M}(\Omega)$ (3.6.3).
$\mathrm{Res}(f, a)$	Residue of the function f at an isolated singularity a (2.5.3).
$\mathrm{Res}(f) := \dfrac{\partial}{\partial \bar{z}} pv(f)$	Residue distribution (3.6.3).
$b_+(f), b_-(f), b(f)$	Boundary values in the sense of distributions; f holomorphic off the real axis (3.6.10).
$\mathscr{H}_b(\Omega_+), \mathscr{H}_b(\Omega_-), \mathscr{H}_b(\Omega)$	Spaces of holomorphic functions of slow growth on Ω (3.6.11).
$m(f, z_0)$	For a holomorphic function f, the multiplicity of z_0 as a zero of f (3.5.5).
$Z(I) := \bigcap_{f \in I} Z(f)$	Zero locus of I ideal $\subseteq \mathscr{H}(\Omega)$ (3.5.3).
$V(I)$	Multiplicity variety of an ideal (3.5.5).
$I(V)$	Ideal of a multiplicity variety (3.5.5).
$\mathscr{H}(V(I))$	Space of holomorphic functions on the multiplicity variety $V(I)$ (3.6.9).
$A(f, z, r)$	Area average, 4.3.1.
$\lambda(f, z, r)$	Circular average, 4.3.1.
$\nu_f(r)$	Number of zeros of f in $\|z\| \leq r$ (4.4.31).
$P_r(\theta)$	Poisson kernel for the unit disk (4.3.4).
$P(f)$	Poisson integral of f (4.3.5).
$\tilde{P}_r(f)$	Perron modification of f (4.7.1).
ω_z	Harmonic measure (4.4.7).
$g(z; z_0, \Omega)$	Green function of Ω with pole at z_0 (4.7.8).
U^{μ}	Logarithmic potential of a measure μ (4.4.24).

U_α^μ	Riesz potential of order α of a measure μ (4.9.6).
$\mathscr{P}(E)$	Space of probability measures on E (4.9.1).
$I(\mu)$	Energy of $\mu \in \mathscr{P}(E)$ (4.9.1).
$C(E)$	Capacity of a compact set E (4.9.1).
$V(E) := \log(1/C(E))$	Robin constant (4.9.1).
$\tau(E)$	Transfinite diameter of a compact set (4.9.14).
$\rho(E)$	Chebyschev constant of a compact set E.
$\mathbb{C}[z]$	Space of polynomials.
$\mathbb{C}\{z\}$	Space of convergent power series (5.6.1).
$\mathbb{C}[[z]]$	Space of formal power series (5.6.1).
$\mathcal{O}_a, \mathcal{E}_a$	Space of germs of holomorphic (resp. C^∞) functions at a (5.6.1).
$\mathcal{O}, \mathcal{E}$	Sheaf of germs of holomorphic (resp. C^∞) functions in $\mathbb{C}$ (5.6.1).
$\mathcal{O}_X, \mathcal{E}_X$	Sheaf of germs of holomorphic (resp. C^∞) functions on a Riemann surface X.
$\mathfrak{S}_F$	Permutation group of a set F (Chapter 5).
$G(p)$	Group of automorphisms of a covering map p (5.4.6).
σ_c, σ_a	Abscissa of convergence (resp. absolute convergence) of a Dirichlet series

Index

Graduate Texts in Mathematics

continued from page ii

97 KOBLITZ. Introduction to Elliptic Curves and Modular Forms.
98 BRÖCKER/TOM DIECK. Representations of compact Lie Groups.
99 GROVE/BENSON. Finite Reflection Groups. 2nd ed.
100 BERG/CHRISTENSEN/RESSEL. Harmonic Analysis on Semigroups: Theory of Positive Definite and Related Functions.
101 EDWARDS. Galois Theory.
102 VARDARAJAN. Lie Groups, Lie Algebras and Their Representations.
103 LANG. Complex Analysis. 2nd ed.
104 DUBROVIN/FOMENKO/NOVIKOV. Modern Geometry--Methods and Applications Vol. II.
105 LANG. $SL_2(\mathbf{R})$.
106 SILVERMAN. The Arithmetic of Elliptic Curves.
107 OLVER. Applications of Lie Groups to Differential Equations.
108 RANGE. Holomorphic Functions and Integral Representations in Several Complex Variables.
109 LEHTO. Univalent Functions and Teichmüller Spaces.
110 LANG. Algebraic Number Theory.
111 HUSEMÖLLER. Elliptic Cruves.
112 LANG. Elliptic Functions.
113 KARATZAS/SHREVE. Brownian Motion and Stochastic Calculus.
114 KOBLITZ. A Course in Number Theory and Cryptography.
115 BERGER/GOSTIAUX. Differential Geometry: Manifolds, Curves, and Surfaces.
116 KELLEY/SRINIVASAN. Measure and Integral, Volume 1.
117 SERRE. Algebraic Groups and Class Fields.
118 PEDERSEN. Analysis Now.
119 ROTMAN. An Introduction to Algebraic Topology.
120 ZIEMER. Weakly Differentiable Functions: Sobolev Spaces and Functions of Bounded Variation.
121 LANG. Cyclotomic Fields I and II. Combined 2nd ed.
122 REMMERT. Theory of Complex Functions.
Readings in Mathematics
123 EBBINGHAUS/HERMES et al. Numbers.
Readings in Mathematics
124 DUBROVIN/FOMENKO/NOVIKOV. Modern Geometry--Methods and Applications Vol. III.
125 BERENSTEIN/GAY. Complex Variables: An Introduction.
126 BOREL. Linear Algebraic Groups.
127 MASSEY. A Basic Course in Algebraic Topology.
128 RAUCH. Partial Differential Equations.
129 FULTON/HARRIS. Representation Theory: A First Course.
Readings in Mathematics.

Printed in the United States
By Bookmasters